ORIGIN AND EVALUATION OF FORMATION PRESSURES

Origin and Evaluation of Formation Pressures

by

BHAGWAN SAHAY

Deputy General Manager,
Oil & Natural Gas Commission, Bombay, India

and

WALTER H. FERTL

Executive Vice President for Atlas,
Western Atlas International, Inc., Houston, Texas, U.S.A.

KLUWER ACADEMIC PUBLISHERS

DORDRECHT / BOSTON / LONDON

Sahay, Bhagwan, Dr.
Origin and evaluation of formation pressures / Bhagwan Sahay, Walter H. Fertl.
p. cm.
Includes index.
ISBN 0-7923-0126-9
1. Reservoir oil pressure. 2. Reservoir oil pressure--India.
I. Fertl, Walter. II. Title.
TN871.18.S24 1988
622'.3382--dc19 89-2755

ISBN 0–7923–0126–9

Published by Kluwer Academic Publishers,
P.O. Box 17, 3300 AA Dordrecht, The Netherlands.

Kluwer Academic Publishers incorporates
the publishing programmes of
D. Reidel, Martinus Nijhoff, Dr W. Junk and MTP Press.

Sold and distributed in the U.S.A. and Canada
by Kluwer Academic Publishers,
101 Philip Drive, Norwell, MA 02061, U.S.A.

Sold and distributed in India, Sri Lanka, Bangladesh, Nepal and Pakistan
by Allied Publishers Private Limited,
P.O. Box 7203, 13/14 Asaf Ali Road, New Delhi-110002, India

In all other countries, sold and distributed
by Kluwer Academic Publishers Group,
P.O. Box 322, 3300 AH Dordrecht, The Netherlands.

First published 1988
by Allied Publishers Private Ltd.

Printed in The Netherlands

Dedicated to
Prof. George Chilingar
and our wives
Shail and Irma

Foreword

This volume, authored by Dr. Bhagwan Sahay, Deputy General Manager of Oil and Natural Gas Commission, India, and Dr. Walter Fertl, Executive Vice President, Western Atlas International, U.S.A., represents a comprehensive study of the origin, detection and evaluation of abnormal pressures. As such, it compiles a host of recent literature on this enormous subject into a splendid treatment of the overall workings of abnormal pressure. The book is unique, in that it includes examples of studies from the Indian subcontinent never before noted in previous books on this subject.

This text is written by two of the foremost authorities on the subject worldwide. Dr. Sahay has published extensively on this topic in Asia, while Dr. Fertl, who has an international reputation in well-logging analysis and evaluation, is the author of an earlier book, *Abnormal Formation Pressures*, published in 1976. This volume has undergone numerous subsequent printings as well as translations into languages other than English.

Since abnormal formation pressures are now encountered with increasing frequency worldwide, due in part to deeper drilling in sedimentary basins where overpressure appears to be the rule, and due in part to geothermal drilling where subnormal pressures are common, this volume will be a welcome addition to the literature. It will certainly serve as a classic reference in those parts of the world where exploration is comparatively recent and experience with these abnormal conditions is limited. Since these pressures can pose severe hazards to the drilling process and introduce complexities to the production and evaluation processes, it is essential that those involved in exploration and production in newly developing basins throughout the world become familiar with the topics covered herein. I believe that this volume will satisfy the needs of the practical geologist, log analyst and petroleum engineer who desires a comprehensive review of this subject, as well as an up-to-date, state-of-the-art description of the techniques associated with detection and evaluation of abnormal pressures.

The manuscript is well organized, and provides an in-depth coverage of the origin detection and management of abnormal pressures. The early chapters outline the factors associated with such pressures, the role of these factors in petroleum exploration and the analytical background associated with pressure gradients. This portion of the text will serve as an introduction to the novice as well as a necessary review for the explorationist and engineer. It is well documented and reads easily. Chapters 4 and 5 cover in some detail the geological factors causing both subnormal and abnormally high pressures, and this comprises *must* reading for those involved in the subject. Portions covering geology are particularly detailed and are explained well.

The literature on this subject is also documented thoroughly. The succeeding chapter focuses on prediction of overpressures prior to drilling, concentrating on geophysical techniques used in such predictions. This is also extremely important, since prior evaluation of prospect, drilling methods and costs and production management are all affected by the initial concept of basin dynamics and factors leading to petroleum occurrence; matters unique to overpressured regions, and not without risk.

Chapter 7 comprises a very comprehensive study of prediction of overpressures while drilling. This section covers specific drilling parameters such as "d" exponent drag, torque and porosity logs to occurrence of overpressure; also covered are drilling fluid parameters including mud logging, mud characteristics such as temperature and weight, pump pressures and others. Finally, well cuttings are related to overpressures and their unusual characteristics in this regime are carefully explained. I found this chapter to be the most complete analysis of overpressure prediction while drilling yet published. It also incorporates state-of-the-art techniques developed in recent years which are often not covered in other works.

Chapter 8, which covers evaluation of overpressure after drilling is specifically aimed at borehole logging techniques, many of which are relatively new. This chapter also presents a comprehensive review of both classic and new methods of overpressure determination including pulsed neutron techniques. Again, new developments in well logging are covered extensively, and the application of many of these to abnormal pressure evaluation is not available in older works.

The next chapter covering reservoir management of overpressured reservoirs details the influence of such factors as rock compressibility, shale water influx and others as drive mechanisms in these unusual reservoirs. Many of these variables are the result of recent field studies in overpressured reservoirs, and will be of great interest to the reservoir engineer.

Chapter 10, an analytical review of abnormal pressures in the Indian subcontinent, represents the first complete representation of this topic in the literature. It details areas of unusual pressure occurrence, and the reasons for such occurrence. This chapter is obviously of immense benefit to those involved in petroleum exploration and management in this developing portion of the world. Actual field examples are included as a portion of the text.

Taken altogether, this volume is a superb accumulation of knowledge on the subject of abnormal pressures. It also represents the first complete study of this matter on the Indian subcontinent, a growing region for new petroleum exploration and production. The text has a very practical approach to subjects covered, and should provide the field engineer and geologist, as well as the academician new to the subject, with a complete primer on abnormal pressures, their location, genesis, evaluation and management, for years to come.

DR. MYRON DORFMAN

University of Texas at Austin
USA

Acknowledgments

The authors are indebted to Col. S.P. Wahi, Chairman of ONGC, and Members of the Commission for encouragement to proceed with this project, for release of pertinent information, and permission to publish the book and also to Mr. S.N. Talukdar, Ex. Member (Exploration) for his encouragement, guidance provided continuously for the work on Overpressures.

Special gratitude is extended to Thomas E. Randall, Gulf Canada Resources Inc., in Calgary, Canada, who contributed Chapter XI, "Reservoir Management of High Pressure Reservoirs". The authors would also like to convey their sincere thanks to Prof. Dr. George V. Chilingar, Department of Petroleum Engineering at the University of Southern California in Los Angeles, California, for his stimulating discussions and valuable suggestions during the preparation of this book.

Compilation of this book could not have progressed so smoothly in such a short time-frame without the cooperation, comments, and useful discussions with colleagues and officials in the Geology section of the Bombay Offshore Project and the Geology and Coordination sections in the Ahmedabad Project. For this active support, the authors convey their sincere appreciation.

The authors also wish to thank the various technical organizations, service companies, and several trade journals for allowing reproduction of illustrations, tables, and other printed material.

Contents

CHAPTER I

Introduction

In the present-day drilling and completion practices, both maximum well control and minimum cost are key factors. In order to successfully complete a well, proper well planning and drilling operations are necessary, to minimize the danger of blowouts, stuck pipes, loss circulation, etc. Hence a basic understanding of the two key formation parameters is a prerequisite, namely the formation pore pressure and the fracture pressure. Both parameters are equally important in cementing, hydraulic fracturing, fluid injection, sand consolidation, etc.

A study of formation pressure gradient sometimes indicates a correlation with size and type of hydrocarbon deposits. Furthermore, distribution of oil and gas is related to regional and local sub-surface pressure and temperature environments. The information on pressure gradient therefore assists in:

- Optimized and safe drilling operations;
- Engineered casing programmes;
- Completion techniques which are effective, safe, and allow for killing of the well without excessive formation damage.

In reservoir engineering, understanding of the formation pressure is useful as it influences compressibility and failure of the reservoir rocks. Water influx into the reservoir from adjacent overpressured shale sections can act as an additional driving mechanism which affects hydrocarbon production behaviour.

In petroleum engineering operations, one often deduces the natural state of stress in rocks while locally imposing a supplementary stress system and observes the displacement and load variations that result from these artificial changes. This concept applies hydraulic fracturing in well stimulation, drilling, casing shoe tests, etc.

Hubbert and Willis (1957) indicated that the general condition of sub-surface stress is characterized by an unequal stress and that hydraulic injection pressures must be nearly equivalent to the next principal compressive stress.

Exploration in the normal pressure environments shows generally predictable trends for time, cost and risk. However, in overpressure environments, especially superpressures, time, cost and risk factors increase considerably.

The various pressures encountered in the sedimentary rocks are broadly divided into three main components, overburden or load pressure, geodynamic pressure, and reservoir pressure.

Reservoir pressures can be further subdivided into hydrostatic, subnormal or subpressures, and abnormally high formation pressures or surpressures.

Since abnormally high pressures are normally associated with well-kicks and blow-outs, which are a big drilling hazard for both men and material, they have attracted great attention. Various methods of their prediction and evaluation both prior to and during the course of drilling have been developed to counteract their detrimental effect. Nevertheless, subnormal pressures are equally important from the point of view of oil and gas exploration. Subnormal and abnormally high formation pressures can co-exist both laterally and vertically provided they are separated by permeability barriers. These pressures occur in formations ranging from Pleistocene to as old as Palaeozoic and are encountered at a level as shallow as a few hundred metres below the surface and even at depths exceeding 6,000 metres. They are present both in shale/sand and massive carbonate and evaporite sections. However, subnormal pressures can be more severe in carbonate sections as compared with clastic section.

Since time is money and the drilling cost of an exploratory well is quite high, it is but natural that effective and systematic formation pressure evaluation is routinely carried out.

It is the objective of the authors of this book to review and discuss the origin and occurrence of all types of pressures so vital at each and every stage of oil and gas exploration, particularly so when dealing with rank wildcats and difficult onshore and offshore areas.

This book presents the synergism of academic, practical and technological factors of several disciplines and their undisputed impact on managerial decision-making. The concepts and methods presented in this book have been compiled from a vast amount of literature, and the authors field experience in various parts of the Indian subcontinent and elsewhere.

The latest technology and methods have been discussed. But in view of the rapid advancements in petroleum industry, new methods and improved techniques constantly evolve. Furthermore, modifications tailored to the local conditions of an area may be necessary, and the petroleum explorer and engineer must keep himself abreast of the latest methods and apply and evaluate them wherever, possible.

CHAPTER II

Role of Different Pressures in Oil and Gas Exploration and Economic Factors in Normal and Abnormal Formation Pressure Environments

The different formation pressures encountered in an area play a vital role both during exploration and exploitation of potential hydrocarbon resources reservoir. Hence it is necessary to clearly understand them. Wireline data in combination with the study of suitable source rock, the basic concepts of the origin of oil and gas, and geological studies on delineating reservoir rocks provide useful information. Synergism of these data serve as a guide in understanding the regional, potential and commercial aspects of the hydrocarbon deposits. These concepts are helpful and can be applied in predicting the physical state of sub-surface hydrocarbon systems such as oil, condensate or gas. These and a multitude of similar studies have attempted to relate the occurrence of oil and gas to single parameters or their combination such as source rock potential, geological time and age, depth, pressure, temperature, sand/shale ratio, shale resistivity ratio, specific geo-chemical and hydrodynamic criteria, etc. (Fig. 2.1).

The effect of depth on reservoir rocks relating to the hydrocarbon properties has been discussed and described by numerous authors for many areas since the early stages of the oil industry. It may be seen that the worldwide average depths of major hydrocarbon deposits are 48 per cent above 1,800 metres, 43 per cent between 1,800 and 2,750 metres, and 9 per cent below 2,745 metres (Klemme, 1972). It should be understood that this statistical decrease in the total reserves of hydrocarbons could be biased due to less exploration activity at greater depths all over the world. The regional distribution of hydrocarbon versus depth also indicates an increase in gas versus oil reserves both for normal and abnormally high pressure environments.

In many areas of the world, a study of the formation pressure gradients indicates a correlation with the size and depth of hydrocarbon deposits. Such information is undoubtedly valuable in evolving operating techniques that will allow safe and economic development of gas and oil-fields in many overpressure regions.

Various authors all over the world have tried to establish a relationship of formation pressure versus depth, e.g., in the United States in the Texas and Louisiana region as early as 1938 (Cannon Craze, 1938). In their study of data on formation pressure versus depth for Texas and Louisiana a maximum pressure gradient approaching 0.177 kg cm^2/m was found. In 1943, Dentan reported maximum gradient of 0.192 kg cm^2/m for

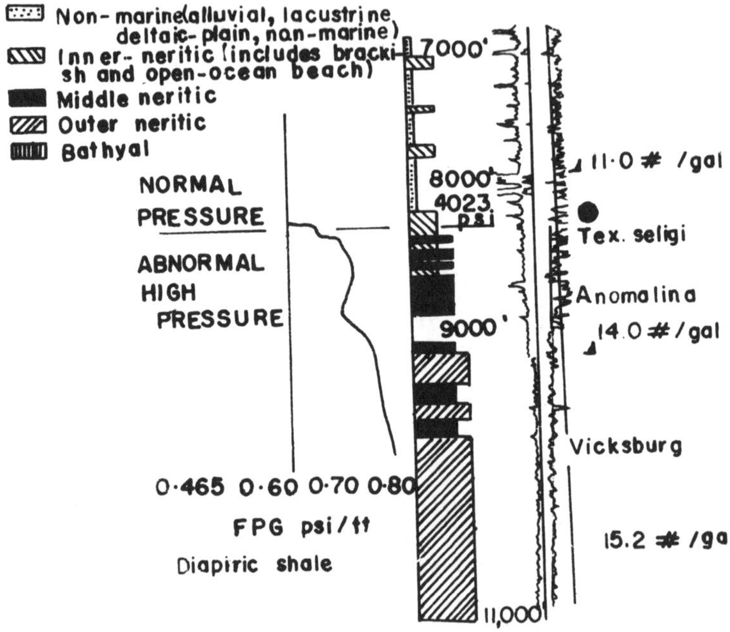

Fig 2.1: Depositional environment and presence of overpressures (after Martin, 1972).

the same region, which was again exceeded in later years. Dickinson (1953) found out values up to 0.872 psi/ft (.201 kg cm^2/m) and suggested this value to approach the upper limit of overpressure gradients in shale/sand sections. He suggested that it was unlikely that 0.90 psi/ft (0.208 kg cm^2/m) would be exceeded. However, later it has been found that higher pressures can be encountered in the U.S. areas. Furthermore, in the Himalayan foothills region of India and in the Potwar plateau region of Pakistan, very high pressure gradients approximating 1 psi/ft in oil/gas bearing areas have been reported. A study of the different pressures encountered in the various sedimentary basins of India, both offshore and onshore, has indicated that the commercial reservoirs occur mostly in the region exhibiting pressure gradients less than 0.8 psi/ft (0.196 kg cm^2/m). Most of the highly overpressured reservoirs in India are devoid of commercial accumulation of hydrocarbons except for fields in the Cambay basin on the western side as well as some fields of the Assam region in the eastern part of India. The production in the Indian subcontinent is mostly from the reservoirs occurring in the near normal hydrostatic pressure environments. In Bangladesh, located in the Indian subcontinent, gas fields occur in the overpressured environments...... The famous Titas field of Bangladesh occurs in the overpressured environments.

In the U.S. Gulf Coast area, several commercial hydrocarbon accumulations have been reported in overpressure environments with pressure gradient as high as 0.9 psi/ft (0.208 kg cm^2/m).

However, at such extremely high pressures, the presence of commercial gas and/or oil reservoirs in shale/sand sequences appear to be the exception rather than the rule.

Such superpressure reservoirs yield water flows with oil and/or gas shows at tremendous flow capacity or have very high but short-lived hydrocarbon production characterized by a rapid pressure decline. Such conditions have been noticed in many parts of the word, e.g. in the U.S. Gulf Coast area and California, the Oligocene reservoir of Burma, the reservoirs in the Himalayan foothills of India, Maikop series (Miocene) of the U.S.S.R., the Potwar plateau of Pakistan, etc.

Fowler (1971) has suggested that the reservoir size decreases with increasing pressure gradient up to a pressure gradient of 0.65 to 0.70 psi/ft (0.15 to 0.16 kg cm^2/m), then it appears to increase with increasing gradient. However, at pressure gradients greater than 0.85 psi/ft (0.196 kg cm^2/m), as per Perry's data (1972), the reservoir size again decreases with increasing gradients.

As the pressure gradient is important for the accumulation of hydrocarbons, both formation temperature and pressure are important factors which significantly affect the physical state of hydrocarbon at depth. Statistical data obtained from various areas of the world show that temperature gradients of the order of 0.3° F/100 ft to 6° F/300 ft have been encountered with gradients of 1° F/100 ft to 1.7° F/100 ft encountered most frequently. In the Indian subcontinent the temperature gradients have been found to range from 2° C/100 m to as high as 5° C/100 m in the areas where oil and gas exploration is actively carried out. However, in India, no direct relationship has been established between high pressure and high temperature gradients, except in some cases. The oil/gas fields in India occur mostly in the temperature regime of 3 to 5°C/100 m with formation pressures in most of the cases being either hydrostatic or slightly above hydrostatic. Notable exceptions are the oil-fields in Cambay and in the eastern part of the country.

The different kinds of reservoir pressures which are usually encountered during the course of drilling are broadly divided into three main components, overburden or load pressure, hydrostatic pressure, and formation pore pressure.

Overburden or Load Pressure

Overburden or load pressures are also sometimes called lithostatic or geostatic pressures, which exceed the average hydrostatic pressure. This pressure originates from the combined weight of the formation matrix (rock) and the fluids (water, oil and gas) in the pore space overlying the formation of interest. Mathematically the overburden pressure (P_0) is expressed as:

$$P_0 = \frac{\text{Wt (rock matrix + fluid)}}{\text{area}}$$

Overburden pressure normally increases with depth. The overburden gradient is normally taken as 1.0 psi/ft. However, the overburden gradient in some cases could be as high as 1.35 psi/ft. (0.312 kg/cm^2). In clastic rocks, corresponding to a force exerted by a formation with an average bulk density of 2.31 gm cm^3. Worldwide observations have resulted in the concept of varying overburden gradient for fracture gradient predictions used in drilling and completion operations. On the basis of sedimentation and combination one may expect overburden gradient at shallow depth to be considerably smaller than 1 psi/ft. This subject will be discussed in detail in the chapter to follow.

Hydrostatic Pressure

Hydrostatic pressure is defined as the pressure which is exerted by a column of water extending from a stratum to a surface. It is the result of the unit weight and the vertical height of a fluid column. The size and shape of a fluid column does not have any effect on the magnitude of this pressure. Mathematically, this can be expressed as:

$$P = p.g.D.$$

where

P = pressure
p = average density
g = gravity value
D = height of the column.

In terms of drilling operations, it can be written as:

$$P_{HY}\,(psi) = C.MW.D.$$

where

D = vertical height of fluid column in feet
MW = mud specific gravity or weight in lb/gal
C = conversion constant (C = 0.052 it MW in lb/gal and C = 0.00695, if MW in lb ft^3).

However, in metric system it is written as:

$$P_{HY} = 0.098\ M.W.D.$$

where

D = vertical fluid column in metres
MW = mud weight.

The hydrostatic pressure gradient is effected by the concentration of the dissolved solids and gas in the fluid column, and different or varying temperature gradients. An increase in the dissolved solids slightly increases the normal pressure gradient, while increasing amount of gases in solution and higher temperature would decrease the normal hydrostatic pressure gradient, e.g. pressure gradient of 0.465 psi/ft (0.107 kg/cm^2) assumes a water salinity of 80,000 ppm NaCl at a temperature of 77°F (25°C).

The average hydrostatic gradient which is encountered during oil and gas exploration in any area is 0.433 psi/ft (0.10 kg/cm^2/m) for fresh and brackish water and

0.465 psi/ft (0.1074) for salt water.

Formation Pore Pressure

Formation pore pressure is the pressure acting upon fluids (water, oil or gas) in the pore space of the formation, expressed either in pound/square inch, atmosphere or kg/cm². Normal hydrostatic reservoir pressures normally correspond to original reservoir pressures, i.e. pressures that existed before the natural pressure equilibrium of formation was disturbed by production. They are described as follows:

$$P_f = hp$$

where

h = depth
p = specific gravity of the fluid taken as 0.433 psi/ft for fresh water gradient and 0.465 psi/ft for salt water gradient
P_f = formation pore pressure.

Formation pressures exceeding the hydrostatic pressure in a geological environment are normally termed as abnormal high formation pressures (superpressures) whereas if the formation pressures are less than the hydrostatic pressure they are called subnormal (subpressures). (Fig. 2.2)

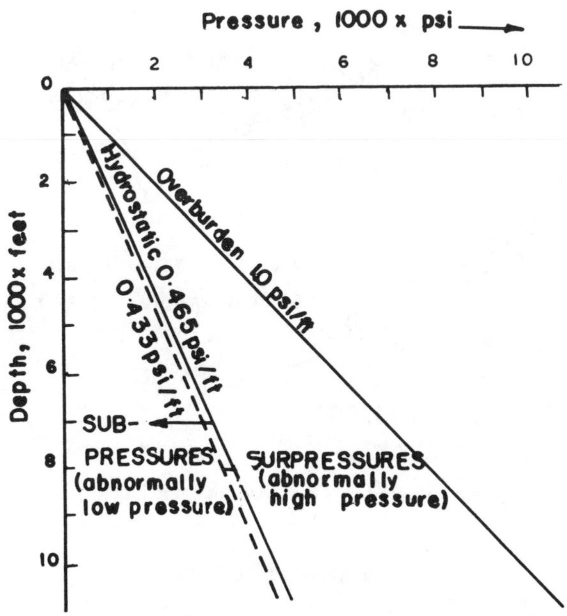

Fig. 2.2: Subsurface pressure concepts

Worldwide exploration for hydrocarbon resources both offshore and onshore has shown that both abnormal formation pressures as well as subnormal formation pressures (subpressures) can co-exist provided they are separated from each other by a permeability barrier, with abnormal pressures occurring more frequently than the subnormal pressure. The permeability barrier separating the two pressure regimes can be of any material or combination of materials present in the earth's crust which prevents movement and passage of substantial volumes of fluids. This pressure seal may be massive shales, carbonates, anhydrite or gypsum. The origin of a pressure seal may be physical or chemical or may be a result of the combination of the two. Several possible types of the formation seals are listed below:

Type of seal	*Nature of trap*
Vertical (e.g. U.S. Gulf Coast; Zechstein, N. Germany, North Sea; Middle East, U.S.S.R.)	Massive shales and silt stones Massive salts Anhydrite Gypsum, Limestone, marl, chalk, dolomite
Transverse (worldwide)	Faults, Salt and shale diapirs
Combination of vertical and transverse	

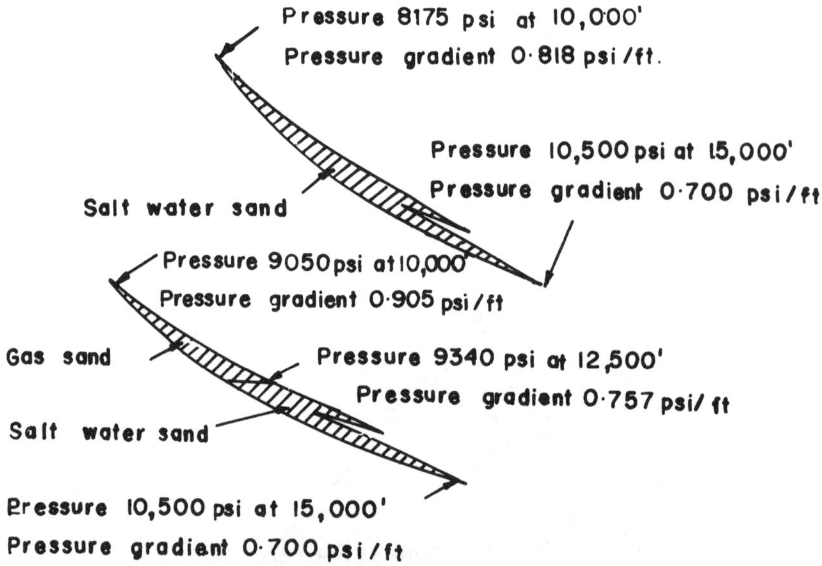

Fig. 2.3: Effect of structure on pressure gradients in sands containing fluids under abnormal pressure (after Dickinson, 1953).

Geodynamic Pressure

These pressures are a result of diastrophism and rock deformation and are present in the rocks of various ages.

The various types of formation pressures with particular reference to subnormal and abnormal formation pressures will be discussed in the subsequent chapters.

REFERENCES

Fertl, W.H., 1976, *Abnormal Formation Pressures*, Elsevier Scientific Publishing Company.

Sahay, B., 1974, *Analytical Studies on Abnormal Pressures in Indian Basin*, manual submitted and distributed to geologists of O.N.G.C.

Sahay, B., et. al., 1983, *Formation Evaluation and Well-Site Geological Techniques.*

Sahay, B., 1982/83, Various lectures delivered at Denver, Los Angeles, and Aberdeen.

CHAPTER III

Analytical Considerations of Overburden, Formation Pressure and Fracture Pressure Gradients

Overburden Pressure

As mentioned previously, overburden or load pressure (sometimes also called lithostatic or geostatic pressure) exceeds the average hydrostatic pressure. It is mathematically derived by the following equation:

$$S = \delta + P$$

where

S = overburden stress
δ = matrix stress
P = pore pressure

Determination of Overburden Stress Gradient (S)

S at any depth = Total rock matrix plus fluid above point of interest
Therefore, S = 0.052 (1-ϕ) pb + 0.052 (ϕ) pn

where

ϕ = Porosity (assumed fluid filled)
pb = Average density of sediments above the point of interest
pn = density gradient of fluid
0.052 = conversion constant

e.g. Assuming ϕ = 12% (.012)-fractional porosity
pb = 20.8 ppg (2.05 sp. gr.)
pn = 9.0 ppg (0.465 psi/ft)
S = 0.052 (1.0-0.012) 20.8 + 0.052 (0.012)
= 1.008 psi/ft.

Hubbert and Wyllis (1957) characterised three principal stresses, whereby hydraulic injection pressures must be nearly equivalent to the least principle compressive stress. Fig. 3.1 shows the generalised trend of three key parameters, namely pore pressure,

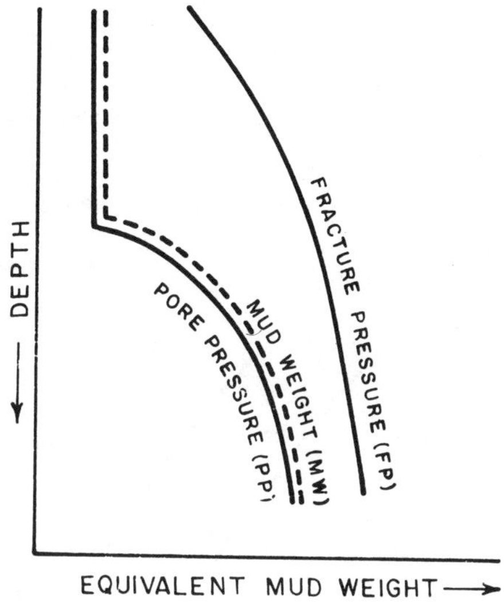

Fig. 3.1: Generalized trends of the three key parameters, formation Pore Pressure, ideal Mud Weight requirements and Fracture Pressure in a High Pressure well. (After Fertl, 1176)

mud weight and fracture pressure v depth.

Formation Fracture Gradient

The formation fracture gradient is defined as the pressure necessary to overcome the formation pressure and strength of rock matrix. The fracture gradient is a function of pore pressure, overburden gradient, and stress rate (Fig. 3.2). This can be calculated at a given depth as follows:

$$P\ frac = Pn + Ps \times 0.0520 \times D$$

where

Pn = equivalent density of fluid column
Ps = surface pressure required for fracturing formation
D = Depth
P frac = formation fracture gradient.

Mathews & Kelly (1967) published a fracture pressure gradient relationship which differs from the Hubbert-Wyllis model in that a variable matrix stress coefficient Ki was introduced. It is mathematically represented by the following equation:

$$FP/D = Pf/Dq + Ki\ \delta/D$$

where

FP = fracture pressure at point of interest, psi
Pf = pore fluid pressure at point of interest, psi
D = depth at point of interest, feet

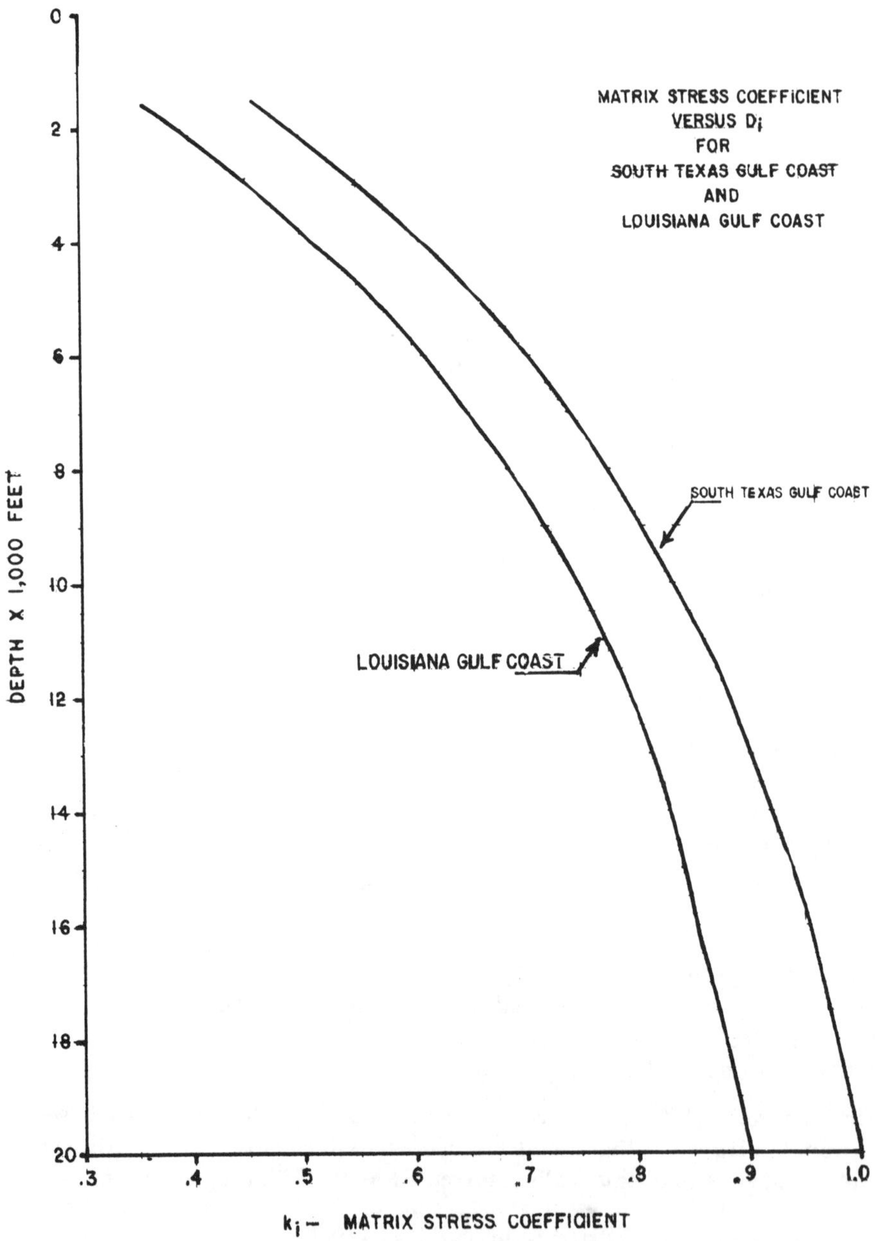

Fig. 3.2: Calculation of fracture gradient (After Mathews & Kelly).

δ = matrix stress at point of interest in psi with $\delta = (Po-Pf)$

Ki = matrix stress coefficient for depth at which the value of δ would be the normal matrix stress, dimensionless.

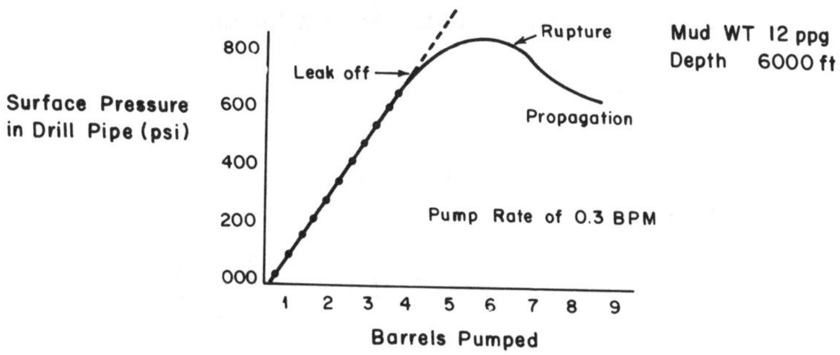

Fig. 3.3: An example of typical Leak Off Test.

The procedure of calculating the fracture pressure gradient by this method is as follows:

(1) Determine pore pressure from well logs, offset well data, etc.
(2) Calculate effective stress δ, where (assuming a constant over-burden load of 1.0 psi/ft) the value of equals (1.0-Pf/D) D.
(3) Determine depth D, in feet for which the δ value would be the normal value, i.e. $\delta =$ (1.0-0.465) Di, or $Di = \delta / 0.535$.
(4) Calculate fracture pressure gradient by the equation given above.

Eaton (1969) has proposed a different equation for the formation fracture gradient, assuming that earth formations are elastic. He related the horizontal stress δH and vertical stress δV by Poisson's ratio. Extending the model by Mathews and Kelly (1967) the Poisson's ratio is introduced into the expression for fracture pressure gradient such as

$$FPG = Pf/D + \left(\frac{V}{1-v}\right) \delta/D$$

where

δ = matrix stress which equals (Po-Pf).
D = depth at point of interest in feet.
Po = 0.4335

All independent variables are assumed to be functions of depths, such as overburden load and Poisson's ratio of rocks.

Eaton's fracture pressure gradient prediction method, one of the best techniques available to oil industry, is used World-wide.

Method of Anderson, Ingram and Zanier (1973)

The world-wide pressure gradients may vary widely at any given depth in the same geologic area (Taylor and Smith, 1979) MacPherson and Berry, 1972).

Anderson et. al. (1973) studied 29 fracture pressure gradient measurements in several U.S. Gulf coast wells during fluid injection for sand consolidation.

$$FPG = \frac{2V}{1-V}\frac{(Po/D)}{1-V} + \frac{1-3V}{1-V}(Pf/D)$$

On the other hand using Biot's stress-strain relationship one derives following fracture-pressure gradient expression

$$FPG = \frac{2V}{1-V}(Po/D) + \alpha\frac{1-3V}{1-V}(Pf/D)$$

where

V = Poisson's ratio and α = rock

compressibility factor in terms of values derived from logs.

Method of MacPherson & Berry

MacPherson and Berry (1972) studied the prediction of fracture pressure gradients from well logs, i.e. correlation of the well bore elastic moduli with fracture pressure gradient data in wells located offshore Lousiana, and drew the following conclusions.

(1) Fracture pressure gradients are higher in shales than in sands.
(2) Fracture pressure gradient versus depth are different in the ten offshore fields studied but appear to be related to the sand/shale ratio depth of overpressures.
(3) At equivalent depth the heavier mud weight used correlated with fracture pressure gradient.
(4) Breakdown pressures in overpressured intervals are only slightly different from normal pressure trends in the direction of higher fracture pressure gradient with increasing overpressure.
(5) Most important, a quantitative relationship exists between the ratio of sand elastic modulus to overburden pressure and the fracture pressure.

Leak Off Tests

Leak off tests are the most accurate method of determining fracture gradients. Unlike pressure tests, which check the formation's ability to withstand a given pressure a leak off test measures the pressure gradient to initiate rock separation. Even the best available charts are sometimes too general to provide the accuracy required on many wells. With the help of the leak off test it is possible to find out exactly which mud weight is needed, how deep the well can be drilled before setting another casing string and define alternatives in case of other well problems.

The following steps serve as a guide to conduct the leak off test.

(1) Close in the well just above the shoe. The leak off test should be run in the first sand below the shoe, However, care needs to be taken that the open hole does not exceed more than 15 metres.
(2) The pumping rate of the mud should vary between 0.25 to 0.5 bbl/min. Faster

rates could cause formation rupture. A small pump, such as a cementing pump, provides better control than a mud pump.

(3) Surface pressure versus volume can be plotted on regular graph paper. Initially this plot should follow a straight line.
(4) Pumping should be continued until the curve slopes away from the straight line and at this point the pump should be shut off.
(5) With the pump off, record the pressure every minute for 10 minute or until pressure readings level off.
(6) Bleed the pressure and measure the volume of fluid released using a trip tank. The volume should be approximate to that pumped into the well bore.
(7) Read the leak off test pressure from the graph where the curve bends away from the straight line portion, and then calculate the fracture gradient.

The above procedure estimates the formation pressure gradient from data thus taken from the leak off test. The leak off pressure is added to the hydrostatic pressure of the mud column from the surface to the point of interest. The resulting fracture pressure is converted to a fracture gradient expressed in mud weights. Constants of 0.052 to 0.00981 are used for US and SI units, respectively.

It is important to have sufficient mud weights to prevent various well hazards in the form of threatened blow-outs and other well problems. However, it is equally important to avoid excessive mud weights which could induce or extend fractures in the formation. Partial or complete mud loss into the formation besides being expensive, reduces the hydrostatic head of the mud and invites well kicks or blow-out conditions. The key to minimizing such problems is the proper selection of casing seats. It is therefore necessary to test formation integrity by carrying out a leak off test in the open hole after drilling about 50 ft (15 m) below the casing shoe. In addition, there are several factors that should be controlled after the casing is set.

(1) Maintain as low a mud weight as possible without exposing the well to a potential kick.
(2) Keep mud viscosities and gel strengths as low as possible to minimize pressure losses.
(3) Keep low circulation rates.
(4) Build up pump pressure slowly particularly when the gel strength of the mud is high.
(5) Calculate the maximum casing pressure on a full column of mud equalling the minimum fracture pressure. Do not exceed this value when the well is initially shut in after the kick.
(6) In critical situation going in the hole (trip-in) should be slowed down. Fig. 3.3 shows a typical example of a leak off test, while Fig. 3.4 shows a Schematic presentation of sub-surface pressure relationship.

This procedure estimates the formation fracture gradient for land and offshore locations. The solution depends on formation pressure gradient, overburden gradient and matrix stress coefficient at the depth of interest.

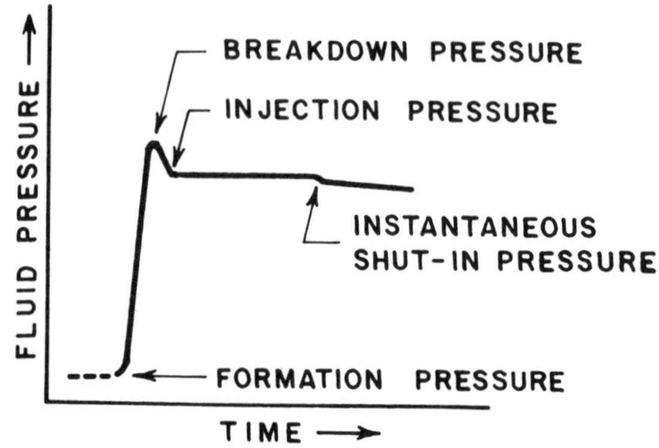

Fig. 3.4: Schematic Presentation of Subsurface Pressure relationships.

REFERENCES

Anderson, R.A., Ingram, D.S. and Zanier, A.M., 1973. Determining fracture pressure gradients from well logs. *J. Pet. Technol.*, **25:** 1259-1268.

Biot, M.A., 1955. Theory of elasticity and consolidation for a porous anisotropic solid. *J. Appl. Phys.*, **26(2):** 115-185.

Chenevert, M.B., Mestare., 1978. How to run casing and open hole pressure tests. *Oil and Gas Journal.*

Christman, S.A., 1973. Offshore fracture gradients. *J. Pet. Technol.*, **25:** 910-914.

Eaton, B.A., 1969. Fracture gradient prediction and its application in Oil field operations. *J. Pet. Technol.*, **21:** 1353-1360.

Fertl, W.H., 1976. Abnormal Formation Pressures, pp. 231-236.

Hubbert, M.K. and Wyllis, D.G., 1957. Mechanics of hydraulic fracturing Trans. AIME, **210:** 153-168.

Kendall, H.A., 1977. Friction pressures, fracture gradients critical offshore well control factors. Ocean resources engineering.

Mac Pherson, L.A. and Berry, L.N., 1972. Prediction of fracture gradients from log derived elastic moduli. *Log Analyst*, **13**(5): 12-19.

Mathews, W.R. and Kelly, J., 1967. How to predict formation pressure and fracture gradient. *Oil and Gas Journal*, **65**(8): 92-106.

Moore, P.L., 1973. Casing Seat testing—why and how. *Oil and Gas Journal*, **71**(32): 73-74.

Rieke, III, H.H. and Chilingarian, G.W., 1974. Compaction of Argillaceous sediments. Elsevier, Amsterdam: pp. 424.

Taylor, D.B. and Smith, R.K., 1970. Improving fracture gradient estimates in offshore drilling. *Oil and Gas Journal*, **68**(15): 67-72.

CHAPTER IV

Subnormal Formation Pressures: Origin and Characteristics

Fluids confined in the pores of the reservoir rock under a certain degree of pressure are referred to as reservoir formation pressure. These pressures are expressed either in pounds per square inch (psi) or in atmospheric pressure (multiple of 14.7 psi) or kg/cm^2 (multiple of 14.2 psi). Reservoir pressure is generally thought of as original or virgin pressure, i.e. the pressure that existed before the pressure equilibrium in the formation has been disturbed by any production. The terms used for reservoir pressure are bottom hole pressure, water pressure, shut-in pressure and rock pressure.

Origin of Normal Formation Pressures

Sediments exhibiting porosity and permeability are subject to the following two stresses:

(a) Compressive and tensile forces transmitted and applied through the rock matrix. In the simplest case, these are dictated by the gravitational weight of the overlying sediments.

(b) Fluid pressure within the pore space due to the head of fluid in the reservoir sequence.

Thus the bulk density of the sediment can be derived from the following relationship:

$$\rho b = \rho g - (\rho g - \rho f)\ \theta$$

where

ρb = bulk density of porous sediment
ρg = density of sedimentary matrix
ρf = density of fluid in pore space
θ = porosity, fraction of void volume

Bulk density varies with depth since the lithology and fluid content do not remain constant. Normal formation pressures are generally classified as those with a pressure corresponding to the hydrostatic head of water. It is, therefore, necessary to allow for a range of gradients to account for variations in salinity. Thus normally pressured reservoires may have a gradient in the range of 0.43 to 0.48 psi/ft.

Influence of Formation Pressures on Drilling Operation

In order to maintain hole stability, i.e. preventing borehole collapse and also the influx of formation fluids into the borehole, it is prudent to maintain the borehole pressure slightly overbalanced to the formation pressure. Significant overbalance of the mud system can lead to:

(a) Reduced penetration rates due to increased 'chip hold' effects.
(b) Lost circulation, i.e. inflow of mud into the formation, due to formation fracturing.
(c) Excesive differential sticking of the drill pipe.

In addition, the pressure gradients of the formation affect the casing string design. Casing is used to "case off" a difficult zone. For instance, a highly pressured zone may exist above a normal (hydrostatic) pressured zone. The mud weight cannot be reduced until the higher zone is sealed off. Furthermore, high mud weight in the lower zone would lead to low penetration rate or in extreme cases even to lost circulation. Hence techniques for predicting pore pressures are helpful for the following reasons:

(a) Assist in selecting adequate mud weights to ensure well safety.
(b) Prevent the use of excessive mud weights to avoid fractued formation and lost circulation problems.
(c) Prevent hole collapse or sloughing of shales.
(d) Assist in the correct design of casing programmes to ensure optimum completion and maximum productivity.

Subnormal Pressures

Subnormal pressures are those, which are less than the hydrostatic gradients. World-wide experience indicates that the frequency of occurrence of sub-normal pressures is less than abnormally high pressures. Nevertheless, sub-normal pressures were encountered in many areas while drilling for hydrocarbons, including Texas, Oklahoma, parts of the Colorado plateau, Un basin in Utah and Appalachean region in the U.S. and areas of Canada, U.S.S.R., etc.

In the Indian sub-continent, sub-normal pressures also occur in various places, both offshore and onshore, in granular as well as carbonate reservoirs.

Characteristics of Sub-Normal Pressures

These pressures generally are of no great importance. However, when they occur, the specific gravity of the mud system has to be kept low and the viscosity high. In case of mud loss in such sub-normal pressure zones, loss circulation material, such as mica, gel flakes, saw dust, rice husk, etc., are added. However, in areas where large losses occur and cannot be stopped, as had happened in some of the wells drilled in India, circulation loss can be stopped by squeezing cement into the formation, provided they are not potentially hydrocarbon bearing. Sometimes, particularly in carbonate sections, even squeezing of cement is no remedy. Problem zones, then, need to be covered by setting casing or liners before drilling deeper.

Sub-normal pressures occur in many depleted oil reservoirs (Brighente, 1967). In some areas of the U.S.A., subnormal formation pressure gradients range from 0.36 psi/ft. to 0.39 psi/ft. Occasionally, surface indication of such in situ pressure reduction is manifested by subsidence which has been observed in several areas.

Today, special subsidence and compaction monitoring devices are available, including (1) Precision casing joint length measuring tools for detection of deep compaction (2) Compaction monitoring wells that measure shallow compaction by means of a cable extensometer, etc. In oil fields, where subsidence is suspected, this data is monitored and reviewed continuously and operating measures such as special water injection are under taken to prevent subsidence. As in production of hydrocarbons withdrawal of ground water may result in sub-pressures and subsidence. For example, a comprehensive study was made of the phenomenon of subsidence occurring in the region of the Po Delta, Italy (Caloi, 1967). It indicated the necessity of shutting in thousand of wells which extracted methaniferous waters to which the severe ground subsidences in the delta were attributed.

In the transition zone from normal to abnormal pressure environments, the temperatures exceed the normal geothermal gradient. Similar conditions are also observed in producing wells drilled into subnormal and subpressure environments. Lower temperature increases are observed in subnormal pressured wells. If one assumes the Morrow formation to be overheated and undercompacted at initial conditions, any formation cooling due to uplift and erosion will reduce the pore pressure. During cooling of an isolated fluid system, which could be due to uplift and/or erosion, the pressure in the isolated fluid system and the pressure in the isolated volume will decrease below the normal hydrostatic value.

Such a temperature decrease may cause a pressure drop in isolated gas zones with the magnitude of reduction depending on the value of pressure change and composition of gas (Fig. 4.1).

Sub-pressures in some of the U.S.S.R. oil/gas fields have been reported in the reservoirs of Middle Miocene age (Fertl, 1976).

Origin of Subnormal Pressures: The origin of subnormal pressures is due to several reasons,

(i) *Erosion and Decompressional Expansion:* Subnormal pressure may result from the exposure of lenticular permeable horizons to erosion, followed by transgressive overlap at a later date by an impermeable stratum, such as salt or anhydrite, and subsequent deep burial due to continuous geosynclinal deposition. In this case, the impermeable seal will prevent fluid flow into this horizon from neighbouring high-pressure strata, when the entire sequence is deeply buried at a later date, and the formation pressures in the lense would then be lower than hydrostatic.

In the wildcat wells drilled in the Ganga basin, India, e.g. in the recent years at Tilhar and Raxaul, mud losses occurred in the low pressure zones which could be due to communication with higher horizons, which exposed that particular formation at an

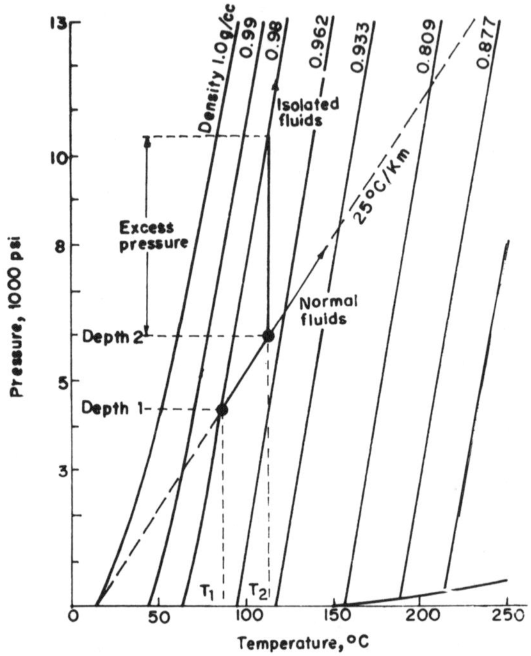

Fig. 4.1: Pressure-temperature-density diagram for water. Excess pressure is the higher pressure due to the temperature increase from T_1 to T_2.

(After Barker, 1972)

abnormally low pressure. Low pressures resulted in heavy mud loss during drilling.

Gas reservoirs in the Appalachain region exhibit sub-surface pressures in reservoirs associated with shales in areas which were eroded. This erosion may have caused reduction of fluid pressure in the shales which was transmitted to the associated reservoir rocks. The pressure drop is caused by absorption of water in clay minerals as overburden pressure and temperature decrease and pore volume increase due to extension of crystal structure.

(ii) *Precipitation*: Subnormal pressures are also caused in very arid areas, such as the Middle-East. The water table in such areas may be hundreds of feet below the surface and hence, underpressured formations can result. Hydrostatic gradient commences at water table only, causing a subnormal gradient from the surface.

(iii) *Temperature Changes*: Reduction of the subsurface temperature causes pore pressure decreases, specially when gas is present. As sediments and pore fluids are buried during sedimentation, temperature rises and if allowed to expand, fluid density would decrease. The magnitude of this effect is very small.

(iv) *Tectonics*: Subnormal pressures do exist in areas of modern tectonic activity, such as along the flanks of the Rocky Mountains. It is suggested that during the compression process, up-warping of upper beds and down-warping of lower beds can result. Intermediate beds must expand to fill the voids created in this process. It is then possible for more competent intermediate beds to have a subnormal gradient.

(v) *Epeirogenic Movements*: Changes in elevation can give rise to abnormal pressures in some formations open to surface laterally but otherwise sealed. Thus, if there is an uplift, formation pressure becomes abnormally high and vice-versa. Pressure changes are seldom caused by changes in elevation alone, since associated erosion and deposition are also significant factors. Loss or gain of water-saturated sediments is equally important.

(vi) *Formation Foreshortening*: This normally occurs in areas of modern tectonic activity specially on the flanks of the rocky mountains. It is expected that during a compression process of up-warping of the upper beds and the down-warping of the lower beds can result and the intermediate beds must expand to fill the void filled by this process. It is therefore, possible for more competent, intermediate beds to have a sub-normal pressure gradient (Fig. 4.2).

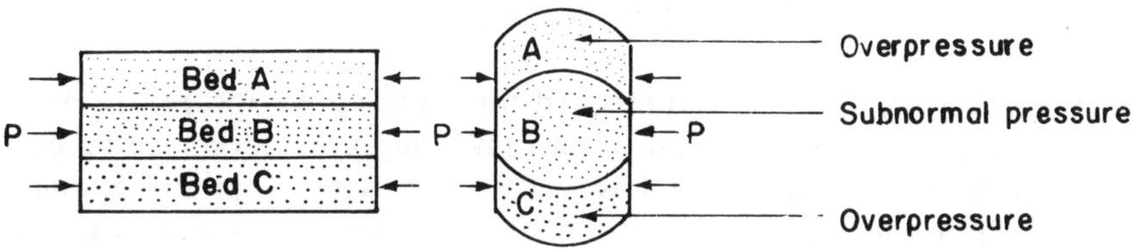

Fig. 4.2: Diagram showing foreshortening of intermediate beds. Shortening of bed B caused by warping of A and C caused pressured problems.

(vii) *Artesian Conditions*: Low pressure may occur in the folded areas of strong topographical relief where the drill bit penetrates the permeable horizon which is exposed on the surface at a lower altitude than the level which is opened by drilling.

(viii) *Artificial Production*: Abnormally low pressures occur when hydrocarbons or water are produced at very high rates and are not compensated by a strong water drive. In cases fresh water acquifers are tapped, reduction in hydrostatic head can cause sub-normal pressures.

REFERENCES

Fertl, W.H., 1976 Abnormal formation pressures, Elseview scientific publishing company.
Gearhart Geodata A manual on overpressure detections.
Sahay, B., 1974 A manual on analytical studies of overpressured zones in Indian basins. (Unpublished).
Sahay, B., Rai, A.,et al. 1988 Wellsite Geological Techniques for Petroleum Exploration.

CHAPTER V

Abnormal High Formation Pressures: Origin and Characteristics

Introduction

Occurrence and evaluation of overpressured formations are critical to exploration concepts, drilling and production operations involving hydrocarbon resources. An interdisciplinary technical approach thus becomes necessary to optimise exploration efforts in such hostile subsurface environments. In the search for hydrocarbon resources, both offshore and onshore, abnormal formation pressures are encountered worldwide in rocks ranging from Caenozoic era (Pleistocene age) to as old as Palaeozoic era (Cambrian age). They are encountered at only few hundred metres below the surface or at depths exceeding more than 6,000 metres, both in clastic as well as carbonate-evaporate sections.

Normal hydrostatic pressured geological environments can be visualised as a hydraulically open system, i.e. permeable fluid—communicating formations allow establishment or re-establishment of hydrostatic conditions. An abnormally high pressure system is a "closed" system preventing or at least greatly restricting fluid communication. In this case the overburden pressure in part is supported by the formation fluid in the pore space (Fertl, 1976).

Abnormally High Formation Pressures (Sur-pressures)

Abnormally high pressured zone is defined as the layer in which the pressure exceeds the vertical hydrostatic pressure column of water (extending from stratum tapped by the well to the surface). Over-pressured zones in view of their association with blow-outs and various other well hazards have attracted more attention in recent years than normal pressured zones.

Sur-pressures are usually common in areas of very rapid deposition, especially in the younger clastic sediments and have the following characteristics:

(1) Represent a very rapid deposition.
(2) Generally fine-grained, partly impermeable sediments.

(3) Sediments exhibiting strong source rock characteristics.
(4) Permeable horizons are few and lenticular in habit.

Common to both normal and over pressures is the sealing mechanism which restricts equalisation of the pressures within the abnormal pressure zone and the rest of geological sequence. Louden (1972) has described two types of permeability barriers or seals.

(a) *Physical seal*: This can be a gravity fault during deposition, deposition of a fine grained material or carbonate, salt or other non-porous material caused by long period of high temperature and low rain fall.
(b) *Chemical seal*: This is pertaining to the chemical deposition of calcium carbonate, for example in warm water, thus restricting average permeability. Another example is the chemical diagenesis during compaction of organic material associated with normal deposition.

The upper limit of such overpressures (sur-pressures) is usually expected to equal the total weight of the overburden which is generally assumed to be equivalent to a gradient of 1 psi/ft. (0.23 kg/cm^{2}m). On a regional basis, overpressures may approach the overburden load, which for reason of stability, cannot surpass. However, in Pakistan, Iran, the U.S.S.R. and in the Himalayan foothills in India, very high pressure gradients approaching the overburden or even exceeding the latter by about 40% have been encountered. Such super pressures can be visualised in the presence of a so-called pressure bridge (Fig. 5.1). In these cases, burial is rather shallow, as laboratory data have shown that the tensile strengths of sedimentary rocks seldom exceed 1000 psi.

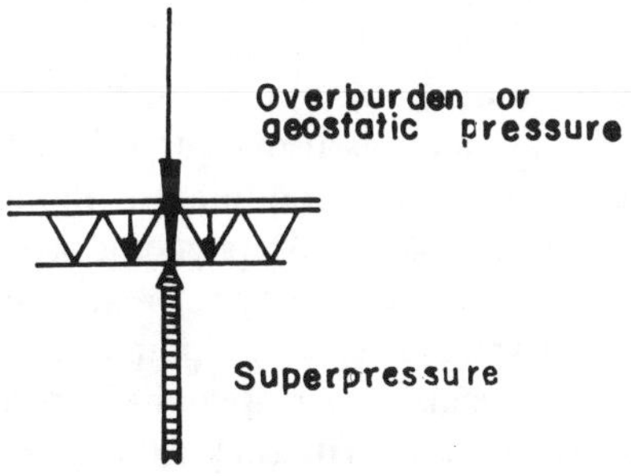

Fig. 5.1: Schematic model of pressure bridge

The internal strength of the rock overlying the pressures assist the overburden load to contain locally these tremendous upward forces.

Reasons for Abnormal High Pressured Horizons

Numerous theories have been propagated in recent years. Frequently the origin can be ascribed to the geological history. The reasons for abnormal high pressure horizons in sediments of various ages include the following:

(1) *Compaction*

Most sediments are deposited in an aqueous medium. At the time of deposition, the sediment consists of mineral grains surrounded by the water. With superincumbent load due to deposition of sediments above, there is a downward directed pressure on the earlier sediments which causes the grains to increase their contact with each other. This causes a reduction in volume caused by expulsion of the water of deposition from around the grains. In case of platy layers of clay minerals, the plates tend to get aligned parallel to each other with increasing pressure and, therefore, permeability decreases rapidly with compaction. Thus, clays which compact to shales yield water of deposition to surrounding porous layers, but, at a certain stage of compaction, begin to act as impermeable barriers. A porous layer completely isolated by shale will take on water squeezed out from the compacting shale until the shale surrounding the isolated porous layer compacts to an impermeable state or the fluid pressure in the porous layer becomes geostatic. In the latter case, no pressure difference between the compacting shale and the porous lenticle will exist longer and fluid flow will stop.

In the first case, fossil pressure will be preserved after the strata are uplifted and the fluid pressure in the porous lens will be higher than normal for the final depth. In the second case, the pressure will be geostatic and, therefore, more than twice the magnitude of hydrostatic pressure appropriate to the depth.

Compaction of thick shale intervals appears to be the main cause for pressuring in the geosynclinal part of the Bengal basin, Gujarat, and Arabian sea shelf. In Bengal, the abnormal pressures encounted approach 1.8 to 1.9 times the hydrostatic pressure.

(2) *Tectonics*

Due to the movement of sedimentary rocks after depositional changes in the physical environment of formation fluids also occur. A fault will vertically displace the fluid bearing layer, either creating new conduits for fluid migration giving rise to pressure changes or establishing updip barriers giving rise to isolation of fluids and preservation of fossil pressure. The barrier may be created by the fault itself or by the movements, bringing the impermeable layer in updip contact with a permeable layer, thus sealing the fluids in the permeable layer. Regional growth faults contribute to the origin of abnormal pressures by redistributing sediments and bringing permeable zones against the impermeable zones, thus inhibiting the flow of fluids to the regions of hydrostatic equilibrium. Geo-pressures occur where a permeable layer being fed with fluids at greater depths. Harkins and Baugher (1969) noted that in the U.S. Gulf Coast, abnormal pressure developed adjacent to the down thrown sides of major growth faults in delta prodelta sequences and these abnormal pressure zones are generally confined to strata in which the particular growth fault is initiated (Figs. 5.2 & 5.3).

In strongly folded areas, there is not only a reduction in pore volume due to compression, but also attenuation of competent layers in limbs and accumulation in cores of anticlinal folds. Additionally, rupturing of layers takes place due to squeezing

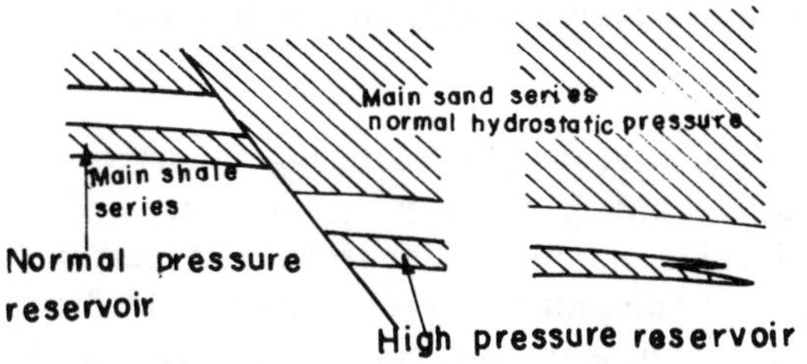

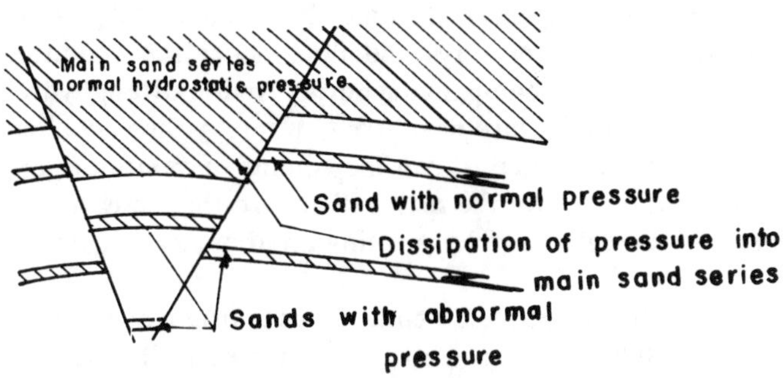

Fig. 5.2: Types of fault seals necessary to pressure abnormal pressure environments.

Fig. 5.3: Schematic section showing stratigraphic rise of overpressures in relation to prograding sedimentation modified by growth faults.

of and stretching beyond the elastic limit. The resulting isolation of blocks of permeable reservoirs causes development of high fluid pressure in them.

In the Surinmastagarh anticline of Jammu in the Siwalik belt of India, overpressures as high as 2.38 times hydrostatic have been encountered during drilling. In the outer folded belt of the Assam-Arakan system, India, pressures 1.8 to 2 times hydrostatic have been encountered while drilling in the Masimpur area of Assam. In the Balh well of Punjab, formation pressures encountered appear to be 2.15 times the hydrostatic. The overpressures encountered in the Potwar plateau of Pakistan also seem to be due to tectonic phenomenon with pressures even 30-40% greater than overburden gradient.

(3) *Diagenesis*

Diagenesis is defined as the chemical and physical changes that sediments undergo during and after accumulation, but before consolidation takes place. Basically diagenesis is the post depositional alteration of a sediment and its constituent minerals. Processes of diagenesis include: (i) formation of new minerals, (ii) redistribution and recrystallization of the substances in sediments, and (iii) lithification. Diagenesis in sand/shale sequences is related to alteration of clay minerals, which contribute to the origin of abnormal fluid pressure by increasing the quantity of free water in the subsurface water system (Fig. 5.4). Montmorillonite, the most predominant mineral in shales, alters to mixed layer clays and illite under the conditions of increasing temperature and pressure accompanying burial. An important hydration stage occurs at depth-dependent temperature near 105°C whereby the clay alteration yields free pore water in an amount equal to half the volume of montmorillonite altered. If the expulsion of free fluid within the rock will absorb the increasing load, the effective stress will slowly increase, thereby creating an abnormal fluid pressure, within the formation.

Diagenesis in carbonate sequence contributes to origin of abnormal fluid pressures by creating permeability barriers in porous sequences. In this way, explusion of fluids from the rocks is inhibited while the static load continues to increase, generating a large potential energy within the pore fluids. Abnormally high pressures in a carbonate sequence are very risky to drilling operations. Such pressures have been encountered in Jurrassic carbonates in closed basins and in Cretaceous-reef trends in the U.S.A., and elsewhere. In India, examples from wells drilled in Arabian Sea off the Bombay Coast can be cited. The porosity in these formations is usually the result of diagenesis. Where the rocks are permeable and formation fluids can escape into the ground water system, the carbonates will be normally pressured. Porous and permeable carbonate members are usually overlain by evaporites. Unlike shales, which often act as semipermeable membrane, the evaporites can be totally impermeable. In this case, pore waters cannot be expelled from porous carbonates into ground water system. The abnormal pressures frequently reach gradients in excess of a 1.98 specific gravity mud.

Volcanic Ash: In the NW region of U.S.A. and South China Sea region (Java, etc.) a rather unusual abnormal formation pressure phenomenon occurs. Original sediments may be fine-grained red bed or sandstone plus large quantity of volcanic ash.

Diagenesis of volcanic ash results in three basic main products—clay minerals, methane and carbon dioxide. Drilling mud is heavily gas cut when penetrating such formations. However, the amount of gas cutting is not directly related to the magnitude of overpressures (Fertl and Timko, 1972).

(4) *Osmosis*

Osmosis is defined as the spontaneous flow of water into a solution or from a more dilute to a more concentrated solution separated by semipermeable membrane. The osmotic pressure differential at constant temperature is almost directly proportional to

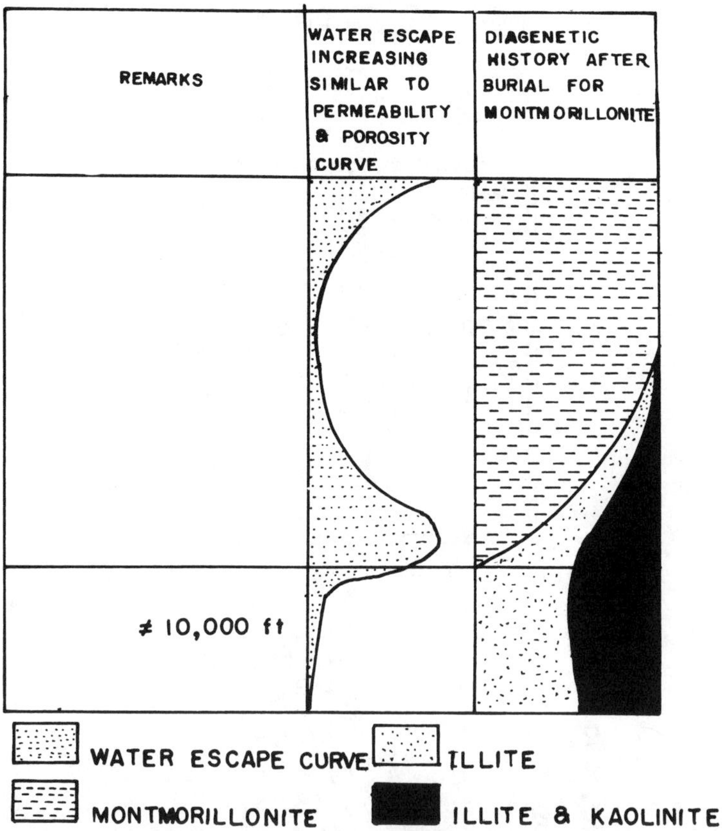

Fig. 5.4: Montmorillonite compaction history.

the concentration differential, and it increases with absolute temperature. Clays act as semipermeable membranes, thereby permitting osmotic and electrosmotic pressures to develop across bedding planes, whenever there is a marked contrast in the concentration of dissolved salts on either side of clay layer. When clay/shale separates two acquifers, fluids from the lower concentration side will pass through clay layers and enter the acquifer of higher concentration. This process continues until fluid pressure in the aquifers of higher concentration is enough to resist such fluid movement. In other words, water moves through the shale to dilute more salty aquifers until it is stopped by the back pressure. The movement of additional water into the aquifer raises the pressure and this is an incremental source of reservoir pressure.

Various studies have shown reverse osmosis or diffusion pressures in some region. It occurs across a membrane wherever the pressure is applied on the more saline side. Water moves towards lesser salt concentration and builds up high pressure in the less saline water. In the Surinsar well, Jammu, India, the reason for abnormal pressure horizon besides tectonics, is also 'Reverse Osmosis', as was revealed by various studies (Figs. 5.5 & 5.6).

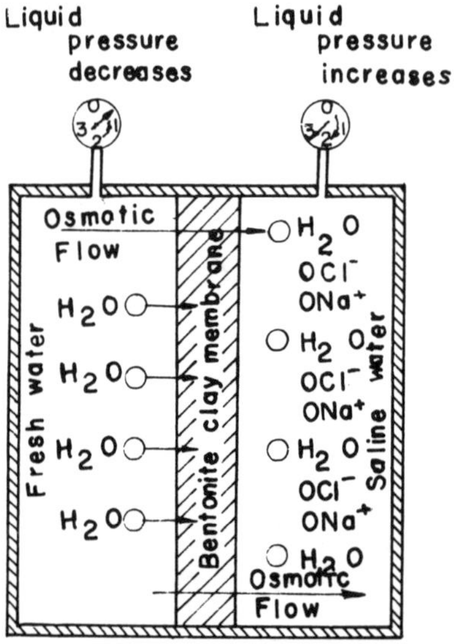

Fig. 5.5: Diagram of Osmotic flow fresh water towards saline water.

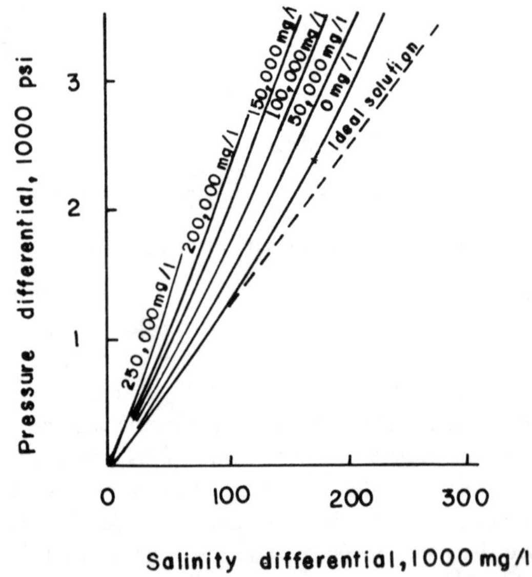

Fig. 5.6: Theoretical values of osmotic pressures across a claymembrane.

(5) *Adsorption*

This phenomenon arises out of the adhesion of the molecules of a liquid or gas to a surface. Under certain conditions, adsorption can inhibit the flow of fluid from compacting shales and thus contributes to the generation of abnormal pressures.

Chennevert (1969) described some lab experiments on different shales which show large internal swelling stresses caused by adsorption. However, the process is temperature-dependent. Under conditions of structural uplift, wherein the displaced sediments are presumably cooler, adsorption could inhibit the expulsion of water and thus extend the duration of abnormal conditions.

(6) *Chemical Caps*

Overpressures can be generated due to chemical conversion of anhydrite to gypsum. This occurs in Mississippi, U.S.A., the Persian Gulf, and many other places. Gypsum will begin to dehydrate below 393 ft of sediments whereby dehydration reduces the volume of formation. In the reverse process, hydration in the epidiagenetic phase anhydrite expands 40%. Classic example of this is Permo Triassic evaporites of England which eat into dolomite. This hydration process can also result from pressure being trapped and causing water to move back into anhydrite. The volume would increase 15-25%, while in the completed state, gypsum volume would increase 30-50%. If this process occurs at depth, tremendous pressures can result. This is seen in central eastern Mississippi, where pressures have resulted in the gradation of anhydrite to gypsum.

(7) *Salt Dome*

Overpressure are frequently present in salt dome areas (Fig. 5.7).

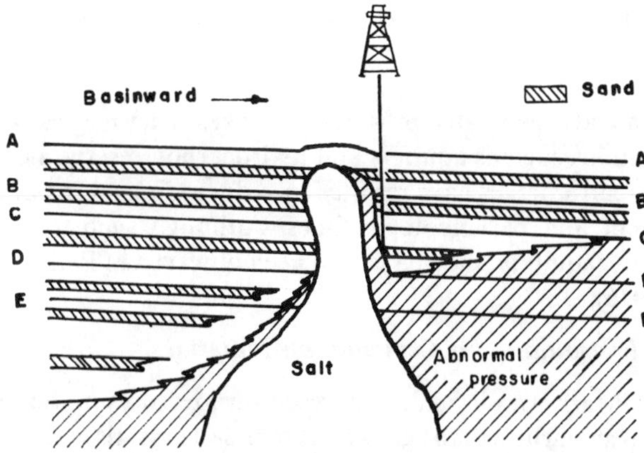

Fig. 5.7: Piercement salt dome showing modification of abnormal pressure surface.

Overpressures zones often occur due to the faulting and folding action associated with diapirism. Furthermore, if salt beds are deposited directly on top of clay, water within the unconsolidated sediments cannot escape through the impermeable salt. As more pressure is exerted due to additional deposition, the clay/shale below the salt becomes overpressured. The water in the clay/shale then carries the entire burden of the salt and any sediment above. Pressure in the formation below the salt is equal to pressure exerted by the total overburden.

Diapiric movement of salt is often accompanied or preceded by diapiric shale movements (Atwater and Forman, 1959). Salt movement results in contemporaneous faulting over intermediate and high relief salt anticlines (Hughes, 1968). The importance of overburden in the generation and continuation of such faulting with local structures in U.S. Gulf Coast area has been emphasised by Hardin and Hardin (1961). Halbouty (1967) has stated that such salt domes are encountered in the U.S. Gulf of Mexico basin during the course of drilling for hydrocarbous.

Salt domes and their association with overpressured zones have been noticed on a worldwide basis, both onshore and offshore.

(8) *Sandstone Dikes*

Sand dikes like shale diapirism are an overpressure phenomenon. Localised loading of sand in relatively mobile conditions causes high pore fluid pressures which are equilibrated by formation of sand dikes. This phenomenon is caused by the loading characteristics of migrating dunes that move over impermeable clayey surfaces, which locally affect pore fluid pressures in underlying water sands. This results in small sand dikes intruding from below into the clayey formation.

Sedimentary beds under lava flows might show extreme internal deformation since lava flow represents an almost classic case of rapid and progressive loading (Gretener, 1969). This is confirmed by field observations (Walton and O'Sultivan, 1950) of sadstone dikes intruding a sill. As the driving mechanism, the authors refer to abnormally high pore fluid pressures caused by the heating of formation water.

(9) *Earthquakes*

Overpressures are also generated by the earthquakes. They may cause compression in subsurface formations such as aquifers and hydrocarbon reservoirs. Such geological phenomenon like earthquakes have resulted in sudden changes in pore fluid pressures throughout present and past geologic times. Although such forces may be more common in shallow aquifers, they have also been observed in Petroleum reservoirs at greater (Fertl, 1976).

(10) *Secondary Precipitation of Cementation Material*

In the Russian Markova oil field, high reservoir pressures in the sediments can be explained by the peculiarities of the growth of rock salt crystals in the more space of the lithified rock skelton. Salt crystals grew at the base rather than at the uper end, gradually filling the voids in the lower part of the deposits, displacing oil. In view of the negligible compressibility of oil, the rock pressure in the closed deposits gradually increased as the volume of voids decreased, reaching abnormally high values. This process was verified by simple calculation of the possible reduction of pore volume under the action of penetrating salt. High rock pressures in the salt bearing carbonate and sulphate carbonate deposits in other fields can be explained by the same processes. Similar increase in pore fluid pressure may also take place due to calcite or silica precipitation.

(11) *Artesian Pressures*

Abnormal pressures are caused by aquifers when a porous horizon outcrops at a much higher elevation than where it is drilled. The effect is, however, usually less in magnitude than that due to a deeply buried permeable lenses (Fig. 5.8).

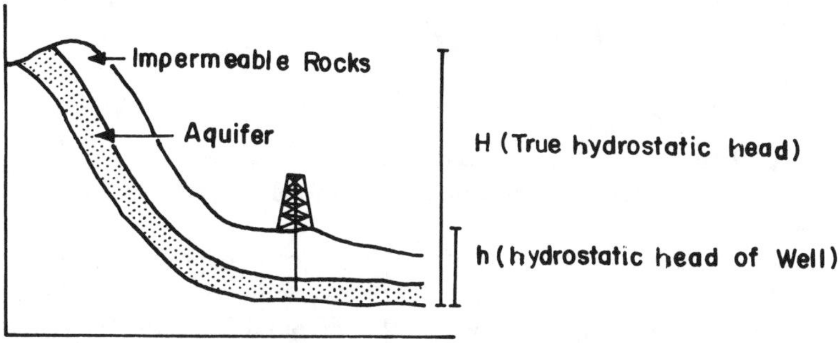

Fig. 5.8: Abnormal pressures caused by Aquifers.

(12) *Hydrocarbon Reservoirs*

The difference in the density between gas and water causes abnormal pressures where hydrocarbon accumulations occur above water irrespective of whether the water is at a normal or abnormal pressure. The magnitude of this pressure depends upon the structural elevation. This usually occurs in lenticular reservoirs, steeply dipping formations and anticlines.

In some areas in the Middle East, such overpressures resulting from density differences in the oil/water system, particularly in gas/water system, are known to approach geostatic pressures of overburden.

(13) *Uplifting*

Formations normally compacted at depth can be uplifted to a shallower depth. In case the original pressures within the formations are retained, such uplift can result in abnormal pressures (Fig. 5.9) provided it is accompanied by other geological processes which reduce the relief between the buried rock and the surface. The magnitude of the pressure is a function of the depth of burial and the degree of uplift. The older the sediments the greater is the possibility that excess reservoir pressures would have dissipated with time through microfissures, cracks., etc. Abnormal pressures are more prevalent and severe in younger rocks than older rocks.

(14) *Thermodynamic and Geochemical Causes*

In areas of high geothermal gradients, e.g. South Texas in the U.S.A., the increase in volume of pore fluid can result in high pressures. These processes may be considered an integral part of the combined effect with other causes which have been discussed earlier. As the system is essentially closed, these factors are incremental addition to the build-up of abnormally high formation pressures. Changes in the formation

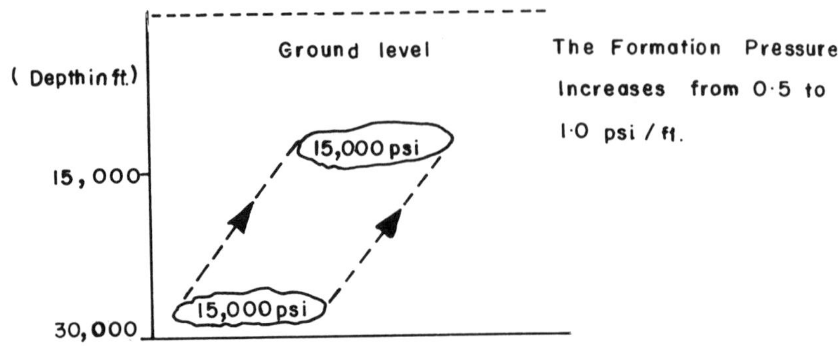

Fig. 5.9: Abnormal pressures caused by uplift.

temperature will also result in change of fluid pressure, either locally or regionally. The pressure-temperature diagram indicates that pressure in an isolated volume increases with rise in temperature more rapidly than the surrounding fluids (Barker, 1972). For higher gradients, small increase in depth is required to cause the same increase in pressure.

Decomposition of organic matter through bacterial action forms pockets of methane under high pressure. This may be observed in drilling shallow formation as well as the wellknown gas seepages in swamps or along beaches.

(15) *Thermal Cracking of Hydrocarbons*

Breakdown of hydrocarbon molecules into simple compounds increases their volume under the effects of temperature and pressure. The volume increase is in the order of 2 or 3 times the original volume and thus causes abnormally high pressures.

(16) *Mud Volcanoes*

Mud volcanoes are overpressure phenomenons caused by the intrusion from depth of mud or a mixture of mud and a more solid rock. As in the case associated with salt diapirism, it refers to the upward movement of a less dense plastic material, e.g. shale (Fig. 5.10).

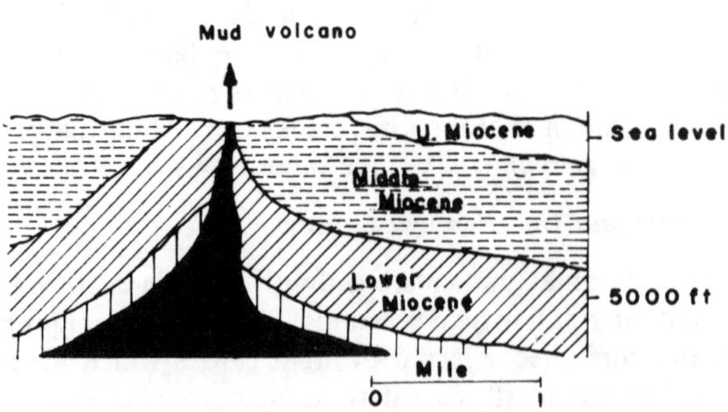

Fig. 5.10: Schematic concept of a mud volcano.

Throughout the world, wherever mud volcanoes are found, there has been rapid Tertiary and or late Cretaceous sedimentation with pore fluid pressures being abnormally high. Pressure gradients of 0.9 psi/ft. (0.208 $kg/cm^2/m$) have been encountered around mud volcanoes in some areas of the U.S.S.R. It is unusual in geology to find a phenomenon associated with rocks of only one particular age. However, geological time is a factor in the escape of abnormally high pore fluid pressure (Ridd, 1970). Nevertheless, mud volcanoes may have erupted in the geological past in association with older sediments, their remains being fossilized. Quiescent mud volcano activity is also believed to be due to normal fluid pressure, except that instead of gradual pressure build-up until eruption, mud gently escapes the surface through fissures. It is postulated that subsurface gas under tremendous pressure is responsible for the driving mechanism responsible for mud volcanoes (Kugler, 1933, 1938; Wilson and Birchwood, 1965; Gansser 1960) listed several criteria common to mud volcanoes:

(1) Generally associated with Teritiary and upper Cretaceous sedimentary strata.
(2) Sedimentary strata are usually of marine origin.
(3) Plastic, pelitic beds predominate.
(4) Gas and connate salt water are invariably present.
(5) Plastic beds are overlain by more competent deposits.
(6) Broad synclines are separated by sharp anticlines in which the deeper plastic sediments push upwards.
(7) Increasing stress mobilizes the plastic clay in the core with salt water, gas, and in many places, oil. The resulting mud is pressed upwards in magma-like fashion and, if the equilibrium of the surface is disturbed, it erupts and forms a mud volcano.
(8) Most eruptive centers consist of several volcanic cones.
(9) Shallow and steep-sided cones can be present together.
(10) The eruptions are in part periodic, but commonly irregular. Many large eruptions have occurred after long periods of quiescence.
(11) Small and large rock fragments are commonly present in the mud, usually originating from older strata.
(12) The life of an individual eruptive center is usually short.
(13) Diapiric zones with mud volcanoes generally coincide with areas of negative gravity anomaly.

(17) *Permafrost Environment*

In the Arctic region, abnormally high formation pressures do occur as a local phenomenon. In permafrost regions, unfrozen areas exist in many places such as under deep lakes. Drastic changes in climate and/or surface conditions cause permafrost encroachment, thereby trapping an unfrozen area in an essentially closed system. As freezing proceeds, build-up of abnormally high formation pressures occur in unfrozen pockets. Due to the proximity to the surface, such pressures may be released, thereby forming giant frost heaves at ground level (Fig. 5.11).

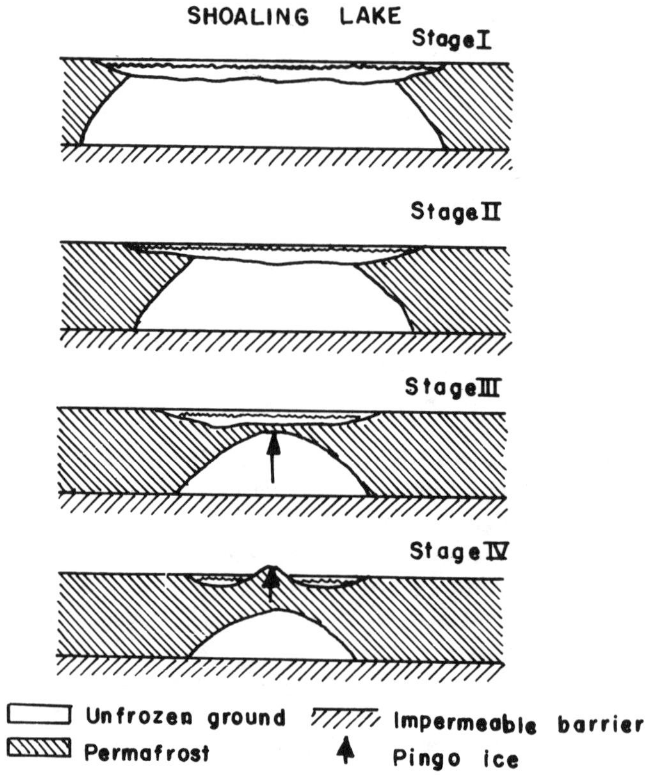

Fig. 5.11: Abnormal pressure conditions in Permafrost area formation of Pingos.

(18) *Other Causes*

Fertl (1976) mentions several other causes responsible for overpressures. Tidal disturbances of shelf regions in oceans, seismic ways caused by sub-marine earthquakes and effects of hurricane may also cause minor, probably innumerable, temporary changes in formations and underlying sediments. As early as 1753, travellers noticed subsurface effects of varying atmospheric pressure in the oil seepage region of the Caspian Sea in Russia. "When the weather is thick and heavy (low pressure), the springs boil up higher and the naphtha often takes fire at the surface". In clearer weather (high atmospheric pressure), the springs do not boil up over 2-3 ft. (0.6-0.91 mts.) (Hanway, 1753).

Although these effects on pore fluid pressures are of minor importance, nevertheless they are present and cause abnormally high formation pressure, either for a short or prolonged period of time.

REFERENCES

Allen, D.R., 1972 Environmental aspects of oil producing operations, Journal of Petroleum Technology.

Fertl, W.H., 1976 Abnormal formation pressures. Elsevier scientific publishing company.

Fertl, W.H., and Timko; 1972 How downhole temperatures, pressures affect drilling.

Kraynine, D.P., 1972 and Judd, W.R., Principle of engineering geology & Geotectonics.

Russell, W.L., 1972 Pressure depth relationships in Appalachian region, Bulletin American Association of Petroleum Geologists.

Sahay, B., 1971 Indication of Abnormal subsurface Pressures on resistivity logs of wells drilled in Bengal basin, ONGC Bulletin.

Sahay B., 1972 Abnormal subsurface pressures, their origin and methods employed for production, society of Petroleum engineering of AIME, Louisiana, USA.

Sahay, B., 1974 Analytical Studies on Abnormal pressures in Indian basins. (unpublished manual).

Sahay, B., 1976 Geopressure, their origin, utility and methods employed for prediction in India. Journal of mining & Metallurgical Society of India.

Sahay, B., et. al. 1983 Formation Evaluation and Wellsite Geological Techniques.

CHAPTER VI

Prediction of Over-Pressures Prior to Drilling

Introduction

Accurate prediction of pore pressures and fracture gradient are essential to the drilling of deep wells. If pore pressures are higher than the normal, their correct prediction not only reduces drilling costs substantially but also prevents many hazards which may result in well-kicks, blow-outs and other potential well complications, which otherwise may lead to the loss or abandonment of the well.

Advantages of proper detection and evaluation of abnormal high pressures include:

1. More effective overall well programming.
2. Maximum penetration rates under controlled drilling mud conditions.
3. Safe and economical selection of casing points.
4. Properly engineered equipment to minimise potential troubles due to circulation loss, kicks, pipe sticking etc.
5. Better understanding of local geology and drilling programmes.
6. Minimising the formation damage.
7. Maintaining high penetration rate.

Prediction

Various techniques have been developed all over the world to predict the overpressure horizon. These methods broadly fall into three categories.

1. Prior to drilling
2. During drilling
3. After drilling

This chapter deals with only such techniques which are concerned prior to drilling in over-pressured areas or into overpressured formations.

Prior to Drilling

(i) Geological correlation of similar areas
(ii) Drilling and mud reports

(iii) General and well-bore seismic data analysis
(iv) Other geo-physical methods

(i) *Geological Correlation*

Overpressured zones are often delineated and thus can be correlated from offset wells. Correlation with the nearby wells thus may (but does not necessarily will) help in the prediction of occurrence of the overpressure zones in a particular well.

(ii) *Drilling and Mud Reports*

These are the studies from offset well in the area to be drilled. Mud weights give a good indication of both the presence and magnitude of the pressures. Any problems associated with well kicks, circulation loss and differential sticking can thus be reviewed and analysed. However, in areas of swelling (heaving, sloughing) clays, mud weights as much as 2 ppg over the actual pore pressure have been used to control such formation problems.

(iii) *General and Well Bore Seismic Data Analysis*

Over the years, numerous developments and refinements in the acquisition, processing and interpretation of geophysical data, such as seismic, gravity, have made it possible for the oil industry to study not only the structural configuration of formations in sedimentary basin, but proper application and interpretation of quality seismic records also allow the investigation of several other parameters, even though attempts to isolate and measure the influence of individual parameters may not be always successful. Such applications are broadly divided into:

(1) Determination of interval velocities
(2) Study of lithological and stratigraphical variations in the geological time scale.
(3) Estimation of geological age and/or average geothermal gradients.
(4) Study of the effects of lateral folding pressure on a regional scale.
(5) Hydrocarbon recognition, especially gas at shallow depth. (Bright and hot spots). Fig. 6.1 shows a typical bright spot.
(6) Detection of natural and/or artificially caused hydrocarbon see pages from the ocean floor.
(7) Investigation of ocean floor, buried ancient glacial channels, etc, which are a definite hazard in the offshore prospecting for oil and gas. Fig. 6.2 shows an ancient glacial channel.
(8) Detection of the presence of abnormal pressure formations and evaluation of the magnitude of such pressures.

This chapter discusses the various applications of geophysical prospecting, particularly, seimic and and gravity, to determine lithological tops and both depth and magnitude of abnormal formation pressure, prior to spudding a well. It is well known that these concepts have a very great impact for planning a wildcat well in order to devise suitable mud casing and drilling policies, so vital for cost factors for exploration in deep water and/or in regions of hostile geological environments.

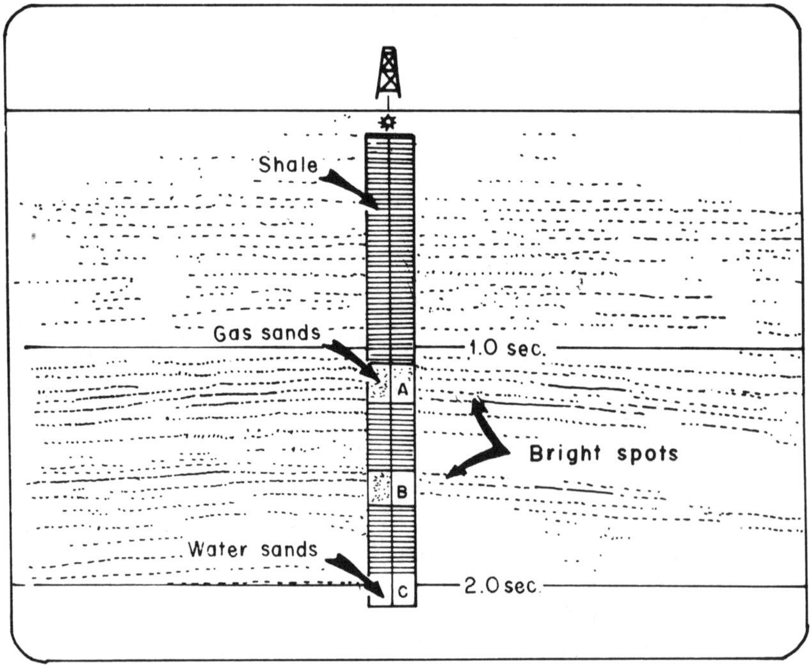

Fig. 6.1: Bright spots show shallow gas zones

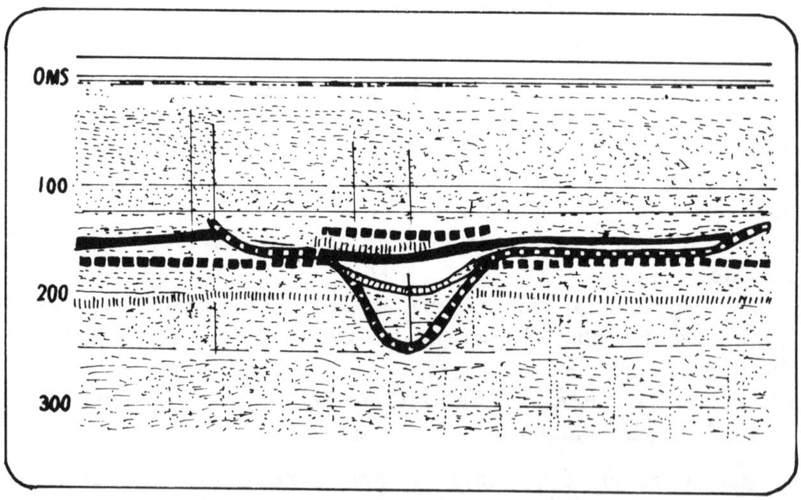

Fig. 6.2: N. sea sparker section shows glaciation

Seismic Data Analysis

This technique focuses on normal compaction with depth. Departures from the normal compaction trend, such as reduction in the velocity of sound waves, assists in the detection and quantitative evaluation of abnormal pore pressures.

Determination of formation velocities from seismic data has been known and utilized for many years based on several methods using high speed digital data handling, (Amenson, Darby, 1969; Yancey, et al 1969, Davis 1973). All these methods are based on elementary reflection analysis as reviewed by Pennebaker (1968).

Since interval velocity, a term used by geophysicists, is the reciprocal of interval travel time as used by the geologists, well log analysis and drilling engineers, one can easily replot velocity data as travel time profiles. These result in an "acoustic log type" presentation in which data are averaged over rather long sections and not as in the case of wireline acoustic logs over a few feet of vertical distance. The most important application of such velocity analysis is the prediction of lithology and formation tops before commencing exploratory drilling in an unknown area (Fig. 6.3). It is a well-known fact that such data is extremely important and greatly assists in properly engineered well-planning. Such predictions can be carried out in both soft rock sand and hard rock carbonate sequences. In the United States, Woeber and Penhollow

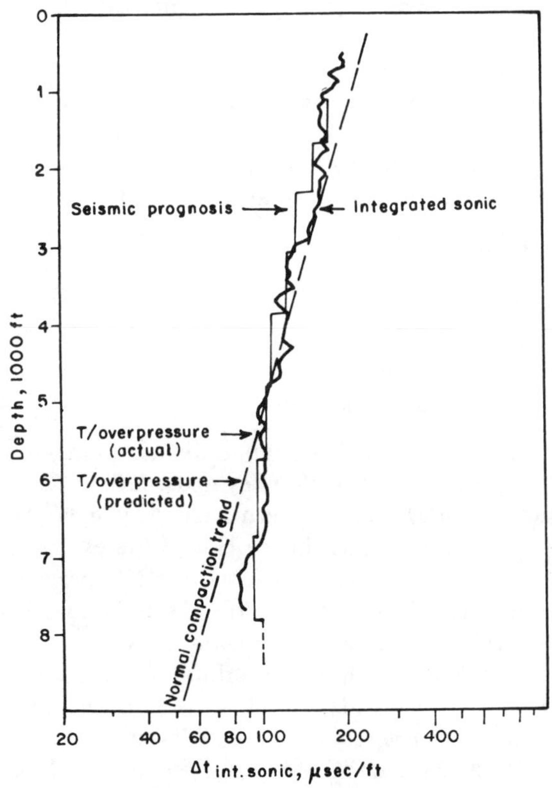

Fig. 6.3: Comparison of seismically derived travel time and actual velocity data in a well (After Reynolds, 1973. Courtesy of the Society of Petroleum Engineers of A.I.M.E.)

(1973) have carried out a stratigraphical study of the accuracy and formation top predictions from seismic velocity profiles in shale-sand sequences. In 80% of the cases, an accuracy of 1—2% was found over the depth range between 1524 mts to 3600 mts. However, accuracy falls off rapidly below 3660 mts. while only 50% of the cases will show accuracy of 2 per cent or more below 4573 mts. (Fertl, 1976).

Applications in the deeper carbonate sections of the andarko basin, Oklahoma has been reported by Louden et al (1971). In specific circumstances, the internal velocity analysis can be used to predict the presence, depth and magnitude of abnormal formation pressures.

Basically, in sand shale sections the interval velocity decreases with depth (West, 1950, Kaufman, 1953, Dix 1955, Sarmento, 1961). Trend lines representative of normal compaction (i.e. hydrostatic formation pressures) can be established. Since abnormally high pressures exhibit higher than normal porosity at a given depth, this is reflected in a travel time increase (i.e. divergence from normal to higher values) in overpressured formations. Degree of departure from the normal trend line indicates the magnitude of the pressure. For quantitative pore pressure evaluation, empirical correlation of the velocity ratio (Pennebaker, 1968) or travel time versus expected pressure gradient and/or mud weight requirements can be used (Hottman and Hohnson, 1965).

The magnitude of the pore pressure can be estimated by comparing the transit time departure at any depth with properly established calibration curves. If no other well test data is available in a rank wildcat area, then rough estimates of pressure magnitude can be made by assuming that the rock matrix stress in an overpressured zone is the same as at the depth on the normal trend at which transit times are the same.

$$PB = SB\ (DB) - CA\ (DA)$$

where

PB = pressure at depth (DB)
SB = Overburden stress gradient at DB
CA = matrix stress gradient at depth DA

This expression has worked well for approximate pressure estimates in rank exploratory basins throughout the world (Reynolds, 1973).

Figure 6.4 shows the various analysis (seismic and logging data) for predicting pore pressures in a rank wildcate, located in the United States. As can be seen in this example, the travel time decrease with depth in the normally pressured section whereas overpressures as designated by the increase in travel time below 3354 mts. As a matter of fact, geophysicists have for sometime been aware of these anomalies which show up on seismic records as regions of rather poor reflections. Such velocity inversions also often coincide particularly with massive shale section and/or large salt masses. Musgrave and Hicks (1966) proposed the use of seismic data supplemented by the gravity information to distinguish between shale section & salt masses.

It is needless to emphaise that in order to interpret the deviations in interval velocities from the normal compaction trend as the function of overpressures, a

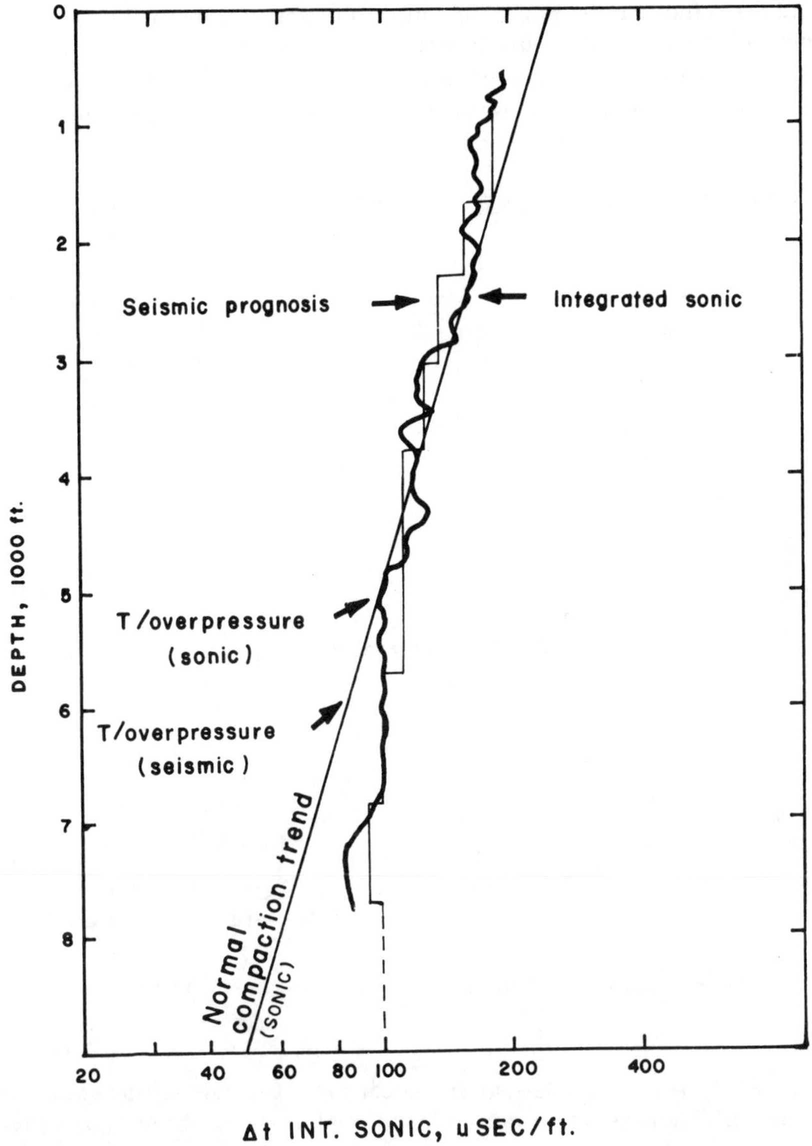

Fig. 6.4: Comparison of seismically derived transit travel time and actual velocity data in well (Courtesy : Gearhart Geodata Services).

thorough knowledge of the lithology of the area is of the utmost importance. In other words, lack of information about major variations in lithology and/or geological age could cause erroneous interpretations. With a proper interpretation of the seismic data, it is possible to predict the top of overpressures within a range of ±550 and pore pressure magnitude can be estimated within±1.01 lbs/gal. (Fertl, 1976).

Figure 6.5 shows overpressure prediction from seismic analysis in a deep well, located in Oklahoma, in comparison with determination from well logging information and the actual mud weights required in the drilling operations. Excellent agreement can be observed.

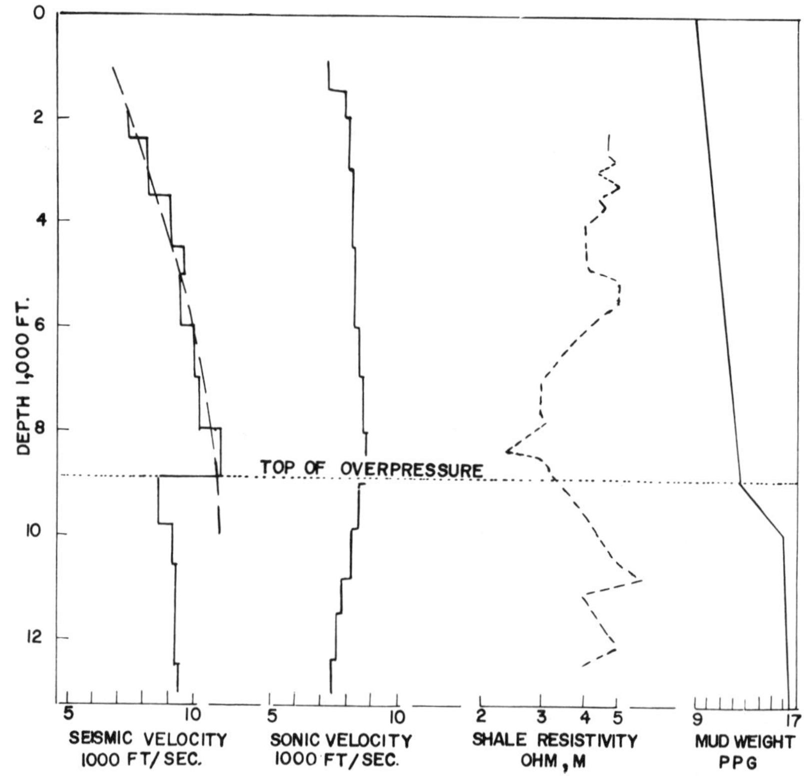

Fig. 6.5: Comparison of predicted top of over pressures with actual well data

With the help of seismic data analysis one can even go one step ahead and attempt to predict the drilling penetration rates prior to spudding-in of the well, based on such formation pressure and lighological data.

In offshore India in Bengal basin on the eastern end of the country, the Natomas Carlsburg group attempted to evaluate interval velocities around well BB-A-1, to see if high pressure zones, if any, could be predicted from the data. The correlation is poor in spite of the fact that the quality of the offshore CDP seismic data of Bay of Bengal area, having 24 fold multiplicity, was very good (Dutta, 1976).

However, in carbonate areas, particularly in Bombay offshore basin, which

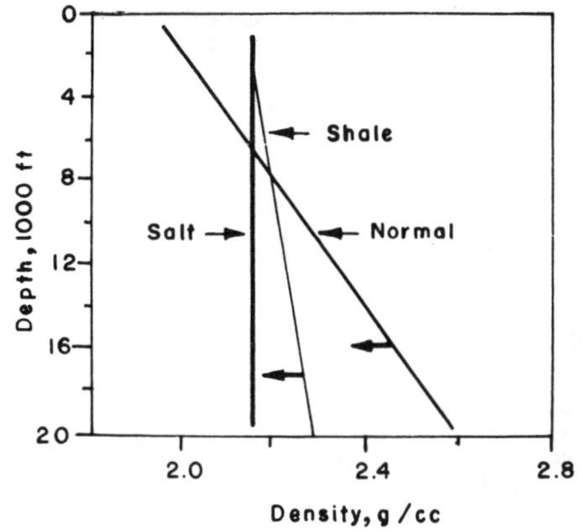

Fig. 6.6: Apparent density control of salt and shale masses; normal hydrostatic section and decreasing slopes in overpressured geological shale/sand sections.

comprises mostly shale and limestone, it has not been possible to predict the occurrence of abnormal high pressures before drilling from seismic section. It is well known that interpretation in the carbonate section can only be speculative because velocities are so high that variations caused by pressures are below the range of resolution of the technique. Hence, forecasting of high pressures prior to drilling in areas, such as Bengal Offshore, India, proves rather difficult.

Inherent constraints in the application of seismic data analysis includes the following conditions.

(a) Highly dipping formation (Levin, 1969)
(b) Normal move out error (Shugart, 1969)
(c) Faulting
(d) Multiple reflections
(e) Static errors (Yancey et al, 1969)
(f) Curved ray path (Dix, 1955)
(g) Anisotropy
(h) Interpretation (Amenson & Darby, 1969)

(iv) *Gravity Data Analysis*

As overpressured zones are characterized by the low bulk desnsities, gravity data thus could be another useful parameter for detection of over pressures (Musgrave and Hicks, 1966). Figure 6.6 shows the apparent density contrast for salt shale and normal hydrostatic shale/sand sequences. However, differentiation of highly pressured shale and salt by gravity data alone may be difficult.

Gravity data thus have to be supplemented with other geophysical surveys in the detection and evaluation of overpressure environments and cannot be used alone for such studies.

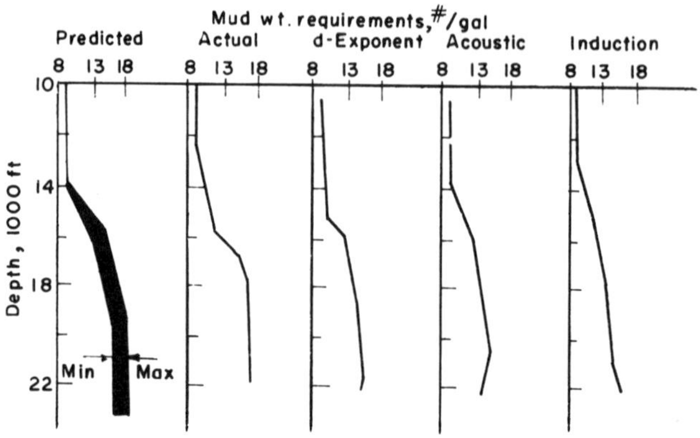

Fig. 6.7: Actual mud weights compared with seismically derived data, values determined from drilling parameters (d-exponent, for description see Chapter 4), and data from acoustic (sonic) and induction logs.

(After Louden et al., 1971. Courtesy of World Oil)

(v) *Other Geophysical Methods*

Besides seismic and gravity analysis generally used for prediction of overpressured horizons, other surveys, like magnetic and possible electrical prospecting methods, could also be included provided the investigation is focussed on a rather shallow depth for prediction of overpressured zones.

(vi) *Wellsite Seismic — Vertical Seismic Profiling (VSP)*

In addition to well logging, seismic and other geophysical measurements, the vertical seismic profiling (VSP) survey provides additional valuable information. Whereas conventional seismeic data are collected with seismic sources and geophones all located on the surface, the VSP survey can be obtained by lowering a geophone down the wellbore, clamping it to the borehole wall, and recording the signal originated by a surface seismic source. The latter includes air-and-water guns, explosives, vibrator (P and S) and weight drop.

While VSP surveys are similar to check-shot surveys, they differ in that the sequential wellbore geophone depth station spacing is usually much closer (20 ft to 100 ft) for VSP applications. Furthermore, whereas check-shots only measure the first break or direct arrival time, VSP surveys not only record direct arrival but also the subsequent events corresponding to energy which reaches the geophone after being simply (i.e. primary went) or multiply (i.e. multiple went) reflected from interfaces in the subsurface. VSP's can be recorded in open and/or cased boreholes and provide good quality data because geophones are in quiet environments, and seismic signals have to pass only once through the so called weathered layer of loose surface rocks and soil, through which traditional seismic energy must pass twice.

While Russian investigators have been actively developing VSP techniques for the last two decades, interest in this method did not pick up in the U.S. until the last few years.

Several types of information can be obtained from VSP, including exact correlation of seismic and well log data, identification of multiple reflections, investigation of effects of attenuation, obtaining usable seismic data in poor signal areas, integrated geophysical interpretation, prediction of subsurface structure, accurate measurement of velocity which allows the calibration of acoustic (sonic) log with certainty (cycle skipping, environmental effects from drilling muds and borehole washouts), estimate of reflection coefficients and consequently acoustic impedence (i.e. velocity times density), mechanical coefficient values (Lame and Poisson's ratio) of geological formations where P and S waves are recorded in the VSP process. Most important to drilling operations, travel times between the bottom of the wellbore and underlying seismic horizon can be determined. This allows "seeing" ahead (beneath) the drill bit for onside decisions in drilling operations. Looking ahead of the drill bit provides three valuable main informations, (1) determining depth to reflectors (formation tops) not yet drilled. (2) compaction curves which yield overpressure information and mud weight requirements in strata not yet drilled, and (3) hardness (penetration rate) of rocks not yet drilled.

CHAPTER VII

Prediction of Overpressures During Drilling

Introduction

Various theories with regard to the causes of overpressures have already been discussed in the preceding chapters. In sand/shale sequences, overpressure environments may be contained by a single cap rock or a sequence of several sealing barriers which may vary greatly in thickness. Hence single or multiple overpressure zones may be present which could be from a few to thousand metres thick. Overpressures due to the tectonic phenomena and faults will further complicate the situation as these may or may not act as overpressure discontinuities. In addition, changes in lithology, such as increasing lime or silt content of the shales, will affect most pressure detection techniques. Problems as mentioned above or similar anomalies vary with local geological conditions, types of the drilling rig and equipment, mud and drilling techniques, capabilities of rig personnel, the condition of the bore nole, whether straight or deviated, etc.

In any case, the objective is to balance the danger of lost circulation and differential pressure sticking by proper mud weights which should be heavy enough to contain any abnormally high pressures encountered in the section as well as light enough to prevent circulation loss problems. All pressure detection concepts are important in drilling and proper mud and casing programs are major cost factors. In order to have an effective and economical mud program, the fracture gradient as well as pore fluid pressure variations need to be clearly understood. If the pressure exerted by the drilling fluid exceeds the strength limit of the formation, mud loss occurs, resulting in the subsequent drop of the hydrostatic pressure head in the well bore, thus causing well kicks or potential blow out conditions.

In oil well drilling, data monitoring on drilling rigs range from a simple standard geologging unit to highly sophisticated digital data collection and analysis systems, both offline geologging and real time online geologging. The rapid advancement in the drilling technology has likewise led to significant developments in the pressure detection techniques, as well as down hole telemetry system are being used for monitoring of the drilling conditions. Today's monitoring equipment is used to keep out of trouble and to reduce the overall cost of drilling operations. Drilling hazards include more than just blow outs and also involve stuck pipe, loss circulation, twisted off pipe and other costly problems.

All the modern geologging technology assist in the accurate monitoring of drilling data untouched by human hands. Important drilling parameters, such as flow rate, pit volume, mud weight, temperature in and out, pipe running speed, torque, weight on bit, etc, can trigger automatic alarms to signal impending danger. The data which is recorded on charts, digital print out, magnetic tapes, etc. are transmitted to headquarters computers on a real time or intermittent basis. It can be manipulated to up-date the drilling program and enters the vast data base to plan future wells. With proper understanding of the relationships involved and the equipment used, the data collected can be used to generate optimum drilling programs, pore pressure estimates at the bit and logs during the course of drilling.

Drilling Parameters

In the past, drilling rate was basically used to distinguish the various lithologies, such as sand, shale, limestone, etc. Changes in penetration rate due to variations in pore fluid pressure has been recognized in the last decade. As is well known, the drilling rate is a function of weight on bit, rotary speed, bit type and size, hydraulics, drilling fluid and formation characteristics (Garnier & Van Lingen, 1959, Somerton, 1959, Maurer, 1962). Under conditions of constant weight on bit, rotary speed, bit types, and hydraulics, the penetration rate in shale decreases uniformly with depth, due to compaction of shales in response to the over burden stress. However, in the case of abnormal pressure zones encountered in the subsurface, the penetration rate will increase in the transition zone. Slower penetration rate is often observed in the cap rock (pressure barrier), which is above the high pressure transition zones. Any variations in lithology, such as silt, sand, limestone, etc., affects the penetration rate.

It is always advisable to establish a normal drilling rate trend while drilling through a normal pressure shale section in order to compare with the section of faster drilling overpressured shales. According to Forgoston (1969), an increase in the drilling rate of shale in excess of twice the normal rate indicates the proximity of overpressure environments. Observed penetration rate increases are due to reduced differential pressure between formation and the mud. Complications can arise due to bit dullness, which can mask the penetration rate variations due to overpressures. A slight fluctuation in the rotary torque can also cause a decrease in the drilling rate.

'd' Exponent

As it is not always possible or feasible to use constant bit weight and rotary speed, an improved method has been developed which facilitates plotting of a normalized penetration rate ('d' exponent) vs. depth (Fig. 7.1).

In 1965, Bingham showed the following relations between several drilling variables:

$$ROP/RPM = K\,(WOB/Dh)^{d}$$

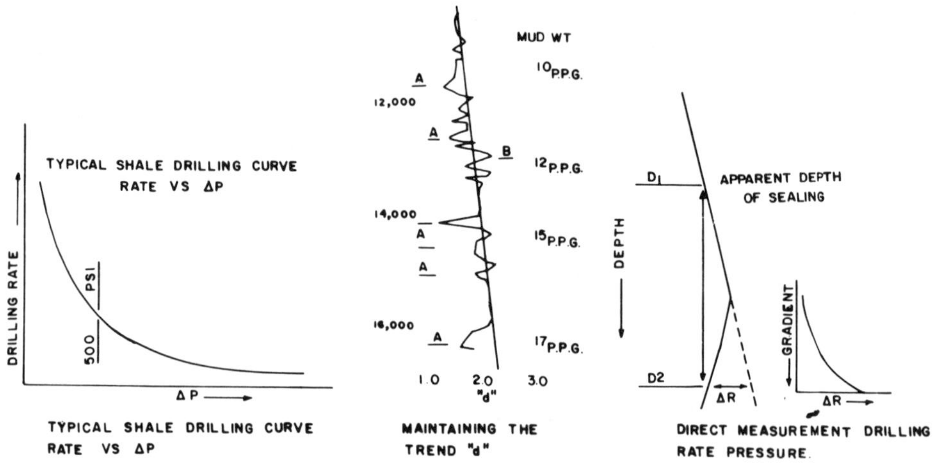

Fig. 7.1: 'D' Exponent Study

(After Bill Rehm & Ray Meclendon Dresser, Swaco, Courtesy, SPE)

Jorden and Shirley (1966) while trying to normalize for the effects of changes in weight on bit, rotary speed and bit diameter on drilling rate, found an excellent correlation between value of 'd' and differential pressure based on drilling data from the U.S. Gulf Coast area. Based on their findings, they proposed the following equation:

$$d = \frac{\log (R/60\ N)}{\log (12^{w}/10^{6}D)}$$

where

R = rate of penetration
N = rotary speed, rotations per minute (RPM)
W = weight on bit in tonnes
D = bit size in cm.
d = 'd' exponent, related to differential pressure between mud column and formation pressure

Nomographs have been developed to solve the above equation easily and quickly (Fig. 7.2). It is recommended to plot 'd' exponent at every 5 mts interval. However, in case of fast drilling the plotting of 'd' exponent should be at every 1 m or as necessary. In order to have a better comparison with other pressure prediction parameters, it is recommended to use the same scale for 'd' exponent plots also.

Quantitative pressure evaluation can then be made using the equivalent depth methods, such as

$$Pf = DA - 0.535\ DE$$

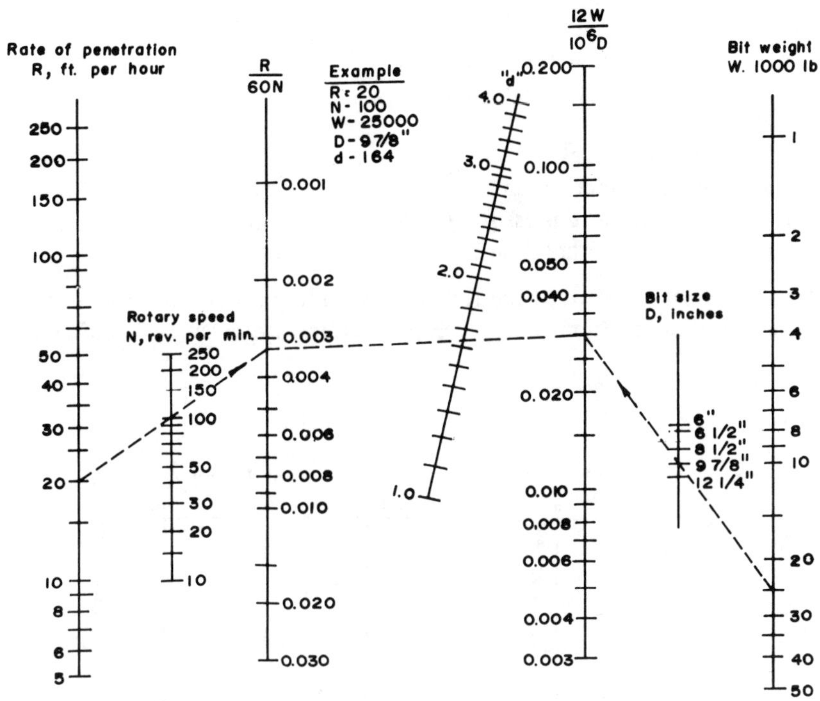

Fig. 7.2: Nomogram for d-exponent determination.
(After Jordon and Shirley, 1966. Courtesy of Journal of Petroleum Technology)

where

Pf = pore fluid pressure in psi
DA = depth of interest in overpressured interval
DE = normal equivalent depth

The equation is valid in areas where the overburden pressure gradient equals 1.0 psi/ft (2.31 kg/cm^2) and hydrostatic gradient equals 0.465 psi/ft (0.107 kg/cm^2).

Since the value of 'd' is affected by any change in the basic input parameters like bit size, weight on bit, type, mud weight, and hydraulics, it was considered necessary to take that into consideration. Hence, Rehm and Maclendon (1971) proposed a correction method to eliminate the effects.

Modified 'd' exponent (dcs): Modified 'd' exponent (dcs) takes into consideration the effects of the effective mud weight as well as the pore pressure gradient. The resulting equation is expressed as:

$$\text{dcs} = \frac{\dfrac{\text{Log (R)}}{(18.19\text{ N})}}{\dfrac{\text{Log (W)}}{(14.9\text{ B})}} \times \frac{\text{Weq}}{\text{ECD}}$$

where

dcs = Corrected 'd' exponent
R = ROP (rate of penetration)
N = RPM
W = Wt. in tonnes
B = Bit diameter in cm.
Weq = Normal pore pressure gradient
ECD = Effective circulating density (Mud weight)

Geoservices (1982) improved the formula by adding a new term to correct for bit wear. The resulting equation thus developed can be expressed at (Fig. 7.3).

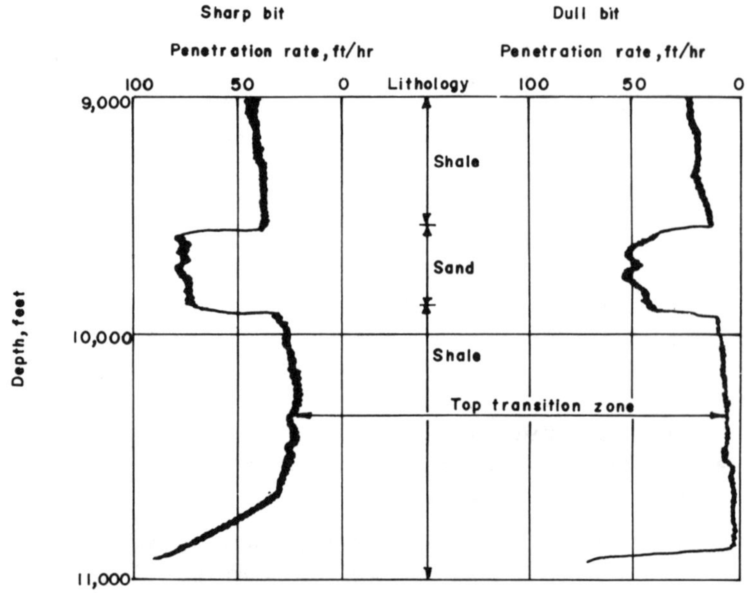

Fig. 7.3: Effect of bit-tooth dullness on penetration rate in pressure transition zone. (After Mathews, 1969. Courtesy Oil & Gass Journal)

$$dcs = \frac{\log\left(\dfrac{ROP}{60\ RPM}.ap\right)}{\log\left(\dfrac{12.\ WOB}{10^6\ BS}\right)} \times \frac{H}{MW}$$

where

ap = bit wear (Tooth wear)
RPM = Rotary speed
ROP = Rate of penetration
WOB = Weight on bit
BS = Bit size
H = Normal hydrostatic gradient
MW = Mud weight

The 'a' term is a function of footage (drilling time) and final bit dullness, while the 'p' exponent varies between 0 and .5 according to the type of drilling bit. The 'a' term may be represented as follows:

$$a = .93\,T^2 + 6\,T + = 1$$

$$\text{With } T = DF\,\frac{(Z-1MO}{L}\;\frac{.31\,Df^2 + 3\,Df + 1}{.31\,Df^2\,\frac{(Z-1M)^2}{L} + 3\,DF\,\frac{(Z-1M)+1}{1}}$$

where

T = current bit dullness in fractional units
DF = final bit dullness in fractional units
1 M = bit run initial depth ft.
L = bit run length or footage
Z = current depth

Zamora (1972) has shown that if 'dcs' is computed in shale intervals (or dense impervious formations) and plotted v/s. depth on a semilog graph, the curve would have a general trend (i.e. a straight line) if the formations are normally pressured. However, in overpressured formations there would be a departure from the extrapolated normal trend line. Hence the formation pressure gradient can be determined as given by Eaton.

$$PF = S - (S - H)\,(dcs/dcn)^{1.2}$$

where

PF = formation pore pressure gradient
S = overburden gradient, lbs/gal
H = normal hydrostatic pressure gradient lbs/gal
dcs = corrected d-exponent
dcn = normal d-exponent extrapolated from trend line

The drawback of this method is that trend line and possible departure from it are known only when all data points have been plotted, i.e., after the well is drilled. In order to eliminate this constraint and to get the running data, a computer program can statistically estimate the best trend line while drilling is going on and to discriminate between new data in order to either improve the trend determination or to compute the overpressure.

This technique, however, does not take into consideration any sudden trend shift due to geological conditions like fault, unconformity, etc., or drilling conditions, poor bit efficiency, drilling practice, etc.

Micro Computer Based Techniques For Calculation of 'd' Exponent

Micro computers assist in 'd' exponent techniques, which estimate formation pressure at any given depth and provide graphical displays of the 'd' exponent versus. depth and the normal trend line involved. Micro computer and printer allow customary transparent overlays to be produced onsite as well. The required input data include penetration rate, RPM, weight on bit, and mud weight at any given depth. This digital method is easy to use and provides greater flexibility since overlays are not required. Mud weight information taken in short intervals is used to correct 'd' exponent which are normalized through Semilog, regression-type analysis (Xling Sun et al, 1984). The calculations make use of local data such as drillability, penetration rate vs. RPM coefficients and normal pressure values and, as such, are applicable in any part of the world. Acceptance of micro computer application in industry has been rapid.

Such digital 'd' exponent determination is quite simple and fast since only few input parameters, such as normal pore pressure gradient, drillability, rotary speed exponent, initial number of normal pressure point and a drilling data file are required. The four major steps used in this technique are:

1. Input drilling information from the data file
2. Calculation of the 'd' and 'dc's'
3. Conduct regression analysis
4. Generate and display final results

This technique can be used on several micro computer systems, e.g., IBM-PC, Apple, HP, etc.

Combs Type Equations

Based on empirical data, Combs (1968) proposed the following mathematical relationship:

$$V = f(\Delta P) = \frac{ROP}{R} \times \left(\frac{WOB}{3500\ Dh}\right) \times \frac{(RPM)}{200} \times \frac{(Q)}{3\ Dh\ Dn} \times \frac{1}{f(Mw) \times f(T)}$$

where

V = value of computed expression
f = function

P = differential pressure
ROP = rate of penetration in ft/hr
RPM = rotary speed
R = factor
WOB = weight on bit
Dh = bit diameter
Q = flow rate, gpm
Dn = nozzle diameter
T = current bit dullness in fractional units
MW = mud weight
ω, μ, σ = exponents

The attached figure is a field example of the use of an equation of this type, in order to compute Δ P. Pore pressure gradient is derived from the differential pressure by the following equation

$$Pf = MW - \frac{1000 \Delta P}{Z}$$

The "R" factor of the Combs equation is dependent upon the geological conditions in a fashion similar to the normal 'd' exponent computation (Fig. 7.4).

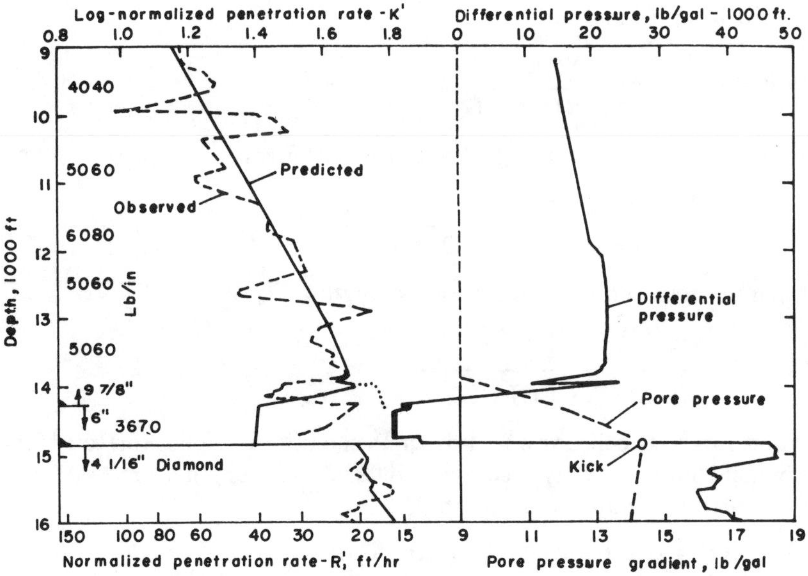

Fig. 7.4: Comparison of observed and predicted penetration rate
(after Combs, 1968. Courtesy of the Society of Petroleum Engineers A.I.M.E.)

The Sigma Log

In the early seventies, AGIP planned a deep drilling project in the PO valley, Northern Italy. Seismic studies of the area, though limited in depth, indicated that abnormal pressures could be expected in lithologies varying from carbonates to argillacaeous sands. Obviously then, the 'd' exponent would not be a satisfactory indicator of overpressure. The basic drilling equation was reworked and a new factor (SIGMA) was introducted which is depth related function of rock strength (Gerard, 1977):

$$ROP = \frac{\text{Depth}}{\sqrt{\sigma t}} (\text{RPM}) \left(\frac{\text{WOB}}{\text{Bit Size}} \right)$$

Where $\sqrt{\sigma t}$ is a function of rock strength. The latter varies due to lithology, compaction/porosity, and the formation pore pressure gradient (FPG).

Primary porosity, of course, is an implicit function of compaction, as compaction increases with depth, so porosity decreases. FPG as a hydrostatic state, does not affect rock strength. But fluid pressures in excess of the normal gradient is a result of some portion of the weight of overlying material being supported by the pore fluids. In this case, the rock will be undercompacted and its strength proportionately diminishes. Thus, if rock strength is plotted against depth under normal conditions, it can be expected to increase with depth.

The step by step presentation of the Sigma Log relationship is given next:

The primary drilling equation is

$$ROP = K (\text{RPM}) \frac{\text{WOB}}{\text{BIT SIZE}} \tag{1}$$

The introduction of a rock strength parameter, σt gives

$$ROP = \frac{\text{Depth}}{\sqrt{\sigma t}} (\text{RPM}) \left(\frac{(\text{WOB})}{\text{BIT SIZE}} \right) \tag{2}$$

Reworking the equation to monitor rock strength σt gives:

$$\sigma t = \frac{\text{WOB}^{0.5} \times \text{RPM}^{0.25}}{\text{BIT SIZE} \times \text{ROP}^{0.25}} \tag{3}$$

Equation 3 is the so called raw rock strength equation and is not corrected for depth — it is in fact considered only accurate at 7000 m. If the depth is greater or less than 7000 m, a correction must be made, thus:

$$\sqrt{\sigma t'} = \sqrt{\sigma t} + 0.028 (7\text{-Depth}/1000) \tag{4}$$

$\sqrt{\sigma t'}$ is simply $\sqrt{\sigma t}$ with a correction for depth that goes to zero at 7000 m. If all conditions were perfect and the drilling fluid used had exactly the same density as the pore fluid column, then it would be sufficient to plot $\sqrt{\sigma t'}$ against depth. This would give a straight line with a slope that reflects increasing compaction and thus rock strength. However, as all bore holes are drilled in over-or near-balanced conditions, and as overbalanced has the effect of slowing down, ROP then to plot $\sqrt{\sigma t}$, against depth without taking into account over balance will result in a curve difficult to interpret. The curve is corrected for overbalance by the following equation, which enables one to plot $\sqrt{\sigma_o}$:

$$\sqrt{\sigma_o} = F \sqrt{\sigma t'} \tag{5}$$

Where

$\sqrt{\sigma_o}$ = Sigmalog
F = Overbalance correction

$$F = 1 + \left(\frac{1 = \sqrt{1 + n^2 p^2}}{n \Delta p} \right) \tag{6}$$

The two unknown parameters are thus ΔP and n, where

ΔP = Differential pressure between mud column and formation pressure

$$\Delta P = (MW\text{-}NG) \left(\frac{DEPTH}{10} \right) \tag{7}$$

Where
MW = Mud weight in SG
NG = Normal Gradient
When plotting the $\sqrt{\sigma_o}$ curve, normal gradient is assumed throughout.

The second unknown is:

n = a function of the time required to equalize the differential pressure that exists between the cuttings and mud weight. This is dependent on the lithology and the porosity of the cuttings themselves, and is therefore reflected in the value of $\sqrt{\sigma t'}$.

if $\sqrt{\sigma t'} > 1$ (high bit weights and slow drilling)

$$n = \frac{1}{\sigma 40} \left(\frac{4\text{-}0.75}{\sqrt{\sigma t'}} \right) \tag{8}$$

if $\sqrt{\sigma t'} < 1$ (low bit weights and fast drilling)

$$n = \frac{3.2\ S}{\sigma 40} \sqrt{\sigma t'} \tag{9}$$

The effect of F is to alter the value of $\sqrt{\sigma t'}$ by a factor proportionate to the overbalance known to exist between the hydrostatic pressure of the formation fluid and the mud weight. A glance at the equation for ΔP (eq. 7) will show you that if mud weight is set to equal the normal gradient the term Δ P becomes zero and F=1 thus $1\sqrt{\sigma t'} = \sqrt{\sigma t'}$.

A Working Example of $\sqrt{\sigma_o}$ follows

Bit Size :	8.5	MW = 1.5 g/cc
WOB :	10 tons	Normal gradient = 1.0 g/cc
RPM :	100	
ROP :	10 m/hr	
Depth :	2000	

$$\sqrt{\sigma t} : \frac{10^{0.5} \times 100^{0.25}}{B.S. \times 10^{0.25}} = \frac{10}{15.11} = 0.66 \text{ (As per equation 3)}$$

$$\sqrt{\sigma t'} = (0.66) +).028\ (7 - 2000/1000) = 0.8 \text{ (As per equation 4)}$$

$$\text{because } \sqrt{\sigma t'} = 1$$

$$n = \frac{3.25}{\sigma 40} (0.8)\ n = 0.00634 \text{ (As per equation 9)}$$

$$\Delta P = (1.5 - 1) \frac{(2000)}{10} = 0.5\ (200) = 100 \text{ (As per equation 7)}$$

$$F = 1 + \frac{\sqrt{1 - 1 + (.00634)^2\ (100)^2}}{(0.00634)\ (100)} \text{(As per equation 6)}$$

$$F = 1 + \frac{(0.184)}{0.634} \quad F = 0.709$$

$$\sqrt{\sigma_o} = F\sqrt{\sigma t'} = (0.709)\ (0.8) = 0.562 \text{ (As per equation 5)}$$

The raw rock strength ($\sqrt{\sigma t'}$) would give a plotted value of 0.8. However, since a pressure difference exists between mud weight (1.5) and formation pressure (1) the true rock strength, taking into account this overbalance, is actually 0.709 and not 0.8. The next step is to use $\sqrt{\sigma_o}$ to get a quantitative measure of the formation pressure, the porosity and, ultimately, the fracture gradient.

With all the drilling parameters successfully normalized and overbalance being taken into consideration, the sigma log ($\sqrt{\sigma_o}$), when plotted against depth, exhibits a slightly slopping trace, the gradient reflecting compaction.

Fig 7.5 shows both the 'd' exponent, dcs and sigma log. Note the superiority of the sigma log over the dcs plot. Any deviation towards the left is indicative of overpressure, while deviation to the right of the trend line shows overcompaction, i.e. possible presence of cap rock or other lithology effects. The deviation of the formation pressure gradient information is exactly the same as it is for 'd' exponent.

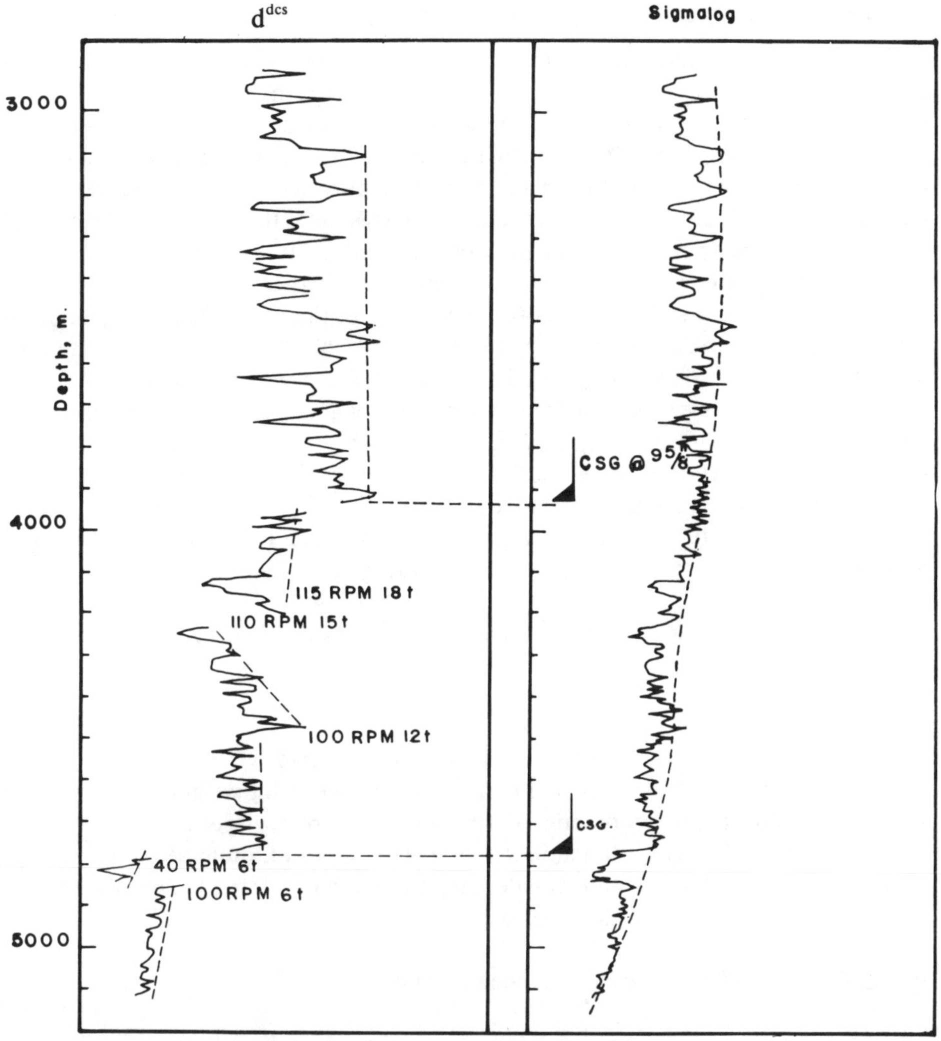

Fig. 7.5: d_{cs} curve is clearly affected by changes in hole sizes and drilling parameters. The sigmalag on the contrary, shows a continuous and regular development which facilitates overpressure detection.

(Courtesy: AGIP)

The significant characteristic of the sigma log is its ability to function as a valid indicator of formation pressure gradient in non-shaly lithologies, i.e. it can be applied in a sequence having silty or tuffaceous and other heterogeneous lithological environments.

Drilling Porosity logs

Several methods have been recently attempted to give a preliminary formation evaluation during the course of drilling a well, thus reducing and/or eliminating the need to pull pipe and running conventional logs. Based on computer models to describe, the relationships between formation characteristics and drilling parameters, it is possible to get an indication of formation porosity and pressure variations (Zoeller, 1970). A review of the method indicates that this can provide drilling optimization, formation evaluation and pressure control. It is necessary that the data collection is accurate and the computer processing effective.

Bonne (1972) discussed the successful application of these specific drilling model in the North Sea area. Sensing devices connected to the drilling rig were used to monitor up to 20 analog and 10 digital signals. The computer monitors the data, as each foot is drilled, prints it on tele typewriter and stores it on magnetic tape. Further analysis of selected variables, for example, may include rate of penetration, bit weight, rotary speed, porosity, pore fluid pressure, bit tooth dullness, bit wear, cost per foot, etc. Many drilling companies have developed their own drilling program models.

Bourgoyne et al (1971) proposed a general drilling model to determine formation porosity, pore fluid pressure, etc., via the following equation:

$$R = K \times f_1 (W/D) \times F_2 (N) \times f_3 (H) \times f_4 (\Delta P)$$

Where

R = rate of penetration (ROP)
K = drilling constant of normalized ROP
$f_1 (W/D)$ = function defining the effect of bit weight
D on ROP, $f_2 (N)$ = function defining effect of rotary speed
N on ROP, $f_3 (H)$ = function defining effect of tooth dullness
H on ROP, $f_4 (\Delta p)$ = function defining the effect of differential pressure across the hole, ΔP on ROP.

Normalized rate of penetration is related to bulk density, by

$$p_b = 2.65 - 1.65 \frac{Sq + \text{Log } K - b}{S_g}$$

Where

b = bulk density in gm/cm^3 in the US Gulf Coast area.
S_g = rock strength parameter

Bulk density then is converted to a formation pseudo porosity which closely agrees with the calculated porosity from density log (Figs. 7.6 & 7.7).

A graphical solution approach of detecting overpressure during drilling for sites without digital facilities was also developed.

Several other interesting techniques have been developed for pore pressure detection, which are discussed briefly. Using historical field data from several Louisiana Gulf Coast wells, equations for predicting pore pressure in both normally and abnormally pressured geologic sections penetrated by the borehole were derived

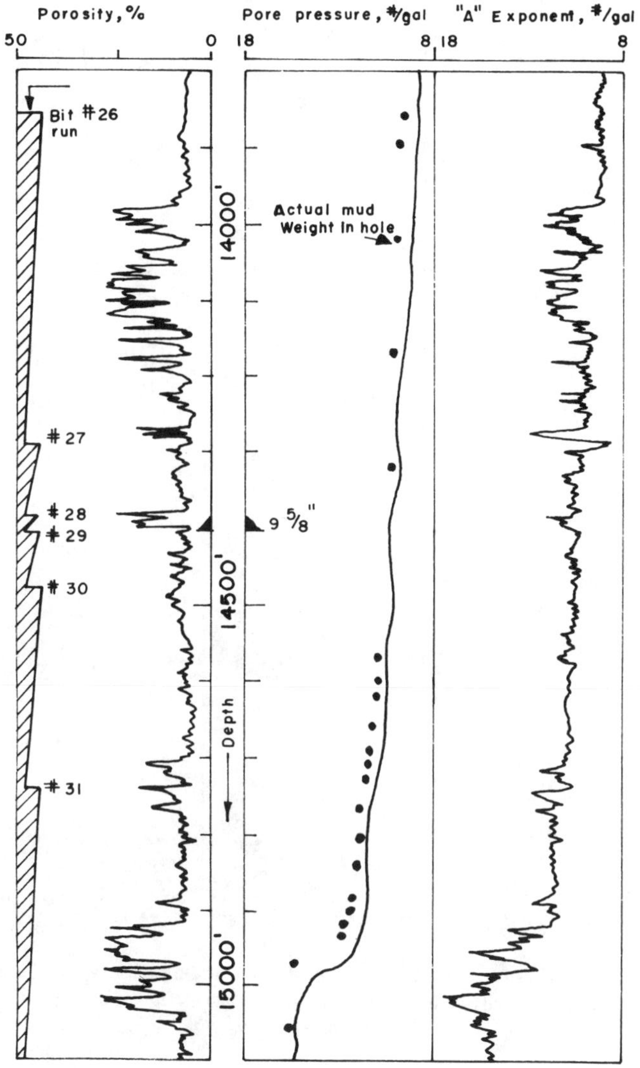

Fig. 7.6: Field example of formation porosity—pressure log in an offshore Louisiana Wildcat, U.S.A. The curve on the right ("A" exponent) gives the value for pore pressure taking porosity and lithology variations into account.

(After Fertl 1976)

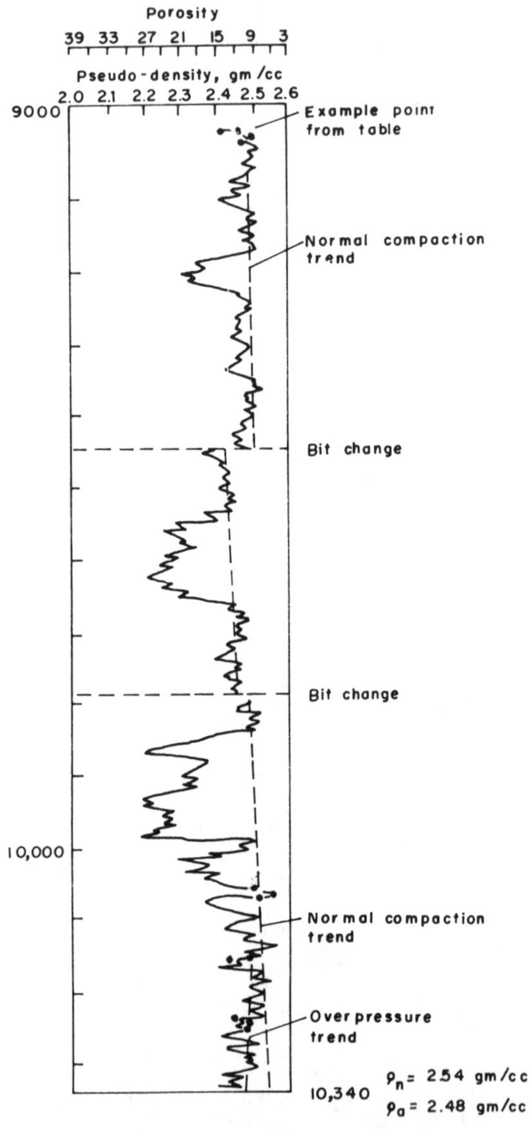

Fig. 7.7: Field example of psudo-density log obtained from drilling data (after Bourgoyne, 1971). ρ_n= density value of normally compacted shale; ρ_n = density value of overpressure shale.
(Courtsy of *Petroleum Engineer)*

by regression analysis (Herbert and Young, 1972). When the results of the analysis are applied to drilling data, the transition from normal to overpressures can be predicted, although, only in a geographically regional basis.

Correlation between well log data and rock drill ability has produced encouraging results (Gstalder and Raynal, 1966, EL-Hadidi, 1970). Results of these studies indicate that transit time data from acoustic logs can be used to predict rock drillability, provided a differentiating lithology parameter is available independently.

Recently, Lutz et al (1972) discussed the SNAP log. Vibrations from the tricone bit are measured at the kelly and interpreted at the derrick floor to give an instantaneous log of the formation while it is being drilled (Fig. 7.7a). In the words of the authors:

"The dynamic theory of drilling presented is based on creation by the drilling of longitudinal vibrations and transmissions of these through the drill stem. A study of the entire mechanical drilling system makes it possible to build instrumentation with which the vibrations received can be compared with the signals from a simulated model of the formation. Adjustment of the simulated signals will give instantaneous data on the hardness of the rock being drilled. Today, the most sophisticated model is not yet available but a simplified treatment unit, associated with sensors and recorders, has been used in the field for several years. An experimental method of instantaneous logging has been developed which allows one to obtain a log (the SNAP log) related to

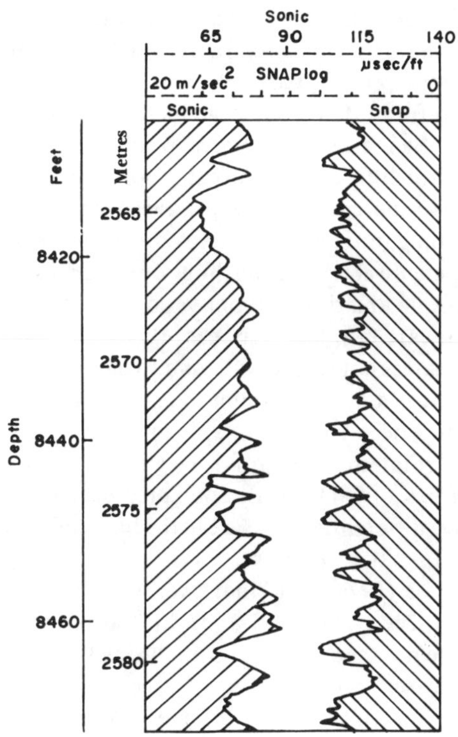

Fig. 7.7a: Detailed correlation between SNAP log and conventional sonic log in 12¼ inch borehole, Aquitanian Basin, France.

(After Lutz et. al., 1972. Courtesy of *Journal of Petroleum Technology.*)

the hardness of the rock. There is an excellent correlation, as would be expected, between the SNAP log and conventional logs, especially the sonic log.

The information received, is of value in formation correlation and in optimizing drilling variables, including making decisions on bit changes. The log is expected to be important in detecting the approach to, and the entrance into, abnormally pressured formations, since it reflects the greatest hardness of the cap rock above them and the lesser hardness and density of the abnormally pressured shales themselves". (Fertl, 1976, pp 135-136).

Drag

Drag is defined as an excess hook load over free handling load. The excess load may be caused by bit balling, dog legs, deviated holes, differential sticking, extra cuttings coming into the well bore when drilling transition zones, junk in the hole and exotic bottom hole assemblies. Drag is noticed at connections or when tripping.

Torque

Undercompacted shales are considered to be plastic in nature. When differential pressure into the well bore exists, these shales will cause a diminished hole diameter that may be smaller than the bit size. Increased torque should be noticed when underbalanced conditions prevail, particularly if full gauge stabilizers are present in the drill string. It is useful in detecting large increases in pore pressure, for example when crossing a fault. Increased torque as a result of underbalanced conditions is virtually unnoticed with differentials into the bore hole less than about one ppg equivalent pressure. Torque can also be produced by an increase in the size and amount of cuttings coming into the wellbore. A drastic increase in torque can result in a twist off, a locked cone on the bit, a washout in the drill string or a change in pore pressure. However, torque and drag are not valid indicators when drilling high angle directional hole (Gearheart Geodata Services, 1980).

McClendon (1977) states that torque will tend to increase in the transition zone with a low density mud because a larger amount of cuttings will enter the bore hole. Shale tends to stick or to impede bit rotation, and bit teeth will take larger bites of the formation as they are rotated. If the mud weight has been weighted to result in a hydrostatic pressure greater than the formation pressure, the torque will be masked. In the presence of negative differential pressure in the bore hole, overpressured shales will tend to flow or heave into the well bore.

Pickington and McKee (1974) state that overpressured shales will tend to slough when drilled underbalanced by 0.5 to 1.0 lb/gal, leading to the increase in torque.

However, special care needs to be taken in analyzing such information, as illustrated on the following field case:

Example: In the subject well, a mud lubricant was added because of high torque during drilling at 7,200 ft offshore Louisiana. A decrease in torque was noted from 7,300 ft. to 7,600 ft.

It was assumed that the lubricant solved the problem. But before Tenneco used the product on other wells, a plot of rotary speed and torque vs. depth was evaluated

(Fig. 7.8).

This analysis showed the torque reduction to be a direct function of slower rotary speeds as evinced by the increased torque caused by higher rotary speeds in the interval from 7,700 ft to 7,900 ft (True, 1984).

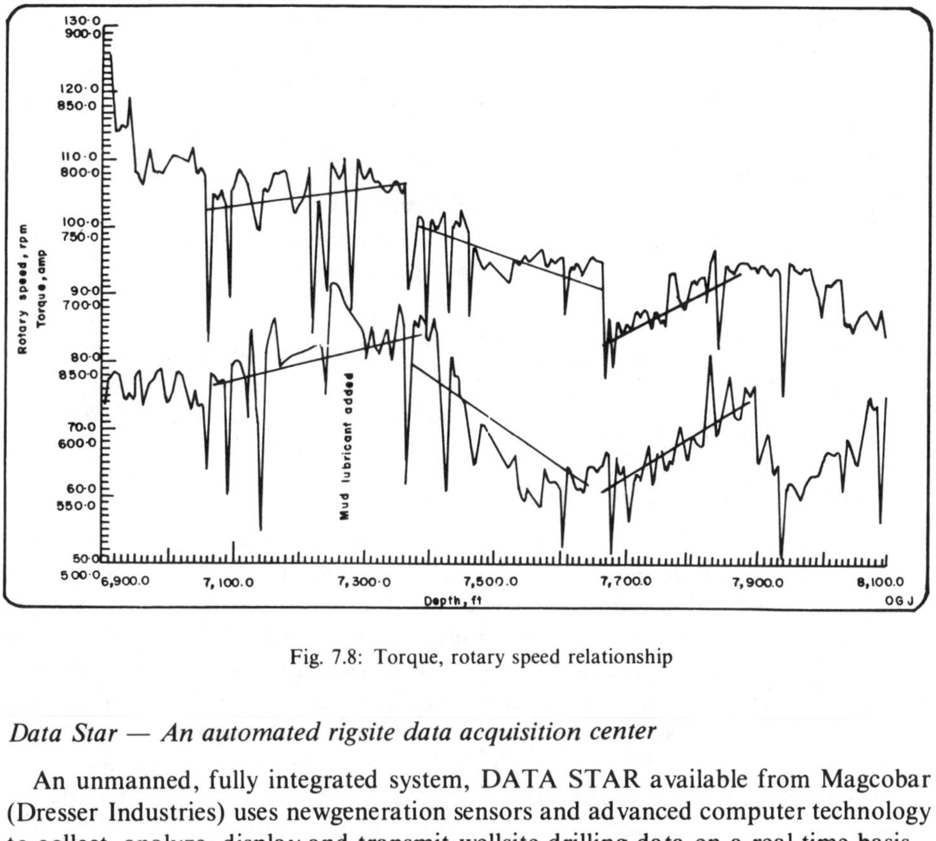

Fig. 7.8: Torque, rotary speed relationship

Data Star — An automated rigsite data acquisition center

An unmanned, fully integrated system, DATA STAR available from Magcobar (Dresser Industries) uses newgeneration sensors and advanced computer technology to collect, analyze, display and transmit wellsite drilling data on a real-time basis.

DATA STAR combines greater reliability and accuracy of acquired wellsite information with ease of operation and modular design. A wide choice of options and total field service help assure minimum downtime. DATA STAR can be used effectively as an aid to drilling operations, whether on land, inland barge or offshore.

The unmanned rigsite data collection center retrieves and displays real-time operational information to help determine the safest and most effective drilling rates, detect high downhole pressures, select the most effective casing seats, and optimize bit weight and rotary speed.

Drilling data displayed by DATA STAR

- Rate of Penetration
- Weight on Bit
- Delta Flow (GPM)
- Accumulative Rotary

- String Weight
- Hook Load
- Pit Volume Total
- Pit Volume (trip tank)
- Flow In (GPM)
- Flow Out (GPM)
- Mud Weight In
- Mud Weight Out
- Temperature In
- Temperature Out
- Peak Pipe Velocity
- Pipe Velocity
- Casing Pressure
- Total Depth
- Rotary Speed
- Torque
- Pump Stroke Totals
- Pump Stroke Rate
- Pump Output (GPM)
- Pump Pressure
- Bit Location
- Bit Time on Bottom
- Bit Time off Bottom
- Military Time
- Fill Strokes
- Fill Total
- Total Combustible Gas
- Ton Miles

By adding optional transmission equipment and the DATA Center, management can track and control drilling operations simultaneously at many wells equipped with DATA STARS.

Figure 7.9 shows the schematic for (a) the basic system at the rigsite and (b) the system layout at company headquarters and for area centers.

Measurement While Drilling (MWD)

These systems have become commercially available during the last five years, with widespread availability during only the last two to three years. Currently there are between 150 and 200 complete systems available for use throughout the world, and these are provided by five major suppliers. Numerous other companies, both in the U.S. and abroad, are developing systems that should reach the market-place in the next few years. MWD systems are finding expanding markets as they demonstrate their reliability and usefulness, and they have proved valuable in monitoring directional drilling and hole deviation, especially offshore. In the near future they will also be used to improve drilling safety and efficiency, though these uses are not currently emphasized.

Three basic Subsystems make up an MWD system:

- Downhole sensor and signal conditioning package.
- Means for transmitting information (from downhole) to surface.
- Surface equipment for receiving, decoding and displaying transmitted information.

Development is proceeding in all three areas. Active research and development of new sensors has provided tools to measure hole inclination and azimuth, tool-face orientation (during periods of non-rotation of drill pipe), formation properties via gamma ray and resistivity tools, downhole weight on bit (WOB), torque and annulus temperature/pressure though seldom are all measurements included in a single package. Work is proceeding to expand the suite of available measurements to include, for example, fluid-in-flux measurements, a turbine rpm indicator and additional

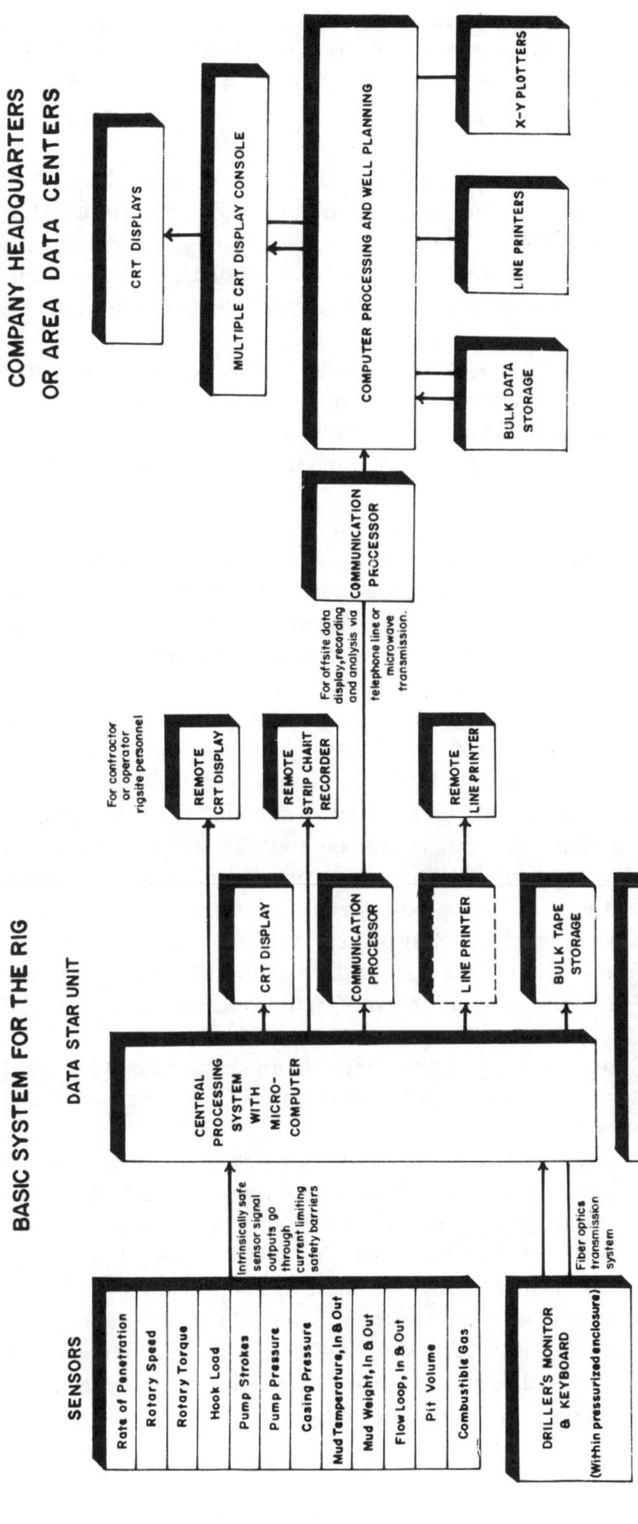

Fig. 7.9: Drilling data display by Data Star

formation evaluation tools. Data-rate limitation of the transmission system is the major bottleneck and will limit the number of sensors in use at any one time, but the list of available sensors will expand as the market and number of suppliers expand. The package employed for any particular portion of a well can thus be determined by the measurements desired (Carson, 1984).

These devices all have sensors located immediately above, or near, the drill bit to monitor drilling variables and, in some cases, formation properties. Information is collected downhole, and sent (telemetered) to the surface through either the drilling fluid, electrical conductors, the drill pipe, or the earth. The systems provide essentially real-time information for monitoring and controlling the drilling operation (Gravley, 1982).

Companies developing new MWD systems, as well as those currently marketing the systems, are working to include downhole computers for signal processing and creative message coding or conditioning. This will partially overcome limitations of the transmission link, by increasing the number of sensors for which an MWD system can give timely information.

Signal transmission for all existing systems and for nearly all of those development is done by "mud-pulse telemetry". In different MWD systems, this is accomplished with three different kinds of signals; a positive pulse, generated by the sudden restriction of a value passage in the mud stream, a negative pulse system, operated by opening a value that reduces pressure in the mud system, thus allowing mud to bypass the motor/bit pressure drops, and siren system that generates a constant, continuous frequency signal and transmits data by changing phase. Other, non-mud-pulse transmission methods have been investigated in the past, and work continues on electromagnetic transmission through the earth, stress wave transmission through the drill pipe, and hardwire transmission via an electrical conductor in the drill pipe. Guaranteed signal propagation is the concern for the first two of these, and expense and possible impact on standard rig operations are concerns for the third.

The final MWD subsystem is surface equipment for receiving and processing signals. Each MWD company develops its own processing and display system based on its sensors, downhole processing and transmission mode. Operating experience indicates that the MWD systems are being accepted in the industry and are providing useful information.

Figure 7.10 & 7.10A shows good agreement between an actual MWD log and the subsequent open hole electric wireline log both run in an US Gulf Coast well (Tanguy and Zoeller, 1981).

Figure 7.11 presents the MWD log in an exploratory well illustrating a transition from normal to overpressure environments. The normal pressure gradient trend line was established from resistivity data higher in the hole and is shown as a solid line. Beginning at approximately 6000 feet (1830 m), the resistivity begins to decrease relative to the established trend indicating the onset of geopressures. At 7500 feet (2290 m), the pore pressure gradient increases to 12 ppg (1.44 g/cm^3) necessitating a corresponding increase in mud density. By using MWD derived pore pressures in real time, the proper mud weight to maintain hydrostatic control can be selected, thus minimizing the risk of taking a kick (Bates and Tanguy, 1983).

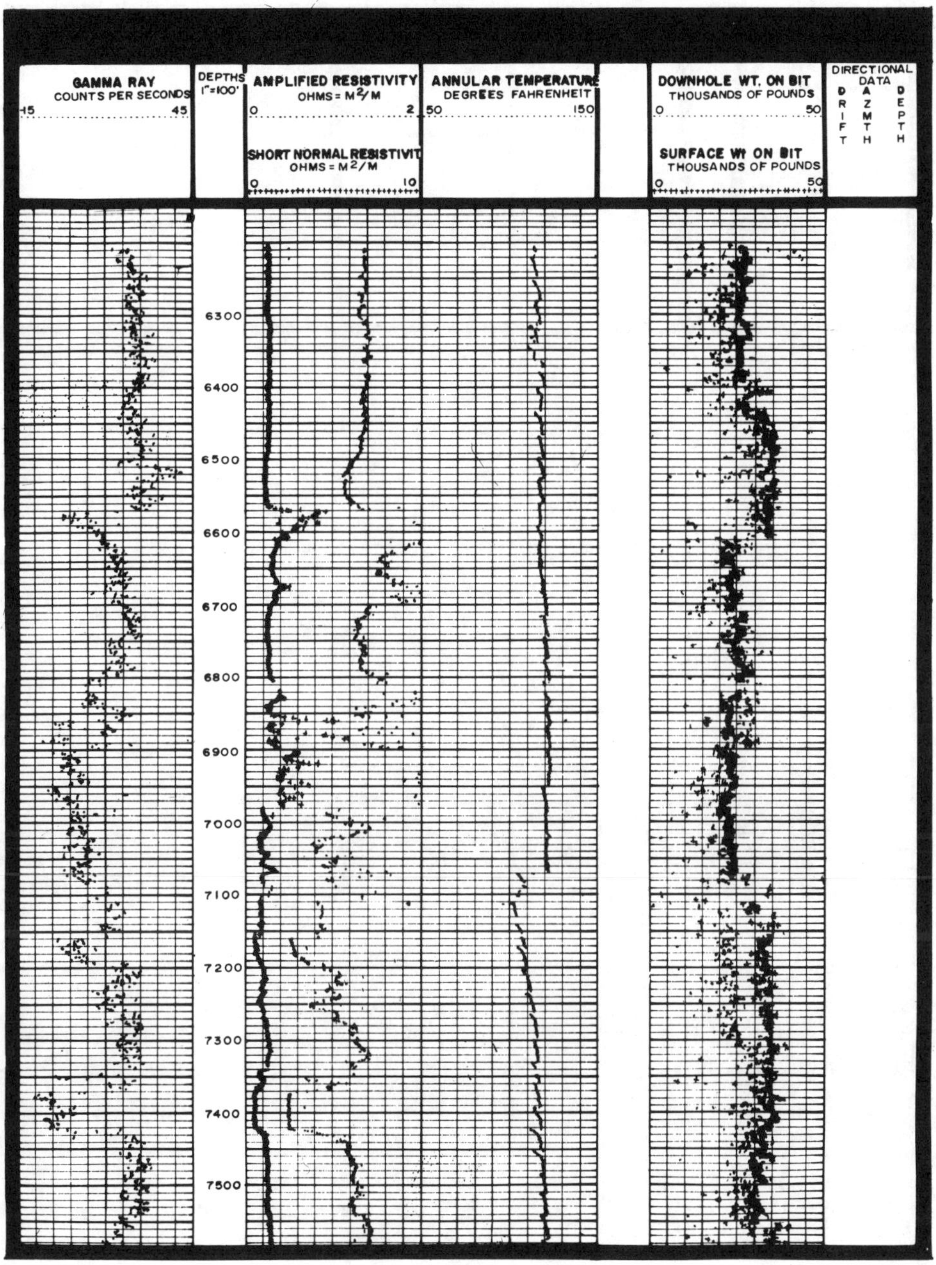

Fig. 7.10: Measurement while drilling

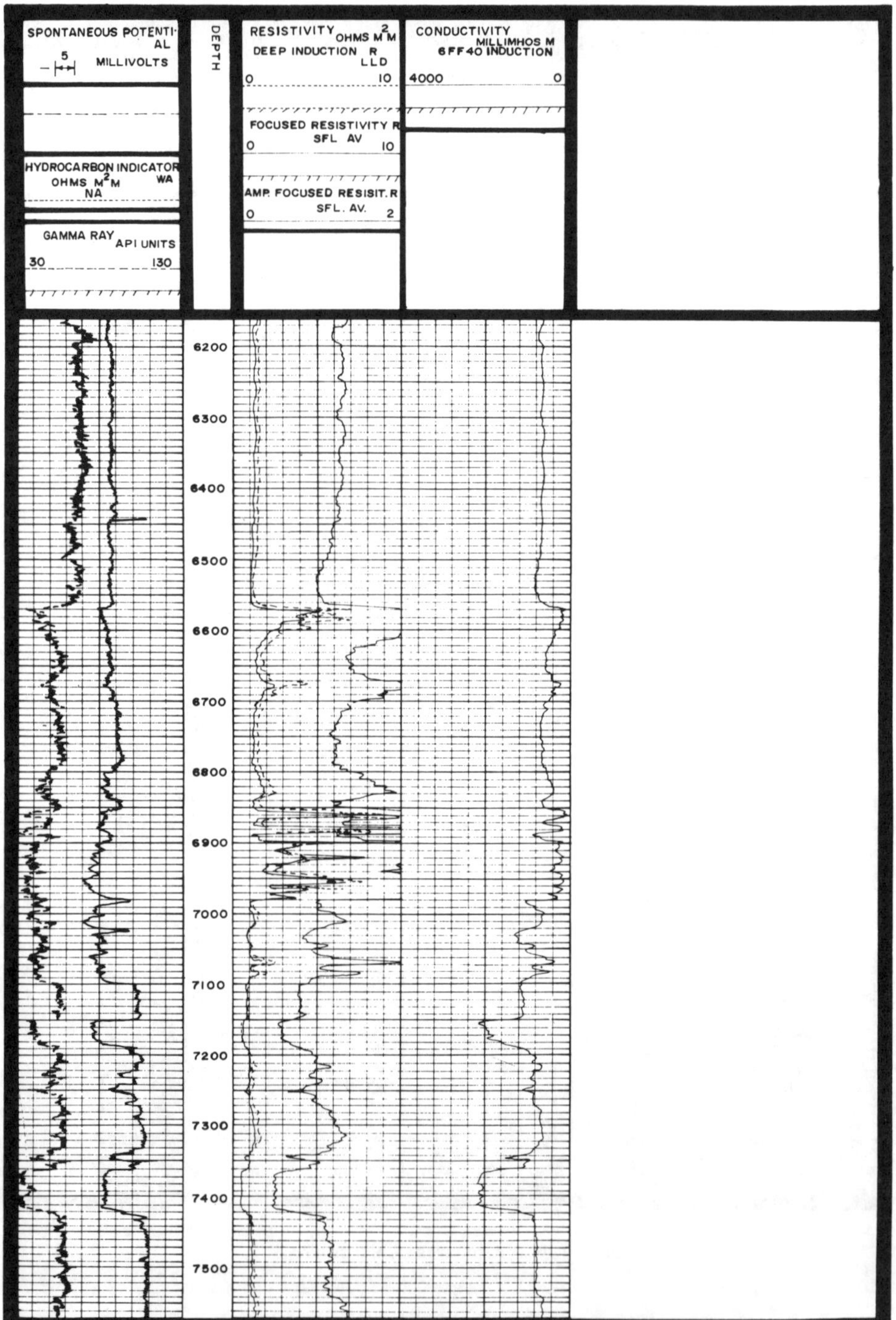

Fig. 7.10A: A Comparison between actual MWD log and Wireline log

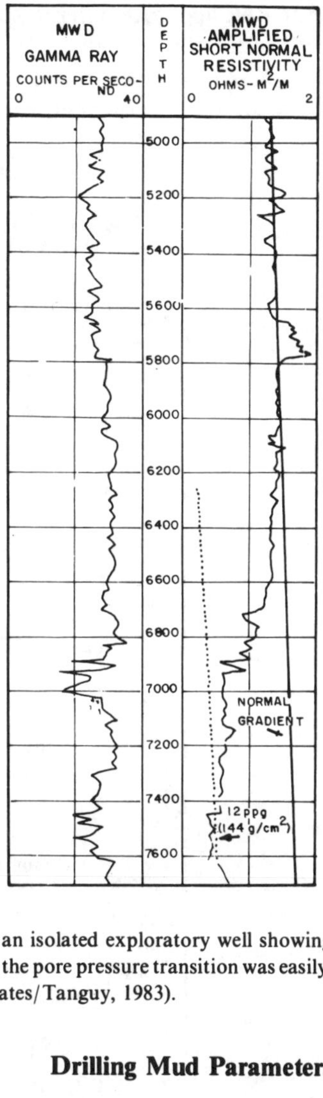

Fig. 7.11: An MWD log from an isolated exploratory well showing the resistivity and gamma ray data. Using the MWD log, the pore pressure transition was easily identified and proper well control was maintained (after, Bates/Tanguy, 1983).

Drilling Mud Parameters

(i) *Gas Logging*

It is customary on both exploratory as well as development wells to have a gas logging device installed to automatically and continuously monitor the presence and/or absence of hydrocarbon gases in the return mud stream. The device thus installed may also be used for detecting the presence of hydrocarbon gas in samples, cuttings, cores, mud and formation water. The reason for having such gas logging device, on both exploratory and development wells is that the presence of dry gas (with little or no distillate) is very difficult to detect by odour, especially in cuttings.

Since mud gas logs are being frequently used to detect overpressures several factors are reviewed here which may complicate application in pressure detection and should be clearly understood prior to interpretation.

(1) *Pay zones*: Presence of high gas readings in potential, permeable pay zones affect such log application for pressure detection. Diffusion of gas into shales lying

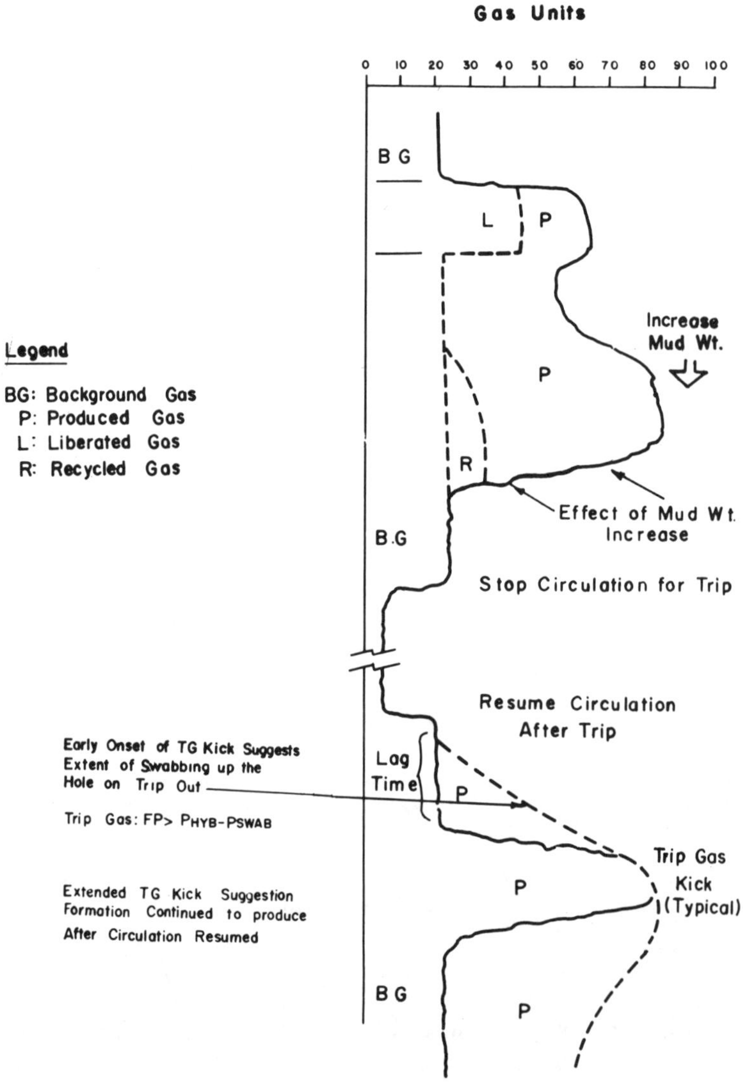

Fig. 7.12: Factors effecting gas shows
(After Mercer, 1974)

immediately above pay zones, can complicate interpretation. Such vertical gas immigration has been encountered above many pools under different geological conditions. A sharp increase in gas contents occurs in these cap rocks with the thickness of the gas-saturated interval depending on gas composition and reservoir pressure within the pool (Fertl, 1976, pp. 138).

(2) *Connection gas/ Trip gas*: Swabbing effect of raising the kelly at the time of pipe connection or during a trip bring small amounts of gas from the previously drilled formations into the mud system, which affects the interpretations required for high pressure predictions (Figs. 7.12 & 7.12a)

(3) *Kelly air*: A simple mechanical operation such as unscrewing the kelly for pipe addition may introduce air into drill pipe. A continuous mud weight recorder or continuous monitoring through gas detector/chromatograph allows its detection at the well head after a complete circulation cycle. However, the effect of kelly air is

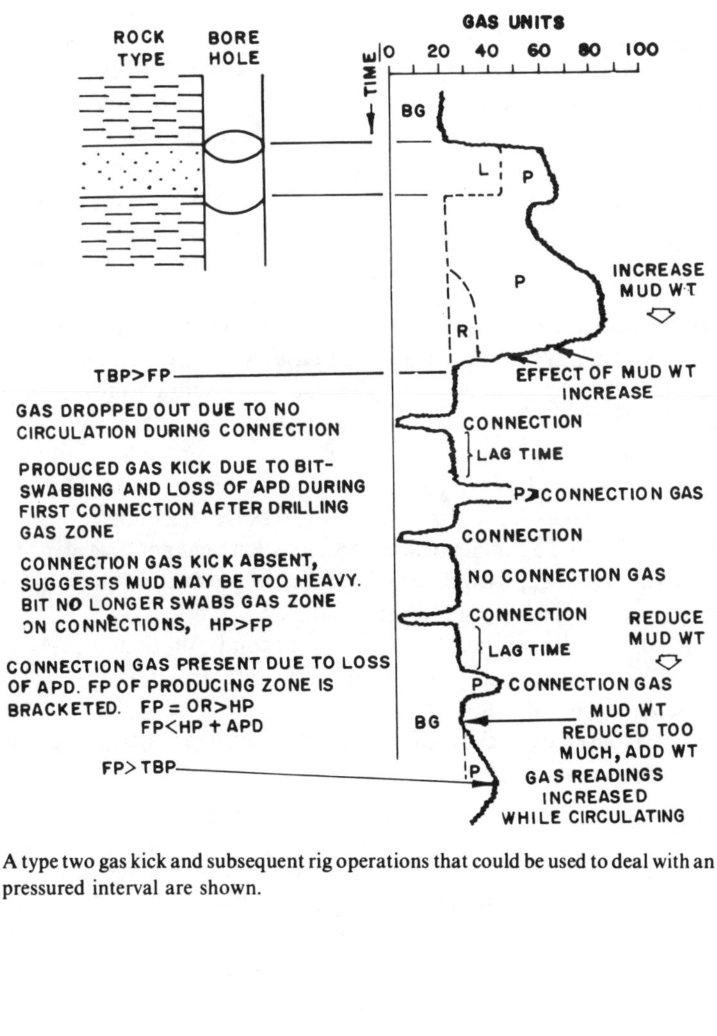

Fig. 7.12a: A type two gas kick and subsequent rig operations that could be used to deal with an abnormally pressured interval are shown.

usually insignificant.

(4) *Gas composition*: The gas composition is also of great importance in the interpretation of gas shows for pressure predictions. H_2S and CO_2 are frequently encountered and these gases being soluble in mud (especially oil muds), their thermodynamic behaviour differs from methane. In deep wells, methane kicks give sufficient warning while H_2S and CO_2 which dissolve at rather low pressures and are dangerous gases for men and material, give very little warning before the bore hole starts to unload. Hence, not enough time would be available to close BOP when necessary (Fertl, 1976, pp. 139).

(5) *Degradation of mud additives*: Certain chemicals in the drilling mud may degradate and may also give H_2S and CO_2 in high temperature wells. For example, degradation of modified lignosulphates at temperature exceeding 400° F may produce H_2S and CO_2 and hence hamper pressure prediction from mud gas logs (Fertl, 1976).

(6) *Gas flushing*: In some parts of the Soviet Union, the total gas readings in drilling mud showed that productive pay zones sometimes indicated only background readings whereas water bearing zones containing dissolved combustible gas showed high total gas values (Rodionov, 1965).

(7) *Volcanic material*: In some parts of the world volcanic beds cause high rate gas flow at shallower depths. Over geologic time, diagenesis of volcanic ash basically resulted in three components, clay minerals, methane and CO_2. Shales cause gas cuttings without directly reflecting formation pressure variation during drilling.

(8) *Deep seated volcanoes*: In some parts of the world, not particularly in the Indian sub-continent, large amounts of solids, fluids and gases expelled by deep-seated volcanic activity may disturb the equilibrium of rocks and thus cause problems on the mud gas log.

(9) *Thermodynamic processes*: Clays exhibiting catalytic activity are present both in formation and drilling fluid. Hydrocarbon gases also originate because of grinding action of the bit, resulting in subsequent increase in the temperature in the presence of catalyst (clay mineral) and organic matter in the rock.

(10) *Lignite zones*: Coal/lignite beds are frequently encountered in the geological section and are associated with high gas readings, thus providing difficulties in the interpretation of mud gas log utilized for overpressure detections.

(11) *Faults*: If faults do not act as barriers, they may channel gas, causing localized gas flows into the wellbore, with the mud becoming gas-cut and posing difficulties in interpretations required for pressure predictions.

(12) *Shale diapirs*: They expel large amounts of solids, fluids and gases which result in frequent gas cutting and can pose problems in interpretations.

Besides, there may be additional but minor factors, like addition of diesel or crude oil in the mud, carbide for checking of the lag time, etc., which can cause interpretation problems for overpressure prediction studies.

Total Gas Detector

A total gas detector detects the air gas mixture from the gas trap and percentage of hydrocarbon gases in that mixture. The reading is plotted on the recorded data as it is calibrated with methane (CH_4) and the reading is said to be methane equivalent. Almost from the inception of the geologging, the hot wire gas detector has been used as a tool to detect the presence of combustible gases separated from the return stream of drilling mud and to make semi quantitative measurements of these gases. Various geologging companies are utilizing more sophisticated analytical instruments to provide qualitative analysis and greatly improved quantitative analysis of the gases. Some of the new instruments have electronic components and thus, in turn, electronic gas analysers have come into use. Regardless of the instruments used for the detection of the presence of hydrocarbon gases in the mud stream, any system used to separate the gas from the return mud stream of the drilling fluid, should be considered every bit as important as the gas analysis instrument itself (Figs. 7.13, 7.14 and 7.15).

As mentioned earlier, various geologging companies which are engaged in the geologging technology have devised different gas detector systems, for example Baroid of the U.S.A. has an automatic recording gas detector, a combination of dual gas detectors, (A) Catalytic, (B) Thermal, while Gearhart Geodata utilizes flame isonisation type.

(ii) *Gas Chromatography*

It is important that the presence of gas as monitored by the total gas detector is

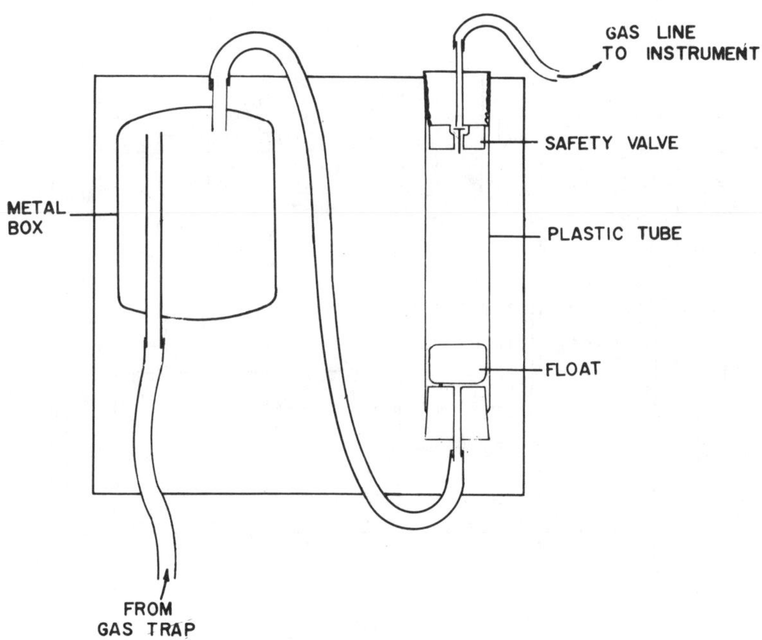

Fig. 7.13: Gas flow system

(Courtesy: Geoservices)

subjected to analytical studies with regard to the type of gas, hence, the necessity of gas chromatography. In this process, the sample of gas obtained from the return drilling mud is mixed with inert gas (Helium, Nitrogen or Air) and flows through a partition column. The column is packed with absorbent material like charcoal which absorbs hydrocarbon gases. These gases are then released to the inert gas system in order of their boiling point. Changes in the composition of gas emerging from the column may

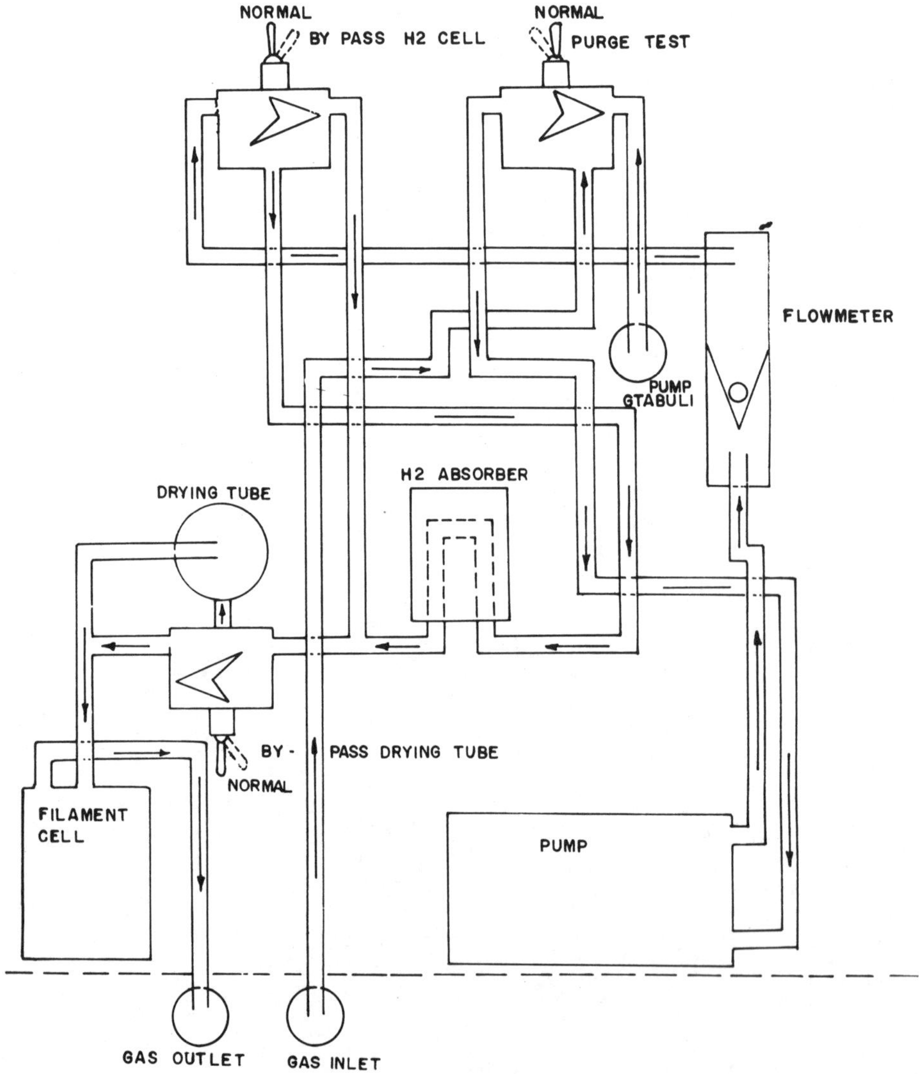

Fig. 7.14: Gas flow pattern for total gas detection

(Courtesy: Geoservices)

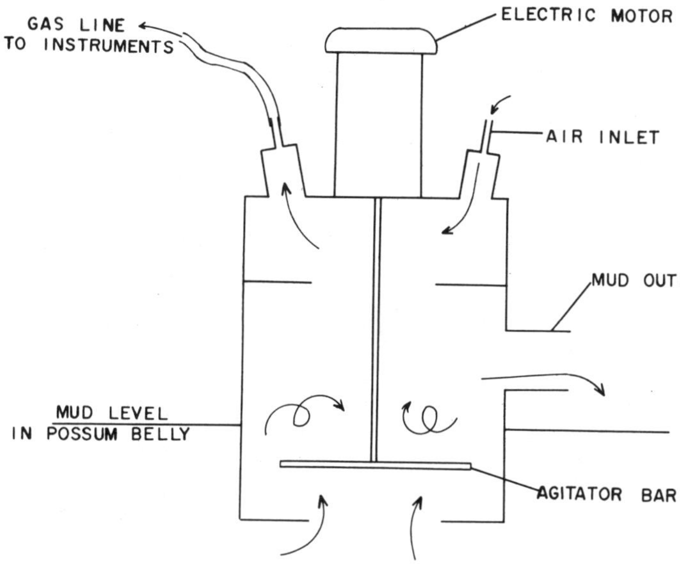

Fig. 7.15: Gas Trap

(Courtesy: Geoservices)

be detected by measuring the thermal conductivity of the gas or by burning the gas in platinum filament. Various geologging companies have devised highly sophisticated gas chromatography instruments to monitor the percentage of each gas in the air-gas mixture during the course of drilling. M/s Geoservices have devised Gal-31 and chromatologger for carrying out the chromatography studies of the gas, monitored with the help of total gas detector, during the course of drilling. Their latest instrument ('chromatologger') automatically identifies the methane, ethane, propane, isobutane and a normal butane. Measured values are kept in memories and are continuously plotted on a 'chromatolog'.

The chromatologger when used with vacuum mud still, a device for quantitative analysis of gases, helps to calculate the gas saturation of the reservoir being drilled which provides an immediate evaluation of the hydrocarbon potential (Fig. 7.16).

Besides the chromatologger of Geoservices, various other companies like Core Laboratories, N.L. Baroid, Gearhart Geodata Services, Samega, etc., have their own chromatography systems for analysing different hydrocarbon gases emitted during the course of drilling and are used in the oil industry under different trade names.

Abnormal Pressure Applications

Connection Gas

Subsequent to elimination of produced gas and if a connection is made, the gas

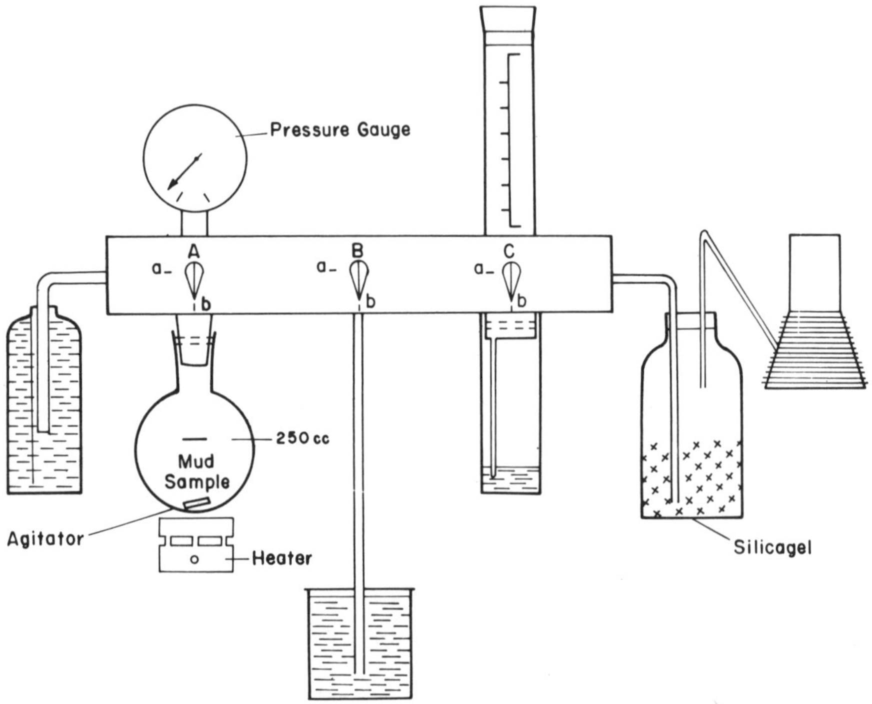

Fig. 7.16: Vacuum mud still connecting diagram

(Courtesy: Geoservices)

detector indicates a decrease in the carried background reading due to no mud circulation during the connection. On resuming circulation after one lag time of mud circulation, a gas kick usually appears which was mistakenly deemed to produce gas, but in reality was related to the connection and was not liberated from the formation. Figure 7.12a shows a typical case of a connection gas.

Because there was no evidence of produced gas in the system while circulating, it was apparent that the mud weight plus the annular pressure drop (APD) were sufficient to create a total bottom hole pressure greater than formation pressure. Thus, the connection gas peak experienced following the first connection after penetrating the gas zone may have been related to swabbing of the zone rather than to insufficient mud weight.

Swabbing may occur when the kelly is raised for a connection. Because annular pressure drop is zero during periods of no circulation, the bottom hole pressure equals the hydrostatic pressure for static mud systems. This decrease of bottom hole pressure may be a factor in magnitude of the connection gas peak. Since the swabbing effect is not measurable, it is difficult to accurately determine the significance of connection gas peaks which result when bit movement on connections extends above the gas zone.

However, on the next connection, and swabbing action being absent, no connection gas peak resulted. This suggests excess mud weight since gas did not enter the wellbore despite the absence of the annular pressure drop. The mud weight was reduced until a moderate connection gas kick occurred with no concurrent increase in background values while circulating.

Subsequent mud weight reduction resulted in measurable quantities of produced gas being observed in the mud system during circulation. This clearly indicates that formation pressure now exceeds total bottom hole pressure. Hence, the mud weight had been reduced too much. The mud weight then was again increased to restore the ideal condition of moderate connection gas peaks with no evidence of produced gas while circulating.

Trip Gas

Trip gas is the general term applied to produced gas which characteristically occurs within the lag time after one round trip circulation is resumed. Three basic factors influence the presence, location, and magnitude of the trip gas kick:

1. Loss of annular pressure drop
2. Bit swabbing the entire hole. This effect is greatly dependent on the speed at which drill pipe is tripped out of the borehole, variations in hole size, configuration of a packed downhole assembly, tripping out with a full core barrel, etc.
3. Time period over which these factors influence the static mud system

Basic principles discussed with regard to connection gas also apply to trip gas. The most significant difference between trip and connection is the extreme impact of the above mentioned parameters during a round trip as compared to their relatively minor effect during a connection. This suggests the importance and prerequisite of ensuring absolute control of already penetrated abnormally pressured zones prior to any trip attempt. It would be foolish to suspend circulation and begin a trip in the midst of a formation gas kick without clear identification of the type of gas kick.

Gas detectors are of limited assistance in providing early warning of zones which blow out immediately upon contact. Other, already proven, methods of well control are available and should be employed in these cases. Statistics show, however, that this type of blowout is the exception rather than the rule. In other words, careful surveillance of all produced gas indications, especially resulting from tripping, becomes a very important rig safety indicator.

A hypothetical situation in which a type-two gas kick occurs, followed by the elimination of produced gas due to increasing the mud weight, is shown in Figure 7.12a Circulation, terminated for a trip, was resumed and a typical trip gas kick was received within one lag time. This kick configuration suggests that the mud weight was sufficient to keep the well under control and no dangerous conditions were experienced during the trip.

If the trip gas kick had occurred in its entirely at a lag time somewhat less than one total lag time, the possibility of gas entry at some shallower depth would have been indicated. Such information is derived from the fact that the initial circulation returns after a trip provides some indication of the various hydrocarbon accumulations penetrated over the length of the uncased hole. If the trip gas kick indicates a very early onset that cannot be attributed to a shallower termination, this may indicate the extent of gas swabbing up the hole following the bit during tripping.

For a normal trip gas kick, formation pressure exceeds hydrostatic pressure. Exhorbitant trip gas peaks and variations in magnitude require proper evaluation. An extended trip gas kick after the main peak suggests a formation that will continue producing gas after circulation is resumed. This suggests a nearly balanced condition and should be treated with utmost care during subsequent trips. Formation pressure in this case may be greater than total bottom hole pressure, until the swabbed trip gas is circulated out of the system.

Occasionally, several factors are superimposed which complicate the proper evaluation of gas detection data, especially if no indications of produced gas are seen. The first trip gas kick after such a zone may indicate the presence of permeability and thus shed greater light on the original interpretation.

Limited Permeability Indications

Occasionally, abnormal pressured intervals are drilled and no indications from other well site parameters (such as penetration rate, sample analysis, etc.) are received. One probable explanation is fractured formations. If such a zone is penetrated, and total gas detection equipment is not on location, the rig crew may be unaware of the danger awaiting the next trip.

In many instances, a kick with large proportions of produced gas may occur while drilling. Attempts to kill such zones and eliminate produced gas by raising the mud weight may prove unsuccessful. Even after building considerable mud weight, produced gas seemingly continues. One must first be certain that the gas is not a continuing build-up of recycled gas. If in fact gas does exist and persist, it suggests an unusually high pressure interval of very low permeability which has apparently established a flow equilibrium with the circulating mud system as a function of deliverability (permeability) rather than formation pressure. Had gas deliverability been greater, the zone would probably have blown out immediately upon contact. Under these circumstances, efforts to completely eliminate all minor produced gas effect may be unwarranted.

Occasionally, evidences of produced gas, such as connection and trip gas, may diminish over extended periods of drilling with no apparent change in mud characteristics. Such evidence may suggest that the source zone was of limited permeability (and/or extent) and has depleted below hydrostatic formation pressure of the mud system.

These factors clearly indicate the importance of using all available gas detection data in interpreting specific hydrocarbon accumulations. This includes the more subtle indications which are observed over long time periods of mud surveillance, especially drill pipe connections and trips.

Proper use of gas detection techniques is beneficial. This can result in better communication among the logging crew, well-site geologist and drilling personnel, and often unnecessary or improper drilling activities may be prevented. Early decisions with respect to already penetrated hydrocarbon accumulations can be made safely. For example, circulation for a trip should not be commenced in the middle of gas kick until it has been clearly determined whether the kick is type-one or type-two. Mud weight increase may be initiated immediately, if necessary, without losing the annular pressure drop during periods of non-circulation. Instructions regarding special care during tripping may be given to the drilling crew immediately after penetration of hydrocarbon accumulations which show characteristics conducive to possible drilling hazards.

Careful application of these principles assists in interpreting well logs and will also assist in planning specific perforating schedule. Because timing is extremely critical when the final total depth has been reached, an in-depth interpretative analysis of the recorded data is a must and should be considered in conjunction with logs and other geological data to influence any final well-testing decisions.

The applications of the gas chromatography are plenty. It is particularly useful in case of low GOR oil pools and high pressure gas fields. The gas chromatography in conjunction with Vacuum Mud Still is helpful for the calculation of gas saturation of the reservoir being drilled and thus in turn provides the evaluation of the hydrocarbon potential of the reservoir. It seems, therefore, that continuous gas logging with the help of total gas detector and the gas chromatography makes it possible to analyse the different gases and to detect overpressured gas reservoirs, the parameters so vital and important, particularly when supplemented with other data either during or after drilling.

(iii) *Mud Weight*

The measurement of mud weight during the course of drilling is a fundamental requirement to ensure safe balance between the hydrostatic and formation pressures. Any deviation in the mud density/weight is indicative of the influx of the formation fluid or gas and is vital for safe and properly engineered drilling practices. It, therefore, becomes absolutely essential to record and monitor the density/weight of the mud, both going in and coming out, during the course of drilling. Deviations towards decreasing mud density is indicative of influx of formation fluids or gas from high pressure reservoirs below. This data need to be properly recorded, reported and acted on to provide safer drilling practices and to avoid well hazards associated with drilling through high pressure reservoirs.

Various instruments for recording of the mud density/weight have been devised by various logging companies. This equipment is kept in the geologging unit along with the other instruments like gas detector, gas chromatograph, for proper formation evaluation. M/s. Geoservices have the Densi mud gamma while other companies have Simpler Densimeters for the density of in and out going mud, for recording of any abnormalities during the course of drilling. Such instruments should be kept on all exploratory and development wells and mud weight/density should be checked every 15 minutes or periodically as necessary during circulation. Abnormalities as and when noticed should be immediately brought to the notice of all concerned.

Mud Viscosity

The viscosity measured during the course of drilling and/or circulation of the mud is also a very useful paramter for overpressure detection. Changes to lower viscosity of the return mud from that of the ingoing mud is indicative of high pressure gas reservoir or influx of salty formation water, while it would increase in case of fresh influx.

Therefore, the viscosity parameters when coupled with the density of return mud along with other overpressure detection technique, can provide meaningful and useful clues.

(iv) *Flow Line Temperature*

Drilling data all over the world are being used to locate depth of logging points at the top of overpressured zone. In such cases, where casing is to be set into the transition zone, this procedure is usually very useful, particularly in such an area where sufficient data are available to provide a degree of stratigraphic control and to permit quantitative log analysis.

Logging and other methods of predicting abnormal pressures with the help of drilling data have a common deficiency, differing only in degree. Most drilling parameters can indicate penetration of the zone, but several methods are only most effective when a very small overbalance exists between the pressure exerted by the drilling mud and pore pressures. It has thus been often necessary in the past to drill a pressure significantly greater than normal before the change could be recognized. The inability to predict the pressures ahead of the bit posed problems in long overpressured sections, which occur at deeper depths. However, the situation becomes critical when abnormal pressures are encountered at shallower depths and the transition between normal and abnormal pressures is very small.

Although greater than the absolute minimum mud weight and multiple casing runs can solve the problem, a limit exists as to the maximum mud weight and the number of casing strings which can be used before one set of difficulties is substituted for another. In cases where pressures occur at shallower depths, other pressure indicators such as penetration rate/conductivity studies, etc., usually do not help. Therefore, a method which is little affected by over-balance or nature of the rock matrix, which does not require penetration far into pressure transition zone and which does not rely on doubtful correlations is of great use not only at shallower depths, but also at deeper depths, particularly when multiple pressure plateaus are encountered. The monitoring of the flow line temperature frequently satisfies the requirements as mentioned above and is of great use for predicting of high pressures (Fig. 7.17).

The theory relating to high temperature and overpressure as discussed by Lewis and Rose (1970) can be compounded with the information recently available during the drilling operations, to reach a greater understanding of temperature behaviour and to increase the usefulness of flow line temperature data. Practically all heat flowing within the earth moves outward. Total heat flowing across all depth increments is constant, but temperature differential across any increment depends upon ability of materials to conduct heat. Below a formation for which conductivity is relatively low,

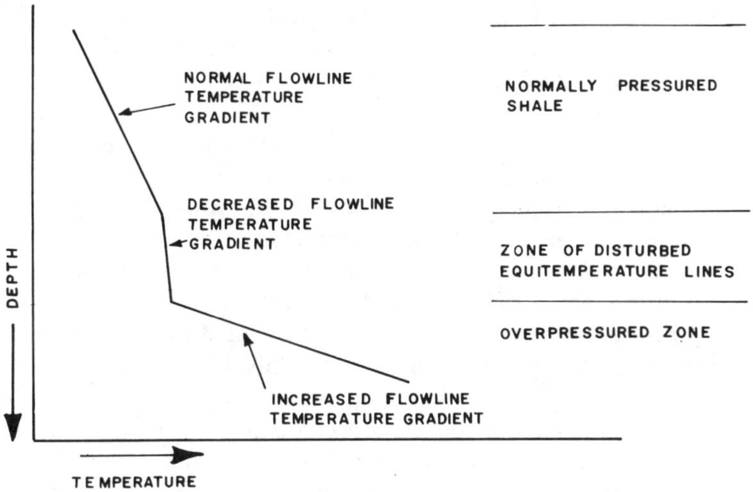

Fig. 7.17: Flowline temperature response to an overpressured shale zone

heat builds up, until temperature differential across the formations allows heat flow through the insulator to equal flow of incoming heat.

Accordingly, higher geothermal gradients exist against insulating formation and a relatively lower temperature gradient is present in formations with relatively high conductivities (Wilson and Bush, 1972).

According to Wilson and Bush (1972), the reason for high geothermal gradient in the transition zone is the greater than normal content of the under compacted formations. Since the thermal conductivity of water is about 1/3rd compared to that of many formation material, the total conductivity of the matrix and the fluid decreases and the geothermal gradient increases as water filled porosity increases. Under normal compaction and pressure, the porosity decreases with depth. However, higher than normal porosity is associated with abnormal pressure. As is known, abnormal pressure is encountered in and below undercompacted shale or other material of some minimum thickness, which is essentially uninterrupted by permeable beds. An increasing geothermal gradient expected in massive overpressure shale is reflected in the increase in flow line temperature, even through mud temperature will be lower than the formation temperature. Therefore, an increase in the flow line temperature above normal is to be interpreted as due to abnormally high pressure zones and the increase in temperature gradient should persist.

It may also be seen that flow line temperature will exhibit changes in the gradient while drilling from shale to sand. This change, as reflected by a decrease in the flow line temperature, is due to the change in the porosities and/or saturating fluids. If one assumes that thermal conductivity of all the formation/matrix material is essentially constant, the basic theory would require higher geothermal gradient to exist across hydrocarbon versus water bearing sands (Wilson & Bush, 1972).

Certain factors make true flow line temperature. These include circulation and penetration rates and tripping effects. Such effects can be minimized by holding circulation rate essentially constant at values dictated by drilling hydraulics. The resumption of drilling immediately after the round trip will show the flow line temperature to be normally less than the temperature observed at the end of the previous bit run. This is to be taken into account while plotting the flow line temperature during the course of drilling. Simply connecting all data points without taking into consideration the time and events would result in an incorrect plot on which temperature gradient increases and decreases in random or cyclic manner. Similarly, temperature decreases are known to be caused by mud addition or down time for the repairs of the equipment on the rig and this should also be discounted. It is, therefore, advisable to take the flow line temperature readings after the circulation is in progress for atleast a couple of hours, especially after a long trip. The flow line temperature to be plotted along with other indicators can thus herald higher pressures several hundred feet above the abnormal pressure zone. It can, therefore, be concluded that the flow line temperature parameter is a valuable addition to the suite of practical indicators available for predicting and defining pressure transition during the course of drilling.

Data available from several of the wells in India and elsewhere show the usefulness of this very important parameter for the prediction of overpressured zones (Fig. 7.18).

(v) *Pit Level and Total Pit Volume*

This important parameter makes it possible to recognize overpressure zones with the help of variations in the total mud volume. Variations in the total mud volume can be monitored with the help of pit level indicators. The mud pit volume recorder can simultaneously measure the mud volume in up to eight pits and thus compute total volume. Continuous recording of this volume enables early detection and evaluation of mud loss/gain and in the event of gas kicks allows an estimate of the kill size and requirements. Since the influx of the formation fluids in the mud column is indicative of overpressure horizons, the mud pit volume recorder, which is usually housed in a geologging unit. This very useful instrument should be used during the drilling of all exploratory and development wells to pin-point anomalies during drilling and circulations and as such help to take remedial measures. It is possible to monitor total volume of the circulating mud simultaneously with the individual volume in each pit with the help of such mud pit volume indicators (sahay, 1983.).

Recently, Dupin de Sainteyr (1973) discussed the development of ultra sonic equipment which accurately measures drilling fluid levels in mud tanks. These measurements are made without any direct mechanical contact with the drilling mud, a method particularly effective on floaters in the offshore areas

(vi) *Hole Fill Up*

During the pulling out the drill string the mud volume needed to fill the hole should

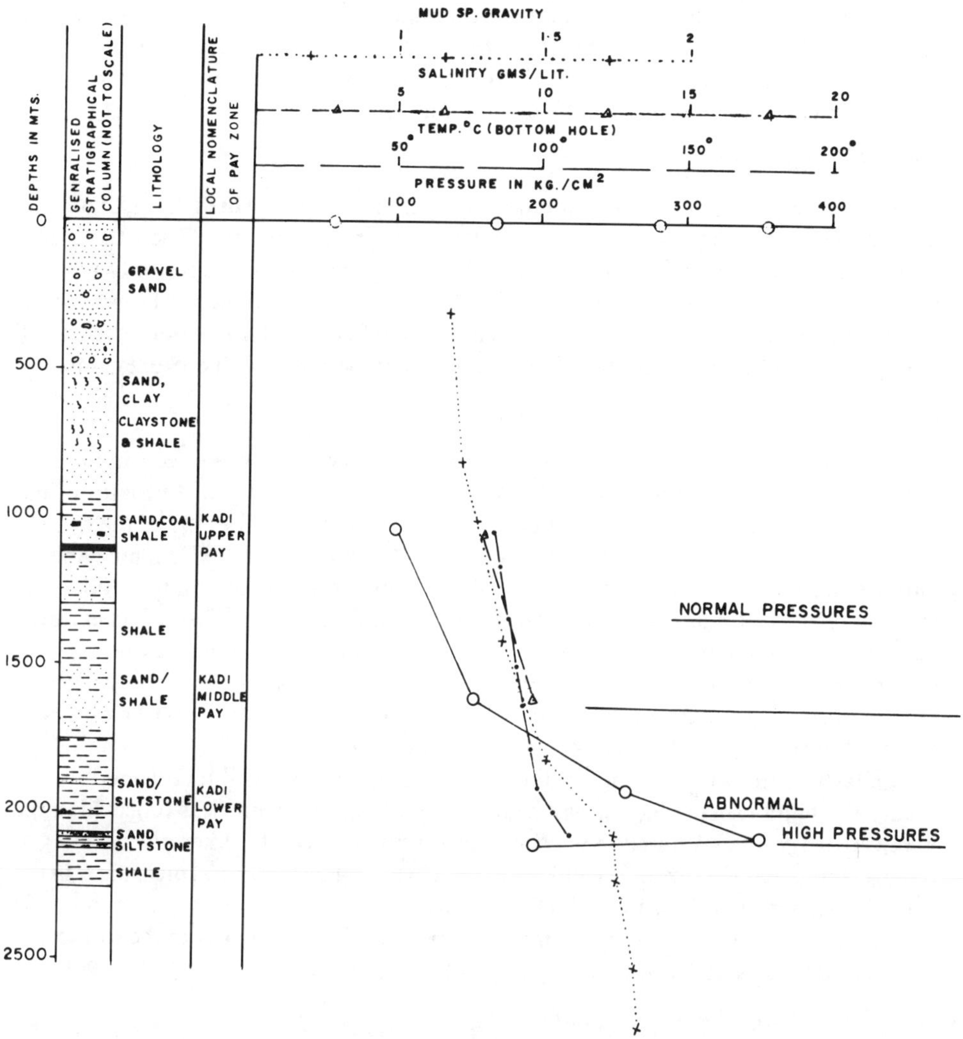

Fig. 7.18Comparative studies on pressure, salinity and bottom hole temperature in Kadi Area, India

equal the pipe volume. Keeping the hole full is even more important at the time the drill collars are pulled since if the same length of collars compared to drill pipe is pulled, the level of drilling mud in the hole will fall four to five times as fast. Furthermore, there could be a temporary pressure reduction while the bit is pulled due to balling of bit, high viscosity of mud, thick mud cake, back pressure valve on string, etc.

In case of influx of gas, oil or formation water into the mud column, the mud volume required to fill the hole will be less than the volume of the pipe pulled and would give the first indication of a kick. Changes in the mud pit level can be known by constantly measuring the mud volume used to fill the hole (Fertl, 1976, pp. 147-150).

(vii) *Mud Flow Rate*

Influx of fluids in the mud column from the formation which is indicated by abnormal rise in pit level will also be reflected in increasing flow rate. The latter can be measured by a standard flow meter. Such flow measurements are superior to pit level checks since small flows can be detected before they become sufficiently large to show up on any pit level device. It would thus always be better to know about such small flows well ahead so that the remedial measures can be taken at the proper time.

(viii) *Mud Pump Pressure/Pump Stroke Counter*

By monitoring pump pressure, it is possible to observe "U" tube effects in the bore hole caused by entry of less dense fluid into the annulus. Entry of formation fluid results in the reduction of hydrostatic pressure in the annulus. Because of the heavier mud in the drill pipe, there will be a tendency to "U-tube" down the drill pipe and up into the annulus provided that the entry is slow enough to allow some of the mud inside the drill pipe to maintain continuity with the mud in the annulus. This influx causes a reduced hydrostatic head which results in a decrease in pump pressure and as such provides an overpressure detection method. However, in case of pore pressure and rock permeability being very large, an increase in drill pipe pressure may be observed rather than the expected decrease (Fertl, 1976, pp. 147-150).

Similarly, influx of formation fluid into the mud column, will tend to reduce the mud weight and it being lighter will have a tendency to pump more as reflected in the pump stroke counter, housed in the geologging unit. This can be taken as a warning of the overpressured horizon and would thus help in taking fast remedial measures. In the Port Canning well in West Bengal, India, such increase in pump strokes from 46 to 51 helped in forewarning of overpressures well ahead in the section before the formation water influx was seen on the other gadgets (Sahay, 1974).

(ix) *Conductivity/Resistivity Studies*

Since overpressure zones show a marked increase in conductivity, it is desirable to have a conductivity/resistivity monitoring device to record the conductivity/resistivity of the mud during the course of drilling. Mathews (1969), proposed chloride content measurements of the mud and stated that "this indicator is valid but is difficult to establish unless there is close control over the mud properties and analysis. It is possible for the mud engineer to make routine chloride, iron tests but the routine test is not adequate to show the minor changes. Additional tests will have to be run".

Many geologging companies are equipped to measure both inlet and flow line mud resistivity/conductivity during the course of drilling. In case of any increase in the conductivity of the return mud above the normal, this could be taken as an indication of the occurrence of overpressures in the section down below. Provided such information is interpreted with other indicators for the prediction of overpressures as discussed earlier, it would provide valuable clues.

Salinities of Interstitial Water

Besides the above indicators, salinities of interstitial water of shale and porous rocks have also provided useful clues in distinguishing a normally pressured section from an abnormally pressured one.

Interstitial waters from shale and sandstone show a contrast in conception and composition. According to Dickey (1969), water in the sandstone from normally compacted sediments increases in total dissolved solids with increasing depth at a rate of 20,000 — 50,000 mg/litre per 1,000 ft. Water from abnormally pressured sandstone has very low total solid contents. The major cations and anions in the sandstone water are sodium and chloride. The sodium and chloride contents increase with depth, except in cases of abnormally pressured zones. The proportion of potassium in the total dissolved solid also increases with depth. The ratio of sodium to potassium is low in the fresher, abnormally pressured water. The ratio is much higher, about (150-400), in more concentrated normally pressured water. The ratio of sodium to magnesium remains constant until the higher pressure zones are encountered. Magnesium decreases rapidly relative to Na^t with freshning of abnormally pressured waters. A similar relation holds true between calcium and sodium.

The bicarbonate concentration of abnormally pressured water increases rapidly to almost 2000 mg/litre when compared with normally pressured water, where it gradually increases from less than 100 to 300 mg/1. Sanstone has high bicarbonates content as compared to the normally pressured water. The sulphate content is usually high in a high pressured zone as compared to the normally pressured zone. Barium is very low or absent in the waters from the highly pressured sanstone.

Interstitial shale waters exhibit salinity trends similar to sandstone waters except that the normally pressured shale water has much lower total dissolved solid concentrations than the adjacent sandstone waters. The sodium and chloride of the interstitial waters in shales show similar trends but at lower concentrations compared to sandstones. The sulphate concentrations in the interstitial water of the shale are unexpectedly high in the abnormally pressured shales and are greater than biocarbonate concentrations which generally exceed the chloride concentrations.

There is a very marked contrast between the composition of the interstitial water of shales and that of sandstone in both normally pressured and abnormally pressured zones. The normally pressured sandstones contain water of high salinity but low bicarbonates and sulphate. The reverse is the case in abnormally high pressured sandstones with bicarbonate concentrations being intermediate between those in the normally pressured sandstone and shale waters (Sahay, 1974).

Thus, a thorough analysis of interstitial waters from both shale and porous reservoir rocks may provide significant clues about the presence of abnormally pressured horizons in a section, as indicated by decreasing salinity of the total dissolved solids compared to the water from a normally pressured section and a contrast in the concentration of other radicals.

Water samples thus obtained are subjected to detailed analysis and basic and acid radicals are worked out and stiff diagrams are plotted, which give a pattern and thus help in distinguishing as to whether the particular pattern is representative of water from a normal or an abnormal pressured zone. Figures 10.10 and 10.11 show several water pattern curves both from abnormal and normal pressured horizons, located in the Bengal Basin, India.

Other Novel Concepts

Several other novel concepts, such as specification variations or redox and pH measurements on the ingoing and outgoing mud have been studied in detail and (Fertl, Timko, 1976) may offer additional possibilities as continuous indicator. However, several factors, such as drilling mud additives, down hole chemical reactions, etc., may significantly affect interpretation of such indicators.

Shale Cutting Parameters

Shale density is a useful tool for determining the time breaks in areas where exploratory drilling, is carried out, which is indicated by the break in the shale density curve. In fine grained clastics, the increase in density is a function of load, time and tectonic stress. Density increases as compaction takes place, the rise being at a very rapid rate in the first few metres of the sediments and below this at a rate which decreases with the depth (Sahay, 1974, 1983).

Under ideal conditions, relatively pure shales will have the same density and porosity at a particular depth. If the shales retain constant composition from the surface upto a great depth, both the porosity and the density values would form smooth curves, when plotted against the depth. In case of any unconformity in the section, a sharp break in the shale density curve from the normal would be noticed. It would thus indicate a method for locating unconformities in a section, thus enabling us to find out the stratigraphic traps in an unknown area. Quite often, the unconformity in a sedimentary sequence is not clearly seen on the basis of lithology, palaeontology, palyonology. The absence of any contrast in lithology, palaeontology, or palyonology and absence of a meagre yield of microfauna/flora in the sediments, account for the above difficulty (Sahay, 1983, pp. 74).

Besides, recognizing unconformities in a section, the shale density is equally useful in locating the high pressure horizons that are likely to be met during the course of drilling. In such cases of high pressures environment, the shale shows lower density. Under the normal conditions, when shale compacts due to overburden pressure, water is squeezed out during the course of sedimentation and the shale porosity decreases and as such the density increases with depth. On the other hand, if the fluids become trapped in the shale formation for some reason, such as lack of enough sands to serve as permeability pipe lines, the fluids support some of the weight of overburden and are under higher than normal pressure. Since fluids are trapped in overpressure shales, the shales exhibit higher than normal porosity and lower than normal density (Sahay,

1974). In the overpressure shale, travel time on the sonic log is higher than normal and resistivity is lower than normal. Normal compaction trends for each type of log are determined from normally pressured shale (Rogers, Oil and Gas Journal, 1966). Plotting of shale density v/s. depth results in a curve which identifies the transition zone between normal and abnormally pressured sections. Locating of the transition zone, as it is called, thus helps the geologists in framing and devising of casing and mud policies and improving of drilling technology and better reservoir evaluations. In the Fig. 10.6, results from the wells drilled in the Bengal Basin would show the difference in densities in a normal pressured section than that of abnormal one. The regions in India were both unconformities and transition zone between normal and overpressured horizon were found out from shale densities are wells in Bengal basin, Arabian Sea. The most suitable method for finding out the densities is to select pure shales from cores, collected at different intervals. In case sufficient cores are not available cutting Samples are also taken. However, while selecting drill, cuttings, proper care should be taken to eliminate the cavings which are generally seen in a cutting sample. Siltyness, calcareous, and organic and Iron matter in shales greatly affect the density measurements. Hence such shales should be eliminated as far as possible. The presence of heavy minerals, sand or silt in a particular shale sample increases density while organic and resinous matter decreases the same.

The method of determining the specific gravity (sp. gr.) is as follows:

(i) *Method*: Cutting samples collected after eliminating cavings should be dried and washed, and the weight in air taken. Use of sp. gr. bottle becomes necessary while dealing with the samples of smaller size and quantity. The weight of the sp. gr. bottle filled with distilled water is taken. Then the shale samples are immersed in the distilled water and the weight of the sample in water is taken. the results may be tabulated and sp. gr. determined by the following formula:

$$\text{Sp. Gr.} = \frac{\text{Weight of the samples in air}}{\text{Wt. of the sp. in air} - \text{Wt. of sp. in water}}$$

In case of cores, it would be desirable to put a coat of varnish and dry them for a couple of hours, before taking the weight in water, so that the pores are all sealed and no water goes inside and affects the density measurements.

In case of cuttings, because of their smaller size, it is not possible to put a coat of varnish on these samples, but due to non-availability of the core samples, the cutting samples too have to be sometimes selected. In such cases, it has been seen that even if some water enters, it would be negligible in quantity and would not substantially affect the results, especially so when the accuracy generally taken for such studies is only up to the second decimal place (Sahay, 1974, 1983).

In some cases, when the bulk density of the clay samples is desired, a different method is suggested.

(ii) *Determination of the Density of the Clays*: The specific gravity bottle is also used in the case of the cutting sample of clay. The following procedure is recommended for determination of the density of clays.

(1) Clean the sp. gr. bottle and find its weight when completely filled with water, when empty and when half filled with dry clay powder.
(2) As clay may be soluble in water, fill the portion of the bottle above the clay with kerosene oil. Place the thumb over the mouth of the bottle and shake the bottle to remove the air bubbles completely. Add more kerosene oil, if necessary, to fill the bottle up to the brim and replace the stopper. Wipe bottle dry on outside and find its weight.
(3) Remove the contents in a beaker and rinse the bottle with kerosene oil to remove any clay powder sticking to the sides. Fill it completely with kerosene oil, replace the stopper. Wipe it dry on outside and find its weight again.
(4) Note the temperature of water and find its density at that temperature from the tables.

Record: Resting point of the balance =

Weight of sp. gr. bottle (empty)	W1 = gms
Weight of sp. gr. bottle + water (full)	W2 = gms
Weight of sp. gr. bottle + clay powder	W3 = gms
Weight of sp. gr. bottle + clay powder + oil	W4 = gms
Weight of sp. gr. bottle + oil (full)	W5 = gms
Temperature of water	t = °c
Density of water t°c	gm/cc

Calculations

(a) *Specific gravity of kerosene oil*

Wt. of oil completely filling sp. gr. bottle = W5 − W1
Wt. of water completely filling sp. gr. bottle = W2 − W1 gms

$$\text{Sp. Gr. of kerosene oil} = \frac{W5-W1}{W2-W1}$$

$$\text{Sp. Gr. of Kerosene oil w.r.t. water at } 4^\circ c = \frac{W5-W1}{W2-W1} = d$$

(b) *Specific gravity of clays*

Wt. of clay powder in bottle = W3 − W1
Wt. of oil filling the space above the clay= W4 − W3
Wt. of oil having the same vol. as the clay = (W5 − W1) − (W4 − W3)

$$\text{Sp. gr. of clay w.r.t. oil} = \frac{W3-W1}{(W5-W1)-(W4-W3)}$$

$$\text{Sp. gr. clay w.r.t. water at } 4^\circ c = \frac{W3-W1}{(W5-W1)-(W4-W3)} \times d$$

Where ‘d’ is sp. gr. of kerosene oil.

(c) *Precautions*

1. All the precautions for the use of a balance should be observed.
2. The sp. gr. bottle should be thoroughly cleaned, rinsed with water and dried before weighing.
3. The bottle should always be held with clean water and never in hand, otherwise the liquid will expand and come out of the bottle due to the heat of hand.
4. The stopper should not be pressed too tight.
5. The bottle being very delicate should never be held by the neck.
6. The air bubbles should be removed completely by inverting the bottle and shaking it gently.
7. The bottle should be wiped clean and dried from outside before each weighing (Sahay, 1974, 1983).

The density logger should take care to select clean shale cuttings from the bottom and uniformly prepare the cuttings for analysis in order to get uniform results.

Shale density studies were carried out in Bengal basin, Surinsar well, Jammu and a well on Tarapur structure and in the well in Bombay High, India. Along with the study of resistivity of the shales, it enabled prediction of the overpressure horizons in the wells in Bengal basin and in another well in India well before they were penetrated by the bit (Sahay, 1974, 1983).

These studies, as such, are strongly recommended where the generation of overpressure in the area is due to the compaction processes in response to stress of overburden. They may, however, also greatly help in the prediction of abnormally high pressure horizon, where such pressures are due to compressional forces (Sahay, 1972, 1974).

Recently, many service companies devised suitable instruments for determining the densities of shales without going through the usual method of finding the specific gravities with the help of a physical or chemical balance. The shale density kit or the Microsol instruments, devised by Baroid and Geoservices are valuable gadgets for undertaking such studies and should be put on all exploratory wells, specially where the high pressures in a section are encountered or anticipated.

(iii) *Baroid Shale Density Kit*: Baroid division of National Lead Co. has developed a convenient method of making density determinations and offers formation density logging as part of its mud logging service (Sahay, 1983, pp. 79).

It provides a formation density kit which includes a 350 ml graduate, two fluids of different densities and four to six calibrated beads of various known densities. The two solutions are poured into the graduate. They are miscible and form a linear density gradient. After the beads are dropped in the graduate, they sink to different levels, dependent on their density. The geologist/logging engineer records the density of each bead and position in millilitre at which each comes to rest in the graduate. These values are plotted on graph paper. A straight line is then constructed using the reference points. When the well passes through a shale section, representative samples of the shale cuttings are taken and dropped into the graduate. The cuttings sink in the column to a depth which is the measure of their density. For example if the cuttings fall to a level of 110 ml in the graduate, the density of the cuttings is 2.55 gm/cc. Temperature changes will affect the calibration of the density gradient. As temperature rises and falls, the calibration will change causing the beads to move up or down the column. Should a noticeable change in ambient temperature occur, the logger checks the position of the beads and if necessary redraws the calibration curve.

(iv) *Microsol*: The Microsol device manufactured by the Geoservices France (Figs. 7.19 and 7.20), is another device to measure shale density. The procedure for determining the density is as follows:

Fill up the Altuglass tube with fresh water until the level is at about 2 cm of the top. Put the floating element in the water. Care should be taken to eliminate air bubbles on the immersed surface. In case some are left, rotate the floating assembly to remove them. When the floating assembly is stabilized, the top of the floater should be at least 1 cm above the water level. If not, remove a little water.

Drop into the water a few drops of teepol or soapy water. The idea is to lubricate the contact between the upper string and the water surface. The graduated scale on the mirror is used to measure the position of the floating assembly without parallel mirror.

To measure, look at the point of the needle so that it is at the same height as its image in the mirror and read the corresponding value on the scale. When there is no sample on the microsol it should read zero. If not, adjust the position of the mirror bean finely to adjust the zero with the knob, right of the mirror. If the room temperature changes, the equilibrium portion of the floating assembly will change too and zero should be reset. Put one or several cuttings on the cup so that the floating assembly descends until ballast weight arrives near the bottom of the tube. Read the graduation L 1.

Remove the cuttings from the cup and put them on the top of the floater inside the water. No air bubble should remain on the cuttings. After the stabilization, read new graduation L 2.

Density is calculated by the following formula:

$$d = \frac{L\,1}{L\,1 - L\,2}$$

Density determinations should be done preferably every 10 metres. In case abnormal shales are expected, the frequency should be increased. Plot the values on the density pressure chart (Fig. 7.21). The points would normally show variations due to small facies differences or could be due to measurement errors. In case, the variations are much larger, it would indicate that either the samples selected are not representative of the formations, not washed properly, dried or there also could be a mistake in measurements. Trace the curve which passes in the middle of the band. In normal formations, it will be a straight line, shown as increase in density with depth (Sahay, 1974, 1983).

The attached density pressure chart (Fig. 7.21) gives an example from well H-12-1, as to how the pressure at 2730 m, which had abnormal high pressures of the order 500 bars (515 atms), was calculated from the shale density curves (Sahay, 1983).

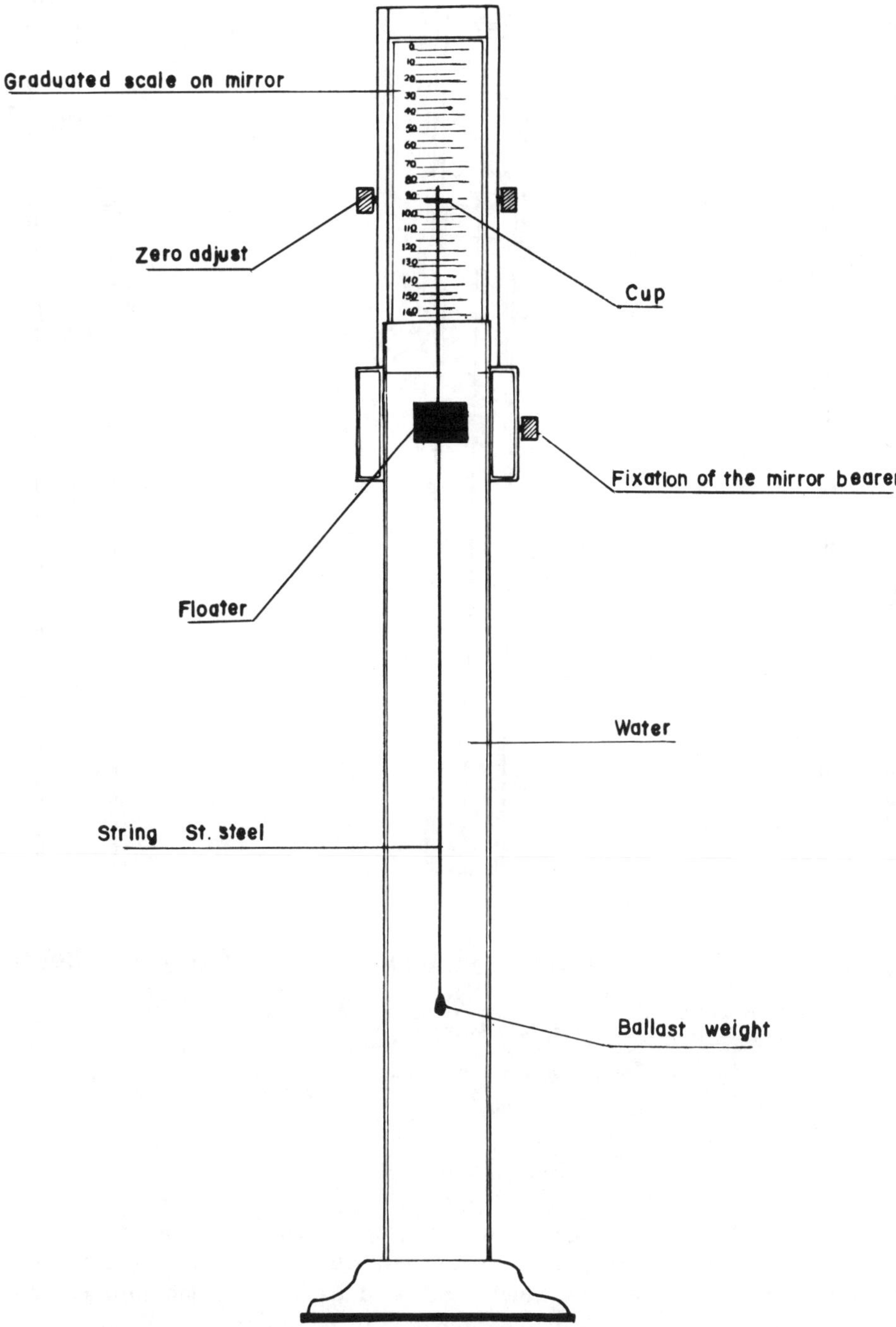

Fig. 7.19: Microsol
(Courtesy: Geoservices, Paris)

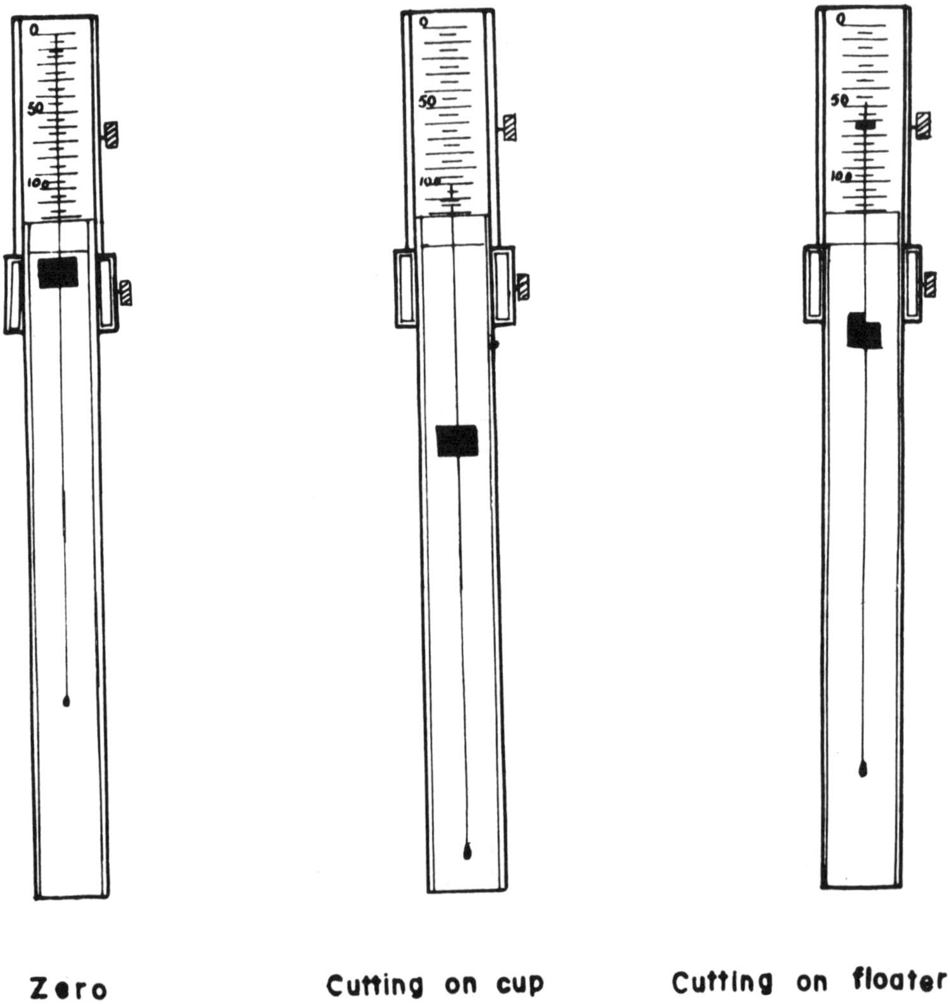

Fig. 7.20: Microsol

(Courtesy: Geoservices, Paris)

Shale Factor

Shale factor analysis is also utilized for over pressure detection. This can be carried out through X-ray diffraction analysis for the investigation of the type and amount of mineral components in the rocks, which is restricted to the regional laboratories, only. However, some simplified rig-site determination of the mineral constituents in the rocks can be carried out. One such method which has proved successful is the

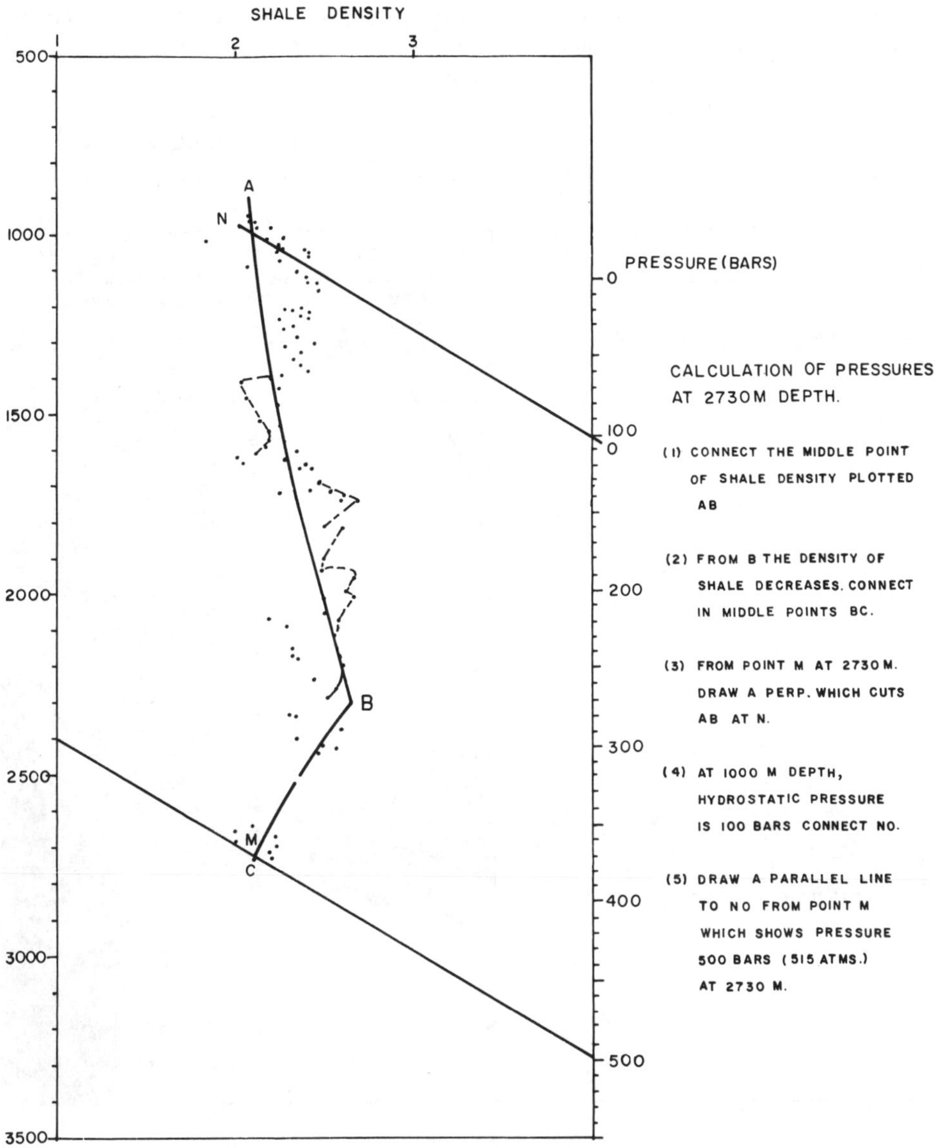

Fig. 7.21: Depth pressure chart, Well A

methylene blue test (Nevins and Weintritt, 1967), better known to geologists and engineers as the "shale formation factor" method (Gill, 1968, Mondshine 1969, Gill and Weintritt, 1970 Gill, 1962). Usually the shale factor may be equated with the cation exchange capacity of solids carried out of the bore hole, which in turn may be related to the montmorillonite content and thus the water holding capacity of the cuttings (Fertl,

1976, pp. 158). As is well known, montmorillonite clays exhibit higher cation exchange capacity than other clays. An increase in the montmorillonite clays will mean a lighter density shale which could mean that it could be interpreted as under-compacted shale. Several authors have associated montmorillonite with deep marine shales in which sands are absent. During the process of diagenesis, montmorillonite is converted to mixed layer and illite clays, hence montmorillonite decreases with depth (Fig. 7.22).

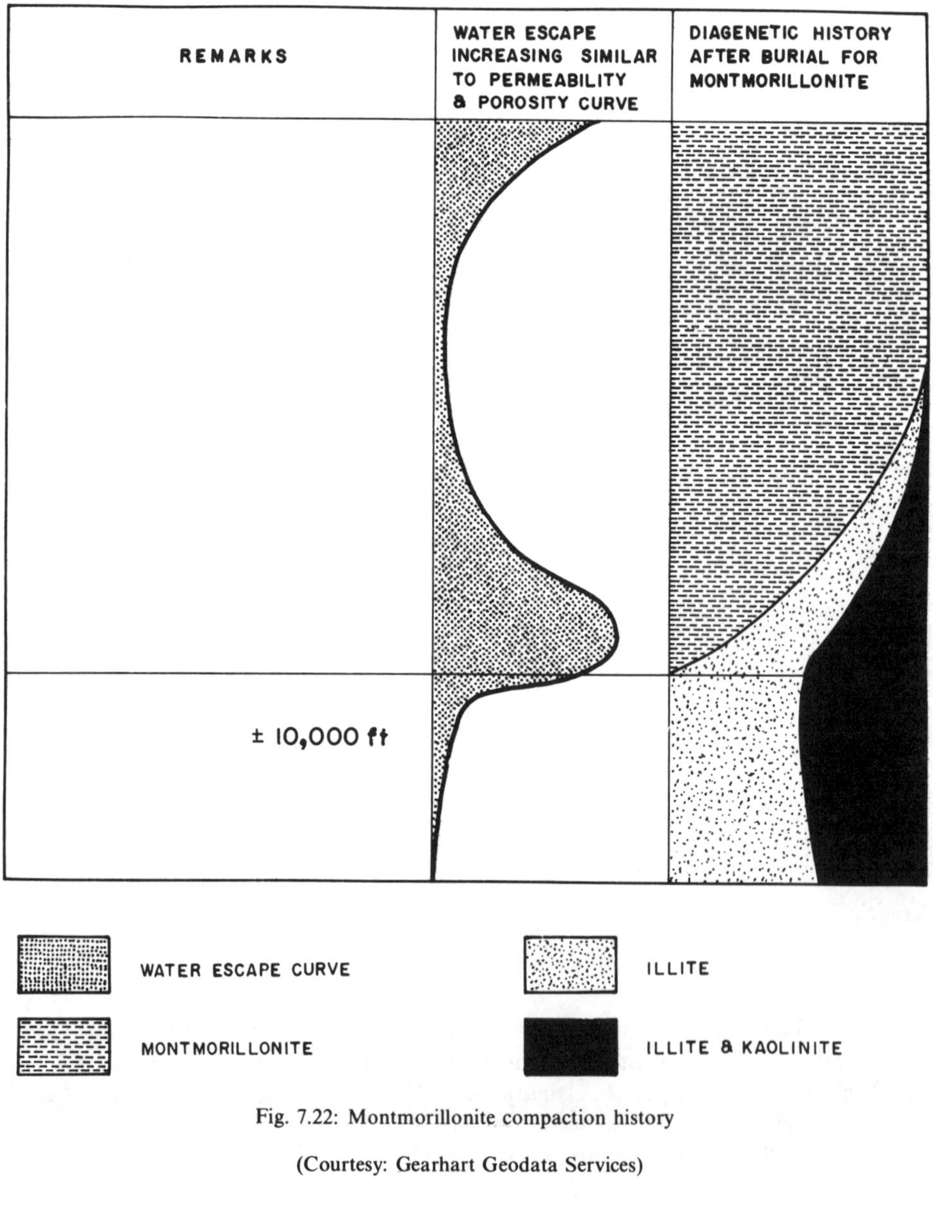

Fig. 7.22: Montmorillonite compaction history

(Courtesy: Gearhart Geodata Services)

Montmorillonite/illite ratio in the sample is measured as the cation exchange capacity in milliequivalents per 100 gms. of samples and termed "shale factor". In essence, shale factor gradually decreases with depth and shows an increase in an abnormally pressured zone.

General Shale Classification (Mondshine, 1969)

Texture	CEC	Water content	Wt. water	Clay content	Wt. % clay	Density Sp. gr.
Soft	20–40	Free & Bound	25–70	Montmorillonite	20–30	1.2–1.5
Firm	10–30	Bound	15–25	Illite and mixed layer montmorillonite illite	20–30	1.5–1.2
Hard	3–10	Bound	5–15	High illite tr. montmorillonite	20–30	2.2–2.5
Brittle	0–3	Bound	2–5	Illite kaolin chlorite	5–30	2.5–2.7
Firm-Hard	10–20	Bound	2–10	Illite and mixed layer montomorillonite illite	20–30	2.3–2.7

Shale factor determination is carried out as follows:

1. Dry the sample
2. Weigh 1 gram of sample and crush to fine powder in a mortar and pestle.
3. Add sample to about 30 ml of distilled water.
4. Wash the suspension into a titrating flask, and acidify with about 2 mls of 2 N Sulphuric acid.
5. Titrate the acidified suspension against methylene blue solution (3.75 gms/lt), vigorously shaking the flask after each 0.5 ml addition of methylene blue solution. After each solution, take one drop of solution from the flask (on a striking rod) and test on a filter paper for the end point.

On the filter paper, each drop will produce a dark blue stained aggregate of clay particles surrounded by a fluid ring. When the fluid ring changes from colorless to blue, the end point is reached.

$$\text{Shale factor in milli equivalents 100 grams} = \frac{100}{\text{Weight of dry sample}} \times \text{Volume of Titrant (mls)} \times \text{Normality of Titrant}$$

Since normality of titrant = 0.01N

$$\text{Shale Factor} = \frac{100}{1} \times \text{Volume of titrant} \times 0.01 \text{ (mls)}$$

Therefore shale factor = Volume of Titrant (in mls). (Gearhart Geodata Overpressure manual, 1980, pp. 198-199).

The shale factor not only helps in the identification of clay composition but it is also representative of the clay/sand abundance ratio in shale cuttings or in general the clay content of the section. It is very important to know the above while drilling carbonate sections with few shale streaks, where density data in such a case could give erroneous interpretations. This also assists even in silty, sandy shales. The shale factor is also useful for detection of the impermeable cap rock often found at the top of high pressure sections.

Recently, shale factor logs have been utilized to locate formations that cause severe bore hole instability. In conclusion, the shale factor analysis, when carried out with other pressure detection techniques, is a useful tool for the prediction of overpressured zones.

(iii) *Volume, Shape and Size of the Cuttings*

As discussed earlier, all overpressure formations are characterized by an increase in the penetration rate, which results in increased volume of cuttings over the shale shaker. It also results in changing the shape and size of cutting samples. In the transition zones, the shape of cuttings overlying overpressure zone is usually angular and sharp rather than rounded, as is the case in the normal pressure environments. In addition, cuttings from overpressure environments are usually large and generally splintery in appearance, hence another clue for the overpressured horizons (Fig. 7.23).

(iv) *Calcimetry Concepts*

As already discussed in Chapter 5, overpressures can be generated due to chemical change of Anhydrite to Gypsum, due to hydration and dehydration effects, which results in a volume increase by about 15-25%. It would, therefore, be essential to understand the carbonate content of the rocks, while carrying out overpressure studies.

While carrying out overpressure studies on cores and cuttings, it would be necessary that the lithology of the samples of the shale cuttings and cores is properly known. In case the shales are calcareous or contain heavy minerals, this will mask the overpressure detection technique. Erroneously high density results because of the carbonate or the iron content of the shale. Geologging companies have devised various instruments as part of their logging services for carrying out the studies on the calcareous content in a sample.

A brief description of the Bernard Calcimeter and Mano Calcimeter is being given (Figs. 7.24 and 7.25).

A graduate tube full of salt-saturated water is connected by its lower extremity to a Mariotte ampoule and by its upper extremity to a 100 cc flask. This apparatus is used for measuring the CO_2 gas volume produced by the reaction of hydrochloric acid on calcite.

Fig. a: These shale samples from just under the pressure seal in a well are mostly small with rounded edges. They were generated by the bit

Fig. b: From 80 ft below the pressure seal of the same well as in Fig. a, samples are fairly large with sharp edges. They were spalled into the well bore

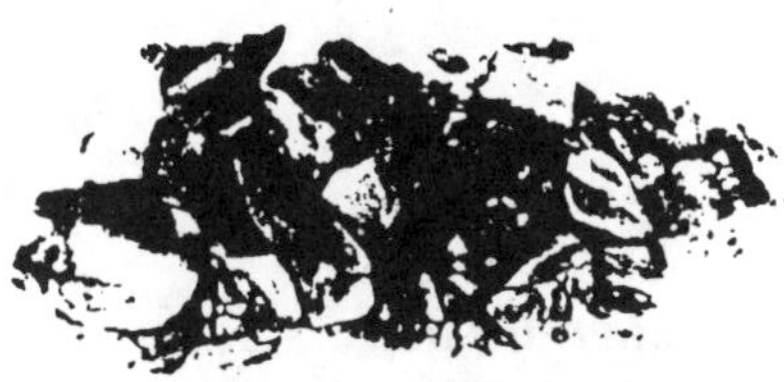

Fig. c: Samples from 150 ft below the pressure seal are mostly very large with sharp edges

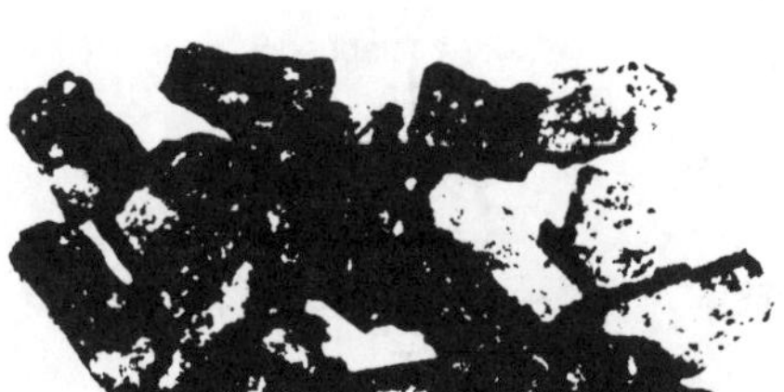

Fig. d: Samples from a normal-pressured well are large with rounded edges

Fig. e: Long slender shale samples with sharp edges indicate the well bore is being destroyed by drill pipe in compression

Fig. 7.23: Shale cuttings in an overpressured environment

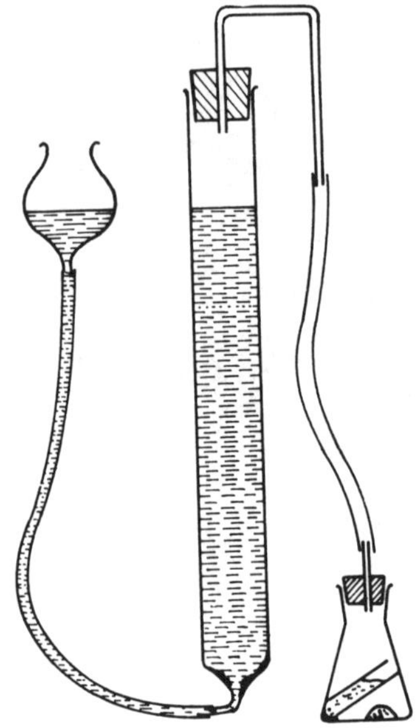

Fig. 7.24: Bernard Calcimeter

(Courtesy: Geoservices)

Directions for use

(i) Weigh 5 g of dried and crushed cuttings on a filter paper.
(ii) Pour 10 cc of hydrochloric acid into the flask.
(iii) Pour pure hydrochloric acid into hemolysis tube and put it inside the flask.

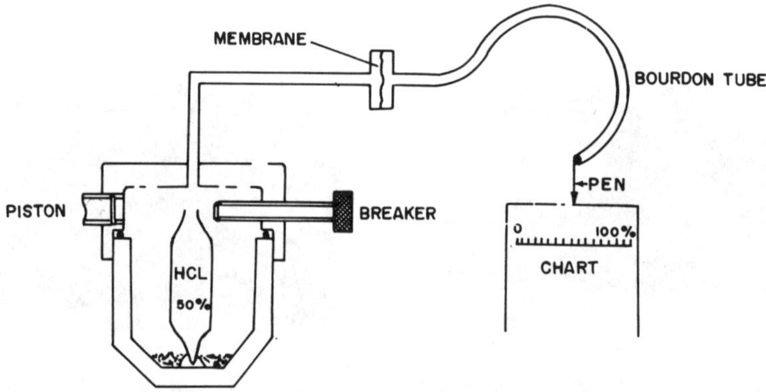

Fig. 7.25: Manocalcimeter

(Courtesy: Geoservices)

(iv) In order to identify two acids, put a few drops of potassium chromate into concentrated acid.
(v) With the help of tweezers, put continuously the filter paper against Enreenmeyer glass wall and be careful that acid does not wet the sample until the flask is connected to graduated tube, lower slightly the Mariotte ampoule to equalize its water level with graduated water level. Read the water level of scale, point zero. Allow the diluted acid to soak sample completely.
(vi) Make the difference between the second and first readings noticed at the room temperature and find out the calcite content with the help of the calcimeter table.

In order to determine the dolomite content proceed as follows:

(i) Proceed again as mentioned before with the concentrated acid.
(ii) After 12 minutes, read the water level on the scale.
(iii) The dolomite content can be found out with the help of calcimeter table.

Remarks

(a) If a calcimeter table is not available, it is sufficient to multiply the reading by '8' for .5 gm. sampels @ 20° C.
(b) Temperature corrections — They are taken from Calcimetry table, if not, use the following formula.

$$Vt = Vo\,(1 + dt)$$
$$Vo = \text{gas vol at } 0°C$$
$$Vt = \text{gas vol at } 0°C$$
$$t = \text{room temperatue} = \frac{1}{173}$$

(a) The carbonate content can also be determined with the help of titremetry, using the method as follows:
(i) 1 gm of dried and crushed cuttings are charged by 20 cc of normal hydrochloric acid.
(ii) For the calcite content operate at normal temperature. For dolomite conent, operate at high temperature.

The following reaction takes place:

$$2\ HCl + CaCO_3 = Ca\ Cl_2 + CO_2 + H_2O$$

2 HCl molecules, neutralized by 1 Ca CO_3 mol, i.e., 100 gm = $CaCO_3$
1 gm $CaCO_3$ is neutralized by 1/50 mol of HCl, i.e., 20 cc of normal HCL. $CaCO_3$ content to be x/100, after reaction HCL vol remaining unneutralized by $CaCO_3$.

$$20\ \frac{(1 - x)}{100}\ cc$$

Filtrate the solution.

Put into a beaker 10 cc of solution, add 2 drops of helianthene. The vol of the HCL remaining unneutralized.

$$10 \frac{(1 - x)}{100}$$

Neutralize with a normal NaOH solution. When the colour of solution changes from red to orange, read the volume of NaOH added.

$$N = \frac{10 - x}{10}$$

$$N = 100 - 10\,n$$

$$CaCO_3 \text{ is } 1 - \frac{n}{10}$$

(b) Wellsite method for determining carbonate content in the rocks.

A simple method of carbonate content determination in the wellsites is discussed below.

Method

(i) The ditch samples at 25 m or any convenient interval are taken for such studies. Approximately 5 gm. of the sample, after being dried, is powdered and exact weight of the sample is measured on a physical balance.

(ii) The sample is then treated with 10% hydrochloric acid and heated till such time as the effervescence ceases.

(iii) It is then treated with distilled water and washed thoroughly with the distilled water till the filtrate stops giving white precipitate with silver nitrate. This ensures that all chlorides present in the sample are removed and the residue left on the filter paper, which is also weighed previously, contains only carbonates.

(iv) After it is ensured that the sample contains only the carbonate and the filtrate does not give any white precipitate, the filter paper with the residue is taken out and dried thoroughly in a drying chamber.

(v) The filter paper along with the dried sample is again weighed on the physical balance and the weight of the residue determined.

(vi) The difference in weight of the sample prior to the treatment with the hydrochloric acid and weight of the residue thus taken after the treatment, gives the carbonate content in a particular sample.

(vii) All precautions are taken for a correct determination of the weight and for preventing the samples from contamination with dust particles, etc.

(c) Manocalcimeter:

Geoservices, France, has devised a gadget for determining the carbonate content of the rock sample with separate determination of limestone and dolomite.

It is carried out by acid attack on a quantity of rock, approximately weighing 1 gm. in a constant volume cell, and recording the amount of CO_2 produced. To use this instrument, the following procedure should be followed.

(i) After putting the manocalcimeter on the table, unscrew the knob to rotate the arm of the reaction cell normally to the right.
(ii) Before using the gadget, the clock should be wound.
(iii) Fill the glass reservoir of pen with ink.
(iv) If necessary, zero should be adjusted.
(v) Unscrew the pot from the reaction cell and put the powdered 1 gm. of sample in it.
(vi) Take an ampoule of acid, cut off one end, insert the other end in the tapered hole of ampoule holder so that the ampoule is held firmly in a vertical position.
(vii) Pull back the button.
(viii) Screw the pot into the cell unit until "o" ring is tightened.
(ix) Connect the flexible hose to the elbow nozzle on top of the cell and place the other end of the hose in a suitable spot for discharge.
(x) Leave the lever of the clock stopper to start the chart driver and push the button to break the ampoule and start the reaction. Pen records the pressure of CO_2.

When pen is stabilized, pull button to vent the CO_2 through the hose and obtain the atmospheric pressure in the cell. Then unscrew the pot.

Novel Miscellaneous Detection Methods

Besides the pressure detection methods as discussed above, several unconventional methods for predicting the presence of overpressures during drilling have been investigated (Fertl and Timko, a,b 1973). These methods include shale cutting resistivity, filtration rate of shale cutting slurry, filtrate (shale water) colour index, shale cuttings, moisture index, Redox and PH potentials of cutting slurry and slurry filtrate. Based on the initial field experiences in various places in the world, several interesting and promising observations have been made. However, additional research and field data will be necessary to study in detail each one of these indicators.

Shale Slurry Resistivity

As is known, analysis of the electrical properties of the shale with wireline logs is one of the best methods for prediction and quantitative formation pressure evaluation. However, this will be discussed in detail in the chapter to follow. In recent years, electrical measurements have also been made on shales at the well-site. Such tests are made on the shale cutting slurry, slurry filtrate, or by measuring minute changes in the mud stream resistivity (salinity) on both in and outgoing mud streams.

Weaver and Beck (1971) carried out shale resistivity measurements on a series of side-wall core samples in the south-west area, Louisiana. Similar tests by Fertl and

Timko also show decreasing shale resistivity values in overpressured zones (Figs. 7.26 and 7.27).

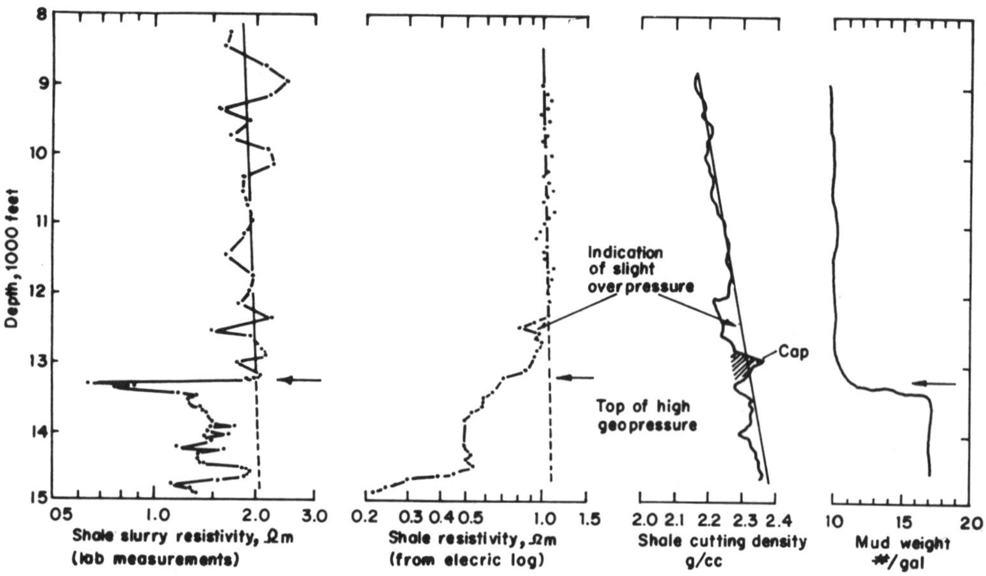

Fig. 7.26: Overpressure indicated by a shale slurry resistivity plot, which compares closely with those of other more commonly used pressure indicators, such as bulk density and log-derived resistivity

(After Fertl and Timko, 1973b, Courtesy of World Oil)

Filtrate Rate of Shale Slurry

As it is well known, in overpressured environments several geo-chemical changes take place in the formations. Similarly a change will occur in filtration rate from shale cutting slurry under constant pressure. Factors affecting the filtration rate include shales and ratios of cutting samples, particles size, type of clay minerals present, pH and electrolite concentration (salinity) of interstitial fluids in shales etc. (Figs. 7.27 and 7.28).

Studies carried out on the wells in the offshore Louisiana, observed filtration times measured on the shale cuttings slurries. Filtration time is the time required to obtain 50 ml of filtrate with filtration pressure of 35 psi (2.46 kg/cm^2). The amount of filtrate can also be measured for a given time. In a normal pressure environment, the time of filtration is fairly constant. However, prior to entering overpressure environment, filtration time decreases, the lowest value coinciding with onset of high pressure as pinpointed by other overpressure detection techniques and mud weight requirements. It would appear that filtration tests may be one of the ways to predict the proximity of overpressure zones, prior to entering them. Such filtration measurements are simple and can be used simultaneously with the colour index indicator.

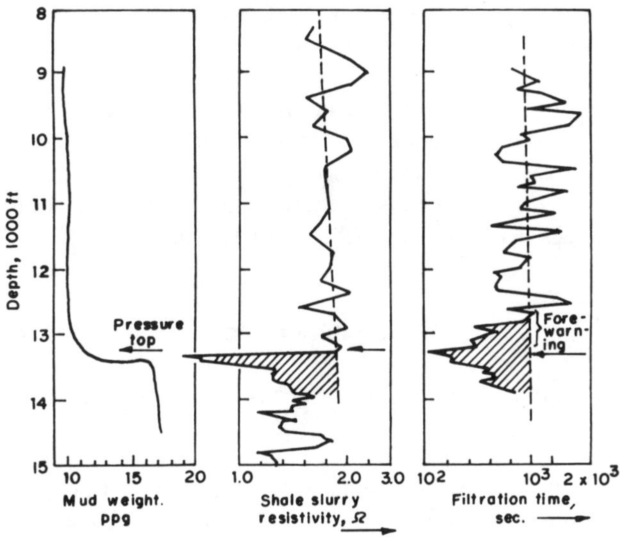

Fig. 7.27: Filtration rate of shale cuttings slurry under constant pressure changes as over-pressure is approached

(After Fertl and Timko, 1973b)

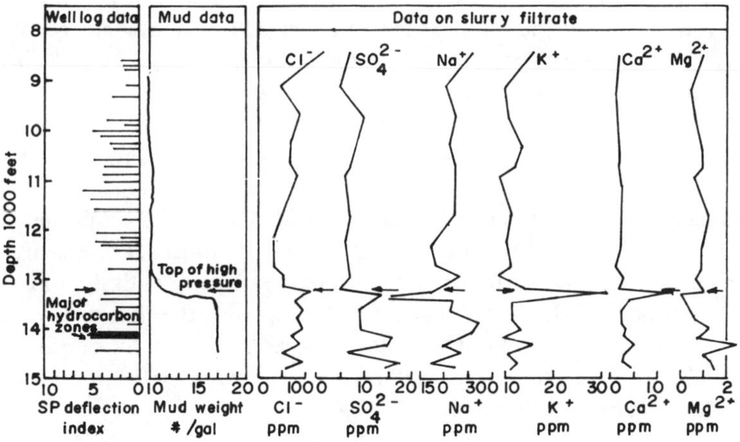

Fig. 7.28: Apparent concentration variations of specific ions in shale cuttings slurries in normal and over pressured zones (after Fertl and Timko, 1973a). SP deflection index gives the maximum value of spontaneous potential (SP) curve for sandstone reservoirs, in mV

(Courtesy of World Oil)

Shale Water Colour

Water squeezed out from shale samples in the abnormal pressure environments has been found to be amber to brown in colour, because of the presence of organic compounds. In several instances, a correlation exists between the colour of the shale slurry filtrate and insitu formation pressure variations. However, the change in the shale water colour is to be taken with caution, as certain mud additives or drilling through lignite zones, will also cause colour variations.

Moisture Index

The shale moisture content (MC) index which is directly related to shale porosity can also be used as pore pressure prediction.

As is well known, abnormal pressures are reflected in higher than normal shale porosity and thus the moisture content at a given depth. The MC-Index of shale cuttings can be found easily at the well-site by measuring the weight loss of wet versus dried samples or by commercially available moisture analysers. Prior to MC-Index measurements, the shale cuttings must be cleaned and surface water removed. The MC-Index method is independent of the mud system used and works equally well in oil base, oil emulsion and water base muds.

Redox and *p*H-Potentials

Recently the development of Redox and pH potential methods that allow prediction of overpressure ahead of the bit has been emphasized (Fertl and Timko, 1973, a). In some of the wells, the redox potential and pH values were measured on the shale cuttings during drilling operations. Redox potential values indicated overpressures much earlier than what was indicated by other parameters, such as shale density measurements, drilling rate, etc. Basically, the Redox potential determines the mobility of an element of variable valence and fixes its dispersions in the earth crust. Today realistic measurements can be made provided certain precautions are taken.

Measurements of the Redox and pH values on the cuttings at a Louisiana well was made. Generally the trend of redox index decreases slightly with depth in normally pressured zone and then drastically increases. It is possible that normal overpressured boundary is marked earlier than what could be found out from the densities of shales (Fig. 7.29). In the overpressured sections, several redox deflections occur which coincide with pay zones (Fertl, 1976, pp. 169).

Based on the experience in both sand/shale and massive carbonate sections, properly measured redox values appear to be related to sub-surface pressure variations, hydrocarbon environments and lithology. However, effects are sometimes superimposed, interpretation may be complicated and hence other indicators, such as drilling parameters, well logs, etc., would still be required. (Fertl, 1976).

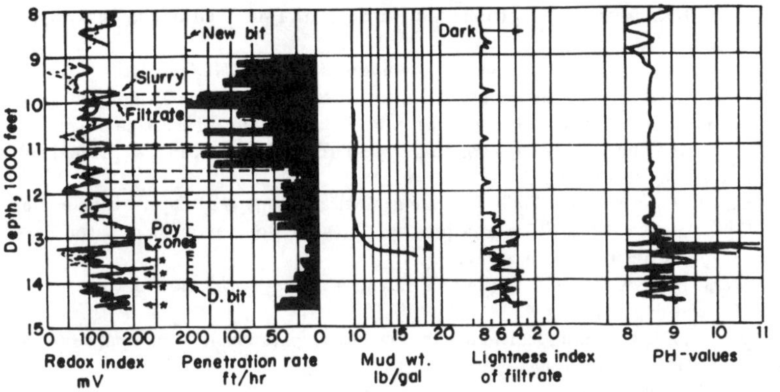

Fig. 7.29: Redox potential measurements showed the pressure top approximately 150ft [45.7 m] earlier than the currently used methods. In effect, pressure was detected ahead of the bit. In addition redox measurements also indicated pay zones (after Fertl and Timko, 1973a).

(Courtesy of World Oil)

However, there are several limitations with regard to the indicators discussed above. These indicators may be affected by the type of drilling mud (water or oil base), down hole environmental changes during drilling, exposure to the mud stream while lifting cuttings to the surface etc. Ion exchange will take place between cuttings and drilling mud and its additives.

Pressure and temperature changes occuring while cuttings are carried to the surface are also important. Magnitude and order of cation exchange processes selectively changes when determinations are made with a dispersed or compacted clay (Fertl, 1976, pp. 172).

REFERENCES

Core Laboratories, 1972-73 Mud logging training manual.

Dupin de Saint cyz, P. 1973. Ultrasonic device monitors mud pit level on floaters. World Oil **177** (7): 57-58.

Fertl, W.H. and Timko, D.J. 1970. How abnormal pressure detection techniques are applied. Oil and Gas Journal, **68** (2): 62-71.

Fertl. W.H. 1973. What to remember when interpreting mud gas cutting. World Oil **177** (4): 67-72.

Fertl. W.H. 1976. Abnormal formation pressures, Elsevier Scientific Publishing Co., New York.

Gearhart Geo-data Services, 1982. Over pressure, a quick look guide.

Hottman and Johnson 1965. Journal of Petroleum Technology.

Jordan, J.R. and Shirley, O.J. 1960. Application of drilling performance data to over pressure detection. J. Pet. Technology **18**: 1387-1394.

Maurer, W.C., 1966. How bottom hole pressure affects penetration rate. Oil and Gas Journal **64** (2): 61-66.

Mondshine, T.C., 1969. New technique determines oil-mud salinity needs in shale drilling. Oil and Gas Journal, **67** (28): 70-75.

Nevins, M.J. and Weitritt, D.J. 1967. Determination of cation exchange capacity by methylene blue absorption. Bulletin of American Ceram Soc. **46**: 587-592.

Overton, H.L. 1970. Resistivity logging from shale alurries. 11th Prof. Well log analysts symp., Los Angeles, California, Paper C.

Sahay, B. and Tikku C.L., 1968. Shale density studies in Bodra-1, ONGC bulletin, Vol. V, pp. 83-85.

Sahay, B., 1972. Abnormal sub-surface pressures, their origin and methods employed for prediction in India. Abnormal sub-surface pressure seminar, S.P.E. Baton Rouge, Louisiana.

Sahay, B., 1974. Analytical studies on abnormal pressures in India basins. A manual submitted and distributed to geologists of ONGC.

Sahay, B., Rai, A., Ghosh, M.K., 1988. Well site geological techniques for petroleum exploration.

Sahay, B., 1984. Analytical studies on over pressured areas in Bombay offshore basin. Paper presented in the 5th Seapex Conference, Singapore.

Weaver, C.E. and Beck, K.C., 1971. Clay water diagnosis during burial; how clay becomes gneiss, Geol. Soc. Am., Special paper, **134**, p. 93.

Wilson and Bush, 1972. Pressure prediction with flowline temp. gradient. Abnormal sub-surface pressure symposium of S.P.E. Batan Rouge, Louisiana.

Zo Bell C.E., 1946. Studies on redox potential of marine sediments. Bull. Am. Asso. Pet. Geol., **30**: 447-514.

Zoeller, W.A., 1970. Drilling Porosity Log. SPE 3066. 45th AIME, Fall meet.

CHAPTER VIII

Evaluation of Overpressure After Drilling

Introduction

As already discussed in the preceding chapters, it is possible to predict the abnormal pressures by several methods. One such method is based on wireline logging methods. However, this method is a postmortem, i.e. it is applicable only after the borehole has been drilled. Nevertheless, it is useful in estimating the formation pressure. The various logs that can detect abnormal high formation pressure are:

1. Electrical surveys such as short normal curve, conductivity or induction log, and spontaneous potential (SP) Curve.
2. Acoustic (Sonic) measurements
3. Bulk density measurements
4. Neutron measurements
5. Pulsed neutron measurements

Except for SP Curve data, parameters recorded in shale formation rather than in sands are plotted vs. depth. The theory behind this prediction technique, is the same as in various other prediction methods, adopted during the course of drilling from the dcs plot or from the shale density studies. As is known, most log interpretations are related either directly or indirectly to porosity. Thus the pressures in the reservoir can be estimated from the pressures of the surrounding shales. Generally speaking porosity decreases exponentially with depth as water is squeezed out from the shales to permeable horizons due to compaction in response to the stress of overburden. In case water from the surrounding shales is not allowed to escape to the sands or to any porous layers, it is retained in the shales, thus increasing the porosity.

It is possible to establish the trend line for normal compaction. The shape and slope of a particular trend line may be characteristic of a formation over an entire geological region or of one field or, sometimes, of a single fault block only.

Interpretation of the resistivity/conductivity versus depth plot depends upon the departure from the normal trend line. Divergence from the normal trend line reflects the degree of abnormality in the formation pressure from the normal hydrostatic pressure at a specific depth. Interpretation methods are often straight forward. However, complications due to variations arising out of changes in lithological characteristics, bore hole diameter, steep and thin formations, salinities of formation waters, etc., can create problems in interpretation.

Electrical Surveys

Resistivity methods: The short normal curve or preferably the amplified short normal curve is considered to be one of the most reliable tools. Hottman and Johnson (1965), the pioneers in this method of pressure detection, observed in their classic paper that under normal compaction the shale resistivity increases with depth as the porosity decreases. However, in case of undercompacted shales, the shale resistivity decreases, with a corresponding increase in porosity. The exact shape and slope of the normal trend line will vary with age and type of shales present (Fig. 8.1). Hence, normal compaction trend will be different from area to area and region to region.

The shale resistivity (Rsh) is plotted on a logarithmic scale vs linear Depth. All data points in normal pressure environments will group along a definite Rsh trend line, Rsh increasing with depth (Fig. 8.2). In the overpressured shale section, the short normal curve will show departure from the normal trend to lower resistivity, due to increasing porosity and higher water content of the shales (Hottman and Johnson, 1965; Ham 1966). The relative salinities of the compacted versus under compacted shales should also be taken into consideration (Chilingar, 1975).

Presence of fresh or brackish water also pose problems and may lead to high resistivity values and will make it extremely difficult, if not impossible to use this resistivity method of pressure detection. It would be thus better to use acoustic logs for pressure detection in such cases.

The procedure followed for evaluating the abnormal pressure zones from the shale resistivities is as follows:

1. Normal compaction trend for the area of interest is established by plotting the logarithm of shale resistivity from the amplified short normal curve vs depth.
2. For selecting the proper shales for such studies care should be taken to see that they are pure and consistent. Siltyness, limyness, sandiness and carbonaceous nature of the shales will be detrimental and should be avoided. For example, calcareous shales show higher resistivity and thus could give erroneous pressure readings. Hence, it is necessary to carry out laboratory analysis of cores and cuttings to quantify their "lithology" effect before considering them for the pressure detection techniques.
3. The thickness of the shale should be 30 ft or more
4. The shale immediately above high pressure pay zone should not be selected.
5. Pressure top at the depth at which the plotted values start to diverge from the normal trend should be noted.
6. The procedure for determining the formation pressure at a specific depth is as follows:

(a) Divergence of observed shale resistivity value Rsh (n) determine the shale resistivity ratio (normal Rsh/observed Rsh).

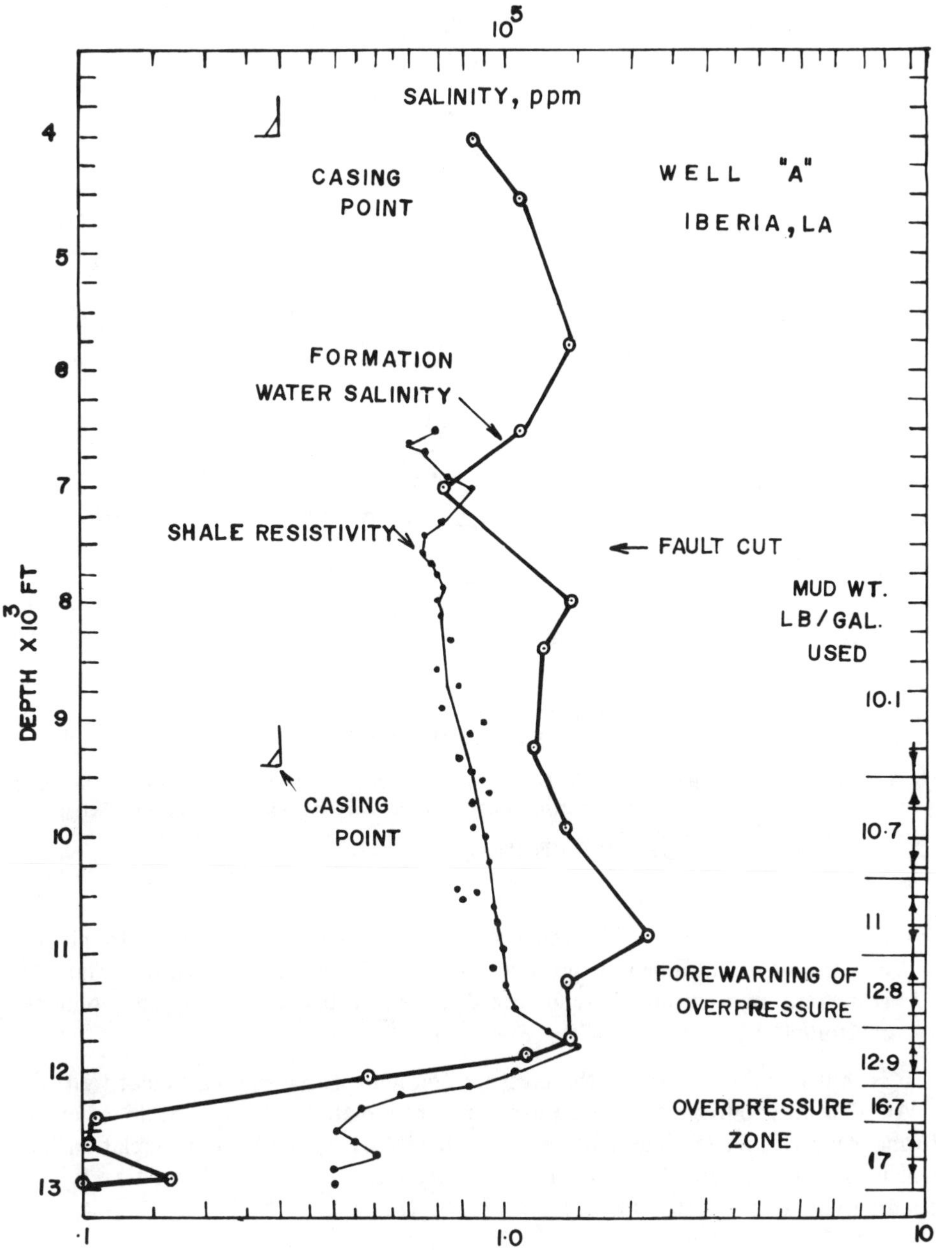

Fig. 8.1: Shale Resistivity

(Courtesy: Abnormal surface pressure, A study group report, 1969-1971)

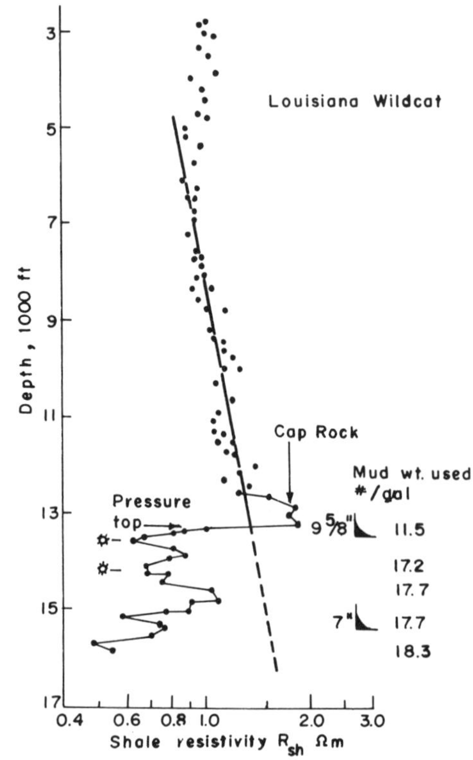

Fig. 8.2: Shale resistivity plot in a U.S. Gulf Coast well. Note limy cap rock (resistivity increase) immediately above the 9⅝ inch casing point and the sharp pressure transition zone below

(After Fertl and Timko, 1971)

(b) Fluid pressure gradient (FPG) corresponding to shale resistivity ratio is found out. FPG Value is multiplied by the depth to obtain the formation pressure. (Fig. 8.3). However, such empirical trends are different in different areas and geological environments.

It is thus possible to predict the pressure gradient profile in a particular well.

Another method, called the equivalent depth method, is utilised to calculate formation pressure. This method is based on the following mathematical relationship, which is valid for any logging parameter, including shale resistivity, travel time, conductivity, density, etc. (Fertl, 1976, pp. 181-182).

(1) Plot the Rsh values and establish the normal trend line
(2) Determine the formation pressure from

$$Pf = Go\ DA - DE\ (Go\text{-}GH),\ Pf = DA - 0.535,\ DE$$

Where Pf= Formation pressure in psi (Fig. 8.4)

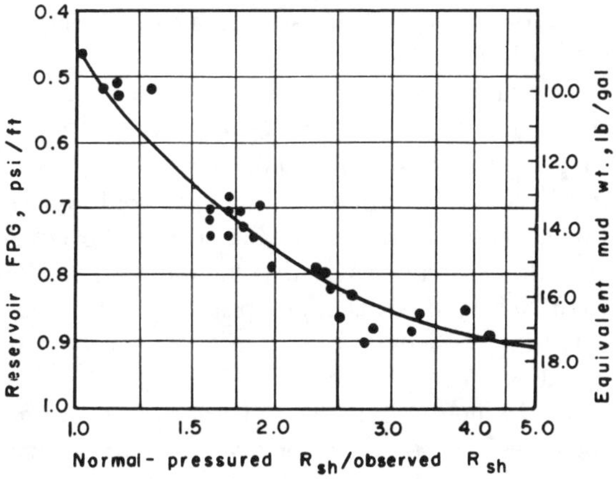

Fig. 8.3: Relationship between shale resistivity ratio ($R_{sh,\ normal}/R_{sh,\ observed}$) and reservoir fluid pressure gradient (FPG) in overpressured Miocene/Oligocene formations, U.S. Gulf Coast area (after Hottman and Johnson, 1965. Courtesy of *Journal of Petroleum Technoloy*)

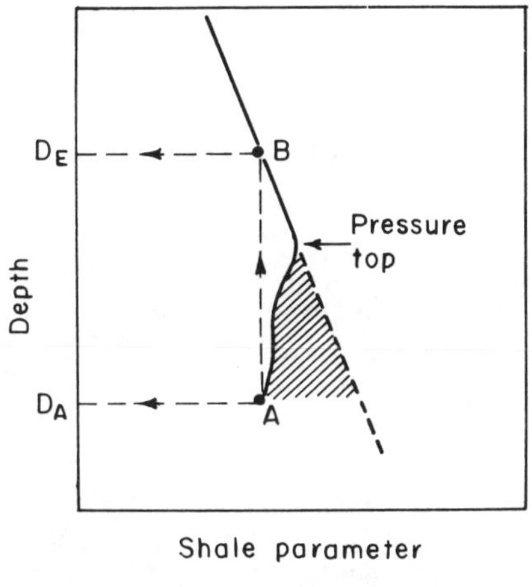

Fig. 8.4: Schematic representation of D_A and D_E used in the equivalent depth method (After Fertl, 1976)

DA = Depth of interest in an abnormal pressure environment in ft/mtrs.
DE = Normal equivalent depth in ft/mtrs. corresponding to DA.
GH = Hydrostatic pressure gradient in psi (equal to 0.433 psi/ft or 1.0 Kg/cm^2 and 0.465 psi/ft or 1.07 Kg/cm^2) for fresh and salt water, respectively.
G_o = Over burden pressure gradient in psi/ft.

Overburden gradient of 1.0 psi/ft (Corresponding to an average bulk density of 2.31 gm/cc) is an initially satisfactory assumption.

The studies for the prediction of overpressures from the study of the shale resistivity vs. depth were carried out successfully in the wells drilled in India in e.g. Bay of Bengal, Port Canning, Bodra in the Mio-Pliocene.

A departure to the left from the normal trend line, to the lower resistivity values just above the top of high pressure sand was seen in the above wells. High pressure-intervals in both the wells contained gas with some water. In the Western part of India, in the South Kadi and Nawagam areas, high pressures were encountered in sediments of Eocene age. Study of the resistivity logs, also indicated the departure towards left from the normal trend.

Conductivity curve: Induction logging devices record formation conductivity (mmhos) and its reciprocal resistivity. If the shale is undercompacted, transition to overpressured shales is reflected by higher than normal conductivity values, resulting from higher than normal water content and porosity (Wallace, 1964, 1965 a,b; MacGregor, 1965).

One of the first applications of the conductivity curve for quantitative pressure evaluation (Wallace, 1964) is shown in (Figs. 8.5 & 8.6). In Fig. 8.5 a departure

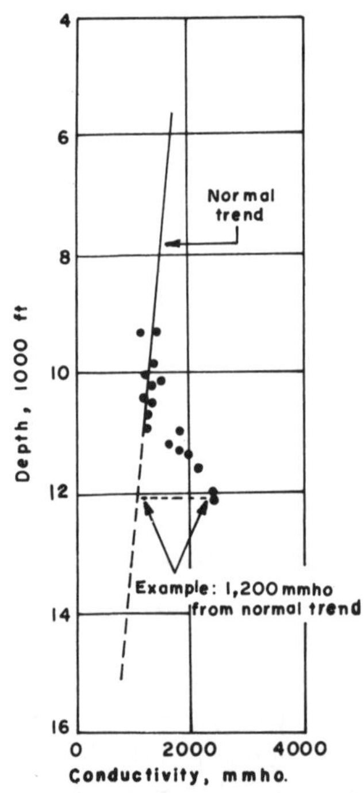

Fig. 8.5: Field case of quantitative formation pressure evaluation in South Louisiana Wildcat (after Wallace, 1965a). There is a departure of 1200 mm/hos at a depth of 12,200 ft.

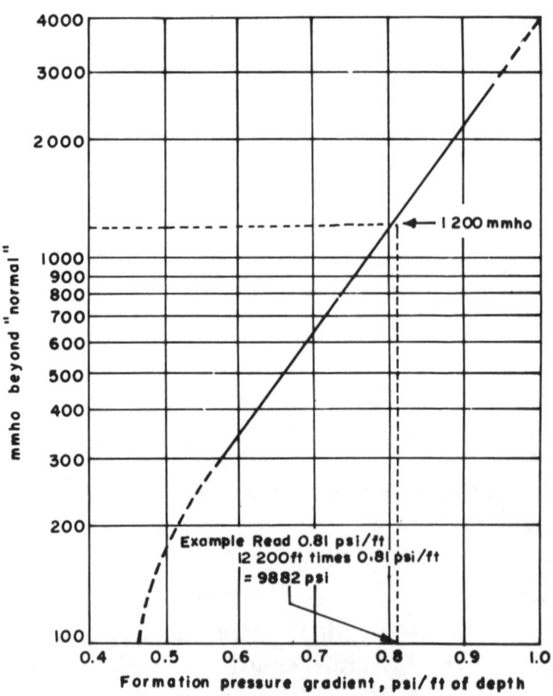

Fig. 8.6: Relationship between the shale conductivity and formation pressure gradients in South Louisiana (after Wallace, 1965)

of 1200 mmhos at a depth of 12,200 ft (3720 m) was obsrved. Using an empirical calibration chart for well log conductivity and formation pressure of 9882 psi (695 Kg/cm^2) is determined. Production testing carried out indicated pressures below the shale zone of 9900 psi (696 Kg/cm^2). Note again that this method is actually based on empirical correlation of actual formation pressure data and shale conductivity values for a specific area.

The equivalent depth method as discussed earlier can also be used (Fig. 8.7). Another version is the development of a transparent overlay of equivalent mud weight lines parallel to normal compaction trend which correspond to hydrostatic conditions. Overlying such calibration charts on the conductivity one allows that to read mud weights and or formation pressures directly from the conductivity curve.

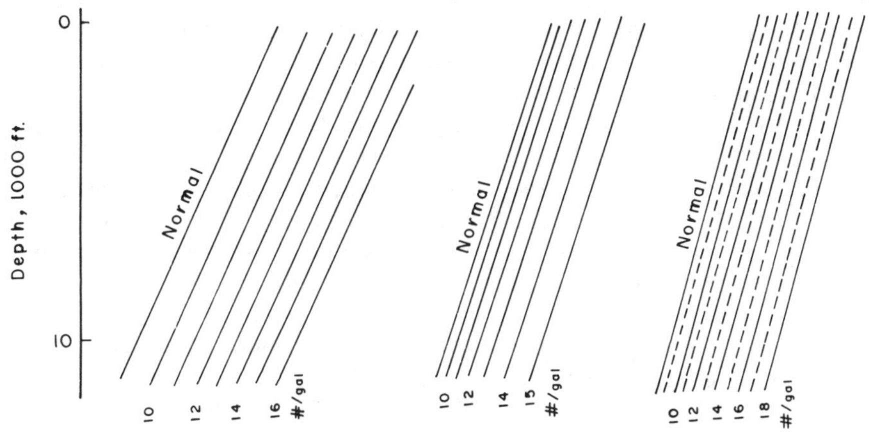

Fig. 8.7: Equivalent mud weight requirements (overlays) for use with conductivity plots in (a) Frio Formation, South Texas; (b) Wilcox Formation, South Texas; and (c) Pleistocene and Pliocene Formation, Itlay (aftr W.R. Mathews, personal communication, 1972)

Shale porosity, temperature, formation water salinity, mineral composition of shales, etc., affect conductivity measurements.

SP Curve

As is known, the SP curve is utilised for (1) detection and correlation of permeable formations such as sands in a shale/sand sequence, (2) determination of formation water resistivity (salinity), (3) estimation of the clay content in potential reservoir rocks, and (4) indication of tectonic stress for use in abnormal formation pressure detection (Fertl, 1976, pp. 190).

Soviet investigators as well as different authors in the Gulf Coast have cited field examples which show decrease of formation water salinities with increasing depth.

Overton and Timko (1969) based on similar observations have proposed salinity variations as found out from SP curve for overpressure studies. The authors suggested a simple relationship, the so-called salinity principle, between salinity of clean sands Cw and porosity of adjacent shales ϕ sh.

$$CW\ \phi\ sh = constant$$

The above suggests that formation water salinity will vary universally with the porosity in adjacent shales. If the shale is normally compacted, porosity decreases with depth, whereas formation water salinity tends to increase. In case overpressured formations occur, divergence from normal trends will be indicated. Apparent shale porosity increases in overpressured zones which in turn suggests a decrease in formation water salinity as is shown by SP curve would show proximity of abnormal pressures. The concept of using the SP curve as a tectonic stress indicator in shale/sand sequences has proved successful in many areas. Often freshening of formation water has been observed prior to actual penetration of overpressured formations. It thus helps in forewarning impending downhole pressure changes.

In several areas of the USA, the Himalayan foothill belt in India (Sahay, 1972) and various other areas, salinity reductions in overpressure environments at greater depths, become so drastic that these salinity variations of the formation water cause even SP reversals, i.e., recorded SP potential changes polarity and SP becomes positive. It is not always possible to straightaway apply the salinity principle and correlations of formation pressure may be limited locally. It could be due to (1) problems inherent to SP measurements (2) the fact that formation water salinities can vary markedly with faulting, salt dome uplift and unconformities and other local or regional geological factors (Fertl & Timko, 1970 b).

Acoustic (Sonic) Logs

This logging device records the transit time of elastic waves, through formations over predetermined distances. The log is used for estimating porosity and measuring sound energy transmission characteristic for use in differentiating fluid content (Hicks and Berry, 1956, Wyllie et al 1958, Pickett, 1963).

The recorded interval transit time is in micro-seconds/ft ($\Delta t/\mu sec/ft$). In a given formation, it depends upon the lithology, degree of compaction, porosity and the fluid content in the pore space. With the exception of air gas-filled holes, it is possible to run this logging device in all drilling mud system. This device has got the advantage of being largely unaffected by changes in hole size, temperature, and formation water salinity. Gas cut mud may cause cycle skipping. The procedure for plotting the acoustic (Sonic) log data is already discussed above. For information pressure evaluation, the transit travel time of shales is plotted vs depth on semi-log paper. In a normal pressure environment, all data points fall along the normal compaction trend line. However, in the case of over pressured shales, the transit time would increase above the normal Δt values. This increase is indicative of an increase in porosity which normally is the case in an overpressured shale section. The method for calculation of the pore pressure from a plot of interval transit time vs depth is as follows:

1. Plot shale transit time Δt and establish normal trend line.
2. It is plotted versus depth in the well. It is necessary to select only pure shales for these studies.
3. The top of abnormal pressure formations is identified by noting the depth at which the plotted points divert from the normal compaction trend.
4. To determine the pressure at the top of over-pressured horizon, the difference between the values of normal Δt (sh) and actual Δt (sh) at depth is found. The pressure gradient is found from these values as shown (Figs. 8.8 & 8.9). Such correlations are prepared from the known pressure data for a given region.

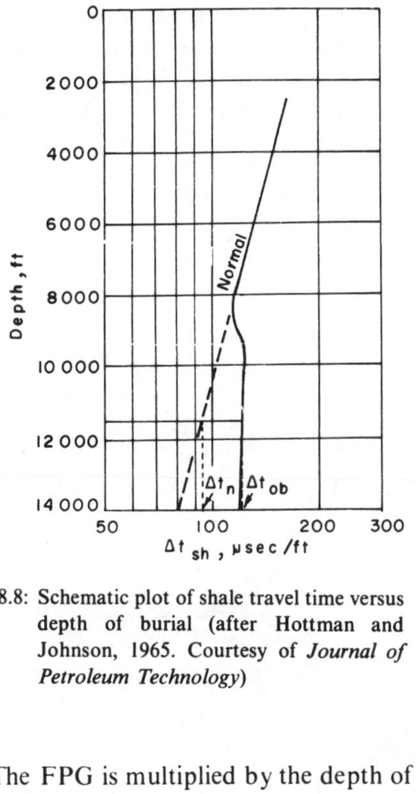

Fig. 8.8: Schematic plot of shale travel time versus depth of burial (after Hottman and Johnson, 1965. Courtesy of *Journal of Petroleum Technology*)

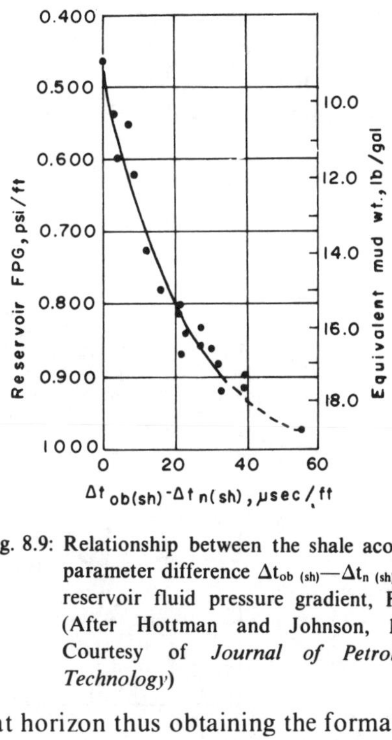

Fig. 8.9: Relationship between the shale acoustic parameter difference $\Delta t_{ob\ (sh)}$—$\Delta t_{n\ (sh)}$ and reservoir fluid pressure gradient, FPG. (After Hottman and Johnson, 1965. Courtesy of *Journal of Petroleum Technology*)

5. The FPG is multiplied by the depth of that horizon thus obtaining the formation pressure. The pressure gradient profile then can be constructed in the well.

Another quantitative method also uses the equivalent depth technique. The step-wise outline for this method has been discussed previously. Once the normal compaction trend has been established for a particular area or a region, then nomographs for a proper evaluation can be constructed for these specific conditions, so that proper and safe drilling precautions can be taken. World wide experiences clearly have shown that formation pressures with this method can be predicted from various logs with reasonable accuracy.

Formation Density Measurements

The density logging device consists of a radioactivity source which radiates the formation with medium energy Gamma rays (Alger et al 1963, Titman & Wahl, 1965). Electron density is related to true bulk density, which in turn depends upon the density of rock matrix, formation porosity and density of the fluids in the pore space.

It is possible to utilise the density measure device for the calculation of pore pressures. In this logging device, the decrease in the bulk density from the normal compaction trend is, therefore, indicative of overpressures. However, in cases of severe hole wash outs in shale sections, as is usually seen in shale sections, the reliability of bulk density plot may decrease drastically. (Fertl, 1976, pp. 203-204).

To calculate pressure from the density log, the equivalent depth method is the most versatile approach. The bulk density in not washed out shale sections, is plotted at linear of logarithmic scale (Fig. 8.10).

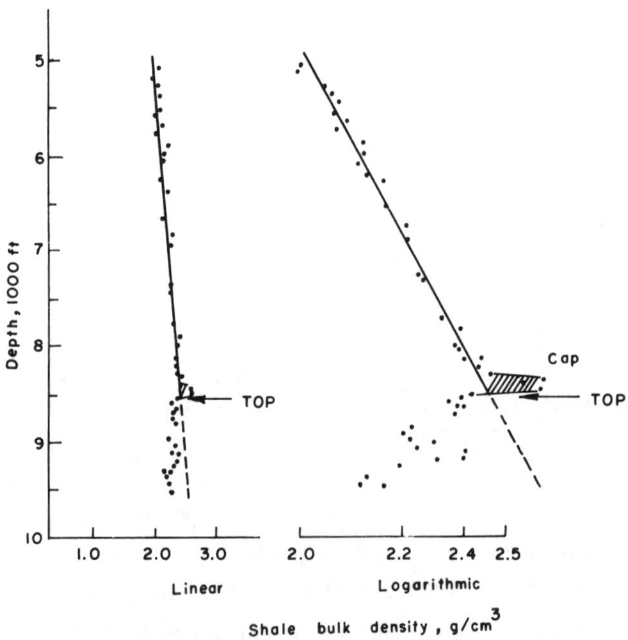

Fig. 8.10: Log-derived shale bulk density plots on a linear and logarithmic scale for the same well

(After Fertl, 1976)

Neutron Measurements

Neutron logs are used mainly for the delineation of porous reservoir rocks and determination of porosity. Since this device measures the porosity of the formations, increase in porosity corresponding to water content of shales in an overpressured environment allows the use of neutron tools for the pressure evaluation.

Pulsed Neutron Measurements

The thermal neutron life time indicates the rate at which thermal neutrons are captured or absorbed by a material. The life time is actually controlled by the microscopic thermal neutron capture cross section (E) of the material. The larger the cross section, the more quickly neutrons are captured. The chlorine ion is the strongest absorber of all commonly occurring elements in sedimentary rocks. (Fertl, 1976, pp. 204-205).

In a normal compacted shale, values decrease in a regular way with depth. However, in overpressured shale sections, divergence from the normal trend to higher values (Fig. 8.11). This log can be run in cased holes bore hole or through the drill string in open hole wells.

The quantitative determination is done by equivalent depth method.

Pulsed neutron logging devices help in investigating special reservoir problems such as communication with and depletion of pay zone by wells in adjacent fault blocks and recharge of aquifers with oil and or their repressuring from a nearby overpressured pay zone. Such data is useful in reservoir engineering and when drilling in partially pressure depleted areas.

General Comments

The short normal or preferentially, its electronically five fold amplications, the amplified short normal curve is frequently used. As mentioned earlier, this tool is affected by large variations of formation temperature and salinity, lithology, presence of steep and rather thin formations, etc. Recently a focussed spherical short normal curve has also become available. Frequently, other wireline logging devices too give very valuable information. However, it appears to be prudent practice not to rely on any of them alone or completely (Fertl, 1976). Fig. 8.12 summarizes how several of these well logging parameters respond to different pressure regimes.

In general, formation pressure plots basically rely on four main considerations, (1) logging sonde was working properly, (2) major detrimental effects on log response, such as wash outs, bit size changes, lithology variations, are either not present, can be avoided in plotting data or at least are recognized, (3) relatively pure and thick shale zones are present, (4) realistic trend lines for normal compaction have been established (Fertl, 1976, pp. 212).

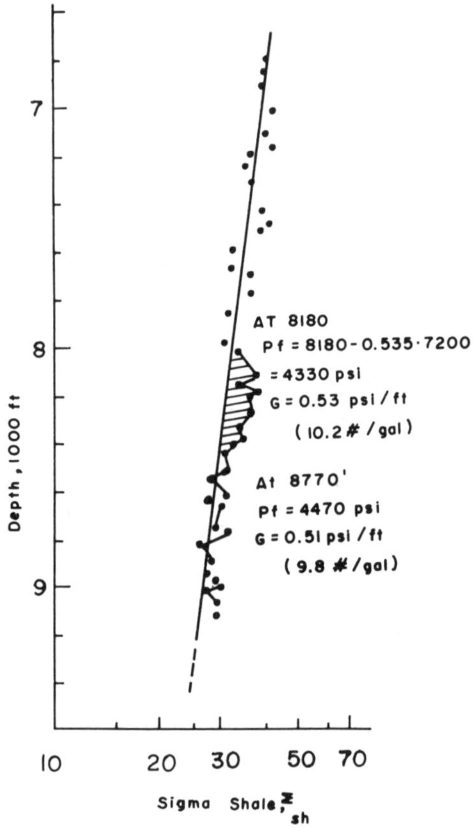

Fig. 8.11: Use of pulsed neutron logging for quantitative formation pressure evalution in cased holes (after Fertl and Timko, 1970b). $\Sigma_{sh}^{\Delta 3\ \Delta 1}$. = neutron capture cross section of shale in 10 cm.

(Courtesy of *Oil and Gas Journal*)

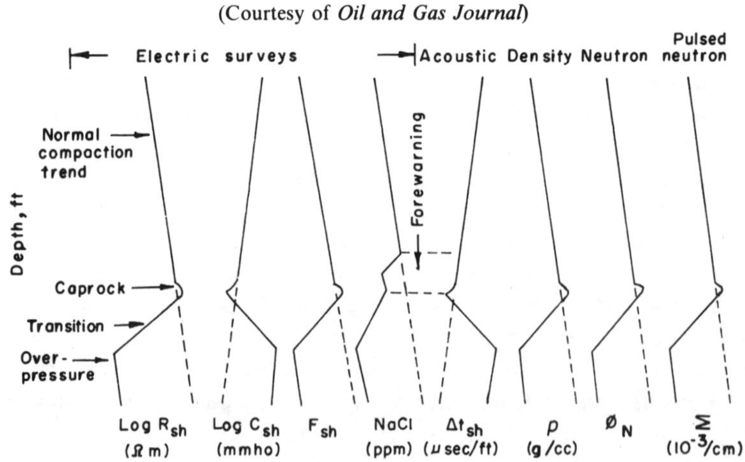

Fig. 8.12: Schematic response of well-logging parameters to normal- and overpressured environments (after Fertl and Timko, 1971a. (Courtesy of the Society of Petroleum Engineer of A.I.M.E.)

Wireline Testers

Open hole logs help in locating and evaluating prospective intervals of commercial significance. After the potential target intervals are identified, it may be ncessary to estimate their potential to justify the additional cost of completing the well. It is possible with the help of wireline formation testing to get a fast and reliable formation evaluation. Depending on the pressure information as obtained with the help of wireline testing, the mud and casing policies can then be modified, accordingly.

The Formation Multi-Tester

Dresser Atlas Formation Multi Tester (FMT) is a sophisticated system of wireline testing equipment which allows multiple pressure recording and a maximum of two separate formation fluid samples are obtained on a single run. It has, therefore, become a major tool in the open hole measurement of reservoir pressure and permeability. High precision pressure data are recorded on surface. The engineer monitors pre-test or sampling operations as it progresses. This offers several important advantages. Pressure and permeability data are immediately available, any flow line blockage or seal failure is readily detected so that the test can be re-run and the best permeable zone can be located by pretest prior to sampling.

Formation Multi Tester (FMT) of Dresser Atlas is run on a multi conductor logging cable and accurately positioned against the desired interval to be tested by reference to either a spontaneous potential or a Gamma ray curve. A packer seal is confirmed with the pre-test prior to opening the fluid sample chamber. The complete cycle of the down hole operation is recorded on film at the surface. Pressure devices are incorporated in the borehole tool to equalise the formation pressure and the hydrostatic head in order to prevent differential sticking of the seal pad. Pressure data recorded on film and analysis of fluids recovered in the sample chambers are used to predict reservoir production characteristics.

The attached figures show a photograph of the formation Multi Tester and the surface equipment used to examine and evaluate the recovered fluids. This sub-surface tool indicates the control electronics, hydraulic system and the test sample chambers. Fig. 8.13 & 8.14 show a diagram of the sub-surface assembly showing the formation seal packer, back up shoe, the control electronics and hydraulic systems at the top and sample chambers at the bottom. The operations of the sub-surface equipment is controlled from the logging track with the complete cycle operation being recorded on film. The measurements recorded are depth, hydrostatic pressure, pre-test, draw down pressures, initial shut in pressure, flowing pressure final shut in pressure and hydrostatic pressure within the tool.

The pressures transducer is calibrated prior to the tool being lowered into the well. The calibration method uses dead weight testers and temperature corrections from the pressure recordings. An after survey calibration repeats the process. The pressures are recorded on the left track of the film in the analog form while the right track gives data

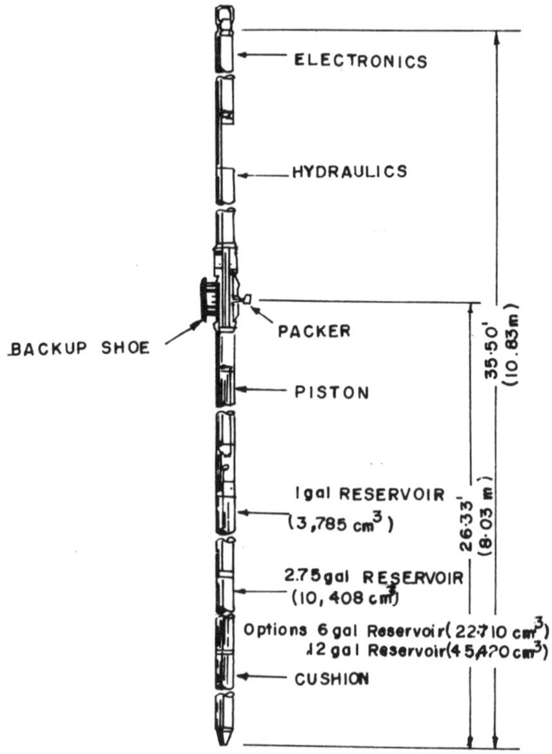

Fig. 8.13: Schematic diagram of FMT borehole assembly

(Courtesy: Dresser Atlas, Dresser Industries Inc.)

in digital form. The fluid sample chambers can be run in combination of one gallon and 2/3/4 gallon capacities or a choice of 2/3/4/ gallon and either 6 gallon or a 12 gallon capacity chamber. In addition to the multiple pre-test pressured recordings, two fluid chambers can be obtained in one trip. Sample chambers are isolated and two samples may be taken from the same formation or two different intervals.

The hydraulic pressure system actuates the rubber formation pad and backup shoe into contact with the borehole wall. The hydraulic pressure is recorded at the surface, indicating the proper setting of the tool. A snorked tube is then forced into or pressed against the formation. This allows formation fluid to enter the sample inlet. The pre-test sample piston is moved. The movement of the piston has the capacity of 10 cm^3 in each stroke operated sequentially. After the pretest chambers are filled, the pressures rise to their initial shut-in-value. The sample pressure transducer transmits the formation shut-in-pressure to the surface recorder.

If the pretest indicates a permeable zone and good seal with the formation, a fluid sample test can be run. Figure 8.14 shows the tool in the process of filling the first sample chamber. The sample chamber contains water in the upper section and air at

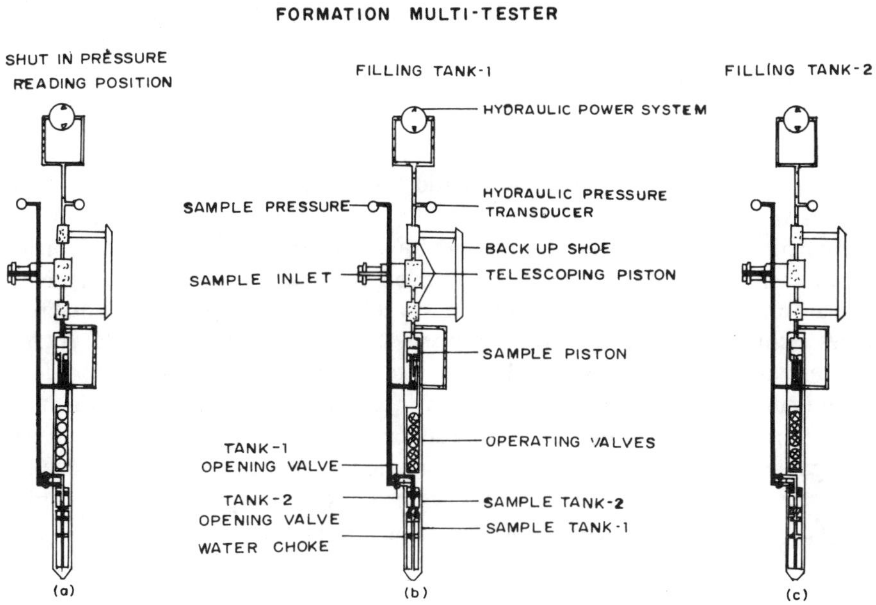

Fig. 8.14: Schematic diagram of formation fluid path during FMT fluid sample test.

(Courtesy: Dresser Atlas, Dresser Industries Inc.)

atmospheric pressure in the lower section. The sections are divided by an insert containing one or more chokes. The choke orifices are pre selected according to the formation pressure anticipated. They restrict the flow of the water cushion into the lower chamber, thus slowing the entry of the formation fluids and maintaining a low negative pressure differential on the formation. As formation pressure is exerted against the floating piston, water is forced into the lower chamber, allowing the upper chamber to fill with formation fluids. The flowing pressure is continuously recorded at the surface under drawdown conditions. After the sample chamber is filled, the valve is closed by hydraulic pressure. The valve is kept in closed position, thereby sealing the fluids in the sample chamber at formation pressure. The pressure transducer transmits the final shut-in pressure which is recorded on film.

If a second segregated sample from the same depth is needed, the valve to the second sample chamber can now be opened. Otherwise, the hydraulic system pressure is released in order to release the pad from the face of the formation. The second fluid sample test is run after the pretest at the desired depth in a series of operations similar to those conducted for two sealed formation fluid samples is returned to the surface. The surface equipment used for extraction and measurement of the recovered samples consists of a pressure regulator, separator and gas meter. The gas recovered is bled from the sample chamber through the separator and measured at the gas meter in cubic feet at surface conditions. Water and oil are drawn off through a valve in the separator into a calibrated vessel where their volumes are measured in cubic centimetres. Further analysis of the recovered fluids can be made in the laboratory. (Formation Multi Tester interpretation Manual, Dresser Atlas, Dreser Industries, Inc., 1980).

Reservoir Fluid Properties

In conjunction with quantitative well log evaluations, the fluid samples and pressure data obtained from FMT can be used to estimate formation pressures, permeabilities, hydrocarbon production rates, depths of oil-water, gas-oil and gas-water contacts. The samples recovered also supply information on the type of formation fluids, gravity of oil, water cut and gas/oil ratio. FMT data interpretation involved considerations of fluid pressure behavior and other physical properties of formation fluids such as density, resistivity and viscosity. The pore space of reservoir rocks may contain water, oil and gas as a single phase for in any combination of these fluids. It is imperative that these properties of reservoir fluids be known or reasonably approximated in order to reliably evaluate the production characteristics of the reservoir rocks.

Pretest Data Interpretation

By recording pressures, while small pretest samples are withdrawn from the formation, hydraulic seal to the formation can be confirmed and indications of formation permeability obtained. With resolution of pressure on the digital recordings being ±1 psi and an accuracy of FMT pressure measurement being ±20 psi, even the small sample pretest provides pressure data which can be used to estimate formation pressure and permeability, when the FMT tool is set against the formation at the desired depth, the pretest is activated to withdraw, in the case of a dual chamber test, 10 cm^3 of formation fluid from the zone in the immediate vicinity of the borehole at a specified flow rate causes a larger pressure drawdown, and if the formation is not competent enough, it may cause plugging of the tool probe. In such cases, it is preferable to run a pretest at a single lower flow rate.

As already mentioned, pretest pressure information tells as to whether or not a fluid sample test should be run. In case the pretest indicates a permeable zone and good packer seal, the fluid sample chamber is opened to recover formation fluids. A fluid sample test may be run with an air cushion in a zone of low permeability or using a water cushion in a high permeability zone. While sampling with a water cushion, the flow rate is controlled by an insert containing a choke of size .02 in, .03 in, or .04 in or a series of .02 in chokes. Fig. 8.15 & 8.16 show typical FMT log, single chamber pretest and'dual chamber pretest.

Sampling Time

An estimate of the time required to fill the sample chamber can be made from the flow rate and pressure drawdown during pretest. The time(t) in minutes needed to fill one gallon chamber is

$$t = \frac{63.1.\ (Ps - Pf)\ \text{pretest}}{\text{Pretest}\ .P}$$

Where

$(Ps - Pf)$ = Pressure drawdown during pretest, psi
P = Formation pressure, psi
63.1 = Conversion factor

Evaluation of Recovered Sample Fluids

The amount of fluids recovered in the sample chamber is not diagnostic of permeability of the formation because the quantities of fluids recovered is a function of time, viscosity, of fluids and pressures in addition to reservoir permeability. Quick sample chamber fill up occurs, in high permeability formations. However, by

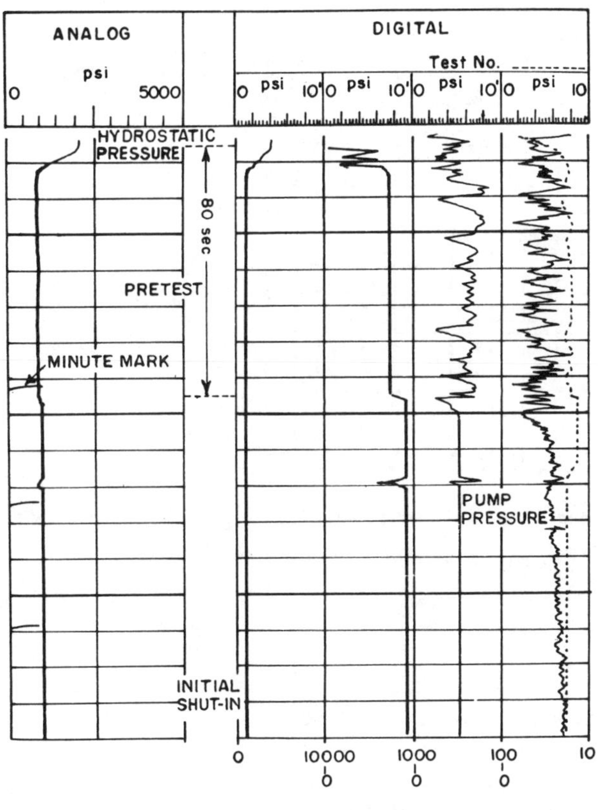

Fig. 8.15: Typical FMT log (single chamber pretest)

(Courtesy: Dresser Atlas, Dresser Industries Inc.)

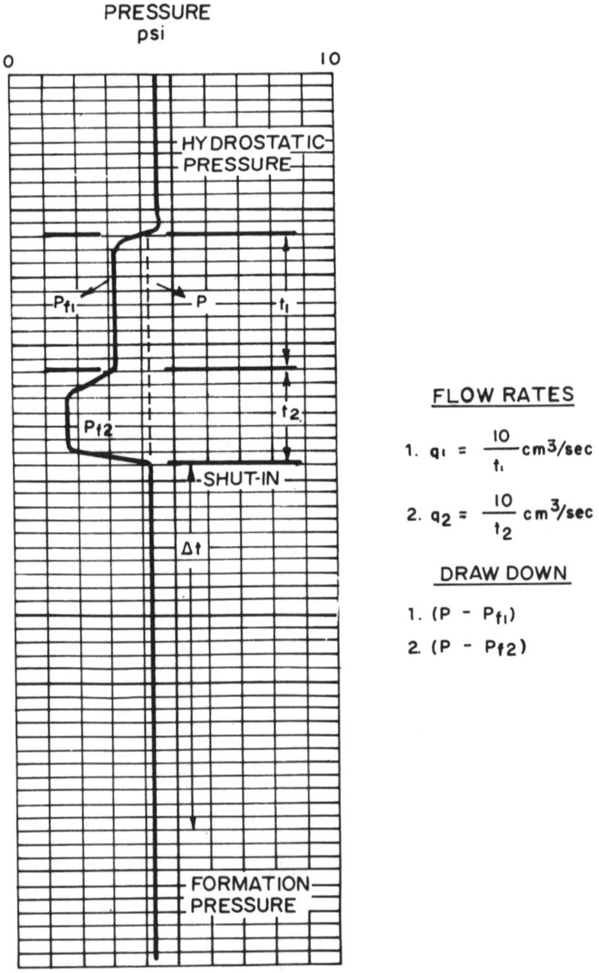

Fig. 8.16: Ideal analog pressure recording for dual chamber pretest

(Courtesy: Dresser Atlas, Dresser Industries Inc.)

extending the sampling time, a sufficient quantity of fluids may be recovered in low permeability formations to analyse the production characteristics.

Sample fluid recovery in excess of 1000 cm^3 allows realistic estimates of (a) Gas/fluid ratio (b) Production prediction. Hydrocarbon versus water, mud filtrate versus formation water, percent water cut, gas solubility in water and or oil. (c) Gravity of oil. API can be measured on the recovered oil sample with a hydrometer (d) Viscosity of recovered fluids.

Flow Rate

The flow rate is defined by the volume of chamber divided by the sampling time. Time required to fill the sample chamber is difficult to determine in high permeability formations where flow rates are high and only small differences exist between the formation shut-in pressure and flowing pressure. If water cushion is used, chokes in sample chamber control flow rate. Formations of low to medium permeability exhibit lower flowing pressures and flow rates. Pressure build up from flowing to shut-in pressure indicates fill-up time. If the test indicates that the sample chamber is full, the volume of sample chamber can be used to derive the flow rate. In case the horizons have low permeability, the sample chamber may not be full at the end of the test, i.e., the test was terminated too early. In such a case, the measured volume of recovered fluids has to be used to determine flow rates. In case sample chamber is found to contain gas, the volume of gas must be corrected to formation conditions. The volume of free gas at reservoir pressure and temperature should be added to the volume of liquid recovered in the chamber.

Permeability

To evaluate permeability from fluid sample, the technique adopted is the same as in the case of pretest. Since the volume of sample chamber is much larger, the resolution on pressure recording during drawdown in high permeability range is considerably improved. The large drawdown is due to very high flow rate, which is possible during fluid sample test. As the resolution of the pressure measurement is greatly enhanced, it allows extension of reliable permeability evaluation to Darcy's range.

Dresser Atlas Formation Multi Tester is a fast and economical means to obtain physical evidence of formation contents. The operation of equipment as mentioned earlier is controlled from surface, through wireline cable, which also serves to transmit pressure data to the surface. After the tool is withdrawn, pressure and volume of the contents of the sample chamber are measured. The pressure recordings allow estimation of formation pressure and pressure gradients. It is possible to detect abnormal pressure zone, Oil/water/gas fluid contacts and reservoir isolation. Fluid samples recovered permit measuring the gravity of oil, predicting type of production, water cut, gas/oil ratio, etc. This information, when combined with pressure analysis, can be utilised to predict permeability and PI of each zone tested.

This tool by virtue of its pinpointing feature, provides individual evaluation of the interbedded formations, besides production characteristics of a massive formation. The former is highly valuable in reservoir studies where producing zones may or may not be laterally interconnected.

Application of FMT Pressure Measures

The application of the FMT is as follows:

(i) *Hydrostatic pressure*: Check on the mud weight and drilling mud homogeneity, overbalanced, balanced, or underbalanced drilling conditions, proper analysis of mud, gas shows.

(ii) *Formation pressures and gradients*: Magnitudes and variation within single interval or sequence, of permeable formations. Overbalanced, balanced, or underbalanced, drilling conditions, proper analysis of mud gas shows.

(iii) *Overpressure detection techniques*: Normalisation of normal (hydrostatic) compaction trend lines used in quantitative formation pressure evaluation based on well logging, or drilling parameters.

(iv) Prediction of pressure regimes in future offset wells penetrating the same formations.

(v) Reservoir isolation or communication tests in single or multi-well studies.

(vi) Reservoir depeltion and drainage studies. Reservoir monitoring is based on time-lapse pressure gradient variations on vertical basis over subject interval and on lateral basis between wells. Also applicable to infill wells.

(vii) Estimates of deepest possible fluid contacts (gas/oil, gas/water, gas/oil/water) in part of formation not penetrated by borehole. These analytical techniques apply to selecting the location of future wells and in preliminary reserve estimates.

(viii) Efficiency of enhanced recovery projects. Monitoring in and/or planning of infill wells. In combination with well logs, high water cut zones are identified which can be avoided in selective completion and injection operations to increase flood efficiency.

Possible Pitfalls in Wireline Methods

Proper calibration of well-logs is a pre-requisite for any realistic formation pressure evaluation. This is clouded with the possibility of malfunctioning, especially in ultra deep and high temperature wells. How can one avoid or at least recognise such malfunctioning of the tools? First of all, tool calibration and the rating of the tools should be compared with the expected down hole conditions (pressure and temperature) in a well where the log is being taken. In case the logs from the nearby wells are available, it is a good idea to have log copies of the corresponding pressure plots on the rig site. However, it is not always the best solution to one's problem either, since adjacent wells may test different fault blocks and faults are known to act either as pressure barriers or conduits. Hence, the pressure profiles encountered, could be identical, similar or completely different.

The other pit fall can be due to the lack of experience of engineers and geologists who are available for this job to decide as to which type of logs to be run for pressure detections dependent upon the formation and bore hole conditions.

(iii) Major lithological or age changes, washouts, or even drastic but unexpected formation pressure variation can also result in unusual pressure profiles. To minimise the effect of such constraints, it is a good idea to have an overlap section of subsequent logging runs. However, shale hydration effects on acoustic (Sonic) logs should not be confused with bad logs.

(iv) Prolonged exposure of shale zones to the drilling mud may alter some of the shale properties. Significant variations in shale resistivity are sometimes observed from one logging run to the next (Fertl and Timko, 1972). These changes appear to be related to the type of drilling mud, its compositions, pH, mud weight, time, magnitude of the formation pressures, downhole temperatures, the presence/absence of shale gas and shale mineralogy.

(v) Reynolds et.al., (1973) have discussed possible shale hydration effects as observed on acoustic (Sonic) logs, to be a constraint on wireline methods.

(vi) Geological age changes, minor unconformities, etc., are often reflected in drastic changes in the shale properties.

(vii) Plotting the data from highly tectonically disturbed shale section should be avoided.

(viii) Shale in the proximity of large salt masses, such as salt domes, etc., have very low resistivity due to increase of salinity in pore water. This may falsely indicate higher than actual formation pressures.

(ix) Extremely fresh formation water, i.e., 3000 ppm sodium chloride (equivalent) or less at deeper depths, increases shale resistivity values and shows formation pressures lower than actual.

(x) The presence of shale gas in the shales, can drastically affect both resistivity and sonic measurements.

(xi) Considerable scattering of acoustic energy occurs in gas cut muds and drilling fluids, containing loss circulation additives. Sometimes attenuation of these muds is so great and variable that no realistic measurements can be obtained.

(xii) Major changes in bit size also affect resistivity parameters.

(xiii) Recognise major lithological changes, i.e., the presence of limy shales. They may occur in the normal and abnormal formation pressure environments and can give erroneous pressure anomalies if not taken into consideration.

(xiv) In thick massive shale sections, it is often observed that the pressure gradient in shales is a function of proximity to the permeable sands.

(xv) If at all possible, avoid shale zones exhibiting severe washouts.

(xvi) Bentonitic and fluid sensitive shales exhibit very low resistivity values, suggesting very high formation pressures which may be non existing.

(xvii) Normal trend line should be realistic since it is the basis for picking pressure top and any quantitative pressure interpretation. They are of the utmost importance in ultra deep wells, since with increasing depth any deviation from correct trend line becomes enhanced, thus aggravating any erroneous quantitative pressure evaluation. They are equally important in boreholes exhibiting major washouts, presence of abrupt age and lithology change, thick and massive pressure caps, etc.

Even after taking these precautions into consideration, if it turns out that the interpretation was wrong, one should reconsider the pressure and all other available data and trace the exact cause of the erroneous interpretation. The experience thus gained will at least help and prevent many of possible technical problems while drilling the next well (Fertl, 1976).

DST & PCT Operations

Besides the latest wire-line methods for overpressure detections like FMT, well bore geophones, etc., it is possible to take pressure measurements along with the fluid samples with the help of drill stem testing.

Drill stem testing (DST) can be done both in open and cased hole. The open hole formation testing procedure permits testing any prospective formation without setting casing. Simplicity and economy of the testing procedure usually permits testing as many different formations as desired, to prevent abandoning the potential productive well, as well as to get an idea about the pressures that may be encountered in the different formations in a similar way as is obtained from the wireline test. Therefore, the information available from the DST, includes samples of the formation fluid, production rate of the formation fluid through various chokes, pressures of the formation and the build-up of the formation, which in turn is helpful in finding out the permeability, bore hole damage, radius of investigation and various other reservoir parameters.

Several factors must be considered in preparing for a formation test in an open hole.

First the clearance between packer seal and supporting member for the packer rubber should be as small as practical to prevent excessive stress on the rubber, as it supports hydrostatic pressure across the clearance gap. The packer should be set over an interval where hole diameter will not be oversized or irregular. In case the packer is to be set in a reduced size drilled hole below the normal well bore, packer size can usually be quite close to the normal size of the drilled hole. There is adequate space between the packer and the hole for most of the running in. Restricted clearance is encountered only while running the packer into the drilled hole.

Prior to running a drill stem test, mud should be circulated and conditioned to remove all cuttings and debris from the well, to prevent plugging fluid passage in the testing equipment. The mud should be conditioned to reduce the geletion or solidification. Viscosity and fluid loss characteristics should be adjusted to guard against sloughing or heaving of the formations. Various types of drill stem tools for open hole, straddle packer testing have been devised by several companies such as Halliburton, Johnston, etc., during the last several years. Open hole test by DST Tool is given in Fig. 8.17. In addition to the different types of drill stem test, Johnston-Schlumberger has devised an offshore test system to compensate for wave motion and thus to provide complete safety for all personnel and equipment involved. Major components of the PCT system are shown in Figures 8.18 & 8.19.

Fig. 8.20 illustrates the different pressures, hydrostate, flowing and shut in pressures obtained during a DST.

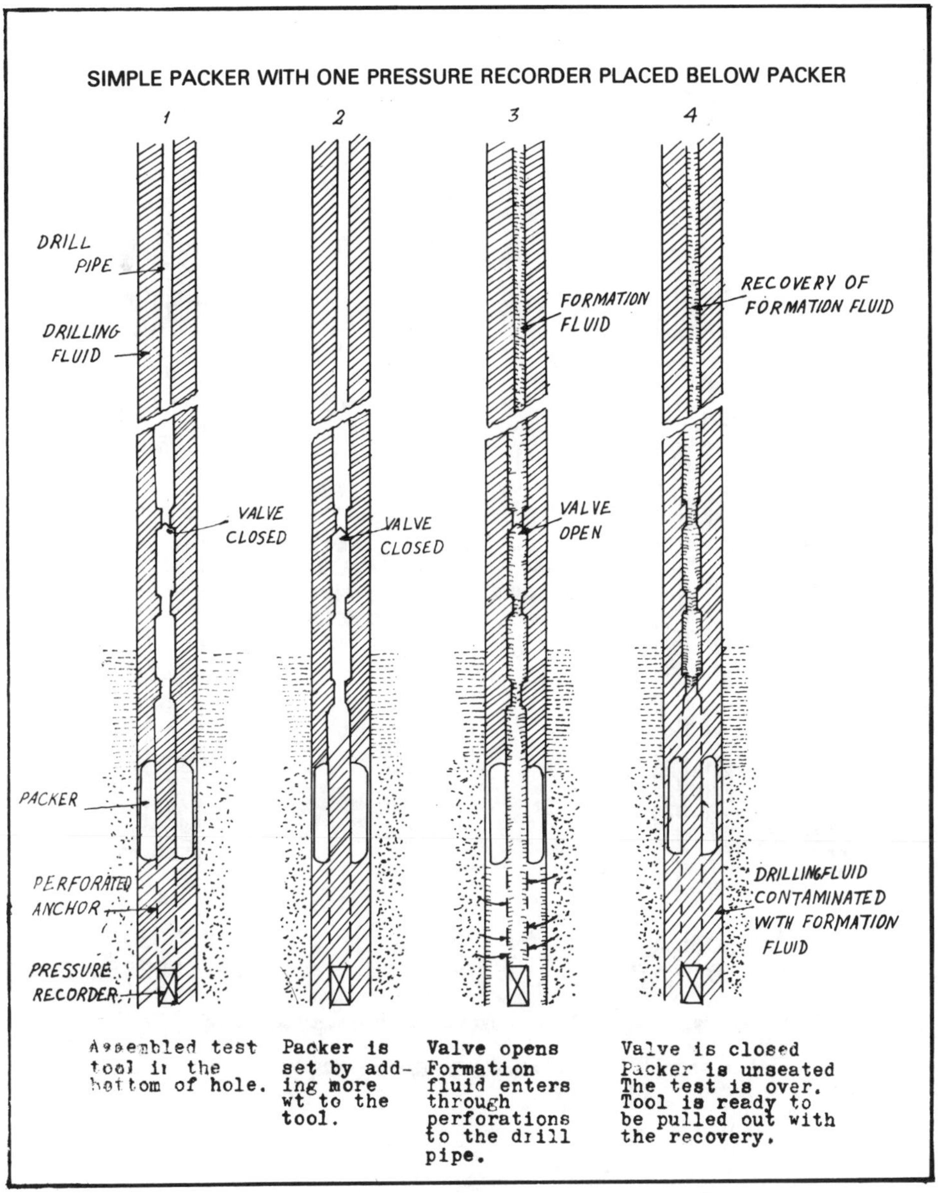

Fig. 8.17: Open hole test by Johnston DST tool

(Courtesy: Johnston Testers)

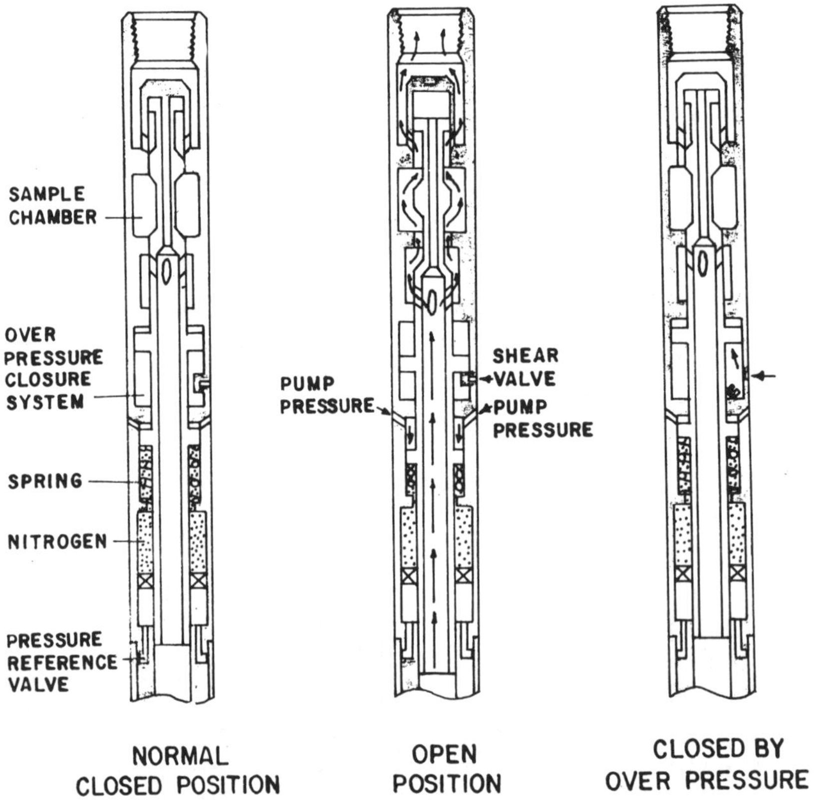

Fig. 8.18: PCT valve and sample operational schematic

*(Courtesy Johnston, Catalog 1978-79)

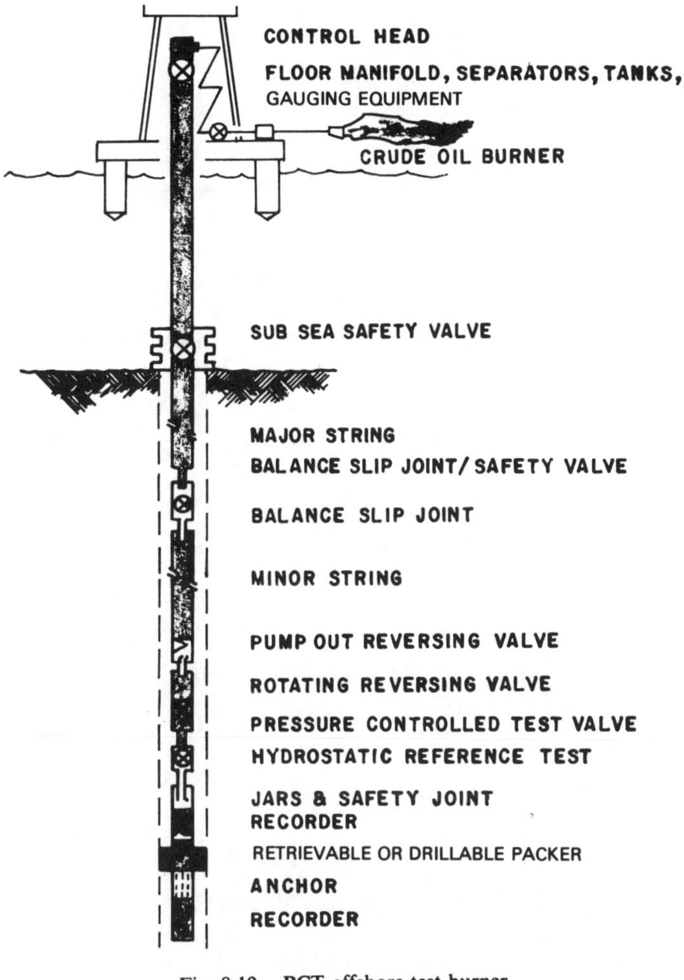

Fig. 8.19: PCT offshore test burner

(Courtsey: Johnston, Catalog 1978-79)

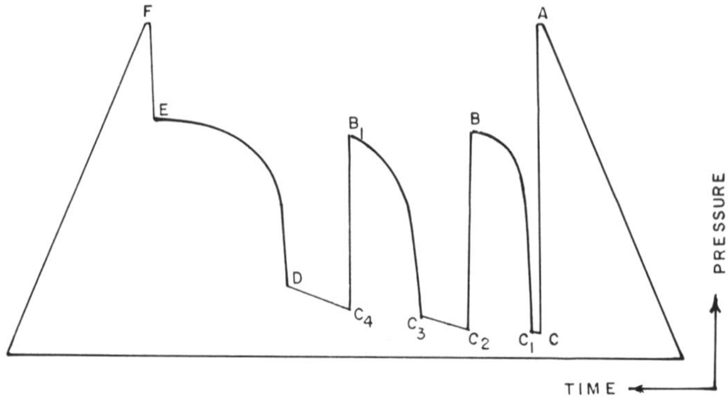

A. Initial Hydrostatic Mud Press
B. Initial Shut-in
C. Initial Flow
D. Final Flow
E. Final Shut-in
F. Final Hydrostatic Mud Press.

The following points are either fluctuating pressures or points indicatng other packer settings (testing different zones).

A-1, A-2, A-3 etc.	Initial Hydrostatic Pressures.
B-1, B-2, B-3 etc.	Subsequent Shut-in Pressures.
C-1, C-2, C-3 etc.	Flowing Pressures.
D-1, D-2, D-3 etc.	Subsequent Final Flow Pressures.
E-1, E-2, E-3 etc	Subsequent Final Shut-in Pressures.
F-1, F-2, F-3 etc.	Final Hydrostatic Mud Pressures.

Fig. 8.20: Guide to identification of DST pressure charts

(Courtesy: DST Manual–P.K. Kulkarni)

REFERENCES

Athy, L.F., 1930a. Density, porosity and Compaction of sedimentary rocks. Bull. Am Assoc. Pet Geol, **14**: 1-24.

Clavier, C., 1969. Quantitative interpretation of TDT logs. SPE 2658 44th AIME. Fall meet, Denver Colo., September.

Dresser Atlas, Formation Multi Tester Interpretation Manual 1980.

Fertl, W.H. and Timko, D.J. 1970a. Occurrence and significance of abnormal pressure formations, I. Oil Gas J., **68**(1): 97-108.

Fertl, W.H. and Timko, D.J., 1970b. How abnormal pressure detection techniques are applied, 2. Oil Gas Journal **168**(2): 62-71.

Fertl, W.H. and Timko, D.J., 1971a. Application of well logs to Geopressure problems in the search, drilling and production of hydrocarbons. Colloq A.R.T.E.P. French Petroleum Institute. Ruell June, paper No. 4.

Fertl, W.H., Timko, D.J., 1970c; Association of salinity variations and geopressures in soft and hard rocks, Transactions SPWLA.

Fertl, W.H., 1976. Abnormal Formation pressures. Elseview Scientific publishing Co.

Foster, J.B. and Whalen, H.E., 1965. Estimation of formation pressures from electrical surveys, Offshore Louisiana. SPE 1200 40th AIME Fall meet. Denver Col., October also J. Pet Technol, **18**: 165-171, 1966.

Ham H.H., 1966. A method of estimating formation pressures from Gulf Coast well logs. Trans Gulf Coast Assoc. **16**: 185-197.

Hicks, W.G. and Berry, J.E., 1956. Application of continuous Velocity logs to determination of fluid saturation of reservoir rocks Geophysics **21**: 739-754.

Hottman, C.E. and Johnson, R.K., 1965. Estimation of formation pressures from log derived shale properties. J. Per. Technol, **17**: 717-723.

Levorsen, A.I., 1967. Geology of petroleum, W.H. Freeman and Co., San Francisco, 2nd ed., **724** pp.

MacGregor, J.R., 1965. Quantitative determination of reservoir pressures from conductivity log. Bull Am. Assoc. Pet Geol, **49**: 1502-1511.

Overton, H.L. and Timko, D.J., 1969. The salinity principles—a tectonic stress indicator in marine sands. 10th Prof well log Analysts Symp, Houstan, Texas.

Pirson, S.J. 1963. Hand book of well log Analysis. Prentice Hall Englewood Cliffs, N.J., **326** pp.

Pickett, G.R. 1963. Acoustic character logs and their applications in formation evaluations. J. Pet Technol **15**: 759-668.

Sahay, B., 1972. Abnormal subsurface pressures: their origin and methods employed for prediction in India. SPE 3900, 3rd Symp on Abnormal Subsurface Pore pressures, Louisiana State Univ, Baton Rouge, La., May.

Sahay, B., et al, 1983. Formation Evaluation and Wellsite Geological Techniques, pp. 229-238.

Sahay, B., 1984. Analytical Studies of Overpressured horizons in Bombay Offshore Basin. 5th Seapex Conference, Singapore.

Schlumberger, C and Leonardon, E.G., 1934. A new contribution to subsurface studies by means of electrical measurements in drill holes. Trans AIME **110**: 237-289.

Seevers, D.O., 1972. A method of determining permeability by means of pulsed NMR and an application of the method to the study of a Gulf Coast Super normally pressured well. SPE 3847, 3rd Symp. On Abnormal subsurface pore pressure, Louisiana State Univ, Baton Rouge, La, May.

Wallace, W.E., 1964. Well induction log yield pressure data Oil Gas Journal **62**(37): 124-126.

Wallace, W.E., 1965(a) Abnormal subsurface pressures measured from Conductivity or resistivity logs 6th Prof well log Analysts, Symp, Dallas, Texas.

Wyllie, M.R.J., De Witte, A.J., and Warren, J.E. 1958(a). On the streaming potential problem Tans AIME, **192**: 3-15.

Wyllie, M.R.J., Gregory, A.R. and Gardner, G.H.F., 1958b. An experimental investigation of factors affecting elastic wave velocities in Porous media. Geophysics, **23**: 459-494.

CHAPTER IX

Reservoir Engineering Concepts in Overpressure Environments

Reservoir engineering, a specialized field of petroleum engineering, describes the reservoir and production behaviour based upon geological and petrophysical data, reservoir fluid properties, reservoir pressure and temperature, etc.

Proper evaluation of hydrocarbon reserves, reservoir behaviour, and production mechanisms affects and controls field operations of abnormally high-pressured reservoir rocks. Generally speaking, such overpressured gas reservoirs are encountered in clastic (sand/shale) sequences, are of limited lateral extent and/or may exhibit limited aquifers, and frequently do not behave as volumetric gas reservoirs. This complicates gas-in-place estimates and may result in rather optimistic hydrocarbon reserve predictions.

Possible mechanisms for overpressured hydrocarbon reservoirs include (a) water influx from adjacent shales into reservoir section ("shale water influx"), (b) formation rock compressibility and associated rock failure, and (c) formation water influx from limited aquifers into the reservoir, etc. Several published studies, however, indicate that any or a combination of these "driving mechanisms" (i.e. energy sources) may be applied successfully to match mathematically the observed field performance data with such numerical simulation studies.

Driving Mechanisms (i.e. Energy Sources) in Reservoir

Shale Water Influx

Gas-in-place can be predicted via a graphical solution for the gas material balance equation. The special case, under specific restrictive assumptions, is the linear (straight-line) relationship between P/Z and the cumulative gas production (P = reservoir pressure, Z = gas deviation factor). Based on the equation of state and corrections for compressibility, the reservoir is considered to be closed and without water encroachment.

The effect of water influx on plots of P/Z versus cumulative gas production has been investigated by Bruns et al. (1965). Results clearly suggest that not only errors in the collected field data but also water influx can be responsible for the typical curved part during the early-life production history of a gas reservoir (Fig. 9.1). In other words, it may be misleading to linearly extrapolate early production data without taking possible water influx into account. Gas-in-place predictions may be overestimated by as much as a factor of two.

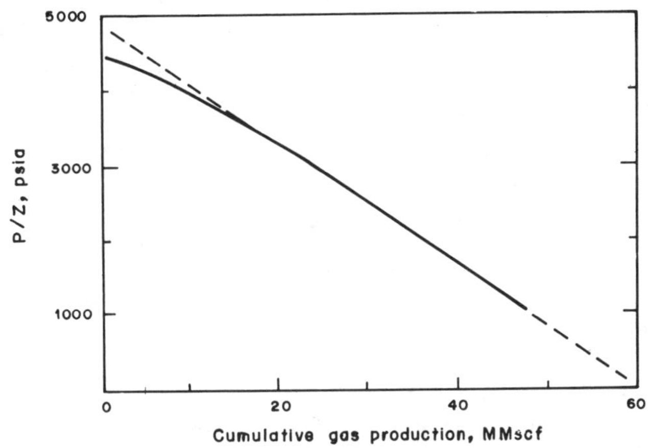

Fig. 9.1: Typical P/Z versus cumulative production behaviour for an overpressured gas reservoir, U.S. Gulf Coast area. Cumulative gas production is in millions of standard cubic feet (M Mscf; standard conditions: 60°F and 14.7 psi)

Wallace (1969) suggested that "shale compaction may be resumed by withdrawal of gas or fluids from the reservoir, thus creating a pressure differential across the shale/sand interface. This initiates a type of abnormally pressured water influx drive that we will call shale water drive. This shale water drive is controlled by the factors involved in Darcy's law, i.e. (1) the permeability of the shale, (2) the pressure differential across the shale/sand interface, (3) the cross-sectional area of the sand interface, and (4) the viscosity of the water."

Field observations and reservoir performance studies (Wickenhauser, 1966; Fowler, 1968; Lavaquial, 1971; Dugga 1972) appear to support the shale water influx concept. Bourgoyne et al. (1972) reviewed shale properties as related to the shale water influx theory in overpressured reservoirs and noted in both model and field case studies that the pressure distribution in the shales should be higher in the vicinity of permeable sands. This concept is supported by independent mathematical model studies (Bredehoeft and Hanshaw, 1968; Smith, 1973) and well-log-derived field observations (Fertl and Timko, 1970).

Fig. 9.2 illustrates "shale water depletion" restricted to the vicinity of the pay sand in an overpressured well, located in South Louisiana (Fertl and Timko, 1970). Two pulsed-neutron logs run a year apart monitor hydrocarbon saturation changes in the pay which is being produced in several adjacent wells. Sigma value changes in the sand are due to the increase in water saturation. Also note the changes in the adjacent shales.

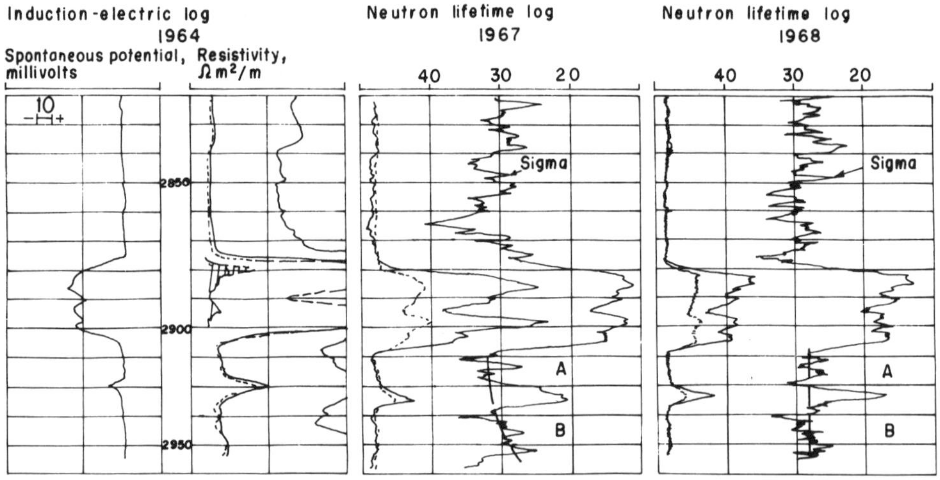

Fig. 9.2: Pulsed neutron logs, run several years following the completion of high-pressured Louisiana well, show effects of pressure depletion (after Fertl and Timko, 1970). The increase in sigma values in the pay is due to increased water saturations caused by production. The sigma decrease in shale A apparently was caused by increased compaction and decreased porosity (sigma is the macroscopic thermal neutron capture cross-section in 10^{-3}/cm. A = shale adjacent to permeable sand, B = shale distant to permeable sand.

Zone A, the shale next to the pay, shows a marked sigma decrease due to increased compaction and porosity decrease (which suggests a shale water influx into the sand), whereas Zone B, the shale, some distance from the permeable sand, shows considerably less, if any, variations due to pressure drawdown.

Rock Compressibility and Failure

Harville and Hawkins (1969) discussed in detail the performance study of the North Ossum gas field, a closed reservoir with a limited aquifer, located in Lafayette Parish, Louisiana. P/Z versus cumulative production based on early-life production (Fig. 9.3) suggests initial gas-in-place reserves as high as 6.23 billion M^3 (i.e. at 15°C and 1 kg/cm²), approximately twice the volumetric gas-in-place estimate.

As reservoir pressure is depleted, the increase in net over-burden pressure causes inelastic rock compaction or rock failure, and as the failure continues with decreasing pore pressure, rock compressibility decreases until it eventually reaches normal values in the range of 6 microsip (microsip = 10^{-6} psi). Fig. 9.4 illustrates rock compressibility for the subject reservoir as a derivative of the pore volume computation as a function of pore fluid pressure. A maximum rock compressibility of 28 microsip occurs during early reservoir depletion when rock failure is presumed to occur. At about 6,000 psi [422 kg/cm²], the rock compressibility again "normalizes" (6 microsip). Low rock compressibility plus higher gas compressibility

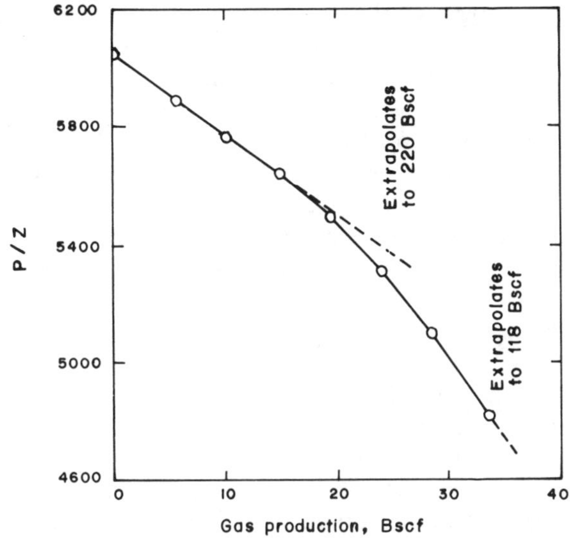

Fig. 9.3: P/Z versus production for the NS 2B reservoir, North Ossum field, Lafayette Parish, Louisiana (after Harville and Hawkins, 1969)

(Courtesy of *Journal of Petroleum Technology*)

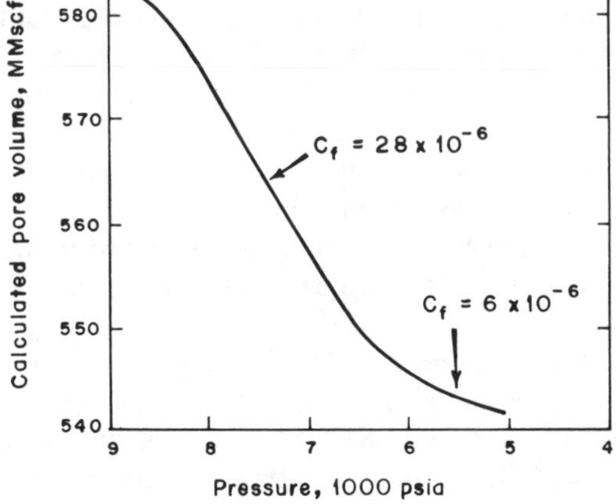

Fig. 9.4: Calculated pore volume of the NS 2B reservoir, North Ossum field. Lafayette Parish, Louisiana, assuming no water influx and gas-in-place of 114 Bcf (after Harville and Hawkins, 1969). c_f = formation compressibility in psi $^{-1}$.

(Courtesy of *Journal of Petroleum Technology*)

(due to decreased pressure) causes the reservoir performance to match closely that of a constant volume system, which should result in reasonably accurate P/Z extrapolation and thus more realistic reserve estimates. This is confirmed by field observations on the subject reservoir since after production of some 0.566 billion m^3 which caused the pressure to decline to 6,500 psi [450 kg/cm^2], the P/Z extrapolation results in a 3.34 billion M^3 gas-in-place estimates, an excellent agreement with the volumetric reserve estimate of 3.2281 billion M^3 (Fig. 9.3).

This study does indicate that the mechanism of rock compressibility and failure can act as a driving force (i.e. energy source) in overpressured gas reservoirs.

Peripheral Water Influx

Numerically generated and actual field performance data of abnormally pressured gas reservoirs were analysed by Bass (1972). Starting with the concept of a combination of rock compressibility, shale water influx, and peripheral water influx as superimposed mechanisms, a generalised mathematical volume balance expression was developed. Gas reservoir performance data were studied to investigate the effect and magnitude of each of these three possible energy sources in overpressured reservoirs. Bass concluded that peripheral water influx makes one of the greatest long-term energy contributions.

General Comments

According to Hammerlindl (1971), supporting previous observations, two distinct slopes are evident on P/Z plots in overpressured reservoirs with the final slope being steeper than the initial one. Consequently, gas-in-place estimates based on earlv production history are erroneously high. The initial slope is attributed to g expansion, reservoir rock compaction, crystal expansion, and water expansion. A soon as pressure approaches normal pressure due to depletion, the reservoir will behave as a normal gas expansion reservoir. At this stage, effective compressibility of the system is about equal to gas compressibility, which causes the second, steeper slope. In other words, reserve estimates using early-life bottom-hole shut-in pressure and cumulative production should be adjusted to allow for pore-volume change and water expansion.

Two prediction methods were proposed for initial gas-in-place estimates from early production history using reservoir rock compressibility, gas production, and shut-in bottom-hole pressure. Method A utilizes the ratio of average values for effective compressibility (Craft and Hawkins, 1959) and gas compressibility, while Method B includes compressibility in the expansion term of the material balance equation.

More recently, Chierici et al. (1978) investigated the behavior of overpressured gas reservoirs in clastic formations with randomly interbedded, thin, non-continuous shale layers using numerical simulation models. Analyses of both "thick-closed" and "thin-edge-water-drive" reservoir geometries lead to the following conclusions:

1. Internal water drive caused by interbedded shale compaction may contribute significantly to reservoir energy and, hence, the maintenance of reservoir pressure. This effect becomes important in low-permeability cases, where the external water drive is weak.
2. Mobile water originated by compaction of interbedded shales may decrease well productivity and increase water production, even in the early life of a well.
3. Partial well penetration into the reservoir may improve the water/gas ratio only when both reservoir thickness and vertical permeability are high enough to allow water segregation toward the bottom of the reservoir.
4. As discussed previously, the internal water drive may have a large effect on the P/Z versus cumulative gas production trend. In other words, the reliability of gas-in-place evaluation from past reservoir performance becomes unreliable.

One of the most recent reservoir engineering studies was performed on an overpressured, retrograde condensate gas field exhibiting a partial water drive, located onshore Louisiana (Sharpe and Vankirk, 1983). The authors concluded that:

1. A material balance analysis not only estimates the original gas-in-place, but can also verify critical parameters used.
2. Very high gas recovery factors (±90%) are possible due to the relatively large amount of original gas-in-place stored per acre-foot of bulk volume.
3. As reservoir pressure declines, the gas in the invaded portion of a partial water-drive reservoir will continue to expand and percolate through to producing wells.
4. Recovery factors can be optimised at final blow-down of a partial water-drive reservoir by increasing field withdrawal rates, thus lowering the abandonment pressure and allowing previously trapped gas to expand farther and percolate through.

REFERENCES

Bass, D.M., 1972. Analysis of abnormally pressured gas reservoirs with partial water influx. SPE 3850, 3rd Symposium on Abnormal Subsurface Pore Pressure, Louisiana State Univ., Baton Rouge, La., May.

Bourgoyne, A.T., Hawkins, M.F., Lavaquial, F.P. and Wickenhauser, T.L., 1972. Shale water as a pressure support mechanism in superpressured reservoirs. SPE 3851, 3rd Symposium on Abnormal Subsurface Pore Pressure, Louisiana State univ., Baton Rouge, La, May.

Bredehoeft, J.D. and Hanshaw, B.B., 1968. On the maintenance of anomalous fluid pressures, 1, Thick sedimentary sequences. Geol. Soc. Am. Bull., **79:** 1097-1106.

Bruns, J.R., Fetkovich, M.J. and Meitzen, V.C. 1965. The effect of water influx on PZ-cumulative gas production curves. J. Pet. Technol., **17:** 287-291.

Craft, B.C. and Hawkins, M.F., 1959. Applied Petroleum Reservoir Engineering. Prentice-Hall, Englewood Cliffs, N.J., 437 pp.

Chierici, G.L., Ciucci, G.M., Sclocchi, G., and Terzi, L., 1978. Water Drive From Interbedded Shale Compaction in Super-pressured Gas Reservoirs-A Model Study. J. Pet. Technol., **30:** 937-946, June.

Dugga, J.O., 1970. The Mobil-David (Anderson "L") field-an abnormally pressured gas reservoir in South Texas, SPE 2938, 45th AIME Fall Meeting, Houston, Texas, October, also J. Pet. Technol., **24:** 132-138, 1972.

Fert, W.H. and Timko, D.M., 1970. Association of salinity variations and geopressures in soft and hard rocks. Trans. Soc. Prof. Well Log Analysts, Paper J., 24 pp.

Fertl, W.H. and Timko, D.J., 1970. How abnormal pressure detection techniques are applied. Oil Gas J., **68(2):** 62-71.

Fowler, Jr., W.A., 1968. Pressures, hydrocarbon accumulation and salinities-Chocolate Bayou field. Brazoria County, Texas, SPE 2226, 43rd AIME Fall Meeting, Houston, Texas, September; also J. Pet. Technol., **22:** 411-423, 1970.

Hammerlindl, D.J., 1971. Predicting gas reserves in abnormally pressured reservoirs. SPE 3479, 46th AIME Fall Meeting, New Orleans, La., October.

Harville, D.W. and Hawkins, M.F., 1969. Rock Compressibility and failure as reservoir mechanisms in geopressured gas reservoirs, J. Pet Technol., **21:** 1528-1530.

Lavaquial, F.P., 1971. Water influx into petroleum reservoirs from adjacent shales. M.S. Thesis, Louisiana State Univ., Baton Rouge, La., 120 pp.

Sharpe, G.F., VanKirk, C.W., 1983. A Reservoir Engineering Study of an Overpressured, Partial Water-Drive Gas Field in Southern Louisiana. SPE 12044, 58th Ann. Fall Meeting. San Francisco, Cal., October 5-8.

Wallace, W.E., 1969. Water production from abnormally pressured gas reservoirs in South Louisiana. J. Pet Technol., **21:** 969-983.

Wickenhauser, T.L., 1966. Shale water as a pressure support mechanism in superpressure reservoirs. M.S. Thesis, Louisiana State Univ., Baton Rouge, La., 115 pp.

CHAPTER X

Overpressure Areas of Indian Sub-Continent: Analytical Study

India

The sedimentary basins of India have a total area of 1,780,000 sq. km. of which approximately 380,000 sq. km. are located offshore the Continental Shelf (Fig. 10.1). Massive sediments have been deposited both on the shelf and the geosyncline region of this basin. Exploration for hydrocarbons carried out so far, both onshore and offshore, has revealed the occurrence of overpressure zones in those areas of massive sediments.

The main onshore areas, where overpressures have been encountered, include the Bengal basin, the Himalayan foot hills, Masimpur and Geleki area in Assam, Tripura, the Cambay basin, on the western side in South Kadi, Nawagam area and Godavari basin. Offshore areas with overpressures include the Bombay Offshore, Godavari and Andaman-Nicobar basins. Super pressures have been encountered in the autochthonous folded belt of the Himalayan foot hills in Jammu and Balh areas in Northern India, Masimpur and several other areas in Assam, Tripura, as well as some part in the Bombay Offshore and Bengal Offshore basins.

The lithology of these overpressure sections is mostly clastic, particularly in the onshore areas, while offshore mostly shales and limestones are present. Overpressures in sediments mostly range from Ececene to Pliocene. Analytical studies carried out to investigate the origin and occurrence of such pressures indicate compaction phenomena of shales to be the primary cause, while tectonic phenomena, prevalent in the Himalayan foot hills and in Masimpur area of Assam, as the secondary reason. In addition, osmosis, reverse osmosis as well as diagenesis of limestone have contributed in part to the generation of overpressures in several areas, both offshore as well as onshore. Overpressures generally range from 1.2 to 2.38 times the hydrostatic.

Fig. 10.2 shows the high pressure areas of India. The overpressure areas, located both onshore and offshore, have been divided into two categories, compaction and tectonic controlled phenomena.

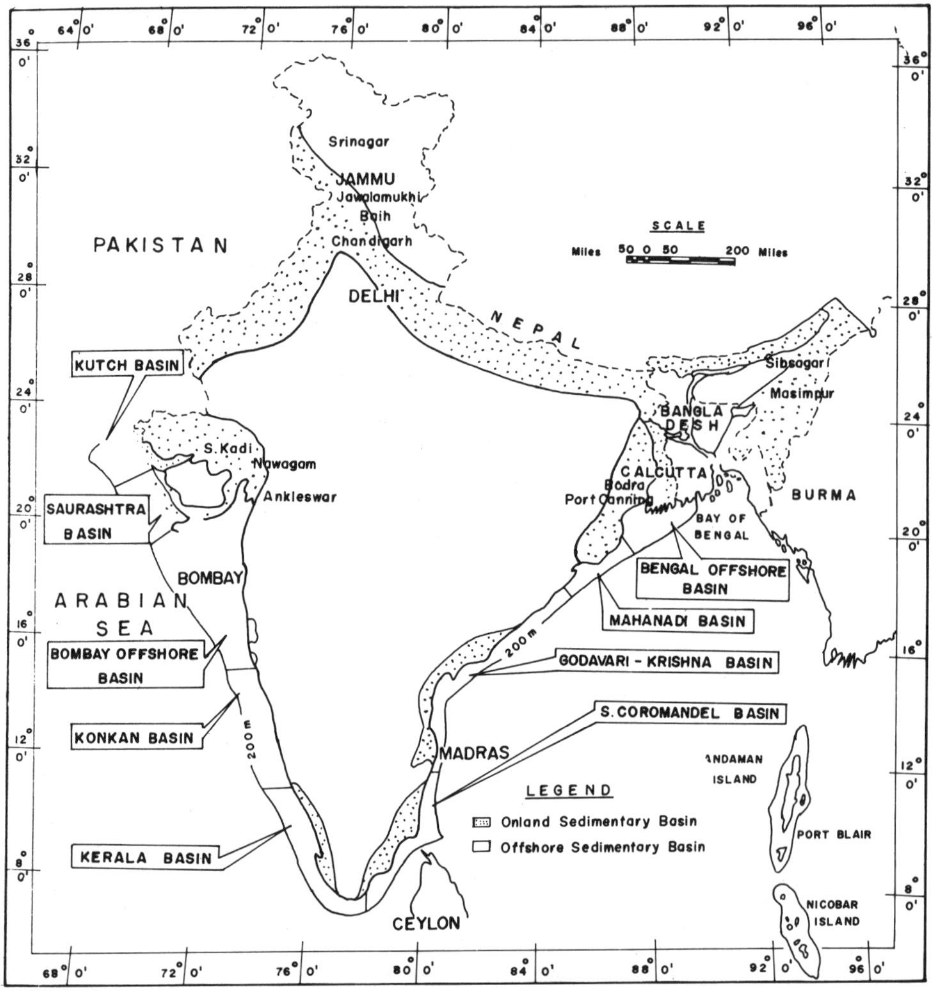

Fig. 10.1: Map of India showing sedimentary basins

(Sahay, 1984, OTC, 4699 Houston Texas)

A. Overpressures Due to Compaction Phenomena

East Coast

(i) *Bengal basin*: The Bengal basin, an area of 30,000 sq. miles, is bounded to the west and north by ancient basement rocks or shield area composed of crystalline rocks and Gondwana sediments, to the north-west by volcanics called Rajmahal trap of Mesozoic age and to the north-east by the Shillong massif with tertiary exposures at its southern flank. The folded belt of Tripura and Chittagong marks the eastern boundary of this basin. Three major structural features, located from west to east, are:

Fig. 10.2: Map of sedimentary basins of India showing high pressure areas

(After Sahay, 1976, P-115. Courtesy: Geological Mining and Metallurgical Society of India)

(a) The basin marginal zone characterised by shallow basement rocks and a system of enechelon scarps and faults;

(b) Shelf zone, characterised by regional monoclinal gentle dips and gradual basinward thickening of the tertiary section;

(c) Eocene hinge zone and deeper basin area, which is characterised by flexure like feature originated as a result of faulting and oscillatory movement of the basement at depth. The hinge zone appears to be a depositional rather than tectonic break, caused by facies change from limestone to shale. The area east of this flexure is characterised by abrupt thickening of strata, regional facies changes from shelf to geosyncline, new deltaic wedges, truncations, etc. (Fig. 10.3). The general stratigraphy and a geological section across Bengal basin showing the facies variations, etc., are given in Figures 10.4 and 10.5.

Occurrence of High Pressures

In the Bengal basin, many exploratory wells have been drilled both by the Indo-Stanvac Petroleum Project and ONGC on the shelf and geosynclinal region. Abnormally high pressures were encountered in the wells drilled in the Port Canning and Bodra areas below the Miocene-Pliocene unconformity in the geosynclinal region. As illustrated in Figures 10.6 and 10.7 shale resistivity and density plots in these wells, indicate a significant decrease in both values around 3800 m in Port Canning and 3730 m in Bodra wells. The formation pressures calculated for these horizons are approximately 1.8 to 1.9 times the hydrostatic. On the basis of this anomaly, observed on the resistivity and density curves of shales around 3700 m in Bodra, it was possible to predict abnormal high pressures well in advance before penetrating deeper, into permeable gas zones and avoid possible drilling and blow out hazards.

Apart from the high pressures encountered in Port Canning and Bodra, where the anomaly from the resistivity and density curves was well pronounced, data obtained from a well drilled close to the hinge zone at Ranaghat-1, also indicated the possible presence of high pressures in a thick monotonous section of Miocene shales below 3650 m. Bottom hole temperature also increased to 147° C at 3650 m.

The salinity of the formation water, obtained from DST, indicates lower values in the first high pressure horizon compared to the salinity in the shallower section in Port Canning-1 (Fig. 10.9). In the Bodra well, drilled by ONGC, the salinity of the formation water obtained via DST operations over a long section (3681-4193 m) is unreliable since several sands were tested simultaneously. However, log derived salinity indicates a decreasing trend above the high pressure formation at 3750 m. This salinity anomaly synchronises with the resistivity and density trends in Bodra-1 (Fig. 10.8).

The water pattern curves for the water obtained in Bodra-1 from DST's at different depths, show a striking change in the bicarbonate and sulphate contents, i.e. they are higher in the water from high pressured horizons at 3800, and 3950 m than that of the normally pressured section above 3600 m (Figs. 10.10 and 10.11).

(ii) *Bengal offshore basin*: The offshore part of Bengal basin has an areal extent of 30,000 sq. km. As a result of seismic surveys, few anticlinal reversals were observed. Exploration in the offshore area was carried out by Carlsburg Natomas Group in 1975-77.

The offshore area has been divided into four tectonic elements which are more or less the continuation of the onshore part of the basin. Two wells were drilled by the above firm in anticlinal structures in the deep basin, east of Hinge zone on 'A' and 'B' features
higher in the water from high pressured horizons at 3800, and 3950 m than that of the

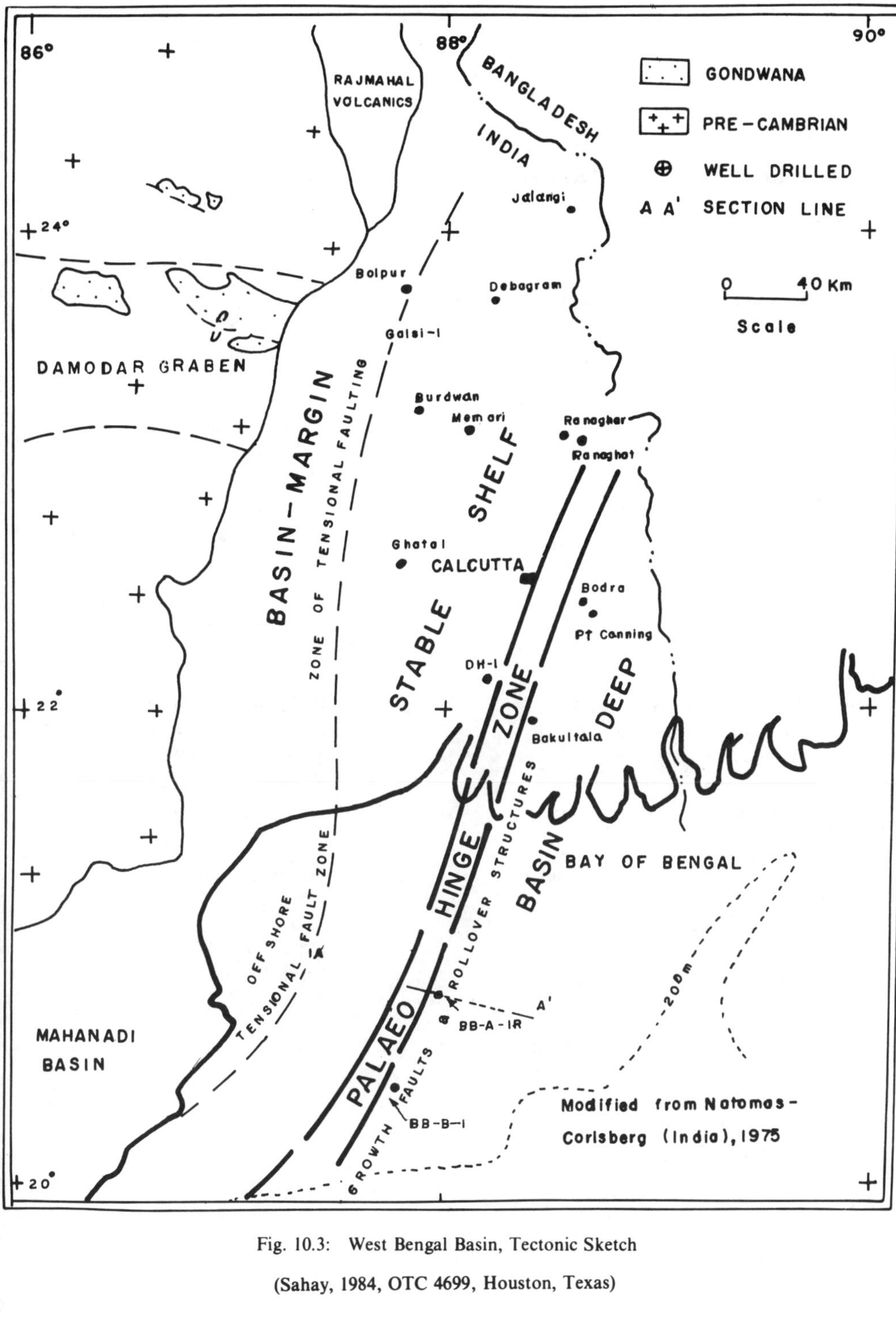

Fig. 10.3: West Bengal Basin, Tectonic Sketch

(Sahay, 1984, OTC 4699, Houston, Texas)

AGE	FORMATION	LITHOLOGY
Pleistocene-Recent	Bengal alluvium	Coarse sand, gravel
Plio-Pleistocene	Debagram formation	Silt, siltstone, clays and minor sands
	. . . Unconformity . . .	
Pliocene-Lr.Miocene	Pandua formation	Grey shale, sand and silty-sandy clay
	. . . Unconformity . . .	
Oligocene	Memari/Burdwan	Calcareous shale, argillaceous limestone and glauconitic sandstone
Upper Eocene	Kopili	Dark grey shale with varying degree of calcareous matter
Middle Upper Eocene	Sylhet	Three limestone bands separated by two clastic inter beds
Lower Eocene-Palaeocene Upper Cretaceous	Jalangi	Dark grey carbonaceous and resinous shales with asphalt, coal, bottom portion sandy
Upper Cretaceous	Ghatal	Silt, siltstone/Sandstone and oolitic limestone
Cretaceous	Bolpur	Ferruginous claystone and sandstone
	. . . Unconformity . . .	
Lr. Cretaceous-Upper Jurassic	Rajmahal	Basltic lava flows

Fig. 10.4: General stratigraphical column of Bengal Basin
(After Biswas, 1961)

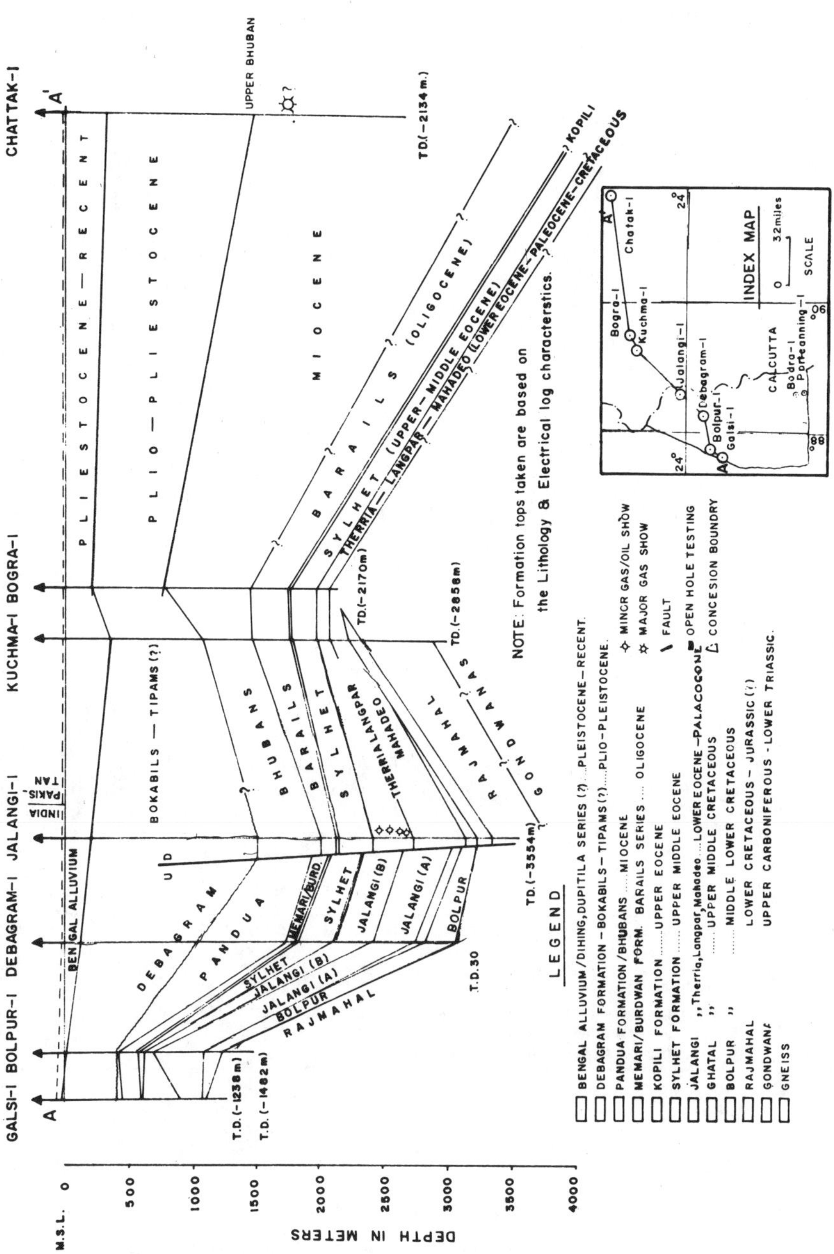

Fig. 10.5: Geological section across Bengal Basin

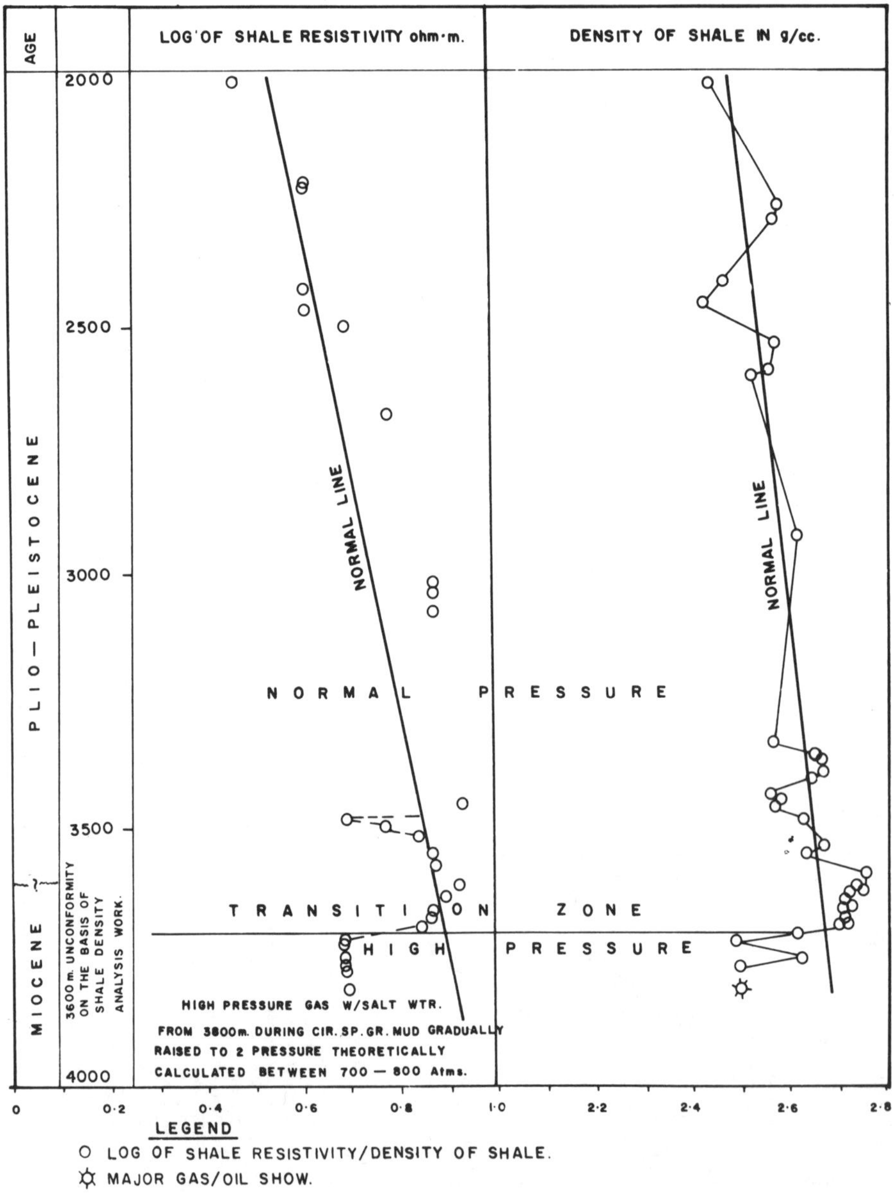

Fig. 10.6: Comparative study between log of shale resistivity and density of shales Bodra—1

(After Sahay, 1972. Courtesy: American Institute of Mining, Metallurgical and Petroleum Engrs., Inc.)

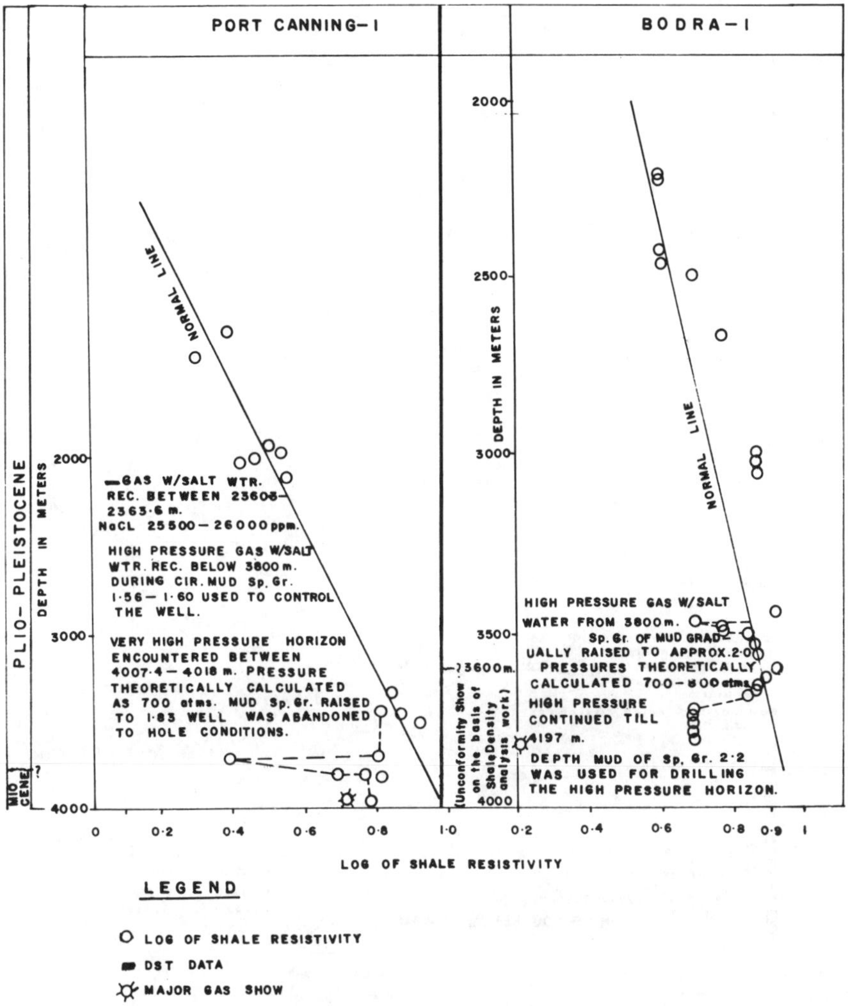

Fig. 10.7: Comparative study of the log of shale resistivities in Port Canning and Bodra wells (After Sahay, 1972, Courtesy American Institute of Mining, Metallurgical & Petroleum Engrs., Inc.)

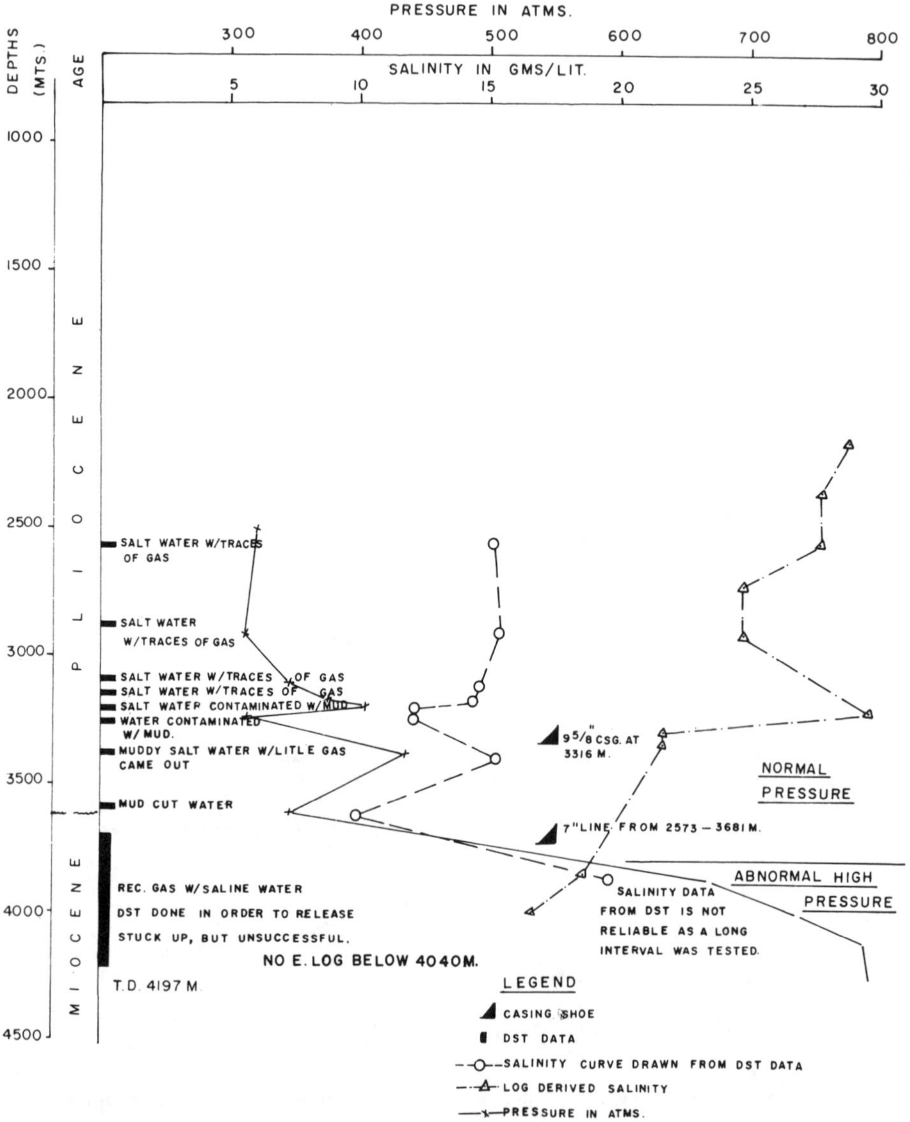

Fig. 10.8: Study on Salinity, Pressures in Bodra-1

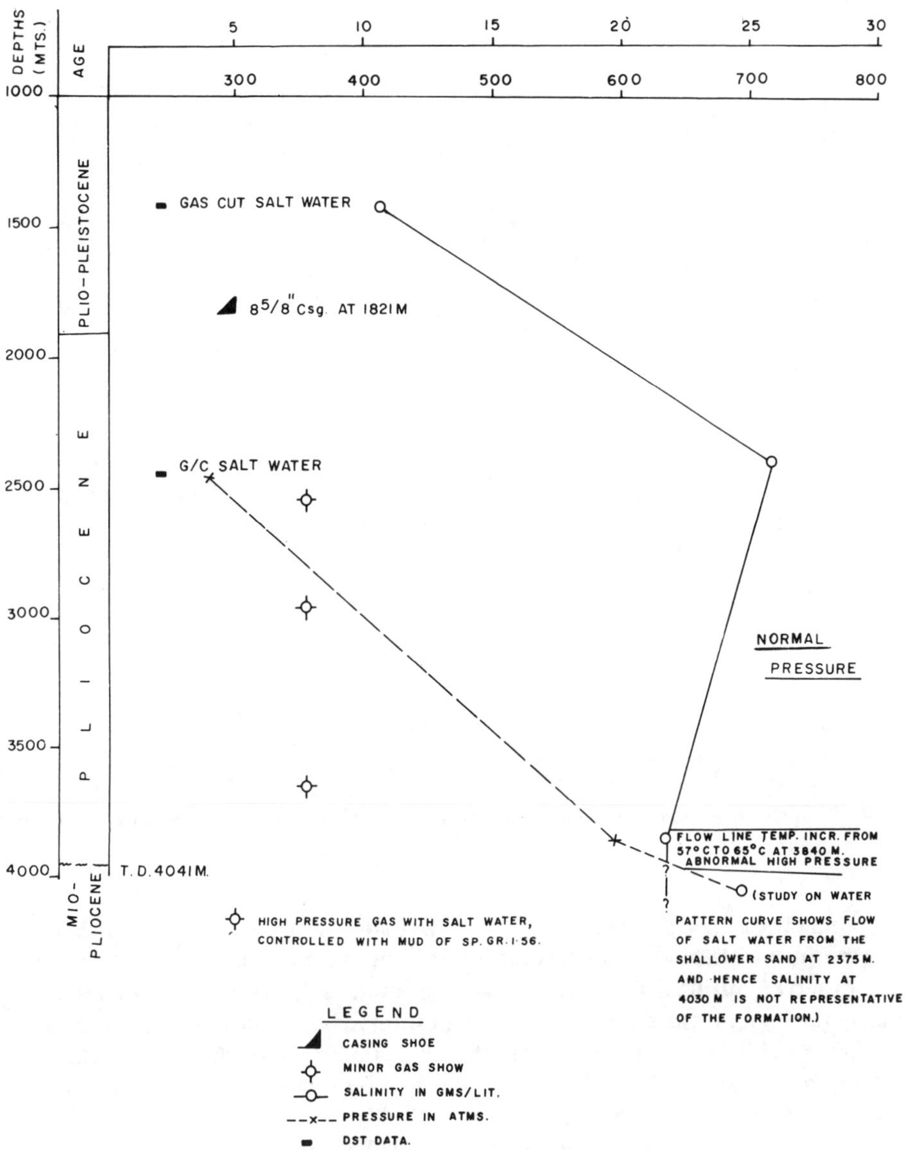

Fig. 10.9: Study on the Salinity and Pressures in Port Canning—1

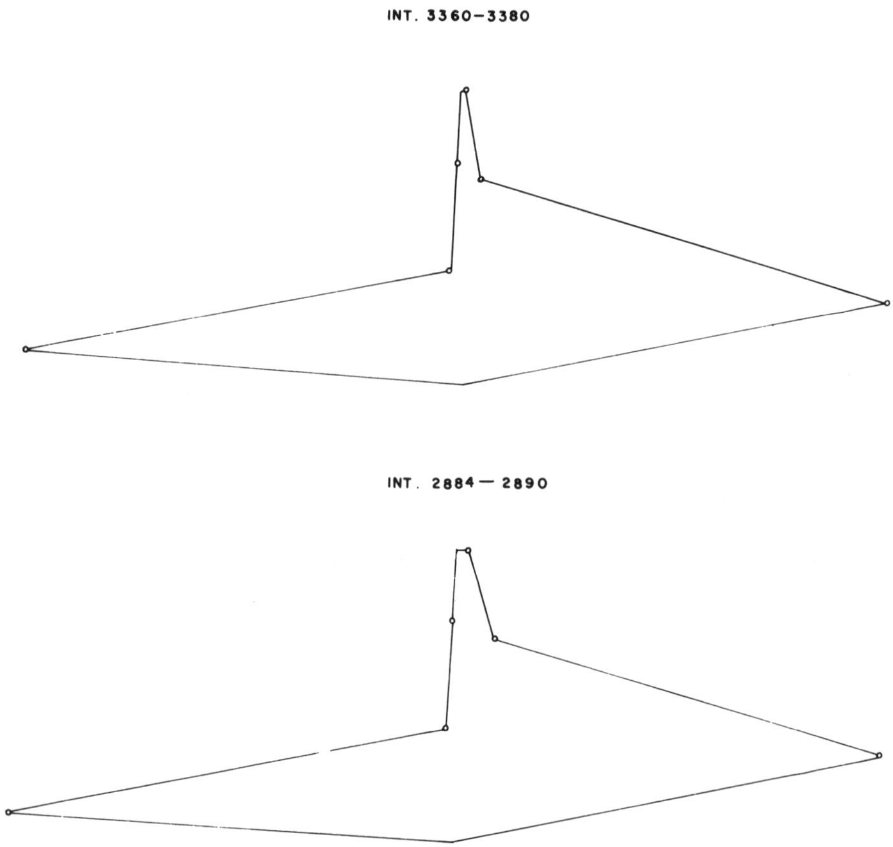

Fig. 10.10: Study on water pattern curves from a normal pressured horizon in Bodra, Bengal Basin, India (Sahay, 1984, 5th Seapex Conference, Singapore)

predominantly a clay/ shale sequence with few sandstone layers. Several sand layers up to 16 m. in thickness were found around 3360 m. No Significant oil and gas shows were observed during drilling. On the basis of mud sp. gravities of 1.8 to 1.9 used to control the high pressures, the pressures are estimated to be aproximately 760 kg/ cm^2 at 3750 m. This would indicate that they are almost twice the hydrostatic. High pressure zones tested water with dissolved gas as was revealed by the FIT's under-taken at various depths below 3200 m.

The other well BB-B-1 on 'B' feature was drilled to 4880 m. The lithology was similar to that of the first well. However, no high pressure zone was encountered in this well. Nor did the well had significant oil and gas shows. The wells were terminated in Miocene, as structure was only mapped up to Miocene. Deeper zones could not be mapped due to absence of reflections. Fig. 10.12 shows a plot of interval transit time versus depth in an offshore well in Bengal.

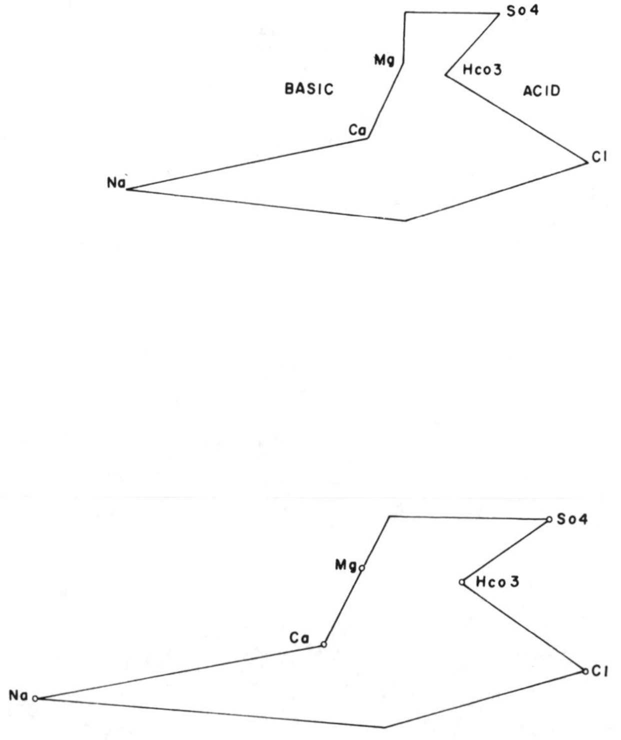

Fig. 10.11: Water pattern curves from abnormal high pressured horizon in Bodra, Bengal Basin (After, Sahay). (Fig. contd. on page 154)

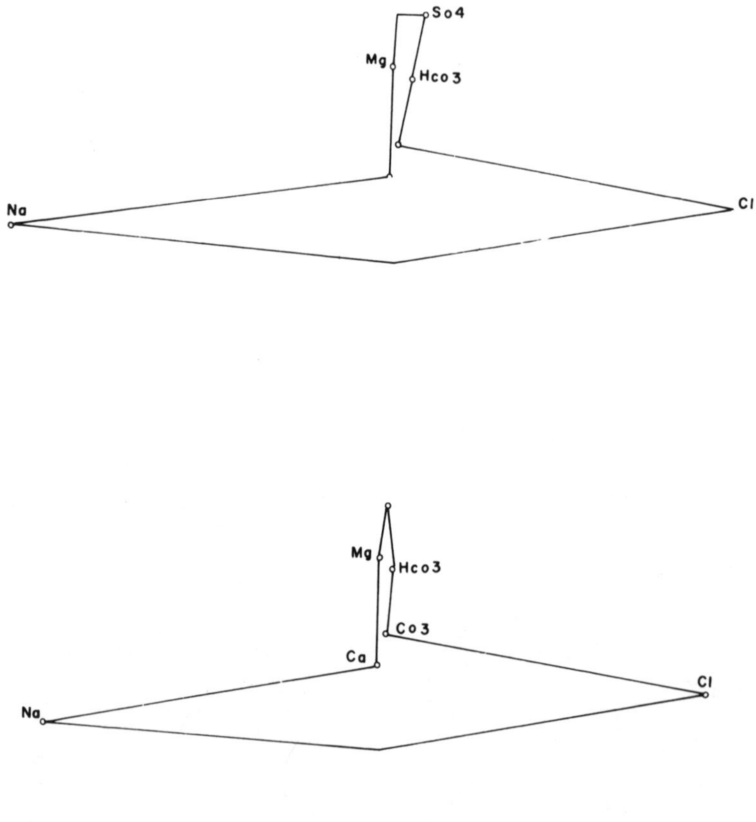

Fig. 10.11: Water pattern curves from abnormal high pressured horizon in Bodra, Bengal Basin (After, Sahay). (Fig. contd. from page 153)

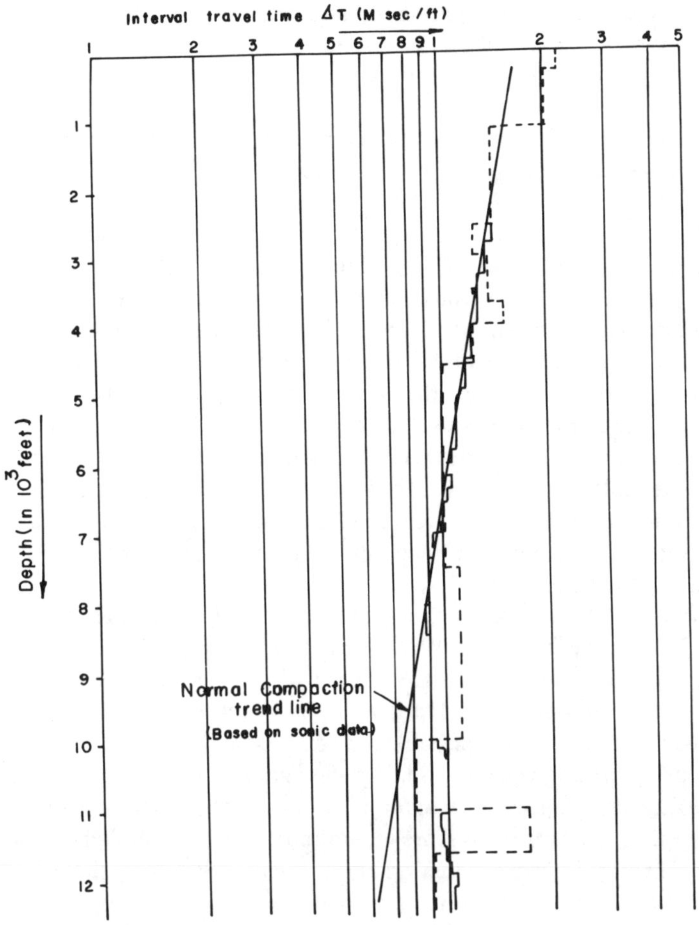

Fig. 10.12: Pressure profile in a well in offshore Bengal showing a plot of interval transit time vs. depth (Courtesy: Data Evaluation Group, Calcutta)

(iii) *Tripura area*: Tripura area which lies in the north-east of the country is the part of southern extension of Naga Lushai hill orogenic belts. The Naga-Lushai belt in Surma Valley Lushai Hills Tripura area is characterised by a wide zone of north-south trending asymetrical anticlines and synclines. The intensity of folding decreases gradually from east to west.

Baramura structure is one such north-south trending anticline with a gentle limb in the west and steep limb in the east, the latter being thrusted and faulted. Several wells have been drilled on this structure, where some indication of gas in Miocene-Oligocene sediments was reported.

Overpressure zones were encountered below 2450-3050 m and 2870 to 3430 m in well nos. 'A' and 'B'. A plot of shale resistivity versus depth for these wells shows the anomaly observed at the depth mentioned above. In view of the occurrence of super pressures observed in these wells, it became necessary to increase the specific gravity of mud to more than 2, indicative of formation pressure being double than the hydrostatic (Fig. 10.13).

In the absence of a definite tectonic phenomenon, such overpressures in the Tripura area appear to be due to compaction of shale in response to the stress of overburden only.

(iv) *Assam area*: The Assam shelf which is comparatively stable and characterised by epiorogenic movements has an areal extent of about 40,000 sq. km. and inlcudes the Shillong plateau and Garo Khasi and the upper Assam valley. Geologically, it is the north eastern prolongation of the Indian peninsular shield with the Garo-Rajmahal hills in between, through which the rivers Ganga and Brahmaputra are deflected southwards to flow into the Bay of Bengal (Fig. 10.14).

The upper Assam valley contains 7000 m of sediments under its alluvial cover. An estimated thickness of about 6400 m Cretaceous andTertiary shelf sediments outcrop in Mikir hills and Shillong plateau. On the southern edge of the platform, lies the mobile belt of the Assam geosyncline with very thick sediments of the same age as those in the shelf. However, the sequence is more complete in the geosynclinal belt (Petroliferous basins of India, 1983, pp. 1 and 2).

The stratigraphy of the Assam areas is shown in the (Figs. 10.15 and 10.16). Various oil and gas fields have been discovered, e.g. Rudrasagar, Lakwa, Geleki, etc. In the Geleki area, formation pressures higher than hydrostatic were encountered.

Shale travel time plotted versus depth indicates a departure from the normal trend at around 3500 m (Fig. 10.17), the possible top of abnormal high pressure zone. This was confirmed during drilling when the mud specific gravitiy had to be raised to 1.45 to 1.5 due to the influx of high pressure water and gas from the sandstone horizon at 3850 m. During the testing, formation pressure was found to be 100 atmosphere more than the hydrostatic. The lithology of the section above the zone is mostly shale with some silty sandy layers. In view of the severely enlarged hole size due to caving, etc., the resistivity logs could not be used properly. As other relevant data like flow line temperature, densities, etc., from Geleki wells were not available, the reasons for overpressures in the Geleki area are not properly understood. However, in view of the absence of any major tectonic phenomena in that area, the pressures appear to be primarily due to the compaction.

Exploration activities carried out in many other areas in Assam to test the various anticlinal features has also indicated the occurrence of overpressures.

(v) *Krishna Godavari basin*: Extensive deltaic plains were formed by the two east coast rivers, Krishna and Godavari, in the State of Andhra Pradesh and the adjoining areas of Bay of Bengal in which these rivers discharge their water. It is known as Krishna Godavari Basin, one of the important petroliferous basins of India. This basin occupies 15,000 sq. km of onland and about 25,000 sq. km of adjoining offshore areas up to the 1,000 m Isobath line. It is a pericratonic basin because it occurs on the margin of Indian craton and consists of metamorphic pre-Cambrian (Fig. 10.18). The general stratigraphic column of this basin is given in Fig. 10.19.

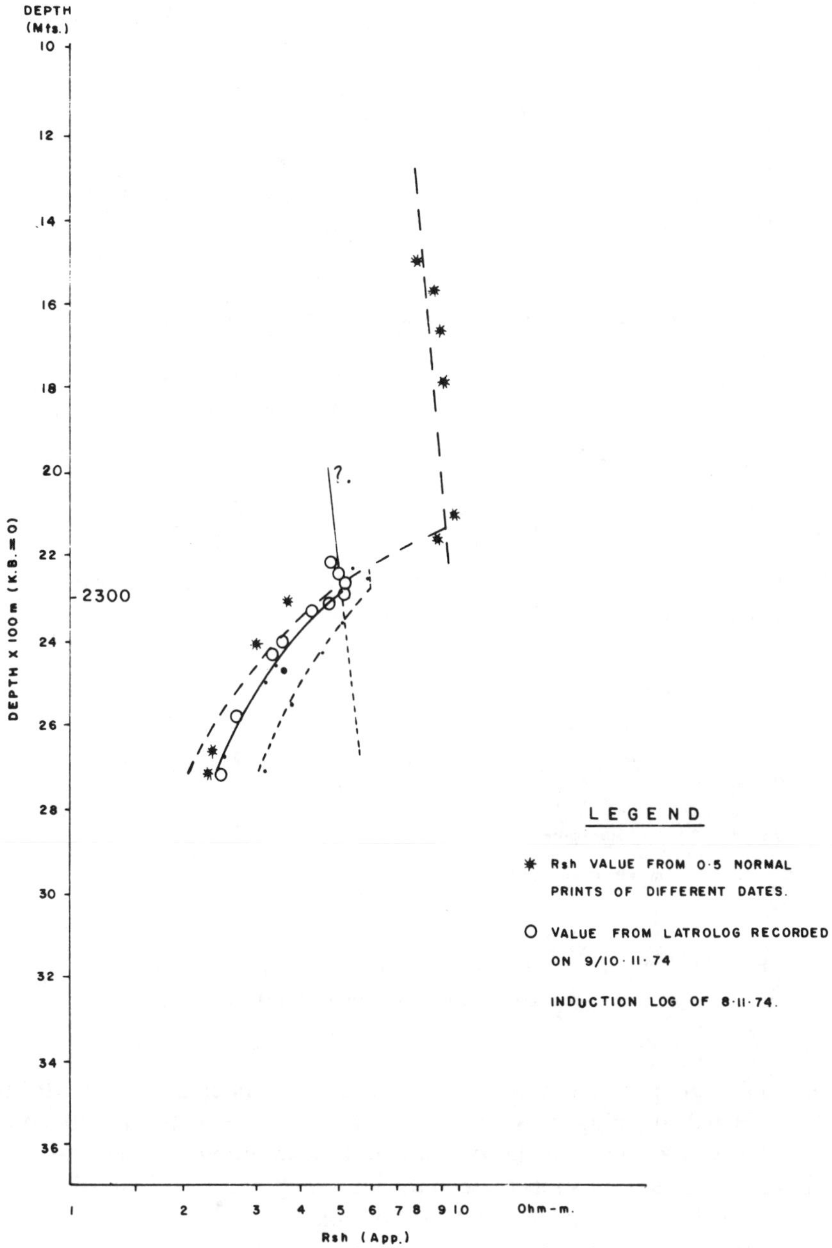

Fig. 10.13: Resistivity vs. depth plot in well—1. Tripura

(Author Logging Section, Assam Project ONGC)

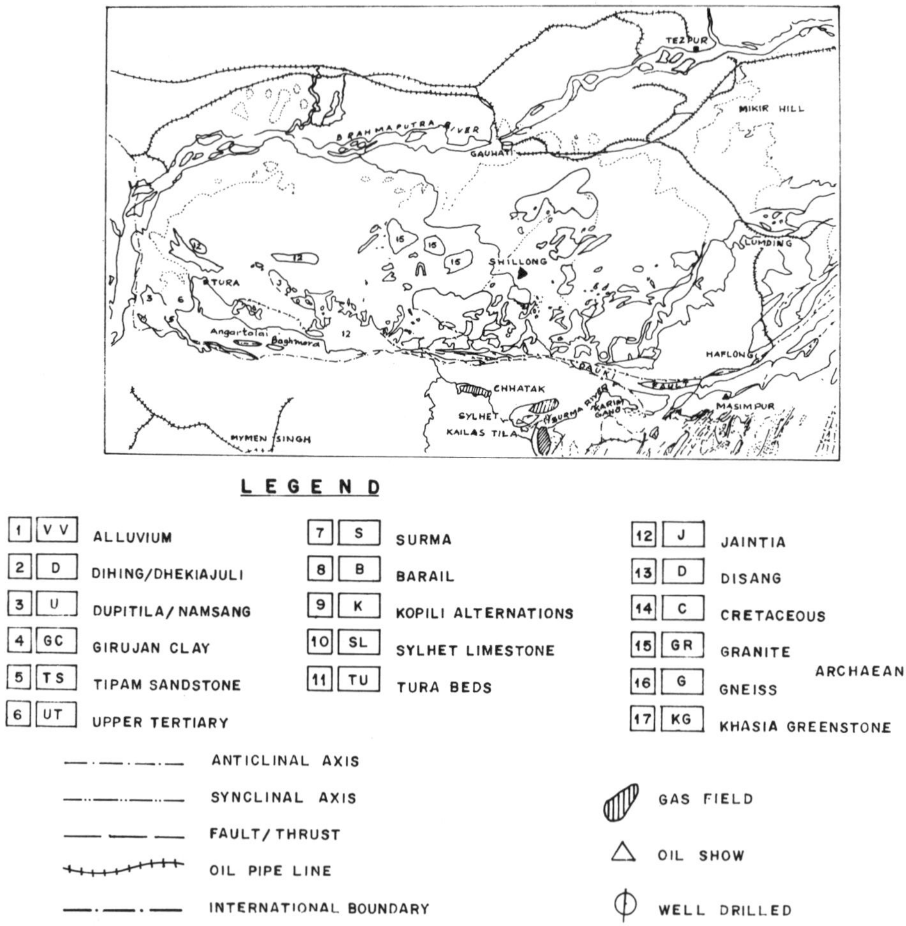

Fig. 10.14: Geological Map of Shillong, plateau and neighbouring areas
(Petroliferous basins of India, 1983)

Abnormal high pressure horizons have been encountered in wells drilled at Amalapuram and adjoining areas, onland as well as offshore. In the first offshore well drilled in the Godavari basin, a high pressure water bearing zone was penetrated below 3100 m which necessitated an increase in the mud weight to 1.51. Indications of high pressure below 3178 m depth were noticed both on dcs and shale density plots.

In addition, some indication of overpressures was also revealed by the decrease of 'd' exponent from 2800 m, corroborated by an increase of flow line temperature from 2800 m. This observed anomaly was confirmed after a sand horizon was penetrated below this depth, which flowed only gas and necessitated increasing the mud specific

AGE (approx)	Series	Stage and local facies: Shelf sediments (Shillong Mikir Hills plateaux and Upper Assam Valley
Pleistocene Recent	Alluvium	Present day alluvium, older alluvium and High Level Terraces
unconformity		
	unconformity	
Plio-Pleistocene	Dihing	Dhekiajulibeds 1800 m in Upper Assam Valley
Mio-Pliocene	Dupi Tila	Upper Dupi Tilas in Garo Hills 1000 m.
		unconformity
		Namsang Beds in Assam Valley 930 m.
	unconformity	
Miocene	Tipam	Girujan Clay 600m. Tipam Sandstone 900m
	Surma	Not subdivided 200m.
	unconformity	
Oligocene	Barail	Not subdivided 1200m.
Eocene	Disangs	Jaintia Series — Kopili Alternations 800m. Sylhet Limestone 600m (including Terria)
Cretaceous Danian Maestrichtian and ? Lower		Langpar stage 200m. Mahadek stage 230m.

Fig. 10.15: Geology of Assam Arakan Basin (After Das Gupta, 1977)

gravity from 1.38 to 1.42. The permeable high pressure zones encountered occur in sediments of Miocene age and are overlain by a thick monotonous shale section.

West Coast

(vi) *Cambay Basin (onland)*: Cambay basin is a flat lying area trending north-south and bounded by the Deccan trap craton on the east and Saurashtra uplift on the west and extending north of Ahmedabad and beyond Surat to the south. Further towards the south it extends below the Gulf of Cambay and co-extensive with Bombay Offshore basin. Most of basinal and marine sediments of Mesozoic and Tertiary ages occur on the eastern and western fringes of the basin. Over a larger part of the basin, the basement of Tertiary sediments appear to be composed of Deccan traps. The top of the trap is generally known to have undergone sub-tropic to tropical weathering.

The tectonic framework of this basin as revealed by various geological and geophysical studies, forms a horst structure bounded by faults at the eastern and western margins. Formation of this basin started in Upper Mesozoic when the sea transgressed from Sind and south Rajasthan to north of Gujarat. In the Tertiary section, the basin has three main synclinal zones, i.e., Ahmedabad, Cambay and Ankleshwar. On their flanks, anticlinal structures occur where oil and gas fields have been discovered. Figs. 10.20 and 10.21 show the structural contour map and stratigraphy of the Cambay basin.

In the Cambay basin, abnormal high pressure horizons were encountered in South Kadi area of Gujarat. Many other structures in this basin also gave indications of high pressures, but cannot be properly defined due to paucity of data.

Probable Age	Group	Formation	Member	General Lithology	Maximum Thickness(m) Borhalla Geleki	Maximum Thickness(m) Dergaon
Oligocene			Barail			
Eocene to Palaeocene	Jaintia	Kopili	D	Sand shale alternations	260	32
			C	Shale	90	25
			B	Sandstones	210	105
			A	Shale with thin sandstones	140	20
		Sylhet	Z	Shale, sandstones, limestones	135	25
			Y	Limestones	110	40
			X	Shale, sands and limestones	40	absent
		Teok-Sandstone	Upper	Sandstones with rare impure lsts.	40	absent
			Middle	Shale and thin sandstones	50	
			Lower	Sandstones	40	
				unconformity		
Cretaceous	Dergaon	Maobund		Shales with sst. intercalations	absent	
				unconformity		
		Bamangaon		Sandstones with minor shales		
				unconformity		
Precambrian				Pre-Cambrian Basement		

Fig. 10.16: Pre-Barail stratigraphical succession in Assam
(Petroliferous Basins of India, 1983)

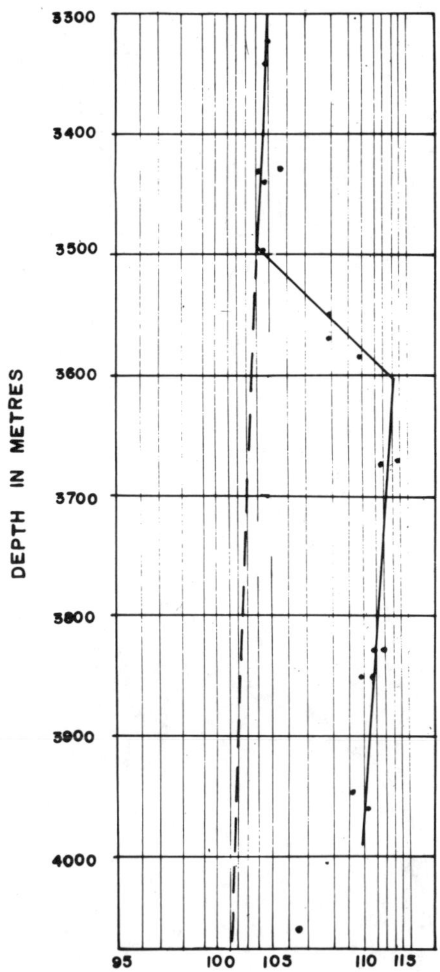

Fig. 10.17: Sonic Transit Time t(μsec/ft). Well No. 'A'
(After, Paul et al...)

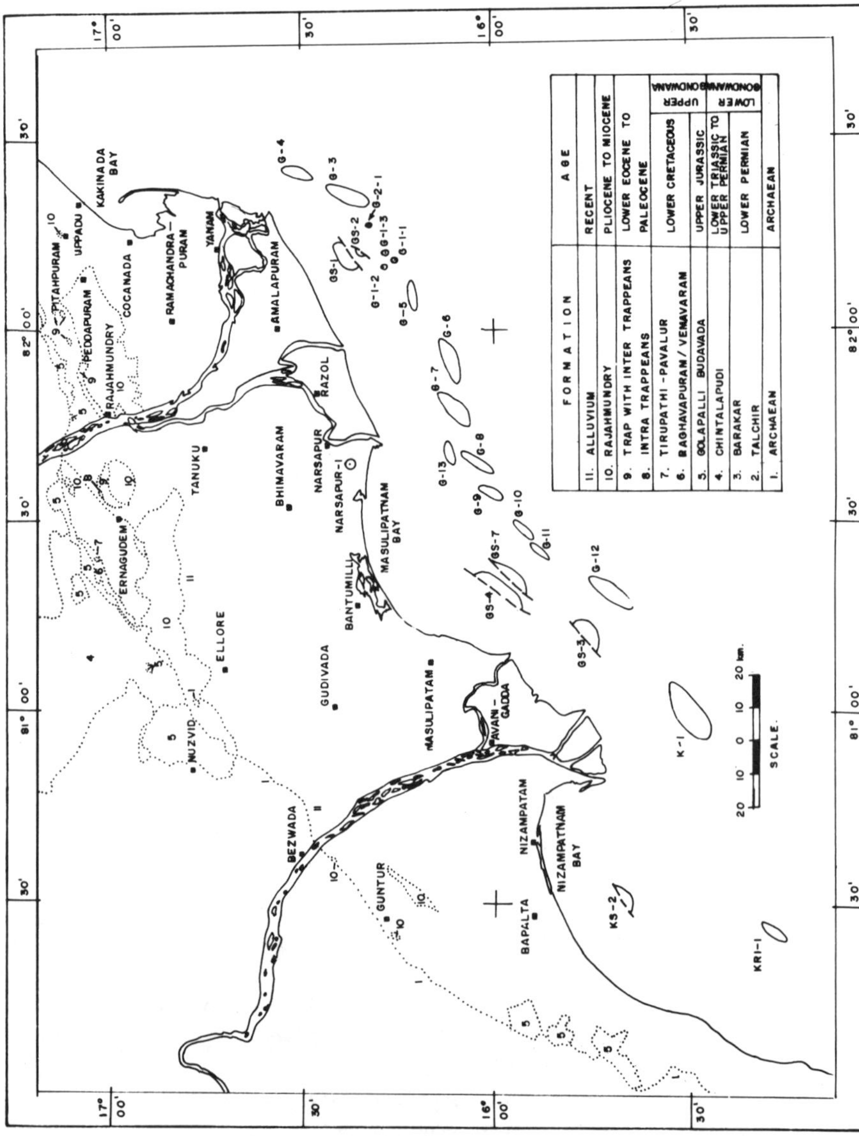

Fig. 10.18: Location map of Krishna-Godavary Basin
(Petroliferous Basins of India, 1983)

Age	Formation	Gross lithology	Maximum thickness (m)		Environment
			Surface	Subsurface	
Pleistocene		Alluvial sands and Kankar; Claystone in sub surface	250	1200	Continental in exposures deltaic and marine in subsurface
		UNCONFORMITY			
Miocene to Pliocene	Rajahmundry	Sandstone in exposures; Claystone and shales with minor sandstone in the subsurface.	600	2000	Continental in exposures marine in subsurface.
		UNCONFORMITY			
Oligocene Late Eocene	–	Claystone and sandstone	–	150	marine
Late Eocene	–	Sandstone with minor clays	–	420	marine
Early Eocene	Traps and intertrappeans	Basalt, marl and limestone in exposures, limestone, sandstone and shales in subsurface.	170	560	Lacustrine in exposures marine in subsurface.
		UNCONFORMITY			
Palaeocene	Intratrappean	Grits, limestone and sandstone in exposures siltstone and shales with basalt flows towards base in subsurface.	70	1100	marine
		UNCONFORMITY			
Late Cretaceous	–	Claystone and silty shales	–	700 +	marine
	Tirupati /Pavalur	sandstone			
	Tirupati /Pavalur	Sandstone ; lateritised	830	–	transition
Early Cretaceous	Raghvdpuram/ Vemayaram/ Budavada	Variegated shales and sandstone.	125	–	marine
	Gollapalli	Gritty and ferruginous sandstone	275	-	Continental
		UNCONFORMITY			
Late Permain	Chintalpudi	Sandstone, ferruginous and conglomeratic.	2000	–	Continental
		UNCONFORMITY			
Archean		metamorphic rocks, charnoc - kites, Khondalites, gneisses and granites.			

Fig. 10.19: Generalised stratigraphy, Krishna-Godavari Basin
(Petroliferous Basins of India, 1983)

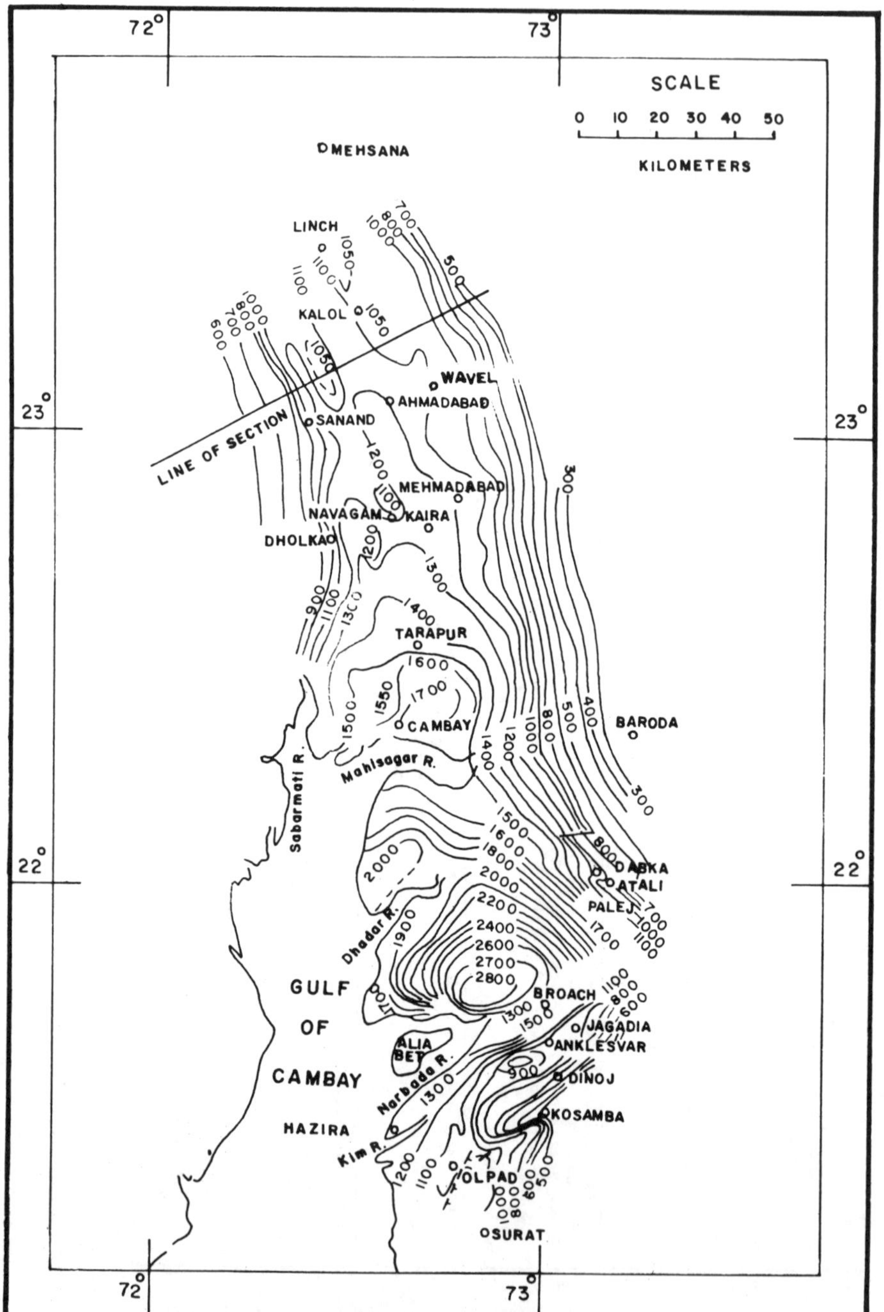

Fig. 10.20: Map of Cambay Basin
(After Mathur and Evans, 1964, P-58. Courtesy: International Geological Congress)

AGE	FORMATION	LITHOLOGY
Pleistocene—Recent	Cambay A	Gravel sand and sandy clay
Pliocene	Cambay B	Claystone, interbedded with sand and shale
Miocene	Cambay C	Variegated shales and sand with shale
	..... Unconformity	
Oligocene	Cambay D	Shale with sandstone
Eocene	Cambay E	Sandy shale and sandstone silty shale, siltstone and carbonaceous shale
Lower Eocene to Upper Cretaceous	Cambay F	Tuffaceous matter—Ash beds
	..... Unconformity	
Upper Cretaceous	Cambay G	Basalt with thin intertrappean flows at places

Fig. 10.21: General stratigraphical column of Cambay Basin

(vii) *South Kadi:* In South Kadi, abnormal high pressures were encountered in a number of wells. As no density determination was done, the overpressure study was confined only to resistivity logs. Two wells, Kadi-4 and Kadi-5, were selected.

As shown in Figures 10.22 and 10.23 the section from the surface down to 1750 m, mainly consists of clays/shales and sandstones and is normally pressured. The abnormal high pressures, 1.6 times the hydrostatic, were encountered in the Kadi Lower Pay below a shale section. The plot of shale resistivity versus depth of wells Kadi-4 and Kadi-5 clearly reveal a significant change in shale resistivity, showing a decreasing trend at 1800 m and 1750 m in both wells, respectively. The abnormal high pressures encountered required the specific gravity of mud to be increased to 1.90 to counter-balance these formation pressures. The abnormal high pressured sands contained gas and oil. Since the salinity data obtained from DST's are scanty, no conclusions could be drawn.

(viii) *Nawagam area*: In the Nawagam field in the Cambay basin in the earlier days abnormal highpressures were also encounterrd in the Kalol Formation (Middle-Upper Eocene) as well as in Nawagam Lower Pay (Lower Eocene-Palaeocene) below a thick section of shale. The formation pressures encountered in the Nawagam Lower Pay were 1.65 times the hydrostatic and occur around 1350 m and at 1860 m in the Nawagam area. Formation water salinity ranges from 4 to 5 gms/lit in the high pressure zone, while the temperature ranges from 110-130° C (Fig. 10.24). Pressures in Nawagam field

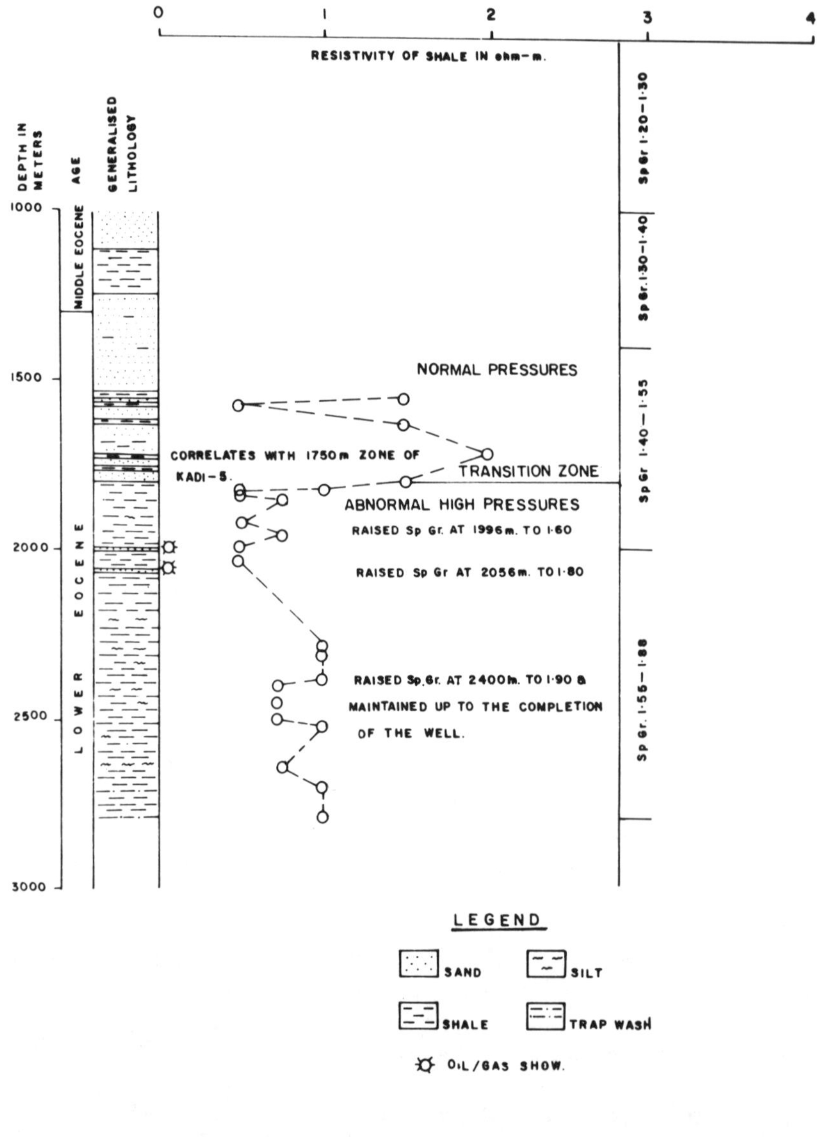

Fig. 10.22: Plot of Resistivity of Shales Kadi Well No. 4

(After Sahay, 1976, P-117. Courtesy: Geological Mining and Metallurgical Society of India)

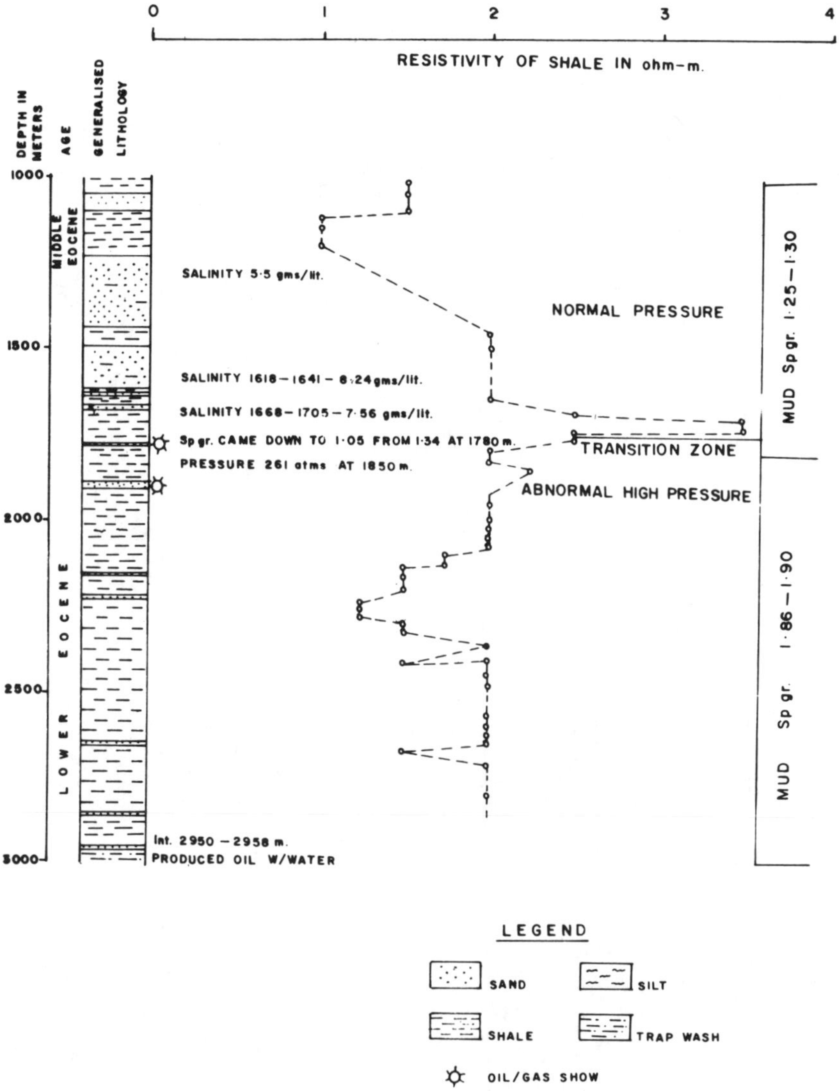

Fig. 10.23: Plot of Resistivity of shales Kadi Well No. 5

(After Sahay, 1976, P-117. Courtesy: Geological Mining and Metallurgical Society of India)

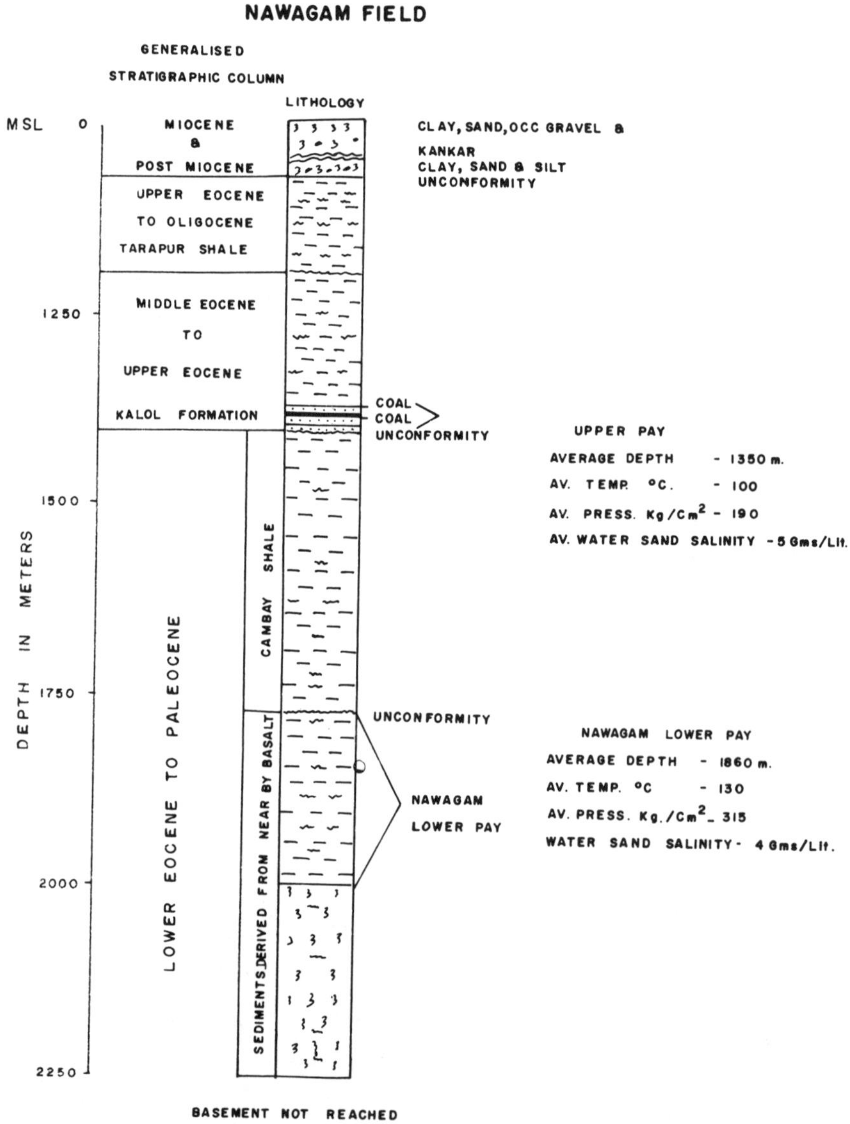

Fig. 10.24: Comparative studies on pressures, salinity and bottom hole temperature in Nawagam area

are now sub hydrostate due to pressure depletion as a result of withdrawal of oil gas from that area.

It is pertinent to point out that the wells in the South Kadi and Nawagam areas were drilled long ago and at a time when the techniques of predicting/delineating of normal and abnormal high pressured horizons were either not known or were poorly understood. Hence, no systematic studies, as carried out in other areas of India, could be undertaken here.

(ix) *Other areas of Cambay basin*: In the wells drilled in Jambusar area sometime back, gas shows in silt layers sandwiched between the thick shale section belonging to the Cambay black shale group of lower Eocene age, were reported frrm 2800 m to 3100 m. Specific gravity of the mud had to be gradually raised to 1.75 to control gas cutting. As no study on shale density and flow line temperature was done, it is not known as to whether the high pressures were actually present or not. However, it is just possible that the shales above these silty layers may also have been overpressured as a mud of almost 1.7 specific gravity had to be used to control the well between 2800 m to 3600 m.

In the Cambay field, e.g. Well No. 40, the mud specific gravity had to be gradually increased to 1.9 around 2750 m in view of too high gas cutting. DST's undertaken showed the zone to be slightly overpressured, with pressure in the 2700 m horizon being 1.2 times the hydrostatic. No studies on density, resistivities of shales and flow line temperatures were carried out for that area above this horizon. In Cambay well No. 45, similar conditions were also experienced.

In well Kalol 236, pressures at 2700 m were observed at 488 atmosphere, which means that they were approximately 1.8 times the hydrostatic.

Offshore West Coast

The continental western margin of India is divided into six sedimentary basins. They include Kutch basin, Saurashtra basin, Bombay Offshore basin, Konkan, Lakshadweep and Kerala basins (Fig. 10.1). Surface geological exposures along the western margin of India are known in Kutch and Saurashtra peninsula and in a thin strip along the coast near Cochin. These basins are garben-type and the stratigraphy of the basins extending into the shelf, is inferred mostly from seismic data.

Exploratory drilling in the various basins has indicated the occurrence of overpressured horizons in the Kutch basin, Bombay Offshore basin, as well as some indications in the Saurashtra basin.

(x) *Kutch basin*: Kutch basin, located to the extreme north, has an area of approximately 28,000 sq. km. (Fig. 10.25). The geology of offshore Kutch is an extension of the onland Kutch basin which forms a part of the fore land shelf of geosyncline. In a thick section Jurassic and Cretaceous (about 2400 m) and a complete Tertiary sequence (about 700 m) separated by 600 m of Deccan trap, volcanics are exposed in the south western part of Kutch peninsula, adjoining the offshore basin. This basin is separated from the Bombay offshore basin by a prominent arch at the level of Eocene and Miocene sediments. Fig. 10.26 shows geological section across shelf in Kutch offshore basin. The difference between the Bombay offshore and the

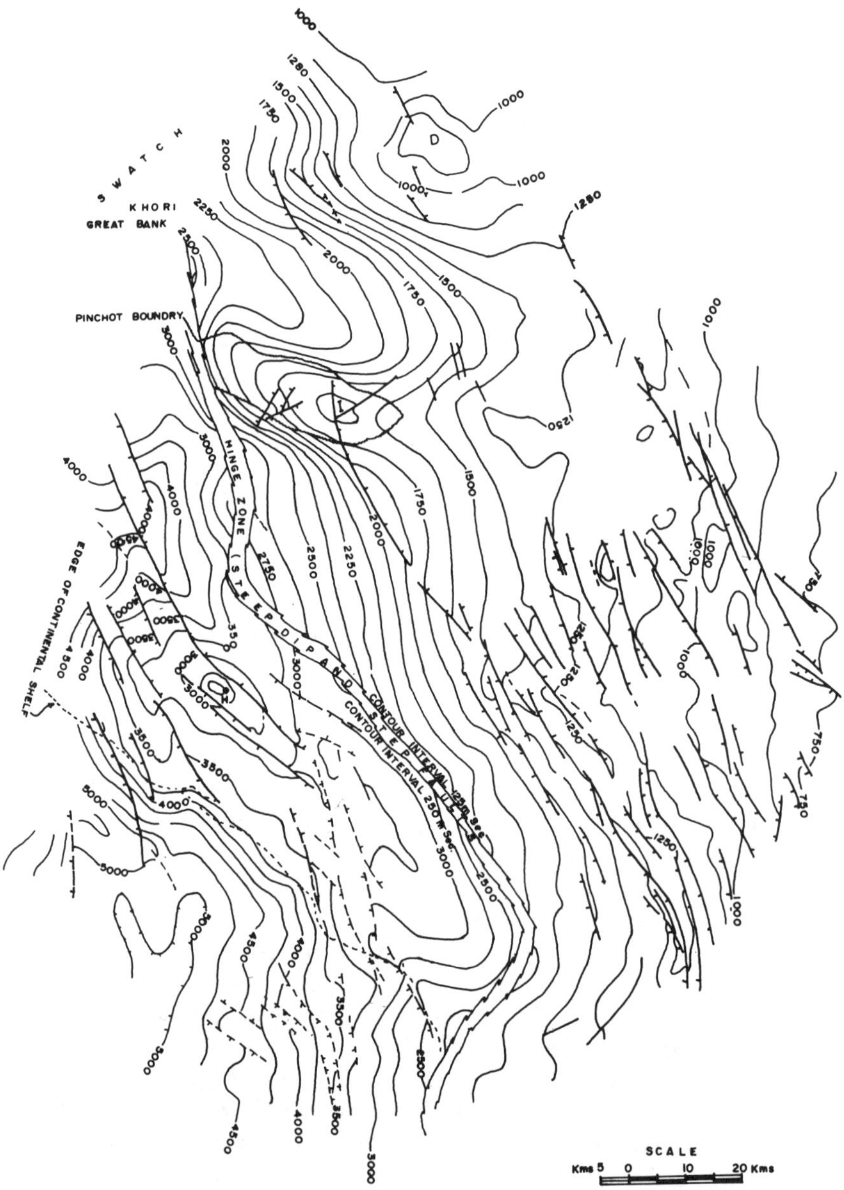

Fig. 10.25: Kutch offshore basin time contour map at the base of Eocene

(Mitra, 1983, Petroliferous Basins of India)

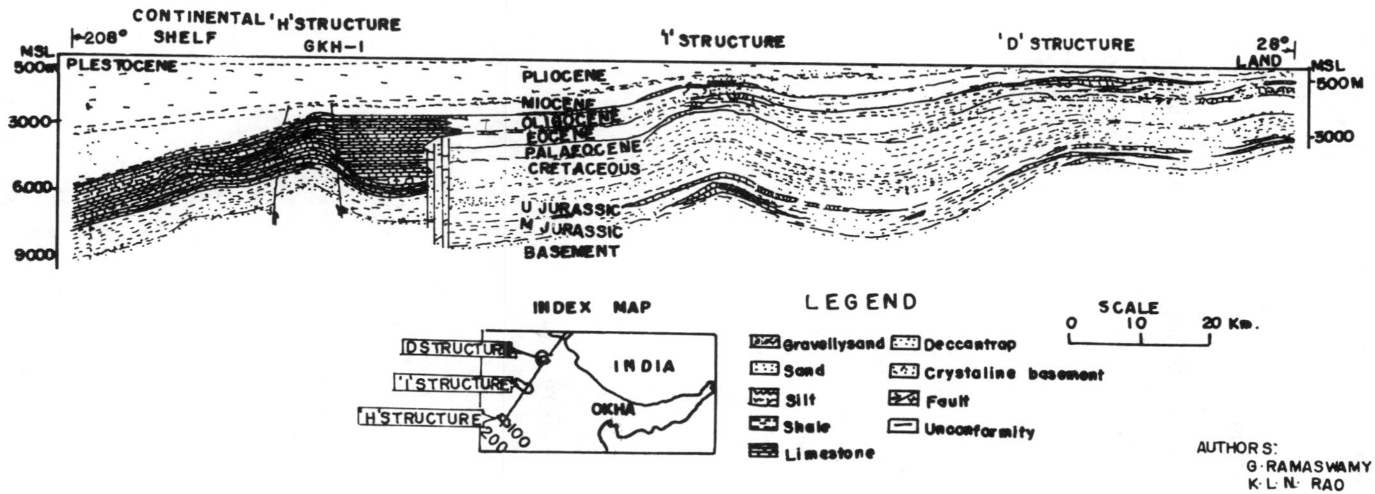

Fig. 10.26: Kutch offshore basin: Geological section across the shelf

(Ramaswamy and Rao, 1978: Petroleum Conference, Calgary)

Kutch basin lies in the absence of Mesozoic rocks in the former while in the latter they are not only present but are also of large areal extent and great thickness.

Exploratory work in the Kutch basin was taken up by the Reading and Bates Group in 1975. A well drilled on 'H' feature by Reading and Bates indicated high pressures ranging between 389 kg/ cm² to 545 kg/ cm² from about 2600 m to the total depth of the well at 4575 m. Formation pressures approximately 1.2 to 1.4 times the hydrostatic were encountered in intervals containing mainly water and dissolved gas. The plot (Fig. 10.27) shows mud weight, formation pressures, temperature and salinity of the water as obtained from DST's, wherever possible. ONGC also drilled wells on the

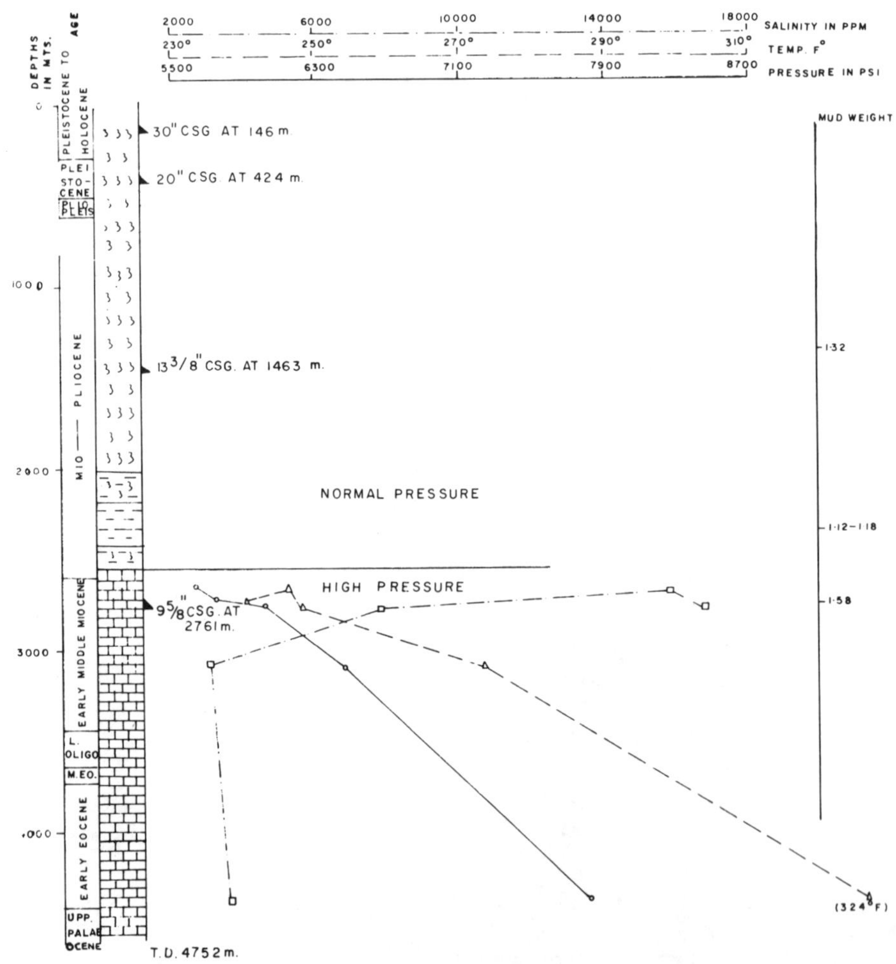

Fig. 10.27: Plot of pressures, salinity and temperature data in offshore Kutch well

other 'D' & 'I' features, but no indications of overpressures were met down to the total depth of the wells drilled.

(xi) *Saurashtra basin*: This basin, adjoining the Kutch basin towards the North, has an areal extent of 28,000 sq. km. The main structural feature is a NW and SE trending arch which plunges north wards the Kutch basin and south towards the Bombay offshore basin. Progressive basinward thickening of the sediments and increase in the carbonate percentage is expected in this basin. A well drilled by Chevron Oil of the U.S.A. on one of the main prospects to the depth of 4400 m showed negative results. A transition zone from normal to overpressure was noticed between 1800-2050 m. Pore pressure indicators called for mud specific gravity to be increased from 1.09 to 1.32. Similarly, on the dcs plot, a normal compaction trend extends down to 1800 m whereas below 1800 to 2800 m, a decrease in the dcs occurs, indicative of a transition to abnormal pressure. No detailed data, however is available for this well.

(xii) *Bombay offshore basin*: Bombay offshore basin occupies an area of 1,20,000 sq. km. It is the southern extension of Cambay basin and encompasses the Gulf of Cambay and the adjoining shelf part of Arabian Sea. It is bounded to the north and north-west by the Saurashtra peninsula, to the east by the Deccan crystalline shield while towards the south the basin may be co-extensive with Kerala and Lakshadweep basins. Towards west and south-west, however, the limits are not well defined (Fig. 10.28).

It is an intra-cratonic basin and came into existence in the Upper Cretaceous time as a result of faulting in the Deccan trap basement. It is divided into several tectonic units, such as the Cambay Gulf shelf to the north, followed to the south by the Dahanu depression, Bombay-Ratnagiri shelf occupying central and southern part of the basin, and the western clastic basin, fringing the Bombay-Ratnagiri shelf.

The full sedimentary section ranging from Paleogene-Cretaceous to Recent has been encountered in this basin. The exception is in the Bombay High itself where no Eocene was deposited. In the adjoining part of Bombay High and the structures east of it, as well as North, North-West and West, a full section has been met. General stratigraphy and geological section illustrating facies variations in this basin, is presented in Figs. 10.29 and 10.30. The Deccan trap of Paleocene to Cretaceous age are usually the basement rocks in the northern part of Bombay High. However, Precambrian basement has been encountered in the southern part of Bombay High as well as in some other structures.

Occurrence of Overpressure Zone

Although several theories with regard to origin of overpressured zones have been proposed by different authors, the occurrence of overpressures in the Bombay offshore basin based on our analytical studies appear to be primarily due to shale compaction phenomena as well as diagenesis of limestone in some cases. Formation pressures encountered range from 1.16 to as high as 2 times the hydrostatic. Geothermal gradients in some of these known overpressure areas range from 3.5 to 4° C per 100 m.

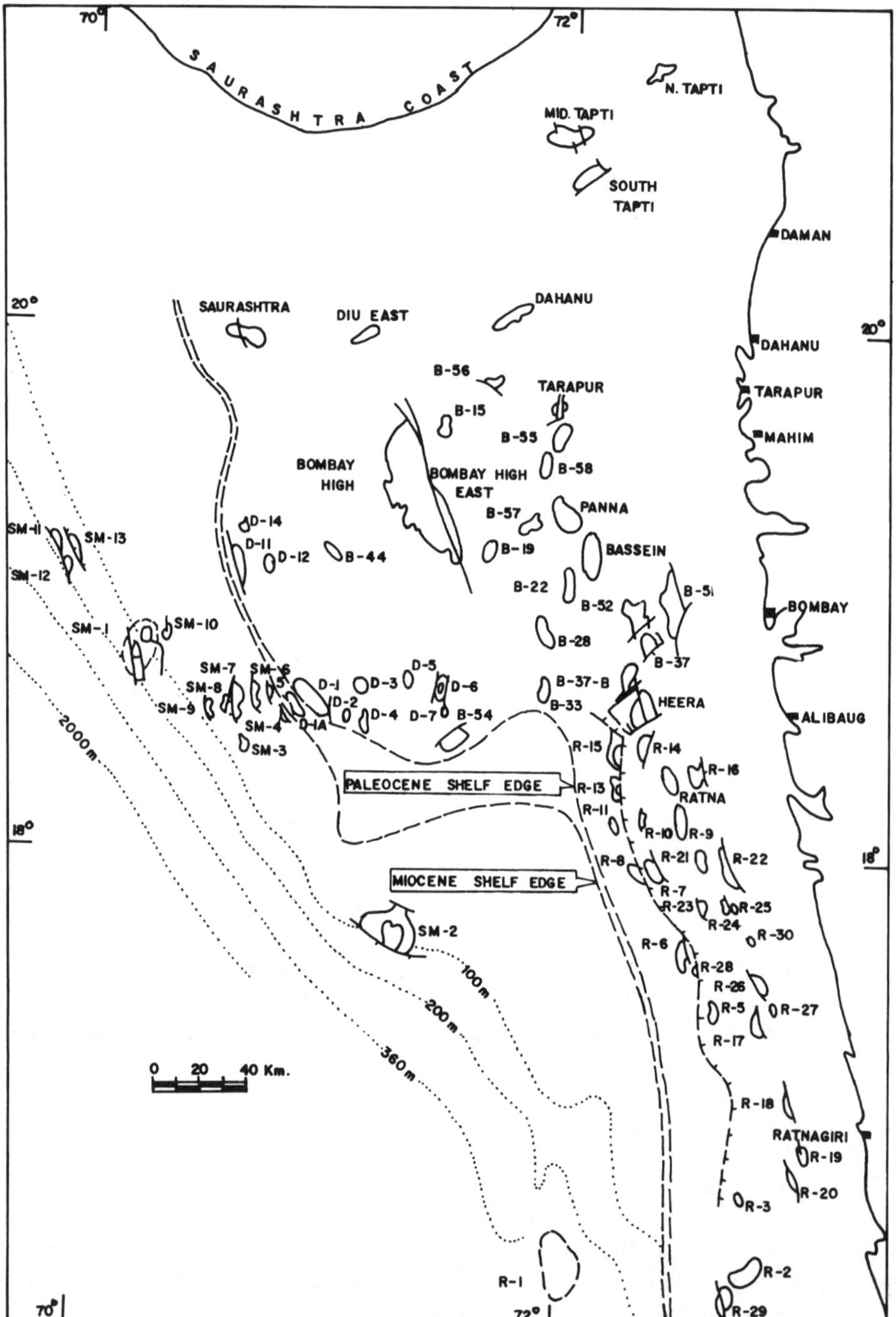

Fig. 10.28: Prospect map of Bombay offshore basin

(After Sahay, OTC, 1984)

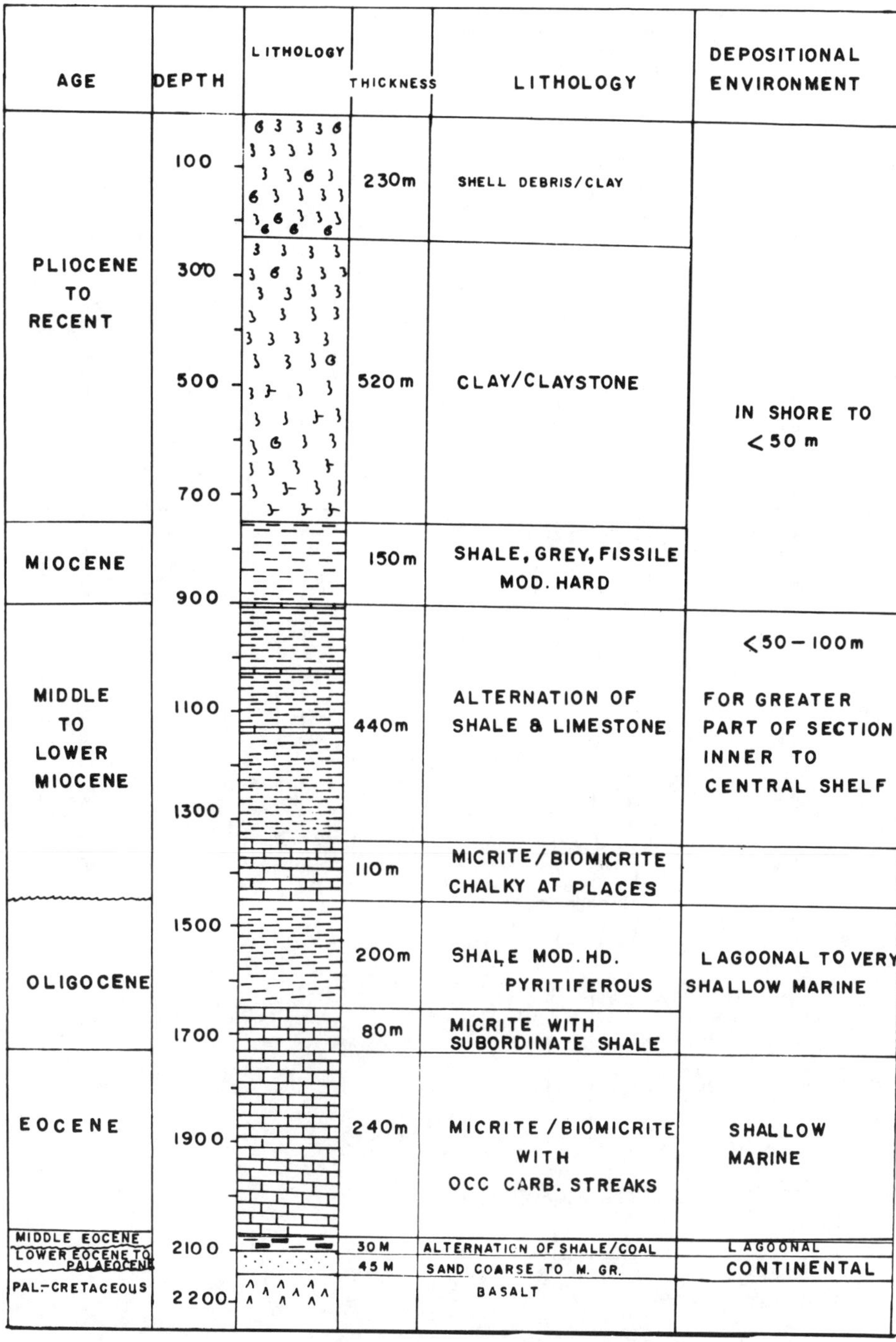

Fig. 10.29: Stratigraphical column of Bombay offshore basin
(After Sahay, 1978 P-4. Courtesy: 2nd offshore South-East Asia Petroleum Conference. Singapore)

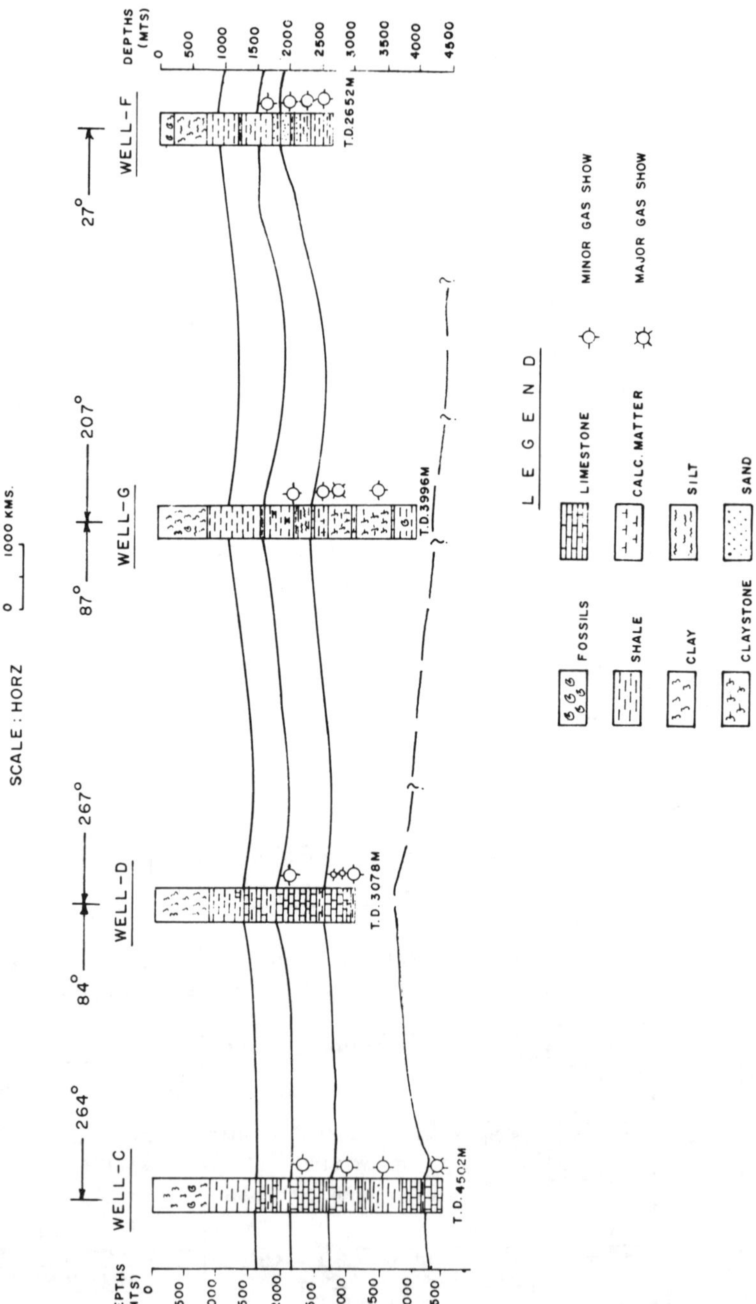

Fig. 10.30: Geological section across Wells C, D, G and F
(Sahay, 1984, 5th Seapex Conference, Singapore)

The main areas of overpressures are confined mostly to the north, north-western region, in the series of east-west trending structures in the western part of DCS area towards north north-east in the Tapti area and in a few sporadic structures east of Bombay High. In the northern Tapti area, shallow overpressure zones were encountered between 200-300 m. Fig. 10.31 shows the different overpressure trends in the Bombay offshore basin.

Figures 10.32 and 10.33 show the geological sections through the wells in which overpressure horizons were encountered. The case histories of the overpressure zones of Bombay offshore basin, are discussed:

Well A

The lithological column of the well on the left side shows mostly a section of shale and limestone of middle Eocene to Recent age. There are two overpressured zones, one at 1600 m, the other around 2700 m. Both zones are overlain by a thick section of shale. Pressure data obtained from drill stem test in the limestone section at 1600 m show the formation pressure to be about 1.16 times the hydrostatic. This limestone section is followed by a thick section of shale. It is normally pressured down to 2550 m. The drill stem test carried out in the limestone section around 2550 m also indicated the pressures to be essentially hydrostatic.

Overpressures encountered below 2700 m were almost superpressures. A plot of shale density, flow-line temperatures carried out during the course of drilling, indicated the anomaly around 2600 m, with a departure towards increasing trend in the flow-line temperatures. Pore pressures calculated from the shale density plot, revealed the pressures to be approx. 1.9 times the hydrostatic. This well had to be finally abandoned around 2782 m depth because of technical complications (Fig. 10.34).

The reasons for abnormal high pressures in this well include compaction phenomena of shales in response to the stress of overburden and diagenesis of limestones.

Well B

Drilled beyond the shelf edge in the clastic basin on its western most limit, overpressures were encountered in Well 'B' on the SM-1 structure below 3800 m. However, no clear indications on shale density and resitivity was observed but an increase in the flow line temperature was corroborated by the RFT pressure data, revealing pressures to be 1.8 times the hydrostatic and the formation fluid to be mostly water. The mud specific gravity had to be gradually raised to 1.9 (Fig. 10.35).

Well C

In Well 'C' drilled on one of the east-west structures, high pressures were encountered below 2100 m depth in a porous limestone, overlain by a thick shale and a tight limestone section. A plot of density versus depth indicates a departure from the

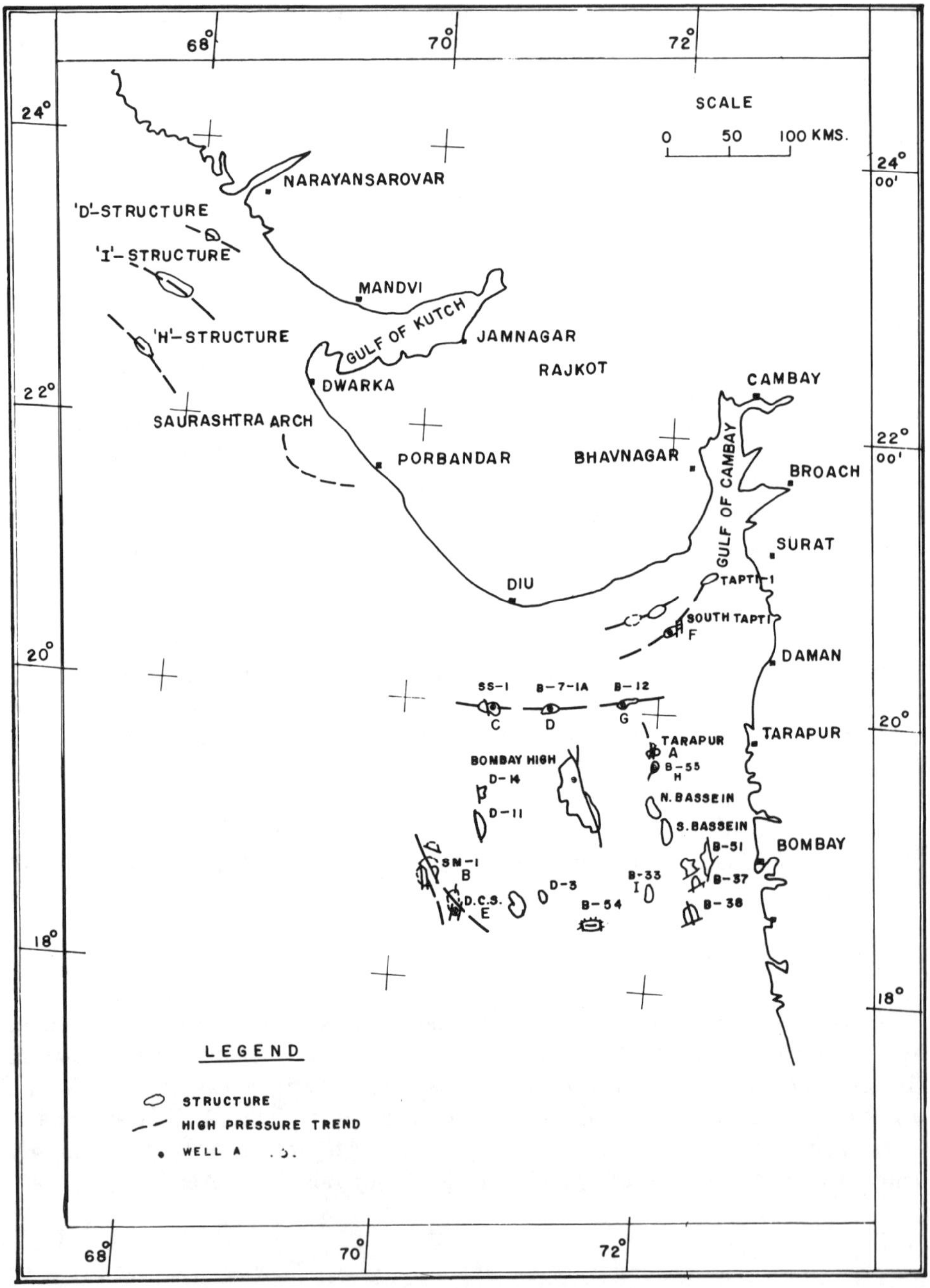

Fig. 10.31: High pressure trends observed in Western Coast of India

(Sahay, 1984, 5th Seapex Conference, Singapore)

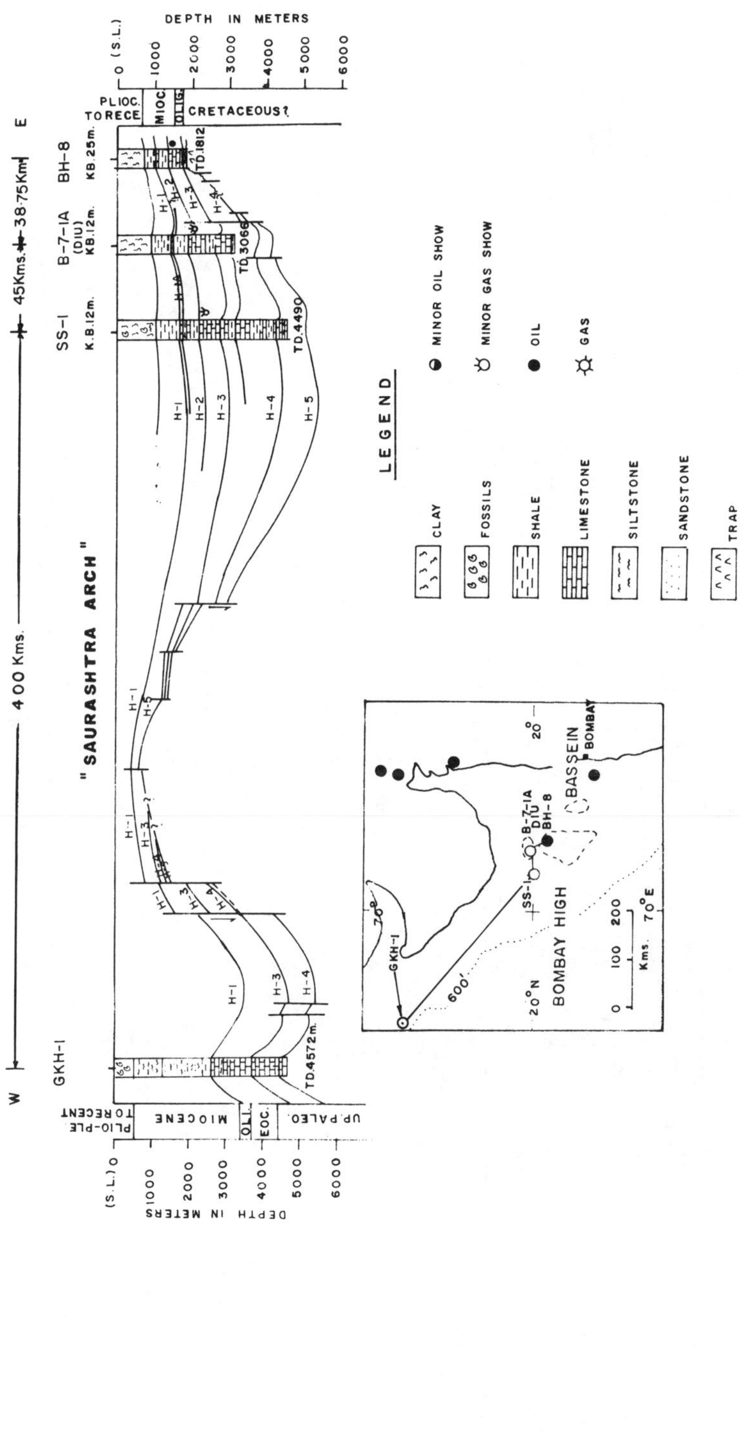

Fig. 10.32: Geological section across Kutch, Saurashtra, Diu and Bombay High (Sahay, 1978, 2nd Seapex Conference, Singapore)

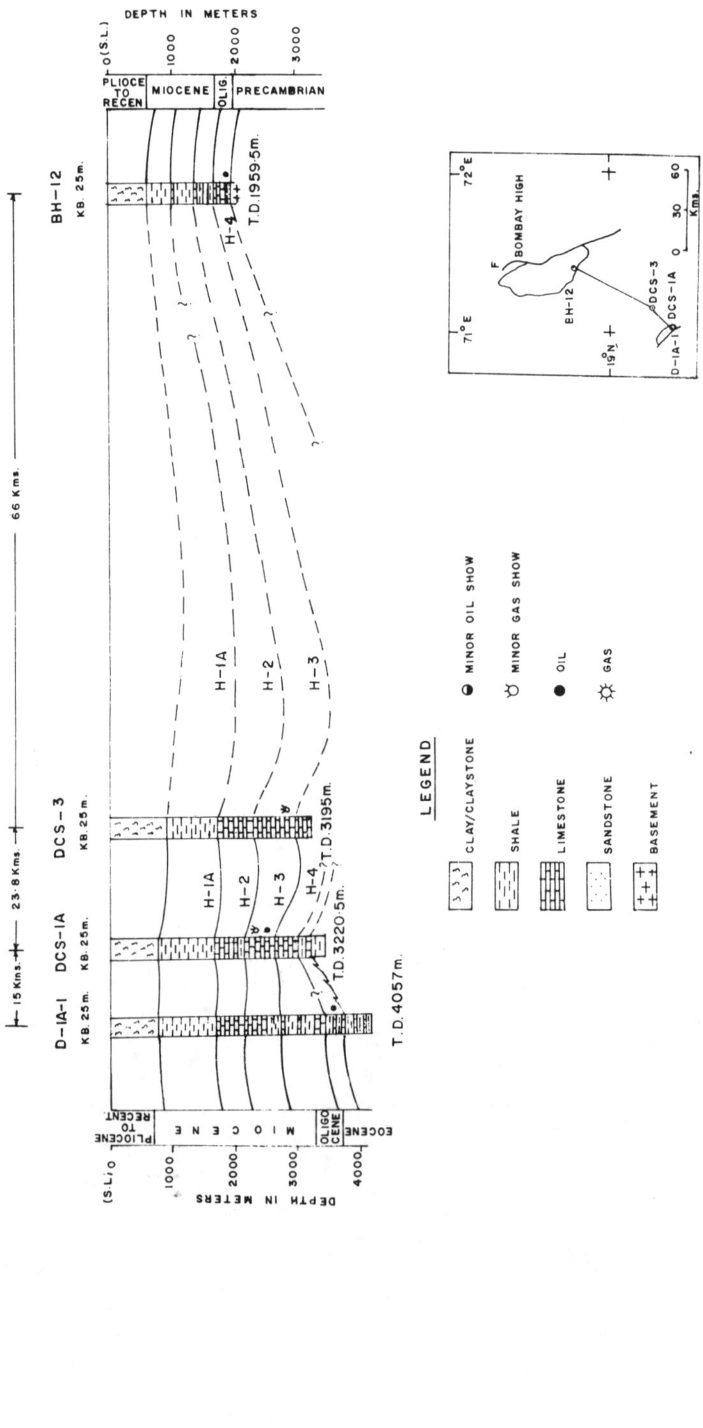

Fig. 10.33: Geological section across D-1A-1, DCS-1A, DCS-3 and BH-12 Wells (After Sahay, Seapex Conference Singapore, 1978)

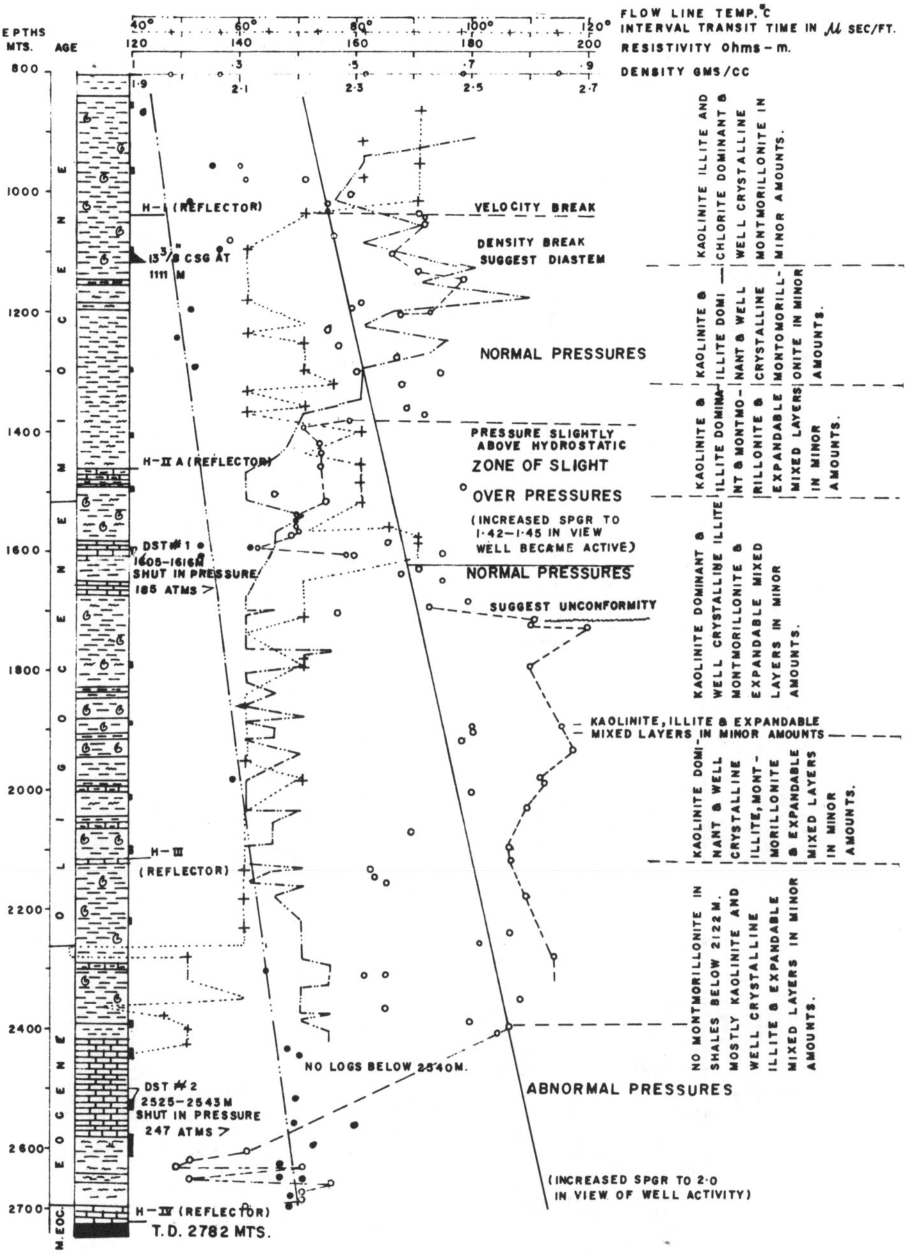

Fig. 10.34: Comparative studies between density, resistivity and interval transit time in Well 'A' for prediction of over pressured zone

(Sahay, 1984, 5th Seapex Conference, Singapore)

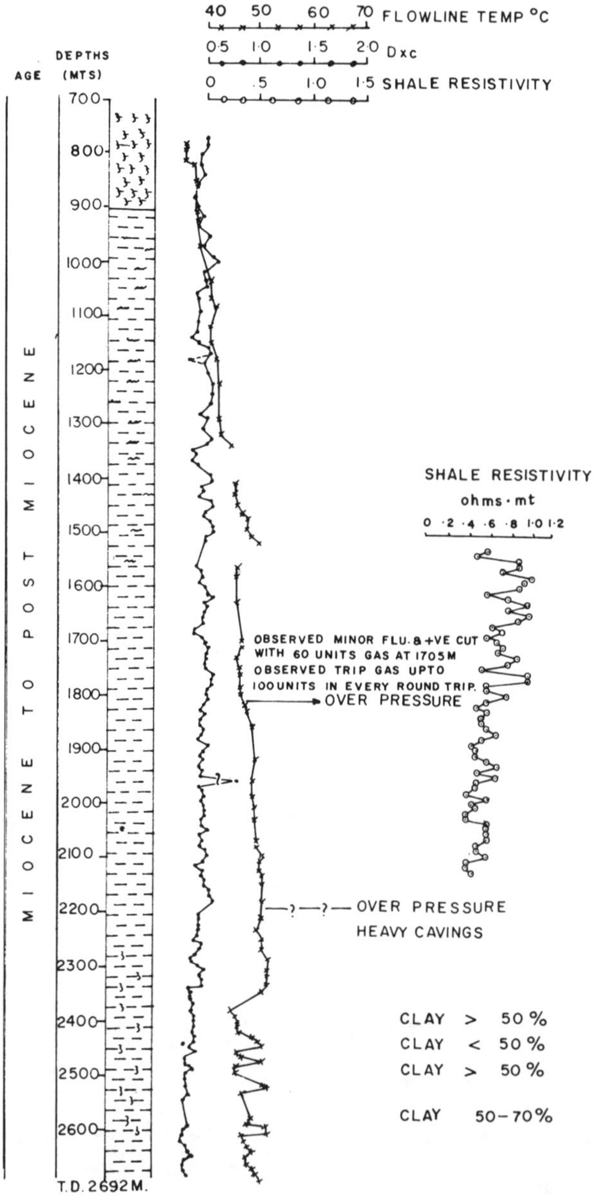

Fig. 10.35: Comparative studies between Dxc, flowline temperature and shale resistivity in Well 'B' (Sahay, 1984, 5th Seapex Conference, Singapore)

normal trend below 1500 m. The flow-line temperature plotted along with the shale density curve also shows a rise in temperature around this depth, thus further confirming the overpressuring. High pressures continue right down to the total depth of the well. An FIT undertaken at 2150 m depth indicated formation pressure of the order of 340 kg/cm^2, i.e. pressures approximately 1.6 times the hydrostatic. The limestone section tested dissolved gas with water (Fig. 10.36).

The reason for this overpressures is mainly compaction of shale in response to the stress of overburden and partly diagenesis of limestone, as discussed earlier, since all other available data rule out tectonic phenomena.

Well D

On the adjoining structure, which also exhibits an east-west trend, Well 'D' encountered high pressure below a thick shale section which were picked on 'd' exponent and shale density curves. Although the indications on 'd' exponent were quite obvious, the shale density curves was not as useful due to the calcareous nature of the shales. The limestone section below 1880 m penetrated high pressure intervals which contained mostly dissolved gas with formation water exhibiting a salinity of 19.8 gms/ litre. The gas composition ranged from Cl to $C5_2$. The FIT's measured formation pressures of the order of 390 kg/cm^2 say 2258 m, i.e., approx. 2.7 times the hydrostatic (Fig. 10.37).

Well E

In another structure located on the western flank of Bombay High in the DCS area, high pressures were encountered in Well 'E' below a shale section at the depth of 3650 m. The specific gravity of the mud during drilling had to be gradually raised to 1.75 and was maintained to total depth of the well. A minor anomaly was observed both on the shale density and 'd' exponent at the depth around 2750 m. The reason for the absence of a sharp observeable anomaly in the shale section is attributed to the calcareous nature of the shales as well as their high iron content, both of which increase the shale density and hence obscure any straight forward overpressure indication. The high pressures were due mostly to water with some dissolved gas. The salinity of formation water was 12.8 gms lit and pressures tested at 3650 m were 665 kg/cm^2, i.e., 1.8 times the hydrostatic (Fig. 10.38).

Well F

In another structure NE of Bombay High, in the Tapti area, high pressures were encountered below 2300 m, which necessitated gradually increasing the specific gravity of the mud to 1.85 below 2350 m. Studies of shale density, dcs and flowline temperature were carried out during the drilling of this well (Fig. 10.39).

Further north-east of Bombay High, in a well drilled in North Tapti structure, a high pressure horizon was met below 200 m. The well kicked since the well could not be

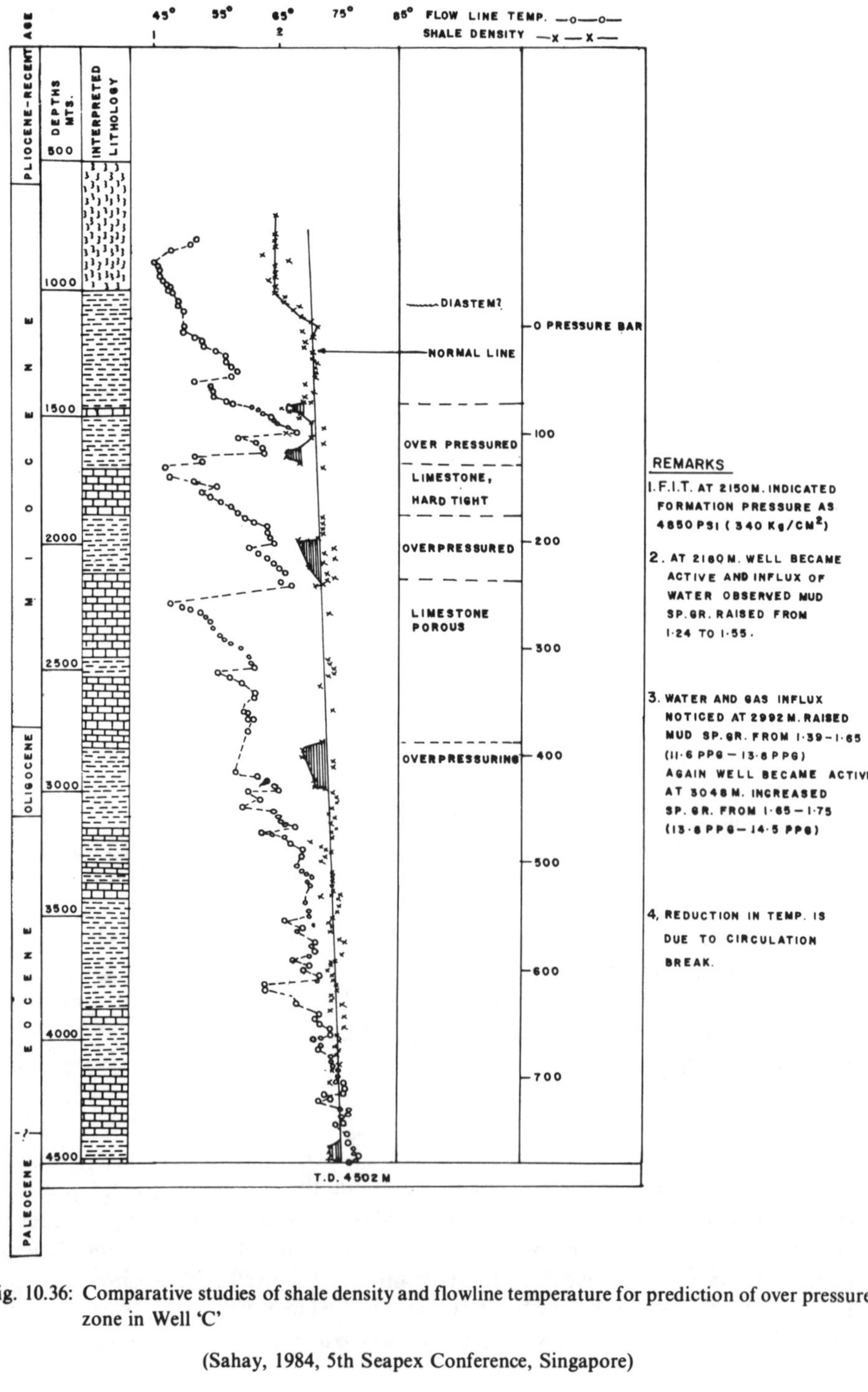

Fig. 10.36: Comparative studies of shale density and flowline temperature for prediction of over pressured zone in Well 'C'

(Sahay, 1984, 5th Seapex Conference, Singapore)

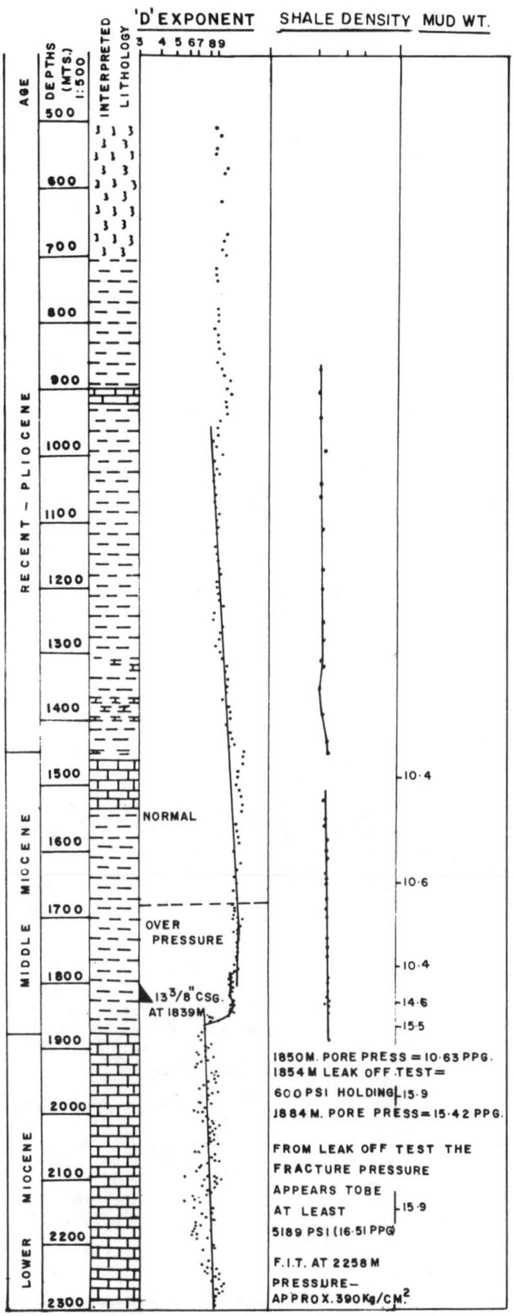

Fig. 10.37: 'D' Exponent and shale density plot for prediction of over pressure zone in Well 'D'

(Sahay, 1984, 5th Seapex Conference, Singapore)

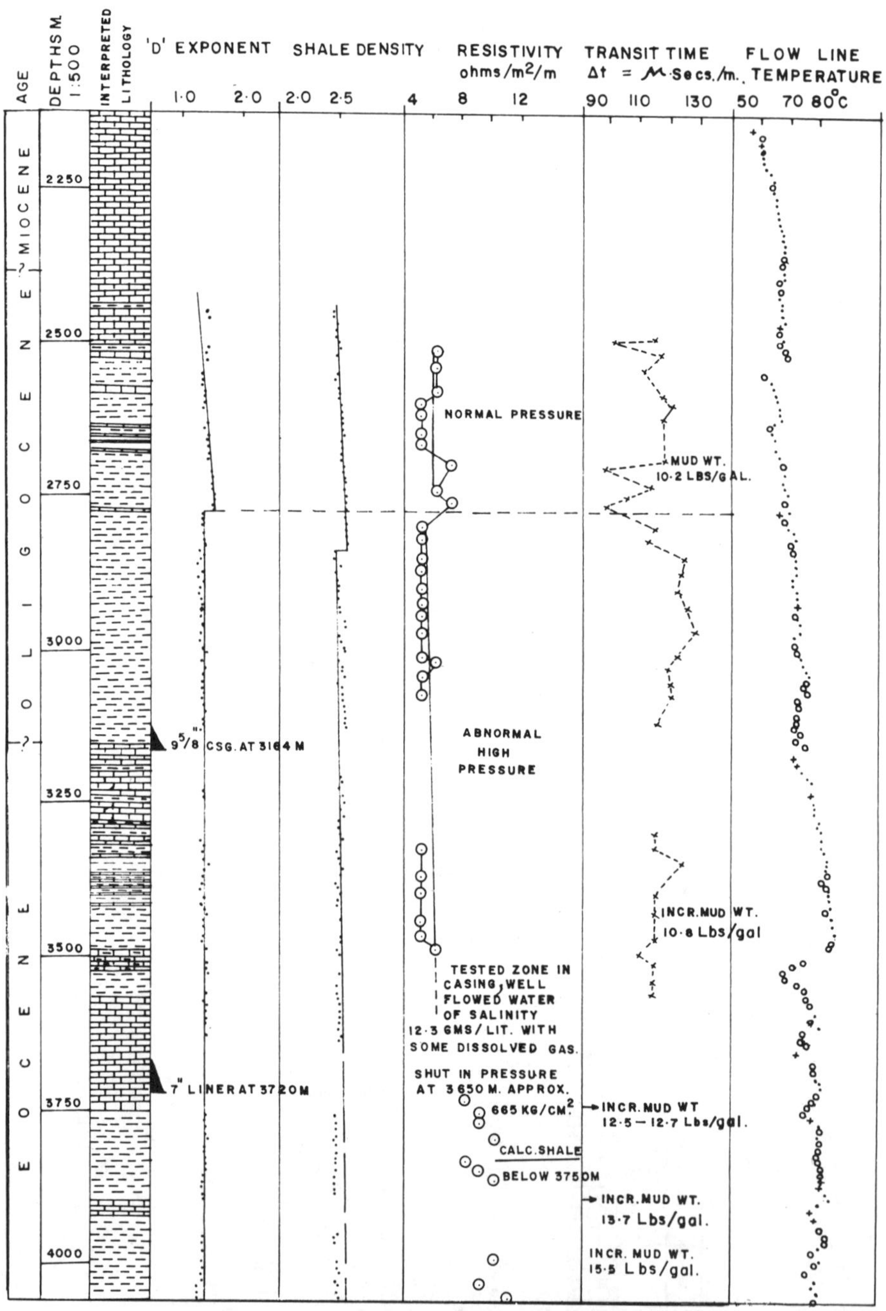

Fig. 10.38: Comparative studies of 'D' exponent, shale density, resistivity, transit time and flowline temperature in Well 'E' for prediction of abnormal high pressure zone
(Sahay, 1984, 5th Seapex Conference, Singapore)

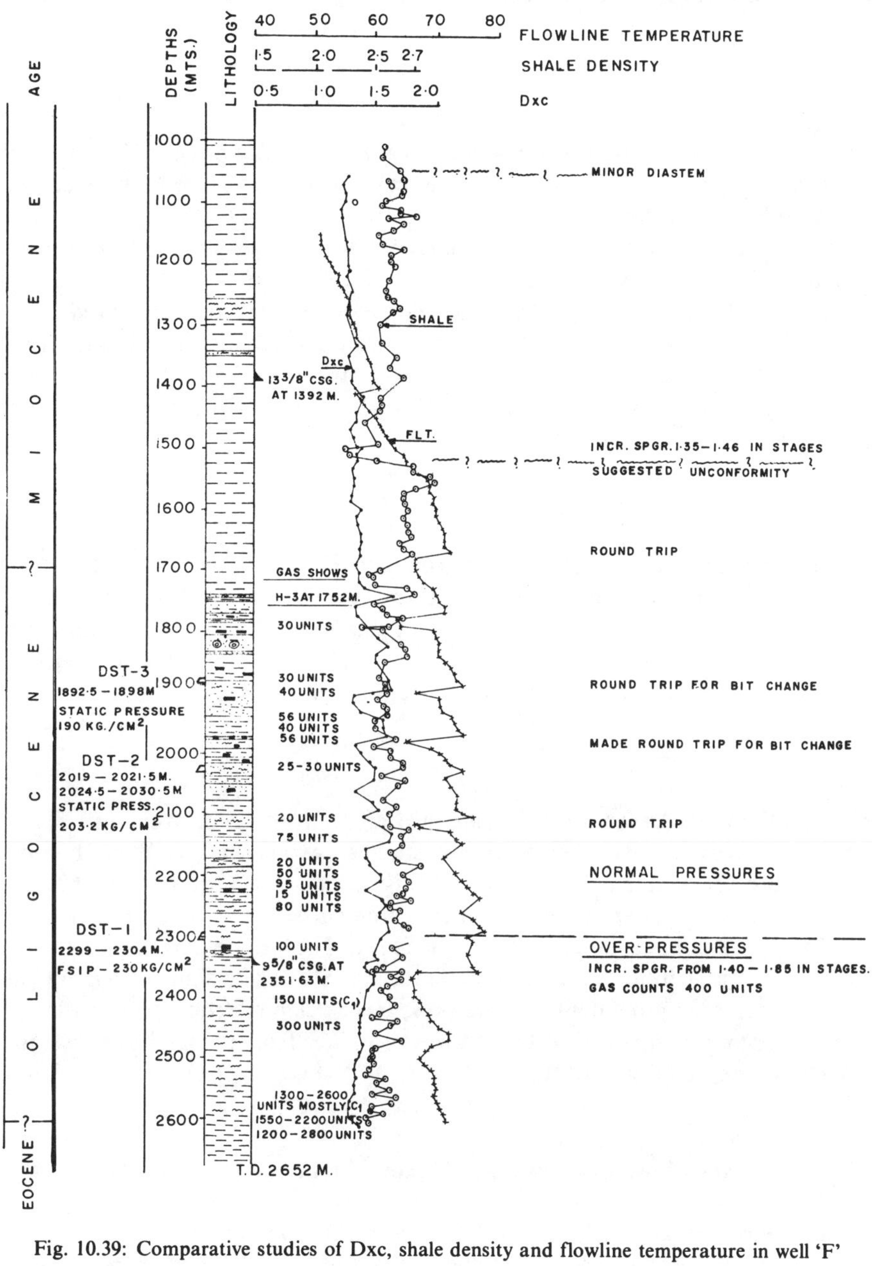

Fig. 10.39: Comparative studies of Dxc, shale density and flowline temperature in well 'F'

(Sahay, 1984, 5th Seapex Conference, Singapore)

controlled because of continuous occurrence of overpressures down to 300 m it had to be abandoned.

Well G

In another adjoining structure in Dahanu area, a well drilled close to Well 'D' in the same structural east-west alignment, indications of high pressures were also encountered below 2300 m.

This necessitated raising of the mud specific gravity in stages and finally to 1.9 at the total depth of 3996 m. An FIT at the depth of 2399 m indicated pressures to be approximately 1.5 times the hydrostatic. Furthermore, the overpressured section continued down below. In view of the overpressure studies carried out, it was possible to predict the occurrence of overpressures well ahead and suitable measures for adjusting both drilling operations and mud weight requirements were taken well in advance (Fig. 10.40).

Well H

In another structure lying north-east of Bombay High and south of Well 'A', high pressure zones were encountered below 2550 m as would revealed by a plot of shale density, dcs, carried out during the drilling of this well. The well was terminated at the depth of 2648 m without reaching the projected depth of 3450 m in view of the interesting horizons obtained in the upper section (Fig. 10.41).

Well I

In one of the wells drilled east of Bombay High, indications of overpressures were found below 2600 m. The overpressure studies carried out in Well 'I' mostly concentrated on flow line temperature and 'Sigmalog'. The Sigmalog was preferred over dcs because the latter applies to an environment which is predominantly shale where this well exhibited heterogeneous lithology. The geopressure log of this well indicates an increase in the flow line temperature at around 2600 m and a drop in the Sigmalog at corresponding depths. Pore pressures calculated from the Sigmalog plot indicated pressures comparable with the values obtained from the RFT's carried out at corresponding depth (Fig. 10.42).

Summary of the Overpressured Zones of Bombay Offshore Basin

As mentioned earlier, overpressure horizons in the Bombay offshore basins occur in sediments of Miocene to Eocene age and are confined mostly to north, north-west in the series of east-west trending structures. In the western most part, in one of the structures in the shelf margin zone in the clastic basin, indications of overpressures have been also encountered.

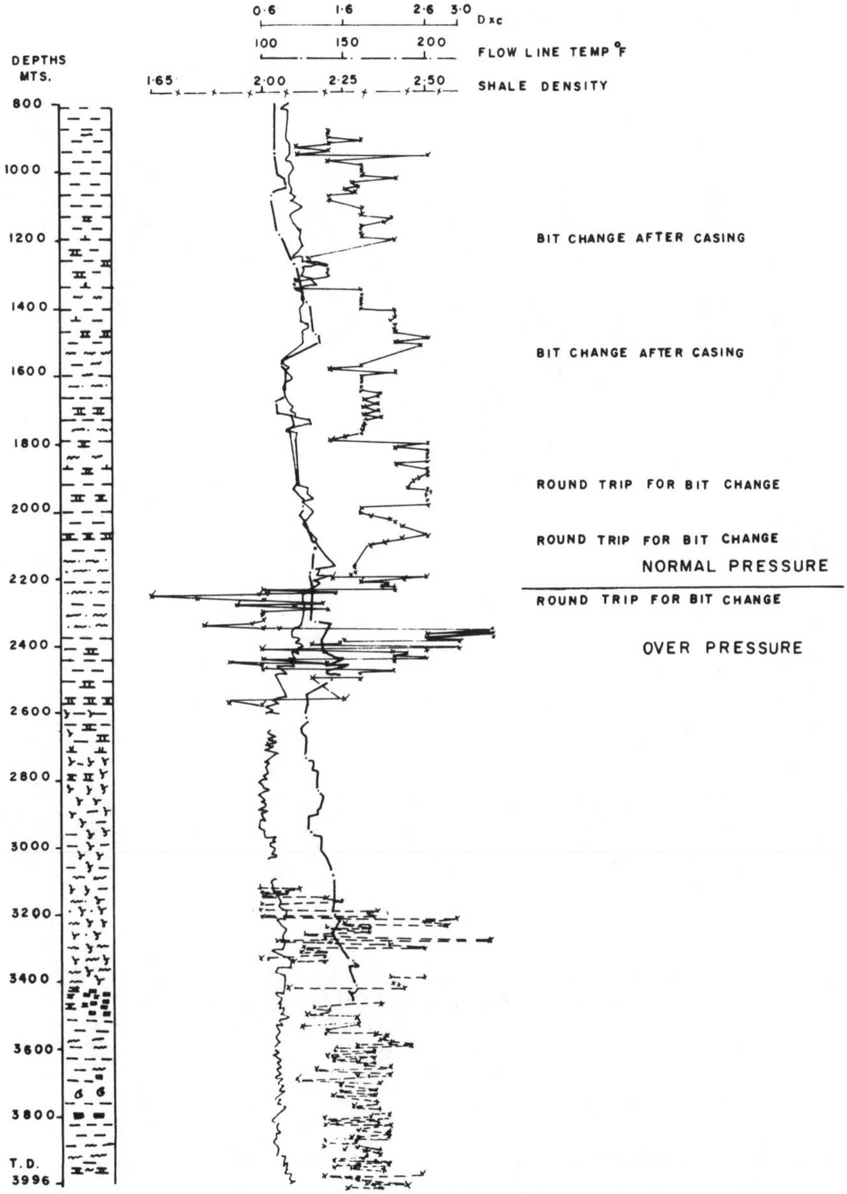

Fig. 10.40: Comparative studies of Dxc, shale density and flowline temperature in Well 'G'

(Sahay, 1984, 5th Seapex Conference, Singapore)

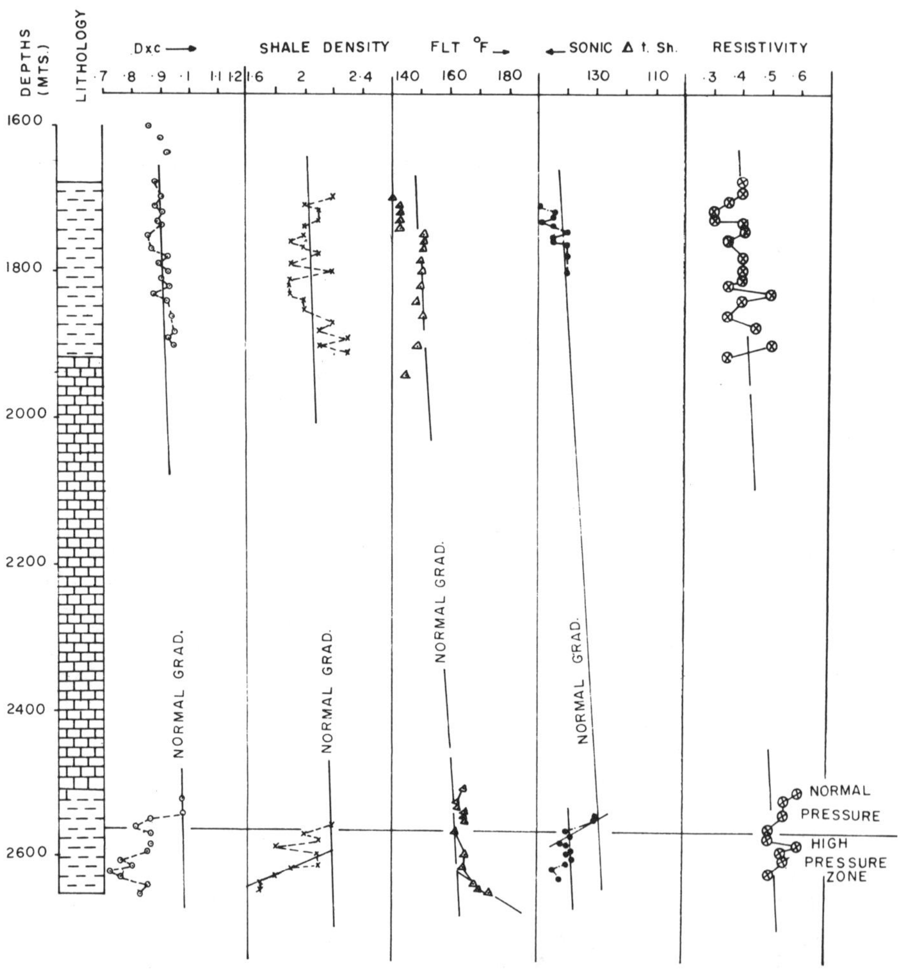

Fig. 10.41: Comparative Studies of Dxc, Shale Density & Flowline Temperature in Well 'H' (Sahay, 1984 5th Seapex Conference, Singapore)

The high pressures in this basin occur mostly at the edge of the shelf margin zone and in the areas which have a huge thickness of sediments, where the rate of compaction probably could not keep pace with the rate of deposition resulting in a few and lenticular type of reservoir rocks devoid of major hydrocarbon accumulation. Such high pressure horizons mostly contain water with dissolved gas. Super pressures, i.e., formation pressures as high as almost double the hydrostatic, ha·· been noticed in mostly such structures which, based on Palaeotectonic studies, are considered of late origin.

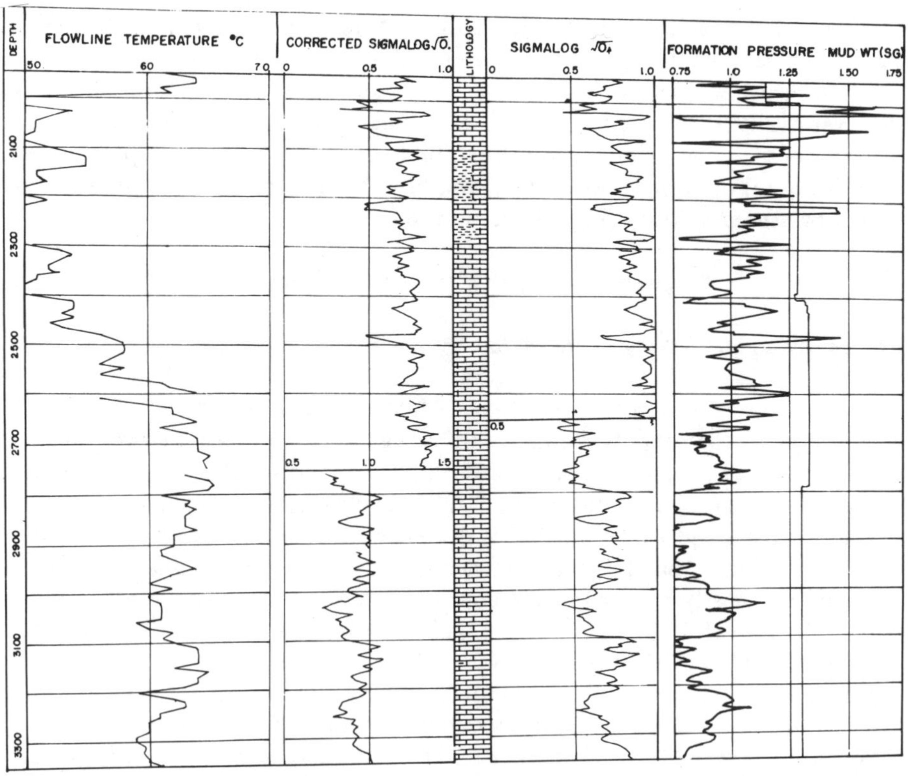

Fig. 10.42: Pressure log showing FLT, corrected sigmalog and formation pressure in Well 'I'

(Sahay, 1984, 5th Seapex Conference, Singapore)

Since the generation of overpressured zones in this basin has been mostly due to the compaction phenomenon, high pressure indicators, such as dcs, shale density, flow line temperatures, etc., have proved quite useful in predicting overpressures during. drilling. Hence, drilling hazards in the form of threatened blow-outs, etc., were always avoided.

Since the geological section in this basin comprises mostly shales and limestones (except in the north east region in Tapti area, where a clastic section has been encountered), it would not have been possible to predict occurrence of such pressures from seismic sections, as a result of interval velocity studies. It is very well known that interpretation in the carbonate section can only be speculative because velocities are so high that variations caused by pressures are below the range of resolution of the technique and hence the difficulty of forecasting of high pressures prior to drilling in such an area, when compared with the adoption of such techniques for clastic sections (Sahay, 5th Seapex Conference, 1984).

In view of the limited data on the salinity of the formation waters obtained, no relationship between the occurrence of overpressures and the salinity variations could be established here except in a few places. However, samples of formation water, whenever available, were analysed and stiff diagrams prepared. The water pattern curves from such overpressured zones exhibit high bicarbonate and sulphte contents; which is typical of such environments, and have been compared to water pattern in normal pressured sections (Figs. 10.43 and 10.44).

B. Overpressures Due to Tectonic Phenomena

General Geological Setting of the Area

In the north-west region of India, there are folded structures with a little faulting. The Punjab foothills thereabouts provide a transitional area between the over-thrusts of the greater part of the foothills belt and the alluvial plains which conceal Siwalik beds that have under-gone but a little folding. Anticlinal structures were delineated as a result of seismic surveys and were evaluated by drilling wells on several important structures, e.g., Surnimastgarh, Balh, Janauri, Jawalamukhi, etc. Lower Siwalik rocks (Lower Miocene) are mostly exposed in the core of these structures. The attached section through the Punjab foothills shows the main geological and tectonic features of the area (Figs. 10.45 and 10.46).

(i) *Surinsar area, Jammu*: In a well drilled in Surinmastgarh anticline in Jammu abnormal high pressures have been encountered in the sediments from lower Siwalik to Murreess (Lower Miocene-Eocene) at the depth of 1200-3665 m. Various studies based on the salinity, temperature, density, resistivity, clay mineralogy and carbonate percentage have revealed that the generation of overpressures there appears to be due to tectonic phenomena. As was observed from the resistivity and density trend curves of the clay/ shales, a decreasing trend was apparent below 2100 m. The reason for higher density and resistivity in the clay/ shale in parts of the upper section (above 2100 m) is attributed to the presence of different clay minerals (i.e. mostly Illite), higher carbonate content in the section above 210 m as compared to the one below that depth, as well as due to the tectonic phenomenon. Salinity data, obtained on water collected during various influxes at different depths, suggest a constant salinity of 5 to 6 gms/ lit from 1200 m up to 2100 m. The log derived salinity plotted, along with the salinity of the mud filtrate as well as the water obtained from the various influx, shows a change in the salinity curve at 2100 m only. Log derived salinities over the section from 100-2100 m is 1 to 2 gms/ lit while at 2100 m the salinity changes to 6 gms/ lit, as was also confirmed during various influx at 1180, 1280, 1527, 2100 m.

The water sample collected at the depth of 2100 m was subjected to detailed chemical analysis and water pattern curves were prepared. This test indicated a high bicarbonate content of the water which is generally observed in formation waters, associated with abnormal high pressure horizons (Fig. 10.47).

The temperature curve of the formation was also plotted along with the resistivity/ density data and indicated a change in the gradient also at 2100 m with a

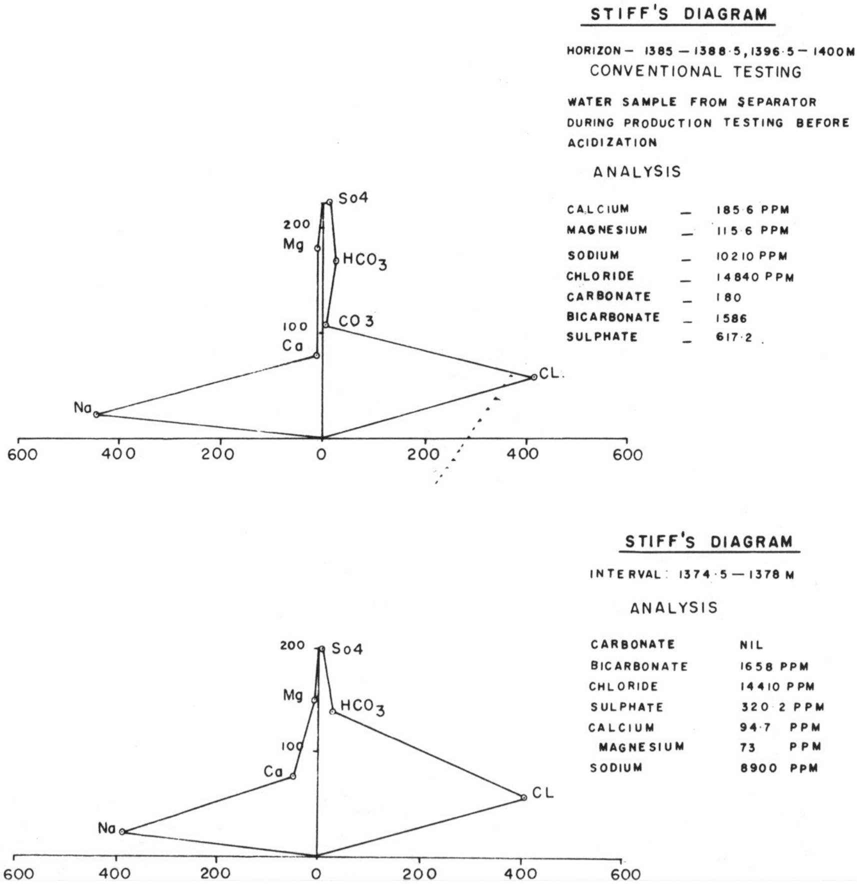

Fig. 10.43: Water pattern curves from wells in a normal pressured area in Bombay offshore basin

(Sahay, 1984, 5th Seapex Conference, Singapore)

slight increase in temperature. The comparative studies undertaken using the various parameters, i.e., salinity, resistivity, density of shales, formation temperatures, clay minerology data, etc., indicated that the abnormal high pressures in the sediments above 2100 m are either due to these being in communication through a fault of fracture connected with deeper horizons or the phenomenon of reverse osmosis is taking place. Flow of water of higher salt concentration from a deeper horizon, perhaps from 2100 m depth, into shallower, less saline, sands separated from each other by shales and clay bands which act as semi permeable membranes may permit the

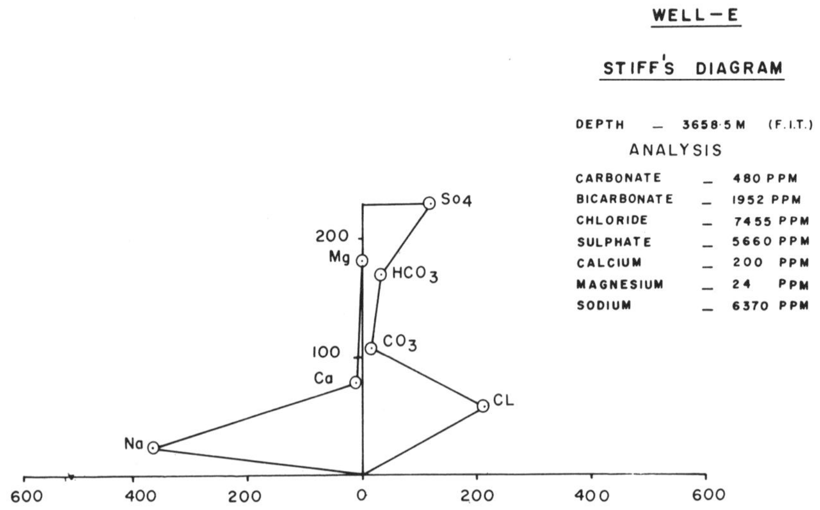

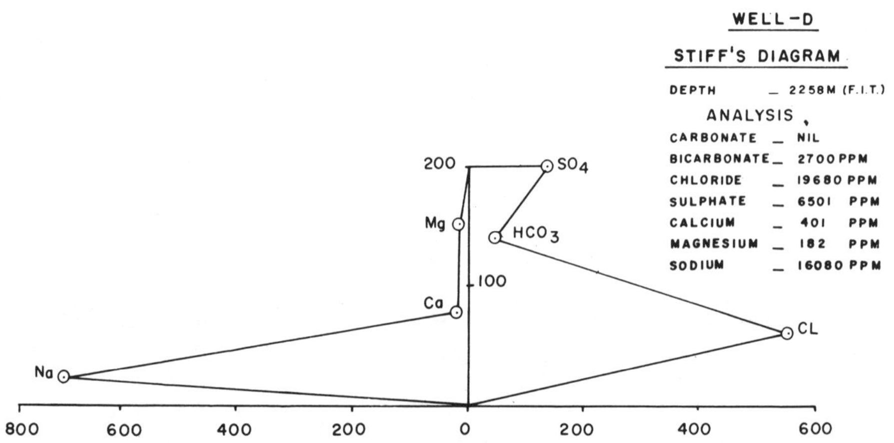

Fig. 10.44: Water pattern curves of Wells 'E' and 'D' in the Bombay offshore basin showing high pressure characteristics

(Sahay, 1984, 5th Seapex Conference, Singapore)

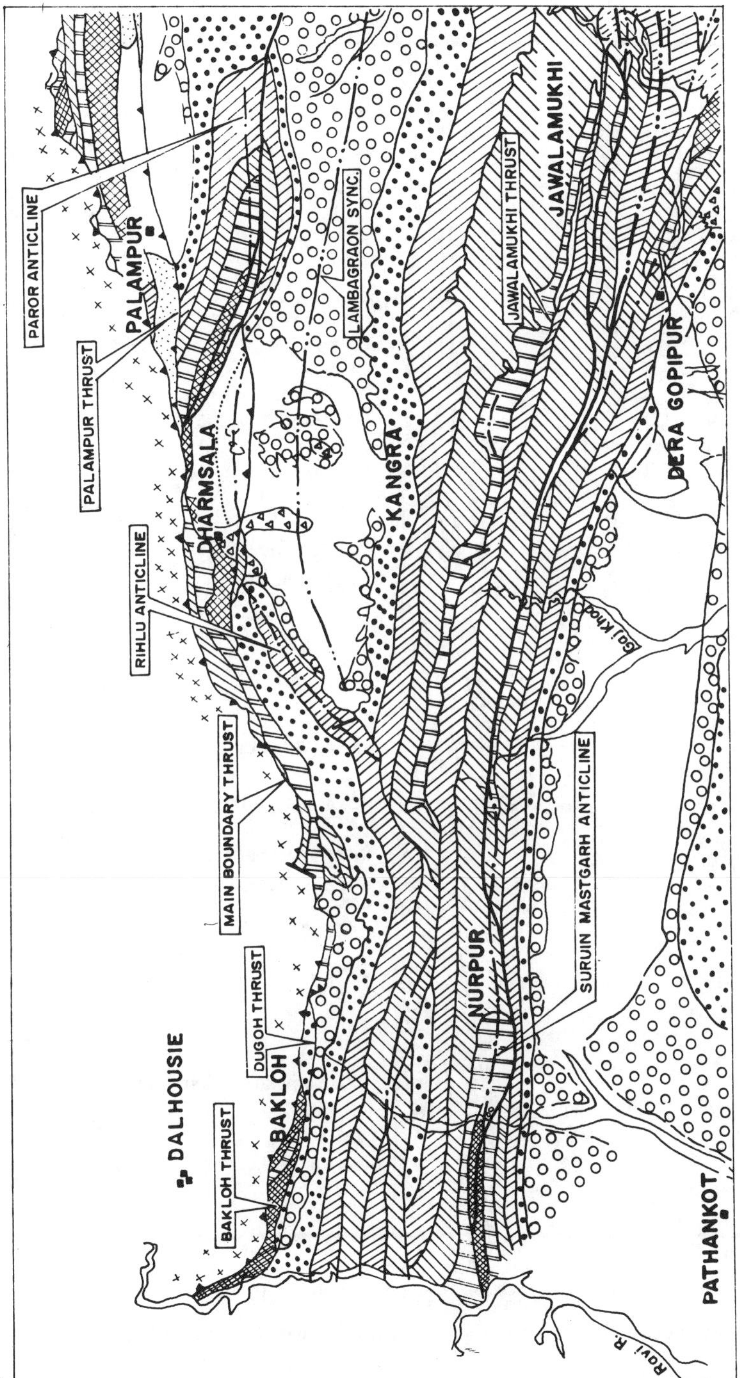

Fig. 10.45: Geological map of foothills of Himachal Pradesh
(Rai Verman, 1983, Petroliferous Basins of India)

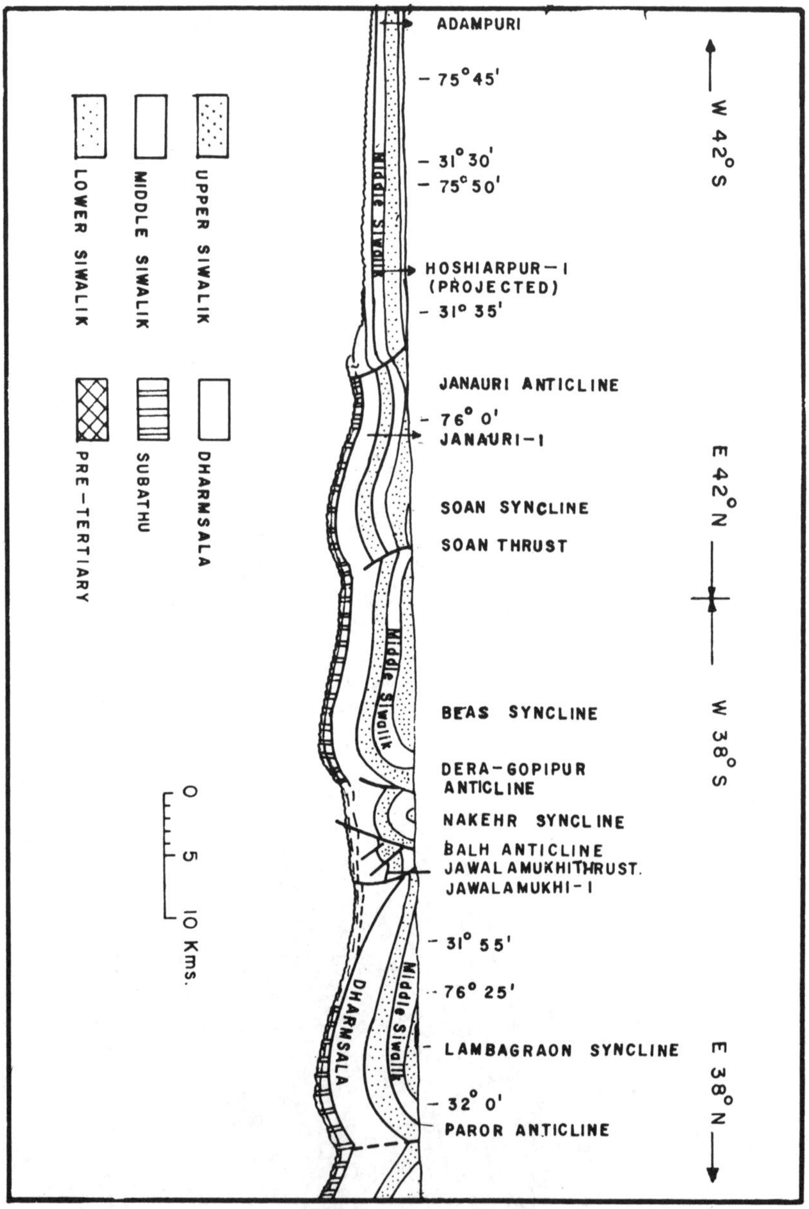

Fig. 10.46: Section through the Punjab foothills.

(After Mathur and Evans, 1964. Courtesy: International Geological Congress—64)

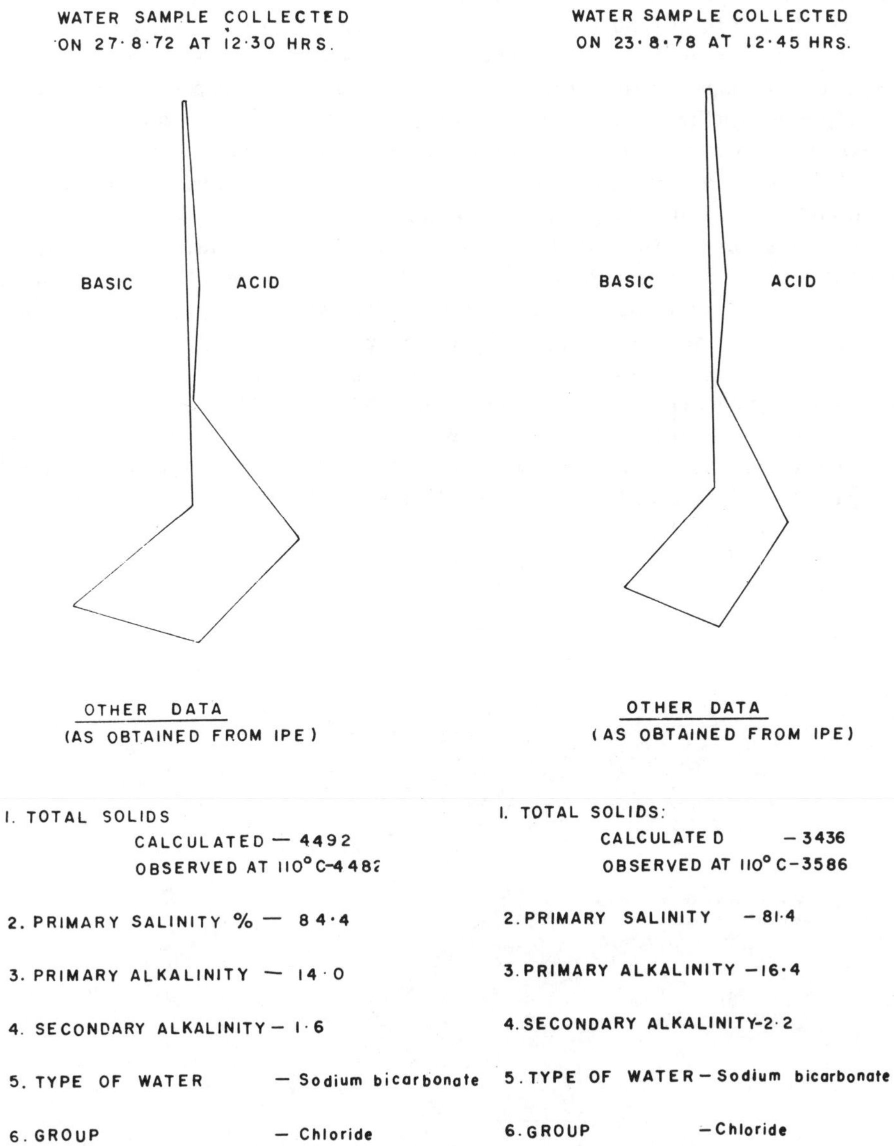

Fig. 10.47: Water pattern curve from an abnormal pressured zone in Surinsar—1, Jammu

process of reverse osmosis. The constant salinity of water observed throughout the section from 1200 to 2100 m depth appears to confirm this. To control the influx of the formation water, the specific gravity of the drilling fluid had to be raised from 1.4 to 1.65 at 1200 m, 1.65 to 2.08 at 1280 m, 2.08 to 2.14 at 1527 m, and from 2.14 to 2.37 at 2100 m. The absence of the anomaly on either density or resistivity curves, which is generally observed above high pressure horizons, at 1200 m depth where such high pressures were first reported, further confirm that generation of overpressures is due to a tectonic phenomenon. It is possible to that reverse osmosis, as discussed above, might have played a role in the generation of high pressures, but this requires confirmation by additional work in this and neighbouring areas.

The formation pressures there postulated based on theoretical calculations are approximately 2.38 times the hydrostatic. A plot of overburden and formation pressures shows that the pressures in the Surinsar well approach overburden. Generally, this is the case in a region of major tectonic activity.

(ii) *Balh well* In a well drilled on Balh anticline in the Himalayan foothills, high pressures were also encountered in the sediments below 2100 m of Murree age. It is interesting to note that in the nearby well, at Jwalamukhi-2, no abnormal high pressures were encountered, although, the same type of formations were penetrated as in the Balh well. Formation pressures in the Balh well are approx. 2.14 times the hydrostatic (Fig. 10.48).

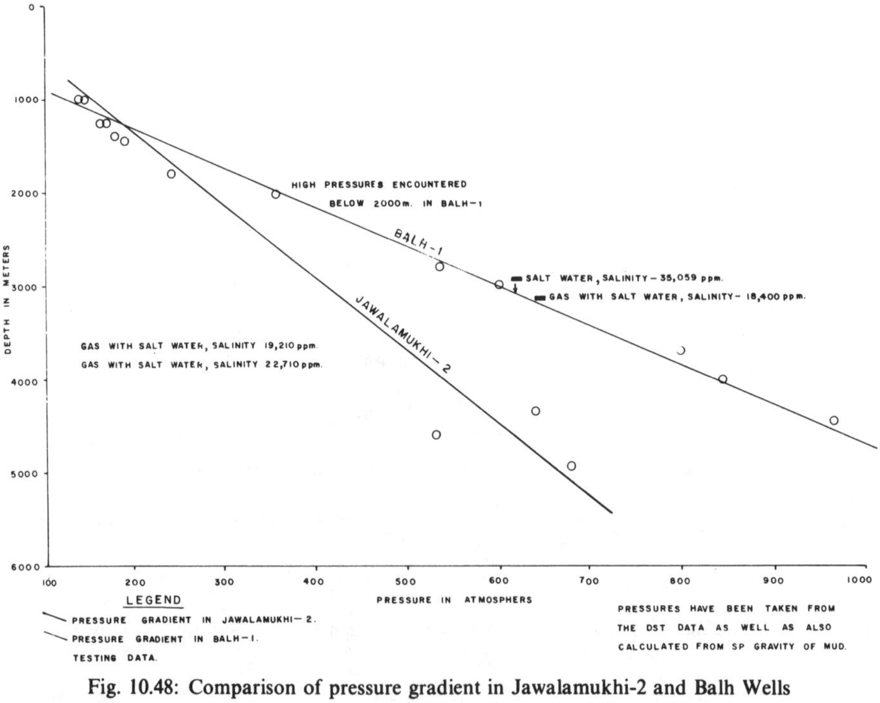

Fig. 10.48: Comparison of pressure gradient in Jawalamukhi-2 and Balh Wells

(After Sahay, 1972. Courtesy: American Institute of Mining, Metallurgical and Petroleum Engineers, IME)

Comparative studies of various parameters such as resistivity, salinity, and flow line temperature data, were undertaken after the well was drilled. A slight increase in the flow line temperature at 2100 m depth just above the top of the abnormal high pressure horizon, was noticed (Fig. 10.49). No indication on the shale resistivity plot was, however, noted as is generally seen in areas where the generation of overpressure is due to shale compaction phenomena, as has been discussed for the Bengal basin. South Kadi and Nawagam area of Gujarat, etc. The studies carried out clearly indicated that the pressures in the Balh well are due to tectonic phenomenon.

(iii) *Masimpur:* In the Masimpur area of Assam, towards the east, no study either on density or resistivity trends of the shales could be done because of the non-availability of data. On the basis of the specific gravity of the mud required to control formation pressures at different depth, pressure magnitudes appear to be oppoximately 1.8 to 2 times the hydrostatic.

As the drilling in the Masimpur high pressure area dates back sometimes to before India's independence, when pressure prediction and evaluation techniques were not available, no study of historical field data could be undertaken.

C. Overpressures Due to Compaction and Tectonics

Andaman and Nicobar Islands—General Geology

The Andaman and Nicobar islands form part of the island chain connecting Burma to the north with the Indonesian archipelago to the south. The arc exhibits consistent morphological features and, from east to west, five elements can be recognized, i.e., block, deep, inner volcanic arc, interdeep, outer non volcanic arc, and fordeep. The Andaman-Nicobar islands chain constitutes the outer non volcanic arc, to the east of which lies Ritchie's Archipelago group of islands flanking the belt of interdeep. The Beta structure (Fig. 10.50). where a well has been drilled, is located between the main Andaman Islands of archipelago group of islands towards east. The islands expose rocks ranging in age from Mesozoic (Middle Cretaceous) to Mio-Pliocene which have been mapped as different geological formations with unconformities separating one from another. The greater bulk of sediments comprise eugeosynclinal flysch which has been separated into two time stratigraphic units viz., Upper Cretaceous to Palaeocene and Oligocene to early Miocene. The Upper Cretaceous to Palaeocene (Baratang Formation) has been affected by ultra basic intrusives and basic volcanics.

The geological section penetrated in this well ranges in age from late Cretaceous to Recent. The general lithology comprises mainly of clastic section with some interbedded limestone bands.

Drilling operations in the Andaman area have encountered high pressures below 1800 m and 3200 m as revealed by plots of dcs, shale density and flow-line temperature. The formation pressures measured by DST's at the depth of 2300 m and around 3375 m were approx. 300 kg/cm^2 and 525 kg/cm^2, respectively. This corresponds to pressures ranging from 1.3 to 1.5 times the hydrostatic.

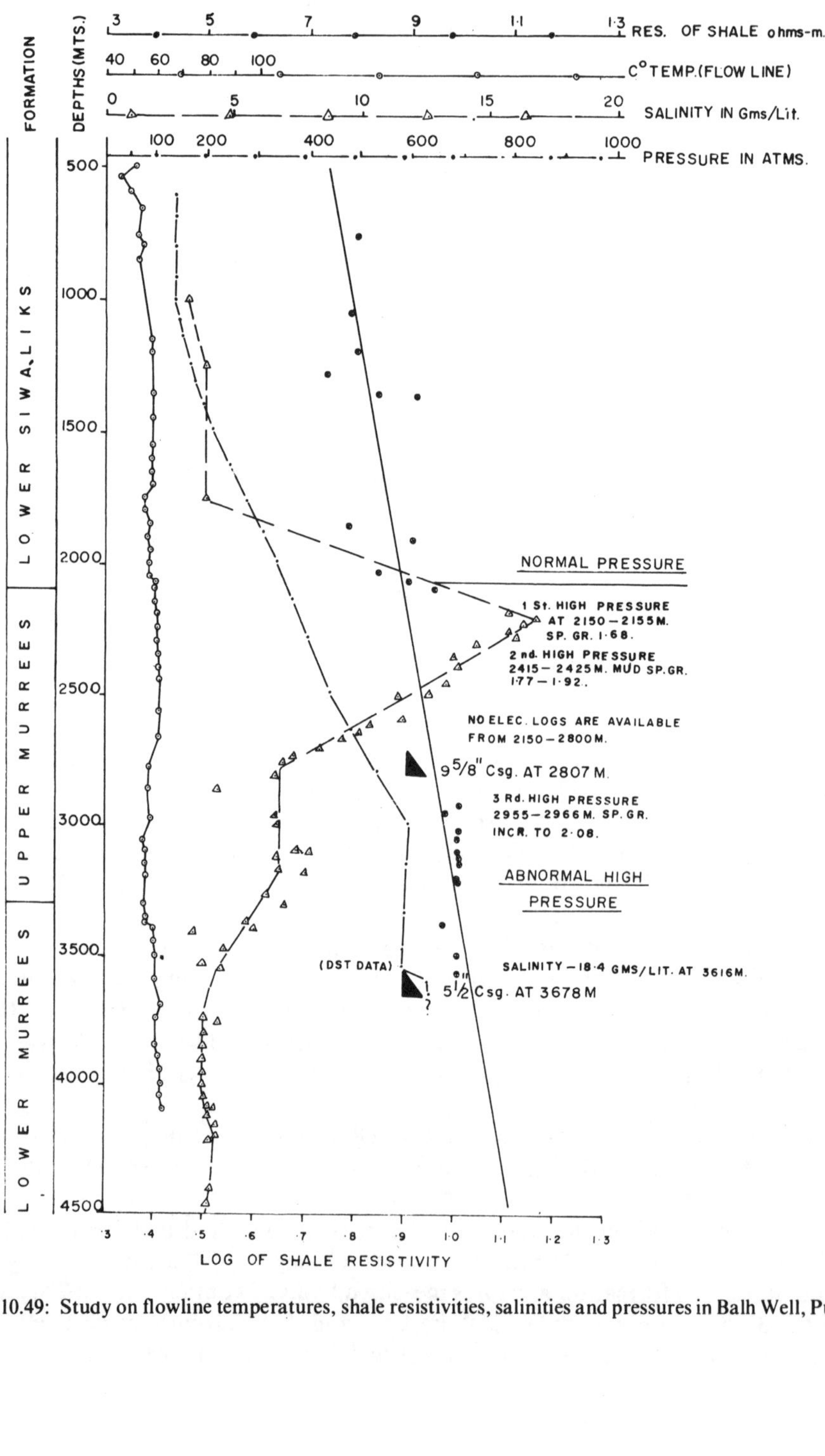

Fig. 10.49: Study on flowline temperatures, shale resistivities, salinities and pressures in Balh Well, Punjab

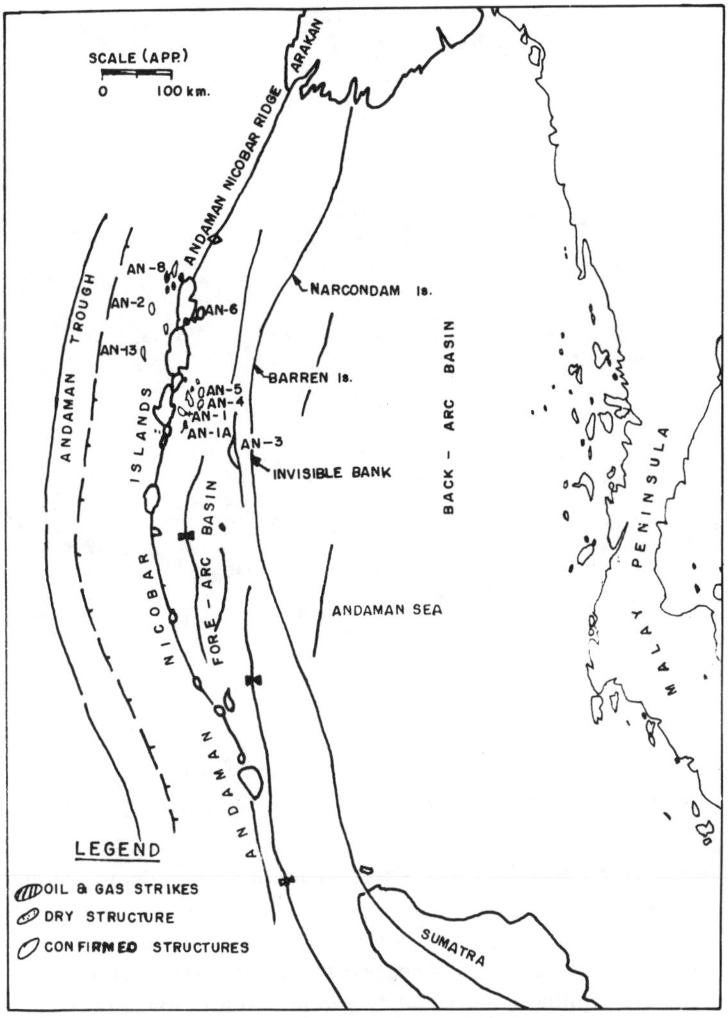

Fig. 10.50: Location map of Andaman-Nicobar basin
(After Sahay, 1984, OTC Houston)

Pakistan

In northern Pakistan's Potwar region (Fig. 10.51), located in the foothills of the Himalayan mountain ranges, and across the border in northern India, extreme overpressure conditions related to tectonic stress and steeply dipping formations caused by the Himalayan folding are found in shallow Pliocene and Miocene sediments. Rapid deposition of fluviatile and lacustrine sand, silt, clay and

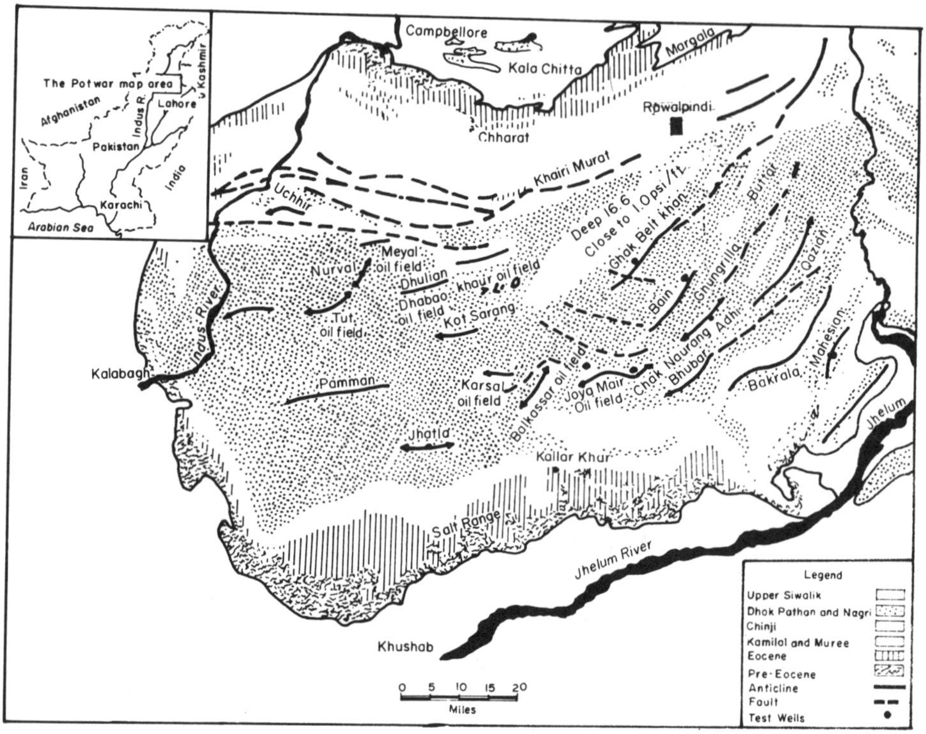

Fig. 10.51: Geological map of Northern Pakistan's Potwar region
(After M. Malick, PPL. Pakistan, Petroleum Engineer International '79)

conglomerates overlie normally pressured Eocene limestones deposited in a shallow marine environment.

Within these post-Eocene Nimadric sands, particularly in the Kamlial and Murree stages, pressures approach and in some cases exceed overburden (geostatic) pressures. These sands show pronounced lenticularity and lateral lithological variation making correlation and internal identification difficult even over short distances within the same structure.

Drilling problems include lost circulation, incursions and flows of water, caving and bridging and other severe hole conditions, stuck pipe and collapsed casing. Many wells have had to be abandoned before reaching their objectives because of drilling problems resulting from overpressure conditions.

The most extreme overpressure conditions were encountered in wells drilled on the Chak Belikan, Baines, and, particularly, the Adhi structure. Well completions became extremely difficult in the first few wells drilled on these structures. When sufficient mud weight was used to control inflows, circulation was lost through fracturing of lower strength formations or other thief zones.

Near the southern limit of the Potwar where tectonic folding is least intense, normal pressures are found in only two fields, i.e., at Jhatla and Kallar Kahur. One well drilled in the Kallar Kahur field encountered several low pressure zones. Recent wells drilled offshore Pakistan have also encountered overpressure conditions and several major well control problems.

A brief review of overpressure zones of the Potwar plateau in Pakistan follows.

The area, commonly known as Potwar, is located in the northern part of West Pakistan, between 1300 m and 518 m above sea level. It is bounded by the Kala Chitta and Margala hills on the north, Salt Range on the south, the Indus river on the west and the Jhelum river on the east. All of Pakistan's producing oil fields are located within this area. Pay zones are mostly in the Eocene and older horizons, although, migrated oil has been found in Lower Miocene sands also.

Stratigraphy and General Geology

The generalized post Eocene succession in Potwar is summarized in Fig. 10.52. The deposition till late Eocene has been mainly in a shallow water shelf environment. Brackish and fresh water lacustrine and estuarine conditions were prevalent locally for intermittent periods only (Mallick, 1982). The Post Eocene Namadric sediments of fluviatile and lacustrine sands, silt, clays and conglomerates, contrast with the older sediments both in composition and rate of deposition. The thickness ranges from 7622-9146 m on the northern side. However, southwards there is thinning of most of the stages, especially the Murrees sedimentation, which showed a significant decrease southwards.

The entire region was affected mainly by epirogenic movements prior to late Tertiary Himalayan Orogeny, resulting in uplift and erosion at certain intervals.

AGE		SYSTEM	SERIES	STAGE	LITHOLOGY
Recent to Pleistocene					Alluvium, loess, clay, gravel, conglomerate
Unconformity					
Pliocene to Miocene	Nimadric	Siwalik	Upper		Light colored, soft sandstone and conglomerates; light red, brown, orange and grey clays.
			Middle	Dhok Pathan	Reddish and orange clays, and subordinate light grey sandstone.
				Nagri	Grey sandstones, and subordinate light red clays.
			Lower	Chinji	Bright red clays with subordinate green-grey sandstones.
				Kamlial	Green-grey and brown, massive sandstones, pseudoconglomerates, and dark red to purplish shales.
		Murree			Grey and greenish massive, fairly hard sandstones, pseudoconglomerates, and dark red to purplish shales with basal conglomerate.
Unconformity					
Middle Eocene and/or Older					

Fig. 10.52: Generalized post-eocene stratigraphic succession in the Potwar

(After M. Malick, PPL. Pakistan, Petroleum Engineer International '79)

Although, a Post-Eocene unconformity is well pronounced throughout the region, there is generally discordance of dip between the Nimadrics and underlying older formations. From early Miocene to Pliocene, the Potwar basin underwent continuous subsidence to the extent of 7622-9146 m in the northern part. Intense folding and faulting movements, caused by the Himalayan uplift, started in the central Himalayas and moved progressively southwards with time. In the Potwar region, the major part of folding and faulting took place in late Pliocene and extended into early Pleistocene. The eastern part of Potwar represents another zone of intense faulting and folding

Numerous oil shows around the Potwar have attracted interest for oil exploration for more than 100 years. The real incentive for intensified exploration, however, started with the discovery of the Khaur oil field in 1915. Additional fields, e.g., Dhulian, Joya Mair, Balkassar, Karsal, Tut and Meyal, etc., have since been discovered. However, Potential pay zones in a number of known structures in the area still await testing. These include structures where attempts to drill and test have already been made, but the objectives could not be reached either because of structural complications or high pressure problems.

Fresh water post-Eocene Nimadric sands over most of the Potwar, particularly in the Kamlial and Murree stages, contain high pressures water. The problem is aggravated because both the lateral as well as vertical extent of the different water sands are very irregular and, therefore, unpredictable. Their pronounced lenticularity and lateral lithological variations make it difficult to identify and correlate the sand even over short distance between wells within the same structure.

The problems faced in the different fields/structures during the course of drilling wells are summarized as follows:

Khaur Field

Formation pressure in some of the Khaur field's water-bearing Muree sands below 4000 ft approached, and in some instances exceeded overburden (geostatic) pressures. This gave rise to serious drilling problems and other complications, such as incursions and flows of water, caving and bridging of the borehole, loss of circulation, and stuck drill pipe, frequently resulting in abandoned holes.

Dhulian Field

In the Dhulian field, mud weights up to 2.01 gms/cc have been required to control chinji water sands in the 610-915 m depth range. High pressure water sands have been also encountered in the lower Nimadric rocks. The first four wells drilled in this structure by the early 1930's, in fact, failed to reach the Eocene and had to be abandoned because of complications resulting from these water sands. In the nearby Meyal oil field, discovered later 1968, mud weights up to 2.13 gms/cc were required to control the Kamlial water sands, while in the other neighbouring field at Tut, mud weights up to 2.26 gms/cc were necessary to safely drill through the Kamlial. Tut was discovered in early 1958.

Kot Sarang

In the Kot Sarang structure, mud weights, ranging from 2.13—2.22 gms/cc reportedly were used to control some of the Kamlial and Murree sands. Further eastward, however, pressure gradients of the Nimadric sands appear to decline gradually. For example, in Karsal, the adjoining structure on the east, only 1.96 gms/cc mud was required to drill through the Kamlial and Murree sands.

Balkassar and Joya Mair Fields

In Balkassar, mud weights of 1.89 gms/cc were needed to balance the pressures of the water sands in Kamlial and Murree. In the joya Mair, the formation pressure in the Kamlial was equivalent to only about 1.53 gms/cc mud. Farther west, however, the declining pressure trend was reversed and mud weights in the adjoining structure of Chak Naurange were again up to 2.14 gms/cc.

Mahesian Structure

High pressure water sands in the Kamlial and Murree Formation were encountered also in a well drilled in the Mahesian structure near the south-eastern edge of the Potwar, requiring mud weights to 2.09 gms/cc. Several low pressure zones in this well resulted in further complications and created difficult problems in reaching the drilling objective in this structure.

It appears that only two locations in the Potwar have not encountered high pressure water sands. Both of them, Jhatla and Kallar Khar, are situated along the southern limit of the Potwar, near the foot hills of the Salt Range. In one of the wells drilled in Kallar, Khar, encountered several low pressure zones in the Nimadric and there was no indication of high pressure water zones even in the Kamlial.

Experience with Nimadric water sands probably has been most severe in wells drilled in the rather tightly folded and faulted structures of the eastern Potwar, namely Chak Beli Khan, Baine, and Adhi.

Chak Beli Khan

In Chak Beli Khan, the first two wells (both located near the surface crest) had to be abandoned at the shallow depths of 322 m and 256 m because of complications resulting from water flows from the Chinji sands at pressures approaching geostatic. In the third well (located downdip on the north-western flank of the displacement on the basis of seismic evidence indicating some lateral displacement of the sub-surface Eocene crest) water sands at high pressures are not encountered in Chinji and Kamlial. However, small but persistent water inflows from some Murree sands in the 3354 to 3659 m range necessitated raising the mud weight to 1.99 gms/cc.

Bains Structure

In the wells drilled on the Bains structure, water sands were penetrated in Chinji, Kamlial and Murree sediments. Mud weights of 2.14 gms/cc were required to control Kamlial water flows and several water sands at geostatic pressure were encountered in Murree requiring mud weights of 2.4 gms/cc.

Adhi Structure

Adhi is a well defined, tightly folded and faulted NE-SW trending anticline. Dips gradually increasing down the flanks are in the range of 10-50°. The first well on this structure, located near the coast, was spudded in December 1950. During drilling at 288 m with 1.22 gms/cc mud, the well encountered the first Chinji water sand, which was controlled by 1.71 gms/cc mud. Mud weight had to be successively raised to 2.25 gms/cc within 30 m. Again at 639 m because of water cut mud returns, mud density was increased to 2.26 gms/cc causing severe loss circulation problems. This required decreasing the mud weight to 2.23 gms/cc which restored equilibrium only briefly. Again at 640 m there was a strong water inflow in excess of 700 bbls/hr. Unable to combat the situation of mud loss and high pressures simultaneously and because of stuck drill pipe, the well had to be abandoned. The flow of water through valves at the well head was allowed to depressurise the sands.

Well No. 2, located 976 m north east of Well-1, along the axis of the fold, was spudded in November 1957. 16 inch conductor casing was cemented at 243 m. In view of the experience in Well-1, mud weight was gradually raised. At the depth of 353 m, when the mud weight reached 2.26 gms/cc as against 2.2 gms/cc in Well-1 at the same depth, there was inflow of water. It was possible to reach an optimum mud weight of 2.33 gms/cc. Loss circulation materials were added to the mud system but again at 599 m there was a severe mud loss. After the mud weight was reduced to 2.30 gms/cc water flow occurred again which was controlled with 2.34 gms/cc mud. Subsequently 11-3/4" casing was cemented at 610 m.

After drilling proceeded with 1.67 gms/cc mud, water influx from the Kamlial sand was noticed at 798 m. It was controlled with 1.73 gms/cc mud but within the next 61 m a series of high pressure water sands were encountered necessitating increasing the mud weight to 2.23 gms/cc. Drilling conditions became extremely difficult because of mud loss as well as high pressure water inflows. At the depth of 1146 m, where the next string of casing was to be cemented, there was an inflow of water from Murree sand with 2.23 gms/cc mud. Mud weight was increased to 2.35 gms/cc to control but again loss circulation occurred. In view of this dual problem and deteriorating well conditions, it became impossible to resume drilling. Unable to combat this situation, the well had to be abandoned and was left to flow water to depressurise the sands.

Adhi, Well-3, spudded in October 1965, is downdip on the NW flank of the anticline about 1402 m west of Well-1. After 20 "conductor casing cementation at 248 m, Chinji water sands were drilled with 1.61 gms/cc mud down to 680 m and 13-3/8" casing was cemented above the expected top of the Kamlial stage. Below this depth, mud weight was gradually raised to 2.01 gms/cc in preparation of penetrating the Kamlial sands. Mud weight was reduced to 1.92 gms/cc in view of slight loss circulation problem around 1012 m but it had to be raised again to 1.96 gms/cc because of a minor water inflow at 1067 m 9-5/8" casing was cemented at 1219 m to isolate these Kamlial sands. While drilling through the Murree sediments, in view of string inflow of water at 2018 m, the mud weight was raised to 2.14 gms/cc. This well also had to be abandoned at the depth of 3237 m because subsurface evidence indicated it was located in a zone of structural complications. Severity of water flow in this well was much less because of (1) partial depressurisation of sands flowing in wells 1 & 2 for about 8-9 years, (2) lenticularity of and lateral lithological variations in the sand bodies, and (3) lower formation pressure gradients because of increased depths of sands in this well.

Adhi, Well-4, was located about 1280 m NE of Well-3, was spudded in April 1967. 20″ casing was cemented at 223 m. Chinji sands were penetrated with 1.61 gms/cc mud to the 13-3/8" casing depth 621 m after which the mud was raised to 1185 gms/cc water inflow from some lower Chinji sands was encountered and mud weight was raised to 2.14 gms/cc. At the depth of 763 m, circulation was completely lost, resulting in deterioration of hole conditions because of cavings and well collapse. While drilling through a Kamlial stage to 9-5/8″ casing depth at 976 m, there were mud losses with 2.14 to 2.21 gms/cc alternating with water inflows. After cementation of 9-5/8″ casing, drilling commenced in Murree with 1.93 gms/cc mud. Inflow of water necessitated raising mud weight to 2.02 gms/cc and then to 2.16 gms/cc. Still another inflow of water at 1328 m was controlled with 2.35 gms/cc mud. From that point onwards, an extremely delicate and precarious balance existed between water bearing high pressure intervals and lost circulation zones. As drilling proceeded, mud weight was gradually reduced. At 1649 m 2.29 gms/cc mud was found to cause slight waterrcut. Heavy intermittent losses of circulation were encountered and were controlled by loss circulation material below 1698 m. There were persistent indications of formation water contamination in the 2.26 gms/cc mud column after tripping the drill pipe.

At 2040 m when a new drill bit was run, the well came in approximately 160 bbls/hr and blew out full bore through the flow line. The well was shut in with 113 to 120 Kg/cm^2 well head pressure and the pressure in the drill pipe was killed with 2.36-2.40 gms/cc mud. Mud up to 2.43 gms/cc was pumped down the annulus but the annulus pressure could not be killed until 2.57 gms/cc mud was pumped in. The bit was pulled back in 9-5/8″ casing but attempts to regain overhead circulation, proved futile. Temperature and cement bond logs indicated the strong possibility of hot formation water from greater depth cycling into some lost circulation zones in the Kamlial sands above the 9-5/8″ shoe. Isolation around the shoe appeared to have broken down.

The thief zone in the Kamlial sands appeared to be accepting fluids from both mud column and water sands below. Attempts to seal off the thief zones with batches of diesel bentonitic junk and plugs of weighted cement proved to be futile. The hole was cleaned to 1890 m because of bad hole conditions. Three attempts were made to set inflatable bridge plugs to isolate high pressure water zones from thief zones but proved unsuccessful. Caliper was attempted but sonde did not go below 1133 m.

After persistent failures over a period of 3 months to restore the borehole to a condition that would permit drilling to continue, the well had to be abandoned.

All the wells drilled on Adhi structure had to be abandoned because of the numerous and severe problems encountered. Chances of overcoming the problems mentioned, availability of techniques that would enable wells to be drilled to pre-Nimadric objectives in all structures of the area would considerably enhance Potwar's petroleum potential. The Pre-Nimadric pay zones in Potwar have been found to be hydrostatic or even sub-hydrostatic. Cementing depth of casing strings in each well has to be determined accurately to isolate high pressure Nimadric formations prior to penetrating low pressure pay zones. Any miscalculation of this critical depth can result not only in the abandonment of a particular well but may also damage the pay zones as well. For example, exposure of heavy mud can plug off fine fractures that provide porosity and permeability in the typical limestone reservoirs of Potwar (Malick, 1979).

Conclusions

It may thus be concluded that most drilling operations all over the Potwar Plateau through post Eocene sediments are abnormally hazardous, time consuming and expensive. The excessive formation pressures of the water-bearing sands are usually responsible for initating a series of events which lead to serious drilling difficulties. They may be due to "sealing-in" of the sands by orogenic movements: residual hydrostatic fossil pressures trapped in lenticular sand bodies dating from the time of deposition of the Nimadrics, partly an expression of the large tectonic stresses and forces of compression during the folding of the anticlines in the sub-Himalayan region, etc. In case some efficient technique capable of providing reliable indication of the proximity of abnormally high formation pressures is available, it could solve lot of problems. Proper prediction techniques for abnormal pressures would result in improved well control and minimize possible drilling hazards.

Prediction techniques as normally employed in the U.S., Gulf Coast area and in India, such as low density, low resistivity, high transit time, high conductivity and flow line temperatures, low dcs values, do not apply here in a straight forward fashion. Furthermore, in the thick Post Eocene sediments of Potwar the conditions are also complicated as shales here are frequently sandy and silty, and alternate very frequently with sand beds.

High pressure water sands usually lead to serious complications, despite modern and improved mud technology which effectively controls formation pressure equivalents to or even higher than geostatic pressure. Problems become acute when severe loss circulation also occurs. This situation usually deteriorates and becomes extremely critical when the simultaneous problem of water inflows and lost circulation is further complicated by caving and heaving of shales, bridging of the borehole, or differential pressure sticking.

Indus Offshore Basin

The Indus offshore basin, located to the south of Karachi trough and Thar slope between Murray ridge and Indian border (23 to 25° N Lat: 66° to 68° E long), has been under exploration for hydrocarbon potential. Nine wells have been drilled, two onshore wells (Karachi 1,2) three wells near Karachi shore (Dabbo creek, Patiani creek and Korangi creek) wells on an offshore platform (Karachi South A-1, B-1, C-1). No oil or gas deposits were found. However, gas shows and hydrocarbon traces were reported. Tectonically the Indus offshore basin can be divided into three units:

(1) An offshore depression on the west.
(2) Offshore Karachi trough platform in the middle between hinge line and Karachi shore line.
(3)Offshore Thar slope platform.

The hinge line zone and the outer margin of the carbonate platform, where there is a possibility of reef build ups appear to be potential areas for hydrocarbons and further investigations and drilling for suitable structural traps, are warranted. No indication of overpressure zones in this part of the basin has so far been reported (Geology and Hydrocarbon potential of offshore Indus basin, Pakistan, AAPG, Vol. 66, 1972 and no information has been made available to the authors, which is more recent and released.

Bangladesh

The geology of Bangladesh is not yet adequately known and whatever is known is primarily based on geophysical studies and exploration drilling.

Bangladesh was the site of vast delta buildups during Eocene to Pliocene times. Subsequent Pliocene-Pleistocene alluvial deposits of the Ganga Brahmaputra river system completely covered the earlier deltaic deposits. The Tertiary strata below the Pleistocene have become compressed into long and narrow folds. Tertiary sediments increase in thickness from west to east and from north to south to a maximum of about 1067 m. Alpine orogeny which started at the end of Oligocene time causes several unconformities, disconformities and tectonic disturbances. The long and narrow, north-trending anticlines are the result of compressional forces of the Alpine orogeny which resulted in numerous thrust faults with gentle east and steep west flanks, deformation decreasing gradually westwards. Figure 10.53 shows petroleum development in Bangladesh as of 1981.

Figure 10.54 shows the stratigraphic sequence of Bangladesh. Tectonically Bangladesh can be divided into three divisions: (1) Bengal foredeep or basin, (2) Arakan magnetic linorium, and (3) Sub-Himalayan foredeep (Sylhet trough, South Shillong Shelf). The Bengal basin or foredeep (basin) is a geographic and geometric feature covering most of the part of the Indian territories of West Bengal and Assam. The basin can be divided into two parts, the western and north-western precambrian platform with a generally reduced sedimentary thickness at places of only 116 to 140 m and the fore deep of the geosynclinal area which exhibits a thickness of sediments in excess of 14,634m to the east and south-east.

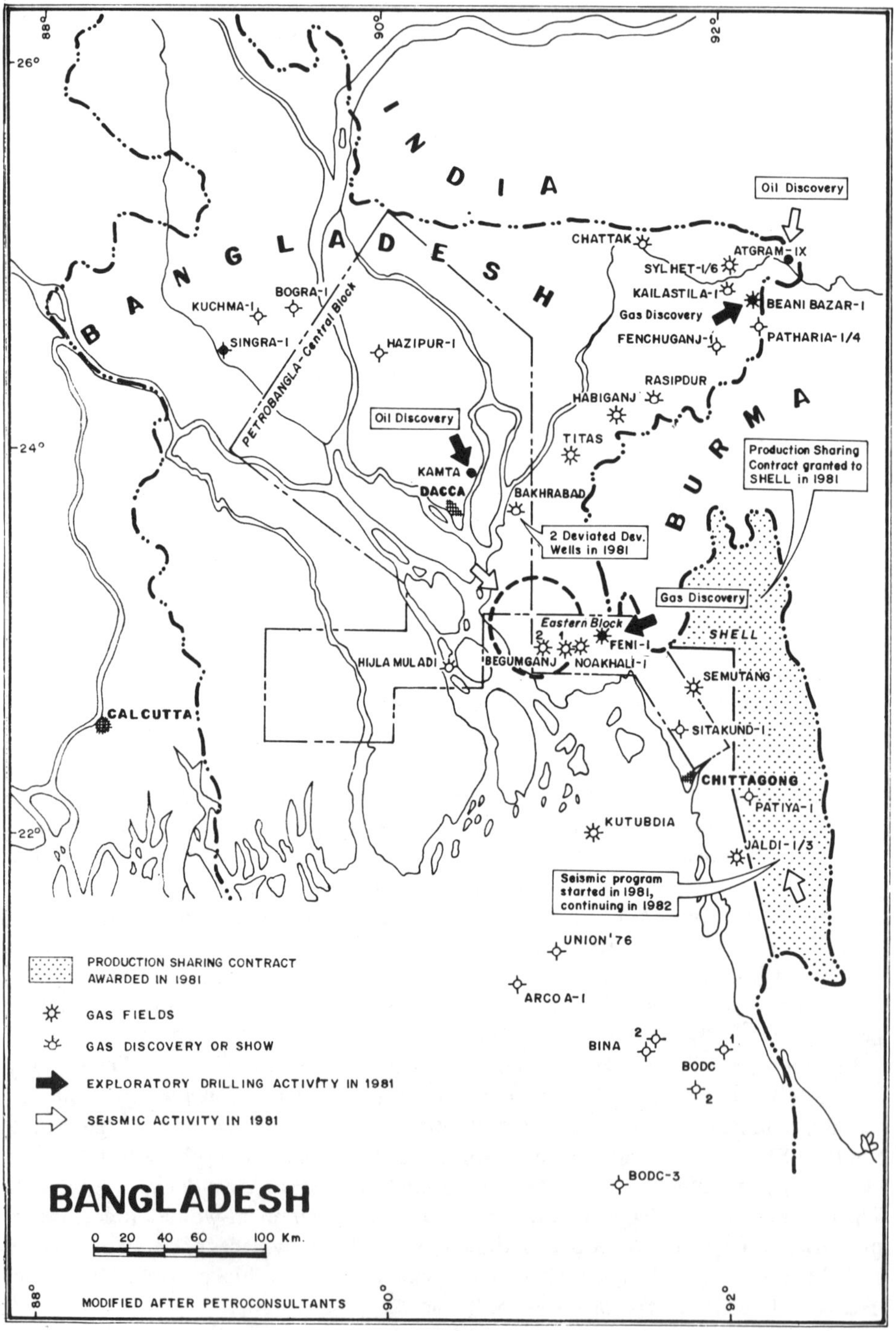

Fig. 10.53: Petroleum developments in Bangladesh, 1981

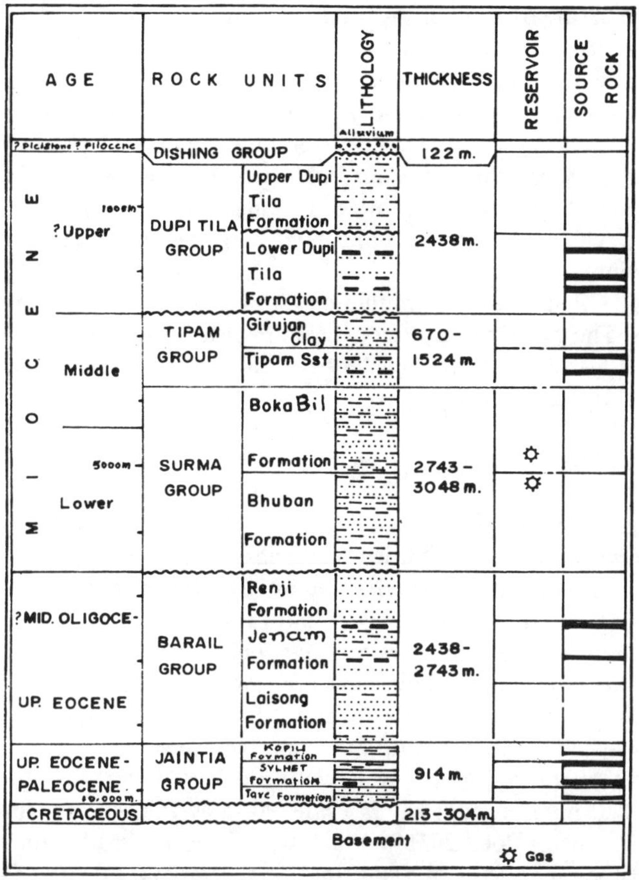

Fig. 10.54 Stratigraphic sequence of Bangladesh

(*Oil and Gas Journal,* 83)

Source : Grunou, et al 1973.

The stable or moderately mobile portion of the basin seems favourable for potential hydrocarbons accumulations. Such areas include the hinge zone, the area between Borga slope of Rangpur saddle and Bengal foredeep. The hinge zone is reflected in the transition of the rock facies from that of shallow shelf to that of deep water deposition facies. Gas and condensate have been generated in this basin, where source rocks are marine and non marine, the temperature gradients are average and slightly lower and the depth of burial exceeds 3049 to 6098 m. Good reservoir and cap rocks condition developed in the shallow marine deltaic environments during Upper Eocene through middle Miocene geologic times. Reservoir porosities range from 20 to 30%, gas saturation averages 75%. While the environment of deposition of sediments is mostly deltaic, gas fields have not been completely delineated and, hence, it is not possible to accurately define the areal extent of any producing reservoir. Areas most promising for potential hydrocarbon accumulations include the Hinge zone, Shillong monocline, Dinajpur high, and anticlinal structures found in the folded belt of Bengal Basin. The gas found is mostly methane, essentially sulphur free and has a high caloric value. Currently the gas is being used in fertilizer and power plants as well as in industrial, commercial and domestic sources. Four fields have been on production since 1981 Titas 3.17 MMm^3/d Sylhet 0.93 MMm^3/d, Habibganj (0.4 MMm^3/d) and Chhatak (0.14 MMm^3/d). The producing sand and sandstones reservoirs are in the Miocene Bokabil Formation.

Recently, Shell Oil has signed a production sharing agreement for a large area in the Chittagong hills. The two 1981 gas discoveries at Feni-land Beanibazar have estimated gas reserves of 1.5 trillion cu. ft. The Japox group holds an exploration assistance contract to develop Bakhrabad gas field.

Offshore Exploration

The sedimentary sequence down to a depth of 4726 m can be broadly divided into three major intervals. Two well stratified successions (the upper 1494 m and the lower one 1006-1098 m, represents sedimentation in low energy environments and an intermediate sequence, approximately 305-356 m thick.

The lower-most succession indicates the source of sediment to come from the east and move westward, whereas the uppermost cross bedded zone suggests the sediments to be transported in a north to south direction. Based on paleontological evidence, older sediments in the offshore are of late Upper Miocene.

The Kutubodia offshore field discovery was drilled by Union Oil of California to a total depth of 3506 m. The main pay is a late Miocene-Pliocene siltstone with 17% porosity and 35% Sw at 2662-2671 m which was tested at 0.51 MMm^3/d. Well log analysis suggests possibilities in two additional gas sands at the depth of 2919-2938 m and 3188-3198 m. Preliminary cumulative reserves for all sandstone reservoirs in this field are less than 28.3 billion m^3. The Kutubodia structure is well pronounced at depth but out only in shallow sedimentary rock sequence. Vertical closure is approx. 198 m. This field has not yet been developed and is not on production.

Bangladesh's gas resources were generated in an "overcooked maturation environment". Therefore, prospects for finding oil are not too promising (Raju, 1968; Gurneu, 1978). Other authors, however, suggest that Palaeo-temperatures and depth of burial for the sediments from which hydrocarbons matured were normal. The source rock for these hydrocarbons are of marine origin, probably Eocene-Miocene shales. There is no supporting evidence that any of the gas found in Tertiary sandstone reservoirs originated in the deeper coal beds, e.g., Permian formations (Woodside, Petroleum Geology of Bangladesh, Oil & Gas Journal, Aug. 8, 1983).

As regards overpressure horizons, wells drilled on Sylhet structure in northern Bangladesh, encountered pressure gradients of up to 0.99 psi/ft in several sands located in massive shale sections. Near the coast and offshore the pressure conditions appear to be similar to conditions across the western borders in India. Similarly, in the

Titas field, overpressure conditions also exist. However, no detailed data on the overpressure areas in Bangladesh is available.

Burma

Petroleum activity in Burma during 1981 was slightly accelerated as compared to previous years. A joint venture company called Burma Petroleum Development Company was formed by Japan National Oil Company (49%), Myanama`(51%), the State controlled petroleum company. The company will explore blocks M-2, M-5, M-6, M-8 and M-9 in the Irrawady delta offshore.

Onshore by means of another Japanese assistance agreement, JNOC is participating in development of oil resources on Ramree island along the Arakan coast. Exploration programmes were carried out in central Burma, along the Irrawady river in the Irrawaddy delta and on Ramree island off the west coast. In 1981, 17 exploratory wells were completed compared with 5 or 6 in 1980. A total of 104 exploratory and development wells were drilled during 1981. Oil and gas indications were seen in approximately 50% of the wells drilled and 31 rigs were active compared to 26 in 1980.

Three major discoveries were reported in 1981. Htanttrin-3 discovered oil and gas. Earlier Kyontani and Pegan-Thyintaong discoveries were reported in early 1981. They appear to be oil fields. Reserves as high as 1.4 billion bbl have been reported. Oil was for the first time encountered in deep limestone formation in the Irrawaddy delta area. Development drilling was confined mainly to the Mann and Htankshabin oil fields. Overall 24 rigs and work over units were active on development and rehabilitation projects. Figure 10.55 highlights the Petroleum Developments in Burma during 1981.

Overpressures in Burma are found in the Tertiary rocks in Irrawaddy and Chindwin river basin and offshore in the gulf of Martaban. In the Chauk field, gradients of 0.86 psi/ft exist and gradients approaching geostatic have been encountered at 1524 m in the Yenangyaung structure. Drilling below 2744 m required 2.27 gms/cc mud. In the gulf of Martaban, severe formation pressures have caused gas blowouts.

Sri Lanka

Sri Lanka has attracted exploration companies since 1980. As a result of various geophysical surveys carried out earlier, three onshore and two offshore wells (up to 2927 m deep) were drilled by the Soviets. The Geological Survey Department of Sri Lanka did some gravity magnetic work in the area of Mannar, Jafna peninsula and islands of Delft and Punguditutivu in 1976-77. Minor gas shows were reported in lower Tertiary, upper Cretaceous and lower Cretaceous formations in three wells drilled by the Russians in Peslai. Reports indicate good porosity in carbonate reservoir rocks, good source rock potential, and possible presence of dry methane in lower and upper Cretaceous zones. On the Indian side in the Cauvery basin, the source rock potential studies indicated sub-commerical shows of hydrocarbon. Organic carbon is abundant in most of the samples, particularly in the lower Cretaceous and Tertiary continental deposits. Geochemical analysis of samples from four wells on the east coast of India

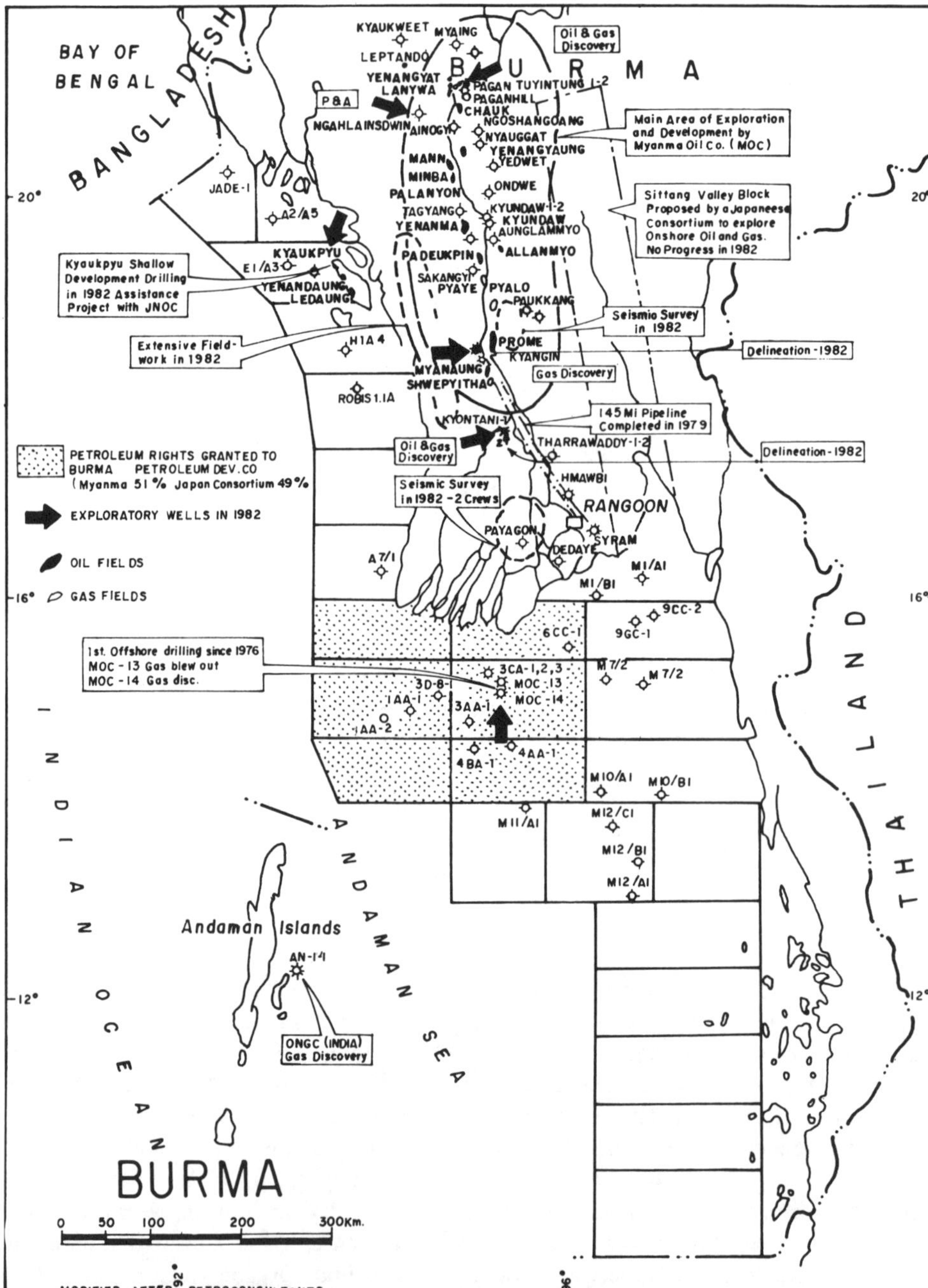

Fig. 10.55: Petroleum developments in Burma, 1982

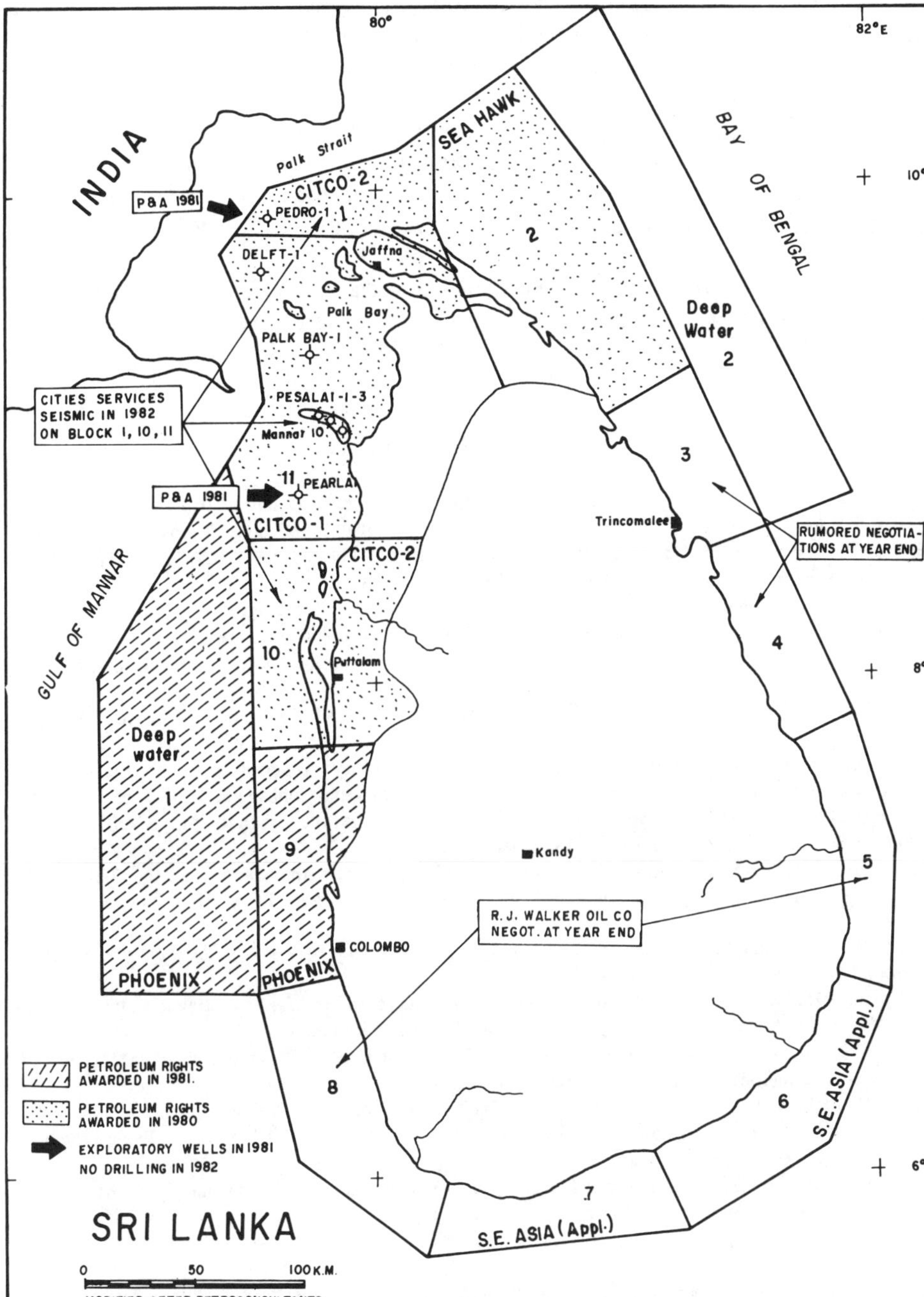

Fig. 10.56: Petroleum developments in Sri Lanka, 1982

has shown no capabilities of light crude or condensate generation. Sub-commercial oil production comes from a well on the Karaikal high and tests in an offshore well near Pondicherry on the Indian side reported a daily gas flow of 68,000m^3/day (Alkakkone, 1985).

Figure 10.56 highlights the petroleum development in Sir Lanka.

Sri Lanka presently has no oil or gas production. However, in view of recent significant oil discoveries in the Cauvery Basin and in the Palk Strait on the Indian side, there exists a reasonable possibility for commercial hydrocarbon deposits in Sri Lanka's territorial waters.

The various wells drilled so far in Sri Lanka as well as in the Cauvery Basin and Palk Strait on the Indian side, have not encountered any overpressures as far as released data, is concerned.

REFERENCES

Alahakkone, R.R., 1985. Sri Lanka needs more exploration, Oil & Gas Journal, July, 28th.

Fertl W.H., 1976. Abnormal Formation pressures, Elsvier Scientific Publishing Company, pp: 346-347.

Fletcher, G.L., 1983, Oil and Gas developments in Far East in 1982. The American Association of Petroleum Geologists Bulletin V. 67, No. 10 (October, 1983) pp: 184-1891.

Malick, A.M., 1978. Problems of drilling through High pressure, water bearing, Post-Eocene sediments of

Malick, A.M., 1979. Pressure plague Pakistan's Potwar, Petroleum engineer international.

the Potwar, West Pakistan. Proceedings of the Fourth Symposium on the Development of Petoleum Resources of Asia and Far East.

Sahay, B., and Tikku C.L., 1968. Shale Density studies in Bodra-1, ONGC Bulletin, August, Vol. 5, pp: 83-85.

Sahay, B., 1969. Comparative studies between logs of shale resistivity and density as a new correlation tool for prediction of high pressures. Paper presented in 2nd Indian Petroleum Conference, Baroda, February, 1969.

Sahay, B., 1971. Indications of Abnormal Sub-surface Pressures on the resistivity logs of wells drilled in Bengal Basin India, ONGC Bulletin, December.

Sahay, B., 1972. Abnormal Sub-surfacepressures, their origin and methods employed for prediction in India, presented in Seminar on Abnormal Sub-surface pressures, SPE of AIME, Louisiana, U.S.A. May.

Sahay, B., 1974. Analytical studies on abnormal pressures in Indian basins. A manual distributed to the geologists of ONGC. (unpublished)

Sahay, B., 1976. Geopressures, their origin, utility and methods employed for prediction in Indian, Journal of Mining and metallurgical society of India, Vol. 48, No. 3, September.

Sahay, B., 1978. Geology of Bombay High and adjoining strictures. Paper presented in 2nd Seapex conference, Singapore. Seapex preceedings Vol. IV, 1977/78.

Sahay, B., etal, 1983. Formation Evaluation and Well-site Geological Techniques, pp:

Sahay, B., 1984. An analytical study of the overpressured area in Bombay Offshore Basin. Paper presented in 5th Seapex Conference, Singapore.

Sahay, B., 1984. A review of Geology and Petroleum possibilities of the Continental margins of India. Paper presented in OTC Conference, Houston.

Sahay, B. and Fertl, W.H., 1984. Occurrence of overpressured formation in and offshore India, Oil and Gas Journal, August.

Shuaib, S.M., 1982. Geology and Hydrocarbon Potential of Offshore Indus Basin, Pakistan, The American Association of Petroleum Geologists.

CHAPTER XI

Reservoir Management of High Pressure Reservoirs

Introductory Remarks

The main thrust of this presentation will be the management of reservoirs with high pressures. No attempt will be made to evaluate why they are high pressured. Once the pressure level is established, however, some special planning for reservoir development, operation and depletion is required. Consideration must be taken of the management environment, including technology, application and reservoir economics. The purpose of this presentation is not to convert readers to reservoir management engineers, but rather to show the steps required and data needed to plan the development of one of these reservoirs (Fig. 11.1).

Caveats

Preparation of this presentation revealed there is no well established body of engineering expertise associated with the depletion of high pressure reservoirs. The literature shows the high pressure, high temperature sour gas reservoirs along the Gulf Coast, U.S.A., produce under pressure depletion, with no apparent attempt to optimize recovery. The high pressure water reservoirs in the same area are being considered for cycling operations to harvest their thermal energy and gas in solution. When they are developed, a more sophisticated reservoir management regime will be required.

The following presentation is, therefore, largely based on the theoretical application of reservoir management techniques learned in conventional reservoirs to high pressure and high temperature situations. Recent interest in the associated problems has resulted in an increasing frequency of technical papers related to reservoir rock and fluid properties in this environment and to drilling and wellbore operations. The list of references compiled may be of assistance in dealing with specific problems.

Occurrence

High pressure and temperature reservoirs have been found throughout the world. Closer to home, of the 29 Alberta reservoirs having discovery pressues over 40 MPa, 19

Alberta High Pressure Reservoirs

Pool	Type	kPa/M	°C/100 m	Depth m	Pressure MPa	Temp °C
Brazeau Nisku K	Gas	18.5	3.0	3844	71	117
Hinton Dunvegan A	Gas	17.1	2.9	3154	54	91
Peco Cardium E	Oil**	16.5	3.1	2482	41	77
Nosehill D-3A	Gas	15.9	2.3	4021	64	94
Brazeau Nisku M	Gas	15.3	3.2	3272	50	104
Brazeau Nisku A	Oil*	15.1	3.4	3114	47	106
Brazeau Nisku E	Oil*	14.4	3.4	3200	46	108
Columbia Nisku B	Gas	14.3	2.7	4123	59	110
Brazeau Nisku H	Oil	14.0	3.4	3134	44	105
Columbia Nisku Un	Gas	14.0	3.0	4214	59	127
Edson Viking A	Gas	13.9	3.0	2803	39	83
Brazeau Nisku F	Oil	13.7	3.4	3353	46	113
Westpem Nisku D	Oil	13.0	3.3	3142	41	104
Westpem Nisku E	Gas	12.9	2.9	3103	40	91
Westpem Nisku Un	Gas	12.7	3.4	3149	40	106
Strachan D-3B	Gas	12.0	3.0	4098	49	124
Strachan D-3A	Gas	11.9	3.0	4122	49	124
Wild River Wab A	Gas	11.9	3.5	3611	43	127
Hanlan Wint B	Gas	10.2	2.9	4133	42	121
Minehead BHL Un	Gas	9.8	3.3	4388	43	146
Waferton Wab B	Gas	9.8	2.4	4191	41	101
Wild River D-3 Un	Gas	9.8	2.6	4172	41	110
Lambert D-3A	Gas	9.7	2.8	4431	43	123
Hanlan BHL A	Gas	9.6	3.1	4596	44	144
Hanlan BHL B	Gas	9.6	2.9	4774	46	138
Blackstone Un	Gas	9.5	3.0	4747	45	144
Harley D-3 Un	Gas	9.5	3.1	4635	44	144
Waterton Rund UN	Gas	9.2	2.3	4441	41	104

*EOR
** Suspended

Some Other High Pressure Reservoirs

Pool Gulf Coast	type	KPa/m	°C/100m	Depth m	Pressure MPa	Tem °C
Thomasville, Miss	Gas				120	185
Piney Woods, Miss	Gas		No Depth Data		125	191
S.W. Piney Woods, Miss	Gas				151	196
Geothermal Aquifer,	Water	18.9	3.5	3650	69	126
Texas & LA	Water					
Clinton, Tx	Water	12.9	3.1	4370	56.5	135
Candaleria, Tx	Water	13.6	2.7	4920	66.8	135
Austen Bayou, Tx	Water	12.8	2.5	6190	79.5	157
Johnson's Bayou, LA	Water	13.2	2.2	4920	65.1	110
Rockfeller, LA	Water	15.3	2.5	6370	97.3	160

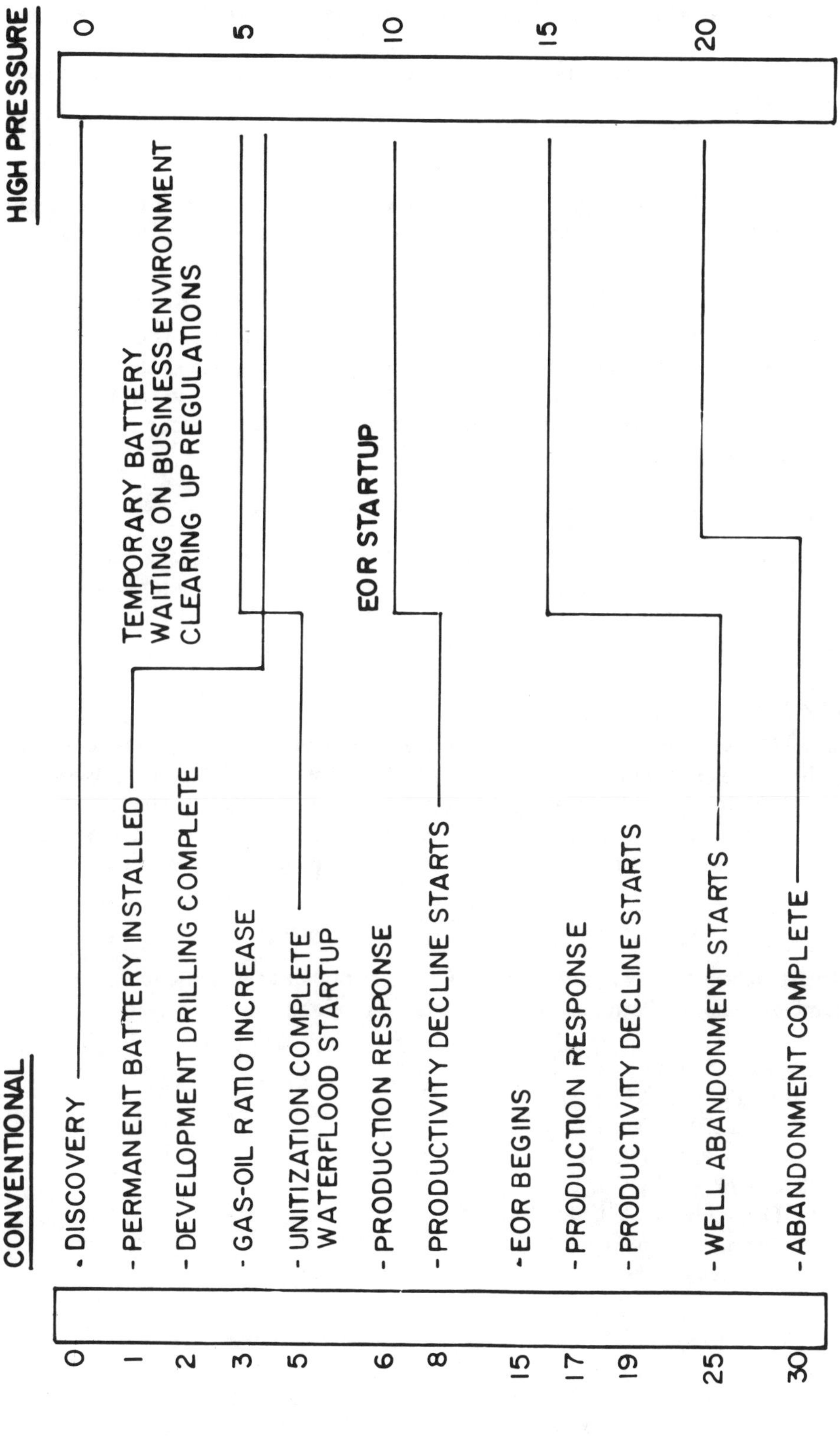

Fig. 11.1: Schematic Timeline Comparison

were found to have pressure gradients exceeding 10 kPa/m. In addition, 23 were found to have temperature gradients exceeding 2.7 deg. C/100 m. Of the six oil reservoirs in this group, two are currently under vertical hydrocarbon miscible flood. One of the gas pools is being cycled.

Expertise

In preparing the following presentation, the importance and properties of high pressure reservoirs were discussed with several engineers from various disciplines.

Development

The most important concern expressed was whether such reservoirs could ever be economically developed. Drilling difficulties are expected to escalate costs, preventing development on small enough spacing to effect optimum depletion. Although the discovery well may indicate considerable potential, reservoir development may not necessarily follow.

Productivity

The number of high pressure reservoirs located in shales or shaly sandstones leads to concern that the high productivity at discovery cannot be maintained for long enough to produce an economic project. These concerns are reinforced by recent studies suggesting permeability to fall rapidly with pressure depletion.

Economics

Development of high pressure oil and gas reservoirs will occur only if the economic cost/benefit relationships are favourable. If high drilling and production costs are not offset by high expected revenues, development will not proceed. The task facing reservoir management is to optimize the costs through efficient placement of wells and design of facilities, and to maximise the recovery by process selection and design.

Units

The Standard International Units (SI) used throughout this presentation may be unfamiliar to some. Pertinent conversions are as follows:

Deg C = (5/9) (Deg F-32)
100 Deg C = 212 Deg F
kPa = (6.85) (psi)
10 kPa/m = 0.43 psi/ft
20 kPa/m = 1.0 psi/ft
1.8 Deg C/100 m = 1 Deg F/100 ft
1 cubic metre = 6.3 barrels
1600 metres = 5000 ft

Meaning in a Reservoir Context

Reservoir Engineering is best defined as a "cradle-to-grave" supervisory responsibility associated with attaining corporate profits. Various other disciplines are involved from time to time in the depletion of an oil or gas reservoir, for example, seismic, drilling, completions, facilities design and installation, operations, accounting, taxation, economics, etc. The economics, taxation and reservoir engineering disciplines are the only ones associated with the full term from discovery through development and operation to abandonment.

High Pressure

The existence of high pressures in the reservoir pore space can have substantial impact on reservoir engineering interpretation of rock and fluid properties, natural drive mechanisms and planning for optimum depletion. The technical risks involved in estimating what has been found (reserves), how much can be recovered (depletion planning), what the recovery will cost (design and estimation), and whether the project is worthwhile (economics) are much greater for reservoirs with pressure over 50 M Pa (7000 psi) than for more conventional 15-30 M Pa (2000-4000 psi) pools.

High reservoir pressure suggests depletion drive to be the dominant reservoir producing mechanism. Plans to inject fluids will probably encounter financial obstacles, if high wellhead injection pressures are required. Conversely, a sizealbe part of the productivity may result from the high pressure itself, as in the case of "expanded" shale reservoirs. Fluid injection may then become part of the "primary" depletion mechanism.

Operating costs for both wells and surface equipment can be expected to be high. If the pressure drop to normal operating conditions is taken across the wellhead, operating costs will result from the need to avoid hydrate formation and other types of plugging. If taken across the separation system, higher yields saleable products can be expected but operation of high pressure battery facilities will attract replacement/-repair costs.

As a result, the optimum depletion programme will probably attempt to achieve maximum recovery in the initial pass. There will probably be no opportunity for the "primary", "secondary" and "tertiary" recovery progression found in more "normal" reservoirs. As a result, gas cycling and solvent flooding with light gases may become "primary" depletion methods. This type of operation is in progress in the recently discovered West Pembina Nisku play.

High Temperature

The data available for high pressure reservoirs suggest the companion problem of high temperature is common. When reservoir temperatures exceed 100°C (212°F) a suite of difficulties develop which complicate depletion plans and facilities design.

High sour gas (carbon dioxide, hydrogen sulfide) concentrations can be expected in such reservoirs. The resulting increase in gas density and compressibility causes the phase behaviour of the reservoir fluid to adversely impact both recovery process design and well productivity.

The need to handle hot fluids on surface further complicates the facilities design, and increases the investment and operating costs. In some cases, the stable operating wellhead temperature may compare unfavourably with the value predicted from pre-design well tests. The result is facility damage resulting from inadequate arrangement of components and expansion loops. Information on high pressure Alberta reservoirs[4] suggests 19 to be "over pressured" (> 10 kPa/km) and 25 to be "over temperature" (> 2.7°C/100 m).

Aquifer Activity

The most probable reason for high pressure in a given reservoir is that the pore volume is disconnected from the pore volume of lower pressure sediments. This may not mean the reservoir is small, although 9 of 12 test wells drilled recently in a Gulf Coast high pressure water zone were suggested to have small volume. All 12 wells encountered reservoirs which had apparent external no-flow boundaries.

As a result, the reservoir depletion program should not count on effective natural water drive unless the closed aquifer contains several billion cubic metres of water-saturated pore space. Any water drive required could be generated by drilling injection wells in an optimum pattern and equipping them suitably with respect to the sour, corrosive, high pressure environment. Since most high pressure reservoirs exceed the normal water gradient, planning should include high wellhead injection pressures.

Management Limitations

As in all projects, the development plans and depletion programme for a high pressure reservoir will be significantly affected by the resulting relationship between costs and benefits for the operating organisation. Questions that must be answered by the reservoir manager are as follows:

(a) Is the productivity of the discovery well high enough to represent an attractive prospect?
(b) Are the reservoir pore volume and potential recovery of saleable product large enough to make development drilling attractive?
(c) Can surface facilities be designed, installed and operated for a reasonable share of the gross revenue?
(d) Can the recoverable reserve be economically improved by using enhanced recovery methods?
(e) What is the supply/demand forecast (volume/price) for the saleable products and injection fluids?
(f) What is the forecast general business environment (regulatory, royalty, tax)?

Although the technology itself is based on physical parameters and is not negotiable, the decisions about how to apply it and how much risk to take require considerable input from financially-oriented management.

Reserve Definition

The discovery well for a reservoir is seldom drilled on an economic basis, but rather is caused by intersection of a number of exploration facets: land, regional geology, play development, rig commitments, etc. The reservoir management may be involved in the decision to set casing, but usually the charisma of attractive open hole logs and tests is the deciding factor. Completion and testing are usually operational responsibilities, and may proceed with little reservoir engineering consultation.

The reservoir management usually begins after initial discovery well tests are complete and authorization is required to spend money on production facilities. At this point, most of the cost/benefit decision questions listed previously are asked and answered based on minimum data. If the cost/benefit relationship is favourable with low risk, facilities construction, development drilling and depletion planning can proceed in parallel. If the risk is moderate, much more attention is paid to the initial depletion plan. If the risk is high, several depletion planning cycles and attempts to improve the business environment may be required before development proceeds.

Management Data

Enlightened reservoir management is impractical at the initial financial decision point. The reservoir engineer usually has little data to work with:

(a) Geologist preliminary estimates of gross pay, net pay, porosity, lithology and relation to the stratigraphic section based on sample cuttings and log correlations.
(b) If a strong structure is present, a seismic gross pay map may be available.
(c) Drill stem test flow rates and pressures, often of questionable stability and depth of investigation.
(d) Core analysis (atmospheric conditions) giving porosity, permeability and lithology.

At this time, initial hydrocarbon in place (OIP) is based on estimated area (A); well pay (H); combined core analysis and log porosity (θ); estimated connate water saturation (S_{wc}); and formation volume factor (B) from correlations.

$$OIP = 10.\ A.\ H.\ \theta.\ (1-S_{wc})/R \qquad (1)$$

Where A = hectares $\quad S_{wc}$ = fraction
H = metres $\quad$ B = Reservoir m^3/surface m^3
θ = fraction

Recoverable reserves (RES) requires an estimates of recovery factor (R) (including plant and fuel shrinkage for gas), usually made on an experience basis.

$$RES = OIP.\ R \tag{2}$$

Once the decision has been made to begin production (usually with temporary facilities) additional information can be derived to characterise the reservoir:

(a) Measured production of reservoir fluids at stable conditions defines initial producing gas-liquid ratio and well productivity.
(b) Associated reservoir pressure measurements permit assessment of completion efficiency, in-situ permeability and production problems.
(c) Samples of hydrocarbons and extraneous fluids at stable conditions identify the reservoir oil and/or gas composition, formation volume factors and the effectiveness of surface facilities in optimising saleable product yields.
(d) Additional samples can be studied in open cells, sand packs and/or reservoir core to identify potential improvement in the reservoir and surface handling processes.

This ongoing data gathering programme influence final facilities design, estimation of recoverable reserves and the overall depletion plan.

Volume Balance

There is a cybernetic principle that states (appropriately amended):

"Operations within a hydrocarbon reservoir can never be defined based on only the input and output."

In other words, infinitely many assumptions about what goes on inside give equally valid performance matches. The "science" in reservoir engineering comprises gathering and interpreting data to describe conditions inside the reservoir and ideal operating mechanisms. The "art" in reservoir engineering is definition of the probable combination of mechanisms operating.

One firm point of reference is the law of conservation of mass and energy. Since mass/energy conversions are unlikely, this can be restated as the conservation of reservoir volume:

$$\text{Expansion} = \text{Withdrawals} - \text{Injection} \tag{3}$$

Withdrawals are the sum over time of

(a) Oil (m^3) × B for oil at subsisting reservoir pressure
(b) Gas (m^3) in excess of that dissolved in the oil at subsisting reservoir pressure × B for gas at subsisting reservoir pressure (4)

(c) Water (m^3) × B for water at subsisting reservoir pressure

Likewise, injection is fun/overtime of

(a) Gas (m^3) × B for injected gas at subsisting reservoir pressure
(b) Water (m^3) × B for water at subsisting reservoir pressure (5)

Expansion is the difference between the initial volume of rock and fluid connected to the wells and their current volume:

(a) Initial oil-in-place (N) × Initial B for oil
(a) Initial oil-in-place (N) × Initial B for oil at subsisting reservoir pressure.
(b) Initial gas-in-place × Initial B for gas
 – Initial gas-in-place × B for gas at subsisting reservoir pressure
(c) Initial water-in-place × Initial B for water (6)
 – Initial water-in-place × B for water at subsisting reservoir pressure.
(d) Initial pore volume for the rock-in-place
 – pore volume for the rock-in-place at subsisting reservoir pressure.

The above description appears to be very complicated, but is a general statement of the foundation on which reservoir engineering methods are built. Even the most complex simulation operates in this way for each small element selected.

The analysis becomes much more meaningful when actual data are applied to calculate the withdrawals, injection and expansion. Each element of the withdrawals and injection uses a measured real value from field reports and a formation volume factor derived from laboratory measurements at reservoir conditions. Near the beginning of the producing life, all of the formation volume factor (B) values are approximate linear functions of pressure (P), for example:

$$Bo = m' \times P + b' \quad (7)$$

By substituting equations of this type into Equations 4 and 5, factoring and recombining, Equation 3 can be rewritten.

$$\text{Expansion} = m'' \times P + b'' \quad (8)$$

Experience also suggests the pore volume change to be a linear function of pressure. The gas and water in place can also be related to the oil-in-place using multipliers. The substitution of linear functions and rearrangement of the expansion terms, in Equation 6, leads to a new version of Equation 3:

$$\text{Oil-in-place} = m \times P + b \quad (9)$$

The solution of Equations 4-6 using the production and pressure measurements obtained during the early depletion period will then yield a succession of oil-in-place estimates. These can be plotted versus pressure, and a best-fit straight line extrapolated to interesect with the initial reservoir pressure to indicate the original hydrocarbon. Extension of the calculations over the early reservoir life can also be useful. Significant deviations from the expected straight line can be used to indicate.

(a) The crude oil bubble point
(b) Onset of water influx
(c) Changes in depletion drive performance

The latter can give interesting data about high pressure reservoirs.

Pulse Tests

The most important void in the information available about a reservoir relates to pay zone continuity. The logs in several wells showing the pay zone to be always in the same structural or stratigraphic location do not provide proof that the wells are in fluid contact through that zone. An almost infinite number of arrangements can yield the same interpretation. The root of the problem lies in the sampling frequency related to a 50 cm^2 wellbore located within 64×10^8 cm^2 (quarter section spacing), or one sample per 100 million. Little statistical significance can be assigned to any interpretation based on such a low sampling frequency.

Information can be gathered in-situ, however, by imposing a step function pulse in a given zone of one well and receiving it at one or more other wells. The step function pulse can be as simple as turning on a well while leaving the others shut-in; changing production rate in one well while leaving the others shut-in or producing at constant rates; or injecting fluid in one well while leaving the others at constant rate. The result is a sharp change in pressure at the test well, repeated in the observation well(s) if any fluid connection exists. Packers must be used above and below any specific interval to be tested, both in the test well and in the receiving well(s). Considerable effort and experience is required to extract the most meaningful information from the location, size, shape and timing of the pulse(s) received.

Similar tests can also be conducted vertically using the injection method to generate the pulse and multiple packers to selectively isolate the pulse and receiver zones. Zones separated by impermeable layers in the wellbore may be found to communicate in the immediate vicinity. Conversely, a reservoir thought to be continuous may be found to really contain several separate pore systems.

Combination of pulse test results with segregated production and pressure data can give new meaning to the initial oil-in-place indicated by volume balance.

Downhole Instrumentation

Measurement devices developed from space program technology are responsible for an increasing capability to analyse reservoir data and define reserves. Ability to measure the rate of decay of thermal neutrons with a TDT log has contributed not only better data on in-situ porosity, but also can estimate the type and amount of fluids present and the location of gas-oil and oil-water interfaces. If waters of different salinities are injected, residual hydrocarbon saturations can be derived. This data indicate the probable recovery value in the reserve calculations.

Another space age device is the crystal pressure gauge, which uses the change in dielectric constant to interpret changes in pressure, usually with very high resolution and accuracy. When protected by a steel jacket, this device can function effectively regardless of downhole environment.

Another method for measuring pressure is to generate a sound pulse within the wellbore and locate the gas-liquid interface from the elapsed time of its return to a microphone pickup. Accurate data is then required to define the pressure gradients in gas and liquid phases.

The repeat formation tester permits capture of fluid samples from various zones in a well. It has built-in packers to isolate the interval sampled. When brought to surface, the samples can be fed into a high-resolution gas chromatograph to give reservoir fluid analysis data, which can then be used to adjust recombinations of the much larger separator samples taken for fluid PVT studies. The tester records the pressure and temperature of each sample at the location taken.

Reservoir Rock Properties

Studies performed with consolidated sandstone cores indicate "abnormal" changes in rock properties at high pressures.

Compressibility

Compressibility of the reservoir rock is most meaningfully expressed as the rate of decrease of pore volume per unit pressure with increasing pressure. In high pressure studies of consolidated cores, the initial rate of change of porosity as pressures increase above atmospheric is low, ($0.5\text{-}7.5\times10$ pv/pw/kPa), suggesting compression of the pore system. At the pressure corresponding to the threshold for plastic deformation of the interpore solids, the slope of the porosity/pressure correlation increases discontinuously.

At yet higher pressures, a mechanical failure threshold is passed, above which the compressibility may not be definable.

Porosity Performance

The results of the rock sample tests at high pressure suggest the pore volume for high pressure reservoirs to comprise three elements:

(a) Pore space defined by the rigid matrix of sand/carbonate grains/crystals and associated cementing minerals.

(b) Incremental pore space caused by elastic deformation and plastic compaction of the inter-pore minerals.
(c) Additional pore space caused by inelastic separation of the solid components resulting from fracturing of the weakest minerals present.

This represents a radical departure from conventional reservoir engineering theory, which encompasses only the first element.

Permeability Performance

The results of high pressure rock tests, particularly the aquifer tests on the Gulf Cost, suggest that pressure depletion results in elimination of the third type of pore spaces. Data presented orally, but not included in the paper, associated this phenomenon with very rapid decline in flow rates from initial conditions, suggesting a significant loss in permeability during the pore collapse phase.

Flow rates then declined much more slowly, suggesting plastic/elastic expansion of the rock solids as per the second porosity type. Depletion was probably insufficient to attain pressures corresponding to Type 1 porosity change.

Fracture Gradient

The pressure/depth ratio required to initiate a fracture in the reservoir rock increases with pore pressure up to the threshold for deformation of inter-pore minerals. Beyond this condition, fracture gradient increases and the fracture initiation pressure tends to coincide with the reservoir pressure.

These considerations are important in drilling high pressure reservoirs, since management of mud weight and rate of drillstring movement is required to prevent mud fracturing and formation damage. The fracture gradient is also important in designing well stimulation treatments.

Production Forecasting

The above core test results suggest very high initial decline rates should be expected in high pressure sandstone/carbonate reservoirs, corresponding to the rapid demise of the Type 3 porosity. This phase should be followed by a period of lower decline rates, perhaps 20-50%/year, corresponding to the loss of porosity, and hence permeability, caused by expansion of the rock solids. Therefore, productivity declines should be more normal (10-20%/year).

Shale reservoirs may exhibit the initial decline performance, but may have very little permeability/productivity once the initial pore collapse phase is completed.

Oil recovery prediction is still quite uncertain. Although the recovery of fluids contained in the Type 3 porosity should be high, no data were found to support this hypothesis. The Berea sandstone work suggests a recovery of 5-7% PV from the Type 2 pores by depletion drive. Recoveries below the threshold deformation pressure will probably be 10-15% by depletion drive, based on conventional reservoir experience. Total depletion recoveries for sandstones/carbonates should range 17-22% and up, depending on the importance of the Type 3 pore volume.

Gas recovery calculations for sandstones/carbonates should include almost 100% of the Type 3/Type 2 pore volume changes and 80-90% depletion drive recovery for the Type I pore space. The nature of high pressure reservoirs suggests low probability of sufficient water drive to significantly reduce the latter values.

These perceptions of probable productivity performance and recovery suggest a strong need to replace reservoir withdrawals from the outset by injecting some relatively worthless substance. This process would prevent collapse of the Type 3 pore volume, and the associated productivity loss. This would be accomplished most inexpensively by injecting very salty water to gain high fluid column weight and reduce surface pumping investment.

Another potentially valuable displacing fluid could be carbon dioxide, if a sufficiently low cost source was nearby. Hydrogen sulfide is yet another possibility. Neither of these fluids is corrosive if no water is present in the injection system. Both exist in the dense phase at least in the upper, low temperature part of the injection string. Low wellhead temperatures could be used to increase fluid column weight and reduce required wellhead pressures. Both fluids, and particularly hydrogen sulfide, act as solvents and could "dry clean" the pore space to maximize hydrocarbon recovery on the first pass.

Hydrocarbon solvents, including ethane, propane, and butane will probably not be required to improve recovery. Methane or nitrogen achieve miscibility directly in most high pressure reservoirs.

Reservoir Fluid Properties

The physical, or pressure-volume-temperature (PVT), characteristics of the fluids encountered in high pressure reservoirs play their own role in reservoir management. Unexpected, non-Newtonian characteristics should be anticipated, since the high pressure affects inter-molecular behavior.

Hydrocarbons

Reservoir hydrocarbon fluids have a range of normally expected properties, including existence of low molecular weight components as gases, separation of low molecular weight materials from high molecular weight materials by well defined interfaces, and ability to identify liquids from gases. In general, this means that reducing pressure on a liquid causes generation of a gas, and reducing pressure on a gas generates a less dense gas. In high pressure reservoirs, some of these concepts are stretched.

Retrograde Condensation: Pressure depletion of reservoir fluid is represented by a vertical straight line on a pressure-temperature (PT) phase behaviour plot at constant volume. Removal of fluid (depletion drive) results in presure reduction. If the reservoir fluid is initially above saturation, a pressure will be reached. (the "dew point pressures") at which the first drop of liquid will form in equilibrium with the remaining gas. As pressure continues to decline, liquid builds up and may exceed 10% of the total pore volume. This is called "retrograde" condensation because it is formed in opposition to the expected behaviour. A maximum is likely to be reached, after which the reservoir liquid content may decline with pressure. If the PT diagram has an appropriate shape, the produced fluid composition may exit this "retrograde" region at the "bubble point" and the reservoir may contain only gas at very low pressures.

To optimize recovery, the reservoir should be kept above the dew point pressure so that no retrograde liquid is lost. The objective is to perform an "end run" around the phase diagram so that the subsequent pressure depletion does not encounter the retrograde region. Since the reservoir temperature cannot practically be altered, the only alternative is to adjust the PT diagram.

The PT diagram can be made to shrink by diluting the reservoir fluid with injected methane or nitrogen. Once the retrograde region no longer intersects the reservoir temperature line, depletion can progress normally. This reservoir process is called "gas cycling".

Injection of carbon dioxide or hydrogen sulfide may tend to aggravate retrograde condensation because these materials favour the dense or liquid phase at high pressures. If they are used as solvents, chase gas (methane or nitrogen) may be injected later to revaporize the residual liquid.

Volatile Oil

In some high pressure reservoirs, it is difficult to tell whether oil or gas is present. The reservoir fluid may be near the two phase region on the PT diagram, so that pressure reduction causes the formation of two phases. If the new phase which forms is gas, the assumption is usually made that the reservoir is an oil pool. If liquid, the reservoir is a gas pool.

Another indicator is the presence of heavy hydrocarbons (C_{20}^{+}) in the liquid phase, and asphalt in particular. These heavy black liquid mixtures can appear in gas pools, but suggest presence of crude oil.

The above distinctions may appear trivial, but can have significant consequences in surface facility design, regulatory procedures and enhanced recovery incentives.

Rich Gas

Once the reservoir is established to be a rich gas pool from PT behaviour, care must be taken to avoid loss of valuable and high-demand natural gas liquids to retrograde condensation. This provides an incentive for gas cycling in high pressure reservoirs.

Crude Oil Source Comparisons

A critical consideration for depletion planning in high pressure reservoirs is the volume of connected pore structure represented. Statistical studies in 1967 indicated that crude oils have a "signature" controlled by the source bed (usually an adjacent shale) from which they originated. If two high pressure wells are sampled and the analysis and PVT behaviour of the recombined reservoir fluid suggest them to be from the same source, the wells are likely to be either in the same reservoir or in reservoirs connected by a common aquifer. This comparison helps in estimating the expansion terms in Equation 3.

Polar Components

The polar components present in a high pressure reservoir are usually as important as the hydrocarbons.

Carbon Dioxide

Carbon dioxide is usually thought of as dense, inert gas, and has many industrial uses relying on these properties. At pressures above 10 MPa (1500 psi) carbon dioxide enters a dense phase region where it has a high affinity for water, can vaporize light and intermediate hydrocarbons from crude oil and can be useful for increasing oil recovery. As pressures and temperatures increase further, the relationship of carbon dioxide to reservoir fluids becomes much more complicated, including the formation of several phases. One of these phases can be dense, pure liquid carbon dioxide, while others have admixtures of hydrocarbons.

Since many high pressured reservoirs have substantial concentrations of carbon dioxide, the latter phase behaviour may become troublesome. Dense liquids that form at wellbore bottom hole conditions tend to aggravate the fluid head in the well, substantially reducing hydrocarbon deliverability. Periodic bottom hole chemical treatments or the installation of liquid lifting equipment may be required, substantially increasing well operating expenses. Unfortunately, unless the well is capable of producing pure carbon dioxide, no offsetting revenue can be realized.

Carbonic acid is formed in the presence of water, and can cause corrosion problems if the produced water contains in unfavourable mixutre of carbonate and sulfate salts.

Hydrogen Sulfide

Hydrogen sulfide is another non-hydrocarbon component common in high pressure reservoirs, often in high concentrations. It can be converted to elemental sulfur, and can form a significant part of the revenue.

Hydrogen sulfide is a colourless, toxic gas at atmospheric conditions, with the characteristic odour of rotten eggs in low concentrations. At higher concentrations it deadens the sense of smell and interferes with the breathing mechanisms, causing unconsciousness and death. As a result, few laboratories have done much work to define its properties, particularly at high pressures. In general, it seems to have essentially the same fluid characteristics as propane.

Hydrogen sulfide contributes beneficially to solvent flood mobilization of reservoir hydrocarbons. It forms part of retrograde condensate when present, indicating its properties to be nearer liquid than gas in high pressure reservoirs. Hydrosulphuric acid is formed when associated with water, which promotes corrosion, particularly in the presence of carbon dioxide and unfavourable carbonate/sulfate mixtures in salt water. Embrittlement of steels is another adverse property.

Water

The water content of hydrocarbon reservoirs can be of several types:

(a) Water associated with the rock structure and its microfractures, usually considered to be part of the rock.
(b) Water located in the connected pore volume but held immobile by the physical forces at pore throats and mineral surfaces, or "connate" water saturation.
(c) Mobile water within the connected pore volume, usually associated with aquifer expansion.
(d) Injected water associated with the drilling and depletion programmes.

The first two varieties have little impact on the reservoir performance, but can cause problems if contact with either of the others causes mineral precipitates to form. The result is usually reduction of porosity, loss of permeability and reduced productivity and recovery.

The water salinity, and the salts of which it is comprised, can be extremely important to surface facility operation. high pressure reservoirs are indicated to have low total dissolved solids but can have much higher carbonate bicarbonate content than conventional brines, leading to scaling problems. As mentioned previously, unfavourable carbonate/sulfate ratio coupled with the presence of acid gases increases corrosion problems. For example, in the SACROC carbon dioxide flood stainless steel wellheads were installed to reduce corrosion.

Reservoir water can also contain significant volumes of hydrocarbon gases, hydrogen sulfide and carbon dioxide. They contribute to the produced gas volume, but a far more important effect is their impact on water compressibility and aquifer energy.

Other Components

The temperatures in high pressure reservoirs tend to cause unusual phase behaviour in materials usually assumed to be inert under more normal reservoir conditions.

Sulphur

Elemental sulphur can become a component of the reservoir fluid in high pressure reservoirs, particularly in equilibrium with hydrogen sulfide. When the fluid enters the well across the pressure drop through the perforations, the phase behaviour favours the solid phase, causing plugging of perforations and wellbore. This can be remedied with periodic wellbore washes with carbon disulfide, an expensive operation. Downhole chemical treatment equipment can be installed to perform this operation continuously.

Sodium Chloride

At high pressures, the phase behaviour relationships between hydrocarbons and reservoir brine become much more complicated. Small volumes of water, including the salt content, may be produced in equilibrium with the hydrocarbon fluids. When this happens, the sodium chloride content may be deposited in the wellbore as a result of the pressure drop, as in the previous description of sulphur deposition.

The cure for this problem is periodic or continuous treatment with fresh alkaline water. If the sodium chloride crystals were to be carried to surface by the flowing hydrocarbons, other erosion and corrosion problems might develop.

Pyrobitumen

Pyrobitumen is the least understood of extraneous compounds found in high pressure reservoirs. It seems to be of asphaltic nature, perhaps with even higher carbon/hydrogen ratios. Another like material is the bitumen derived from coal. In most reservoir rocks, this material is considerrd to be part of the matrix, and can cause porosity reduction in pores and permeability reduction in pore throats.

In high pressure reservoirs, particularly those with low molecular weight hydrocarbons, the pyrobitumen can be plastic at reservoir temperature and pressure. It then is mobilized by high pore channel velocities and enters the wellbore, where it attaches to any solid available. This leads to the conclusion that it must flow as a sticky, high viscosity material.

The only sure remedy is to remove the bottomhole equipment and mechanically clean it. Research is underway to develop solvents to improve in-situ treatment capability. The inevitable result is higher operating costs.

Surface Handling of Fluids

From a hydrocarbon phase behaviour viewpoint, the surface handling of fluids requires several stages of separation to optimise the saleable liquid yield. The equipment installed then depends on the objective. If the reservoir fluid has low molecular weight and the gas processing plant is close, wellhead heating and insulated lines may keep the fluid single phase until it enters the plant. If gathering distances are greater, separation and separate liquid and vapour lines may be required.

Progression through the continuum of reservoir fluids to higher molecular weights reveals no clearcut differentiation between "gas" and "oil" pools. If the higher molecular weight liquids are to be delivered directly to an oil pipeline, multistage separation/treating stages must be designed to maximize liquid yield within the specified vapour pressure limits.

Corrosion and scaling can bear significantly on surface facility design, particularly if the cocktail of hydrocarbons, carbon dioxide, hydrogen sulphide and saline water is very corrosive. Water disposal below surface is mandatory, and is usually into the zone of origin. If replacement with some other fluid is in progress, it is advisable to find a shallower low-pressure zone for disposal of water at minimum expense.

Injection of fluids to maintain pressure and enhance recovery poses little problem from a surface equipment viewpoint. Procedures and equipment are well established to handle high pressure gases and water. Carbon dioxide and hydrogen sulfide are not corrosive as long as they are anhydrous. Nitrogen is readily obtainable from air.

Recovery Predictions

The intent here is not to get deeply involved in all of the considerations required to design, install and operate a recovery project for a high pressure reservoir, but rather to touch upon the concepts required to plan to do so.

Volumes

The recovery fraction of the oil discovered will be determined by the balance between depletion operations and voidage replacement operations.

Depletion operations (Fig. 11.2) represent a "run down" philosophy, with little commitment to conservation or good reservoir management practices. This is not necessarily a bad philosophy, since management may perceive any other course to be too risky. The volume recovered over time usually totals 10-20% of the initial oil-in-place and 80-90% of initial gas-in-place in conventional reservoirs. As discussed above, recoveries from the pore volume attributed to grain separation and to elastic/plastic deformation of the pore space may be somewhat higher. Conversely, the degradation of permeability, and accompanying reduction of reservoir volume that can be drained by each well, may operate to severely reduce recovery.

The other alternative is to inject a relatively worthless fluid to replace voidage and recover the valuable hydrocarbons before permeability degradation occurs.

Displacement Efficiency

The fraction of hydrocarbons displaced from a given pore volume through complete contact with the displacing fluid is termed the displacement efficiency. If the displacing fluid is immiscible with the hydrocarbons (eg. water displacing oil), the remaining hydrocarbon space can be entered only by a different, miscible fluid. Immiscible displacement efficiencies (water displacing oil or hydrocarbon gas) usually range 40-60%. Miscible displacement efficiencies usually range 85-100%.

There is no substitute for data in this area. Risk reduction requires work on rock and fluid samples to define the flow and displacement behaviour of high pressure pore

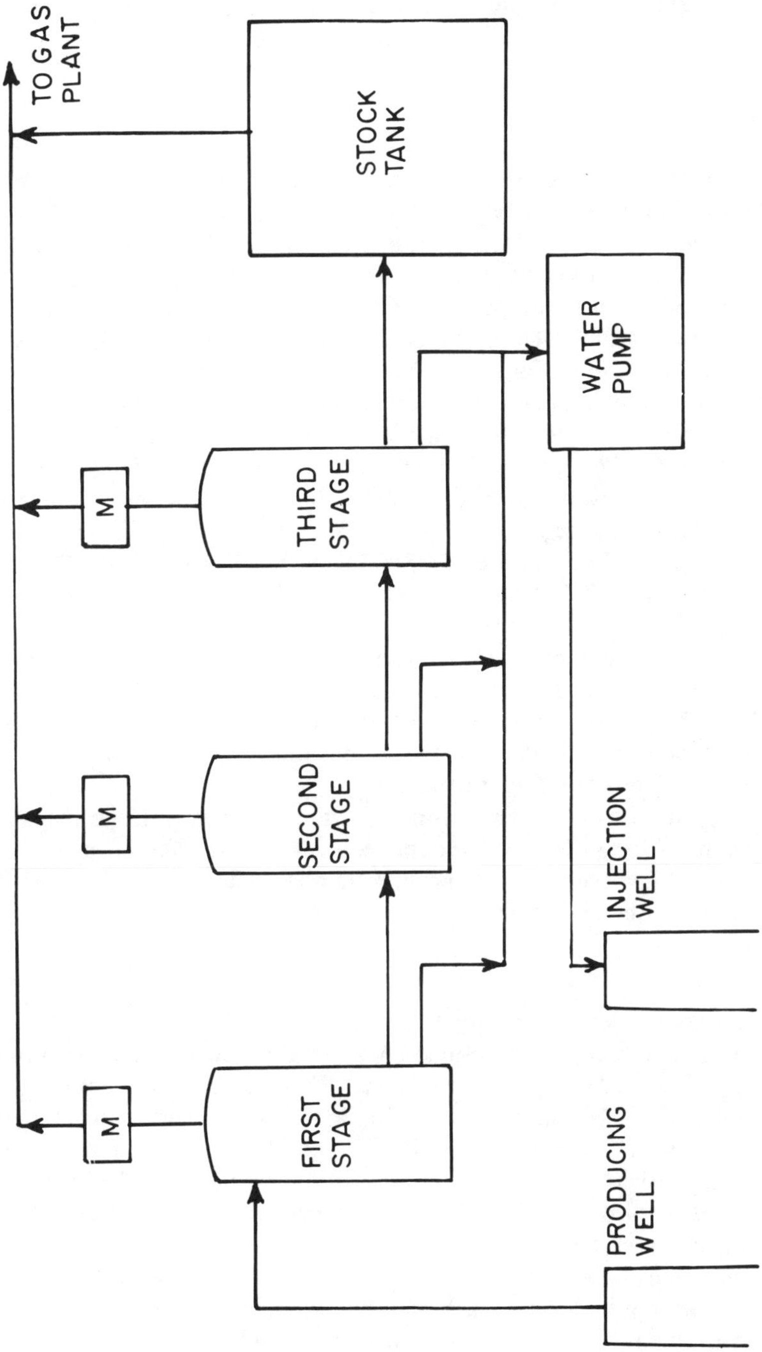

Fig. 11.2: Depletion Operation

systems under operating conditions. The above limits, however, suggest enough benefit over the depletion recoveries to warrant the work.

Areal Sweep

The portion of the areal cross section of a reservoir contacted by displacing fluid is called the "Areal Sweep Efficiency" or "Areal Conformance". The principle that matter and energy cannot be created or destroyed, discussed previously in volume balance form, applies universally to all pore space. The physical principles of "continuity" and fluid distribution within pore space apply equally to conventional and high pressure reservoirs. The solutions for ideal arrangements of injectors and producers in incompressible flow can be found in most text books. Since the fluid is already at maximum compression in high pressure reservoirs, these ideal solutions may be directly applicable. Otherwise simulation is required.

The areal sweep efficiency for a given reservoir is defined by the relative location of the injection and withdrawal points, and ranges from very low values for a one injector/one producer arrangement to 90-100% for tightly spaced "five spot" (one injection well surrounded by four producers) arrangements. Reservoirs usually cannot be developed to contain only full patterns, and some inefficiencies are associated with the edges.

Another factor affecting areal sweep efficiency is mobility, which is the ratio of permeability to viscosity for any given fluid flowing. The mobility of the displacing fluid divided by that of the displaced fluid is termed the "mobility ratio". This parameter has "favourable" impact on the sweep efficiency if it is near 1.0, and "unfavourable" if it is much different from 1.0.

Areal sweep efficiencies of reservoirs with favourable mobility ratios usually range 75-95%. If the mobility ratio is unfavourable, it can sometimes be made more favourable by modifying the effective injection fluid (eg. alternately injecting water and gas to counteract the high mobility of the gas).

Vertical Sweep

The fraction of the vertical cross section swept by the displacing fluid is termed the "vertical sweep efficiency" or "vertical coverage". This factor is influenced to some extent by mobility ratio, because the most mobile fluids will tend to penetrate the reservoir layers fastest, causing instability in the displacement process. This phenomenon emphasizes the impact of vertical variations in the horizontal permeability, and vertical sweep efficiency is most commonly correlated with statistical permeability variation.

The above problems of frontal stability are further aggravated by gravity, which tends to make the lowest density, highest mobility fluids prefer to travel in the uppermost permeable layers. Vertical permeability restrictions caused by reservoir layering may counteract this problem.

In reservoirs of moderate permeability variation (0.6-0.8) and with appropriate mobility ratio control, vertical sweep efficiencies usually range 70-90%.

The calculated recoverable oil volume can be derived by multiplying the initial oil-in-place by the product of the three efficiencies.

Rates

The usual procedure for predicting the rates at which reservoirs will produce is to develop some set of performance equations, or "model", tune it to reality with past performance of the reservoir in question or one similar, and use it to predict the future.

One approach is to use a "stochastic" model, based on assumptions about initial buildup of reservoir productivity as development drilling proceeds, followed by decline of productivity after development ends. Constraints can then be built in, including market restrictions, rate limits imposed by regulatory bodies, producing equipment limitations, the impact of gas conservation orders, etc. The impact of changes in operations can be imposed, followed by return to normal decline rates. The production forecasts generated from such a model tend to become more objective as history progresses, and may initially reflect only the preconceived ideas about success or failure held by their author. Care must be taken in setting up the model, since only a few of the variables are truly independent, and inability to properly define inter-relationships of parameters may give misleading results.

A more deterministic approach is to use a "simulation" model, developed from the continuity equation through the simultaneous solution of many reservoir elements. The response of producing and injection elements to changes in saturation and pressure, however, must still be derived from experience in the pool under study or one similar. The assumption must also be made that the implicit inter-relationships between parameters in the model agree with those in the real world. The potential pitfalls are little different from those of the stochastic model, since the simulator can only quantify the impact of the assumptions made by the author of the forecast.

The above discussion suggests rate forecasting is really an art form. As experience is developed, a system of quality control tests that can be run on critical variables usually develops for a given area. Confidence is attained by sensitivity testing, or the "what if?" process.

Cost/Benefit Relationships

Investments are seldom made for mineral resource development if they are not perceived to be of benefit to those making the investment. A large part of the planning for development of high pressure reservoirs will, therefore, hinge on whether the product marketplace, regulatory environment and business environment permit benefits to be perceived by the investors.

The marketplace comprises the world oil, NGL and sales gas supply, demand, and price structure. The regulatory environment includes such facets as conservation regulations, environmental protection guidelines and import/export controls. The business environment includes provincial royalties, federal income and sales taxes, PGRT, IORT, provincial and federal incentives, land sale bonuses, etc., and is by far the least predictable.

DRILLING
- DEVELOPMENT
- INCENTIVES

PRESSURE MAINTENANCE
- WATERFLOOD
- CARBON DIOXIDE
- HYDROCARBONS
- NITROGEN

FEDERAL INCENTIVES
- INCOME TAX
  - CANADA
  - USA
- WINDFALL PROFITS TAX (USA)
- PGRT (CANADA)

PROVINCIAL INCENTIVES
- ALBERTA ROY. LTY
- SASKATCHEWAN NET PROFITS

Fig. 11.3: Cost/Benefit relationships

The limit for development viability can be assessed from the viewpoint of "social benefit", defined as the value of goods and services generated less the cost to generate them, all exclusive of payments to government and in constant dollars. Ideally, the costs should include every expense, from the cost of exploration rights through discovery, development drilling and the depletion programme, to abandonment. In practice, exploration "ups" are expected to counteract "downs", and all costs up to the creation of the cased discovery well have already been written off against successes.

Drilling

The initial development planning decision is usually whether to complete the high pressure discovery well for production, including installation of a rudimentary battery (perhaps rented). A deliberately conservative depletion forecast (no voidage replacement) is usually developed to define revenue sources, and some rule-of thumb investment and operating cost estimates are made. If social benefit calculations (usually sub-conscious) at this point are negative, the well is usually semi-abandoned and left in the hope that someday it will be worthwhile to produce. If they are positive, the investor's perceptions about the business and regulatory environments are applied, sensitivities investigated, and the well put on production if its ability to return investment on an after tax basis within reasonable time has moderate risk.

The decision to drill followup wells to develop the reservoir is often much more complex. The need to have an overall game plan for the reservoir emerges. Drilling of stepout wells to delineate pool extent way defer the decision to install permanent, full scale battery facilities. Economics calculations at this stage are often performed in reverse, to identify how much reserve and productivity are required to warrant full scale operations. The remoteness of the discovery from other operations is also important, and the perceived probability of other developments in the immediate area.

Pressure Maintenance

Voidage replacement for pressure maintenance and recovery improvement appears to be a cornerstone of depletion planning in high pressure reservoirs. The injection fluid alternatives have been mentioned previously, and are compared below in terms of reservoir depletion.

Waterflood

Water is the most readily available "relatively worthless substance" for use in reservoir voidage replacement. It is non-toxic, has a low vapour pressure and is easily handled with conventional fluid handling equipment. Contaminants are usually unimportant, provided they do not react chemically with the formation or its connate water to form solids that reduce injection well permeability. Most reservoirs eventually produce some water anyway, and the injection/disposal facilities will then be required. The water injection fluid column yields high bottom-hole pressure, enabling injection at substantial rates with low wellhead pressure.

Water is capable of only immiscible displacement, limiting the potential hydrocarbon recovery to 35-50%. It cannot be used in low molecular weight reservoirs, since it operates to reduce the recovery. Many high pressure reservoirs are closely associated with shales, which may be damaged by contact with water.

Carbon Dioxide

Carbon dioxide is a very compressible substance, and at bottomhole conditions in a high pressure reservoir will have an even higher density than water. It tends to extract water and light hydrocarbons from the porespace it contacts, provided it does not become contaminated with either nitrogen or methane. In low molecular weight reservoirs, it may aggrevate retrograde condensation and decrease NGL recoveries.

The main problem with carbon dioxide is that it is expensive to purchase and inject. Its operation as a vaporizing gas drive usually results in considerable breakthrough, at which time it appears in the produced gas and increases the processing cost required to make it saleable. Thus, it costs money to inject and increases operating costs again when it returns. No information is available on performance of high pressure carbon dioxide floods, but there is reason to believe it will probably exist in the porespace as a dense, immiscible pure phase, which will have properties similar to water. There will probably be associated carbon-dioxide-rich hydrocarbon and hydroarbon-rich carbon dioxide phases. These complexities suggest high recoveries will be difficult to attain.

Hydrocarbons

Most high pressure reservoirs contain hydrocarbon fluids which are soluble in methane. The injection fluid is readily available as pipeline sales gas, for about half the cost of carbon dioxide. Methane injection is really a storage operation, since it is saleable when it breaks through and reappears on the surface. Phase behaviour suggests the method to be equally applicable to vertial miscible floods, horizontal miscible floods and retrograde condensate recovery (gas cycling). Recoveries can be expected to be high (70-90%) provided sufficient wells are drilled to assure reservoir sweep. Water can be injected alternatively for mobility control in horizontal floods.

Nitrogen

Most high pressure reservoirs contain fluids which are soluble in nitrogen. The injection fluid is readily obtainable on site from air for about 1/3rd the cost of methane, so requires little pipeline investment. When nitrogen breaks through and reappears in the produced gas, it must be rejected before the gas can be processed for sale. Nitrogen suppliers claim the rejection cost will be low, but the fact remains that the nitrogen will increase operating costs. Water can be injected alternatively for mobility control.

Federal Incentives

Although, some companies still evaluate development drilling and pressure maintenance operations on a before-tax basis, incentives estimates are required in comparing costs to true benefits.

Income tax credits are the most important federal incentive to drill additional wells. The intangible costs are written off in the year incurred, and tangible costs are credited according to a diminishing balance in the corporate account. The well and battery operating costs are also credited in the year incurred, and are deducted from gross revenue before calculating PGRT.

In addition, PIP grants and other encouragements to develop in the frontiers have been offered in Canada. The underlying rationale is that federal incentives should reduce net federal take from the resource development, allowing investors to perceive potential benefits.

The equipment required to inject voidage replacement fluids is treated mainly as tangible, and subject to the diminishing balance calculation. All operating costs, however, including the value of the fluids injected, are given credit for both income tax and PGRT in the year incurred.

Provincial Incentives

The western provinces have, from time to time, offered incentive programs to stimulate the drilling of wells. If the incentive amount is paid in cash, there tend to be compensating changes in investment accounts for federal income tax. Benefits then turn out to be less than superficially apparent. If the incentive is paid as a reduction in royalty payments, no compensating change in federal income tax can occur because it is calculated independent of royalty.

A separate incentives program under Section 4.2 of the Royalty Regulations of Alberta was put in place 77-01-01 to encourage recovery improvement by enhanced recovery methods. Waterflood and immiscible gas flood projects were excluded. The guidelines for administering this incentive permitted amorization of capital investment; injected fluid costs; and injection facility fuel costs to be accumulated using asset of formulas and deducted from gross lease revenue before calculating royalty. This driving force has led to development of over 30 new enhanced recovery schemes in Alberta.

Saskatchewan is finalizing guidelines for enhanced recovery scheme incentives, which will minimize royalty rates for a project until payout. After payout, project royalties will be based on revenue less direct operating costs.

REFERENCES

Beggs, H.D.: Gas Production Operations, OGCI Publications, Oil & Gas Consultants International Inc., Tulsa, Oklahoma, pp. 208-211.

Brennan, R.M. and Annis, M.R.: A New Fracture Gradient Prediction Technique That Shows Good Results in Gulf of Mexico Abnormal Pressure, Society of Petroleum Engineers Paper SPE 13204, Sept. 17-19, Houston, Texas.

ERCB Report ST84-18: Alberta's Reserves of Crude Oil, Oil Sands, Gas, Natural Gas Liquids, and Sulphur at December 31, 1983, Sections 2 and 4.

Fertl, W.H.: Abnormal Formation Pressures, Implications to Exploration, Drilling and Production of Oil and Gas Resources, Elsevier Scientific Publishing Company, New York, 1976, p. 382.

Hammerschaimb, E.C. and Kuuskra, V.H.: The Economic Potential of Reinjection into Geopressured Aquifers, Journal of Petroleum Technology, July, 1984, p. 1164.

Henderson, L.J.: Deep Sucker Rod Pumping for Gas Well Unloading Society of Petroleum Engineers Paper SPE 13199, Sept. 17-19, Houston, Texas.

Hamouz, M.A. and Mueller S.L.: Some New Ideas for Well Log Pore-Pressure Prediction, Society of Petroleum Engineers Paper SPE 13204, Sept. 17-19, Houston, Texas.

Muskat, M.: Physical Principles of Oil Production, International Human Resources Development Corporation, Boston, 1981, p. 922.

Morita, N.: Gray, K.E.; Srouji, F.A.A.: Jogi, P.N.: Rock Property Change During Reservoir Compaction, Society of Petroleum Engineers Paper SPE 13090, Sept. 17-19, Houston, Texas.

Osif, T.L.: The Effects of Salt, Gas, Temperature, and Pressure on the Compressibility of Water, Society of Petroleum Engineers Paper SPE 13174, Sept. 17-19, Houston, Texas.

Place, M.C. Jr. and Smith, J.T.: An Unusual Case of Salt Plugging in a High-Pressure Sour Gas Well, Society of Petroleum Engineers Paper SPE 13246, Sept. 17-19, Houston, Texas.

Podio, A.L.; Weeks, S.G.; and McCoy, J.N.: Low Cost Wellsite Determination of Bottomhole Pressure in High Pressure Gas Wells, Society of Petroleum Engineers Paper SPE 13254, Sept., 17-19 Hauston, Texas.

Randall, T.E. and Gemeroy, C.H.: PVT Analysis Comparison. An Aid to Understanding Reservoir Performance, Annual Technical Meeting, Petroleum Society of CIM, 1971.

Swanson, R.K.; Bernard, W.J.; and Osoba, J.S.: Gulf Coast Geopressure: Old Questions, New Answers, Society of Petroleum Engineers Paper SPE 13090, Sept. 17-19. Houston, Texas.

Glossary

A

ABC Method: A weathering computation method whereby the effect of irregular weather thickness determined from shots above the base of weathering.

Abnormal events: Coherent events which are not reflections but refractions, reflected refractions, etc.

Abyssal: Belonging to the environment below 4,000 m water depth in which the only sediments are red clay and the deep-sea oozes (Abyssal deposits). At this depth the temperature does not exceed 4 degrees C and only specialised animals are found (Bathyal).

Abyssal deposits: Sediments formed at depths greater than 2,000 m in the deeper part of the bathyal zone and the abyssal zone itself.

Abyssal rocks: Rocks formed at very great depth. A synonym of plutonic rocks.

Accessory minerals: Minerals present in a rock in such small amounts that their presence or absence is not significant when considering for classification purposes.

Accommodation structures: Small structures such as folds and faults, which permit beds to occupy completely the available space during major folding or faulting. It is impossible for holes to develop in the structure and any such tendency is prevented by the development of accomodation structures. Similarly, in the cores of tight folds structures develop which allow the excess material in the core to be accommodated.

Accicular: (mineralogy) Needle-like.

Acid rock: An igneous rock with 10% or more free quartz. The term arises from the concept of the silica as an acidic oxide, i.e., in theory, together with water, it can form a range of 'silica acids' and thus the minerals forming the rocks were regarded as salts of these acids. If therefore a rock contained excess silica, it was regarded as having an excess of the acidic 'principle'. The name persists although the theory behind it is certainly untenable. Acid rocks were formerly defined in terms of silica percentage—any rock containing more than 60% Silica being termed acid (Granite, Granodiorite, Microgranite, Pitchstone, Rhyolite).

Acoustic impedence: Seismic velocity multiplied by density. Reflection coefficient depends on changes in acoustic impedence.

Aeration: The technique of injecting air or gas in varying amounts into a drilling fluid for the purpose of reducing hydrostatic head.

Aeromagnetic: Magnetic measurements made from an aircraft.

Aerolites: A class of meteorite.

Aeolian deposits: Sediments deposited after transport by wind (Dunes).

Agglomerate: Pyroclastic rocks consisting mainly of fragments larger than 2 cm in diameter.

Airwave: Energy from the shot which travels in the air at the velocity of sound.

Airborne magnetometer: Device used to measure variations in the earth's magnetic field, while being transported by an aircraft.

Alkali basalt: The term is here used to cover all igneous rocks of generally basaltic character which contain feldspathoids.

Allochem: Transported particles, pellets, skeletal detritus, oolites etc. which represent the framework of the rock.

Allochthonous: A term applied to the material forming rocks which have been transported to the site of deposition.

Allotriomorphic: A term applied to mineral grains showing no development of crystal form whatsoever.

Alluvial deposits (alluvium): Deterital material which is transported by a river and deposited usually temporarily at points along the flood plain of a river. Commonly composed of sands and gravels. (River terrace; rejuvenation; alluvial deposits. Many important ore locally found concentrated in alluvial deposits.

Analog: A continuous physical variable which bears a direct linear relationship to another variable such as earth motion so that one is proportional to another. Continuous as opposed to discrete or digital.

Angoumian: A stratigraphic stage name for the European Upper Turonian (Cretaceous system).

Anhydrite: Evaporite mineral found in sedimentary rocks associated with gypsum.

Anisotropic: Applied to a substance having different physical properties when measured in different directions.

Annulus: The space surrounding the drill pipe or tubing suspended in the hole. The outer wall of the annulus may be the wall of the open hole or inner wall of a casing string.

Annular velocity: The Velocity of a fluid moving in the annulus.

Anticline: A fold or fold system in the form of an arch.

Antistress minerals: Minerals formed under conditions of metamorphism where stress was entirely absent.

API gravity: The gravity of crude oil or other related fluids as measured by a system recommended by American Petroleum Institute. It is related to specific gravity by the following formula.

$$\text{API degree} = \frac{141.5}{\text{Sp. gr. } 60^\circ\text{F}} \qquad 131.5$$

Aquifer: A water bearing bed of strata either by its virtue of porosity or because it is pervious.

Aquitanian: A stratigraphic stage name for the top of the Oligocene to the base of the Miocene.

Arenaceous rocks: A group of detrital sedimentary rocks typically sandstone in which the particles range in size from 1/16 mm to 2mm.

Argillaceous rocks: A group of detrital sedimentary rocks commonly clays, shales mudstones and marls.

Anchor: Portion of Drill Stem Test tool assembly below the packer, extending to the bottom of the hole.

Artesian Structures: A series of sedimentary rocks disposed in such a way that an aquifer holds water under a pressure head between two layers of impermeable strata.

Artificial lift: The pumping of a well to bring oil to the surface when natural pressure is not sufficient.

Assimilation: A process by which material is incorporated into an igneous rock by a process of melting and solution.

Authigenic: Generated on the spot. Refers to sediments consisting of minerals deposited directly out of solution by chemical process.

Authochthonous: A term used to imply not transported; in situ.

Automorphic: A term used to describe the texture of a rock in which the mineral grains display fully developed crystal form.

B

Bailer: A long cylindrical container with a valve at the bottom, which is used for removing the water, cuttings and mud from bottom of a cable tool well.

Bank: In situ limestone deposit formed by organisms which do not have the ecological potential to erect a rigid wave resistant structure.

Bar: A more or less linear deposit of sand and or gravel generally found in the sea which is parallel or subparallel to the coast line.

Barrier: An emerged ridge of sand, extending parallel to the coast.

Basalt: A fine grained, glassy basic igneous rock.

Base of Weathering: The boundary between low velocity surface layer and an underlying comparatively high velocity layer. May or may not correspond to the geological base of weathering or to the water table.

Base level: The surface to which long profiles of rivers are related. The base level of a stream is usually the sea.

Basin: A depression of large size which may be structural or erosional origin. The sediments and the associated fauna which is found in the deeper part of the basin is called basinal fades.

Bathyal: A term used to describe the environment of the continental slope between 200 and 400 m.

Bays: Minor disturbances in the geomagnetic field that are regional in character.

Bayon: An ox-bow lake developed in a delta.

Benthos: Those animals which live on the sediments of the sea floor including both mobile forms.

Bentonite: A special assemblage of clay minerals, used in the drilling fluid for making the main body of the fluid.

Biofacies: A rock unit which contains an assemblage of charactristic of a particular environment.

Biogenic: A term applied to material produced by the action of living organism.

Bioherm: Synonymous with large organic reef. The term applies for organisms in situ.

Biostrome: Stratified deposits such as shell beds, crinoid beds and coral beds consisting of and built mainly by organisms mostly sedimentary biostromes which do not have mound like or lense like forms.

Bit: The drilling tool that actually cuts the hole in the rock.

Bitumen: Originally, native mineral pitch tar or Asphalt.

Bituminous: Containing oil or yielding oil on distillation.

Blanket sand: A thin layer of a sandstone horizontally extensive in two dimensions.

Bonebed: A sedimentary layer characterised by a high proportion of fossil beds, scales, teeth etc.

Brachyanticline: A long narrow anticline as distinguished from the circular dome or anticline.

Bradenhead: First fitting or flange attached to the top of surface casing.

Brea: A viscous asphalt formed from the evaporation of oil seepage.

Bromerite: A frogenic rock with microcrystalline ooze matrix. Signifies either that fossils were sedimentary or that currents were calm in depositional area (Folk 49).

C

Cainozoic: The division of geological time which succeeds the Mesozoic and ends at the Quaternary. The duration is approx 63 m.y. to 2 m.y. It is commonly used as a synonym for Tertiary.

Calcarenite: Clastic limestones of grain-size 1/16 mm 2 mm.

Calcilutite: Limestone composed of 50% or more clay size calcareous particles.

Calcirudite: Limestone having grain size 2mm or more.

Calcisiltite: Mechanically deposited limestone composed of 50% or more of silt sized calcareous particles.

Canyon: A deep valley with vertical sides excavated by a river generally associated with rejuvenation.

Caprock: An impervious formation overlying directly a reservoir rock which acts like a seal.

Carbonisation: The reduction of organic tissue to a carbon residue.

Casing pressure: It is the pressure at the top of the well within the tubing and casing annulus.

Cation: The positively charged particle in the solution of an electrolyte which under the influence of an electrical potential moves towards the Cathode (negative electrode) examples are Na, H, NH, Ca, Mg, Al.

Chorology: The study of geographical distribution of organisms.

Christmas tree: The assemblage of valves, pipes and fittings assembled at the top of the well to control the flow of oil and gas.

Clay gall: When a patch of clay or mud dries out, the upper surface cracks and peels away from underlayers. These thin layers of hardened clay may occassionally be transported a short distance and deposited in a sand or similar sediment in the form of flat or lens shaped clay galls, generally oriented parallel to the bedding.

Closed basin: A basin not connected to larger body of water.

Closure: It is the vertical distance between the highest part of the fold and lowest continuous encircling structural contour.

Coal: The general name given to a stratified accumulations of carbonaceous material derived from vegetation.

Columnar structure: A structure seen mainly in lava flows and sill but most commonly in basic rocks.

Compaction: Compaction is a reduction in bulk volume of sediments caused mainly by vertical forces, exerted by an increasing overburden. It is conveniently expressed as a change in porosity brought out by tighter packing of the grains.

Competent bed: A rock layer during folding flexes without appreciable flow or internel shear.

Concordant: A term applied to intrusive igneous rocks where the margins of the intrusion do not cut across the bedding or foliation of the country rock.

Concretions: Certain constituents of sediments tend to concentrate in certain parts of the rock, often accumulating around nucleous. The masses so formed may be rounded or irregular and are called concretions.

Condensed sequence: A stratigraphic term used to describe a series of beds which show a relatively much thinner development than the equivalent beds elsewhere.

Conductivity: A measure of the quantity of electricity transferred across unit area per unit potential gradient per unit time. It is the reciprocal of resistivity.

Conformable: A sequence of beds are said to be conformable when they represent an unbroken period of deposition.

Conglomerate: Rudaceous rocks consisting of rounded or subrounded fragments implying rather more transport than breccias.

Coning: Excessive rates of oil production may bring about a local rise of oil-water. contact gushing of water from bottom section of the stratum called coning. The process of coning is either limited or impossible where penetrated stratum has poorly permeable layers.

*Connate water:*Water trapped in the sediments at the time of deposition.

Continental: A term commonly used in geology in the sense of land, implies condition on a land mass as opposed to oceans.

Continental drift: A hypothesis put forward to account for various major features of the earth surface e.g., the distribution of the continents and ocean basins in which the continent as we now see them are believed to have been formed by the break-up of one large land mass and to have drifted into their present positions. Amongst the evidence which has been put forward in support of the theory is the distribution of the permo-carboniferous glaciation in the southern hemisphere, the apparent wandering of the poles, as revealed by place—magnetic studies, the morphological 'fit' of South America and Africa and the distribution of certain plants and animals, both as fossils and at the present time.

Continental margins: Base of the continental slope at approximately 200 m water depth marks the approximate edge of continents. Both continental shelf and slope referred to as the continental margin structurally part of the continents even though covered by sea water.

Continental Shelf: It is that area of the sea or ocean floor between that mean low water line and that sharp change in the inclination of the floor that marks the begining of the continental slope. The sharp change in inclination, from about 1/8th to 10 to more than 3 degrees ocean at various depths but usually between about 130 and 200 meters. The width of the shelf ranges from less than a mile to 800 miles.

Continental rise: Aprons (built of sediment from the continents) formed at the mouths of many canyons, coalesce to form a thick sediment deposit, called continental rise, The sediment cover is thick enough to bury pre-existing bottom irregularities for many nearly featureless bottom. They are common in North Pacific Ocean, Atlantic and North Indian Oceans.

Continental Slope: It is usually from 10 to 20 miles wide, extends from outer edge of the continental shelf to the abyssal ocean floor. The inclination of the slope, varies from 30 to over 45 degrees.

Coqunna: Carbonates consisting wholly or almost entirely of mechanically sorted fossil debris.

Constorast: Concretionary limestones generally formed under arid conditions, typical of the Devonian and Permo-Trias of Great Britain.

Crater: A depression which is more or less circular in plan.

Crest plane: The plane connecting the crest lines of successive beds in an anticlinal structure.

Cross bedding: A series of inclined bedding planes having some relationship to the direction of current flow, the angle of rest of the sediment and rate of supply of sediment.

Cryptocrystalline: A term used to describe a very finely crystalline aggregate in which crystals are so small as to be indistinguishable except under powerful magnification.

Culmination: The result of coincidence of two anticlines belonging to different sets of folds in cross folding.

Cycle of erosion: A concept in which the various stages in the erosion of a region are regarded as parts of a cycle process.

Cyclothem: A series of beds formed during a sedimentary cycle and showing many repetitions of roughly the same sequences in the stratigraphic column.

D

Decollement: The plane of dislocation caused by an upper series of rocks folding and in the process sliding over a large series of rocks, which may be unfolded or only slightly folded. It has been likened to the crumpling of a table cloth when pushed over a table cloth.

Delomitisation: A process whereby a dolomite bearing rock is converted to a calcite bearing rock either by metamorphism or diagenesis.

Deep sea deposits: Abyssal deposits.

Deformation: A structural term used to describe any change in the attitude, shape or volume of a bed or layer after its formation.

Delta: An alluvial deposit at the mouth of river usually of triangular in shape.

Deltaic sediments: Formed in a river delta sediments found are in general sands of various sorts, clay materials, silts together with a certain amount of debris.

Denudation: The sum total of the process which result in general lowering of the land surface. It is normally taken to consist of the processes of weathering, transportation erosion.

Dendrite: Refers to a valley drainage pattern that resembles the branches of a tree or the lines of broad leaf.

Depression: The result of coincidence of an anticline and syncline.

Detrital: A term applied to any particle of minerals or more marely rocks, which have been derived from preexisting rocks by process of weathering erosion.

Diachronous: A term applied to lithological units (e.g., a bed of sandstone, a reef limestone) which appear to be continuous beds but which, in fact, represent the development of the same facies at different places at different times, i.e., the lithological unit transgresses the absolute time planes.

Diagenesis: Those processes affecting a sediment while it is at or/near the earth's surface, i.e., at low temperature and pressure. Lithification is the term used to describe those changes which result in the formation of massive rock from a loose sediment. Two important processes in diagenesis operate (1) compaction (2) Cementation.

Diapire: An intrusion which domes the overlying cover after piercing lower layers. It may develop narrow neck at depth, giving a balloon-like form.

Diastem: A term often used for an unconformity produced by a period of non-deposition which is localised and/or short in duration.

Diastrophism: The large-scale deformation of the earth's crust which produces continents, ocean basins, mountain ranges, etc.

Differential erosion: The features develop in rocks which have varying resistance to the agencies of erosion and/or weathering so that parts of the rock are removed at a greater rate than others. A typical example is the removal of soft beds from between harder beds in a series of sedimentary rocks.

Digital recording unit: The basin information through this is collected in terms of digits by using different mathematical filters. We can discard and accept the appropriate events. It is ideal for mapping out sand lenses, pinch outs, deeper horizons, poor reflecting horizons, by using different filters.

Dip: The true dip of a plane is the angle that it makes with a horizontal plane—the angle being measured in a direction perpendicular to the strike of the plane. Apparent dip is the angle measured in any other direction. Given the strike and an apparent dip or two apparent dips, it is possible to obtain the true dip.

Dip slope: A topographic surface which dips in the same direction as the underlying beds and is often more or less parallel to them.

Discordant: A term used to show the sudden variation in the dip.

Dismicrite: Consists predominantly of microcrystalline calcite but contains irregular patches, lenses of sparry calcite.

Dispersion: Violet light is refracted slightly more than red light and hence refractive index of a transparent material differs for red and violet lights. The difference between the refractive indices is termed as dispersion.

Dog leg: The 'elbow' caused by a sharp change of deviation the well bore.

Dolomitisation: The process by which an original Calcium carbonate rock is converted into the double calcium magnesium carbonate, either wholly or in part.

Drift: As applied to movements of continents, (1) continental drift. (2) Any slow movement of surface water, sand, etc., or any deposit which has been transported by wind. (3) Specifically, all the glacial and fluvio-glacial deposits left after the retreat of glaciers and ice sheets.

Drill pipes: A string of drill pipe, screwed together and extending from the rig floor to drill collar and bit at the bottom of the hole.

Drill stem: A cylindrical bar of steel or iron screwed into the cable tool bit to give it weight.

Drill stem test: A test of the productive capacity of a well when it is still full of drilling mud. The testing tool is lowered in the hole attached to the drill pipe and placed opposite the formation to be tested. Packers are set to shut off the weight of the drilling mud and tool is opened to permit the flow of any formation fluid into the drill pipe, where the flow can be measured.

Dunes: In areas of loose, unconsolidated surface deposits, e.g., sandy deserts, a well-established prevailing wind will tend to heap up the sand into regular accumulations which are known as dunes.

Dyke: A sheet like body of igneous rock which cuts across the bedding or structural planes of the host.

E

Earth:	It has radius	6357 Km towards North pole
		6378 Km towards equator
	Mean density	5.517 gms/ cc
	Mass	5.977×10^{27} gms
	Age	4-5 billion years

Earthquake: A series of shock waves generated at a point (The focus) within the earth's crust or mantle. The point on the surface of the earth above the focus is called the epicentre. The three main types of wave motion generated by an earthquake are (A) L-waves (B) S-waves (C) P-waves.

Electro telluric currents: Electric currents generated by a sub-terraneous accumulation of electro chemically active minerals, such as hydrocarbons, sulphide ores, geothermic electric effects, redox potential contrasts, etc.

Ellectrotelluric surveys: Surface exploration procedure from which vertical directions and relative intensities of electro telluric currents are determined and mapped at various perspected depths.

Ellipsoid: A solid figure of which all plane sections are either ellipses or circles used for illustrating the variation of a property. The major axis of such a solid represents the maximum value of a variable, defines a plane. The intermediate axis is chosen so as to lie perpendicular to this plane.

Eluvial deposits: Weathered material which is still at, or near its point of formation. The term is especially applied to deposits of economic substances. (Placer deposits, Alluvial deposits, Colluvial deposits).

Embayment: A deep depression in a shoreline forming a large open bay.

Environment: (1) The place in which an animal or plant lives, e.g., in burrows in sand, in shallow marine or tropical waters. (1) The environment may be continental (environments, continental) or marine, (Littoral, Continental shelf and continental slope, Bathyal, Abyssal) (Adaptive radiation, Habitat) (2) The term has come to be commonly applied also to the condition of deposition of a sediment.

Eperic Sea: A shallow inland sea.

Epirogeny: The uplift or depression of continental or subcontinental land masses as a result of widespread adjustments of land. Epeirogenic movements are generally even in character, producing little more than tilting, slight warping and minor faulting of the rocks.

Epicontinental: Situated within the limits of a continental mass, e.g., an epicontinental sea is one covering past of a continent, one covering the continental shelf.

Epigenesis: Includes all processes at low temperature and pressure that affect sedimentary rocks after diagenesis and upto metamorphism.

Equipotential surveys: Surveys exploration procedure from which horizontal distribution and relative intensities of electro telluric currents are determined and mapped in subsoil.

Equivalent circulating density: For a circulating fluid, the equivalent circulating density in lb/gal equals the hydrostatic head (psi) plus the total annular pressure drop (psi) divided by the depth (ft) and by 0.052.

Erosion: That part of the process of denudation which involves the wearing away of land surface by the mechanical action of transported debris.

Escarpment: A more or less continuous line of cliffs or steep slopes facing in one general direction and due to erosion or faulting. The abbreviated form 'Scarp' is sometimes used as a synonym but is more commonly limited to cliffs produced by faulting.

Estuarine: Pertaining to an esturary or the mouth of a river.

Eithinic: A term used to describe an environment characterised by the presence of large volumes of stagnant water which is de-oxygenated, giving rise to reducing conditions. The characteristic sediments are black carbonaceous pyritic muds. Such conditions exist at the present in the deeper parts of the Black Sea. Perhaps parts of the British. Rhaetic deposits represent a fossil example.

Evaporite: A sediment resulting from the evaporation of saline water. Most evaporites are derived from bodies of sea water, but under special conditions lakes may give rise to bodies of borate minerals, sodium carbonate minerals, and other similar types. There are two views as to the origin of thick beds of evaporities.

One suggests formation in a Sabkha environment, the other suggests formation by the evaporation of an enclosed body of saline water. Evaporites of the second kind begin to form when sea water is concentrated to about 50% of its original volume.

The minerals are formed in the reverse order of their solubilities, i.e., the least soluble first. It is probable that the original character of most evaporite deposits has been destroyed by replacement through the agency of circulating fluids, e.g., polyhalite (K_2 So_4 Mg So_4. $2CaSo_4$ $2H_2o$) is not a primary evaporite as it is formed by the reaction of appropriate solutions with Ca, So_4 minerals. Experimental work shows that the temperature at which crystallisation takes place and the concentration of the various ions present control the nature of the actual substances deposited, e.g., pure solutions containing calcium and sulphate ions will deposit gypsum ($CaSo_4$. $2H_2o$) below 42 degrees and anhydrit ($CaSo_4$) above this temperature. The presence of quite a small proportion of Sodium Chloride (NaCl) depresses this temperature considerably.

A typical evaporate sequence is as follows:

4 Potash and magnesium salt (kayanite, Carnallite, Sylvite, Kieserite)
3 Rock Salt (Halite)
2 Gypsum of anhydrite
1 Calcite and dolomite

Extinction: When the vibration directions of the ordinary and extraordinary rays of an anisotropic mineral are parallel to the vibration directions of the polars in petrological microscope. No light reaches the eye and the mineral is said to be in extinction. This is because light which can pass the polariser (vibrating E-W also passes the mineral (Following one of its vibration directions) but is stopped by the analyser, which has a N—S vibration direction. This phenomenon occurs every 90 degrees of rotation of the mineral.

F

Facies: The sum total of features such as sedimentary rock type mineral content sedimentary structures bedding characterises fossil content etc. Which characterise a sediment as having been deposited in a given environment. Facies which are particularly characterised by rock type are referred to as lithofacies whereas those especially characterised by their fauna are called biofacies. It should be noted that a particular lithofacies may be diachronous and this may be only detectable if the fossil evidence is adequate. Study of the distribution of facies leads to reconstruction of past geographies.

Fault breccia: During the process of faulting, the rocks are commonly broken upto a greater or lesser extent. If the rocks on either side of the fault plane are hard, the fragments produced will be large and angular and are termed fault breccia.

Fingering: The process of separating suspended solids from the liquid by forcing the later through a porous medium. Two types of fluid filtration occur in a well, dynamic filtration while circulating and static filtration when not circulating.

Fingering: Due to different permeability of the rock over the area of stratum and different productivity of wells, edge water penetrates to the centre of oil reservoir through most permeable sections of formation. Such water sections look like fingers running from oil drainage boundary and this phenomenon is callad fingering. Such fingers can divide the oil bearing section of a field into different blocks.

Fishing: Searching for a piece of drill pipe or drilling equipment broken off from the drilling tools and left in the hole.

Flocculates: Groups of aggregates or particles in suspension subject to being up by normal shaking and stirring and reforming on standing.

Flocculation: The change which takes place when the dispersed phase of a colloid forms a series of discrete particles which are capable of settling out from the dispersion medium.

Flow Structure: The structure arising when directional movement of a liquid containing crystals causes these crystals to take up a parallel orientation. It is not confined to volcanic rocks.

Fluviatile: Sedimentary deposits laid down by a river or stream are termed as Fluviatile deposits.

Flysch: This term originated in Switzerland and is applicable only to sediments associated with Alpine orogeny. This is also termed as greywacke suite because it is typical of many geosynclines.

Foliation: Parallel orientation of platy minerals or mineral banding inrocks.

Foredeep: A deep oceanic trench on the convex side of an island arc.

Formation Volume Factor: The factor applied to convert a barrel of gas free oil in stock tank at the surface into an equivalent amount of oil in the reservoir. It ranges between 1.14 and 1.16.

G

Gas Cap: An accumulation of gas above an oil pool.

Geanticline: The broad uplifted area bordering geosyncline which supplies sediments for its infilling.

Geochemistry: The primary purpose of geochemistry is on the one hand to determine the composition of the Earth and its parts, and on the other to discover the laws which control the distribution of individual elements. The term was introduced by the Swiss Chemist Schobein in 1838. With the more refined techniques of modern analytical chemistry, detailed studies can be made of the distribution of many elements often present in quantities of only few parts per million. This has resulted in information becoming available concerning not only the distribution of elements in rock typcs but also their distribution amongst the major crystal units on the one hand, and in various latter types (Silicates) on the other. Refinements of technique have also made it possible to establish the relative abundance of the individual isotopes of the elements.

Geophysical prospecting: The study of variations in the values of the physical parameters of the crust of the Earth is done with the object of gaining information for economic purposes, about sub-surface structure or the presence of mineral bodies. Methods employed involve the study of (a) ambient fields and (b) the effects of applied fields. Examples of (a) are, gravitational, magnetic, electrical (both large scale telluric currents and localised currents due to the spontaneous potential associated with certain ore bodies). Natural radioactivity, temperatures gradients and heat flow. Examples of (b) are, seismic (by the flow application or shock waves), electrical (both the direct application of DC or AC currents-resistivity methods and by electro-magnetic induction), radioactivity (by excitation by a suitable source). Techniques include aerial surveys, both land and marine surface surveys, and surveys in bore holes.

The objectives of geophysical prospecting include the search for ores of economic importance, the delineation of potential oil and gas-bearing structure, the location and nature of underground water supplies, the estimation of reserves of natural construction materials, the investigation of shallow structural conditions affecting civil engineering works and the determination of the physical properties of foundation materials.

Geosynclines: Major structural and sedimentational units of the crust of the earth. They consist of elongated basin which become filled with very great thicknesses of sediment and because of this fact progressive subsidence of the basin floor must have occurred. Volcanic rocks are found intercalated with sediments. The accumulated pile of sediments is subsequently strongly deformed by orogenic forces into a fold-mountain chain, and during this process the lower portions of the sedimentary pile may become highly metamorphosed and granite emplacement may take place. It has been suggested that the Gulf of Mexico is a modern example.

Geothermal gradient: A measure of rise of geo-temperature with depth below the earth's surface.

Graben: A downthrown block between two parallel faults (Rift valley).

Gravel: This term is used in geology with a precise meaning of a grain size, and not in implying loosely compacted coarse sediment. In general, the term is applied to grains larger than coarse sand and finer than pebbles, i.e., 2 mm to 4 mm. The term however, is also used stratigraphically and by civil engineers in a much looser sense, e.g., river gravels and glacial gravels which range from medium sand to boulders (particle size).

Gravel packing: Placing of gravel opposite a producing horizon to prevent or retard the movement of loose sand grains into the well bore along with the oil.

Gravity (gliding) tectonics: The operation of gravitational forces to produce faulting, thrusting and folding by the movement of the cover down the slope of an unlifted region.

Gravity transport: Rock material may move under the influence of gravity either as a movement of weathering products down a slope or as a mass movement of rock along joint planes, bedding planes etc.
Gumbo: Any relatively sticky formation such as clay, encountered in drilling.
Guyot: A submarine table mountain. They occur in the deep oceans, especially in Pacific, and may perhaps be the foundation of coral atolls.

H

Heaving: The partial or complete collapse of the walls of a hole resulting from internal pressures due primarily to swelling from hydration or formation gas pressures.
Hydraulic limestone: A limestone containing sufficient siliceous and alluminous impurities to yield a true cement (one which will set under water).
Hydrogeology: The study of the geological factors relating the Earth's water.
Hydrology: The study of all waters in and upon the Earth. It includes underground water, surface water, and rainfall, and embraces the concept of the hydrological cycle.

I

Imbricate structure: Fault (Thrust fault) (2) A fabric found in conglomerates and pebble beds when these consist of fragment having a noticeable elongation, deposited under the influence of a powerful current, the long axes of the pebbles tend to be more or less parallel with one another leaving in the direction of current flow. It has been suggested that under certain circumstances sand grains might show a micro-imbricate structure.
Impregnation: The infilling of pores in a rock by mineral matter, or the replacement of existing pore filling material. Not equivalent to cementation (Diagenesis) the term is usually used with the sense of a later event, especially the introduction of an ore mineral or liquid, e.g., oil.
Incompetent bed: A rock layer that, during folding, flexes with appreciable flow and shear and hence develops slaty cleavage. (Competent bed)
Inconsequent drainage: A drainage system which does not conform to the structural pattern of the region, although, it may show local small adjustments to it. Two types may be recognised.
1) Superimposed drainage
2) Antecedent drainage.

Index fossil: A fossil species which characterises a specific horizon by its abundance, although, not restricted to it in time.
Index mineral: A mineral, occuring in a series of metamorphic rocks, whose first occurrence marks the outer limit of metamorphic zone as one moves in the direction of increasing metamorphic grade.
Induration: A process by which soft sediment becomes hard rock. Originally the term implied hardening due to baking, i.e., proximity to an igneous rock, but nowadays it

is used more generally to include hardening by pressure and cementation (Diagenesis)

Inhibited Mud: A drilling fluid having an aqueous plane with a chemical composition that tends to retard and even prevent appreciable hydration or dispersion of formation clays and shales through chemical and or physical means.

Inlier: A limited area of older rocks completely surrounded by younger rocks. It may be produced by erosion-faulting, or few folding or a combination of any two, or all three, of these agencies.

In situ (Latin, 'in place'): Geologists use the term frequently to distinguish material, e.g., rocks, minerals, fossils, etc., found in their original position of formation, deposition or growth, as opposed to loose, disconnected, or derived material.

Interbedded: Literally between two layers, e.g., a lava flow may occur interbedded between two layers of sediments, a limestone may be interbedded between two layers of shale.

Intergrowth: A textural term applied to inter-locking arrangements of two minerals, usually due to simultaneous crystallisation of the phases or the exsolution of one phase from another (Graphic texture, Perthite).

Intraclast: Weakly consolidated carbonate sediment that have been eroded from adjoining parts of sea bottom and redeposited to form a new sediment.

Intraformational: A term applied to rocks, e.g., conglomerates and breccias, or structural features which occur between two sets of defined strata. An informational conglomerates or breccias is one formed by the contemporaneous erosion and redeposition of a bed of sediment previously deposited. It implied temporary change in the conditions of sedimentation. Intraformational slumping or folding is also possible.

Intrusion: A body of igneous rock which has forced itself into pre-exisiting rocks, either along some definite structural feature, e.g., bedding planes-joints cleavages etc. or by deformation and cross cutting of the invaded rocks. (Batholith, Boss, Dyke, Laccolith, Lopolith, Phacolith, sill, stock)

Island arc (Festoon): A arcuate chain of islands e.g. Japan, Aleutian Islands. They are found around the margins of the Pacific, Malaya, and Indonesia and in the Carribean. Their abundance and distribution suggest that they have a common origin. The following are some of their main characteristics: (1) They are associated with areas of strong seismic activity and deep-focus-earthquakes (2) deep oceanic trenches (fore deeps) occur on the convex side and deep basins on the concave side (3) They are regions of intense gravitational and magnetic anomalies. (4) When a double arc occurs, the inner one is volcanic, the outer one non-volcanic. (Plate tectonics).

Isobath: A line joining points of equal depth of a horizon below the surface or a datum plane.

Isochore: A line joining points of equal vertical interval between two datum planes. (Cf-Isopach)

Isopach, Isopachyte: A line joining points of equal bed thickness.

Isostasy: The tendency of the earth's crust to maintain a state of near equilibrium, i.e., if anything occurs to modify the existing state, a compensating change will occur to maintain a balance. The most obvious example of this is the deep roots of mountain chains, as erosion of the mountains occurs, reducing their height compensation in the form of renewed uplift occurs. Continental blocks may be depressed under the load of an ice sheet, recovering when the ice melts. In this way variations in the relative levels of the land and sea may be explained. (Rejuvenation). The concept of isostasy depends upon the model of the earth's crust in which lighter, continental masses 'float' on a denser sub-stratum.

K

Killing a Well: Bringing a well under control that is blowing out. Also the procedure of circulating water and mud into a completed well before starting well service operations.

Klippe (Klippen): A tectonic-outlier resulting from the erosion of a thrust sheet or nappe, so that an isolated mass of rock is separated from the underlying 'basement' by a thrust plane. On a map this appears as a continuous closed outcrop of the thrust plane. In the Swiss Alps, klippen occur in which older rocks rest upon younger ones.

L

Lamination: The development of thin, discrete layeıs of rocks. The individual layers may differ from one another or may be homogeneous. The layers being separated by a physical discontinuity, e.g., bedding planes. (For igneous lamination, layered igneous structures)

Lava: The material extruded by a volcano which consists of molten or part-molten silicate material (Pyroclastic rocks, volatiles).

Lineation: (Liner Structure) lineation is any one dimensional feature in a rock or shown on a rock surface. It may arise in any of the following ways: (1) By a linear parallel arrangement of minerals, either by growth or mechanical orientation, (2) by the intersection of a planar structure with the rock surface, (3) by the intersection of two planar structures, e.g., bedding and-cleavage, (4) by the development of a series of small parallel puckers in a planar structure, e.g., microfolds (Phyllitic lineation), (5) by the deposition of elongate particles (sometimes fossils, especially elongate ones) in a preferred orientation during sedimentation and (6) as—striae or slickensides.

Lithification: The process which result in the formation of a massive rock from a loose sediment.

Lithoglastic: Carbonate detrities mechanically formed and deposited, derived from previously formed carbonates.

Lithofacies: A facies particularly characterised by its rock type (CF-Biofacies)

Lithographic stone: An extremely fine grained compact-limestone slabs of which were formerly used as lithographic plates. The best known example is the Kimmeridgian stone from Solenhofen, Bavaria, famous for its well-preserved fauna.

Lithology: A term usually applied to sediments, referring to their general characteristics. It is used in several rather loose senses, e.g., lithological variation, referring to variation in composition and texture lithology of a formation meaning the rock types present in a stratigraphic sub-division. Lithology generally relates to descriptions based upon hand specimens and outcrops rather than microscopic or chemical features.

Lit-par-Lit-injection: The forcible intrusion of igneous material usually granitic in character, along bedding planes foliation planes (Cleavage rock) etc. to give a close-spaced, more or less parallel series of veinlets. If the process is extensive and regular gneiss like rock may be produced. Reaction between the invading magma and the country rock produces a magmatite.

Littoral: The environment between the highest and lowest levels of spring tides. This is practically synonymous with beach more precisely with 'forebeach' (The area between the cliff line and high water marks is the 'back-beach' which is usually where a storm beach is developed). Littoral deposits are mainly sands and or pebbles and sometimes boulders, more rarely mud. Long-shore drift and the immediate coastal rocks will largely control the character of such deposits. Some littoral zones are bare rock, not covered by a veneer of sediment littoral faunas and floras have special characteristics to enable them to survive the continuously changing environment.

Loadcast: When a bed of sand is deposited on a layer of plastic mud, there is a tendency for the sand to press downwards into the mud in the form of buttons, owing to unequal loading from above. These projections (which may superficially resemble—ripple marks or ripple casts) are known as loadcasts. Occasionally they may be accentuated by later deformation (collapse) or by post diagenetic deformation.

Loess: Accumulations of wind-born dust (mainly siliceous). The dust is derived originally from desert area or from vegetation free areas around ice sheets. Loess is well graded and has a particle size of between 1/16 and 1/32 mm. It is commonly unstratified and shows a well developed vertical joining. Loess reworked by river action is called brickearth.

Longshore drift: The process by which material is transported along the shore. On many coasts the dominant wave-direction is oblique to the shore line and material is thus carried obliquely up the beach, but is then moved straight down the beach by the back-wash. The resulting action is that material is carried side way. This action is not confined to the immediate coastline but may take place down to the lower limit of wave action. The drift may in some cases threaten to remove sand from the beaches, but this may be arrested by the erection of groynes across the beach.

M

Magneto Electric Surveys: Surface or air borne exploration technique that determines distribution, directions and relative intensities of magnetic field vertices associated with vertical part of electrotelluric currents.

Magnetotelluric Surveys: A method of exploration of earth's crust and of the mantle to very great depths based on relation divisional variations in earth's magnetic field and association magnetotelluric currents. It has no relation to ET Surveys.

Meander: Any naturally occuring bend which a river or stream may develop along its course, often making a series of reentrant curves. The precise cause of meandering is unknown, but recent work suggests that meanders are produced as a result of the river taking a preferred route, determined by gradient cross-sectional area of channel, and ratio of rate of discharge to load.

Meteoric Water: A term applied to water which penetrates the rock from above, i.e., rain, dew, hail, snow, and also the water of river and streams. (e.g., Connate water; juvenile; ground water).

Micrite: A consolidated or unconsolidated ooze or mud of either chemical or mechanical in origin. It is about 0.005 mm or similar in diameter.

Mineralisation: This term is used almost exclusively for the introduction of ore-minerals and Gangue minerals into pre-existing rocks, whether by veins; replacement (5 and 6) or in a disseminated fashion, mineralisation can be considered on any scale from a hand specimen to region of several thousand square miles. The term can also be used in a stratigraphic sense, e.g., one can refer to a "post-carboniferous mineralisation", meaning the formation of the mineral deposits associated with the Hercynian (Armorican) granites (e.g., metallogentic—mineral deposit, Ore body).

Molasse: The term originated in Switzerland, and is used to describe sediments produced by the erosion of mountain ranges after the final phase of an-orogeny (post-tectonic) (e.g., flysch). In Switzerland, these post-tectonic sediments of the Alpine orogeny include arkoses. (Arenaceous rocks) (Polymict-Conglomerates and breecias, and reddish-brown shales). (Argillaceous rocks). They appear to have developed in intermontare basins, which are commonly non-marine. Other deposits which have been claimed as molasse are the Nework Sandstone. (Triassic) of the eastern United States derived from the Appalachians, and the Old Red Sandstone. (Devonian of Britain, derived from the Caledonian mountains).

Mud volcano: A hot spring in a volcanic region in which is incorporated sufficient argillaceous material or volcanic ash to produce a boiling mass of mud. A conical mound which mimics a normal volcano formed as a result of earth quake activity during which masses of semiliquid mud are forced up through fissures.

N

Nappe: A fold in which the axial plain is horizontal or sub-horizontal.

Natural Gas: It is restricted to gaseous hydrocarbons which may occur in association with oil accumulations.

Negative area: An area where deposition is taking place

Neogene: The name given to the Miocene and Pliocene periods of the Teritary, when grouped together. It is almost obsolete.

Neritic: Sub-littoral (Environment)

O

Offshore: The zones extending from the lowest limit of low spring tides to the edge of the continental shelf.

Oil Shale: A fine grained black or dark grey-argillaceous rock containing-Kerogen, which yields liquid hydrocarbons, i.e., oil on distillation. Oil shales do not contain free oil. They appear to form in a special kind of anaerobic environment and are not uncommonly found in association with coal-bearing strata. (Boghead coal).

Outcrop: (1) Often used as synonym of exposure, but should strictly be used only with the sense of (2) below: (2) The total area over which a particular rock unit occurs at the surface whether visibly exposed or not. The term is especially used for the delineation of such an area on a geological map.

Outlier: A limited area of younger rock completely surrounded by older rocks. It may be produced by-erosion-faulting, or folding or a combination of any two, or all three, of these agencies.

Overburden: (1) Useless material which overlies a bed of useful material, e.g., shale overlying a coal seam. (2) Any loose unconsolidated material which rests upon solid rock.

Over lap: A term used to describe the relationship of the beds in an unconformity where progressively younger members of an upper series rest upon an older series.

Over step: An unconformity which develops during a marine transgression when the younger series rests upon progressively older members of the underlying series.

Ox-bow-lake: Meander loops tend to enlarge themselves by the erosion of their outer banks, which gradually narrows the neck between successive loops. In time of flood the neck may be breached and the ends of the loop become silted up the river taking the new 'preferred' straighter course. The out-off loop is variously called an Ox-bow lake, a mort lake, billabong, or a bayov. The lake subsequently becomes a marsh and in time dries out.

P

Palaeoclimatology: The study of the climatic conditions obtaining in past geological eras. Detailed study of sedimentary rocks and their enclosed fossils have made possible estimates of such climatic factors as wind direction, rainfall, atomspheric and oceanic temperature and effects of atmospheric changes. The most obvious palaeoclimatic determinations are the recognition of-ice ages, and the hot, dry periods in which evaporites form. Estimates of climatic conditions become less and less reliable as they are projected further and further back in time. The Pleistocene climates are relatively well known whereas climates for lower Palaeozoic periods are probably little better than intelligent guesswork. In exceptional circumstances, e.g., deposition of varved clays, seasonal variations from year to year can be detected.

Palaeoecology: Fossils were once animals and plants and were therefore governed by precisely the same ecological factors as now govern living animals and plants. Palaeoecology is therefore a study of these factors and how they affected the mode of life of organisms in the past. A palaeoecological study is designed to obtain as much information as possible about the environment of the fossil or fossils, particularly their relation to the sediment.

Palaeogeography: A reconstruction of the presumed geography, especially the relative positions of land and water, at some particular period in the past. It is generally more difficult to reconstruct the geographies of the older periods and in many cases published maps are no more than aproximations. It should be noted that features, e.g., coastlines, shown in a particular position super-imposed on the present geography probably did not occupy this position relative to other modern features and to other palaeogeographical features, owing to the fact that palaeogeogrphical maps are rarely plainspastic. (Continental drift)

Palaeomagnetism: It is well known to physicists that the positions of the earth's magnetic poles change over a period of time. The investigation of such changes during geological time is the field of palaeomagnetic studies. Some rocks acquire a permanent record of the earth's magnetic field by virtue of certain iron bearing minerals which they contain.

Paleoslope: The inclination of an ancient land surface as existing in some period of geological history.

Paragenesis: The relationship of minerals expressed in terms of a time sequence, i.e., if conditions are changing progressively along a time axis, then a particular mineral will appear as a derivative of an earlier mineral, and may ultimately pass into yet another mineral, the process taking place begin recrystallisation. The term is sometimes used with the specific meaning of the order of crystallisation of the minerals forming a rock, and it is also used loosely in the sense of the mode of origin of a rock or mineral.

Paralic: A term applied to sedimentary basins developed in marginal marine environment e.g., lagoons, littoral basins, etc.,

Parallel growth: During the formation of crystals it commonly happens that, owing to local controls, the crystals of a particular substance develop the same habit and have one crystallographic direction in common. Parallel growth should be carefully distinguished from twinning (Twinned crystal). It is an accidental feature of crystal growth, with an element of randomness, compared with the strictly regular laws governing twinning. The commonest type of parallel growth involves crystals having a prismatic habit, the axes of which are all parallel.

Parautochthonous: A term used to describe a feature formed at great distance from the place in which it is found. It is most commonly used to describe nappe structure in the Alps. (Granite series)

Peat: A partially decomposed mass of vegetation which has grown in a shallow lake or marsh. Generally peat is dark brown or black and contains recognisable vegetable fragments, but very little mineral material (coal). The main plants which make-up peat are the 'Peat mosses'. Sphagnum and Hypnum; with some contributions from rushes, sedges, horse-tails, etc. Peat deposits may be some tens of metres thick and cover large areas. For large scale peat formation, rapid plant growth, high moisture content and the development of anaerobic conditions are essential. High land and low land (or fen) peats are recongnised, the former being spongy and mainly borns from mosses, the latter being more compact and homogeneous mainly formed from sedges, rushes etc. Diagenesis of peat lead to lignite and coal formation, under the right conditions.

Pediment: A plain of eroded bed rock (which may or may not be covered by a thin veneer of alluvium) in an arid region developed between mountain and basin areas. (Cycle of erosion)

Pelagic: A term used to describe the mode of life of those animals which live in the open sea but not on the sea floor, including forms which are free-swimming (nektonic), e.g., fish and forms which float passively in the surface waters (planktonic) such as jellyfish.

Pellets: Ovoid particles of sediments commonly composed only of calcium carbonate, which range in size from 0-25 mm in length. They are usually considered to be faecal pellets which may have been excerted by molluscs, echinoderms, and possibly some other groups of invertebrate animals.

Pisolite: A grain type similar to oolite and generally 2.0 mm or more in diameter.

Penecontemporaneous: Almost at the same time

Pennsylvanian: The younger of the two sub-systems into which the Carboniferous period in North America is divided.

Permafrost: Permanent frost or permanently frozen ground. It extends from 2000 to 3500 Km south of north pole. For permafrost to form, heat loss from the ground must counterbalance insulation and the interior heat of earth.

Permeability: A rock is said to be permeable if water or other liquid in contact with its upper surface tends to pass through the rock more or less freely to the lower surface.

Photogeology: Strictly speaking, the term may be used for the geological interpretation of any photograph. In practice it is used almost exclusively for the geological interpretation of aerial photographs, mainly vertical photograph, although 'high obliques' can also be used. It is usual to study vertical photograph in Stereoscopic pairs, using a stereoscope, so that relief may be fully appreciated.

Phreatic Water: Ground water particularly below water table (Vadose water).

Plate-tectonics: Modern study of the major architectural features of the Earth's crust mid oceanic ridges, major tear faults, oceanic trenches, continental blocks, mountain ranges, earthquake belts volcanic zones etc., suggests that these may be used to define a series of major regions of the Earth's crust (termed plates). The typical plate includes the continental shelf, sea, and oceanic areas. The theory of continental drift requires relative movement between continental blocks. It is suggested that in fact the movement is between plates, and features such as rear-faulting (Transcurrent faulting) represent lateral movement of plates, mid oceanic ridges result from plates moving apart (with creation of crust at the 'Join'), and trenches (Island ores) occur where one plate moves under another. Other features, such as mountain ranges, earthquakes, etc., can also be related to plate movements. The whole concept seems to unify a number of major observations; and there is little doubt that plate tectonics will be much used to explain both recent and more remote events in Earth history.

Polymer: A substance formed by the union of two or more molecules of the same kind linked into another compound having the same elements in the same proportion but a higher molecular weight and different physical properties.

Porosity: A rock is said to be porous, if it possesses cavities between the mineral grains making up the rock which can contain liquid.

Pressure: Physicists define pressure as the force per unit area, i.e., stress. However the geologist commonly uses pressure as a loose, all embracing term to include concepts of stress strain and shear arising from external sources. He generally uses the phrase 'hydrostatic pressure' (load pressure) as that force which acts upon a rock due to the mass of superincumbent material.

Protozoa: A unicellular animal which may or may not secrete a test or skeleton ranging in size from 0-1 mm, to 8 mm. Modern protozoa are divided into a number of groups, but only two of these are of geological importance—the Foraminifera and the Radiolaria.

Provenance: The source area or areas of the material making up a sediment more especially the nature of the rocks from which the material has been derived.

Pseudomorph: A pseudomorph is one mineral occuring in the crystal form of another.

Pyrometamorphism: An intense form of thermal metamorphism developed at or near a contact with a high temperature igneous intrusion.

R

Radiactive dating: (Radiomotric dating): The most reliable method of obtaining a data for a rock depends upon observation that the rate of decay of a radoactive element is a constant (Age of the earth). The earliest methods, utilising uranium and thorium minerals.

Raised beach: A wave cut platform with or without a covering of beach deposits, which is now raised above the present sea-level.

Recrystalisation: The process whereby a mass of crystals passes through a solution phase in developing a new set of crystals of the same kind. The change requires the presence of a suitable energy level. Probably small amounts of solvent circulating in pore spaces can bring about recrystallisation.

Rejuvenation: Rejuvenation takes place with the relative uplift with respect to sea level of a region which has developed a mature drainage system (cycle of erosion). Streams whose-base level has fallen cut down very rapidly and become emerged (entrenched) into the land in an attempt to re-create, the previous long-profile of equilibrium. An uplift of the land (or fall in sea level) caused rivers to reach the sea by means of waterfalls. In time these regress upstream adjust themselves to the new profile. Break such as waterfalls in the long-profile of a river are known as knick (nick) points, and represents the junction of the new, adjusted profile and the old, now mat adjusted one.

Relict structure: A near synonym of a Palimpsest structue.

Replacement: A term widely used in geology to describe the process whereby one constituent in a system is progressively situated by another.

Residual deposits: Material left behind when a part of a rock is removed by chemical weathering processes, usually solution or leaching. Laterite and bauxite are types of residual deposits as is the terra rossa of limestone regions, which consists of the insoluble residue after the carbonate has been removed in solution.

Retrograde condensation: The phenomenon of appearance and disappearance of the liquid by pressure reduction at constant temperature is called retrograde condensation.

Reverse circulation: Normally drilling fluid is circulated down through the drill pipe and up through the Annulus. Sometimes this normal circulation is reversed when drilling fluid returns to surface through the drill pipe after being pumped down through the annulus.

Rhythmic Sedimentation: The term applied to a sequence of sediments which changes their character progressively from one extreme type to another, a change which is followed directly by a return to the original type. A typical example would be the coal measure rhythm.

Marine	
Swamp	
Deltaic	Rhythmic unit
Lagoonal	
Marine	
Swamp	

The essential features of the rhythmic unit is a steady sequence of changes followed by an abrupt return to the starting point. This may be symbolically expressed.

d
c
b
a

---

d
c
b
a

Rift Valley: An elongated trough bounded by faults; in all cases the tectonic depression and a geomorphological valley coincide. The term 'graben' appears to be synonymous. Taphrogeosyncline is also a synonym.

Ripple marks: Undulations produced by fluid movement over sediments. Oscillatory currents produce symmetric ripples whereas a well defined current direction produces asymmetrical ripples. The crest lines of ripples may be straight or sinuous. The characteristic features of ripples depend upon current velocity, particle size, persistence of current direction and whether the fluid is air or water.

River capture: The abstraction of the head-waters of one river by a more actively eroding stream-usually by the breaching of local watershed.

River terrace: A river terrace marks the position of the former flood plain which developed when the river flowed at a higher level. It usually consists of a bench on the side of the valley, covered with the usual flood plain deposits of clay, sand and gravel.

S

Salt domes: Diapiric masses of salt. Under high pressure, salt deforms plastically and behaves like an intrusive-magma, deforming and piercing the overlying sediments. Occasionally salt domes reach the surface giving rise to salt glaciers. Salt domes are of considerable economic importance as they may develop oil traps, while in Texas and Louisiana sulphur occurs in association with them.

Saturation, Saturated: The principle of saturation derives from the application of phase rule studies to igneous and metamorphic rocks. It has been shown that if there is an excess or a deficiency of a certain component in the rock, particular minerals can or cannot develop. An alternative statement of this is that certain minerals are stable in the presence of an excess of a component while other minerals are stable only if this component is not in excess.

Secondary enrichments: Secondary (supergene) enrichment is a term used especially of ore deposits and is applied to parts of the ore body in which the metal content of the ore (or, in the case of non metallic material, the valuable mineral content) has been increased as a result of the downward percolations of water carrying material in solution.

Sedimentary by-passing: The laws of selective transportation show that some sedimentary particles may travel faster or farther than others, owing to sizes, shape or density difference which favour continued movement. Eaton applied terming by passing to instances where one particle passes another simultaneously transported or continues its motion after other has come to rest.

Sedimentary rocks: Rocks formed from material derived from pre-existing rocks by processes of denudation, together with material of organic origin. The term includes both consolidated and unconsolidated material. The latter is also referred to as sediment. They are classified on the basis of the character of the material and the process which leads to its deposition.

Sedimentary structures: In recent years, a voluminous vocabulary of terms relating to the structural features of sedimentary rocks has grown up. In general, the terms used to relate to processes taking place during, or immediately after, deposition of a sediment prior to diagenesis. The following outline classification of sedimentary structures has been adapted from the work quoted at the end of the entry.

Seepage: Prolonged erosion of the formations overlying a petroleum accummulation may permit gas and/or oil to seep to surface.

Shelf facies: A term applied to those sediments and their associated fauna which are found in the shallower marginal parts of a basin or geosyncline.

Shield: A major structural unit of the Earth's crust, consisting of a large mass of precambrian rocks, both metamorphis and igneous, which have remained unaffected by later — orogenies (positive area). Well-known examples are the Canadian and Baltic Shields. The term is practically synonymous with kraton.

Shoe-String sand: A body of sediment which is thin and extensive in one horizontal dimension.

Shutter valley: A valley which has been affected by a tear-fault at right angles to the axis of the valley so that the two halves no longer coincide.

Sial: The upper portion of the crust of the Earth composed predominantly of silica and aluminium.

Sill: A sheet-like body of igneous rock which conforms (conformable) to bedding or other structural planes. The original sill is the great thin sill of Northern England, which has a very wide extent underlying much of North under Land and Durham. Sills may be composite or multiple. Sills which have a small area may be confined to a single plane, but those which have a larger geographical extent can commonly be found occupying different planes in different places. Sills of this kind are said to be transgression. Sills may be differentiated and it is been suggested that laccoliths may grade into sills by increasing lateral extent and decreasing convexity upwards. This has also been claimed for lopoliths and phacoliths. Sills are normally medium grained but large ones may be coarse enough to be described as plutonic. Basic sills (Dolerite) are by far the commonest.

Sinistral: A term applied to tear (strike) faults to describe the apparent direction of movement, in this case to the left.

Sinistral coiling: Mollusca (Cephalopoda)

Skin effect: It is a measure of the additional pressure drop necessary to overcome the flow resistance of the reduced permeability zone around the well bore.

Slickenside: When one surface of a rock moves over another surface in close contact under pressure, the two surfaces develop a kind of polish, with linear grooves and ridges parallel to the direction of movement. This is termed slickensiding and is commonly seen on fault surfaces. It should be noted that slickensides may develop with surprisingly little movement if the rock type is suitable.

Slope: There are four recognised elements of hillside slopes. (1) A convex or waxing slope (2) A free face or outcrop of bare rock (3) A uniform slope of talus or detrital material. (4) A concave or waning slope the pediment. It is possible for one or more of these elements to be missing, most commonly the free face, since it depends for its existence on a hard resistant band of rock. If both the free face and the constant slope are missing the landscape is said to be concaves-convex, with the waxing slope Merging into the waning slope. The concept of parallel retreat of slopes is embodied in Pencks scheme of the cycle of erosion.

Sparite: Any transparent of translucent cyrstalline calcite and argonite. It is larger than .02 mm in diameter.

Spudding in: The very beginning of drilling on a well location.

Squeeze job: Usually a secondary cementing job where cement slurry is pumped into the formation through the bottom of the casing or through the perforations in the casing to obtain a shut off of undesirable fluids.

Stalactite: A pendant accumulation of calcium carbonate found in limestone caves.

Stalagmite: A general term used for the massive calcium carbonate deposits found in limestone caverns, but more specially for the upward projection corresponding to stalactite.

Stand pipe: Two or three or sometimes four joints of drill pipes fastened together called a double, trible or favourable respectively.

Stress minerals: Minerals formed under conditions of metamorphisim when stress was present to a greater or lesser degree.

Strike: The direction in which a horizontal line can be drawn on a plane. In geologic usage the strike is important in determining the direction in which to measure the true dip.

Structural bench: A flattening of dip without any change in direction.

Sub-littoral: The area between the lowest limit of the litoral zone and the cotinental shelf. A synonym is neritic.

Swabbing: Operation of a lifting device to bring well fluids to the surface when the well does not flow of its own. During drilling or pulling out of drill pipe from the hole due to coming out of the mud in the hole there is a drop in the hydrostatic pressure which is also called swabbing.

Syntexis: A term which is used to include both pure melting and assimilation of country rock.

T

*Taphrogenesis:*Essentially vertical movements in the Earth's crust which result in major high angle faulting such as commonly associated with rift valleys.

Taphrogeosyncline: A synonym of rift valley.

Tectogene: A hypothetical deep tightly folding downward below an orogenic belt, suggested as an explanation of the greater thickness of sialic material.

Tectonics: It is used to relate a particular phenomenon to a structural or orogenic concept.

Tepes: Minor non tectonic anticlines are called as tepes which are abundant in the carbonate rocks of Eastern Guadalupe mountains.

Tethys: A Mesozoix geosyncline which developed between Laurasia and Gondwanaland and covered Southern Europe, the Mediteranean, North Africa, Iran and Himalayan region, possibly extending into Burma and South-East Asia. It was the site of the Alpine orogeny.

Thixotropic: A term applied to certain types of solid/liquid systems which are effectively solid when stationary but which become mobile liquids when subjected to shearing stresses.

Three point problem: A geometrical method of finding the dip and strike of a plane surface, given the altitudes above a datum plane at three known positions.

Transgression: The invasion of a large area of land by the sea in a relatively short space of time. It is generally considered that the sea advances over a more or less peneplained surface.

Transition series: A group of beds linking to series of sediments formed in contrasting environments.

Transportation of sediments: The main agencies by which sedimentary materials are moved are gravity, running water, ice, wind and the sea.

Tubing: The pipe through which oil or gas both are brought from the reservoir to the surface. It is generally placed inside the casing and is 2 to 4 inches in diameter.

Turbidite: It is a matter of common observation and an easily demonstrated fact that slurries of sediment and water behaves as discrete fluid phase when poured into fresh or sea water. Such slurries can be generated in large quantity in various types of basin and will flow down a slope at remarkably high speeds, covering distances of tens of kilometres.

Turbidity limestones: Some of the litho calcarlnites and biocalcarlnites show certain features which indicate their transportation and re-sedimentation by turbidity currents.

Twist off: To twist a joint of pipe in two by excessive force applied by the rotary table.

U

Ukrainian: A stratigraphic division of the Russian Precambrian Pre Saamian.

Uniformitarianism: The concept fundamental to geology that processes which operate at present also operated in the past and produced the same results. These processes need not have operated at the same rate nor at the same intensity.

Unsaturated: A term applied to minerals which cannot develop in stable equilibrium in the presence of silica. Such minerals are olivine, feldspathoids, corundum, garnets, etc.

W

Wash out: A channel out into sediment and filled by later material.

Weathering: The process by which rocks are broken down and decomposed by the section of external agencies such as wind, rain, temperature changes, plants and bacteria. It is an initial stage in the process of denudation.

Whipstock: A long slender steel wedge with a groove along one side that is placed at the bottom of the hole and is used for deflecting the bit during drilling directional holes.

Wildcat: A borehole sunk in the hope of finding oil in a region where deposit of oil or metallic ores have not been recorded.

Wildflysch: It consists of clastics of varying size, some of them enormous, in a matrix of fine-grained, typically dark-coloured marine shale, siltstone or mudstones. They are found in ancient geosynclinal tracts.

Window: An area where erosion has penetrated thrust or recumbent fold to expose the rocks lying beneath.

Work over: To perform one or more of a variety or remedial operations, on a producing well with the hope of restoring or increasing production.

X

Xenocryst: The term applied to crystals often apparently porphyritic, in an igneous rock which have not formed in the place in which they are now found.

Conversion Factors and Tables

TABLE 1

(a) Units

British Units		*Metric Units*	
12 inches	= 1 foot	10 millimetres	= 1 centimetre
3 feet	= 1 yard	10 centimetres	= 1 decimeter
220 yards	= 1 furlong	10 decimetres	= 1 metre
8 furlongs	= 1 mile	10 metres	= 1 decametre
6 feet	= 1 fathom	10 decametres	= 1 hectometre
120 fathoms	= 1 cable length	10 hectometres	= 1 kilometre
6080 feet	= 1 nautical mile		

(b) Conversion Equivalents

	Inches	*Feet*	*Yards*	*Miles*	*Centi-metres*	*Metres*	*Kilo-metres*
1 Inch	1	0.0833	0.0278		2.54	0.0254	—
1 Foot	12	1	0.333		30.48	0.3048	—
1 Yard	36	3	1	0.00057	91.44	0.9144	0.000914
1 Mile	63360	5280	1760	1	160934	1609.3	1.609
1 Centi-metre	0.3937	0.0328	0.0109	—	1	0.01	—
1 Metre	39.37	3.2808	1.0936	0.0006	100	1	0.001
1 Kilo-metre	39370	3280.8	1093.6	0.62136	100.000	1000	1

TABLE 2
AREA

(a) Units

British Units		*Metric Units*	
144 sq. inches	= 1 sq. foot	100 sq. millimetres	= 1 sq. centimetre
9 sq. feet	= 1 sq. yard	100 sq. centimetres	= 1 sq. decimetre
4840 sq. yards	= 1 acre	100 sq. decimetres	= 1 sq. metre
640 acres	= 1 sq. mile	100 sq. metres	= 1 acre
484 sq. yards	= 1 sq. chain		= 1 sq. decametre
10 sq. chains	= 1 acre	100 acres	= 1 hectare
			= 1 sq. hectometre
		100 hectares	= 1 sq. kilometre

(b) Conversion Equivalents

1 sq. inch	=	6.45 sq. centimetres	1 sq. centimetre	=	0.155 sq. inch
1 sq. foot	=	9.29 sq. decimetres	1 sq. decimetre	=	15.5 sq. inches
1 sq. yard	=	0.836 sq. metres	1 sq. metre	=	10.76 sq. feet
1 acre	=	0.4047 hectares		=	1.196 sq. yard
1 sq. mile	=	259 hectares	1 sq. hectare	=	2.47 acres

TABLE 3

WEIGHTS

(a) Units

British Units			*Metric Units*		
Avoirdupois Units			10 milligrams	=	1 centigram
16 drams	=	1 ounce	10 centigrams	=	1 decigram
16 ounces	=	1 pound	10 decigrams	=	1 gram
28 pounds	=	1 quarter	10 grams	=	1 decagram
4 quarters	=	1 cwt	10 decagrams	=	1 hectogram
20 cwt	=	1 ton (long)	10 hectograms	=	1 kilogram
14 pounds	=	1 stone	100 kilograms	=	1 quintal
2000 pounds	=	1 short ton	1000 kilograms	=	1 metric tonne
2240 pounds	=	1 long ton			

(b) Conversion Equivalents

	Pounds	*Kilograms*	*Short tons*	*Metric tons*	*Long tons*
1 Pound	1	0.45359	0.0005	0.00045359	0.00044643
1 Short cwt	100.0	45.359	0.05	0.04536	0.4464
1 long cwt	112.0	50.802	0.056	0.05080	0.05
1 Kilogram	2.2046	1	0.0011023	0.001	0.0009842
1 Short ton	2000	907.185	1	0.90718	0.89286
1 Metric ton	2204.6	1000	1.1023	1	0.98421
1 Long ton	2240	1016.06	1.12	1.01606	1
1 Maund	82.28	37.32	0.41	0.037	0.037

*Indian Units**	=	1 tola	1 seer	=	2.057 pounds (avoirdupois)
80 tolas	=	1 seer		=	2.5 Pounds (apothecaris)
40 seers	=	1 maund		=	0.9331 kilograms

*As defined in the Indian Standards Weight Act 1939 but now replaced by Metric Units.

TABLE 4

CAPACITY

(a) Units

British Units			*Metric Units*		
60 minims	=	1 fluid drachm			
8 fluid drachms	=	1 fluid ounce	10 millilitres	=	1 centilitre
20 fluid ounces	=	1 pint	10 centilitres	=	1 decilitre
2 pints	=	1 quart	10 decilitres	=	1 litre
4 quarts	=	1 Imperial gallon	10 litres	=	1 decalitre
8 gallons	=	1 bushel	10 decalitres	=	1 hectolitre
8 bushels	=	1 quarter	10 hectolitres	=	1 kilolitre

(b) Conversion Equivalents

1 IMPERIAL GALLON		1 LITRE	= 1000.03 cu. centimetres
	= 277.42 cubic inches		= 61.026 cubic inches
	= 4.54596 litres		= 1.7598 pints
	= 0.160544 cubic foot		= 0.264178 U.S. gallons
	= 1.20094 U.S. galls.		= 0.219975 Imperial gallons
	= 0.028594 U.S. barrel		= 0.035316 cubic foot
	= 0.0045461 cubic metre		= 0.00629 U.S. barrel
1 U.S. GALLON			= 0.01 hectolitre
	= 231 cubic inches		= 0.00100003 cu. metre
	= 3.78533 litres	1 CUBIC FOOT	
	= 0.83268 Imperial gallon		= 28.316 litres
	= 0.133681 cubic foot		= 7.4805 U.S. gallons
	= 0.0238095 U.S. barrel		= 6.2288 Imperial gallons
	= 0.0037854 cubic metre		= 0.17811 U.S. barrel
			= 0.028317 cu. metre
1 U.S. BARREL			
	= 9702 cubic inches	1 CUBIC METRE	
	= 158.984 litres		= 35.315 cu. feet
	= 42.00 U.S. gallons		= 999.97 litres
	= 34.9726 Imperial gallons		= 264.17 U.S. gallons
	= 5.6146 cubic feet		= 219.97 Imperial gallons
	= 0.15899 cubic metre		= 6.2898 U.S. barrels

TABLE 5

VOLUME

British units			*Metric units*		
1728 cubic inches	=	1 cu. foot	1000 cu. millimetres	=	1 cubic centimetre
27 cubic feet	=	1 cu. yard	1000 cu. centimetres	=	1 cubic decimetre
			1000 cu. decimetres	=	1 cubic metre

TABLE 6

Composition, Volume, Weight and Calorific Value of Dry Gases

Gas		*Molecular wt*	*Weight per cu. ft. (dry) lb*		*Cu. ft. per lb.*		*Calorific Value in B.Th.U. per cu.ft. (dry)*	
							At 0° C and 760 mm	*At 60° F and 30 in.*
			At 0° C and 760 mm	*At 60° F and 760 mm*	*At 0° C and 30 in.*	*At 60° F and 30 in.*	*Gross*	*Gross*
Oxygen	O_2	32.00	0.0892	0.0846	11.22	11.82	—	—
Nitrogen[1]	N_2	28.17	0.0785	0.0745	12.74	13.42	—	—
Air		(29.0)	0.0807	0.0766	12.39	13.05	—	—
Carbon dioxide	CO_2	44.00	0.1226	0.1163	8.15	8.60	—	—
Hydrogen	H_2	2.02	0.0056	0.0053	178.50	188.70	343.2	325.6
Carbon monoxide	CO	28.0	0.0780	0.0740	12.82	13.52	341.1	323.6
Methane	CH_4	16.03	0.0447	0.0424	22.38	23.58	1067.0	1012.0
Ethane	C_2H_6	30.05	0.0837	0.0794	11.95	12.59	1856.0	1761.0
Ethylene	C_2H_4	28.03	0.0781	0.0741	12.80	13.49	1673.0	1587.0
Propylene	C_3H_6	42.05	0.1171	0.1111	8.54	9.00	2467.0	2340.0
Benzene	C_6H_6	78.05	0.2174	0.2063	4.60	4.85	4011.5	3805.0
Hydrogen sulphide	H_2S	34.08	0.0950	0.0901	10.50	11.10	675.7	640.7

[1]Atmospheric Oxygen in air taken at 21.0 V percent

TABLE 7

TEMPERATURE

(a) Scales

Fahrenheit scale: A thermometer scale in which water freezes at 32°F and boils at 212°F at sea level atmospheric pressure.

Centigrade scale: A thermometer scale based upon the freezing point (0 degree C) and boiling point (100 degree C) of water.

(b) Conversion Factors

(i) Degree Fahrenheit (°F) to degree Centigrade (°C)

$$°C = 5/9\ (t^{*}-32)$$

(ii) Degree Centigrade (°C) to degree Fahrenheit (°F)

$$°F = 9/5\ T^{\dagger}+32$$

***Is temperature in Fahrenheit.**

† T is temperature in Centigrade.

1 B.T.U.	= 252 calories	= 0.252 kilogram calories	
1 Therm	= 100,000 B.T.U.	= 25,200 kg. cal	= 29.32 kwh@
1 k.w.h.@	= 3600,000 Joules	= 1.34 Horse Power hours	= 3412 B.T.Us.

@K.W.H.=Kilowatt hour is an electrical unit

TABLE 8

(a) Specific Gravity and Metric Tonne Equivalents of Petroleum Products

Products	*Specific Gravity (range)*	*Barrels per Metric Tonne (range)*
Crude oils	0.80 — 0.97	8.0 — 6.6
Aviation gasolines	0.70 — 0.78	9.1 — 8.2
Motor gasolines	0.71 — 0.79	9.0 — 8.1
Kerosenes	0.78 — 0.84	8.2 — 7.6
Gas oils	0.82 — 0.90	7.8 — 7.1
Diesel oils	0.82 — 0.92	7.8 — 6.9
Lurbricating oils	0.85 — 0.95	7.5 — 6.7
Fuel oils	0.92 — 0.99	6.9 — 6.5
Asphaltic bitumens	1.00 — 1.10	6.4 — 5.8

(b) Specific Gravity and A.P.I. Gravity of Petroleum Products at 60° F

Products	*Specific Gravity (range)*	*A.P.I. Gravity (range)*
Crude petroleums	0.770 — 1.000	52 — 10
Aviation gasolines	0.700 — 0.780	70 — 50
Motor gasolines	0.710 — 0.790	68 — 48
White spirits	0.760 — 0.820	55 — 41
Kerosenes	0.780 — 0.840	50 — 37
Gas oils	0.820 — 0.900	41 — 26
Diesel oils	0.820 — 0.902	41 — 22
Lubricating oils	0.870 — 0.960	31 — 16
Fuel oils	0.920 — 1.010	22 — 9
Bitumens	0.995 — 1.000	11 and below

TABLE 9

Volumetric Equivalents of different Crude Oils by Origin

Origin	*Average U.S. Barrels per*	
	Long ton	*Metric tonne*
Abu Dhabi	7.342	7.460
Albania	6.567	6.672
Algeria	7.633	7.756
Angola	7.132	7.246
Argentina	6.881	6.992
Austria	6.783	6.892
Bahrein	7.219	7.335
Bolivia	7.968	8.096

(Contd.)

Origin	*Average U.S. Barrels per*	
	Long ton	*Metric tonne*
Brazil	7.199	7.315
British Borneo	7.346	7.464
Bulgaria	7.150	7.265
Burma	7.346	7.464
Canada	7.311	7.428
Chile	7.679	7.802
China (Mainland)	7.151	7.266
Colombia	6.977	7.089
Cuba	6.547	6.652
Czechoslovakia	6.675	6.782
Ecuador	7.372	7.490
Egypt	6.814	6.923
France	7.087	7.201
Gabon	6.921	7.032
Germany (East)	7.016	7.129
Germany (West)	7.132	7.247
Hungary	7.509	7.630
India	7.389	7.508
Indonesia	7.253	7.369
Iran	7.298	7.415
Iraq	7.336	7.454
Israel	7.171	7.286
Italy & Sicily	6.705	6.813
Japan	7.077	7.191
Kuwait	7.145	7.260
Libya	7.145	7.620
Mexico	6.992	7.104
Morocco	7.413	7.532
Netherlands	6.865	6.975
Neutral Zone	6.717	6.825
New Guinea (Western)	7.350	7.468
Nigeria	7.184	7.299
Pakistan	6.995	7.107
Peru	7.398	7.517
Poland	7.302	7.419
Qatar	7.565	7.686
Rumania	7.335	7.453
Saudi Arabia	7.275	7.392
Senegal	6.921	7.032
Syria	7.382	7.500
Taiwan	7.302	7.419
Trinidad	6.879	9.989
Turkey	6.789	6.898
United Kingdom	7.164	7.279
United States	7.301	7.418
USSR	7.151	7.266
Venezuela	6.889	7.000
Yugoslavia	7.027	7.070

TABLE 10

Volumetric Equivalents of Products

Products	*Cu. Metres*	*Litres*	*U.S. Barrels*	*Imperial Gallons*	*U.S. Gallons*
Motor spirit	1.427	1427	8.98	312.5	375.00
Superior kerosene	1.284	1284	8.08	281.2	337.44
Inferior kerosene	1.222	1222	7.69	267.6	321.12
Aviation turbine fuel	1.291	1291	8.12	282.7	339.56
High speed diesel oil	1.216	1216	7.65	266.3	319.56
Light diesel oil	1.172	1172	7.37	256.7	308.04
Fuel oil	1.068	1068	6.72	233.9	280.68
Power alcohol	1.286	1286	8.09	281.6	337.92
Lubricating oils	1.114	1114	7.01	244.0	292.80

TABLE 11

Conversion of Feet to Metres

(1 inch = 25.4 millimetres)

Feet	*Metres*	*Feet*	*Metres*	*Feet*	*Metres*	*Feet*	*Metres*	*Feet*	*Metres*
1	0.3048	21	6.4008	41	12.4968	61	18.5928	81	24.6888
2	0.6096	22	6.7056	42	12.8016	62	18.8976	82	24.9936
3	0.9144	23	7.0104	43	13.1064	63	19.2024	83	25.2984
4	1.2192	24	7.3152	44	13.4112	64	19.5072	84	25.6032
5	1.5240	25	7.6200	45	13.7160	65	19.8120	85	25.9080
6	1.8288	26	7.9248	46	14.0208	66	20.1168	86	26.2128
7	2.1336	27	8.2296	47	14.3256	67	20.3736	87	26.5176
8	2.4384	28	8.5344	48	14.6304	68	20.7264	88	26.8224
9	2.7432	29	8.8392	49	14.9352	69	21.0312	89	27.1272
10	3.0480	30	9.1440	50	15.2400	70	21.3360	90	27.4320
11	3.3528	31	9.4488	51	15.5448	71	21.6408	91	27.7368
12	3.6576	32	9.7536	52	15.8496	72	21.9456	92	28.0416
13	3.9624	33	10.0584	53	16.1544	73	22.2504	93	28.3464
14	4.2672	34	10.3632	54	16.4592	74	22.5552	94	28.6512
15	4.5720	35	10.6680	55	16.7640	75	22.8600	95	28.9560
16	4.8768	36	10.9728	56	17.0688	76	23.1648	96	29.2608
17	5.1816	37	11.2778	57	17.3736	77	23.4696	97	29.5656
18	5.4864	38	11.5824	58	17.6784	78	23.7744	98	29.8704
19	5.7912	39	11.8872	59	17.9832	79	24.0792	99	30.1752
20	6.0960	40	12.1920	60	18.2880	80	24.3840	100	30.4800

TABLE 12

Conversion of Metres to Feet

(1 inch = 25.4 millimetres)

Feet	*Metres*	*Feet*	*Metres*	*Feet*	*Metres*	*Feet*	*Metres*	*Feet*	*Metres*
1	3.280	21	68.898	41	134.514	61	200.131	81	265.748
2	6.561	22	72.178	42	137.795	62	203.412	82	269.029
3	9.842	23	75.459	43	141.076	63	206.693	83	272.310
4	13.123	24	78.740	44	144.357	64	209.974	84	275.591
5	16.404	25	82.021	45	147.638	65	213.255	85	278.871
6	19.685	26	85.302	46	150.919	66	216.535	86	282.152
7	22.966	27	88.583	47	154.199	67	219.816	87	285.433
8	26.247	28	91.864	48	157.480	68	223.097	88	288.714
9	29.528	29	95.144	49	160.761	69	226.378	89	291.995
10	32.808	30	98.425	50	164.042	70	229.659	90	295.276
11	36.089	31	101.706	51	167.323	71	232.940	91	298.556
12	39.370	32	104.987	52	170.604	72	236.220	92	301.837
13	42.651	33	108.268	53	173.885	73	239.501	93	305.118
14	45.932	34	111.268	54	177.165	74	242.782	94	308.399
15	49.213	35	114.549	55	180.446	75	246.063	95	311.680
16	52.493	36	118.110	56	183.727	76	249.344	96	314.961
17	55.774	37	121.391	57	187.008	77	252.625	97	318.241
18	59.005	38	124.672	58	190.289	78	255.906	98	321.522
19	62.336	39	127.953	59	193.570	79	259.186	99	324.803
20	65.617	40	131.234	60	196.850	80	262.467	100	328.084

TABLE 13

Conversion of Cubic Feet to Cubic Metres

(1 inch=25.4 millimetres)

Cubic Feet	*Cubic Metres*	*Cubic Feet*	*Cubic Metres*	*Cubic Feet*	*Cubic Metres*	*Cubic Feet*	*Cubic Metres*	*Cubic Feet*	*Cubic Metres*
10	0.2836	210	5.9465	410	11.6099	610	17.2733	810	22.9366
20	0.5663	220	6.2297	420	11.8931	620	17.5564	820	23.2198
30	0.8495	230	6.5129	430	12.1762	630	17.8396	830	23.5030
40	1.1327	240	6.7960	440	12.4594	640	18.1228	840	23.7862
50	1.4158	250	7.0792	450	12.7426	650	18.4060	850	24.0693
60	1.6990	260	7.3624	460	13.0257	660	18.6891	860	24.3525
70	1.9822	270	7.6455	470	13.3089	670	18.9723	870	24.6357
80	2.2654	280	7.9287	480	13.5921	680	19.2555	880	24.9188
90	2.5485	290	8.2119	490	13.8753	690	19.5386	890	25.2020
100	2.8317	300	8.4951	500	14.1584	700	19.8218	900	25.4882
110	3.1149	310	8.7782	510	14.4416	710	20.1050	910	25.7683
120	3.3980	320	9.0614	520	14.7248	720	20.3881	920	26.0515

(Contd.)

Cubic Feet	Cubic Metres	Cubic Feet	Cubic Metres	Cubic Feet	Cubic Metres	Cubic Feet	Cubic Metres	Cubic Feet	Cubic Metres
130	3.6812	330	9.3446	530	15.0079	730	20.6713	930	26.3347
140	3.9644	340	9.6277	540	15.2911	740	20.9545	940	26.6178
150	4.2475	350	9.9109	550	15.5743	750	21.2376	950	26.9010
160	4.5307	360	10.1941	560	15.8574	760	21.5208	960	27.1842
170	4.8139	370	10.4772	570	16.1406	770	21.8040	970	27.4673
180	5.0970	380	10.7604	580	16.4238	780	22.0871	980	27.7505
190	5.3802	390	11.0436	590	16.7069	790	22.3703	990	28.0337
200	5.6634	400	11.3267	600	16.9091	800	22.6535	1000	28.3168

TABLE 14

Conversion of Cubic Metres to Cubic Feet

(1 inch=25.4 millimetres)

Cubic Metres	Cubic Feet	Cubic Metres	Cubic Feet	Cubic Metres	Cubic Feet	Cubic Metres	Cubic Feet	Cubic Metres	Cubic Feet
1	35.36	21	741.61	41	1447.90	61	2154.19	81	2860.49
2	70.63	22	776.92	42	1483.22	62	2189.51	82	2895.90
3	105.94	23	812.24	43	1518.53	63	2224.82	83	2931.12
4	141.26	24	847.55	44	1553.83	64	2260.14	84	2966.43
5	176.57	25	882.87	45	1589.16	65	2295.45	85	3001.75
6	211.86	26	918.18	46	1624.47	66	2330.77	86	3037.06
7	247.20	27	953.50	47	1659.79	67	2366.08	87	3072.38
8	282.56	28	988.81	48	1695.10	68	2401.40	88	3107.69
9	317.83	29	1024.13	49	1730.42	69	2436.71	89	3143.00
10	353.15	30	1059.44	50	1765.73	70	2472.03	90	3178.32
11	388.46	31	1094.75	51	1801.05	71	2507.34	91	3213.63
12	423.78	32	1130.07	52	1836.36	72	2542.66	92	3248.95
13	459.09	33	1165.38	53	1871.68	73	2577.97	93	3284.26
14	494.41	34	1200.70	54	1906.99	74	2613.28	94	3319.58
15	529.72	35	1236.01	55	1942.31	75	2648.60	95	3354.89
16	565.03	36	1271.33	56	1977.62	76	2683.91	96	3390.21
17	600.35	37	1306.64	57	2012.94	77	2719.23	97	3425.52
18	635.66	38	1341.96	58	2048.25	78	2754.54	98	3460.84
19	670.98	39	1377.27	59	2083.57	79	2789.86	99	3496.15
20	706.29	40	1412.59	60	2118.88	80	2825.17	100	3531.47

TABLE 15

Conversion of Degrees Centigrade (°C) to Degrees Fahrenheit (°F)

°C	°F	°C	°F	°C	°F	°C	°F	°C	°F
1	33.8	21	69.8	41	105.8	61	141.8	81	177.8
2	35.6	22	71.6	42	107.6	62	143.6	82	179.6
3	37.4	23	73.4	43	109.4	63	145.4	83	181.4
4	39.2	24	75.2	44	111.2	64	147.2	84	183.2

(Contd.)

°C	°F	°C	°F	°C	°F	°C	°F	°C	°F
5	41.0	25	77.0	45	113.0	65	149.0	85	185.0
6	42.8	26	78.8	46	114.8	66	150.8	86	186.8
7	44.6	27	80.6	47	116.6	67	152.6	87	188.6
8	46.4	28	82.4	48	118.4	68	154.4	88	190.4
9	48.2	29	84.2	49	120.2	69	156.2	89	192.2
10	50.0	30	86.0	50	122.0	70	158.0	90	194.0
11	51.8	31	87.8	51	123.8	71	159.8	91	195.8
12	53.6	32	89.6	52	125.6	72	161.6	92	197.6
13	55.4	33	91.4	53	127.4	73	163.4	93	199.4
14	57.2	34	93.2	54	129.2	74	165.2	94	201.2
15	59.0	35	95.0	55	131.0	75	167.0	95	203.0
16	60.8	36	96.8	56	132.8	76	168.8	96	204.8
17	62.6	37	98.6	57	134.6	77	170.6	97	206.4
18	64.4	38	100.4	58	136.4	78	172.4	98	208.4
19	66.2	39	102.2	59	138.2	79	174.2	99	210.2
20	68.0	40	104.0	60	140.0	80	176.0	100	212.0

TABLE 16

Conversion of Degrees Fahrenheit (F) to Degrees Centigrade (C)

°F	°C	°F	°C	°F	°C	°F	°C
1	−17.2	54	12.2	107	41.7	150	71.1
2	−16.7	55	12.8	108	42.2	161	71.7
3	−16.1	56	13.3	109	42.8	162	72.2
4	−15.6	57	13.9	110	43.3	163	72.8
5	−15.0	58	14.4	111	43.9	164	73.3
6	−14.4	59	15.0	112	44.4	165	73.9
7	−13.9	60	15.6	113	45.0	166	74.4
8	−13.3	61	16.1	114	45.6	167	75.0
9	−12.8	62	16.7	115	46.1	168	75.6
10	−12.2	63	17.2	116	46.7	169	76.1
11	−11.2	64	17.8	117	47.2	170	76.7
12	−11.1	65	18.3	118	47.8	171	77.2
13	−10.6	66	18.9	119	48.3	172	77.8
14	−10.0	67	19.4	120	48.9	173	78.3
15	− 9.4	68	20.0	121	49.4	174	78.9
16	−8.9	69	20.6	122	50.0	175	79.4
17	−8.3	70	21.1	123	50.6	176	80.0
18	−7.8	71	21.7	124	51.1	177	80.6
19	−7.2	72	22.2	125	51.7	178	81.1
20	−6.7	73	22.8	126	52.2	179	81.7
21	−6.1	74	23.3	127	52.8	180	82.2
22	−5.6	75	23.9	128	53.3	181	82.8
23	−5.0	76	24.4	129	53.9	182	83.3
24	−4.4	77	25.0	130	54.4	183	83.9

(Contd.)

°F	°C	°F	°C	°F	°C	°F	°C
25	−3.9	78	25.6	131	55.0	184	84.4
26	−3.3	79	26.1	132	55.6	185	85.0
27	−2.8	80	26.7	133	56.1	186	85.6
28	−2.2	81	27.2	134	56.7	187	86.1
29	−1.7	82	27.8	135	57.2	188	86.7
30	−1.1	83	28.3	136	57.8	189	87.2
31	−0.6	84	28.9	137	58.3	190	87.8
32	0.0	85	29.4	138	58.9	191	88.3
33	0.6	86	30.0	139	59.4	192	88.9
34	1.1	87	30.6	140	60.0	193	89.4
35	1.7	88	31.1	141	60.6	194	90.0
36	2.2	89	31.7	142	61.1	195	90.6
37	2.8	90	32.2	143	61.7	196	91.1
38	3.3	91	32.8	144	62.2	197	91.7
39	3.9	92	33.3	145	62.8	198	92.2
40	4.4	93	33.9	146	63.3	199	92.8
41	5.0	94	34.4	147	63.9	200	93.3
42	5.6	95	35.0	148	64.4	201	93.9
43	6.1	96	35.6	149	65.0	202	94.4
44	6.7	97	36.1	150	65.6	203	95.0
45	7.2	98	36.7	151	66.1	204	95.6
46	7.8	99	37.2	152	66.7	205	96.1
47	8.3	100	37.8	153	67.2	206	96.7
48	8.9	101	38.3	154	67.8	207	97.2
49	9.4	102	38.9	155	68.3	208	97.8
50	10.0	103	39.4	156	68.9	209	98.3
51	10.6	104	40.0	157	69.4	210	98.9
52	11.1	105	40.6	158	70.0	211	99.4
53	11.7	106	41.1	159	70.6	212	100.0

TABLE 17

Conversion of Pounds per Square Inch to Kilograms per Square Centimetre

1 Inch = 25.4 millimetres
1 Pound = 0.4535924 kilogram

Pound per sq. inch	*Kilograms per sq. cm.*	*Pound per sq. inch*	*Kilograms per sq. cm.*	*Pound per sq. inch*	*Kilograms per sq. cm.*	*Pound per sq. inch*	*Kilograms per sq. cm.*	*Pound per sq. inch*	*Kilograms per sq. cm.*
1	0.0703	21	1.4765	41	2.8826	61	4.2887	81	5.6949
2	0.1406	22	1.5468	42	2.9529	62	4.3590	82	5.7652
3	0.2109	23	1.6171	43	3.0232	63	4.4293	83	5.8355
4	0.2812	24	1.6874	44	3.0935	64	4.4997	84	5.9058
5	0.3515	25	1.7577	45	3.1638	65	4.5700	85	5.9761
6	0.4218	26	1.8280	46	3.2341	66	4.6403	86	6.0464
7	0.4922	27	1.8983	47	3.3044	67	4.7106	87	6.1167

(Contd.)

Pound per sq. inch	Kilograms per sq. cm.	Pound per sq. inch	Kilograms per sq. cm.	Pound per sq. inch	Kilograms per sq. cm.	Pound per sq. inch	Kilograms per sq. cm.	Pound per sq. inch	Kilograms per sq. cm.
8	0.5625	28	1.9686	48	3.3747	68	4.7809	88	6.1870
9	0.6328	29	2.0389	49	3.4450	69	4.8512	89	6.2573
10	0.7031	30	2.1092	50	3.5154	70	4.9215	90	6.3276
11	0.7734	31	2.1795	51	3.5857	71	4.9918	91	6.3979
12	0.8437	32	2.2498	52	3.6560	72	5.0621	92	6.4682
13	0.9140	33	2.3201	53	3.7263	73	5.1324	93	6.5385
14	0.9843	34	2.3904	54	3.7966	74	5.2027	94	6.6089
15	1.0546	35	2.4607	55	3.8669	75	5.2730	95	6.6792
16	1.1249	36	2.5311	56	3.9372	76	5.3433	96	6.7495
17	1.1952	37	2.6014	57	4.0050	77	5.4136	97	6.8198
18	1.2655	38	2.6717	58	4.0778	78	5.4839	98	6.8901
19	1.3358	39	2.7420	59	4.1481	79	5.5543	99	6.9604
20	1.4061	40	2.8123	60	4.2184	80	5.6246	100	7.0307

TABLE 18

Conversion of Kilograms per Square Centimetre to Pound per Square Inch

Kilograms per sq. cm.	Pound per sq. inch	Kilograms per sq. cm.	Pound per sq. inch	Kilograms per sq. cm.	Pound per sq. inch	Kilograms per sq. cm.	Pound per sq. inch	Kilograms per sq. cm.	Pound per sq. inch
0.1	1.422	2.1	29.869	4.1	58.316	6.1	86.762	8.1	115.209
0.2	2.845	2.2	31.291	4.2	59.738	6.2	88.185	8.2	116.631
0.3	4.267	2.3	32.714	4.3	61.160	6.3	89.607	8.3	118.054
0.4	5.689	2.4	34.136	4.4	62.583	6.4	91.029	8.4	119.476
0.5	7.112	2.5	35.558	4.5	64.005	6.5	92.452	8.5	120.898
0.6	8.534	2.6	36.981	4.6	65.427	6.6	93.874	8.6	122.321
0.7	9.956	2.7	38.403	4.7	66.850	6.7	95.296	8.7	123.743
0.8	11.379	2.8	39.825	4.8	68.272	6.8	96.719	8.8	125.165
0.9	12.801	2.9	41.248	4.9	69.694	6.9	98.141	8.9	126.588
1.0	14.223	3.0	42.670	5.0	71.117	7.0	99.563	9.0	128.010
1.1	15.646	3.1	44.092	5.1	72.539	7.1	100.986	9.1	129.432
1.2	17.068	3.2	45.515	5.2	73.961	7.2	102.408	9.2	130.855
1.3	18.490	3.3	46.937	5.3	75.384	7.3	103.830	9.3	132.277
1.4	19.913	3.4	48.359	5.4	76.806	7.4	105.253	9.4	133.699
1.5	21.335	3.5	49.782	5.5	78.228	7.5	106.675	9.5	135.122
1.6	22.757	3.6	51.204	5.6	79.651	7.6	108.097	9.6	136.544
1.7	24.180	3.7	52.626	5.7	81.073	7.7	109.520	9.7	137.966
1.8	25.606	3.8	54.049	5.8	82.495	7.8	110.942	9.8	139.389
1.9	27.024	3.9	55.471	5.9	83.918	7.9	112.364	9.9	140.811
2.0	28.447	4.0	56.893	6.0	85.340	8.0	113.787	10.0	142.233

TABLE 19

Density of Oil-field Waters

Specific Gravity of Water at 60° F	Approximate Total Solids in Parts per Million (mg/litre)	Hydrostatic Pressure Gradient (psi/ft)[1]
1.000	none	0.433
1.010	13,500	0.437
1.020	27,500	0.441
1.030	41,400	0.445
1.040	55,400	0.450
1.050	69,400	0.454
1.060	83,700	0.459
1.070	98,400	0.463
1.080	113,200	0.467
1.090	128,300	0.471
1.100	143,500	0.476
1.110	159,500	0.480
1.120	175,800	0.485
1.130	192,400	0.489
1.140	210,000	0.493

Source: From C.E. Reistle and C.E. Lone, Tech. Paper 432, U.S. Bur. Mines (1928). See also O. & G. Jour., February 3, 1945, p. 69, and December 23, 1948, p. 75.

[1]Water compressibility and the presence of gas in solution have been ignored.

TABLE 20

Densities and Specific Volumes of Petroleum
(all measurements at 60° F)

Degrees API Gravity	Specific Gravity	Gallons per Pound	Pounds per Gallon	Pounds per Barrel	Barrels per Short Ton	Barrels per Metric Ton	Barrels per long Ton
0	1.076	0.1116	8.962	376.40	5.31	5.86	5.95
10	1.000	0.1201	8.328	349.78	5.71	6.30	6.40
15	0.9659	0.1243	8.044	337.85	5.92	6.53	6.63
18	0.9465	0.1269	7.882	331.04	6.04	6.66	6.77
20	0.9340	0.1286	7.778	326.68	6.12	6.75	6.86
22	0.9218	0.1303	7.676	322.39	6.20	6.84	6.95
24	0.9100	0.1320	7.578	318.28	6.28	6.93	7.04
26	0.8984	0.1337	7.481	314.20	6.37	7.02	7.13
28	0.8871	0.1354	7.387	310.25	6.45	7.11	7.22
30	0.8762	0.1371	7.296	306.43	6.53	7.19	7.31
32	0.8654	0.1380	7.206	302.65	6.61	7.28	7.40
34	0.8550	0.1405	7.119	299.00	6.69	7.37	7.49

(Contd.)

Degrees API Gravity	Specific Gravity	Gallons per Pound	Pounds per Gallon	Pounds per Barrel	Barrels per Short Ton	Barrels per Metric Ton	Barrels per long Ton
36	0.8448	0.1422	7.034	295.43	6.77	7.46	7.58
38	0.8348	0.1439	6.951	291.94	6.85	7.55	7.67
40	0.8251	0.1456	6.870	288.54	6.93	7.64	7.76
42	0.8155	0.1473	6.790	285.18	7.01	7.73	7.85
44	0.8063	0.1490	6.713	281.95	7.09	7.82	7.94
46	0.7972	0.1507	6.637	278.75	7.17	7.91	8.04
48	0.7883	0.1524	6.563	275.65	7.26	8.00	8.13
50	0.7796	0.1541	6.490	272.58	7.34	8.09	8.23
55	0.7587	0.1583	6.316	365.27	7.54	8.31	8.44
60	0.7389	0.1626	6.151	258.34	7.74	8.53	8.67
65	0.7201	0.1668	5.994	251.75	7.94	8.76	8.90
70	0.7022	0.1711	5.845	245.49	8.15	8.98	9.12
75	0.6852	0.1753	5.703	239.53	8.35	9.20	9.35
80	0.6690	0.1796	5.568	233.86	8.55	9.43	9.58
85	0.6536	0.1838	5.440	228.48	8.75	9.65	9.80
90	0.6388	0.1881	5.316	223.27	8.96	9.87	10.03
95	0.6247	0.1924	5.199	219.36	9.16	10.10	10.26
100	0.6112	0.1966	5.086	213.61	9.36	10.32	10.49

Source: Adopted from *Petroleum and Figures 9th ed.* American Petroleum Institute.

(b) Conversion Equivalents

	U.S. gallons	Imperial gallons	Cubic feet	Barrels	Cubic centimetres	Litres	Cubic metres
1 U.S. gallon*	1	0.83268	0.13368	0.02381	3785.4	3.7853	0.0037854
1 Imperial gallon†	1.201	1	.16054	.028594	4546.04	4.5460	.004546
1 Cubic foot	7.4805	6.22888	1	.17811	28317.01	28.316	.028317
1 Barrel‡	42	34.972	5.6146	1	158987.55	158.98	.15899
1 Cubic centimetre	.00026417	.00021996	.000035315	.0000062895	1	.00099997	.000001
1 Litre	.26418	.219976	.035316	.0062899	1000.027	1	.00100027
1 Cubic metre	264.17	219.97	35.314	6.2898	1000000	999.97	1

*U.S. gallon the volume occupied by 231 cu. inches

†1 Imperial gallon the volume occupied by 10 lb. of water at 62° F when weighed against brass in air at 30 inch barometric pressure

‡1 barrel = 42 U.S. gallons

Name Index

Subject Index

254
Advances in Polymer Science

For further volumes:
http://www.springer.com/series/12

Aims and Scope

The series Advances in Polymer Science presents critical reviews of the present and future trends in polymer and biopolymer science. It covers all areas of research in polymer and biopolymer science including chemistry, physical chemistry, physics, material science.

The thematic volumes are addressed to scientists, whether at universities or in industry, who wish to keep abreast of the important advances in the covered topics.

Advances in Polymer Science enjoys a longstanding tradition and good reputation in its community. Each volume is dedicated to a current topic, and each review critically surveys one aspect of that topic, to place it within the context of the volume. The volumes typically summarize the significant developments of the last 5 to 10 years and discuss them critically, presenting selected examples, explaining and illustrating the important principles, and bringing together many important references of primary literature. On that basis, future research directions in the area can be discussed. Advances in Polymer Science volumes thus are important references for every polymer scientist, as well as for other scientists interested in polymer science - as an introduction to a neighboring field, or as a compilation of detailed information for the specialist.

Review articles for the individual volumes are invited by the volume editors. Single contributions can be specially commissioned.

Readership: Polymer scientists, or scientists in related fields interested in polymer and biopolymer science, at universities or in industry, graduate students.

Special offer:
For all clients with a standing order we offer the electronic form of Advances in Polymer Science free of charge.

P.K. Dutta • Joydeep Dutta
Editors

Multifaceted Development and Application of Biopolymers for Biology, Biomedicine and Nanotechnology

With contributions by

A.K. Anal · S. Banerjee · A. Bhowmick · Y.M. Chabre ·
J. Dutta · P.K. Dutta · J. Gopikrishna · M.K. Gupta ·
C. Haldar · R. Jayakumar · R. Kaur · N. Kottari · R. Kumar ·
P.P. Kundu · H.T. Lee · P. Maiti · S. Maya · D. Menon ·
S.V. Nair · D. Narayanan · S.K. Pandey · D.K. Patel ·
P. Pushp · R. Roy · M. Sabitha · R. Sharma · R. Srivastava ·
A. Tuladhar

Springer

Editors
P.K. Dutta
MN National Institute of Technology
Department of Chemistry
Allahabad, Uttar Pradesh
India

Joydeep Dutta
Institute of Engineering & Technology,
ITM University
Department of Applied Sciences & Humanities
Raipur, Chattrisgarh
India

ISSN 0065-3195 ISSN 1436-5030 (electronic)
ISBN 978-3-642-40122-0 ISBN 978-3-642-40123-7 (eBook)
DOI 10.1007/978-3-642-40123-7
Springer Heidelberg New York Dordrecht London

Library of Congress Control Number: 2013949496

Printed on acid-free paper

Springer is part of Springer Science+Business Media (www.springer.com)

Preface

Polymers today have a very important role in the technological advancement of our society. Of the polymers, biopolymers are particularly attractive materials because they are generally nontoxic and biodegradable, thus combining excellent functional properties with environmental friendliness and prospects for sustainable development. Furthermore, the input materials used for the production of these polymers can be either renewable, based on agricultural plant or animal products, or synthetic. In this regard, synthetic biopolymers are more costly than natural biopolymers, but both "types" of biopolymers possess interesting properties that make them suitable for a wide repertoire of biomedical applications. In addition, multifunctional entities can be created through decoration of the polymer with specific molecules such as proteins, peptides, drugs, antibodies, biomimetic ligands, transfection agents, cells and various ligands. The advent of nanotechnology has given a new momentum towards the development of biopolymer-based nanomaterials, which can undoubtedly change the shape of our lives.

The present volume entitled "Multifaceted Development and Application of Biopolymers for Biology, Biomedicine and Nanotechnology" is an attempt to provide the scientific community and entrepreneurs with a thorough understanding and knowledge of the combination of nanotechnological, biotechnological, technological and medical aspects of the recent advances in biopolymer applications.

The chapter "Functionalized Nanoparticles and Chitosan-Based Functional Nanomaterials" summarizes the main advances in design and preparation of nanoparticles published over the last 10 years in terms of polymeric nanomaterials, emphasizing functionalized metal nanoparticles, carbon nanotubes, graphene, fullerene, liposomes, quantum dots and nanocomposites, and the current developing interest in functionalization of chitosan derivatives in the form of nanomaterials to provide new strategies for a wide range of applications. The chapter "Nanoparticles for Gene Delivery into Stem Cells and Embryos" describes the available nanoparticle-based gene delivery systems and their use in stem cells for maintaining self-renewal, pluripotency and/or targeted differentiation into specific cell types for cell-based therapy and/or gene therapy. The chapter also discusses and reviews the progress and future of nanoparticles for the generation of transgenic animals via gene delivery into embryos – a research area that is yet to be fully explored. The chapter "Engineering of Polysaccharides via Nanotechnology" deals with various sources of chitosan and their physical, chemical and biological properties; synthesis of chitosan-based nanoparticles, nanospheres and nanogels; characterization; and biomedical applications. The chapter "Hydroxyapatite-Packed Chitosan-PMMA Nanocomposite: A Promising Material for Construction of Synthetic Bone" is focused on natural bone structure, polymer blends and recent advances in polymer/(nano)hydroxyapatite nanocomposites for application as synthetic bone. The chapter "Biodegradable Polymers for Potential Delivery Systems for Therapeutics" will provide support to research scientists and clinical physicians who are interested in the development and application of biodegradable polymeric nanoparticles as potential delivery systems for therapeutics. The chapter "Phytomedicine-Loaded Polymeric Nanomedicines: Potential Cancer Therapeutics" explains the design and development of nanoformulations and their application as potential anticancer agents. The chapter "Proteins and Carbohydrates as Polymeric Nanodrug Delivery Systems: Formulation, Properties and Toxicological Evaluation" embodies an in-depth discussion of various carbohydrate- and protein-based nanomedicines with respect to their formulation, properties of the nanoconstructs, manipulation and utilization of these properties towards developing a better drug delivery device and also the toxicological interactions of these nanocarriers with host physiological systems. The chapter "Biopolymeric Micro- andNanoparticles: Preparation, Characterization and Industrial Applications" reviews the preparation, characterization and applications of biopolymeric micro- and nanoparticles. The chapter "Applications of Glyconanoparticles as Glycobiological Therapeutics and Diagnostics" provides an overview of the most recent syntheses and applications of glyconanoparticles in glycoscience that have deepened our understanding of multivalent carbohydrate–protein interactions. Together with sensitive detection devices, inhibitors of bacterial adhesion to host tissues, cancer vaccines and therapeutic systems (including photosensitizers for photodynamic therapies), these novel bionanomaterials are finding widespread relevance. The recent development of glyconanoparticle-based chemotherapeutic drug delivery agents is also addressed, together with their molecular imaging properties for MRI or X-ray contrast agents, as platforms for selective immunolabelling and imaging of cells, and as colorimetric devices in various bioassays.

Last but not least, we would like to thank all contributors for their generous support, the publisher for accepting our book and the administrative heads of both MNNIT Allahabad and IET Raipur, India, for their encouragement and cooperation, without which it would have been extremely difficult to complete this task in time.

Allahabad, India P.K. Dutta
Raipur, India Joydeep Dutta
April 2013

Contents

Adv Polym Sci (2013) 254: 1–50
DOI: 10.1007/12_2012_200

Published online: 3 January 2013

Functionalized Nanoparticles and Chitosan-Based Functional Nanomaterials

P.K. Dutta, Rohit Srivastava, and Joydeep Dutta

Abstract Newly developed nanomaterials offer unique opportunities in the fields of industry and medical sciences that are complementary to current technology. Nanomaterials can be obtained through physicochemical processes from various inorganic and organic substances. The properties and functions of materials can be tuned through controlling the composition, structure, and morphology of the nanoparticles. Chitosan is the principle derivative of chitin, which is the second-most naturally occurring polysaccharide after cellulose. Chitosan has an amino group in the C-2 position and OH groups in the C-3 and C-5 positions of each repeat unit and can react with functional nanomaterials through various kinds of reaction mechanisms. There have been several reports on the preparation of nanoparticles and functional nanomaterials and their uses. This chapter summarizes the main advancements in the design and preparation of nanomaterials over the last 10 years, with an emphasis on functionalized metal nanoparticles, carbon nanotubes, graphene, fullerene, liposomes, quantum dots, and nano-composites, and outlines the current developing interest in functionalization of chitosan derivatives in the form of nanomaterials to provide new strategies for a wide range of applications.

Keywords Carbon nanotube · Chitosan · Nanomaterials · Tissue engineering

---

P. Dutta (✉)
Department of Chemistry, MN National Institute of Technology, Allahabad 211004, India
e-mail: pkd_437@yahoo.com

R. Srivastava
Department of Chemistry, MN National Institute of Technology, Allahabad 211004, India

Department of Applied Chemistry, Birla Institute of Technology, Deoghar Campus, Deoghar 814142, India

J. Dutta
Department of Applied Sciences & Humanities, Institute of Engineering & Technology, ITM University, Raipur, Chattrisgarh 493661, India

Contents

1 Introduction

Nanosized or nanostructured materials, commonly known as nanomaterials, are the backbone of nanotechnology and nanoscience. They have unique physicochemical properties compared with the bulk materials of the same composition and have been demonstrated to be capable of changing their properties and applications. Nanomaterials are a "minute discrete entity with at least one dimension being 100 nm or less" [1]. Nanomaterials are mostly found as: (1) carbon-based nanomaterials that consist of carbon atoms and can possess different nanostructures, e.g., single-walled or multiwalled carbon nanotubes (CNT), graphene, and fullerene. (2) Metallic nanomaterials, which can consist of nanomaterials of metals, transition metals and their compounds or composites, e.g., gold, iron oxide, silver, silica, and quantum dots. (3) Silicon nanomaterials, which are mainly composed of silicon and its compounds, e.g., silicon or silica nanoparticles. (4) Organic nanomaterials, which are formed via the agglomeration or assembly of organic molecules, e.g., dendrimers, polymers biomolecules, or biomacromolecules Nanomaterials with size <50 nm can enter most cells without any difficulty. When nanomaterials are <20 nm, they can enter the human body and can be used in the treatment of various diseases, including cancer, human immunodeficiency syndrome, viral infection, and central nervous system disorders, because most disease processes occur at the molecular and cellular levels. Further, the ideal imaging resolution is in nanometers because most biological processes take place on this length scale. Therefore, the functionalization of nanoparticles and nanomaterials and their use in biomedical and pharmaceutical industries is of great interest. Functionalization involves the design and development of novel functional nanomaterials such as multifunctional liposomal [2], functionalized fullerenes [3, 4], functionalized nanotubes [5], functionalized nanoparticles [6], polymeric micelles [7], dendrimers [8], nanoshells [9], and polymeric microspheres [10]. Both academic communities and the public worldwide have already benefited from the unique application of these nanomaterials.

Polymer-based functional nanomaterials are mostly used as nanosized drug carriers. When the drug are dissolved or encapsulated in the nanoparticles some interactions occur both physically and chemically. The polymeric nanoparticles have some advantages for drug delivery. Nanoparticles can be made from polymers (polymeric nanoparticles, micelles, or dendrimers), lipids (liposomes), viruses, and even from organometallic compounds. The use of polymeric nanoparticles as drug delivery systems has many advantages [11–14]. Two types of polymers are used as drug conjugates, i.e., natural and synthetic polymers. Various types of polymers such as albumin, chitosan, and heparin occur naturally and are used to deliver oligonucleotides, DNA, proteins, and drugs.

Of all the natural polymers, chitosan has received increased attention in recent years, especially for biomedical applications [15]. Chitosan is the second-most abundant biopolymer in nature, just after cellulose. It is a linear polysaccharide consisting of β-(1,4) linked D-glucosamine residues and *N*-acetyl-glucosamine groups [16] and is obtained from chitin by deacetylation. Chitin is commonly found in the shells of insects and crustaceans, as well as in the cell walls of some fungi. Chitosan has some unique properties, including biocompatibility, biodegradability, hydrophilicity, nontoxicity and nonantigenicity, as well as bioadherence and cell affinity [17]. It is the only positively charged naturally occurring polysaccharide and can interact strongly with negatively charged entities. Thus, chitosan itself has several inherent characteristics and no doubt its use in functional nanomaterials [18] will draw the attention of the scientific community. However, the work reported on chitosan-based functional nanomaterials is limited and not well documented. In this chapter, an effort has been made to develop an interest in chitosan-based functional nanomaterials, looking at its versatility and end-use applications for the benefit of the readers, researchers and workers involved in this field of research.

2 Functionalized Nanomaterials

Although nanomaterials have a wide range of applications, these application are limited. However, when these nanomaterials are functionalized with different organic or inorganic ligands and various type of biomolecules then the number of applications increases rapidly. Although some nanomaterials have excellent physical and chemical bulk properties, they do not posses surface properties suitable for specific applications. Consequently, it may be necessary to modify or functionalize the surface of such materials. There are the several advantages to the surface modification or functionalization of nanomaterials, e.g. to stabilize nanomaterials against agglomeration [19], to render them compatible with another phase, to use modified inorganic nanofillers in organic polymers, and to enable their self-organization. Functional organic groups on the particle surface may allow deliberate interaction of the nanoparticles with molecules, other nanoparticle surfaces or solids. The functionalized nanoparticles with combination of unique

Table 1 Physicochemical properties of metal nanoparticles

Physical state	Characteristics
Shape and size	Potential for various applications such as catalysis [32], biosensing [33, 34], recording media [35], and optics [36]
Noble metal nanoparticles	Nanoparticles in the form of colorants [37], metal coatings [38], electronics [39], optics [40], chemical catalysts [41], and medicines [42]
Assembled nanoparticles	Used in diverse field such as photodetectors, nanoelectronics, and chemical and biological sensors [43, 44]
Nanoparticles of different geometrics and techniques	Nanowires in the microchip industry and as nanowaveguides for electromagnetic radiation, for solvent evaporation of hydrophobic nanoparticle molecular crosslinking in colloidal aggregates and templates [45–47], and in assemblies using biomacromolecules [48] such as DNA [49] and bacterial S-layer proteins [50]

structural characteristics and integrated functions will attract increasing research interest and could lead to new opportunities in biomedical applications [20].

3 Metal Nanoparticles

Nanoparticles can be composed of any substance, including metals [21, 22], semiconductors [23, 24], core–shell composite architectures [25–27], and organic polymers [28]. These particles often display properties intermediate between quantum and bulk materials because of their intermediate size [29] and large surface area to volume ratio [30]. Nanoparticle of different sizes and shapes exhibit different absorbance and fluorescence features and reveal polymerization effects [31]. Some of the characteristics of metallic nanoparticles are shown in Table 1.

Several types of NPs have been employed for various applications. We have emphasized three types of functionalized metal nanoparticles, i.e., gold, iron oxide, and silica nanoparticles.

3.1 Gold Nanoparticles

Functionalized gold nanoparticles (GNPs) have been the subject of intense research during the past decade due to their potential applications in gene delivery, drug delivery, sensing, imaging, and chemotherapy [51–55]. Mostly, GNPs are used for biomedical application because of their very good oxidation resistance, easy synthesis and optical properties. The ultimate goal in functionalization is to preserve the properties of both the GNP and the bound biological molecule. In other words, the biological molecule should be stable and able to retain its biorecognition properties, and the GNPs should be able to retain their unique properties such as strong plasmon

Table 2 Preparation of functional GNPs by various workers and their observations

Preparation method	Observations	References
GNPs reduced and stabilized by citrate have a negative surface charge and preferentially bind to thiol, amine, cyanide, or diphenylphosphine functional groups	Binding via α-amino groups at low pH and suppressed at neutral and high pH, due to electrostatic repulsion between the surface of gold and the charged carboxyl groups	Zhong et al. [58]
Fluorescent 3D superlattices of dansyl glutathione-protected GNPs, with potential applications in molecular detection	Functional nanoparticle-based solids make it possible to use them in novel applications like selective detection of bovine serum albumin (BSA) based on the selective binding of the naphthalene ring of the dansyl moiety with site I of BSA	Edakkattuparambil et al. [59]
GNP/CNT hybrid platform with horseradish peroxidase-functionalized GNP label for sensitive detection of human IgG as a model protein	This approach provided a linear response range between 0.125 and 80 ng/mL, with a detection limit of 40 pg/mL and showed good precision, with acceptable stability and reproducibility. Used for the detection of human IgG in real samples, which provides a potential alternative tool for the detection of protein in the clinical laboratory	Cui et al. [60]
Rapid and sensitive colorimetric sensor for the detection of Hg^{2+} using L-cysteine-functionalized GNPs induced by ultraviolet radiation	The L-cysteine-modified GNPs aggregate quickly in the presence of Hg^{2+}. Functionalized GNPs can be used for delivery of various payloads into cells, such as small drug molecules or large biomolecules like DNA and siRNA	Chai et al. [61]
Charge-reversal functional GNPs prepared by layer-by-layer technique	To deliver small interfering RNA (siRNA) and plasmid DNA into cancer cells. Polyacrylamide gel electrophoresis measurements of siRNA confirmed the occurrence of the charge-reversal property of functional GNPs	Guo et al. [62]
Functionalized GNPs with branched 2 kDa polyethyleneimine (PEI) for transfection vectors	The transfection efficiency into monkey kidney (Cos-7) cells varied with the PEI:gold molar ratio in the conjugates. The study indicates that increasing the hydrophobicity of the transfection agent enhances cellular internalization. Positively charged synthetic materials are widely used as transfection vectors	Klibanov et al. [63]

(continued)

Table 2 (continued)

Preparation method	Observations	References
GNPs functionalized with single-stranded oligodeoxynucleotides for gene therapy	Incubating cells with fluorophore-labeled DNA conjugated to nanoparticles resulted in efficient uptake of the particles. Protein expression was controlled by conjugation of antisense DNA with GNPs. EGFP-expressing C166 cells incubated with antisense DNA-functionalized nanoparticles (ASNP) resulted in reduction of fluorescence intensity in cells incubated with ASNP	Mirkin et al. [64]
GNPs coated with a short peptide to promote intracellular delivery of membrane-impermeable proteins	Microscopy and enzyme assays showed that the particles were able to transport functional enzymes into a variety of cell lines. Significantly, the transported proteins were able to escape from endosomes and the particles showed no apparent cytotoxicity	Ghosh et al. [65]

absorption bands [56], light scattering [57], etc. For biomedical applications, surface-functionalization of GNPs is essential in order to target them to specific disease areas and allow them to selectively interact with cells or biological molecules. In general, functionalization of GNPs can be performed by using either chemical functional groups or biological molecules. Colloidal GNPs are normally stabilized against aggregation by using long hydrocarbon ligand chains consisting of various functional groups. Table 2 describes the preparation and evaluation of functional GNPs by various workers.

It has also been seen that two nanomaterials can be functionalized with each other. One example is the synthesis of fullerene (C_{60})-functionalized GNPs by the coupling of the fullerene molecules with peripheral amine moieties on the particle surface [66]. On the other hand, GNPs were selectively attached to chemically functionalized surface sites on nitrogen-doped carbon nanotubes. A cationic polyelectrolyte was adsorbed on the surface of the nanotubes by electrostatic interaction between carboxyl groups on the chemically oxidized nanotube surface and polyelectrolyte chains.

Functional gold nanomaterials have been used for various types of biolabeling and life-related molecular diagnostics due to their unique optical, catalytic, and biocompatible properties. Recently, attention has been paid to chemiluminescent functionalized gold nanomaterials as bioprobes for bioassays [67].

3.1.1 GNPs in Polymers

The use of various functionalized polymers as stabilizers in the design of metal core–organic shell hybrid nanoparticles architectures has attracted increasing interest for different applications. Polymer chemistry allows for many variations, whereby polymeric nanoparticles can be easily manipulated without the loss of their desired physical, chemical, and biological properties. The conjugation of GNPs with functional polymer is the first step towards fabricating functional gold nanocomposites. Many investigations involving the fabrication of GNPs functionalized with polymer have been reported, such as "grafting from" fabrication (Table 3), "grafting to" fabrication (Table 4), and post-modification.

Chemically bonded scaffolds are superior to physically adsorbed scaffolds in terms of the robustness of the resulting gold nanocomposites. Although most polymers introduced by the "grafting from" method are artificial polymers [80, 81], representative biopolymers such as oligonucleotides and peptides can also be introduced.

A marked advantage of the covalent "grafting to" technique is the possibility of achieving a high surface graft density of polymer brush on the GNP surface.

3.1.2 Post-modification

Conjugation of as-prepared GNPs with as-prepared polymers is the most common and simplest method for preparing gold nanocomposites because mixing of the as-prepared materials can eliminate uncertainty factors such as the dispersion of GNP size and molecular weight.

3.2 Iron Oxide Nanoparticles

Iron-oxide nanoparticles (IONPs) represent a significant class of inorganic nanomaterial that is contributing to the current revolution in nanomedicine [82, 83]. Their unique physical properties, including high surface area to volume ratios and superparamagnetism, confer useful attributes for medical applications such as magnetic resonance imaging (MRI), drug and gene delivery, tissue engineering, and bioseparation [84]. Superparamagnetic iron-oxide nanoparticles (SPIONs) have been prepared with a mean particle diameter of 50–100 nm, and ultrasmall superparamagnetic iron-oxide nanoparticles (USPIONs) with a size below 50 nm. These two classes of IONPs have been studied widely for medical applications, particularly as the next (potential) generation of MRI contrast agents. They are also seen as potential vectors for drug and gene delivery. The biodistribution of these nanoparticles can be altered by the application of an external magnetic field; they also have potential applications in hyperthermia therapy because some magnetic

Table 3 "Grafting from" fabrication of GNPs in polymers

Preparation method	Observations	References
Derivatization of tiopronin-protected GNPs with ethylenediamine and bis (3-aminopropyl)-terminated PEG and their functionalization with the GRGDSP peptide sequence	Particles subsequently tested in vitro with a human fibroblast cell line to determine the biocompatibility and the cell–particle interactions, using fluorescence and scanning electron microscopies	De la Fuente et al. [68]
Water-soluble thioether polymers introduced as attractive ligands to produce monodisperse nanoparticles in aqueous media	A very useful system for restricting the size distribution and for facile functionalization with other ligands for a variety of applications. This technique was successful in the preparation of monodisperse fluorescent GNPs with diameters of 1.1–1.7 nm	Wang et al. [69] and Baker et al. [70]
Facile preparation of size-controlled GNPs using versatile and end-functionalized thioether polymer ligands	Multidentate thioether polymeric ligands (PTMP–PVAc) lead to formation of smaller but special "multimer" morphology in the organic phase. Fairly uniform nanoparticles were produced using monodentate thioether functionalized ligands (DDT–PVAc). Further modification of such polymer ligands to introduce hydrophilic functionalities results in the phase transfer of GNPs from organic to aqueous media. Hybrids are used to trap and encapsulate other nanoparticles, biomolecules, dyes, or drugs by a temperature-introduced "breathing" process. Fabrication of pH-responsive nanocomposites of GNPs and poly (4-vinylpyridine) are used as smart supports to entrap transition metal ions. The ions were reduced in situ to construct novel bimetallic nanocomposites, regarded as intelligent catalysts with activity regulated by environmental stimuli	Haung et al. [71]
Core–shell GNPs and a slightly crosslinked poly(*N*-isopropyl acrylamide) hybrid	Water-soluble GNPs carry primary amino groups at the solvent-exposed interface that are used for further conjugation of biologically active molecules	Li et al. [72]

(continued)

Table 3 (continued)

Preparation method	Observations	References
Quantification and reactivity of functional groups in the ligand shell of PEGylated GNPs via a fluorescence-based assay	The "grafting from" technique used with atom transfer free-radical polymerization (ATRP) to grow polymer chains from the surface of GNPs (~20 nm) enables dense, uniform, and homogeneous coverage of polymer chains on the surface of GNPs. Other advantages like growth of polymer chains without appreciable chain termination or chain transfer, and the presence of an active initiator site at the end of the growing polymer chain, facilitates synthesis of surface-grafted block copolymers	Maus et al. [73]
Thermoresponsive polymer brushes on 20 nm colloidal gold formed through ATRP of *N*-isopropyl acrylamide in aqueous media	The system was found to exhibit thermoresponsive behavior above and below the LCST	Chakraborty et al. [74]
End-functionalized 3D self-assembled monolayers (SAMs) on GNPs by living cationic ring-opening polymerization reaction directly on GNP surfaces	Dense polymer brushes were prepared in a "one-pot multistep" reaction. Gold/polymer nanocomposites were very stable, "brush-type" shells of linear macromolecules and terminal mesogen prepared by means of a quantitative termination reaction. The result was an amphiphilic core–shell material with a well-defined hydrophilic–lipophilic balance	Jordan et al. [75]
Preparation of hydrogels through the chemical crosslinking of poly(vinyl alcohol) (PVA) and functionalized GNPs	Carboxylic group-functionalized GNPs were synthesized, dispersed in a PVA matrix, and allowed to react with the hydroxyl groups of PVA at high temperature	Moreno et al. [76]

particles can heat up under the influence of a localized high-frequency magnetic field.

There are many chemical methods that can be used to synthesize magnetic nanoparticles for biomedical applications, i.e., microemulsification [85], sol–gel syntheses [86], sonochemical reactions [87], hydrothermal reactions [88], hydrolysis and thermolysis of precursors [89], flow injection syntheses [90], and electrospray syntheses [91].

Biofunctionalization of IONPs with biological molecules is an alternative functionalization approach, which displays unique characteristics. Biofunctionalized

Table 4 "Grafting to" fabrication of GNPs in polymers

Preparation method	Observations	References
Post-modification route for preparation of smart hybrid GNPs with poly (4-vinylpyridine) (P4VP) based on RAFT and click chemistry	A new azide-terminated ligand was first synthesized to modify GNPs by ligand exchange reaction, and then click reaction used to graft alkyne-terminated P4VP, which was prepared by RAFT, onto the surface of GNPs	Zhang et al. [77]
A sensitive mediatorless sensor of H_2O_2. Gold electrode was immersed in thiol-functionalized poly(styrene-*co*-acrylic acid) (St-*co*-AA) nanosphere latex prepared by emulsifier-free emulsion polymerization of St with AA and functionalized with dithioglycol to assemble the nanospheres. Then, GNPs were chemisorbed onto the thiol groups and formed monolayers on the surface of poly(St-*co*-AA) nanospheres. Finally, horseradish peroxidase was immobilized on the surface of the GNPs	The sensor displayed an excellent electrocatalytical response to reduction of H_2O_2 without the aid of an electron mediator	Xu et al. [78]
Polymer terminated by 2-(2,4-dinitrophenylsulfanyl) ethanol was synthesized by ring-opening bulk polymerization of ϵ-caprolactone initiated by the reaction product of aluminum isopropoxide and 2-(2,4-dinitrophenylsulfenyl) ethanol	The corresponding thiolyated PCL was obtained after removal of the protecting group under slightly basic conditions. A decrease in polydispersity after deprotection was observed. PCLS-H was grafted to the GNPs either by direct incorporation or ligand exchange with the existing undecanethiol on the GNPs	Aryal et al. [79]

IONPs have unique interaction with biological entities and magnetic nanoparticles, assisted by biological recognition elements attached to their surface, can access the target specifically. Methods of preparation for functional IONPs are enumerated in Table 5.

3.2.1 Iron Oxide Nanoparticles in Polymers

The preparation of wholly organic polymeric materials possessing high magnetic movements and susceptibilities comparable to inorganic systems remains a challenge. To overcome this inherent problem, the synthesis of composites combining organic polymers and inorganic magnetic particles has been conducted to prepare

Table 5 Preparation of functional iron-based nanoparticles

Preparation method	Observations	References
Production of IONPs, ranging from iron oxide to iron and iron carbide, was carried out by systematically modifying the degree of reduction during flame spray synthesis under a controlled atmosphere	At a laboratory scale, continuous production yields iron-based particles of 20–50 nm at a production rate of >10 g/h. Carbon-encapsulated iron carbide (C/Fe_3C) combines exceptionally high saturation magnetization (140 emu/g), air stability (up to 200°C), and resistance against acidic dissolution (1 week in 24% HCl). Top graphene-like carbon layer can be covalently functionalized with various linkers to chemically design the particle surface	Herrmann et al. [92]
Biocompatible, hydrophilic, magnetofluorescent nanoparticles with surface-pendant amine, carboxyl, and aldehyde groups were designed by using *O*-carboxymethyl chitosan (OCMC)	The free amine groups of OCM-stabilized magnetite nanoparticles on the surface allow covalent attachment of a fluorescent dye, rhodamine isothiocyanate, to develop a magnetofluorescent nanoprobe for optical imaging. In order to impart specific cancer cell targeting properties, folic acid and its aminated derivative can be conjugated onto these magnetofluorescent nanoparticles. The potential of these nanoconjugates as T2-weighted negative contrast MRI agent was evaluated in folate-overexpressed HeLa and normal L929 fibroblast cells	Bhattacharya et al. [93]
Immobilization of BSA on surface-modified SPIONs was performed by two different double-step immobilization approaches. The first approach consists of preparation of SPIONs by controlled chemical co-precipitation in the presence of BSA solution. The second approach includes preliminary surface modification of SPIONs with an amine group using a coupling agent of 3-aminepropyltrimethoxysilane	Both procedures were followed by EDC hydrochloride activation, with sequential immobilization of the layer of BSA. TEM shows that the particle size varies in the range 10–15 nm and does not change significantly after the coating process. In vitro tests performed after the functionalization of SPIONs with L-aspartic acid (LAA) and BSA. BSA-coated SPIONs incubated with cells demonstrated a cell response similar to that of control cells, with no adverse cell damage and no endocytosis, whereas LAA-coated SPIONs show partial endocytosis without cytoskeletal disorganization	Mikhaylova et al. [94]

(continued)

Table 5 (continued)

Preparation method	Observations	References
Simple one-pot strategy was used to prepare surface-functionalized, water-dispersible IONPs. Small organic molecules that required functional groups such as amines, carboxylics, and thiols as capping agents were injected into the reaction medium at the end of the synthesis	A successful surface modification of IONPs with selected functionalities was studied. The functionalized nanoparticles were very stable and mostly present as individual units in buffer solutions. The pedant functional groups of the capping ligand molecules were very reactive, and their availabilities investigated by covalently linking fluorescent dyes to the nanoparticles through the crosslinking of EDC hydrochloride. The quenched quantum yield and shortened lifetime of the dyes strongly indicated a direct bonding between the functional group of the nanoparticles and the fluorescent molecules	Qu et al. [95]
Design of smart MRI contrast agent based on SPIONs and aptamers is described for the detection of human α-thrombin protein	The contrast agent was based on the assembly of aptamer-functionalized nanoparticles in the presence of thrombin. A detectable change in MRI signal was observed with 25 nM thrombin in human serum. Changes were not observed with control analytes, streptavidin, or BSA, nor with inactive aptamer-functionalized nanoparticles	Yigit et al. [96]
Preparation and characterization of new magnetic fluorescent nanoparticles (Fe_2O_3) to label living cells	Bifunctional magnetic nanoparticles were prepared, each particle consisting of a magnetic oxide core (Fe_2O_3) covered by a dimercaptosuccinic acid (DMSA) ligand and a covalently bonded fluorescent dye as a surface modification. The technique attaches biological entities (proteins, DNA, and enzymes) to synthetic materials to biofunctionalize IONPs	Bertorelle et al. [97]

hybrid materials with modular organic composition and functionality. Several methods have been developed to modify the magnetite nanoparticles. The most commonly used method is to encapsulate magnetic nanoparticles in a polymeric matrix, such as polylactic acid (PLA) or polylactic-*co*-glycolic acid (PLGA) microspheres [98], Therefore, many approaches have been focused on the encapsulation of magnetic nanoparticles with biocompatible materials such as dextran, chitosan, gelatin, poly(ethylene glycol) (PEG), PLGA, poly(D,L-lactide) (PLA),

Table 6 Use of iron oxide nanoparticles in a polymer matrix

Work and observations	Findings	References
Maghemite (γ-Fe_2O_3) magnetic nanoparticles (MNPs) coated with poly(3-thiopheneacetic acid) were prepared by an oxidative polymerization method using $KMnO_4$. IONPs with a size range of 10–20 nm were also prepared by modified controlled chemical co-precipitation from a ferrous/ferric mixed salt solution in alkaline medium and the IONPs covalently modified by biodegradable polymers such as PEG and poly(ethylene glycol)-*co*-poly(D,L-lactide) (PELA)	Used as an immobilization matrix for the fabrication of a solid-state tris (2,2-bipyridyl)ruthenium(II) $(Ru(bpy)^{2+})_3$ electrogenerated chemiluminescence (ECL) sensor. The poly(3-thiopheneacetic acid)-coated MNPs formed clusters with average diameter of 200–500 nm. The multilayer films of poly(3-thiopheneacetic acid)-coated MNPs were uniformly formed on the surface of a Pt electrode using an external magnet	Lu et al. [99]
A series of ethylenediamine (EDA)-functionalized magnetic polymers were prepared via suspension polymerization using varying amounts of the functional monomer glycidyl methacrylate during the suspension polymerization	EDA-functionalized Fe_3O_4 magnetic polymers for removal of Cr(VI) in wastewater	Yong-Gang et al. [100]
Surfaces of MNPs were coated with (3-aminopropyl) triethoxysilane (APTES) by a silanization reaction and then linked with PEG diacid via the reaction between $-NH_2$ and $-COOH$ to form well-dispersed surface-functionalized biocompatible MNPs	The magnetite (Fe_3O_4) nanoparticles were coated with PEG diacid via covalent bonds prepared for MRI. Experimental results imply that PEG diacid was covalently modified on the surface of the MNPs via at least one of their carboxylic groups, leaving the second carboxylic group available for further chemical reaction	–
Chemical synthesis and in vitro drug delivery response of PEG-functionalized MNPs, activated with a stable ligand, folic acid, and conjugated with an anticancer drug, doxorubicin	Functionalization and conjugation steps in the chemical synthesis confirmed the drug-release behavior of PEG-functionalized and folic acid–doxorubicin-conjugated MNPs, characterized by two stages involving an initial rapid release, followed by a controlled release	Zhang et al. [101]
Synthesis of poly(*N*-isopropyl acrylamide-acrylamide-allylamine)-coated MNPs using silane-coated MNPs as a template for radical polymerization of *N*-isopropylacrylamide, acrylamide, and allylamine	Properties of nanoparticles such as size, biocompatibility, drug loading efficiency, and drug release kinetics were evaluated in vitro for targeted and controlled drug delivery	Rahimi et al. [102]

and poly(glycolide) (PGA) because these materials have biocompatible and biodegradable properties as well as low toxicity. Table 6 describes the use of IONPs in a polymer matrix.

3.3 Silica Nanoparticles

Silica nanoparticles have been widely used as fillers in the manufacture of coatings [103], rubber [104], plastics [105], binders [106], functional fibers [107], etc. In recent years, the preparation of organic–inorganic hybrid materials composed of polymers and functionalized silica nanoparticles has been widely investigated. The combination of organic polymer components with nanometer-sized silica fillers in a single material has extraordinary significance for the development of hybrid materials with unique properties.

Recently, the straightforward synthesis and easy modification of silica nanoparticles has been demonstrated, making them well suitable for biomedical applications such as bioseparation, cell recognition, bacterium and DNA detection, gene delivery, and so on. Silica nanoparticles are biologically inert, have very low toxicity, and are amenable to surface functionalization with a wide variety of molecules and functional groups [108–111]. In Table 7, the synthetic methods for preparation of silica nanoparticles in polymers are shown.

3.4 Carbon-Based Nanomaterials

3.4.1 Carbon Nanotubes

Among all the carbon nanostructures (e.g., fullerene and graphene), the carbon nanotubes (CNTs) are probably the most studied and used in different applications. CNTs can be classified according to their superstructures into two types: (a) multiwalled carbon nanotubes (MWCNTs) or double-walled carbon nanotubes [126], which are composed of multiple layers of concentric cylinders with a spacing of about 0.34 nm between the adjacent layers; and (b) single-walled carbon nanotubes (SWCNTs), which consists of single layers of graphene sheets seamlessly rolled into cylindrical tubes.

CNTs are very difficult to dissolve or disperse in most organic or aqueous solutions. Thus, it is necessary to modify or functionalize CNTs. The functionalization of CNTs can be carried out by covalent functionalization or by noncovalent functionalization.

Meso–meso linked diporphyrins ($[H_2Por]_2$) covalently functionalized soluble single-walled carbon nanotubes ($[H_2Por]_2$-SWCNTs) have been successfully prepared [127]. As a light-harvesting chromophore, meso–meso linked diporphyrins have been incorporated into a photosynthetic electron-transfer model with

Table 7 Synthesis of silica nanoparticles in polymers

Preparation method	Experimental details and findings	References
Synthesis and characterization of a new NO-releasing scaffold prepared from amine-functionalized silica nanoparticles	Inorganic–organic hybrid silica was prepared via co-condensation of tetraethoxy- or tetramethoxysilane (TEOS or TMOS) and aminoalkoxysilane with appropriate amounts of ethanol (or methanol), water, and ammonia. The amine functional groups in the silica were converted to *N*-diazeniumdiolate NO donors via exposure to high pressures of NO (5 atm) under basic conditions. Control over both the structure and concentration of the silane precursors (i.e., tetraalkoxy- and aminoalkoxysilanes) and specific synthetic conditions allowed for the preparation of NO donor silica particles of widely varying sizes (20–500 nm), NO payloads (50–1,780 nmol/mg), maximum amounts of NO released (10–5,500 ppb/mg), half-lives (0.1–12 h), and NO release durations (up to 30 h)	Shin et al. [112]
Size-tunable silica crosslinked micellar core–shell nanoparticles (SCMCSNs) from a Pluronic nonionic surfactant (F127) template system with organic swelling agents such as 1,3,5-trimethylbenzene (TMB) and octanoic acid at room temperature	Size and morphology of SCMCSNs was directly evidenced by TEM imaging and DLS measurements (up to ~90 nm). Pyrene and coumarin 153 were used as fluorescent probe molecules to investigate the effect and location of swelling agent molecules. Papaverine as a model drug was used to measure the loading capacity and release property of nanoparticles. The swelling agents can enlarge the nanoparticle size and improve the drug loading capacity of nanoparticles. Moreover, the carboxylic acid group of fatty acids adjusts the release behavior of the nanoparticles	Chi et al. [113]
Novel nanformulation of a photosensitizer for photodynamic therapy (PDT) of cancer, whereby the photosensitizer molecules are covalently incorporated into organically modified silica (ORMOSIL) nanoparticles	The covalently incorporated photosensitizer molecules retained spectroscopic and functional properties and robustly generated cytotoxic singlet oxygen molecules upon photoirradiation. The synthesized nanoparticles of ultralow size (20 nm) were highly	Ohulchanskyy [114]

(continued)

Table 7 (continued)

Preparation method	Experimental details and findings	References
	monodispersed and stable in aqueous suspension. The advantage offered by this covalently linked nanofabrication is that the drug is not released during systemic circulation. These nanoparticles are also avidly uptaken by tumor cells in vitro and demonstrate phototoxic action, thereby highlighting their potential in diagnosis and PDT of cancer	
Uniformly sized silica-coated magnetic nanoparticles (magnetite-silica) synthesized in a simple one-pot process using reverse micelles as nanoreactors	Core diameter of the magnetic nanoparticles is easily controlled by adjusting the ratio of polar solvent to surfactant concentrations in the reverse-micelle solution. The thickness of the silica shell is easily controlled by varying the amount of tetraethyl orthosilicate added after synthesis of the magnetite cores. Several grams of monodisperse magnetite-silica nanoparticles can be synthesized without going through any size-selection process. When crosslinked, enzyme molecules form clusters on the surfaces of the magnetite-silica nanoparticles and the resulting hybrid composites are magnetically separable, highly active, and stable under harsh shaking conditions for more than 15 days. Conversely, covalently attached enzymes on the surface of the magnetite-silica nanoparticles are deactivated under the same conditions. Used for the convenient, rapid, and efficient separation of trace aromatic compounds	Lee et al. [115]
Preparation, characterization, and application of magnetic silica nanoparticle-functionalized MWCNTs	Pyrene, a fluorophore that displays excimer emission specifically at higher density/concentration was chosen for highlighting the fidelity of surface functionalization. The UV–visible absorbance of the pyrene chromophore capping the nanoparticle was used to determine the amount of pyrene units present on the surface of silica nanoparticles	Deng et al. [116]

(continued)

Table 7 (continued)

Preparation method	Experimental details and findings	References
Tuning the surface functionality of silica nanoparticles through a "click" chemistry-based protocol	Aggregation studies, using SEM, DLS, and zeta potential analysis, indicate that severe aggregation between amine-modified silica nanoparticles can be reduced by adding inert functional groups, such as methyl phosphonate, to the surface	Chandran et al. [117]
Design of nanoparticle surfaces for optimum balance of the use of inert and active surface functional groups to achieve minimal nanoparticle aggregation and reduce nanoparticle nonspecific binding. Silica nanoparticles were prepared in a water-in-oil microemulsion and subsequently surface-modified via co-hydrolysis with tetraethylorthosilicate (TEOS) and various organosilane reagents. Nanoparticles were produced with different functional groups, including carboxylate, amine, amine/phosphonate, PEG, octadecyl, and carboxylate/ octadecyl groups	To determine the effect of various surface modification schemes on nanoparticle nonspecific binding, the interaction between functionalized silica nanoparticles and a DNA chip was studied using confocal imaging and fluorescence microscopy. Dye-doped silica nanoparticles functionalized with octadecyl and carboxylate groups showed minimal nonspecific binding. Using these surface modification schemes, fluorescent dye-doped silica nanoparticles more readily conjugated with biomolecules and were used as highly fluorescent, sensitive, and reproducible labels in bioanalytical applications	Bagwe et al. [118]
Preparation of a hybrid material consisting of various sizes of silica nanoparticles with MWCNTs	Poly (acrylic acid) oligomer reacted with hydroxyl groups on acid-treated MWCNTs leading to a grafted encapsulation of the MWCNTs. These subsequently reacted with 3-aminopropyltriethoxysilane (APTES), resulting in sub-grafting of APTES on the MWCNTs. Such siloxane-MWCNTs were further hydrolyzed to make MWCNTs indirectly bearing Si–OH groups. Finally, a bud-like MWCNT/silica hybrid was obtained by forming a number of silica nanoparticles from Si–OH groups on the surface of MWCNTs by introducing siloxane-MWCNTs into a solution of TEOS, ammonia, and ethanol. The average size of the silica nanoparticles on the surface of the MWCNTs was controlled by adjusting the concentration of ammonia and the reaction time	Zhou et al. [119]

(continued)

Table 7 (continued)

Preparation method	Experimental details and findings	References
Emulsion polymerization of ethylene from mesoporous silica nanoparticles (MSNs) with vinyl (V)-functionalized monolayers	V-MSNs were synthesized via deposition of vinyl monolayers on the pore walls, and the relative surface coverage of the vinyl monolayers was 74%. A fluorinated P–O–chelated nickel catalyst was coordinated to the vinyl groups. These V-MSNs hosting catalysts fully dispersed in water with the assistance of an ultrasonic processor in the presence of surfactants. After addition of ethylene, polyethylene chains grew from the pores of V-MSNs, forming stable nanocomposite lattices with solid content of up to 17.3%	Wei et al. [120]
Mesoporous silica nanoparticles were synthesized and characterized to develop a drug delivery system by loading them with hydralazine and functionalizing them with PEG	These agents restored damaged cell membranes and ameliorated abnormal mitochondrial behavior induced by the endogenous toxin acrolein. Such a formulation shows potential as a novel therapeutic agent	Cho et al. [121]
Surface functional monomer-directing strategy for the highly dense imprinting of 2,4,6-trinitrotoluene (TNT) molecules at the surface of silica nanoparticles	Vinyl functional monomer layer of the silica surface not only directs the selective occurrence of imprinting polymerization at the surface of silica through the copolymerization of vinyl end groups with functional monomers, but also drives TNT templates into the formed polymer shells through the charge transfer complexing interactions between TNT and the functional monomer layer	Gao et al. [122]
Functionalization of silica nanoparticles via the combination of surface-initiated RAFT polymerization and click reactions	A functional monomer with a pendant azide moiety, 6-azidohexyl methacrylate (AHMA) was polymerized on the surface of silica nanoparticles via surface-initiated RAFT polymerization, with considerable control over the molecular weight and molecular weight distribution. The kinetics of AHMA polymerization mediated by 4-cyanopentanoic acid dithiobenzoate (CPDB)-anchored nanoparticles was investigated and compared with that of AHMA polymerization mediated by free CPDB under similar conditions.	Li et al. [123]

(continued)

Table 7 (continued)

Preparation method	Experimental details and findings	References
	The subsequent postfunctionalization of PAHMA-grafted nanoparticles was demonstrated by reacting with various functional alkynes via click reactions. Kinetic studies showed that the reaction of surface-grafted PAHMA with phenylacetylene surface-grafted PAHMA was much faster than that of free PAHMA with phenylacetylene. In the case of high molecular weight alkynes, surface-grafted PAHMA showed lower reaction rates than free PAHMA	
On-line synthesis of enzyme-functionalized silica nanoparticles in a microfluidic reactor using polyethyleneimine (PEI) polymer and R5 peptide	A simple microfluidic reactor system for the effective synthesis of enzyme-functionalized nanoparticles offers many advantages over batch reactions, including excellent enzyme efficiencies. Better control of the process parameters in the microfluidic reactor system over batch-based methodologies enables the production of silica nanoparticles with the optimum size for efficient enzyme immobilization with long-term stability. The synthetic approach used glucose oxidase and two different nucleation catalysts of similar molecular mass: the natural R5 peptide, and PEI polymer	He et al. [124]
RAFT polymerization used to graft polystyrene onto silica nanoparticles	A novel route was used to prepare the RAFT agent, 2-butyric acid dithiobenzoate (BDB), by substitution of dithiobenzoate magnesium bromide with sodium 2-bromobutyrate under alkali condition in aqueous solution. Epoxy groups were covalently attached to silica nanoparticles by condensation reaction of 3-glycidyloxypropyltrimethoxysilane (GPS) with the hydroxyl on the silica particle surface. RAFT agent-functionalized nanoparticles produced by ring-open reaction of the epoxy group with the carboxyl group of BDB	Liu and Pan [125]

Table 8 Preparation of functional carbon nanotubes

Preparation method	Observations and evaluation	References
Thionine was employed as a functional molecule for the noncovalent functionalization of CNTs because it shows a strong interaction with either SWCNTs or MWCNTs	Attachment of thionine molecules onto the sidewalls of CNTs improves the solubility and lowers the thermal stability of the original CNTs. More importantly, it functionalizes the surface of CNTs with rich NH_2 groups and therefore opens up more opportunities for the surface chemistry of CNTs. The modification of small thionine molecules allows other kinds of species such as cytochrome *C* and TiO_2 nanoparticles to be easily and selectively introduced onto the surface of CNTs. With this approach, SWCNTs or MWCNTs can be tailored with desired functional structures and properties	Li et al. [128]
A novel cascade of chemical functionalization of MWCNTs through chemical modification by a carbohydrate, D-galactose. Galactose-conjugated or galactosylated. MWCNTs were synthesized involving the sequential steps of carboxylation, acylation, amine modification, and finally, galactose conjugation	Modification of MWCNTs with galactose was characterized at every sequential step of functionalization. That galactosylation improved the dispersibility of MWCNTs in aqueous solvents was confirmed by investigation of their dispersion characteristics at different pH values. Galactosylated MWCNTs were used for delivery of different bioactive(s) as well as active ligand (galactose)-based targeting to hepatic tissue	Jain et al. [129]

SWCNTs as an electron-acceptor. Some additional methods for preparing functional CNTs are depicted in Table 8.

3.4.2 Dispersion of CNTs in Polymer Matrix

The performance of CNT-based polymer composites include the extent to which the CNTs can be wetted by a given polymer and the resultant adhesion between the nanotube and the surrounding polymer matrix material. Therefore, the interface between the polymer matrix and the reinforcement plays a crucial role in the physical properties of the composites. Several techniques like physical processing and melt compounding methods (ultrasonication, milling, or grinding) are known [130] to be efficient at dispersing CNTs into polymer matrices (Table 9).

Table 9 Application of dispersing CNTs in polymer matrices

Work	Observations	References
Study of processed MWCNT–polystyrene nanocomposites in a twin-blade mixer	The effect of shear mixing energy was studied and it was found that the tube length diminishes with increasing mixing energy, and that the rate of tube breakage reduced as the tubes were better dispersed	Mamedov et al. [131] and Andrews et al. [132]
Series of studies on melt processing of nanocomposites based on CNTs	The interesting relation between nanotube connectivity and the onset of non-Newtonian nanotube–polymer behavior was investigated	Potschke et al. [133–134]
Study involving melt compounding polydimethylsiloxane with MWCNTs	Viscosity changes were measured as a function of nanotube–polymer mixing time and gave some quantitative understanding of CNT dispersion in the matrix polymer. Shear stresses in the mixer were far below the ultimate tensile strength of MWCNTs, indicating that the direct scission of MWCNTs is unlikely to occur during mixing	Huang et al. [135]
Incorporation of MWCNTs into shape memory polyurethane fiber by in situ polymerization, with treatment of MWCNTs in concentrated HNO_3 and H_2SO_4	Mechanical stirring, ultrasonic vibration, melt blending, extraction, and melt spinning of the MWCNTs resulted in them being distributed homogenously and preferentially aligned along the fiber-axial direction	Meng et al. [136]
Premixing CNT with poly(ethylene terephthalate) (PET) in a solvent followed by melt spinning after drying the mixture	Tensile strength of the composite fibers increased by 36.9% and the tensile modulus increased by 41.2% by adding 0.02 wt% of acid-treated MWCNTs	Shen et al. [137]

3.4.3 Graphene

In recent years, graphene has become a very hot topic of research in the field of nanomaterials. In graphite, each graphene layer is arranged such that half of the atoms lie directly over the center of a hexagon in the lower graphene sheet, and half of the atoms lie directly over an atom in the lower layer. A number of researchers have developed a range of approaches for synthesizing graphene, including micromechanical cleavage, liquid-phase exfoliation of graphite, epitaxial growth, and reduction from graphite oxide (the precursor of graphene) by chemical or thermal exfoliation. Recently, the biological applications have also started to be investigated [138, 139].

Table 10 Graphene-based functional nanomaterials

Work	Finding	References
Studies on PEG-functionalized nanographene sheets (NGS)	High solubility and stability in physiological solutions. Used for drug carriers for controlled loading and release of antitumor drugs and for biological imaging	Sun et al. [140] and Liu et al. [141]
PEG-functionalized NGS were labeled with Cy7, a commonly used NIR fluorescent dye, and the in vivo behavior of graphene studied by in vivo fluorescence imaging	Compared with CNTs with one-dimensional structure, the 2D nanographene has unique shape and size as well as various intriguing physical and chemical properties, thus making it an interesting material for nanomedicine	Zhang et al. [142] and Yang et al. [143]
Study on the long-term in vivo biodistribution of ^{125}I-labeled NGS functionalized with PEG	Uncovered the highly efficient tumor passive targeting ability of PEGylated NGS and used graphene as a powerful photothermal therapy agent for effective photo-ablation of tumors in a mouse model	Yang et al. [144]
A low-cost technique via simple rapid-mixture polymerization of aniline using graphene oxide (GO) and graphene papers as substrates, respectively, to fabricate free-standing, flexible GO-polyaniline (PANI) and graphene-PANI hybrid papers	The potential toxicity of graphene was studied over time. The results showed that PEGylated NGS mainly accumulate in the reticuloendothelial system, including liver and spleen, after intravenous administration and gradually cleared, probably by both renal and fecal excretion. PEGylated NGS does not cause appreciable toxicity at the tested dose (20 mg/kg) in treated mice over a period of 3 months. This work greatly encourages further studies of graphene for biomedical applications	Zhang et al. [145]
A homogeneous aqueous colloidal suspension of chemically crosslinked GO sheets was generated by addition of polyallylamine to an aqueous suspension of GO sheets followed by sonication of the mixture	Nanostructural PANI was deposited on the surfaces of GO and graphene papers, forming thin, lightweight, and flexible paperlike hybrid papers	Yan et al. [146]
Production of a homogeneous colloidal suspension of crosslinked GO sheets	"Paper" material samples, made by simple filtration from a colloidal suspension of crosslinked GO sheets, showed excellent mechanical stiffness and strength, thus demonstrating the utility of such colloidal suspensions	Park et al. [147]

The functionalization of graphene makes it a promises candidate for various applications. Several approaches have been developed for noncovalently and covalently functionalizing graphene (Table 10).

3.4.4 Fullerene

In recent years, studies on functionalized fullerenes for various applications in the field of biomedical sciences have seen a significant increase (Table 11).

The most common technique used is to react pristine C_{60} with a functionalized polymer. Three examples are: (a) azido-functionalized polymers, which result in the formation of aza[60]fulleroids and aziridino[60]fullerene-containing polymers [157] when reacted with C_{60}; (b) polymers appended with aldehyde functional groups have been used to produce fulleropyrrolidine-containing polymers [158]; and (c) the fullerene is functionalized in such a way so as to form a monomer, which can then be polymerized. Polymerization of C_{60} monomers has usually been done by condensation, although examples of addition polymerization do exist. The wide-ranging reactivity of the remaining double bonds on the C_{60} cage is one potential drawback of this approach because further reaction of these double bonds will result in the loss of the required properties of the mono-fullerene adduct.

3.5 Liposomes

The main advantages of liposomes are their biocompatibility, nontoxicity, adjustable size [159], the possibility to functionalize their surface to make them targeting or stealthy [160], and their ability to encapsulate hydrophilic molecules in their water pool [161]. All these advantages have attracted scientists and researchers across the world to work in liposomal chemistry. The great issue in liposome research is the design of hybrid liposomes containing functional nanoparticles for use as nanoscale therapeutics [162, 163]. Liposomes, which have a lipid bilayer shell of less than 5 nm thickness surrounding an aqueous core, can provide a means of dispersing and concentrating nanoparticles via encapsulation or binding, thus shielding them from biomolecular adsorption and delivering them through established liposome targeting strategies. When loaded with drugs and nanoparticles, they resemble classic liposome–drug formulations with additional functionality owing to the nanoparticles. This functionality could be related to imaging [164], biosensing [165], or heating through external activation using lasers [166] or alternating current electromagnetic fields. Nanoparticles include inorganic nanocrystals such as iron oxide, gold, and quantum dots; natural nanoparticles such as viruses, lipoproteins, and apoferritin; and hybrid nanostructures composed of inorganic and natural nanoparticles. Of these, the most investigated nanoparticles platforms for biomedical purpose are lipidic aggregates such liposomal nanoparticles, micelles, and microemulsions. Their relative ease of preparation and functionalization, as well

Table 11 Preparation of functionalized fullerenes

Preparation method	Observations	References
Three-point hydrogen bonding assembly between a conjugated poly (*p*-phenylene vinylene) (PPV) and a functionalized fullerene	A new self-assembly system was demonstrated between PPV derivatives and an organo-fullerene through a three-point hydrogen bonding interaction. Two new compounds, U-PPV bearing an uracil moiety (electron donor) and DAP-C_{60} containing a 2,6-diacylamidopyridine unit (electron acceptor), were prepared and used to build a supramolecular system	Fang et al. [148]
Advanced preparation and reactivity of a surface-functionalized, fullerene-enriched liposome	A C70-enriched liposome, LMIC70, was obtained by exchange from an unstable water-soluble complex of fullerene (C70-CDx). The biological activity of LMIC70 toward DNA was assayed under visible-light irradiation. The results showed that LMIC70 can serve as a superlative system for DNA photocleavage	Ikeda et al. [149]
A range of fullerenes functionalized with long alkyl chains was synthesized and inserted into SWCNTs	The impact of the alkyl chain length and of the type of linker between the addend and the fullerene cage on the geometry of molecular arrays in nanotubes was studied by high-resolution TEM. In the presence of functional groups, the mean inter-fullerene separations significantly increased by 2–8 nm, depending on the length of the alkyl chain, but the periodicity of the fullerene arrays was disrupted due to the conformational flexibility of the alkyl groups	Chamberlain et al. [150]
Mono- and multiple-functionalized C_{60} derivatives were synthesized and studied for their photophysical properties	Fluorescence quantum yields of the C_{60} derivatives were quantitatively determined. For the mono-functionalized C_{60} derivatives, the compound with a (5–6)-open fulleroid addition pattern on the fullerene cage appeared to be considerably less fluorescent than those with a [6,6]-closed cage addition pattern. Despite the disturbance of the electronic structure via multiple additions to the fullerene cage, the multiple-functionalized C_{60} derivatives exhibited no dramatic changes in fluorescence quantum yields in	Sun et al. [151]

(continued)

Table 11 (continued)

Preparation method	Observations	References
	comparison with the mono-functionalized C_{60} derivatives. The fluorescence lifetimes of the C_{60} derivatives, were obtained using the time-correlated single	
Cis- and *trans*-functionalized bis-18-crown-6 porphyrins self-assembled with fullerene functionalized with pyridine or alkyl ammonium cation entities	Photo-induced electron-transfer processes reported	Souza et al. [152]
A self-assembled photoactive antenna system containing GNP as the central nanocore and appended fullerene moieties as the photoreceptive hydrophobic shell was designed by functionalizing a GNP with a thiol derivative of fullerene	Upon suspension of fullerene-functionalized GNPs (Au–S–C_{60}) in toluene, formation of 5–30 nm diameter clusters was observed. The quenching of fluorescence emission as well as decreased yields of triplet excited state suggests the participation of excited singlets in the energy transfer to the gold nanocore. Application of electrophoretically deposited Au–S–C_{60} nanoassemblies on optically transparent electrodes in the photoelectrochemical conversion of light energy was demonstrated	Sudeep et al. [153]
The Diels–Alder reactions of masked *o*-benzoquinones (MOBs) with [60] fullerene afforded novel and highly functionalized bicyclo-[2.2.2]-octenone-fused [60]fullerene derivatives	MOBs reactive with cyclohexa-2,4-dienones undergo Diels–Alder cycloadditions with [60]fullerene to produce hitherto unknown, stable, and highly functionalized bicyclo-[2.2.2]-octenone derivatives. In light of the mild conditions and considerable generality, these reactions are certainly noteworthy	Yen et al. [154]
A variant of the Huisgen 1,3-dipolar cycloaddition reaction provides a new and convenient functionalization of fullerenes	This method complements the widely used Prato and Bingel–Hirsch reactions. The derived, highly functionalized cyclopentenone and cyclopentenamine fullerene compounds are suitable for further functionalization upon hydrolysis and may serve well in the synthesis of new C_{60} derivatives possessing uncommon and interesting properties	Zhou and Magriotis [155]
One-step preparation of polyfunctionalized fullerene derivatives by regioselective penta-addition of an organocopper reagent	A functionalized aryl iodide was first converted to the corresponding Grignard reagent and then to a copper reagent and finally allowed to react with C_{60}. The method allows	Zhong et al. [156]

(continued)

Table 11 (continued)

Preparation method	Observations	References
	introduction of five functional groups to the C_{60} skeleton in a convergent manner. The shuttlecock-like molecules crystallize into a columnar packing structure. If the R group could bear a wide variety of functional groups, this reaction would provide a uniquely short, general approach to the synthesis of densely functionalized molecules such as polysaccharides that would tightly bind to proteins. It is reported that a mild iodine/magnesium exchange procedure pioneered by Knoche removes some of the limitations and allows one- or two-step syntheses of functionalized fullerene compounds and their derivatives in good to high yield	

as the synthetic ability to combine multiple amphiphilic moieties, are the most important reasons for their popularity. Lipid-based nanoparticle platforms allow the inclusion of a variety of imaging agents, ranging from fluorescent molecules to chelated metals and nanocrystals.

Liposomes are lipid bilayers encapsulating an aqueous core. Due to their ease of preparation in a size-controlled manner and their biocompatibility, they are widely used as lipid-based nanoparticles for drug delivery. Incorporation of PEG-containing lipid reduces opsonization and renders the liposomes long-circulating in the blood stream. After intravenous administration, long-circulating liposomes passively accumulate in the tumor tissues due to an enhanced permeation and retention effect. For active targeting, antibodies, peptides, and small molecules have been conjugated to the surface of the liposomes. The side effects of anticancer drugs are considerably reduced by encapsulating them in the liposomes. Numerous clinical trials are currently in progress using liposome-encapsulated anticancer drugs. Table 12 shows the efficient preparation methods for functional lyposomes.

Bionanotechnology involves the use of biomolecules to control both the structure and properties of nanomaterials. One of the most studied examples is DNA-directed assembly of inorganic nanoparticles such as gold nanoparticles (GNPs). However, systematic studies on DNA-linked soft nanoparticles, such as liposomes, are still lacking.

Table 12 Preparation of functional lyposomes

Preparation method	Observations	References
Study of the formation, characterization, and release properties of bilayer-decorated magnetoliposomes (dMLs) that were prepared by embedding small hydrophobic SPIONs at different lipid molecule to nanoparticle ratios within dipalmitoylphosphatidylcholine (DPPC) bilayers	The dML structure was examined by cryo-TEM and DSC, and release examined by carboxyfluorescein leakage. Nanoparticle heating using alternating current electromagnetic fields (EMFs) operating at radio frequencies provided selective release of the encapsulated molecules at low nanoparticle concentrations and under physiologically acceptable EMF conditions	Chen et al. [167]
Programmable assembly and systematic characterization of DNA-linked liposomes	Parameters such as liposome size, charge, and fluidity have little effect on the DNA melting temperature. Cryo-TEM studies showed that programmable assemblies were obtained and that the majority of the liposomes maintained a spherical shape in the assembled state. Although liposome and GNP systems are similar in many aspects, the important differences are explained by their respective physical properties. Cycloaddition is a novel generic chemical tool for the facile in situ surface modification of liposomes. Fluorescence resonance energy transfer was used to demonstrate that the reaction takes place at the surface and a colorimetric assay was developed to follow the reaction in time without the need for any equipment	Dave and Liu [168]
Chemical modification of liposome surfaces via a copper-mediated [3+2] azide–alkyne cycloaddition reaction based on click chemistry and monitored by a colorimetric assay	These vesicles were subjected to physical characterization and stability analysis at varying pH values. The enzymatic activity of alkaline phosphatase was evaluated following exposure to simulated gastric pH. The alginate-loaded liposomes were typically of the order of 10 μm; however, there was evidence of vesicle aggregation thought to be due to alginate present on the surface of the vesicles. The typical size of the aggregated vesicles was ~30 μm. Targeted delivery of imaging agents and therapeutics to tumors would provide early detection and increased therapeutic efficacy against cancer	Cavalli et al. [169]

(continued)

Table 12 (continued)

Preparation method	Observations	References
Potential of alginate-loaded liposomes as a vehicle for the oral delivery of bioactive proteins. The vesicles were prepared from the phospholipid dipalmatoyl phosphatidylcholine using a simple dry film hydration technique. Alkaline phosphatase (ALP) was used as a model bioactive protein. ALP activity following 2 h of exposure to simulated gastric pH was maintained at a significantly higher level (80%) when encapsulated in the alginate-loaded liposomes as compared to ALP loaded in conventional liposomes (55%)	Evaluation of individual phage clones after screening revealed that a phage clone displaying the CSNIDARAC peptide bound to H460 lung tumor cells at higher extent than other phage clones. The CSNIDARAC peptide-targeted and doxorubicin-loaded liposomes inhibited tumor growth more efficiently than untargeted liposomes or free doxorubicin. In vivo imaging of fluorescently labeled liposomes demonstrated selective homing of the CSNIDARAC-liposomes to tumor. Nanomedicines and luminescent QDs can be loaded into immunoliposomes for cancer diagnosis and treatment	Smith et al. [170]
A phage-displayed peptide library was screened to identify peptides that selectively bind to lung tumor cells. The synthetic CSNIDARAC peptide strongly bound to H460 cells and was efficiently internalized into the cells	Pharmacokinetic properties and in vivo imaging capability of QD-conjugated immunoliposome-based nanoparticles (QD-ILs) was investigated. Nontargeted QD-ILs showed minimal binding and uptake in these cells. Doxorubicin-loaded QD-ILs showed efficient anticancer activity, while no cytotoxicity was observed for QD-ILs without chemotherapeutic payload. In athymic mice, QD-ILs significantly prolonged circulation of QDs, exhibiting a plasma terminal half-life of ~2.9 h, as compared to free QDs. In MCF-7/HER2 xenograft models, localization of QD-ILs at tumor sites was confirmed by in vivo fluorescence imaging	He et al. [171]
Synthesis, biophysical characterization, tumor cell-selective internalization, and anticancer drug delivery of QD-ILs. QD-ILs prepared by insertion of anti-HER2 scFv exhibited efficient receptor-mediated endocytosis in HER2-overexpressing SK-BR-3 and MCF-7/HER2 cells but not in control MCF-7 cells, as analyzed by flow cytometry and confocal microscopy	The C-type lectin asialoglycoprotein receptor (ASGPR) is particularly suitable for liver-specific delivery due to its exclusive expression by parenchymal hepatocytes. For targeting of ASGPR, rhodamine-B-loaded β-cyclodextrins (βCDs) were functionalized with glycodendrimers. Liposomes were equipped with synthetic glycolipids containing a terminal *N*-acetyl-D-galactosamine (D-GalNAc) residue to mediate binding to ASGPR. Uptake studies in the human hepatocellular carcinoma cell line HepG2 demonstrated that βCDs and liposomes displaying terminal D-Gal/D-GalNAc residues	Weng et al. [172]

(continued)

Table 12 (continued)

Preparation method	Observations	References
	were preferentially endocytosed. In contrast, uptake of βCDs and liposomes with terminal D-Man or D-GlcNAc residues was markedly reduced. The D-Gal/D-GalNAc-functionalized βCDs and liposomes enable hepatocyte-specific targeting. Gal-functionalized βCDs are efficient molecular carriers to deliver doxorubicin in vitro into hepatocytes and can induce apoptosis	
Design, synthesis, and biological evaluation of carbohydrate-functionalized cyclodextrins and liposomes for hepatocyte-specific targeting. Targeting of glycan-binding receptors is an attractive strategy for cell-specific drug and gene delivery	DLS measurements showed that the size of noncoated liposomes remained stable upon raising the temperature from 25°C to 46°C. Polymer-coated liposomes aggregated around 43°C. The study demonstrates that liposomes surface-modified with *N*-(2-hydroxypropyl)methacrylamide (HPMA) mono/dilactate copolymer are attractive systems for achieving temperature-triggered content release	Bernardes et al. [173]
Thermosensitive HPMA mono/dilactate polymer was prepared and the temperature-triggered release of content from polymer-coated liposomes studied. HPMA mono/dilactate polymer was synthesized with a cholesterol anchor suitable for incorporation in the liposomal bilayers and with a cloud point temperature of the polymer slightly above normal body temperature (42°C)	Controlled polymerization conditions allow preparation of copolymer lipids with accurate degrees of polymerization, varying the lengths of both blocks while keeping polydispersity low. Adjusting the length of the PEG block permits tailoring the shielding properties of the polyether lipid further. The use of different initiators broadens the fields of application of these materials, ranging from commercially available cholesterol to different aliphatic glyceryl ethers as the anchors for incorporation in the liposomes. MALDI-ToF mass spectrometry confirmed complete incorporation of the respective initiators	Paasonen et al. [174]
Synthesis of linear-hyperbranched lipids for liposome preparation based on linear PEG and hyperbranched polyglycerol	Conditions for copper-catalyzed azide–alkyne cycloaddition were optimized for rapid attachment of azides with controlled composition onto the liposomes	Hofmann et al. [175]
"Clickable", polymerized liposomes as a general liposomal platform for rapid attachment of functional moieties to their peripheries via click chemistry. The platform was based on polydiacetylene lipids terminated with alkynyl groups	–	Kumar et al. [176]

3.6 Quantum Dots

Among various nanomaterials developed, quantum dots (QDs) represent one of the most successful new biological probes. Compared to conventional organic fluorophores, QDs have advantageous properties that include tunable emission, photostability, and high brightness [177, 178]. These unique properties make QDs highly desirable for certain biological applications and bring new possibilities for biological imaging [179]. These applications range from solar cells, light-emitting diodes, laser technologies, and chemical sensing to bioimaging. This has been motivated by a desire to reach a fundamental understanding of several of their unique properties and by the wealth of potential applications involving the use of these materials, ranging from electronic devices to in vivo cellular imaging.

The functionalization of nanoparticles is very important for improving their application. To date, most of the QDs are prepared in the organic phase and their surfaces are functionalized with hydrophobic moieties such as tri-*n*-octyl phosphine oxide (TOPO) and fatty acids. These hydrophobic QDs are not dispersible in aqueous or biological fluids [180]. Thus, surface modification of biocompatible water-dispersible QDs has received tremendous attention in the last decade [181]. Various approaches to the preparation of QDs (Table 13) have been reported.

3.7 Nanocomposites

Polymer nanocomposites are composites with a polymer matrix and filler with at least one dimension of less than 100 nm. Organic–inorganic nanocomposites are generally organic polymer composites with inorganic nanoscale building blocks. They combine the advantages of the inorganic material (e.g., rigidity, thermal stability) and those of the organic polymer (e.g., flexibility, dielectric properties, ductility, and processability) [192]. Moreover, they usually also contain special properties due to the nanofillers, which leads to materials with improved properties. A defining feature of polymer nanocomposites is that the small size of the fillers leads to a dramatic increase in interfacial area as compared with traditional composites. Even at low loadings, this interfacial area creates a significant volume fraction of interfacial polymer with properties different from the bulk polymer [193, 194]. Organic–inorganic nanocomposite systems can be prepared by various synthesis routes, reflecting the various methods available to introduce each phase. The organic component can be introduced as: (a) a precursor (monomer or oligomer), (b) a preformed linear polymer (in molten, solution, or emulsion states), or (c) a polymer network that is physically (e.g., semicrystalline linear polymer) or chemically (e.g., thermosets, elastomers) crosslinked. The inorganic part can be introduced as a precursor (e.g., TEOS) or as preformed nanoparticles. Organic or inorganic polymerization generally becomes necessary if at least one of the starting moieties is a precursor. There are various types of nanostructural materials, i.e.,

Table 13 Preparation of functional QDs

Preparation method	Observations	References
Synthesis of TOPO-functionalized CdSe QDs by pyrolysis of organometallic reagents and characterization of their optical properties	Controlling the relative amount of precursors injected into the coordinating solvent, e.g. TOPO, enabled modification of the size and quality of nanocrystals, the spectra of which show fairly sharp excitonic spectral features even in a limited size range. Passivation with ZnS over-coating provides a large enhancement of the fluorescence intensity, i.e., quantum efficiency reached over 50% at room temperature in the best case	Hashizume et al. [182]
Different sizes of glutathione (GSH)-functionalized CdTe QDs were prepared directly in aqueous solution and their use in imaging of cells investigated	The QDs have tunable fluorescence in the range of 510–670 nm, and also have high photoluminescence (PL) quantum yield without any postpreparative treatment. Furthermore, the QDs have strong resistance to photobleaching, and are also considered as cytocompatible. In addition, for the first time, folic acid was covalently conjugated to the GSH-functionalized CdTe QDs for imaging of cancer cells, demonstrating potentially broad application as biolabels	Xue et al. [183]
Simple strategy for the synthesis of water-soluble, luminescent, citrate-functionalized CdSe QDs and their application for selective detection of silver ions	PL spectra show a single, narrow emission band at ca. 554 nm without any contribution from the trap states. The effect of various ions, including physiologically important metal ions, on the PL intensity of citrate-capped CdSe QDs was studied. Selective luminescence quenching with Ag^{+} ion was found to be predominant	Ingole et al. [184]
Water-soluble, biologically compatible CdSe QDs with L-cysteine as capping agent were synthesized in aqueous medium	Fluorescence spectra, absorption spectra, and TEM studies showed that both the molar ratio of Se:Cd and the reaction time are determining factors for the size distribution of CdSe/L-cysteine QDs. The interaction of QDs bioconjugated to BSA was	Liu et al. [185]

(continued)

Table 13 (continued)

Preparation method	Observations	References
	also studied by absorption and fluorescence titration experiments	
L-Cysteine-functionalized CdTe QD-based sensor was developed for simple and selective detection of trinitrotoluene (TNT) based on the formation of a Meisenheimer complex between TNT and cysteine	Fluorescence of QDs quench because electrons of the QDs transfer to the TNT molecules via the formation of a Meisenheimer complex. TNT was detected with a low detection limit of 1.1 nM. Studies on the selectivity of this method show that only TNT generates an intense signal response. The synthesized QDs are excellent nanomaterials for TNT detection	Chen et al. [186]
NH_2-functionalized CdSe/ZnS QD-doped SiO_2 NPs with both imaging and gene carrier capabilities were developed and characterized	QD-doped SiO_2 NPs were internalized by primary cortical neural cells without inducing cell death in vitro and in vivo. Moreover, the ability to bind, transport, and release DNA into the cell allows GFP-plasmid transfection of NIH-3T3 and human neuroblastoma SH-SY5Y cell lines. QD-doped SiO_2 NPs properties make them a valuable tool for future nanomedicine application	Bardi et al. [187]
Investigation of surface-modified QDs as a sensing receptor for Cu^{2+} ion detection by an optical approach. Water-soluble L-cysteine-capped ZnS QDs were used as a fluorescence sensor for Cu(II) ions involved in fluorescence quenching	The optimum fluorescence intensity was found to be at pH 5.0 with a nanoparticle concentration of 2.5 mg/L. The effect of foreign ions on the intensity of ZnS QD fluorescence showed a low interference response towards other metal ions except Ag^+ and Fe^{3+} ions. The quenching mechanism was studied and the results show the existence of both static and dynamic quenching processes	Koneswaran and Narayanaswamy [188]
Functionalized core–shell CdSe/ZnS QDs with short-chain 3-mercaptopropionic acid (3MPA) to render these nanocrystalline semiconductor water-soluble	Ligand-exchange reaction was significantly improved with the use of an organic base to first remove the thiolic hydrogen. Non-bound 3MPA was removed from the colloid by dialysis. The pH of the reaction medium has an important impact on the successful attachment of functional DNA on the QDs	Pong et al. [189]

(continued)

Table 13 (continued)

Preparation method	Observations	References
Fluorescence enhancement of amine-capped CdSe/ZnS QDs by thiol addition; addition of alkyl thiols drastically reduces the fluorescence of CdSe core QDs	Thiol addition enhances the emission properties of already highly fluorescent amine-capped CdSe/ZnS core–shell QDs	Aguilera-Sigalat et al. [190]
Preparation of a compact, functional QD-DNA conjugate, where the capturing target DNA is directly and covalently coupled to the QD surface	This enables control of the separation distance between the QD donor and dye acceptor to within the range of the Förster radius. Moreover, a tri(ethylene glycol) linker was introduced to the QD surface coating to effectively eliminate the strong, nonspecific adsorption of DNA on the QD surface. As a result, this QD-DNA conjugate hybridizes specifically to its complementary DNA with a hybridization rate constant comparable to that of free DNA in solution	Zhou et al. [191]

metal nanoparticles, carbon-based materials (CNTs, graphene, fullerene), and QDs. Here we will discuss nanocomposites based on these nanostructural materials.

SnSb nanoparticles were coated onto the surface of MWCNTs by reductive precipitation of metal chloride salts within a CNT suspension [195]. The composites have a possible application in lithium-ion batteries. The SnSb–CNT nanocomposite showed a high reversible capacity of 480 mAh/g and stable cyclic retention until the 50th cycle. The improvement of reversible capacity and cyclic performance of the SnSb–CNT composite is attributed to the nanoscale dimension of the SnSb alloy particles (<50 nm) and the structural advantages of CNTs as a framework material. The CNTs could be pinning the SnSb alloy particles on their surfaces so as to hinder the agglomeration of SnSb particles, while maintaining electronic conduction as well as accommodating drastic volume variation during electrochemical reactions.

Wang et al. [196] reported synthesis of MnO_2/CNT nanocomposites. The MnO_2 nanoparticles were coated on the CNTs by a facile direct redox reaction between $KMNO_4$ and CNT without any other oxidant or reductant addition. Xia and coworkers [197] synthesized poly(ϵ-caprolactone) (PCL)/functionalized MWCNT nanocomposites and their electroactive shape memory properties were demonstrated. The crosslinking reaction of the pristine PCL was realized by using benzoyl peroxide as an initiator. The raw MWCNTS were prefunctionalized by acid-oxidation processes and covalent grafting with PEG.

4 Chitosan-Based Functional Nanomaterials

Chitosan is the principle derivative of chitin, which is the second-most naturally occurring polysaccharide after cellulose. Chitosan has an amino group in the C-2 position and OH groups in the C-3 and C-5 positions and can react with functional nanomaterials by various kinds of mechanisms. The current developing interest is the functionalization of chitosan derivatives in the form of nanomaterials to provide new strategies for a wide range of applications like multifunctional QD-based magnetic chitosan beads [198], synthesis and characterization of monodispersed chitosan nanoparticles with embedded QDs [199], synthesis of green CdSe/chitosan QDs using a polymer-assisted γ-radiation route [200], water-soluble chitosan QDs hybrid nanospheres for bioimaging and biolabeling [201], etc.

The authors' laboratory has demonstrated some of these unique functional nanomaterials, particularly those based on chitosans [202]. The preparation of QDs, chitosan-GNP derivatives, and fullerene-based chitosan derivatives, and the immobilization of gold nanoparticles onto thiol-functionalized chitosans are shown in Fig. 1.

5 Applications of Chitosan-Based Functionalized Nanomaterials: Biomedical and Industrial Applications

It is assumed that nanotechnology will provide new tools for medicine using the new techniques of material design and uses. It could radically change the way that surgery is done. It will make it possible to do molecular-scale surgery to replace defective cells and to repair and rearrange cells. Since disease is the result of physical disorder, i.e., misarranged molecules and cells, medicine at this level should be able to cure most diseases. Mutations in DNA could be repaired and cancer cells, toxic chemicals, and viruses could be destroyed through the use of medical nanodevices. Nanotechnology has the power to radically change the way cancer is diagnosed, imaged, and treated. Currently, there is a lot of research going on to design novel nanodevices capable of detecting cancer at its earliest stages, pinpointing its location within the body and delivering anticancer drugs specifically to malignant cells. In recent years, nanotechnology has found innumerable applications in the field of medicine: from drug delivery systems, nanorobots, and cell repair machines to imaging, nanoparticles, and nanonephrology. Owing to the extensive use of nanomaterials in medical equipment and devices, nanomedicine has become a significant branch of nanotechnology. Some of the potential applications are tissue engineering, wound healing, cancer diagnostics, and drug delivery.

The primary objective of the drug delivery system is to make the life-saving drug available in that part of the body where it is required the most. Most of the time, these systems fail to work efficiently because the particles of the drug are too large

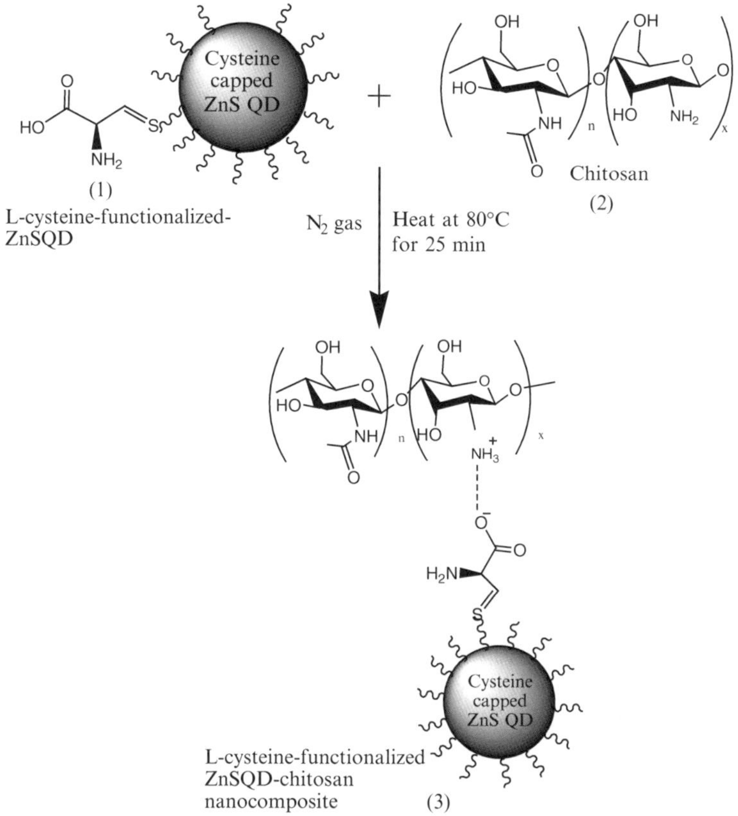

(a) Preparation of quantum dot (QD)

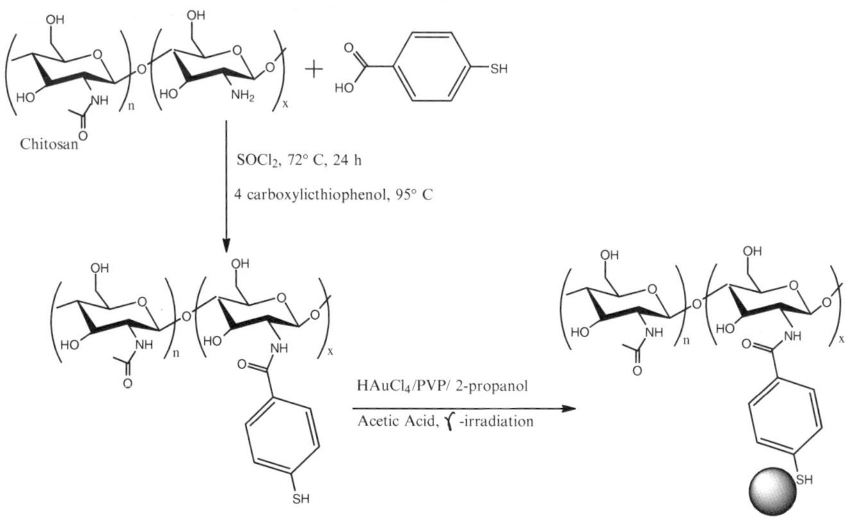

(b) Chitosan-GNP derivative

Fig. 1 (continued)

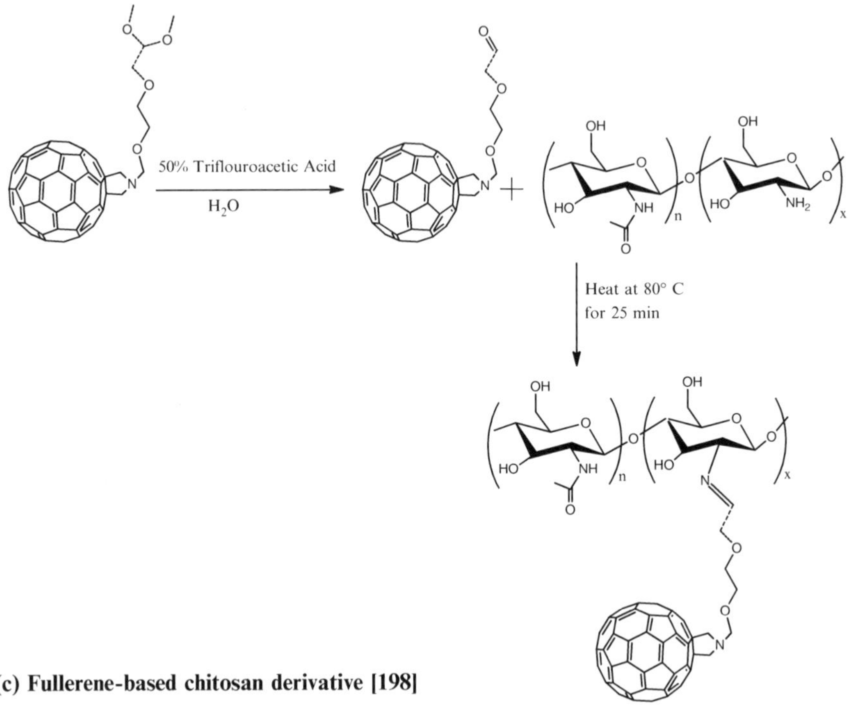

(c) Fullerene-based chitosan derivative [198]

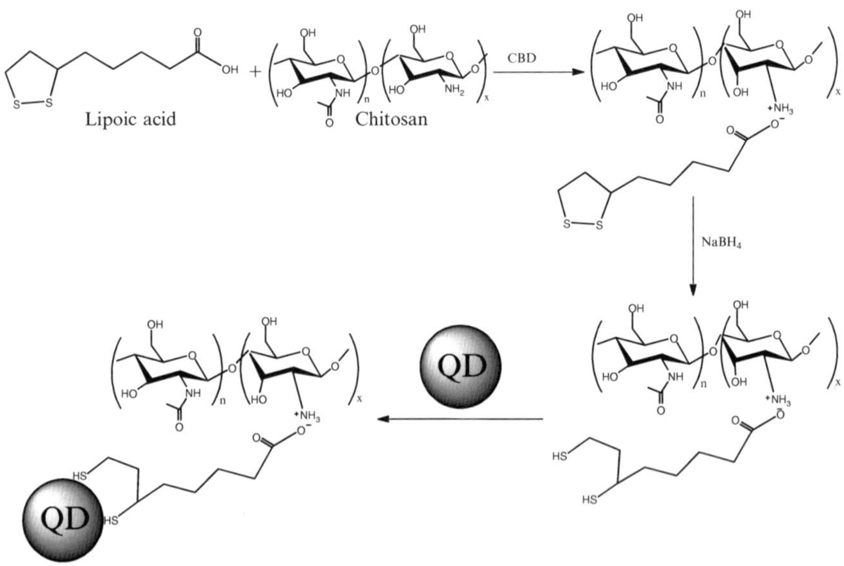

(d) Immobilization of gold nanoparticles onto thiol-functionalized chitosans

Fig. 1 Preparation of chitosan-based functional nanomaterials. (**a**) Preparation of quantum dot (QD). (**b**) Chitosan-GNP derivative. (**c**) Fullerene-based chitosan derivative [198]. (**d**) Immobilization of gold nanoparticles onto thiol-functionalized chitosans for the cells to absorb, or they are insoluble, or they have the potential to cause tissue damage. On the other hand, nanoparticles are easily taken up by the cell due to their exceedingly small size. Moreover, they are completely soluble and do not damage the tissues. Hence, the efficiency of the drug delivery system can be increased several times by integrating nanoparticles within them. Similarly, the effectiveness of biopharmaceuticals can be increased several times by coupling them with nanoparticles, which will proficiently deliver the peptides or proteins at the tumor site and, in this manner, cure cancer without causing extensive damage to the adjacent tissues and organs. Owing to their size and properties, nanomaterials are extensively used for the treatment of a number of diseases. Nanoparticle contrast agents are being developed for tumor detection purposes. Labeled nanoparticles and nanolabeled particles are used for MRI. QDs can be used to measure levels of cancer markers because these are robust and very stable light emitters. Their photochemical stability and the ability to tune broad wavelengths make QDs extremely useful for biolabeling [203].

Template synthesis strategies provide excellent control over both the internal and external dimensions of nanotubes. Because these methods also allow internal and external surfaces to display disparate functional groups, template-synthesized nanotubes are very attractive starting points for drug delivery vehicles. Several key advances have recently been made that bring this goal closer to reality. First, nanotubes can now be prepared from a much broader range of materials that includes biocompatible and biodegradable materials such as amino acid polymers, DNA, alginate, and chitosan. Of those, chitosan is a very promising material for various kinds of effective applications. In addition, a number of capping strategies have been developed to retain drug payloads within nanotubes. Finally, nanotubes have been successfully taken up by target cells, where their payload has been delivered. In the near future, these advances will be combined for drug-delivery studies in cell culture and, more importantly, in whole animals.

Drug-delivery strategies using template-synthesized nanotubes [204] can overcome the blood–brain barrier (BBB), which limits the transport of therapeutic molecules from the blood compartment into the brain, thus greatly reducing the species of therapeutic compounds that can be efficiently accumulated in the central nervous system (CNS). Various strategies have been proposed for improving the delivery of drugs to the brain, and numerous invasive and noninvasive methods have been proposed by different scientists in an attempt to circumvent the BBB and to increase the delivery of drug compounds into the brain. An interesting alternative, which solves this problem and also that of reaching a suitable target in the CNS, has recently been provided through the use of nanoparticulate colloidal devices as a noninvasive technique for drug delivery to the brain. These systems offer diverse advantages over invasive strategies, because (a) they are designed using biocompatible and biodegradable materials; (b) they avoid the disruption and/or modification of the BBB; and (c) they modulate the biopharmaceutical properties of the entrapped drugs. Moreover, the possibility of targeting specific brain tissue, thanks to ligands linked to the surface of the nanoparticulate colloidal devices, confers the necessary characteristics for the treatment of CNS pathologies to these drug carriers.

This section emphasizes recent research on different aspects of chitosan-based functional nanomaterials, including applications of chitosan-based functional metal nanoparticles (gold, iron oxide, and silica), carbon-based nanomaterials (CNTs, graphene and fullerene), quantum dots, and liposomes for tissue engineering, wound dressing, drug delivery, cancer diagnosis, and some industrial applications.

Multifunctional nanocarriers based on chitosan/gold nanorod (chitosan–AuNR) hybrid nanospheres have been successfully synthesized [205]. The anticancer drug cisplatin was subsequently loaded into the obtained hybrid nanospheres, utilizing the loading space provided by the chitosan spherical matrix. In vitro cell experiments demonstrated that the chitosan–AuNR hybrid nanospheres can be utilized not only as contrast agents for real-time cell imaging but also as a near-infrared thermotherapy nanodevice to achieve irradiation-induced cancer cell death owing to the unique optical properties endowed by the encapsulated gold nanorods. An amperometric glucose biosensor was synthesized by co-deposition of glucose oxidase with chitosan–gold nanoparticles (chitosan–AuNP) on glassy carbon electrodes modified with gold-Prussian blue nanoparticles [206]. By reducing $HAuCl_4$ with the chemiluminescent reagent luminol in the presence of hydrophilic polymer chitosan, three-dimensional (3D) flowerlike gold nanostructures were synthesized via a convenient one-pot method [207]. Due to their shape-dependent surface plasmon resonance (SPR) properties and specific surface structures, these gold nanostructures might also have great potential for applications in biomedicine and surface-enhanced Raman scattering. Hortiguela et al. [208] reported a simple synthetic route to induce chitosan gelation by the in situ formation of AuNPs and its application in the processing of macroporous scaffolds. A novel glucose biosensor based on immobilization of glucose oxidase in thin films of chitosan containing nanocomposites of graphene and AuNPs at a gold electrode was developed [209]. The graphene/AuNs/glucose oxidase /chitosan composite film shows a prominent electrochemical response to glucose, which makes it a promising application for electrochemical detection of glucose. Gee et al. [210] reported synthesis of novel fluorescent chitosan-coated magnetic nanoparticles and their use in high-efficiency cellular imaging. They evaluated the feasibility and efficiency of labeling cancer cells (SMMC-7721) with these nanoparticles. Liu et al. [211] synthesized magnetic chitosan nanocomposites on the basis of amine-functionalized magnetite nanoparticles for the application of heavy metal ion removal from water. A tyrosinase biosensor based on a Fe_3O_4-chitosan nanocomposite has been developed for the amperometric detection of dopamine by the biocatalytically liberated dopaquinone at −0.25 V versus a saturated calomel electrode [212]. Kaushika et al. [213] reported urease and glutamate dehydrogenase co-immobilized onto a SPION–chitosan-based nanobiocomposite film deposited onto an indium-tin oxide (ITO)-coated glass plate via physical adsorption, and its use for urea detection. Liu et al. [214] synthesized nanosized silica particles with sulfonic acid groups as a crosslinker for chitosan to form chitosan–silica complex membranes, which were applied to pervaporation dehydration of ethanol–water solutions.

Qui et al. [215] designed functionalized MWCNTs incorporated into a chitosan membrane for separation of ethanol–water mixtures by pervaporation. Venkatesan et al. [216] synthesized chitosan with natural hydroxyapatite (HAp) derived from Thunnus Obesus bone and chitosan grafted with functionalized MWCNTs in addition to HAp (f-MWCNT-*g*-chitosan/HAp) scaffolds for bone tissue engineering applications. A bionanocomposite film consisting of glucose oxidase/Pt/functional graphene sheets/chitosan for glucose sensing was reported [217]. With the electrocatalytic synergy of functional graphene sheets and Pt nanoparticles to hydrogen peroxide, a sensitive biosensor with a detection limit of 0.6 μM glucose was achieved. Biofunctionalization and manipulation of graphene nanosheets are important for biomedical research and application. In this connection, chitosan-modified graphene nanosheets were successfully prepared under microwave irradiation in *N*,*N*-dimethylformamide medium, which involved the reaction between the carboxyl groups of the graphene oxide nanosheets and the amido groups of chitosan, followed by the reduction of graphene oxide nanosheets into graphene nanosheets using hydrazine hydrate [218]. Fan et al. [219] reported graphene/chitosan films prepared by a solution-casting method and the biocompatibility of graphene/chitosan composite films was checked by tetrazolium-based colorimetric assays in vitro.

Lin et al. [201] prepared CdSe/ZnS QD-encapsulated chitosan hybrid nanospheres. They demonstrated that these hybrid nanospheres can be internalized by tumor cells and, hence, act as a labeling agent in cell imaging by optical microscopy. Further, the nanospheres can be used for imaging of tumors in tumor-bearing mice via intratumoral administration and can accumulate at the tumor site via the blood circulation after intravenous injection. Jayshree et al. [220] prepared chitosan–ZnS QDs conjugated with mannose ligand for targeted cancer imaging. Their study highlighted the applicability of polysaccharide-protected and mannosylated fluorescent ZnS nanoprobes for active targeting of cancer cells. Yuan et al. [221] reported research on the synthesis of blue-light emitting ZnO QDs combined with biodegradable chitosan (*N*-acetylglucosamine) for tumor-targeted drug delivery.

Wang et al. [222] prepared cholesterol succinyl chitosan-anchored liposomes (CALs) and investigated their characterization, physical stability, and drug release behavior in vitro. Compared with plain liposomes and chitosan-coated liposomes, CALs had larger sizes, higher zeta potentials, and better physical stability after storage at $4 \pm 2°C$ and $25 \pm 2°C$. Wang et al. [223] prepared folate–PEG-coated polymeric liposomes formed from octadecyl-quaternized lysine-modified chitosan and cholesterol. They demonstrated that the folate–PEG-coated polymeric liposomes could be a useful drug delivery system.

Functional nanomaterials are a powerful tool for bottom-up techniques in addition to the simplicity of large-scale preparation and the low manufacturing cost. Within a short time span, nanomaterials will be used in multiple examples of electronic, sensor, optical, and other devices. One of the great benefits of self-organized systems is that they can easily traverse different scales and be integrated with microscale devices. From that point of view, the chitosan-based functional nanomaterials will find more opportunities and success.

6 Concluding Remarks

In order to successfully prepare and functionalize nanomaterials for a given biomedical application, a wide range of physical, chemical, biological, and physiological factors and conditions must be taken into account. Strategies for synthesis of functional nanomaterial provide excellent properties. Because these nanomaterials also allow internal and external surfaces to display disparate functional groups, template-synthesized materials are very attractive starting points for drug delivery vehicles, bioimaging, biolabeling, etc. Several key advances have recently been made that bring this goal closer to reality. Functional nanomaterials can now be utilized from a much broader range of materials that include biocompatible and biodegradable materials such as amino acids, polymers, DNA, chitosan, and alginate. This chapter has focused especially on chitosan-based functional nanomaterials as a current developing interest. In the near future, these advances will be combined for drug-delivery, bioimaging, and other studies in cell culture and, more importantly, in whole species.

Acknowledgments The authors gratefully acknowledged the financial assistance from University Grants Commission and CSIR, New Delhi in the form of major research projects.

References

1. Nalwa HS (ed) (2004) Encyclopedia of nanoscience and nanotechnology. American Scientific Publishers, New York
2. Torchilin VP (2006) Multifunctional nanocarriers. Adv Drug Deliv Rev 58:1532–1555
3. Brettreich M, Burghardt S, Bottcher C, Bayerl T, Bayerl S, Hirsch A (2000) Globular amphiphiles: membrane-forming hexaadducts of C60. Angew Chem Int Ed 39:1845–1848
4. Burghardt S, Hirsch A, Schade B, Ludwig K, Bottcher C (2005) Switchable supramolecular organization of structurally defined micelles based on an amphiphilic fullerene. Angew Chem Int Ed 44:2976–2979
5. Klumpp C, Kostarelos K, Prato M, Bianco A (2006) Functionalized carbon nanotubes as emerging nanovectors for the delivery of therapeutics. Biochim Biophys Acta 1758:404–412
6. He X, Wu X, Cai X, Lin S, Xie M, Zhu X, Yan D (2012) Functionalization of magnetic nanoparticles with dendritic–linear–brush-like triblock copolymers and their drug release properties. Langmuir 28:11929–11938
7. Nishiyama N, Kataoka K (2006) Current state, achievements, and future prospects of polymeric micelles as nanocarriers for drug and gene delivery. Pharmacol Ther 112:630–648
8. Villalonga-Barber C, Micha-Screttas M, Steele BR, Georgopoulos A, Demetzos C (2008) Dendrimers as biopharmaceuticals: synthesis and properties. Curr Top Med Chem 8:1294–1309
9. Lal S, Clare SE, Halas NJ (2008) Nanoshell-enabled photothermal cancer therapy: impending clinical impact. Acc Chem Res 41:1842–1851
10. Barratt G (2003) Colloidal drug carriers: achievements and perspectives. Cell Mol Life Sci 60:21–37
11. Eaton M (2007) Nanomedicine: industry-wise research. Nat Mater 6:251–253

12. Rapoport N (2007) Physical stimuli-responsive polymeric micelles for anti-cancer drug delivery. Prog Polym Sci 32:962–990
13. Mitragotri S, Lahann J (2009) Physical approaches to biomaterial design. Nat Mater 8:15–23
14. Ydens I, Degee P, Nouvel C, Dellacherie E, Six JL, Dubois P (2005) Surfactant-free stable nanoparticles from biodegradable and amphiphilic poly(ε caprolactone)-grafted dextran copolymers. e-Polymers 46:1–11
15. Csaba N, Koping-Hoggard M, Fernandez-Megia E, Novoa-Carballal R, Riguera R, Alonso MJ (2009) Ionically crosslinked chitosan nanoparticles as gene delivery systems: effect of PEGylation degree on in vitro and in vivo gene transfer. J Biomed Nanotechnol 5:162–171
16. Dutta J, Dutta PK (2005) Chitosan a material for 21st century. In: Dutta PK (ed) Chitin and chitosan: opportunities and challenges. SSM Intl. Publication, Contai, pp 1–34
17. Kickelbick G, Schubert U (2003) Organic functionalization of metal oxide nanoparticles. In: Baraton MI (ed) Synthesis, functionalization and surface treatment of nanoparticles. American Scientific Publishers, Stevenson Ranch, p 91
18. Grancharov SG, Zeng H, Sun S, Wang SX, O'Brien S, Murray CB, Kirtley JR, Held GA (2005) Bio-functionalization of monodisperse magnetic nanoparticles and their use as biomolecular labels in a magnetic tunnel junction based sensor. J Phys Chem B 109:13030–13035
19. Doty RC, Tshikhudo TR, Brust M, Fernig DG (2005) Extremely stable water-soluble Ag nanoparticles. Chem Mater 17:4630–4635
20. Gao J, Gu H, Xu B (2009) Multifunctional magnetic nanoparticles: Design, synthesis, and biomedical applications. Accounts of Chemical Research 42:1097–1107
21. Trindade T, O'Brien P, Pickett NL (2001) Nanocrystalline semiconductors: synthesis, properties, and perspectives. Chem Mater 13:3843–3858
22. Grieve K, Mulvaney P, Grieser F (2000) Synthesis and electronic properties of semiconductor nanoparticles/quantum dots. Curr Opin Colloid Interface Sci 5:168–172
23. Nedeljkovic JM (2000) Nanoengineering of inorganic and hybrid composites. Trends Adv Mater Processes Mater Sci Forum 352:79–85
24. Farmer SC, Patten TE (2001) Photoluminescent polymer/quantum dot composite nanoparticles. Chem Mater 13:3920–3926
25. Hodak JH, Henglein A, Hartland GV (2001) Tuning the spectral and temporal response in Pt/Au coreshell nanoparticles. J Chem Phys 114:2760–2765
26. Hughes MP (2002) Dielectrophoretic behavior of latex nanospheres: low-frequency dispersion. J Colloid Interface Sci 250:291–294
27. Xu XJ, Chow PY, Gan LM (2002) Nanoparticles of latexes from commercial polystyrene. J Nanosci Nanotechnol 2:61–65
28. Patra A, Koenen JM, Scherf U (2011) Fluorescent nanoparticles based on a microporous organic polymer network: fabrication and efficient energy transfer to surface-bound dyes. Chem Commun 47:9612–9614
29. Brust M, Kiely CJ (2002) Some recent advances in nanostructure preparation from gold and silver particles: a short topical review. Colloid Surf A 202:175–186
30. Lue JT (2001) A review of characterization and physical property studies of metallic nanoparticles. J Phys Chem Solids 62:1599–1612
31. Bönnemann H, Richards RM (2001) Nanoscopic metal particles—synthetic methods and potential applications. Eur J Inorg Chem 10:2455–2480
32. Sun S, Murray CB, Weller O, Falks L, Moser A (2000) Monodisperse FePt nanoparticles and ferromagnetic FePt nanocrystal superlattices. Science 287:1989–1992
33. Kamat PV (2002) Photophysical, photochemical and photocatalytic aspects of metal nanoparticles. J Phys Chem B 106:7729–7744
34. Zheng Z, Yang M, Zhang B (2008) Reversible nanopatterning on self-assembled monolayers on gold. J Phys Chem C 112:6597–6604

35. Rowe MP, Stienecker WH, Zellers ET (2007) Exploiting charge-transfer complexation for selective measurement of gas-phase olefins with nanoparticle-coated chemiresistors. Anal Chem 79:1164–1172
36. Rao CNR, Kulkarni GU, Govindaraj A, Satishkumar BC, Thoms PJ (2000) Metal nanoparticles, nanowires and carbon nanotubes. Pure Appl Chem 72:21–36
37. Xu S, Hartvickson S, Zhao JS (2008) Engineering of SiO2−Au−SiO2 sandwich nanoaggregates using a building block: single, double, and triple cores for enhancement of near infrared fluorescence. Langmuir 24:7492–7499
38. Mohr C, Hofmeister H, Radnik J, Claus P (2003) Identification of active sites in gold-catalyzed hydrogenation of acrolein. J Am Chem Soc 125:1905–1911
39. Bruchez M Jr, Moronne M, Gin P, Weiss S, Alivisatos AP (1998) Semiconductor nanocrystals as fluorescent biological labels. Science 281:2013–2016
40. Tang Z, Kotov NA (2005) One-dimensional assemblies of nanoparticles: preparation, properties and promise. Adv Mater 17:951–962
41. Shipway AN, Katz E, Williner I (2000) Nanoparticle arrays on surfaces for electronic, optical and sensor applications. Chem Phys Chem 1:18–52
42. Kalsin AM, Fialkowski M, Pazewski M, Smoukov SK, Bishop KJM, Grzybowski BA (2006) Electrostatic self-assembly of binary nanoparticle crystals with a diamond-like lattice. Science 312:420–424
43. Leunissen ME, Christova CG, Hynninen AP, Royall CP, Campbell AI, Imhof A, Dijkstra M, Roij R, Blaadern A (2005) Ionic colloidal crystals of oppositely charged particles. Nature 437:235–240
44. Jackson AM, Hu Y, Silva PJ, Stellacci F (2006) From homoligand to mixed-ligand-monolayer-protected metal nanoparticles: a scanning tunneling microscopy investigation. J Am Chem Soc 128:11135–11149
45. Li H, Park SH, Reif JH, LaBean TH, Yan H (2004) DNA-templated self-assembly of protein and nanoparticle linear arrays. J Am Chem Soc 126:418–419
46. Nath N, Chilkoti A (2001) Interfacial phase transition of an environmentally responsive elastin biopolymer adsorbed on functionalized gold nanoparticles studied by colloidal surface plasmon resonance. J Am Chem Soc 123:8197–8202
47. Wang G, Murray RW (2004) Controlled assembly of monolayer-protected gold clusters by dissolved DNA. Nano Lett 4:95–101
48. Gyorvary E, Schroedter A, Talapin DV, Weller H, Pum D, Sleyter UB (2004) Formation of nanoparticle arrays on S-layer protein lattices. J Nanosci Nanotechnol 4:115–120
49. Storhoff JJ, Lazarides AA, Mucic RC, Mirkin CA, Letsinger RL, Schatz GC (2000) What controls the optical properties of DNA-linked gold nanoparticle assemblies? J Am Chem Soc 122:4640–4650
50. McConnell WP, Nowak JP, Brousseau LC, Fuierer RR, Tenent RC, Feldheim DL (2000) Electronic and optical properties of chemically modified metal nanoparticles and molecular bridged nanoparticle arrays. J Phys Chem B 104:8925–8930
51. Prabaharan M, Grailer JJ, Pilla S, Steeber DA, Gong SQ (2009) GNPs with a monolayer of doxorubicin-conjugated amphiphilic block copolymer for tumor-targeted drug delivery. Biomaterials 30:6065–6075
52. Gibson JD, Khanal BP, Zubarev ER (2007) Paclitaxel-functionalized GNPs. J Am Chem Soc 129:11653–11661
53. Hong R, Han G, Fernandez JM, Kim BJ, Forbes NS, Rotello VM (2006) Glutathione-mediated delivery and release using monolayer protected nanoparticle carriers. J Am Chem Soc 128:1078–1079
54. Aryal S, Grailer JJ, Pilla S, Steeber DA, Gong SQ (2009) Doxorubicin conjugated GNPs as water-soluble and pH-responsive anticancer drug nanocarriers. J Mater Chem 19:7879–7884
55. Kim CK, Ghosh P, Pagliuca C, Zhu ZJ, Menichetti S, Rotello VM (2009) Entrapment of hydrophobic drugs in nanoparticle monolayers with efficient release into cancer cells. J Am Chem Soc 131:1360–1361

56. Nikoobakht B, El-Sayed MA (2003) Surface-enhanced Raman scattering studies on aggregated gold nanorods. J Phys Chem A 107:3372–3378
57. Jans H, Liu X, Huo Q, Austin L, Maes G (2009) Dynamic light scattering as a powerful tool for gold nanoparticle bioconjugation and biomolecular binding studies. Anal Chem 81:9425–9432
58. Zhong Z, Patskovsky S, Bouvrette P, Luong HT, Gedanken A (2004) The surface chemistry of Au colloids and their interactions with functional amino acids. J Phys Chem B 108:4046–4052
59. Shibu ES, Muhammed MAH, Kimura K, Pradeep T (2009) Fluorescent superlattices of GNPs: a new class of Functional nanomaterials. Nano Res 2:220–234
60. Cui R, Huany H, Yin Z, Gao D, Zhu JJ (2008) Horseradish peroxide-functionalized GNPs label for amplified immunoanalysis based on GNPs/carbon nanotubes hybrids modified biosensor. Biosens Bioelectron 23:1666–1673
61. Chai F, Wang C, Wang T, Ma Z, Su Z (2010) L-cysteine functionalized GNPs for the colorimetric detection of Hg^{+2} induced by ultraviolet light. Nanotechnology 21:025501
62. Guo S, Huany Y, Jiang Q, Sun Y, Deng L, Liang Z, Du Q, Xing J, Zhao Y, Wang PC, Dong A, Liang XJ (2010) Enhanced gene delivery and siRNA silencing by GNPs coated with charge-reversal polyelectrolyte. ACS Nano 4:5505–5511
63. Thomas M, Klibnove AM (2003) Conjugation to gold nanoparticles enhances polyethylenimine's transfer of plasmid DNA into mammalian cells. Proc Natl Acad Sci USA 100:9138–9143
64. Rosi NL, Giljohann DA, Thaxton CS, Lytton-Jean AKR, Han MS, Mirkin CA (2006) Oligonucleotide-modified gold nanoparticles for intracellular gene regulation. Science 312:1027–1030
65. Ghosh P, Yang X, Arvizo R, Zhu ZJ, Agasti SS, Mo Z, Rotello VM (2010) Intracellular delivery of a membrane-impermeable enzyme in active form using functionalized GNPs. J Am Chem Soc 132:2642–2645
66. Deng F, Yang Y, Hwany S, Shon YS, Chen S (2004) Fullerene-functionalized GNPs: electrochemical and spectroscopic properties. Anal Chem 76:6102–6107
67. Tian D, Zhang H, Chai Y, Cui H (2011) Synthesis of N-(aminobutyl)-N-(ethylisoluminol) functionalized gold nanomaterials for chemiluminescent bio-probe. Chem Commun 47:4959–4961
68. De la Fuente JM, Berry CC, Riehle MO, Curtis ASG (2006) Nanoparticle targeting at cells. Langmuir 22:3286–3293
69. Wang Z, Tan B, Hussain I, Schaeffer N, Wyatt MF, Brust M, Cooper AI (2007) Design of polymeric stabilizers for size-controlled synthesis of monodisperse gold nanoparticles in water. Langmuir 23:885–895
70. Baker CO, Shedd B, Tseng RJ, Martinez-Morales AF, Ozkan CS, Ozkan M, Yang Y, Kaner RB (2011) Size control of gold nanoparticles grown on polyaniline nanofibers for bistable memory devices. ACS Nano 5:3469–3474
71. Huang X, Li B, Zhang H, Hussain I, Liang L, Tan B (2011) Facile preparation of size-controlled gold nanoparticles using versatile and end-functionalized thioether polymer ligands. Nanoscale 3:1600–1607
72. Li D, He Q, Cui Y, Wang K, Zhang X, Li J (2007) Thermosensitive copolymer networks modify gold nanoparticles for nanocomposite entrapment. Chemistry 13:2224–2229
73. Maus L, Spatz JP, Fiammengo R (2009) Quantification and reactivity of functional groups in the ligand shell of PEGylated GNPs via a fluorescence-based Assay. Langmuir 25:7910–7917
74. Chakraborty S, Bishoni SW, Perez-Luna VH (2010) GNPs with poly(N-isopropylacrylamide) formed via surface initiated atom transfer free radical polymerization exhibit unusually slow aggregation kinetics. J Phys Chem C 114:5947–5955
75. Jordan R, West N, Ulman A, Chou YM, Nuyken O (2001) Nanocompositees by surface-initiated living cationic polymerization of 2-oxazolines on functionalized GNPs. Macromolecules 34:1606–1611

76. Moreno M, Hernandez R, Lokez D (2010) Crosslinking of poly(vinyl alcohol) using functionalized GNPs. Eur Polym J 46:2099–2104
77. Zhang T, Wu Y, Pan X, Zheng Z, Ding X, Peng Y (2009) An approach for the surface functionalized gold nanoparticles with pH-responsive polymer by combination of RAFT and click chemistry. Eur Polym J 45:1625–1633
78. Xu S, Tu G, Peng B, Han X (2006) Self-assembling GNPs on thiol-functionalized poly (styrene C-CO-arylic acid) nanospheres for fabrication of a mediatorless biosensors. Anal Chim Acta 570:151–157
79. Aryal S, Remant Bahadur KC, Bhattarai N, Lee BM, Kim HY (2006) Stablization of GNPs by thiol Functionalized poly (ϵ-caprolactone) for the labeling of PCL biocarrier. Mater Chem Phys 98:463–469
80. Zhao H, Kang X, Liu L (2005) Comb–coil polymer brushes on the surface of silica nanoparticles. Macromolecules 38:10619–10622
81. Luo S, Xu J, Zhu Z, Wu C, Liu S (2006) Phase transition behavior of unimolecular micelles with thermoresponsive poly(N-isopropylacrylamide) coronas. J Phys Chem B 110:9132–9139
82. De M, Ghosh PS, Rotello VM (2008) Applications of nanoparticles in biology. Adv Mater 20:4225–4241
83. Jun YW, Lee JH, Cheon J (2008) Chemical design of nanoparticle probes for high-performance magnetic resonance imaging. Angew Chem Int Ed 47:5122–5135
84. Laurent S, Forge D, Port M, Roch A, Robic C, Vander Elst L, Robert N (2008) Muller magnetic iron oxide nanoparticles: synthesis, stabilization, vectorization, physicochemical characterizations, and biological applications. Chem Rev 108:2064–2110
85. Chin AB, Yaacob II (2007) Synthesis and characterization of magnetic iron oxide nanoparticles via w/o microemulsion and Massart's procedure. J Mater Process Technol 191:235–237
86. Albornoz C, Jacobo SE (2006) Preparation of a biocompatible magnetic film from an aqueous ferrofluid. J Magn Magn Mater 305:12–15
87. Kim EH, Lee HS, Kwak BK, Kim BK (2005) Synthesis of ferrofluid with magnetic nanoparticles by sonochemical method for MRI contrast agent. J Magn Magn Mater 289:328–330
88. Wan J, Chen X, Wang Z, Yang X, Qian Y (2005) A soft-template-assisted hydrothermal approach to single-crystal Fe_3O_4 nanorods. J Cryst Growth 276:571–576
89. Kimata M, Nakagawa D, Hasegawa M (2003) Preparation of monodisperse magnetic particles by hydrolysis of iron alkoxide. Powder Technol 132:112–118
90. Alvarez GS, Muhammed M, Zagorodni AA (2006) Novel flow injection synthesis of iron oxide nanoparticles with narrow size distribution. Chem Eng Sci 61:4625–4633
91. Basak S, Chen DR, Biswas P (2007) Electrospray of ionic precursor solutions to synthesize iron oxide nanoparticles: modified scaling law. Chem Eng Sci 62:1263–1268
92. Herrmann IK, Grass RN, Mazunin D, Stark WJ (2009) Synthesis and covalent surface functionalization of nonoxidic iron core-shell nanomagnets. Chem Mater 21:3275–3281
93. Bhattacharya D, Das M, Mishra D, Banerjee I, Sahu SK, Maiti TK, Pramanik P (2011) Folate receptor targeted, carboxymethyl chitosan functionalized iron oxide nanoparticles: a novel ultradispersed nanoconjugates for bimodal imaging. Nanoscale 3:1653–1662
94. Mikhaylova M, Kim DK, Berry CC, Zagorodni A, Toprak M, Curtis ASG, Muhammed M (2004) BSA immobilization on amine-functionalized superparamagnetic iron oxide nanoparticles. Chem Mater 16:2344–2354
95. Qu H, Caruntu D, Liu H, O'Connor CJ (2011) Water-dispersible iron oxide magnetic nanoparticles with versatile surface functionalities. Laungmuir 27:2271–2278
96. Yigit MV, Mazumdar D, Lu Y (2008) MRI detection of thrombin with aptamer functionalized superparamagnetic iron oxide nanoparticles. Bioconjug Chem 19:412–417

97. Bertorelle F, Wilhelm C, Roger J, Gazeau F, Ménager C, Cabuil V (2006) Fluorescence-modified superparamagnetic nanoparticles: intracellular uptake and use in cellular imaging. Langmuir 22:5385–5391
98. Bazile D, Prud'homme C, Bassoullet MT, Marlard M, Spenlehauer G, Stealth Me VM (1995) PEG–PLA nanoparticles avoid uptake by the mononuclear phagocytes system. J Pharm Sci 84:493–498
99. Lyu YK, Kyung KJ, Lee WX (2009) Functionalized magnetic nanoparticles with poly (3-thiophenaacetic acid) and its application for electrogenerated chemiluminescence sensor. Synth Met 159:571–575
100. Gang ZY, Yu SH, Dong PS, Qin HM (2010) Synthesis, characterization and properties of ethylenediamine-functionalized Fe_3O_4 magnetic polymers for removal of Cr(vi) wastewater. J Hazard Mater 182:295–302
101. Zhang J, Rana S, Srivastava RS, Mishra RDK (2008) On the chemical synthesis and drug delivery response of folate receptor or activated, polyethylene glycol-functionalized magnetic nanoparticles. Acta Biomater 4:40–48
102. Rahimi M, Wadajkar A, Subramanian K, Yousef M, Cui W, Hsieh JT, Nguyen KT (2010) In vitro evaluation of novel-polymer coated magnetic nanoparticles for controlled drug delivery. Nanomedicine 6:672–680
103. Zhou SX, Wu LM, Sun J, Shen WD (2000) The change of the properties of acrylic-based polyurethane via addition of nano-silica. Prog Org Coat 45:33–42
104. Arrighi V, McEwen IJ, Qian H, Serrano PMB (2003) The glass transition and interfacial layer in styrene-butadiene rubber containing silica nanofillers. Polymer 44:6259–6266
105. Arkhireeva A, Hay JN (2004) Synthesis of organically-modified silica particles for use as nanofillers in polymer systems. Polym Polym Compos 12:101–110
106. Von Hohenesche CF, Unger KK, Eberle T (2004) Agglomerated non-porous silica nanoparticles as model carriers in polyethylene synthesis. J Mol Cat A Chem 221:185–199
107. Kim YK, Lewis AF, Patra PK, Warner SB, Mhetre SK, Shah MA, Nam D (2002) Nanocomposite fibres. Mater Res Soc Symp - Proc 740:441–446
108. He XX, Wang KM, Tan WH et al (2003) Concentration of trace amounts oligonucleotide using super-paramagnetic DNA nano-en-richer. Chem J Chin Univ 24:40–42
109. Bruce IJ, Sen T (2005) Surface modification of magnetic nanoparticles with alkoxysilanes and their application in magnetic bioseparations. Langmuir 21:7029–7035
110. He XX, Wang KM, Tan WH et al (2001) A novel fluorescent label based on biological fluorescent nanoparticles and its application in cell recognition. Chin Sci Bull 46:1353–1356
111. Wang L, Yang CY, Tan WH (2005) Dual-luminophore-doped silica nanoparticles for multiplexed signaling. Nano Lett 5:37–43
112. Shin JH, Metzger SK, Schoenfisch MH (2007) Synthesis of nitric oxide-releasing silica nanoparticles. J Am Chem Soc 129:4612–4619
113. Chi F, Guo YN, Liu J, Liu Y, Huo Q (2010) Size-tunable and functional core-shell structured silica nanoparticles for drug release. J Phys Chem C 114:2519–2523
114. Tymish Y, Ohulchanskyy, Roy I, Lalit N, Goswami, Chen Y, Bergey EJ, Pandey RK, Oseroff AR, Prasad PN (2007) Organically modified silica nanoparticles with covalently incorporated photosensitizer for photodynamic therapy of cancer. Nano Lett 7:2835–2842
115. Lee J, Lee Y, Youn JK, Na HB, Yu T, Kim H, Lee SM, Koo YM, Kwak JH, Park HG, Chang HN, Hwang M, Park JG, Kim J, Hyeon T (2008) Simple synthesis of functionalized superparamagnetic magnetite/silica core/shell nanoparticles and their application as magnetically separable high-performance. Biocatalysts. Small 4:143–152
116. Deng Y, Deng C, Yang D, Wang C, Fu S, Zhang X (2005) Preparation. characterization and application of magnetic silica nanoparticle functionalized multi-walled carbon nanotubes. Chem Commun 44:5548–5550
117. Chandran SP, Hotha S, Prasad BLV (2008) Tunable surface modification of silica nanoparticles through 'click' chemistry. Curr Sci 95:1327–1333

118. Bagwe RP, Hilliard LR, Tan W (2006) Surface modification of silica nanoparticles to reduce aggregation and non-specific binding. Langmuir 22:4357–4362
119. Zhou H, Zhang C, Li H, Du Z (2011) Fabrication of silica nanoparticles on the surface of functionalized multi-walled carbon nanotubes. Carbon 49:126–132
120. Wei L, Zhang Y (2009) Emulsion polymerization of ethylene from mesoporous silica nanoparticles with vinyl functionalized monolayers. J Polym Sci A Polym Chem 47:1393–1402
121. Cho Y, Shi R, Borgenes R, Ivanisevic A (2008) Functionalized mesoporous silica nanoparticles-based drug delivery system to rescue acrolein-mediated cell death. Nanomedicine 3:507–519
122. Gao D, Zhang Z, Wu M, Xie C, Guan G, Wang D (2007) A surface functional monomer-directing strahegy for highly dense imprinting of TNT at surface of silica nanoparticles. J Am Chem Soc 129:7859–7866
123. Li Y, Benicewicz BC (2008) Functionalization of silica nanoparticles via the combination of surface-Initited RAFT polymerization and click reactions. Macromolecules 41:7986–7992
124. He P, Greenway G, Haswell SJ (2008) The on-line synthesis of enzyme functionalized silica nanoparticles in a microfluidic reactor using polyethylenimine polymer and R5 peptide. Nanotechnology 19:315603–315610
125. Liu CH, Pan CY (2007) Grafting polystyrene on to silica nanoparticles via RAFT polymerization. Polymer 48:3679–3685
126. Bandow S, Takizawa M, Hirahara K, Yudasaka M, Iijima S (2001) Raman scattering study of double-wall carbon nanotubes derived from the chains of fullerenes in single-wall carbon nanotubes. Chem Phys Lett 337:48–54
127. Hea L, Zhua YZ, Zhenga JY, Mab YF, Chen YS (2010) Meso-meso linked diporphyrin functionalized single-walled carbon nanotubes. J Photochem Photobiol A Chem 216:15–23
128. Li Q, Zhang J, Yan H, He M, Liu Z (2004) Thionine-mediated chemistry of carbon nanotubes. Carbon 42:287–291
129. Jain AK, Dubey V, Mehra NK, Lodhi N, Nahar M, Mishra DK, Jain NK (2009) Carbohydrate-conjugated multiwalled carbon nanotubes: development and characterization. Nanomedicine 5:432–442
130. Xia HS, Wang Q, Li KS, Hu GH (2004) Preparation of polypropylene/carbon nanotube composite powder with a solid-state mechanochemical pulverization process. J Appl Polym Sci 93:378–386
131. Mamedov AA, Kotov NA, Prato M, Guldi DM, Wicksted JP, Hirsch A (2002) Molecular design of strong single-wall carbon nanotube/polyelectrolyte multilayer composites. Nat Mater 1:190–194
132. Andrews R, Jacques D, Qian DL, Rantell T (2002) Multiwall carbon nanotubes: synthesis and application. Acc Chem Res 35:1008–1017
133. Potschke P, Fornes TD, Paul DR (2002) Rheological behavior of multiwalled carbon nanotube/polycarbonate composites. Polymer 43:3247–3255
134. Potschke P, Bhattacharyya AR, Janke A, Goering H (2003) Melt mixing of polycarbonate/multi-wall carbon nanotube composites. Composite Interfaces 10:389–404
135. Huang YY, Ahir SV, Terentjev EM (2006) Dispersion rheology of carbon nanotubes in a polymer matrix. Phys Rev B 73:125422–125431
136. Meng QH, Hu JF (2008) Self-organizing alignment of carbon nanotube in shape memory segmented fiber prepared by in situ polymerization and melt spinning. Compos Part A Appl Sci Manuf 39:314–321
137. Shen LM, Gao XS, Tong Y, Yeh A, Li RX, Wu DC (2008) Influence of different functionalized multiwall carbon nanotubes on the mechanical properties of poly(ethylene terephthalate) fibers. J Appl Polym Sci 108:2865–2871
138. Mohanty N, Berry V (2008) Graphene-based single-bacterium resolution biodevice and DNA transistor: interfacing graphene derivatives with nanoscale and microscale biocomponents. Nano Lett 8:4469–4476
139. Zhang R, Hummelgard Lu G, Olin H (2011) Real time monitoring of the drug release of rhodamine B on graphene oxide. Carbon 49:1126–1132

140. Sun X, Liu Z, Welsher K, Robinson JT, Goodwin A, Zaric S, Dai H (2008) Nanographene oxide for cellular imaging and drug delivery. Nano Res 1:203–212
141. Liu Z, Robinson JT, Sun XM, Dai HJ (2008) PEGylated nanographene oxide for delivery of water-insoluble cancer drugs. J Am Chem Soc 130:10876–10877
142. Zhang L, Xia J, Zhao Q, Liu L, Zhang Z (2009) Functional graphene oxide as a nanocarrier for controlled loading and targeted delivery of mixed anticancer drugs. Small 6:537–544
143. Yang X, Zhang X, Liu Z, Ma Y, Huang Y, Chen Y (2008) High-efficiency loading and controlled release of doxorubicin hydrochloride on graphene oxide. J Phys Chem C 112:17554–17558
144. Yang K, Zhang S, Zhang G, Sun X, Lee ST, Liu Z (2010) Graphene in mice: ultrahigh in vivo tumor uptake and efficient photothermal therapy. Nano Lett 10:3318–3323
145. Yang K, Wan J, Zhang S, Zhang Y, Lee ST, Liu Z (2011) In vivo pharmacokinetics, long-term biodistribution and toxicology of PEGylated graphene in mice. ACS Nano 5:516–522
146. Yan X, Chen J, Xang J, Xue Q, Miele P (2010) Fabrication of free-standing electrochemically active and biocompatible graphene oxide-polyaniline and graphene-polyaniline hybrid papers. ACS Appl Mater Interfaces 2:2521–2529
147. Park S, Dikin DA, Ngyyen ST, Ruoff RS (2009) Graphene oxide sheets chemically cross-linked by polyallylamine. J Phys Chem 113:15801–15804
148. Fang H, Wang S, Xiao S, Yang J, Li Y, Shi Z, Li H, Liu H, Xiao S, Zhu D (2003) Three-point hydrogen bonding assembly between a conjugated PPV and a functionalized fullerene. Chem Mater 15:1593–1597
149. Ikeda A, Doi Y, Hashizume M, Kikuchi JI, Konishi T (2007) An extremely effective DNA photocleavage utilizing functionalized liposomes with a fullerene-enriched lipid bilayer. J Am Chem Soc 129:4140–4141
150. Chamberlain TW, Camenisch A, Champness NR, Briggs GAD, Benjamin SC, Ardavan A, Khlobystov AN (2007) Toward controlled spacing in one-dimensional molecular chains: alkyl-chain-functionalized fullerenes in carbon nanotubes. J Am Chem Soc 129:8609–8614
151. Sun YP, Guduru R, Lawson GE, Mullins JE, Guo Z, Quinlan J, Bunker CE, Gord JR (2000) Photophysical and electron-transfer properties of mono and multiple-functionalized fullerene derivatives. J Phys Chem B 104:4625–4632
152. Souza FD, Chitta R, Gadde S, McCarty AL, Karr PA, Zandler ME, Sandanayaka ASD, Araki Y, Ito O (2006) Design, syntheses and studies of supramolecular porphyrin-fullerene conjugates, using bis-18-crown-6 appended porphyrins and pyridine or alkyl ammonium functionalized fullerenes. J Phys Chem B 110:5905–5913
153. Sudeep PK, Ipe BI, Thomas KG, George MV (2002) Fullerene-functionalized GNPs. A self-assembled photoactive antenna-metal nanocore assembly. Nano Lett 2:29–35
154. Yen CF, Peddinti RK, Liao CC (2000) Highly functionalized bicyclo[2.2.2]octenone-fused b [60] fullerenes from masked o-benzoquinones and C_{60}. Org Lett 2:2909–2912
155. Zhou Z, Magriotis PA (2005) A new method for the functionalization of [60] fullerene: an unusual 1,3-dipolar cycloaddition pathway leading to a C_{60} housane derivative. Org Lett 7:5849–5851
156. Zhong YW, Matsuo Y, Nakamura E (2006) Convergent synthesis of a polyfunctionalized fullerene by regioselective five-fold addition of a functionalized organocopper reagent to C_{60}. Org Lett 8:1463–1466
157. Goh HW, Goh SH, Xu GQ (2002) Synthesis and miscibility studies of [60] fullerenated poly (2-hydroxyethyl methacrylate). J Polym Sci Part A: Polym Chem 40:1157–1166
158. Zhang F, Svensson M, Andersson MR, Maggini M, Bucella S, Menna E, Inganas O (2001) Soluble polythiophenes with pendant fullerene groups as double cable materials for photodiodes. Adv Mater 13:1871–1874
159. Charrois GJR, Allen TM (2003) Rate of biodistribution of STEALTH liposomes to tumor and skin: influence of liposome diameter and implications for toxicity and therapeutic activity. Biochim Biophys Acta 1609:102–108
160. Martina MS, Fortin JP, Menager C, Clement O, Barratt G, Gabrielle-Madelmont C, Gazeau F, Cabuil V, Lesieur S (2005) Generation of superparamagnetic liposomes revealed as highly efficient MRI contrast agents for in vivo imaging. J Am Chem Soc 127:10676–10685

161. Al-Jamal WT, Al-Jamal KT, Bomans PH, Frederik PM, Kostaleros K (2008) Functionalized-quantum-dot-liposome hybrids as multimodal nanoparticles for cancer. Small 4:1406–1415
162. Al-Jamal WT, Kostarelos K (2007) Liposome nanoparticle hybrids for multimodal diagnostic and therapeutic applications. Nanomedicine 2:85–98
163. Soenen SJH, Hodenius M, Cuyper MD (2009) Magneto liposomes: versatile innovative nanocolloids for use in biotechnology and biomedicine. Nanomedicine 4:177–191
164. Elbayoumi TA, Torchilin VP (2009) Tumor-targeted nanomedicines: enhanced antitumor efficacy in vivo of doxorubicin loaded, long-circulating liposomes modified with cancer specific monoclonal antibody. Clin Cancer Res 15:1973–1980
165. Sau TP, Urban AS, Dondapati SK, Fedoruk M, Horton MR, Rogach AL, Stefani FD, Radler JO, Feldmann J (2009) Controlling loading and optical properties of GNPs on liposome membranes. Colloids Surf A 342:92–96
166. Wu GH, Milkhailovsky A, Khant HA, Fu C, Chiu W, Zasadzinski JA (2008) Remotely triggered liposome release by near-infrared light absorption via hollow gold nanoshells. J Am Chem Soc 130:8175–8177
167. Chen Y, Bose A, Bothun GD (2010) Controlled release from bilayer- decorated magnetoliposomes via electromagnetic heating. ACS Nano 4:3215–3221
168. Dave N, Liu J (2011) Programmable assembly of DNA functionalized liposomes by DNA. ACS Nano 5:1304–1312
169. Cavalli S, Tipton AR, Overhand M, Kros A (2006) The chemical modification of liposome surfaces via a copper-mediated [3+2] azide–alkyne cycloaddition monitored by a colorimetric assay. Chem Commun 30:3193–3195
170. Smith AM, Jaime-fonseca MR, Grover LM, Bakalis S (2010) Alginate-Loaded liposomes can protect encapsulated alkaline phosphatase functionality when exposed to gastric ph. J Agric Food Chem 58:4719–4724
171. He X, Na MH, Kim JS, Lee GY, Park JY, Hoffman AS, Nam JO, Han SE, Sim GY, Oh YK, Kim LS, Lee BH (2011) A novel peptide probe for imaging and targeted delivery of liposomal Doxorubicin to lung tumor. Mol Pharm 8:430–438
172. Weng KC, Noble CO, Sternberg BP, Chen FF, Drummond DC, Kirpotin DB, Wang D, Hom YK, Hann B, Park JW (2008) Targeted tumor cell internalization and imaging of multifunctional quantum dot-conjugated immunoliposomes in vitro and in vivo. Nano Lett 8:2851–2857
173. Gonc ALO, Bernardes JL, Kikkeri R, Maglinao M, Laurino P, Collot M, Hong SY, Lepenies B, Seeberger PH (2010) Design, synthesis and biological evaluation of carbohydrate-functionalized cyclodextrins and liposomes for hepatocyte-specific targeting. Org Biomol Chem 8:4987–4996
174. Paasonen L, Romberg B, Storm G, Yliperttula M, Urtti A, Hennink WE (2007) Temperature-senstive poly (N-(2-hydroxypropyl) methacrylamide mono/dilactat)-coated liposomes for triggered contents release. Bioconjug Chem 18:2131–2136
175. Hofmann AM, Wurm F, Huhn E, Nawroth T, Langguth P, Kney H (2010) Hyperbranched polyglycerol-based Lipids via oxyanionic polymerization: toward multifunctional stealth Liposomes. Biomacromolecules 11:568–574
176. Kumar A, Erasquin UJ, Qin G, Li K, Cai C (2010) Clickable, polymerized liposomes as a versatile and stable platform for rapid optimization of their peripheral compositions. Chem Commun 46:5746–5748
177. Medintz IL, Uyeda HT, Goldman ER, Mattoussi H (2005) Quantum dot bioconjugates for imaging, labeling and sensing. Nat Mater 4:435–446
178. Michalet X, Pinaud FF, Bentolila LA, Tsay JM, Doose S, Li JJ, Sundaresan G, Wu AM, Gambhir SS, Weiss S (2005) Quantum dots for live cells, in vivo imaging, and diagnostics. Science 307:538–544
179. Han M, Gao XH, Su JZ, Nie SM (2001) Quantum-dot-tagged microbeads for multiplexed optical coding of biomolecules. Nat Biotechnol 19:631–635
180. Yong KT, Roy I, Swihart MT, Prasad PN (2009) Multifunctional nanoparticles as biocompatible targeted probes for human cancer diagnosis and therapy. J Mater Chem 19:4655–4672

181. Smith AM, Duan H, Mohs AM, Nie S (2008) Bioconjugated quantum dots for in vivo molecular and cellular imaging. Adv Drug Deliv Rev 60:1226–1240
182. Hashizume K, Matsubayashi M, Vacha M, Tani T (2002) Individual mesoscopic structures studied with sub-micrometer optical detection techniques: CdSe nanocrystals capped with TOPO and ZnS-overcoated system. J Lumin 98:49–56
183. Xue M, Wang X, Wang H, Tang B (2011) The preparation of glutathione-capped CdTe quantum dots and their use in imaging of cells. Talanta 83:1680–1686
184. Ingolea PP, Abhyankara RM, Prasa BLV, Haram SK (2010) Citrate-capped quantum dots of CdSe for the selective photometric detection of silver ions in aqueous solutions. Mater Sci Eng B 168:60–65
185. Liu P, Wang Q, Li X (2009) Studies on CdSe/L-cysteine quantum dots synthesized in aqueous solution for biological labeling. J Phys Chem C 113:7670–7676
186. Chen Y, Chen Z, He Y, Lin H, Sheng P, Liu C, Luo S, Cai Q (2010) L-cysteine-capped CdTe QD-based sensor for simple and selective detection of trinitrotoluene. Nanotechnology 21:125502–125507
187. Bardi G, Malvindi MA, Gherardini L, Costa M, Pompa PP, Cingolani R, Pizzorusso T (2010) The biocompatibility of amino functionalized CdSe/ZnS quantum-dot-Doped SiO_2 nanoparticles with primary neural cells and their gene carrying performance. Biomaterials 31:6555–6566
188. Koneswaran M, Narayanaswamy R (2009) L-cysteine-capped ZnS quantum dots based fluorescence sensor for Cu^{2+} ion. Sens Actuators B 139:104–109
189. Pong BK, Trout BL, Jim-Yang LEE (2002) Preparation of DNA-functionalised CdSe/ZnS Quantum Dots. Curr Opin Biotechnol 13:40–46
190. Aguilera-Sigalat J, Rocton S, Galian RE, Prieto JP (2011) Fluorescence enhancement of amine-capped CdSe/ZnS quantum dots by thiol addition. Can J Chem 89:359–363
191. Zhou D, Ying L, Hong X, Hall EA, Abell C, Klenerman D (2008) A compact functional quantum dot-DNA conjugate: preparation, hybridization, and specific label-free DNA detection. Langmuir 24:1659–1664
192. Zou H, Wu S, Shen J (2008) Polymer/silica nanocomposites: preparation, characterization, properties and applications. Chem Rev 108:3893–3957
193. Balazs AC, Emrick T, Russell TP (2006) Nanoparticle polymer composites: where two small worlds meet. Science 314:1107–1110
194. Krishnamoorti R, Vaia RA (2007) Polymer nanocomposites. J Polym Sci, Part B: Polym Phys 45:3252–3256
195. Park MS, Needham SA, Wang GX, Kang YM, Park JS, Dou SX, Liu HK (2007) Nanostructured SnSb/carbon nanotube composites synthesized by reductive precipitation for lithium-ion batteries. Chem Mater 19:2406–2410
196. Wang H, Peng C, Peng F, Yu H, Yang J (2011) Facile synthesis of MnO2/CNT nanocomposite and its electrochemical performance for supercapacitors. Mater Sci Eng B 176:1073–1078
197. Xiao Y, Zhou S, Wang L, Gong T (2010) Electro-active shape memory properties of poly (ϵ-caprolactone)/functionalized multiwalled carbon nanotube nanocomposite. Appl Mater Interfaces 2:3506–3514
198. Tan WB, Zhang Y (2005) Multifunctional based QD based magnetic chitosan beads. Adv Mater 17:2375–2380
199. Nie Q, Tan WB, Zhang Y (2006) Synthesis and characterization of monodispersed chitosan nanoparticles with embedded QDs. Nanotechnology 17:140–144
200. Kang B, Chang SQ, Dai YD, Chen D (2008) Synthesis of green CdSe/chitosan quantum dots using a polymer-assisted g-radiation route. Radiat Phys Chem 77:859–863
201. Lin Y, Zhang L, Yao W, Qian H, Ding D, Wu W, Jiang X (2011) Water soluble chitosan QD hybrid nanospheres towards bioimaging and biolabeling. ACS Appl Mater Interfaces 3:995–1002
202. Dutta PK, Kumar H, Tiwari DK, Archana D, Rizvi KS, Kumar A, Singh BK, Srivastava R (2011) The glimpses of chitosan nanoparticles. Asian Chitin J 7:103–106

203. Laxmi K (2011) Nanotechnology and nanoscale devices applications in the treatment of cancer. Everyman's Science 45:301–307
204. Perry JL, Martin CR, Stewart JD (2011) Drug-delivery strategies by using template-synthesized nanotubes. Chemistry 17:6296–6302
205. Guo R, Zhang L, Qian H, Li R, Jiang X, Liu B (2010) Multifunctional nanocarriers for cell imaging, drug delivery, and near-IR photothermal therapy. Langmuir 26:5426–5434
206. Kulys J, Stupak R (2008) Glucose biosensor based on chitosan-gold and Prussian blue-gold nanoparticles. Open Nanosci J 2:34–38
207. Wang W, Cui H (2008) Chitosan-Luminol reduced gold nanoflowers: from one-pot synthesis to morphology-dependent SPR and chemiluminescence sensing. J Phys Chem C 112:10759–10766
208. Hortiguela MJ, Aranaz I, Gutiérrez MC, Ferrer ML, del Monte (2011) Chitosan gelation induced by the in situ formation of gold nanoparticles and its processing into macroporous scaffolds. Biomacromolecules 12:179–186
209. Shana C, Yanga H, Hana D, Zhanga Q, Ivaskab A, Niua L (2010) Graphene/AuNPs/chitosan nanocomposites film for glucose biosensing. Biosens Bioelectron 25:1070–1074
210. Gee Y, Yu Z, He S, Nie F, Teng G, Gu N (2009) Fluorescence modified chitosan-coated magnetic nanoparticles for high-efficient cellular imaging. Nanoscale Res Lett 4:287–295
211. Liu X, Hu Q, Fang Z, Zhang X, Zhang B (2009) Magnetic chitosan nanocomposites: a useful recyclable tool for heavy metal ion removal. Langmuir 25:3–8
212. Wang Y, Zhang X, Chen Y, Xu H, Tan Y, Wang S (2010) Detection of dopamine based on tyrosinase-Fe_3O_4 nanoparticles-chitosan nanocomposite biosensor. Am J Biomed Sci 2:209–216
213. Kaushika A, Solankia PR, Ansaria AA, Sumanaa G, Ahmadb S, Malhotra BD (2009) Iron oxide-chitosan nanobiocomposite for urea sensor. Sens Actuators B Chem 138:572–580
214. Liu Y-L, Hsu C-Y, Su Y-H, Lai J-Y (2005) Chitosan-silica complex membranes from sulfonic acid functionalized silica nanoparticles for pervaporation dehydration of ethanol-water solutions. Biomacromolecules 6:368–373
215. Qiu S, Wu L, Shi G, Zhang L, Chen H, Gao C (2010) Preparation and pervaporation property of chitosan membrane with functionalized multiwalled carbon nanotubes. Ind Eng Chem Res 49:11667–11675
216. Venkatesana J, Qianb Z-J, Ryua B, Kumar NA, Kima S-K (2011) Preparation and characterization of carbon nanotube-grafted-chitosan–natural hydroxyapatite composite for bone tissue engineering. Carbohydr Polym 83:569–577
217. Wua H, Wanga J, Kanga X, Wanga C, Wanga D, Liua D, Aksay IA, Lin Y (2009) Glucose biosensor based on immobilization of glucose oxidase in platinum nanoparticles/graphene/chitosan nanocomposite film. Talanta 80:403–406
218. Hua H, Wanga X, Wanga J, Liua F, Zhanga M, Xu C (2011) Microwave-assisted covalent modification of graphene nanosheets with chitosan and its electrorheological characteristics. Appl Surf Sci 257:2637–2642
219. Fan H, Wang L, Zhao K, Li N, Shi Z, Ge Z, Jin Z (2010) Fabrication, mechanical properties, and biocompatibility of graphene-reinforced chitosan composites. Biomacromolecules 11:2345–2351
220. Jayasree A, Sasidharan S, Koyakutty M, Nair S, Menon D (2011) Mannosylated chitosan-zinc sulphide nanocrystals as fluorescent bioprobes for targeted cancer imaging. Carbohydr Polym 85:37–43
221. Yuan Q, Hein S, Misra RDK (2010) New generation of chitosan-encapsulated ZnO quantum dots loaded with drug: synthesis, characterization and in vitro drug delivery response. Acta Biomater 6:2732–2739
222. Wang Y, Tu S, Li R, Yang XY, Liu L, Zhang Q, Wang Y, Tu S, Li R, Yang XY, Liu L, Zhang Q (2010) Cholesterol succinyl chitosan anchored liposomes: preparation, characterization, physical stability, and drug release behaviour. Nanomedicine 6:471–477
223. Wang H, Zhao P, Liang X, Gong X, Song T, Niu R, Chang J (2010) Folate-PEG coated cationic modified chitosan–cholesterol liposomes for tumor-targeted drug delivery. Biomaterials 31:4129–4138

Adv Polym Sci (2013) 254: 51–86
DOI: 10.1007/12_2012_194

Published online: 5 February 2013

Nanoparticles for Gene Delivery into Stem Cells and Embryos

Pallavi Pushp, Rajdeep Kaur, Hoon Taek Lee, and Mukesh Kumar Gupta

Abstract Gene delivery is an important issue in embryo and stem cell studies for transgenic animal production, cell fate regulation, gene therapy, generation of patient-specific stem cells for cell-based therapy, cell tracing and imaging. Gene delivery has been classically achieved by a variety of methods that use a viral or a non-viral vector packaged with the nucleic acid of interest. In the last decade, several newer approaches to gene delivery have emerged that utilize nanomaterials to provide an efficient, safe and targeted gene delivery, both in vitro and in vivo. These nanomaterials, including polymeric nanoparticles, ceramic nanoparticles, magnetic nanoparticles, polymeric micelles and dendrimers, modify the kinetics, distribution and release of the genes into the cells and, thereby, increase the efficiency of gene delivery. This chapter describes the available nanoparticle-based gene delivery systems and their utility in stem cells for maintaining self-renewal, pluripotency and/or targeted differentiation into specific cell types for cell-based therapy and/or gene therapy. The chapter further discusses and reviews the progress and future of nanoparticles for generation of transgenic animals via gene delivery into embryos – a research area that is yet to be fully explored.

Keywords Embryo · Gene delivery · Genetic engineering · Nanoparticle · Reprogramming · Stem cells · Transgenesis

---

P. Pushp, R. Kaur, and M.K. Gupta (✉)
Department of Biotechnology and Medical Engineering, National Institute of Technology, Rourkela, India
e-mail: pushppallavi88@gmail.com; rajdeepkaur.10@gmail.com; guptam@nitrkl.ac.in

H.T. Lee
Department of Animal Biotechnology, Bio-Organ Research Center, Konkuk University, Seoul, South Korea
e-mail: htl3675@konkuk.ac.kr

Contents

1 Introduction

Genes can be introduced into mammalian cells primarily by two methods: non-viral and viral vector methods. The viral vector method is very efficient in gene delivery but carries safety risks and could be a biohazard. Integration and accidental re-activation of viral vectors into the host genome may result in aberrant gene expression, immune reaction and carcinogenesis. Possible recombination of viral vector with endogenous retroviral sequences may also form a self-replicative virus to cause diseases or induce mutations. Non-viral gene delivery methods are therefore, attractive. Electroporation and lipid-based gene delivery (lipofection) have been a common methodology for years. However, in recent years, nanoparticle (NP)-based gene delivery methods have aroused intense interest among researchers owing to their convenience in manufacturing, handling and use (Fig. 1). The versatility of the nanotechnology platform has allowed tailoring of NPs for their size, contents and surface electronic properties by relatively simple physical and/or

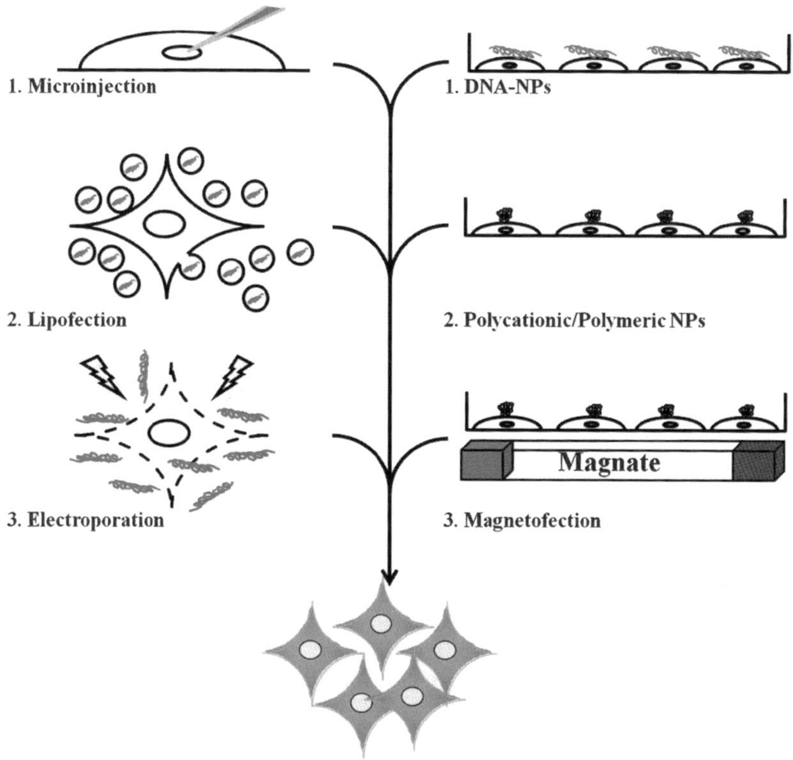

Fig. 1 Non-viral gene-delivery options for stem cells

chemical methods in order to use them as gene delivery vehicles. NP-based gene delivery methods can also be designed to have additional functionality such as cell-specific gene targeting, multigene delivery, controlled gene delivery, enhanced cellular uptake of genes and environment-sensitive degradability. Several types of NPs, including magnetic NPs, carbon nanotubes and several well-known synthetic cationic polymers and copolymers such as polyethyleneimine (PEI), poly (L-lysine) (PLL), polyamidoamine dendrimer, dextran, derivatized chitosan, poly [(2-methylamino) ethyl methacrylate], polylactide (PLA), polylactide-*co*-glycolide (PLGA), etc., have been explored for their commercial utilization for effective gene delivery in vitro as well as in vivo.

2 Non-viral Options for In Vitro Gene Delivery into Cells

2.1 Naked DNA

Naked DNA possesses little-to-no ability to transfect mammalian cells, with the exception of muscle cells [1]. Physical methods such as direct microinjection and electroporation are, therefore, commonly incorporated into naked DNA delivery methods. These methods induce temporary holes in the plasma membrane, which allows easy entry of the naked DNA into the cells. Among the two commonly used methods, direct microinjection requires special equipment (cell injector) and training but is suitable for non-dividing cells such as neurons and for large-sized cells such as oocytes and embryos. Electroporation, on the other hand, is an efficient method of naked DNA delivery but can result in a high rate of cellular death due to the electrical stimulation. Unfortunately, despite efficient delivery into the cells, gene integration into host genome and subsequent gene expression efficiency has been very low in both the methods. This is primarily due to their degradation in cytoplasm by nuclease enzymes, sequestration by DNA-binding proteins and cytoskeletal elements in the cytoplasm and their inability to cross the nuclear membrane.

2.2 *Compacted DNA (DNA NPs)*

DNA is negatively charged and, therefore, can be condensed by polycations that range from inorganic polycations to organic polyamines to polypeptides such as polylysine and protamine. Because condensed DNAs are compacted, they are less accessible to nuclease degradation in the cytoplasm and can cross the nuclear membrane at an increased efficiency. Studies have shown that naked DNA (~1,200 nm [2]) can acquire toroidal (~50 nm [3]), ellipsoidal (~22 × 50 nm [4]), rod-shaped (~8–11 × 200 nm [4]) and spherical (~130 nm [2]) configuration upon compaction by polycations such as spermine, CK30-PEG trifluoroacetate, CK30-PEG acetate and PEG-POD, respectively, in the nanometre size range and, accordingly, may technically be called DNA NPs.

The shape and size of the DNA NPs and the percentage of the DNA that condenses can be controlled by controlling the length of the polycation, the concentration of the salt (e.g. NaCl), the molecular charge ratio of polycation: DNA and the type of counterion [5–7]. The polycation may also impart additional functionality such as cell-type-specific gene delivery. For example, DNA compacted with galactosylated polylysine can specifically target hepatocytes, which express the asialoglycoprotein receptor [7]. Similarly, polylysine can be PEGylated to avoid the likelihood of particle aggregation [7]. When properly compacted and processed, DNA NPs are homogeneous in size and shape, consist only of compacted DNA, do not form aggregates, are colloidally stable in physiological salt concentrations and protect the DNA from digestion by cytoplasmic nuclease. Nonetheless, problems associated with electroporation and microinjection methods of introducing the DNA or DNA NPs into the cells cannot be solved by DNA compaction alone.

2.3 Liposomes and Lipoplexes

Negatively charged DNA can also be condensed by cationic lipids. Further, cationic lipids can form clusters of aggregated vesicles (liposomes) to encapsulate the DNA within a lipid bilayer to form a lipoplex. Because liposomes can interact and fuse with the cell membrane, DNA can be delivered directly across the plasma membrane. Consequently, liposome-mediated gene delivery (lipofection) has become a common protocol for gene delivery in a variety of cell types, and several commercial products are now available on the market. Cationic liposomes can be formed from a variety of cationic lipids including *N*-[1-(2,3-dioleoyloxy)propyl]-*N*,*N*,*N*-trimethylammonium-methyl sulfate (DOTAP) and *N*-[1-(2,3-dioleoyloxy) propyl]-*N*,*N*,*N*-trimethylammonium chloride (DOTMA). A neutral lipid such as 1,2,-dioleoyl-3-phosphatidylethanolamine (DOPE) is often included in the formulation to facilitate membrane fusion and to destabilize the liposomes for DNA release in the cytoplasm.

Liposomes and lipoplexes are usually self-assembling, easy to prepare and biodegradable. They allow increased uptake of naked DNA and DNA NPs. They can also be combined with polycations to form lipid–DNA NPs. Caracciola et al. [8] observed that lipid–protamine–DNA (LPD) NPs were more efficient than lipoplexes for gene delivery in CHO (Chinese hamster ovary cells), HEK293 (human embryonic kidney cells), NIH 3T3 (mouse embryonal cells) and A17 (murine cancer cells) cells. Unfortunately, cationic liposomes exhibit significant variability in gene delivery efficiency and are often toxic to cells.

2.4 NP-Based Gene Delivery

Gene delivery NPs can be formulated from diverse materials with unique architectures and loaded with DNA by condensation, encapsulation, surface attachment or entrapment. They offer multifold advantages over other methods of gene delivery:

- Due to their small size, NPs can efficiently penetrate across the cell membrane barrier to increase the efficiency of gene delivery.
- NPs can modify the condensation and physico-chemical state of the loaded DNA to protect them against cytoplasmic nuclease.
- Unlike many viral vectors, use of NPs is not limited by DNA size. They are capable of delivering large-sized DNA having multiple regulatory sequences. They can also be used to deliver multiple genes simultaneously.
- Functionality of the NPs can be tailored for specific or multiple bioactivities, such as controlled release of genes, cell-type-specific gene delivery and environmentally sensitive degradability, etc. Use of NPs may also allow the delayed release of DNA into cells until the cells enter mitosis and dissolve their nuclear

envelope to allow increased chances of interaction with host genome and, hence, increased success of gene integration.
- NPs are relatively easier to make, less expensive and more stable during long-term storage. Furthermore, they can be tested for the absence of endotoxins and other harmful ingredients and, therefore, are safe to use.
- Because NPs can be synthesized chemically, free of animal-derived components, they are ideal for use when absence of animal-derived components is a priority, for example, in biopharmaceutical applications. Absence of animal-derived components may also facilitate regulatory compliance.
- The absence of lipids in non-lipid-based NPs makes them suitable in lipid or signal transduction research.
- In the field of transgenesis, NP-mediated gene delivery might be useful in species for which conventional methods of gene delivery are not effective (e.g. chicken).

3 Types of NPs Used for In Vitro Gene Delivery into Cells

3.1 Inorganic NPs

3.1.1 Metal and Metal-Based NPs

Gold (Au) and silver (Ag) NPs are the two most frequently commercialized NPs for variety of uses. Of the two, AuNPs have been at the forefront of metal NP research investigating gene delivery applications, owing to their well-established surface chemistry and physico-chemical properties. AuNPs are biocompatible, nontoxic and are relatively easy to synthesize in a range of sizes by simple, cheap and reliable methods. DNA loading and release from AuNPs are governed, for the most part, by hydrogen bonding and Au-thiol chemistry [9–11]. Consequently, gene delivery using AuNPs has been well demonstrated by several researchers [12, 13]. AgNPs were, however, found to have cytotoxic and genotoxic potential in mammalian cells such as human mesenchymal stem cells (MSCs) [14] and, hence, are not normally used.

Several authors have also functionalized AuNPs for controlling the manner, place and timing of DNA release. When AuNPs were functionalized with polyethylene-glycol–orthopyridyl-disulfide (PEG–OPSS), the loaded DNA could be released from the AuNPs by laser irradiation at a power density value of 80 mJ/pulse without any fragmentation of DNA [15]. Chen et al. [16] attached thiol-modified DNA to the surface of Au nanorods through Au–S bonds. When femtosecond NIR irradiation was applied to the Au nanorod–DNA conjugates, a change of shape from rod to sphere was observed, which induced the release of DNA. A similar phenomenon was described by Takahashi et al. [17]. Wijaya et al. [18] further showed that Au nanorods can be used to selectively release multiple DNA. Electroporation is yet another external stimulus that can be used to release genes from AuNPs. Kawano et al. [19] showed that electroporation can be used to release the DNA from AuNPs modified with mPEG-SH5000. Bimetallic NPs, made up of alloyed combination of Au and Ag in the form of nanorods, have also been used for gene delivery [20]. Thus, metallic NPs offer an opportunity to remote-control gene delivery.

3.1.2 Magnetic NPs (Magnetofection)

Magnetic NPs (MNP) have recently gained great interest as non-viral carriers for gene delivery [21]. In this system, DNA can be attached to MNPs (normally in suspension) and introduced into the cell culture medium. The DNA–MNP complexes (called "Magnetoplex") are then focused to the target cells by applying a high-field or high-gradient magnetic force produced by rare earth magnets (or electromagnets) placed below the cell culture to increase the sedimentation of the complex. Upon binding to the cell surface, the Magnetoplex can be taken up by endocytosis and the DNA is released from the MNPs intracellularly. The transfection efficiency can be further increased by using an oscillating magnetic force [22, 23] and MNP heating [24, 25]. The technique (called magnetofection) promotes rapid transfection with increased gene delivery efficiency and, consequently, many static-field magnetofection systems are now available commercially.

The MNPs generally consist of superparamagnetic iron oxide NPs (SPION) (magnetite, Fe_3O_4, or maghemite, Fe_2O_3), which magnetize strongly under an external magnetic field but retain no permanent magnetism upon removal of the magnetic field at room temperature. This property prevents aggregation or clumping and, thereby, helps in easy dispersal. However, SPION can readily agglomerate to form large particles in aqueous solutions of ~pH 7. Thus, for use as MNPs, they are either encapsulated within a polymer (PEG, poly-L-lactic acid) or metallic (gold, silver) shell or are dispersed within a polymer matrix (silica, polyvinyl alcohol, polyvinylpyrrolidone or dextran). Pre-coating of the MNPs also makes them biostable, biodegradable and nontoxic. The shell or matrix can be functionalized by attaching carboxyl groups, amines, biotin, streptavidin, antibodies, etc. to promote uptake by the target cells, prevent aggregation and increase transfection efficiency and reduce cytotoxicity. Several types of coating agents have been used, including anionic surfactants (oleic acid, lauroyl sarcosinate), nonionic water-soluble surfactant (Pluronic F-127), fluorinated surfactant (lithium 3-[2-(perfluoroalkyl) ethylthio]propionate), polymers (PEG, PLL, poly(propyleneimine) dendrimers), carbohydrates (chitosan, heparan sulfate), silica particles (MCM48), proteins (serum albumin, streptavidin), hydroxyapatite, phospholipids, cationic cell-penetrating peptide (TAT peptide), non-activated virus envelope (HVJ-E), transfection reagent (Lipofectamine 2000) and viruses (adenovirus, retrovirus). See [23, 26] for a detailed review. These coating agents are often used in conjunction with PEI, which not only binds with DNA to the MNPs but also serves as a NP dispersant [27].

3.1.3 Ceramic NPs

Ceramic materials such as silica, zirconium phosphate, cerium oxide (CeO_2, ceria), aluminium oxide (Al_2O_3, alumina), yttrium oxide (Y_2O_3, yttria), etc. have received very little attention for gene delivery applications. Of the various ceramic materials, silica NPs were shown to protect the loaded DNA against denaturation induced by changes in the external pH and temperature and, thus, have potential for use as non-viral gene delivery vectors. Consequently, several authors have used surface-modified (multifunctional) silica NPs to deliver DNA [28–31]. Kim et al. [32] have used silicon nanowires to deliver GFP-encoding plasmid DNA (pDNA). Organically modified silica (ORMOSIL) NPs have also been used as a non-viral vector for gene delivery [28, 33]. Unfortunately, silica NPs showed cytotoxicity that increases with increase in dose, exposure duration and metabolic activity of the cell [34]. Exposure of cells to silicon oxide resulted in increased activity of reactive oxygen species (ROS) and reduced glutathione levels, indicating an increased oxidative stress [35].

3.1.4 Carbon Nanofibers and Nanotubes

Carbon nanofibers and nanotubes have also shown great promise for non-viral gene delivery. Cai et al. [36] used a technique called "nanotube spearing" wherein DNA can be attached to nickel-embedded, elongated, magnetic nanotubes that can be aligned parallel, like spears, to penetrate the cell membrane along the lines of a magnetic flux. The penetration of cell membrane helps in delivery of the genes into the cells. With this method, nearly 100% cell viability was reported with high transfection efficiency [36, 37]. Vertically aligned carbon nanofibers have also been used to deliver multiple genes into the cell [38, 39]. However, others have shown potential toxicity of carbon nanofibers and nanotubes.

3.2 *Organic NPs*

3.2.1 Polyionic Bioreducible Polymers

Polycationic polymers having disulfite linkages in their polymeric structures have been extensively investigated for use as non-viral gene delivery systems. These polycations not only polyplex the negatively charged DNA to condense and protect them against nuclease digestion but also release the loaded DNA intracellularly upon breakage of disulfite linkages by the reducing environment of the cytoplasm. These polycationic bioreducible polymers show reduced cytotoxicity and controlled intracellular release of DNA, leading to increased transfection efficiency. Examples of bioreducible polymers include PEI, polyion complex (PIC) micelles, polyamidoamine (PAA) and polypeptides.

PEI and PEI Conjugates

PEI is polycationic polymer that can polyplex DNA at low PEI:DNA ratios and is easily taken up by cells through endocytosis. Within the endosome, the amine groups of PEI can buffer the protons to undergo mechanical swelling (proton sponge effect), increase intra-endosomal osmotic pressure and, thereby, promote endosomal disruption that can lead to efficient endosomal escape of the polyplex [40]. Thus, PEI can result in increased success of gene delivery. Unfortunately, PEI was shown to have high cytotoxicity that increased with increase in its molecular weight.

Studies have shown that conjugation of PEI with disulfite linkages in their polymeric structure can impart bioreducible properties, increase intracellular release of the DNA and reduce cytotoxicity. Crosslinking of low molecular weight PEI with a homo-bifunctional and amine-reactive crosslinker such as dithiobis(succinimidylpropionate) (DSP) and dimethyl 3,30-dithiobispropionimidate.2HCl (DTBP) [41, 42], cystamine bisacrylamide (CBA) [43] or methylthiirane (thiolation) [44] significantly reduced cytotoxicity and improved gene delivery efficiency. In a comparative study, Breunig et al. [45] reported that disulfide crosslinked low molecular weight linear PEIs (polycationic bioreducible PEIs) had higher transfection efficiency and lower cytotoxicity than commercial transfection reagents such as PolyFect, SuperFect, Lipofectamine, FuGENE6 or JetPEI. These beneficial effects were observed with both linear [46] and branched PEIs [47].

Acetylation [48] and PEGylation [49, 50] were also shown to influence the gene delivery efficiency of PEI. Hosseinkhani et al. [48] reacted PEI with acetic anhydride to acetylate 80% of the primary and 20% of the secondary amines. This acetylated PEI was shown to have enhanced gene delivery efficiency over unmodified PEI for MSCs. Chen et al. [51] showed that a PEG–PEI copolymer had better gene delivery efficiency than cationic liposomes and did not affect the bionomics, proliferation and differentiation potential of MSCs.

Others have used PEI to coat the biopolymers to form NPs. The PEI coated on biopolymers caused polyplexing of DNA [52, 53] while the biopolymers increased the cellular uptake [54] and reduced the cytotoxicity [55] by modifying the surface charge and dispersing the stability and buffering capacity of the resulting NPs [55]. The PEI coated on biomaterials such as hyaluronan (HA) also helped in controlled, sustained and prolong release of the DNA [55]. Park et al. [56] polyplexed four genes (SOX5, SOX6 and SOX9 genes fused to GFP, YFP or RFP marker) with PEI coated onto PLGA NPs and obtained ~80% transfection efficiency in human MSCs. By polyplexing with PEI, the cell-uptake ability of the DNA-loaded NPs was enhanced for both in vitro and in vivo culture systems, including human MSCs [56]. Jeon et al. [54] achieved co-delivery of DNA and siRNA into human MSCs by complexing them with PEI coated on

PLGA NPs. Mahor et al. [52] used branched PEI as a transfecting agent for DNA encapsulated in HA biomaterials and obtained significantly higher expression levels than for naked DNA.

PIC Micelles

PIC micelles are self-assembling co-polymers consisting of a core of hydrophobic blocks (e.g. PLL, PEI) stabilized by a corona of hydrophilic polymeric chains (e.g. PEG). They have a core–shell structure with high water-solubility and colloidal stability and have polycation properties that are capable of condensing and compacting the negatively charged DNA. Attachment of disulfite linkages to the PIC micelles impart bioreducible properties, with reduced cytotoxicity and increased ability to release the loaded DNA inside the cells. Kakizawa et al. [57] showed that thiolated PEG–PLL micelles could successfully encapsulate the oligonucleotides, enter the cells by endocytosis and efficiently release the loaded oligonucleotides in response to the reducing intracellular environment of the cells. Oishi et al. [58] used thiolated PEG–PEI to take advantage of the "proton sponge effect" of the PEI in endosomal release. These PIC micelles showed higher gene delivery efficiency than those of thiolated PEG–PLL. When oligonucleotides were conjugated to PEG via disulfide linkages and complexed with PEI to form polyelectrolyte complex (PEC) micelles, a further enhancement was observed due to more effective endosomal escape [59]. The potential of complexes formed with naturally occurring biomaterials, protamine and HA conjugates via a disulfide linkage has also been reported as a safe and effective non-viral gene delivery option [60].

Polyamidoamine

PAAs can be synthesized by Michael reaction of amine monomers and acrylamide monomers. Lin et al. [61] reported a series of novel bioreducible PAAs by Michael-type polyaddition of various primary amines [4-amino-1-butanol (ABOL), 5-amino-1-pentanol (APOL), *N*,*N*-dimethyl-1,3-ethylenediamine (DMEA), 2-(2-aminoethoxy) ethanol (AEEOL), 3-methoxypropylamine (MOPA), 3-morpholinopropylamine (MPA) or histamine (HIS)] with disulfide bond-containing cystamine bisacrylamide (CBA). These bioreducible PAAs had higher buffer capacities than PEI in the endosomal pH range and, therefore, contributed to the greater endosomal escape of the polyplexes. Of the above, bioreducible PAAs containing amino alcohol pendant groups (pAPOL, pABOL) exhibited the highest gene delivery efficiency [62]. In another study, Lin et al. [63] reported the syntheses of bioreducible PAA consisting of bioreducible CBA and two amino groups with distinctly different basicity, HIS and 3-(dimethylamino)-1-propylamine (DMPA). The copolymers at a HIS:DMPA ratio of 70:30 were shown to combine optimal DNA condensation ability and buffer capacity and, thereby, resulted in high gene delivery efficiency and lower cytotoxicity than observed with homopolymers.

Polypeptides

Polypeptides having bioreducible linkages (mainly disulfide bonds formed between free thiol groups of cystein residues) and a net positive charge can also be used for complexing the DNA into NPs that can release the loaded DNA intracellularly. These peptides can also be conjugated to polymers such as PEG for enhanced gene delivery [64, 65]. Several reductively degradable polycations (RPCs) consisting of HIS and PLL residues have also been developed as gene delivery carriers by oxidation of terminal cysteinyl-thiol groups [66, 67].

The polypeptides can also be designed to ascribe a specific function. For example, inclusion of a nuclear localizing signal (NLS), DNA binding proteins such as histones and a high mobility group (HMG) protein sequence in the polypeptide can enhance the intranuclear entry and gene integration of the DNA. Manikam and Oupicky [68] reported the synthesis of novel reducible copolypeptides (rCPP) by an oxidative copolymerization of a histidine-rich peptide and a NLS peptide. The rCPPs exhibited minimum cytotoxicity, enhanced intracellular release of the DNA and high gene integration rate. Lo and Wang [69] also designed novel polypeptides incorporating a Tat sequence, which is a cationic cell-penetrating peptide known to enhance the cellular uptake of various drugs and proteins. Nearly 7,000-fold improvement in gene transfection efficiency was observed. Similar effects were also observed by incorporation of nona-arginine (D-R9), which is a cell-penetrating peptide with protein transduction domains [70].

3.2.2 Biodegradable Polymeric NPs

Several biodegradable and biocompatible polymers exhibit good potential for surface modification and functionalization and are good candidates for non-viral gene delivery. However, most of these complexes are too large to pass through the plasma membrane and the nuclear pores to be effective for gene delivery. In recent years, generation of nanoscale polymeric NPs, nanospheres and nanocapsules have revolutionized their utility as a gene delivery system. Nanospheres have a matrix-like structure wherein DNA can be firmly adsorbed at their surface, entrapped or dissolved in the matrix. Nanocapsules, on the other hand, have a polymeric shell and an inner core wherein DNA is usually dissolved in the core but can also be adsorbed at their surface. One advantage of using polymeric NPs is that many of the polymers (PLGA, PLA, etc.) are already FDA-approved for the delivery of some drugs, which should facilitate their approval for gene delivery applications.

Poly(beta-amino esters)

Poly(beta-amino esters) are cationic, hydrolytically degradable polymers that can be produced as polymeric NPs for gene delivery. Green et al. [71] developed small (~200 nm), positively charged (~10 mV), polymeric NPs by the self assembly of poly(beta-amino esters) and DNA. These NPs had four times greater gene delivery efficacy than those observed for Lipofectamine 2000 in human embryonic stem (ES) cells. These materials exhibited minimal toxicity and did not adversely affect the colony morphology or cause nonspecific differentiation of the ES cells.

Polylactide-*co*-Glycolide

PLGA is biodegradable, biocompatible and FDA-approved biomaterial that has aroused considerable interest among researchers developing biodegradable NPs for gene delivery. PLGA NPs complex with DNA at a low PLGA:DNA ratio and have allowed robust gene expression in several cell types, including human MSCs [72]. They can be modified with other polymers such as PEI. Polyplexing with PEI enhanced the cellular uptake of DNA complexed to PLGA NPs both in vitro and in vivo [72, 73]. Park et al. [56] polyplexed four genes (SOX5, SOX6 and SOX9 genes fused to GFP, YFP or RFP marker genes) with PEI coated onto PLGA NPs and obtained ~80% transfection efficiency in human MSCs.

Polyethylene-Glycol

PEG as such is not used as an NP. Repeating PEG moieties are usually added to polymers to alter electrostatic binding properties and increase hydrophilicity of NPs. The bulky nature of PEGylated polymers can also protect the NPs from degradation by cellular enzymes, increase their stability and prevent aggregation. PEGylation also enhances the transfection efficiency of NPs [74].

Chitosan and Chitosan Derivatives

Chitosan is a biodegradable, biocompatible, nontoxic, natural polysaccharide consisting of repeating units of glucosamine and *N*-acetyl-glucosamine, the proportions of which determine the degree of deacetylation and, hence, the polymer properties including solubility, hydrophobicity and the ability to interact with polyanions. Chitosan can bind to the minor groove of DNA to condense and protect it against nuclease degradation without affecting the native conformation. Furthermore, chitosan NPs are stable during storage and their preparation does not require sonication and organic solvents, which minimizes possible damage to DNA during complexation. Thus, it is a good candidate for non-viral gene delivery.

Chitosan–DNA NPs (~20–500 nm) can be readily formed by coacervation between the positively charged amine groups on the chitosan and negatively charged phosphate groups on the DNA. The size of the NP and the degree of DNA condensation depends upon the nitrogen:phosphate (amine group on chitosan: phosphate groups on DNA) ratio, molecular weight of chitosan and the degree of deacetylation [75, 76]. Generally a low molecular weight, highly deacetylated chitosan at high N:P ratio results in small sized NPs with highly condensed DNA. Chitosan–DNA NPs have been used for gene delivery in a variety of cell types including human MSCs and were shown to have lower cytotoxicity than lipoplexes [77, 78]. Unfortunately, results so far have only been modest. Furthermore, gene delivery efficiency is dependent on cell type [77, 78].

Several researchers have attempted to improve the gene delivery efficiency of chitosan–DNA NPs by attaching endosome-disrupting molecules or targeting ligands such as transferrin [78], KNOB (C-terminal globular domain of the fibre protein) [75], lactose and lactobionic acid [79, 80], galactose and PEG [79, 81], PEI [82], trimethyl groups [83], deoxycholic acid [84], pH-sensitive polymer poly (propyl acrylic acid) (PPAA) [76, 85], etc. to the reactive amine groups on chitosan. These modifications were successful in improving the gene delivery efficiency and/or cell-type-specific gene delivery. However, improvement in gene delivery efficiency was only modest and was generally lower than achieved with standard gene delivery agents such as Lipofectamine.

Hyaluronan

HA has been used to modify the surface charge, dispersing stability and buffering capacity of polymers such as PEI and chitosan to form NPs for non-viral gene delivery [86]. NPs made of HA and chitosan showed lower cytotoxicity and induced a higher rate of gene integration in neural stem cells and spinal cord slice tissue compare to those obtained with PEI [86]. Similar results were also obtained with PEI-introduced chitosan NPs for rat MSCs [55].

Gelatin

Gelatin has also been used as a nanocarrier of DNA for transfecting HeLa cells, chicken cells and chicken embryos. Tseng et al. [87] encapsulated the DNA into gelatin to produce NPs by a water–ethanol solvent displacement method. The DNA–gelatin NPs (~300 nm) were nontoxic to cells and effectively induced transgene expression 24 h after cell transfection. Direct injection of the DNA–gelatin NPs in the area opaca of the chicken egg resulted in transgenic embryos without affecting their embryonic development and hatching.

3.2.3 Serum Albumin

Serum albumin has also been tested as NPs for gene delivery. Mo et al. [88] encapsulated the DNA into human serum albumin (HSA) by a desolvation-crosslinking method to produce DNA–HSA NPs having a mean size of 120 nm and zeta potential of −44 mV. The DNA–HSA NPs were easily taken up by the cells via receptor-mediated endocytosis that involved primarily caveolae pathways. Within the cells, DNA–HSA NPs protected the DNA against nuclease attack and showed sustained release of DNA over 6 days without significant cytotoxicity. The overall transfection rate was found to be fivefold higher than obtained with Lipofectamine.

3.2.4 Dendrimers

Dendrimers are polymeric molecules composed of multiple branched monomers radially emanating from a central core. Dendrimers based on polymers such as polyamidoamines (PAMAMs) and poly(propylene imine) can form compact polycations under physiological conditions and, therefore, have been of interest to researchers for use in gene delivery both in vivo and in vitro [78, 89, 90]. They can also be functionalized with other molecules such as α-cyclodextrin [91], PEG [92], etc. to enhance gene integration and expression. However, the use of dendrimers for in vitro gene delivery into somatic and stem cells has not yet been demonstrated.

3.3 Composite NPs and Other NPs

Composite NPs such as calcium phosphate NPs have also gained interest for non-viral gene delivery. Cao et al. [93] used calcium phosphate (CP) nanocomposite particles to encapsulate DNA for their delivery into MSCs. The CP–DNA NPs (~100 nm) were reported to be less cytotoxic and more efficient in gene delivery into MSCs than those of Lipofectamine or a standard calcium phosphate transfection kit.

4 Application of NPs in Stem Cells

NP-based gene delivery has plethora of applications in: (1) cellular reprogramming of somatic cells to derive induced pleuripotent stem (iPS) cells, (2) genetic engineering for clinical application of stem cells, (3) creating a three-dimensional (3D) in vitro niche for in vitro culture and/or directed differentiation of stem cells, (4) tracking the transplanted stem cells in vivo and (5) investigation of gene function in vitro. NPs, loaded with or without marker genes, can also be used for isolation, purification and enrichment of adult stem cells such as MSCs, and germ-line stem cells. One such approach is to load the NPs with a cell-type-specific cell surface marker, which can then be used to label the stem cells in a mixed population of cells and to purify,isolate and enrich by fluorescence-activated cell sorting (FACS) or magnetic-activated cell sorting (MACS). MNPs loaded with anti-CD34 antibody have been used to label CD34-positive stem cells, which were then enriched by magnetic sorting [94]. A similar approach has been used extensively for isolation and enrichment of male germ-line stem cells, which have otherwise been very difficult to isolate and purify by conventional means [95].

NPs, loaded with or without genes, can also be used to create a nanostructured 3D scaffold or cell culture substrate to mimic the in vivo stem cell niche or nano-environment for stem cell self-renewal, proliferation and/or targeted differentiation. Several nanostructured NPs based on chitosan, PLGA, gelatin, etc. have been shown to be safe and of potential application in stem-cell-based tissue engineering applications. Carbon nanotubes have recently been gaining potential interest as a promising nanomaterial because they have dimensions similar to the 3D structure of proteins found in extracellular matrices [96, 97]. Mooney et al. [97] found that small carbon nanotubes promoted the adhesion of human MSCs without noticeable differentiation, whereas large carbon nanotubes led to a dramatic stem cell elongation, inducing cytoskeletal stress and selective differentiation into osteoblast-like cells. Unfortunately, carbon nanotubes have also been reported to be genotoxic and, therefore, further improvement is required prior to their safe and effective use [98]. A biocompatible, self-assembling peptide nanofiber scaffold (SAPNS) that mimics the structure of extracellular matrix has also been developed and has been demonstrated to mimic a 3D nano-environment for the migration and differentiation of neural stem cells and the growth of blood vessels and axons in the scaffolds [99–101]. Another recent study has established a culture system to expand and maintain mouse ES cells using MNPs, creating the magnetic field-MNP culture system without affecting the pluripotency [102].

4.1 NP-Based Gene Delivery for Stem Cell Isolation and Culture

NP-based gene delivery is a relatively recent concept in stem cell engineering. Among non-viral methods, lipofection and electroporation have been optimized for several stem cell types and were shown to give an acceptable level of transfection. A need for improvement was felt for stem cell types that are resistant to gene introduction (e.g. germ-line stem cells) or grow as clump (e.g. ES cells) or when multiple genes need to be introduced simultaneously (e.g. iPS cells). Subsequently, with advancements in the nanotechnology, several NP-based gene delivery options have been explored, tested and commercialized

(Fig. 1). Many of these NP-based gene delivery systems have been shown to be superior to commercially available lipofection systems [53, 93]. Qiagen has developed NanoFect Transfection Reagent, based on a chemically synthesized, lipid-free reagent that has now been shown to be efficient in DNA delivery in a broad range of cell types. However, it is MNPs that have grabbed maximum acceptance. System Biosciences has developed a MNP suspension, LentiMag, that effectively binds DNA (and viral) vectors and very quickly concentrates them onto target cells by use of a magnetic plate. Using LentiMag, higher transduction efficiencies have been achieved compared with transductions performed with Polybrene (hexadimethrine bromide). Commercially available or in-house produced MNPs have now been used for several cell types, including both embryonic and adult stem cells [103–105].

NP-based gene delivery systems have gained fresh impetus with the identification of iPS cells. The iPS cells offer several advantages over other existing cell types such as ES cells (see [106, 107] for a detailed review on iPS cells). iPS cells are derived by cellular reprogramming of patient-derived somatic cells through introduction of one or more pluripotency genes (Oct4, Sox2, Klf4 and cMyc OR Oct4, Sox2, Nanog and Lin14) [108]. Introduction of multiple genes necessitates the use of retroviral or lentiviral vectors, which raises safety issues. Furthermore, despite using viral vectors, the efficiency of cellular reprogramming has been very low (in the range of 0.001–0.01%). Accordingly, intense research has been focused on either reducing the number of genes required for cellular reprogramming [109], and/or using non-viral vectors [110], non-integrating episomal vectors [111], mRNAs [112], proteins [113], novel culture methods [114–116], pluripotency-inducing proteins [117], etc. Furthermore, Lee et al. [104] demonstrated that MNPs were efficient in simultaneous delivery of four genes (Oct4, Sox2, Klf4 and cMyc) into somatic cells to reprogram them into iPS cells at an improved efficiency. Similarly, Ruan et al. [118] obtained efficient generation of iPS cells by introduction of four genes (Oct4, Sox2, LIN28, and Nanog) into somatic cells using polyamidoamine dendrimer-modified MNPs as the delivery system. Thus, magnetofection provided safe, virus-free and exogenous DNA-free iPS cells.

Because stem cells can be grown long-term in vitro, their genetic modification prior to transplantation provides a unique opportunity for correcting genetic defects such as ADA severe combined immunodeficiency (ADA-SCID), Shwachman–Bodian–Diamond syndrome (SBDS), Gaucher disease (GD), Duchenne muscular dystrophy (DMD), Becker muscular dystrophy (BMD), Parkinson disease (PD), Huntington disease (HD), Lesch–Nyhan syndrome (HPRT), Diabetes mellitus (JDM) and Down syndrome [119]. Indeed, several genetic diseases were shown to be curable using stem-cell-based genetic engineering in animal models [119]. Considering the several advantages of NP-based gene delivery systems over the viral vector methods, it can be envisioned that NPs will find potential application in stem-cell-based genetic correction of diseases.

4.2 NP-Based Gene Delivery for Inducing Differentiation of Stem Cells

Prior to their transplantation into patients, stem cells need to be differentiated into the desired cell type (e.g. cardiomyocyte). Targeted differentiation of stem cells can be achieved by culturing them in the presence of specific growth factors, providing specific cell culture substrate, modifying cell surface properties and through up- or downregulation of specific genes. In recent years, genetic engineering has become an important strategy for the induction and regulation of the targeted differentiation of stem cells into a specific cell type. At least two strategies (described below) have been applied to use NPs in inducing and controlling the differentiation of stem cells.

4.2.1 Delivery of DNA- or siRNA-Loaded NPs

Introduction of genes and/or introduction of siRNA to up- and downregulate specific genes involved in signalling pathways controlling the cell phenotype can induce specific differentiation of stem cells into specific cell types. For example, Kim et al. [72] showed that PLGA NPs loaded with SOX9 genes induced chondrogenesis in human MSCs both in vitro and in vivo. Park et al. [56] further showed that introduction of the SOX trio (SOX5, SOX6, and SOX9) complexed with PEI-modified PLGA NPs led to a dramatic increase in the chondrogenesis of human MSCs in in vitro culture systems. NPs containing siRNAs for silencing Bcl2l2 and Trib2 were shown to enhance osteogenic and adipogenic differentiation, respectively, of MSCs [120]. Use of NPs also allows simultaneous introduction of DNA vector and siRNA and, thereby, enhanced and efficient differentiation [54].

4.2.2 Incorporation of Oligonucleotide- or DNA-Loaded NPs and Differentiating Agent into Scaffolds

DNA- or siRNA-loaded NPs can be combined with 3D tissue-engineered scaffold impregnated with or without bioactive molecules. Such systems can provide a combination of differentiation-inducing gene delivery (by NPs), physical support and surface properties (by scaffold) and differentiation stimulants (by bioactive molecules) and, hence, might be a better alternative for target differentiation of stem cells into a specific cell types and production of specific tissue constructs for tissue engineering. Cao et al. [121] developed a 3D NP gene delivery system (3D-NGDS) based on collagen/chitosan scaffolds, in which pTGFβ1/calcium phosphate NPs mixed with fibronectin were used to transfect MSCs. They observed that 3D-NGDS could successfully transfect the MSCs and induce chondrogenic differentiation in vitro without dexamethasone. The transfection efficiency was higher than obtained with the Lipofectamine 2000 method.

DNA- or siRNA-loaded NPs immobilized on the surface-coated ECM may also allow the controlled release of DNA, leading to long-term expression of the desired protein for enhanced differentiation [122]. Hosseinkhani et al. [48] observed that simple mixing of plasmid DNA (encoding BMP-2) and acetylated PEI solutions and their encapsulation within scaffolds (collagen sponges reinforced by incorporating of poly(glycolic acid) fibres) led to homogenous bone formation throughout the sponges. This strategy could be of particular importance for delivery of siRNAs, which are known to have a short half-life. Furthermore, adhering NPs containing different DNAs or siRNAs into nanostructured 3D scaffolds could allow spatial retention of the DNA or siRNA within nanopores until their cellular delivery. Different NPs localized to spatially distinct locations within a single implant might allow two different tissue types to develop in controllable areas of an implant. Thus, complex tissues and organs can be engineered by the in situ development of multiple cell types guided by spatially restricted NPs [120].

4.3 Stem Cells as Carriers of NPs or DNA NPs

NPs, loaded with or without genes, can also be used to track the cellular distribution, differentiation and fate of stem cells after their in vivo transplantation. NPs such as quantum dots, MNPs and magnetic carbon nanotubes can be easily loaded into stem cells and visualized by imaging techniques such as magnetic resonance imaging (MRI) or fluorescent imaging for monitoring the fate of the transplanted stem cells. These NPs have better photostability and longevity than chemical dyes and, hence, are advantageous. Ruan et al. [118] labelled iPS cells with MNPs and found them suitable for long-term observation and tracking of stem cells through fluorescent microscopy and MRI [118]. Alternatively, fluorescent markers such as EGFP, YFP, CFP, RFP, etc. can be introduced into the stem cells by NP-based gene delivery methods for tracking the fate of the transplanted cells [56].

Because stem cells have the intrinsic ability to "home" into transplanted organs, they can also be used as a delivery vehicle to deliver therapeutic genes and/or track NPs into a target organ. Tang et al. [123] introduced SPION into therapeutic MSCs to act as contrast enhancers for tracking the transplanted cells by MRI. Kim et al. [124] used SPION to transfer genes into MSCs and found the method to be safe and effective. However, SPION/PLL labelling of C17.2 neural stem cells was shown to result in altered gene expression as an early cellular response and, therefore, further improvement may be necessary [6].

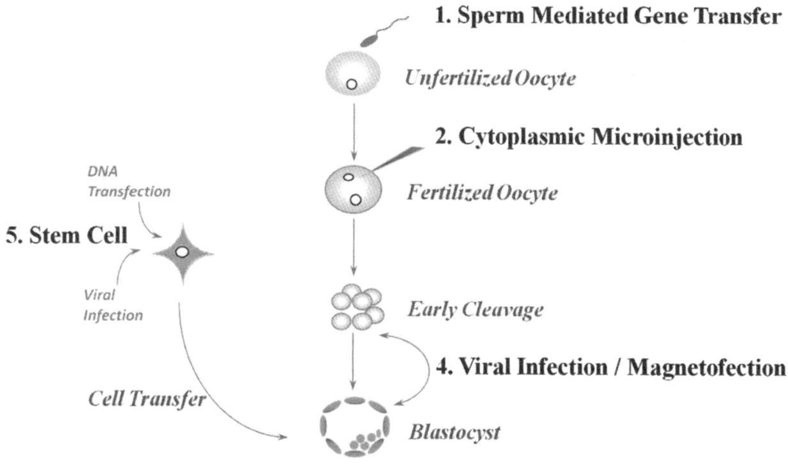

Fig. 2 NP-based transgenic strategies. NP-loaded DNA may be transfected into sperm for sperm-mediated gene transfer (*1*), microinjected into unfertilized or fertilized oocytes (*2*) or magenofected into early cleavage-stage to blastocyst-stage embryos (*4*). Alternatively, DNA-loaded NPs may be delivered into stem cells (*5*) for production of transgenic embryos by morula aggregation of blastocyst injection

5 NP-Based Gene Delivery for Transgenesis

Transgenic animals can be produced by introduction of DNA vector into sperm (sperm-mediated gene transfer or SMGT), metaphase oocytes (MII transgenesis) or pronucleus-stage zygote (PN microinjection) or by introduction of viral vector into cleavage-to-blastocyst stage embryos (Fig. 2). Alternatively, genes can be introduced into somatic or stem cells and transfected cells used to produce embryos via somatic cell nuclear transfer (SCNT), morula aggregation of stem cells or blastocyst-injection of stem cells (Fig. 3). Clearly, NPs can be utilized to produce transgenic animals via NP-based gene delivery into somatic and stem cells. Recently, however, attempts have been made to use NP-based gene delivery system to introduce DNA into sperm and oocytes but not into zygote or embryos.

5.1 NP-Based Gene Delivery into Sperm (nanoSMGT)

SMGT offers several advantages over other methods of transgenesis, not least of which is its ease in methodology. Unfortunately, despite several successful reports on efficient uptake of DNA by sperm through electroporation [125], lipofection [126] and DMSO–DNA complex [127], generation of offspring remains low [128]. Furthermore, except for a very few selected laboratory, most researchers have failed to produce viable offspring via artificial insemination of transfected sperm

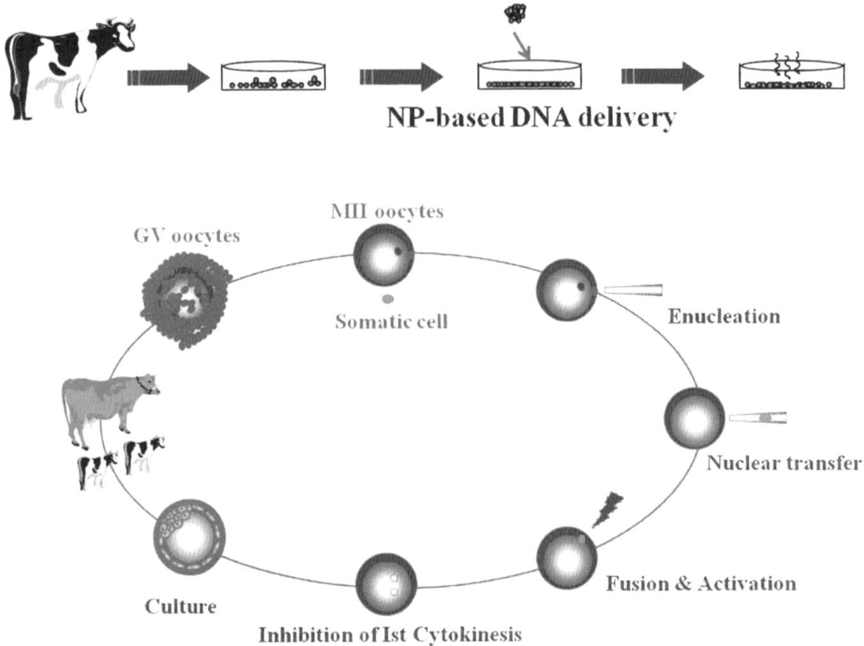

Fig. 3 Somatic cell nuclear transfer strategy for transgenesis. NPs can be used to delivery DNA into somatic cells, which can then be selected for gene expression and microinjected into enucleated oocytes for production of transgenic animals

[129]. Transfection of sperm almost always resulted in significant loss of motility and necessitated artificial fertilization through intracytoplasmic sperm injection. Thus, there is a need to improve the transfection method of sperm without using electroporation or lipofection.

The successful use of NPs to introduce foreign DNA into somatic and stem cells has brought new perspectives for production of transgenic embryos via SMGT. In a recent study, Kim et al. [130] showed that MNPs can be successfully used for introducing genes into pig sperm. The DNA-loaded MNPs bound ejaculated spermatozoa at a higher efficiency than those obtained by using DNA alone or lipofection. Clusters of MNPs were detected both in the sperm nucleus and at the inner surface of the plasma membrane. In vitro fertilization (IVF) of oocytes with transfected sperm resulted in successful production of transgenic embryos – a method named nanoSMGT. Similar results were also obtained by Campos et al. [131] for cattle sperm. Campos et al. [131] used NanoFect Transfection Reagent (Qiagen) for nanoSMGT in cattle sperm and observed that nanopolymer efficiently introduced exogenous DNA into the sperm and resulted in successful production of transgenic embryos. Interestingly, unlike other methods, nanoSMGT was not affected by DNA preparation methods and parameters such as the linear-to-circular DNA ratio. The ratio of linear-to-circular DNA can influence exogenous DNA uptake by sperm when electroporation or lipofection is used. However, during nanoSMGT, the ratios of linear-to-circular plasmid did not influence the uptake by sperm cells and none of the tested treatments affected sperm motility and viability after nanotransfection [130, 131]. Campos et al. [131] also used halloysite clay nanotubes (HCN) for nanoSMGT. They observed that the mean number of plasmids taken up by cattle sperm was higher in HCN-based gene delivery than by using lipofection. IVF of oocytes with HCN-transfected sperm successfully resulted in transgenic embryos with higher efficiency but, unfortunately, the transgene did not express.

5.2 NP-Based Gene Delivery into Oocytes

Transgenic animal production via introduction of genes into oocytes have not been very fruitful by conventional methods due to the high cytoplasmic content of nuclease enzymes, cytoplasmic sequestration of injected DNA by DNA binding proteins and cytoskeletal elements and the lack of DNA to transport across the nuclear membrane. Some studies have shown that simultaneous introduction of DNA along with the sperm can lead to successful production of transgenic mice [132]. However, this method, called MII transgenesis, required intracytoplasmic sperm injection to be performed and was not very fruitful in non-rodent species. Use of lentriviral vectors, on the other hand, has resulted in successful production of transgenic animals via gene introduction in oocytes.

MII-stage oocytes lack a nuclear membrane and offer a unique opportunity for DNA to interact with the host chromatin to produce transgenic animals. However, all efforts to produce transgenic animals via direct injection of DNA into oocytes have failed. We have recently developed a NP-based gene delivery method for mammalian oocytes, called oocyte-mediated gene transfer (OMGT) (Fig. 4). Pig oocytes recovered from abattoir-derived prepubertal porcine ovaries were matured in vitro for 42–44 h and microinjected with DNA NP solution (10 ng/μL) using a femtojet microinjector (Eppendorf, Hamburg, Germany). The DNA (4.7 kb) was derived from the pEGFP-C1 plasmid (Clontech Laboratories, CA, USA), which contains EGFP-encoding transgene under the control of cytomegalovirus (CMV) promoter, and linearized with *Apa*LI restriction enzyme. Injected oocytes were then in vitro fertilized using fresh epididymal sperm obtained from abattoir-derived porcine testis and cultured in NSCU23 medium supplemented with 0.4% BSA. The efficiency of transgenesis was monitored by visualization of green florescence under UV illumination using an EGFP filter set. Results showed that the cleavage rate of injected oocytes (68.7 $\pm$ 0.5%) was similar to that of uninjected control oocytes (67.8 $\pm$ 0.4%) although a high percentage of injected oocytes showed developmental block at the 2–4 cell stage. The EGFP expression rate at the 2–4 cell stage, when expressed as proportion of injected oocyte, was 17.2 $\pm$ 0.1%. Interestingly, mosaicism was not observed. The EGFP expression rate increased to 26.7 $\pm$ 0.1% by increasing the DNA concentration to 40 ng/μL. Injecting the DNA

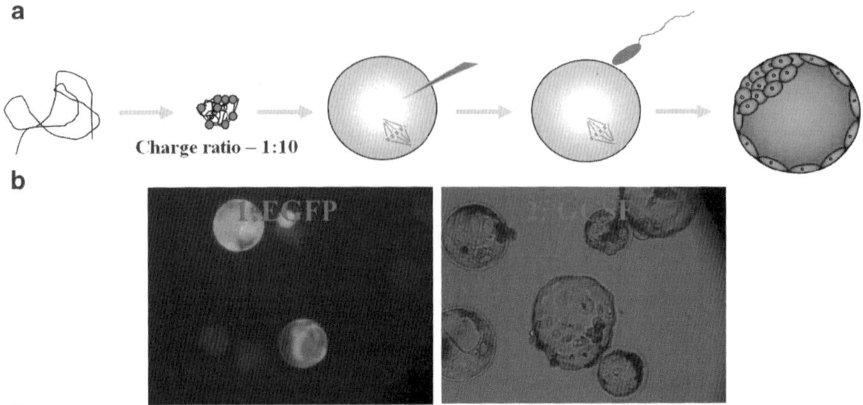

c

- ➢ Require less skill
- ➢ Suitable for species such as pig
- ➢ No mosaicism
- ➢ Multigene insertion
- ➢ Injection of large size vector is possible

Fig. 4 OMGT strategy of transgenesis and its advantages. (**a**) DNA vector is complexed with a polycationic polypeptide at a charge ratio of 1:10, microinjected into MII-stage oocytes and fertilized in vitro. (**b**) Transgenic pig blastocysts produced by OMGT of EGFP or GCSF genes. (**c**) Advantages of OMGT

solution near to the metaphase plate of the oocyte did not improve ($P < 0.05$) the EGFP expression rate (22.2 ± 0.1%). We further show that complexion of DNA with polypeptide having four NLS and 22 basic amino acids in its sequence, at a charge ratio of 1:10, improves the transgene expression efficiency to 100% at blastocyst stage. Thus, our results suggest that OMGT is a promising tool for producing transgenic livestock. The OMGT method of mammalian transgenesis has several advantages over existing methods such as PN injection. It requires less skill to learn, allows microinjection of large number of oocytes in a relatively small time (>200 embryos in 10 min), allows multigene transgenesis and eliminates the chances of mosaicism. Use of pipettes with ~100-fold larger tip aperture in our methodology also facilitates the handling of large constructs such as yeast or mammalian artificial chromosomes.

5.3 NP-Based Gene Delivery into Embryos

Although promising, NP-based gene delivery methods have not yet been applied to mammalian embryos. However, a few attempts have been made to use NP-based gene delivery methods in species in which conventional methods of gene introduction have not been successful. In particular, transgenic chicken has proven to be difficult to produce by conventional transgenesis owing to their unique reproductive system and hard shell around the egg. Consequently, retroviral or lentiviral injection into the blastoderm layer of Stage X embryos are generally used. We and others have shown that Polybrene (hexadimethrine bromide) can increase the infection rate of viral vectors to increase the transgenesis rate [133–136]. Thus, other polycations might be of use in improving the success of transgenesis and need to be explored.

In a recent study, Tseng et al. [87] used gelatin as a nanocarrier of plasmid DNA for transfecting chicken embryos. The plasmid DNA was encapsulated in gelatin to produce NPs (~300 nm) by a water–ethanol solvent displacement method. The NPs were nontoxic to cells, and its direct injection in the area opaca of the egg resulted in the highest hatching rate without affecting embryo development. Gene expression in embryo sections was observed 4 days after injection.

6 Factors Affecting Gene Delivery Efficiency of NPs

6.1 Cell Type

The gene delivery efficacy of several NPs are known to be cell-type-dependent and they preferentially transfect certain cell types over the others [77]. In certain cases, NPs are intentionally modified to allow the transfection of specific cell types by attaching a ligand that specifically identifies a particular cell surface property of the target cell. Some cell types are also relatively resistant to gene transfection (e.g. germ cells).

6.2 Cell Cycle Stage

The nuclear envelope is one of the major cellular barriers in the intranuclear delivery of DNA [137]. In a non-dividing cell, the nuclear enclosure of NPs is dependent on size (with 100 and 200 nm particles being better included than 500 nm particles) and charge (with positively charged particles being better included than negatively charged particles) on the NPs [138]. However, nuclear membrane breakdown during mitosis and meiosis facilitates the access of NPs to the chromatin and it is highly plausible that at least few of them are included by chance in the nuclei of the daughter cells. Thus, cell division has a positive influence on the efficiency of gene delivery. Conversely, transfection efficiency of NPs is higher in dividing (mitotic) cells than in non-dividing (non-mitotic) cells.

6.3 Cell Culture Conditions

Certain ingredients, such as electrolytes and macromolecules, and the pH of the cell culture medium may alter the surface and physico-chemical properties of the NPs and, thereby influence the gene delivery efficiency. In particular, presence of serum or serum proteins is often reported to lower the transfection efficiencies of several NPs. Even if the NPs are stable in the presence of serum or serum proteins, batch variations in serum quality can lead to differences in transfection efficiency. Thus, it is often advisable to test a small lot of serum from a reputable supplier in a control experiment. Once a given lot has yielded satisfactory and reproducible results, sera from this lot should be used for further experiments.

Cell culture substrate can also the influence gene delivery by affecting the occurrence of endocytosis and, thereby, the uptake of DNA-bound NPs by the cells. Hsu et al. [139] observed that the culture of MSCs on chitosan or HA-modified chitosan membranes increased the intracellular uptake of iron oxide NPs (~5 nm) as well as naked DNA (3.3 kb, ~5 nm) by more than fivefold. The increased internalization of NPs was associated with an increase in clathrin-mediated endocytosis on chitosan (~50%) and in caveolae-mediated endocytosis on chitosan-HA (~30–40%). In the case of naked DNA, but not iron oxide NPs, macropinocytosis also occurred on both substrates.

Microbial contamination by bacteria, fungi and mycoplasma during in vitro cell culture are additional factors that may influence the efficiency of gene transfection by modifying the growth behaviour of the infected cells. Variation in the growth behaviour of infected cells may lead to different transfection efficiencies between replicate experiments.

6.4 Cell Density and Passaging

In most cases, the optimal confluency for gene transfection for adherent monolayer cells is ~40–80%. If cell density, at the time of adding NP–DNA complexes, is not optimal, it can lead to insufficient uptake of complexes into the cells. Furthermore, cells that have been passaged a large number of times tend to change their growth behaviour, morphology and potential for transfection. When cells with high passage numbers are used for replicate experiments, decreased transfection efficiencies may be observed in later experiments. Using cells with a low passage number (<50 splitting cycles) is, therefore, recommended.

6.5 DNA (Vector Design) Quality and Quantity

The DNA backbone of a vector as such does not influence the gene delivery efficiency of NPs. However, the type of promoter, regulatory sequences, coding sequence, etc.

used to design the vector sequence has a strong influence on the gene expression rate. For example, DNA vectors harbouring NLS sequences, DNA binding proteins such as histones and HMG proteins, Simian Virus 40 (SV40) promoter and origin of replication have been shown to increase the intranuclear delivery and expression of genes. Furthermore, the presence of impurities (e.g. endotoxins) in the DNA can also lower the transfection efficiency.

The optimal quantity of DNA used for transfection also needs to be titrated to regulate the gene copy number in the transfected cells to obtain optimal gene expression and to avoid post-integrative gene silencing due to high copy number or overexpression of exogenous genes. In certain situations, the non-viral vector does not integrate into the nuclear genome and remains episomal. The tendency of non-viral vectors to stay episomal can be considered beneficial for cellular reprogramming of somatic cells into stem cells (iPS cells) [111, 140]. However, when used for transgenesis and stable expression of genes, the episomal form of non-viral vectors is not desirable because they are not passed on to daughter cells. The development of self-replicating vectors, vectors without regions prone to epigenetic silencing and vectors containing scaffold or nuclear matrix attachment regions (S/MARs) to keep them in transcriptionally active regions are some of the approaches that have shown promise in increasing the persistence of expression for episomal vectors.

Apart from mosaic and variable expression of genes, NP-mediated gene delivery methods also suffer from integration-mediated activation or inhibition of other nearby genes. Random genomic integration may also lead to insertional mutagenesis and/or trans-activation of cellular proto-oncogenes, resulting in cellular transformation to cancerous cells. This can be addressed by proper design of DNA vector to include homologous sequences for locus-specific gene targeting, Sleeping Beauty transposon-transposase or piggyBac transposition for non-random preferential integration at microsatellite repeats [141–143] or by using a ϕC31 integrase system [144, 145].

Post-integrative gene silencing can occur for a variety of reasons, such as high copy number or overexpression of exogenous genes, random integration into heterochromatin regions and episomal silencing due to heterochromatin spreading [146]. Although optimizing the DNA concentration can reduce the chances of high gene copy number, random integration into heterochromatin regions can be overcome by incorporating insulator sequences into the vectors [140]. Several regulatory sequences that have insulating properties have been described [147], including S/MARs [148]. Evidence suggests that inclusion of an S/MAR region can provides an insulating effect by inhibiting promoter region methylation and silencing, as seen for CMV and HAAT promoters [140]. S/MAR-containing vectors have been used to drive transgene expression in hematopoietic stem cells [149].

6.6 NP Size

The size of the NP–DNA complexes is of crucial importance for their cellular uptake through endocytosis and/or pinocytosis and subsequent transfer into the nucleus through the nuclear pore complexes (NPCs) in the nuclear envelopes [150]. Using the *Xenopus* nuclear envelope reassembly (XNER) assay, Symens et al. [138] found that the nuclear enclosure of NPs was dependent on the size (with 100 and 200 nm NPs being better included than the 500 nm NPs) and charge (with positively charged NPs being better included than negatively charged or PEGylated NPs) of the DNA–NP complexes. Accordingly, smaller NPs (generally <25 nm) are considered to be more efficient for NP-based gene delivery systems. Interestingly, irrespective of the size of the DNA–NP complex, combining an NP-based gene delivery system with physical methods such as electroporation does not seem to have any beneficial effect on the gene delivery efficiency.

6.7 NP:DNA Ratio (Nitrogen:Phosphate Ratio), Concentration and Incubation Period

The NP:DNA ratio influences the N:P ratio, which in turn determines the shape and size of the DNA–NP complexes and the degree of DNA compaction. Furthermore, if the ratio of NP:DNA is suboptimal, the overall charge of the complexes might be negative, neutral or strongly positive, which can lead to inefficient adsorption to the negatively charged surface of mammalian cells. Thus, the NP:DNA ratio needs to be optimized to obtain an optimal rate of gene transfection. Furthermore, the optimal concentration of a given NP–DNA complex should be evaluated for each cell type because they are known to be dependent on cell type and cell density. Different cell types may also require different incubation periods with the DNA–NP complex to achieve maximal transfection level. This should be kept in mind when determining the length of incubation after transfection. If the time-point of maximal expression is not known for a particular cell line, a time-course experiment may be necessary.

6.8 Controlled Intracellular Release of DNA

Besides minimizing the cytotoxicity, surface coating and functionalization of NPs are also carried out occasionally to allow slow or delayed release of DNA until the cells enter mitosis and dissolve their nuclear envelope. Under such circumstances, DNA has an increased chance of coming in contact with the host genome and, hence, of increased transfection rate. Mahor et al. [52] observed that encapsulation of DNA in HA increased transgene expression when it was delivered into MSCs using branched PEI as a transfecting agent. DNA-loaded NPs can also be immobilized on the surface-coated ECM (a process termed reverse transfection), which not only provides a surface for cell attachment but also sustains the release of DNA from the surface, thereby inducing transgene expression for a prolonged period of time. Consequently, long-term expression of the desired protein can be achieved with a smaller amount of required DNA, as compared with bolus delivery.

6.9 Cytotoxicity

Several NPs have been reported to have inherent cytotoxicity, which might be augmented by excessive exposure and/or high concentrations of NP–DNA complexes and by stress due to temperature shifts or long periods without medium, etc. Full attention, therefore, must be given to minimize the cytoxicity. Interestingly, although most NPs are reported to cause oxidative stress by increasing the levels of reactive oxygen species (ROS) and reducing the glutathione levels [34], some NPs were demonstrated to have antioxidants action by blocking ROS production or scavenging the ROS [151, 152]. NPs can also cause mitochondrial damage. Thus, a biodegradable, biocompatible NP with minimal cytotoxicity should be chosen and properly exposed to cells to achieve the maximal number of transfected cells. Several approaches such as use of biodegradable polymers and/or surface coating of NPs with biodegradable polymers have, therefore, been used.

6.10 Stability, Storage and Shelf-Life of NPs

In order to maintain the structural and functional integrity of the entrapped DNA, the NP preparation process needs to be optimized for molecular weight, crosslinking method, crosslinking time and N:P ratio. Tzeng et al. [153] developed polymer-DNA NPs that remained stable in normal serum and could also be stored for at least 3 months in ready-to-use form with no measurable decrease in efficacy, thus expanding their potential in a practical setting.

7 Conclusions

NP-based gene delivery into stem cells provides an unprecedented opportunity for isolation, in vitro culture, differentiation and post-transplantation tracking of stem cells. It has also enabled the fabrication of controlled and high-throughput in vitro culture methods for culturing stem cells that were otherwise difficult to grow and differentiate in vitro. However, a variety of factors influence the application of NPs in stem cells and must be suitably addressed. Despite tremendous improvements, most NP-mediated gene delivery systems still suffer from low transfection efficiency and further research is needed on tailoring the size, content and surface electronic properties through chemical and physical methods. Furthermore, before NP technology can be used to deliver genes, several issues regarding the fate, toxicity and safety of NPs must be addressed. A pipeline of assays is needed to select the most efficient NPs and may include investigation of parameters crucial for efficient cellular uptake and retention; molecular analysis of uptake mechanisms, intracellular trafficking and degradation pathways of NPs; cellular tests for the effects of NPs on cellular physiology, proliferation and differentiation; and toxicity assays for genotoxicity, mutagenesis and oncogenesis. Finally, NP-based gene delivery offers new opportunities for transgenesis that need to be explored to utilize its full potential.

Acknowledgements This work was partly supported by grants from the BioGreen 21 Program (#PJ0080962012 and PJ0090142012), Rural Development Administration, Republic of Korea. The authors acknowledge the financial assistance to Pallavi Pushp in the form of an Institute Research Fellowship from NIT, Rourkela.

References

1. Wolff JA, Malone RW, Williams P et al (1990) Direct gene transfer into mouse muscle in vivo. Science 247:1465–1468
2. Read SP, Cashman SM, Kumar-Singh R (2010) A poly(ethylene) glycolylated peptide for ocular delivery compacts DNA into nanoparticles for gene delivery to post-mitotic tissues in vivo. J Gene Med 12:86–96
3. Wilson RW, Bloomfield VA (1979) Counterion-induced condesation of deoxyribonucleic acid. A light-scattering study. Biochemistry 18:2192–2196
4. Farjo R, Skaggs J, Quiambao AB et al (2006) Efficient non-viral ocular gene transfer with compacted DNA nanoparticles. PLoS One 1:e38
5. Fink TL, Klepcyk PJ, Oette SM et al (2006) Plasmid size up to 20 kbp does not limit effective in vivo lung gene transfer using compacted DNA nanoparticles. Gene Ther 13:1048–1051
6. Kedziorek DA, Muja N, Walczak P et al (2010) Gene expression profiling reveals early cellular responses to intracellular magnetic labeling with superparamagnetic iron oxide nanoparticles. Magn Reson Med 63:1031–1043
7. Liu G, Molas M, Grossmann GA et al (2001) Biological properties of poly-L-lysine-DNA complexes generated by cooperative binding of the polycation. J Biol Chem 276:34379–34387
8. Caracciolo G, Pozzi D, Capriotti AL et al (2011) Factors determining the superior performance of lipid/DNA/protammine nanoparticles over lipoplexes. J Med Chem 54:4160–4171
9. Han G, Chari NS, Verma A et al (2005) Controlled recovery of the transcription of nanoparticle-bound DNA by intracellular concentrations of glutathione. Bioconjug Chem 16:1356–1359
10. McIntosh CM, Esposito EA 3rd, Boal AK et al (2001) Inhibition of DNA transcription using cationic mixed monolayer protected gold clusters. J Am Chem Soc 123:7626–7629
11. Niidome T, Nakashima K, Takahashi H et al (2004) Preparation of primary amine-modified gold nanoparticles and their transfection ability into cultivated cells. Chem Commun (Camb) 1978–1979
12. Rosi NL, Giljohann DA, Thaxton CS et al (2006) Oligonucleotide-modified gold nanoparticles for intracellular gene regulation. Science 312:1027–1030
13. Seferos DS, Giljohann DA, Rosi NL et al (2007) Locked nucleic acid-nanoparticle conjugates. Chembiochem 8:1230–1232

14. Hackenberg S, Scherzed A, Kessler M et al (2011) Silver nanoparticles: evaluation of DNA damage, toxicity and functional impairment in human mesenchymal stem cells. Toxicol Lett 201:27–33
15. Niidome Y, Niidome T, Yamada S et al (2006) Pulsed-laser induced fragmentation and dissociation of DNA immobilized on gold nanoparticles. Mol Cryst Liq Cryst 445:201–206
16. Chen CC, Lin YP, Wang CW et al (2006) DNA-gold nanorod conjugates for remote control of localized gene expression by near infrared irradiation. J Am Chem Soc 128:3709–3715
17. Takahashi H, Niidome Y, Yamada S (2005) Controlled release of plasmid DNA from gold nanorods induced by pulsed near-infrared light. Chem Commun (Camb) 2247–2249
18. Wijaya A, Schaffer SB, Pallares IG et al (2009) Selective release of multiple DNA oligonucleotides from gold nanorods. ACS Nano 3:80–86
19. Kawano T, Yamagata M, Takahashi H et al (2006) Stabilizing of plasmid DNA in vivo by PEG-modified cationic gold nanoparticles and the gene expression assisted with electrical pulses. J Control Release 111:382–389
20. Huang YF, Sefah K, Bamrungsap S et al (2008) Selective photothermal therapy for mixed cancer cells using aptamer-conjugated nanorods. Langmuir 24:11860–11865
21. Boyer C, Priyanto P, Davis TP et al (2010) Anti-fouling magnetic nanoparticles for siRNA delivery. J Mater Chem 20:255–265
22. Kamau SW, Hassa PO, Steitz B et al (2006) Enhancement of the efficiency of non-viral gene delivery by application of pulsed magnetic field. Nucleic Acids Res 34:e40
23. McBain SC, Griesenbach U, Xenariou S et al (2008) Magnetic nanoparticles as gene delivery agents: enhanced transfection in the presence of oscillating magnet arrays. Nanotechnology 19:405102
24. Ito A, Shinkai M, Honda H et al (2001) Heat-inducible TNF-alpha gene therapy combined with hyperthermia using magnetic nanoparticles as a novel tumor-targeted therapy. Cancer Gene Ther 8:649–654
25. Tang QS, Zhang DS, Cong XM et al (2008) Using thermal energy produced by irradiation of Mn–Zn ferrite magnetic nanoparticles (MZF-NPs) for heat-inducible gene expression. Biomaterials 29:2673–2679
26. Kami D, Takeda S, Itakura Y et al (2011) Application of magnetic nanoparticles to gene delivery. Int J Mol Sci 12:3705–3722
27. Huth S, Lausier J, Gersting SW et al (2004) Insights into the mechanism of magnetofection using PEI-based magnetofectins for gene transfer. J Gene Med 6:923–936
28. Bharali DJ, Klejbor I, Stachowiak EK et al (2005) Organically modified silica nanoparticles: a nonviral vector for in vivo gene delivery and expression in the brain. Proc Natl Acad Sci USA 102:11539–11544
29. Gemeinhart RA, Luo D, Saltzman WM (2005) Cellular fate of a modular DNA delivery system mediated by silica nanoparticles. Biotechnol Prog 21:532–537
30. Li Z, Zhu S, Gan K et al (2005) Poly-L-lysine-modified silica nanoparticles: a potential oral gene delivery system. J Nanosci Nanotechnol 5:1199–1203
31. Radu DR, Lai CY, Jeftinija K et al (2004) A polyamidoamine dendrimer-capped mesoporous silica nanosphere-based gene transfection reagent. J Am Chem Soc 126:13216–13217
32. Kim W, Ng JK, Kunitake ME et al (2007) Interfacing silicon nanowires with mammalian cells. J Am Chem Soc 129:7228–7229
33. Roy I, Ohulchanskyy TY, Bharali DJ, Pudavar HE, Mistretta RA, Kaur N, Prasa PN (2005) Optical tracking of organically modified silica nanoparticles as DNA carriers: a nonviral, nanomedicine approach for gene delivery. Proc Natl Acad Sci USA 102:279–284
34. Chang JS, Chang KL, Hwang DF et al (2007) In vitro cytotoxicitiy of silica nanoparticles at high concentrations strongly depends on the metabolic activity type of the cell line. Environ Sci Technol 41:2064–2068
35. Lin W, Huang YW, Zhou XD et al (2006) In vitro toxicity of silica nanoparticles in human lung cancer cells. Toxicol Appl Pharmacol 217:252–259
36. Cai D, Mataraza JM, Qin ZH et al (2005) Highly efficient molecular delivery into mammalian cells using carbon nanotube spearing. Nat Methods 2:449–454

37. Gao L, Nie L, Wang T et al (2006) Carbon nanotube delivery of the GFP gene into mammalian cells. Chembiochem 7:239–242
38. Mann DG, McKnight TE, McPherson JT et al (2008) Inducible RNA interference-mediated gene silencing using nanostructured gene delivery arrays. ACS Nano 2:69–76
39. Yu Z, McKnight TE, Ericson MN et al (2012) Vertically aligned carbon nanofiber as nano-neuron interface for monitoring neural function. Nanomedicine 8:419–423
40. Boussif O, Lezoualc'h F, Zanta MA et al (1995) A versatile vector for gene and oligonucleotide transfer into cells in culture and in vivo: polyethylenimine. Proc Natl Acad Sci USA 92:7297–7301
41. Gosselin MA, Guo W, Lee RJ (2001) Efficient gene transfer using reversibly cross-linked low molecular weight polyethylenimine. Bioconjug Chem 12:989–994
42. Neu M, Sitterberg J, Bakowsky U et al (2006) Stabilized nanocarriers for plasmids based upon cross-linked poly(ethylene imine). Biomacromolecules 7:3428–3438
43. Sun YX, Zeng X, Meng QF et al (2008) The influence of RGD addition on the gene transfer characteristics of disulfide-containing polyethyleneimine/DNA complexes. Biomaterials 29:4356–4365
44. Peng Q, Zhong Z, Zhuo R (2008) Disulfide cross-linked polyethylenimines (PEI) prepared via thiolation of low molecular weight PEI as highly efficient gene vectors. Bioconjug Chem 19:499–506
45. Breunig M, Lungwitz U, Liebl R et al (2007) Breaking up the correlation between efficacy and toxicity for nonviral gene delivery. Proc Natl Acad Sci USA 104:14454–14459
46. Lee Y, Mo H, Koo H et al (2007) Visualization of the degradation of a disulfide polymer, linear poly(ethylenimine sulfide), for gene delivery. Bioconjug Chem 18:13–18
47. Koo H, Jin GW, Kang H et al (2010) Biodegradable branched poly(ethylenimine sulfide) for gene delivery. Biomaterials 31:988–997
48. Hosseinkhani H, Hosseinkhani M, Gabrielson NP et al (2008) DNA nanoparticles encapsulated in 3D tissue-engineered scaffolds enhance osteogenic differentiation of mesenchymal stem cells. J Biomed Mater Res A 85:47–60
49. Neu M, Germershaus O, Behe M et al (2007) Bioreversibly crosslinked polyplexes of PEI and high molecular weight PEG show extended circulation times in vivo. J Control Release 124:69–80
50. Ahn CH, Chae SY, Bae YH et al (2002) Biodegradable poly(ethylenimine) for plasmid DNA delivery. J Control Release 80:273–282
51. Chen XA, Zhang LJ, He ZJ et al (2011) Plasmid-encapsulated polyethylene glycol-grafted polyethylenimine nanoparticles for gene delivery into rat mesenchymal stem cells. Int J Nanomedicine 6:843–853
52. Mahor S, Collin E, Dash BC et al (2011) Controlled release of plasmid DNA from hyaluronan nanoparticles. Curr Drug Deliv 8:354–362
53. Park JS, Na K, Woo DG et al (2010) Non-viral gene delivery of DNA polyplexed with nanoparticles transfected into human mesenchymal stem cells. Biomaterials 31:124–132
54. Jeon SY, Park JS, Yang HN et al (2012) Co-delivery of SOX9 genes and anti-Cbfa-1 siRNA coated onto PLGA nanoparticles for chondrogenesis of human MSCs. Biomaterials 33 (17):4413–4423
55. Pimpha N, Sunintaboon P, Inphonlek S et al (2010) Gene delivery efficacy of polyethyleneimine-introduced chitosan shell/poly(methyl methacrylate) core nanoparticles for rat mesenchymal stem cells. J Biomater Sci Polym Ed 21:205–223
56. Park JS, Yang HN, Woo DG et al (2011) Chondrogenesis of human mesenchymal stem cells mediated by the combination of SOX trio SOX5, 6, and 9 genes complexed with PEI-modified PLGA nanoparticles. Biomaterials 32:3679–3688
57. Kakizawa Y, Harada A, Kataoka K (2001) Glutathione-sensitive stabilization of block copolymer micelles composed of antisense DNA and thiolated poly(ethylene glycol)-block-poly(L-lysine): a potential carrier for systemic delivery of antisense DNA. Biomacromolecules 2:491–497

58. Oishi M, Hayama T, Akiyama Y et al (2005) Supramolecular assemblies for the cytoplasmic delivery of antisense oligodeoxynucleotide: polyion complex (PIC) micelles based on poly (ethylene glycol)-SS-oligodeoxynucleotide conjugate. Biomacromolecules 6:2449–2454
59. Kim SH, Jeong JH, Lee SH et al (2006) PEG conjugated VEGF siRNA for anti-angiogenic gene therapy. J Control Release 116:123–129
60. Mok H, Park JW, Park TG (2007) Antisense oligodeoxynucleotide-conjugated hyaluronic acid/protamine nanocomplexes for intracellular gene inhibition. Bioconjug Chem 18:1483–1489
61. Lin C, Zhong Z, Lok MC et al (2007) Novel bioreducible poly(amido amine)s for highly efficient gene delivery. Bioconjug Chem 18:138–145
62. Namgung R, Brumbach JH, Jeong JH et al (2010) Dual bio-responsive gene delivery via reducible poly(amido amine) and survivin-inducible plasmid DNA. Biotechnol Lett 32:755–764
63. Lin C, Zhong Z, Lok MC et al (2007) Random and block copolymers of bioreducible poly (amido amine)s with high- and low-basicity amino groups: study of DNA condensation and buffer capacity on gene transfection. J Control Release 123:67–75
64. McKenzie DL, Kwok KY, Rice KG (2000) A potent new class of reductively activated peptide gene delivery agents. J Biol Chem 275:9970–9977
65. McKenzie DL, Smiley E, Kwok KY et al (2000) Low molecular weight disulfide cross-linking peptides as nonviral gene delivery carriers. Bioconjug Chem 11:901–909
66. Oupicky D, Parker AL, Seymour LW (2002) Laterally stabilized complexes of DNA with linear reducible polycations: strategy for triggered intracellular activation of DNA delivery vectors. J Am Chem Soc 124:8–9
67. Read ML, Singh S, Ahmed Z et al (2005) A versatile reducible polycation-based system for efficient delivery of a broad range of nucleic acids. Nucleic Acids Res 33:e86
68. Manickam DS, Oupicky D (2006) Multiblock reducible copolypeptides containing histidine-rich and nuclear localization sequences for gene delivery. Bioconjug Chem 17:1395–1403
69. Lo SL, Wang S (2008) An endosomolytic Tat peptide produced by incorporation of histidine and cysteine residues as a nonviral vector for DNA transfection. Biomaterials 29:2408–2414
70. Won YW, Kim HA, Lee M et al (2010) Reducible poly(oligo-D-arginine) for enhanced gene expression in mouse lung by intratracheal injection. Mol Ther 18:734–742
71. Green JJ, Zhou BY, Mitalipova MM et al (2008) Nanoparticles for gene transfer to human embryonic stem cell colonies. Nano Lett 8:3126–3130
72. Kim JH, Park JS, Yang HN et al (2011) The use of biodegradable PLGA nanoparticles to mediate SOX9 gene delivery in human mesenchymal stem cells (hMSCs) and induce chondrogenesis. Biomaterials 32:268–278
73. Andersen MO, Lichawska A, Arpanaei A et al (2010) Surface functionalisation of PLGA nanoparticles for gene silencing. Biomaterials 31:5671–5677
74. Ziady AG, Gedeon CR, Miller T et al (2003) Transfection of airway epithelium by stable PEGylated poly-L-lysine DNA nanoparticles in vivo. Mol Ther 8:936–947
75. Mao HQ, Roy K, Troung-Le VL et al (2001) Chitosan-DNA nanoparticles as gene carriers: synthesis, characterization and transfection efficiency. J Control Release 70:399–421
76. Kiang T, Bright C, Cheung CY et al (2004) Formulation of chitosan-DNA nanoparticles with poly(propyl acrylic acid) enhances gene expression. J Biomater Sci Polym Ed 15:1405–1421
77. Corsi K, Chellat F, Yahia L et al (2003) Mesenchymal stem cells, MG63 and HEK293 transfection using chitosan-DNA nanoparticles. Biomaterials 24:1255–1264
78. Sato N, Kobayashi H, Saga T et al (2001) Tumor targeting and imaging of intraperitoneal tumors by use of antisense oligo-DNA complexed with dendrimers and/or avidin in mice. Clin Cancer Res 7:3606–3612
79. Gao S, Chen J, Xu X et al (2003) Galactosylated low molecular weight chitosan as DNA carrier for hepatocyte-targeting. Int J Pharm 255:57–68
80. Kim TH, Park IK, Nah JW et al (2004) Galactosylated chitosan/DNA nanoparticles prepared using water-soluble chitosan as a gene carrier. Biomaterials 25:3783–3792

81. Park IK, Kim TH, Park YH et al (2001) Galactosylated chitosan-graft-poly(ethylene glycol) as hepatocyte-targeting DNA carrier. J Control Release 76:349–362
82. Kim TH, Kim SI, Akaike T et al (2005) Synergistic effect of poly(ethylenimine) on the transfection efficiency of galactosylated chitosan/DNA complexes. J Control Release 105:354–366
83. Thanou M, Florea BI, Geldof M et al (2002) Quaternized chitosan oligomers as novel gene delivery vectors in epithelial cell lines. Biomaterials 23:153–159
84. Kim YH, Gihm SH, Park CR et al (2001) Structural characteristics of size-controlled self-aggregates of deoxycholic acid-modified chitosan and their application as a DNA delivery carrier. Bioconjug Chem 12:932–938
85. Cheung CY, Murthy N, Stayton PS et al (2001) A pH-sensitive polymer that enhances cationic lipid-mediated gene transfer. Bioconjug Chem 12:906–910
86. Gwak SJ, Jung JK, An SS et al (2012) Chitosan/TPP-hyaluronic acid nanoparticles: a new vehicle for gene delivery to the spinal cord. J Biomater Sci Polym Ed 23(11):1437–1450
87. Tseng CL, Peng CL, Huang JY et al (2012) Gelatin nanoparticles as gene carriers for transgenic chicken applications. J Biomater Appl (in press). doi:10.1177/0885328211434089
88. Mo Y, Barnett ME, Takemoto D et al (2007) Human serum albumin nanoparticles for efficient delivery of Cu, Zn superoxide dismutase gene. Mol Vis 13:746–757
89. Dutta T, Burgess M, McMillan NA et al (2010) Dendrosome-based delivery of siRNA against E6 and E7 oncogenes in cervical cancer. Nanomedicine 6:463–470
90. Luo D, Li Y, Um SH et al (2006) A dendrimer-like DNA-based vector for DNA delivery: a viral and nonviral hybrid approach. Methods Mol Med 127:115–125
91. Arima H, Kihara F, Hirayama F et al (2001) Enhancement of gene expression by polyamidoamine dendrimer conjugates with alpha-, beta-, and gamma-cyclodextrins. Bioconjug Chem 12:476–484
92. Ofek P, Fischer W, Calderon M et al (2010) In vivo delivery of small interfering RNA to tumors and their vasculature by novel dendritic nanocarriers. FASEB J 24:3122–3134
93. Cao X, Deng W, Wei Y et al (2011) Encapsulation of plasmid DNA in calcium phosphate nanoparticles: stem cell uptake and gene transfer efficiency. Int J Nanomedicine 6:3335–3349
94. Jing Y, Moore LR, Williams PS et al (2007) Blood progenitor cell separation from clinical leukapheresis product by magnetic nanoparticle binding and magnetophoresis. Biotechnol Bioeng 96:1139–1154
95. Kanatsu-Shinohara M, Takashima S, Ishii K et al (2011) Dynamic changes in EPCAM expression during spermatogonial stem cell differentiation in the mouse testis. PLoS One 6:e23663
96. Lee W, Parpura V (2009) Chapter 6 – Carbon nanotubes as substrates/scaffolds for neural cell growth. Prog Brain Res 180:110–125
97. Mooney E, Dockery P, Greiser U et al (2008) Carbon nanotubes and mesenchymal stem cells: biocompatibility, proliferation and differentiation. Nano Lett 8:2137–2143
98. Lindberg HK, Falck GC, Suhonen S et al (2009) Genotoxicity of nanomaterials: DNA damage and micronuclei induced by carbon nanotubes and graphite nanofibres in human bronchial epithelial cells in vitro. Toxicol Lett 186:166–173
99. Ellis-Behnke RG, Liang YX, You SW et al (2006) Nano neuro knitting: peptide nanofiber scaffold for brain repair and axon regeneration with functional return of vision. Proc Natl Acad Sci USA 103:5054–5059
100. Guo J, Su H, Zeng Y et al (2007) Reknitting the injured spinal cord by self-assembling peptide nanofiber scaffold. Nanomedicine 3:311–321
101. Tysseling-Mattiace VM, Sahni V, Niece KL et al (2008) Self-assembling nanofibers inhibit glial scar formation and promote axon elongation after spinal cord injury. J Neurosci 28:3814–3823
102. de Freitas ER, Soares PR, de Santos RP et al (2011) Magnetic field-magnetic nanoparticle culture system used to grow in vitro murine embryonic stem cells. J Nanosci Nanotechnol 11:36–44

103. Lee CH, Kim EY, Jeon K et al (2008) Simple, efficient, and reproducible gene transfection of mouse embryonic stem cells by magnetofection. Stem Cells Dev 17:133–141
104. Lee CH, Kim JH, Lee HJ et al (2011) The generation of iPS cells using non-viral magnetic nanoparticle based transfection. Biomaterials 32:6683–6691
105. Pickard MR, Barraud P, Chari DM (2011) The transfection of multipotent neural precursor/stem cell transplant populations with magnetic nanoparticles. Biomaterials 32:2274–2284
106. Okita K, Yamanaka S (2011) Induced pluripotent stem cells: opportunities and challenges. Philos Trans R Soc Lond B Biol Sci 366:2198–2207
107. Yoshida Y, Yamanaka S (2010) Recent stem cell advances: induced pluripotent stem cells for disease modeling and stem cell-based regeneration. Circulation 122:80–87
108. Takahashi K, Yamanaka S (2006) Induction of pluripotent stem cells from mouse embryonic and adult fibroblast cultures by defined factors. Cell 126:663–676
109. Huangfu D, Osafune K, Maehr R et al (2008) Induction of pluripotent stem cells from primary human fibroblasts with only Oct4 and Sox2. Nat Biotechnol 26:1269–1275
110. Okita K, Nakagawa M, Hyenjong H et al (2008) Generation of mouse induced pluripotent stem cells without viral vectors. Science 322:949–953
111. Yu J, Hu K, Smuga-Otto K et al (2009) Human induced pluripotent stem cells free of vector and transgene sequences. Science 324:797–801
112. Warren L, Manos PD, Ahfeldt T et al (2010) Highly efficient reprogramming to pluripotency and directed differentiation of human cells with synthetic modified mRNA. Cell Stem Cell 7:618–630
113. Kim D, Kim CH, Moon JI et al (2009) Generation of human induced pluripotent stem cells by direct delivery of reprogramming proteins. Cell Stem Cell 4:472–476
114. Han DW, Tapia N, Hermann A et al (2012) Direct reprogramming of fibroblasts into neural stem cells by defined factors. Cell Stem Cell 10:465–472
115. Yang JH, Lee SH, Heo YT et al (2010) Generation of insulin-producing cells from gnotobiotic porcine skin-derived stem cells. Biochem Biophys Res Commun 397:679–684
116. Yang JH, Shim SW, Lee BY et al (2010) Skin-derived stem cells in human scar tissues: a novel isolation and proliferation technique and their differentiation potential to neurogenic progenitor cells. Tissue Eng Part C Methods 16:619–629
117. Shi Y, Do JT, Desponts C et al (2008) A combined chemical and genetic approach for the generation of induced pluripotent stem cells. Cell Stem Cell 2:525–528
118. Ruan J, Shen J, Wang Z et al (2011) Efficient preparation and labeling of human induced pluripotent stem cells by nanotechnology. Int J Nanomedicine 6:425–435
119. Park IH, Arora N, Huo H et al (2008) Disease-specific induced pluripotent stem cells. Cell 134:877–886
120. Andersen MO, Nygaard JV, Burns JS et al (2010) SiRNA nanoparticle functionalization of nanostructured scaffolds enables controlled multilineage differentiation of stem cells. Mol Ther 18:2018–2027
121. Cao X, Deng W, Wei Y et al (2012) Incorporating pTGF-beta1/calcium phosphate nanoparticles with fibronectin into 3-dimensional collagen/chitosan scaffolds: efficient, sustained gene delivery to stem cells for chondrogenic differentiation. Eur Cell Mater 23:81–93
122. Nguyen YT, Kim HK, Kwon JS et al (2010) Efficient transfer of reporter gene-loaded nanoparticles to bone marrow stromal cells (D1) by reverse transfection. J Nanosci Nanotechnol 10:3170–3174
123. Tang C, Russell PJ, Martiniello-Wilks R et al (2010) Concise review: nanoparticles and cellular carriers-allies in cancer imaging and cellular gene therapy? Stem Cells 28:1686–1702
124. Kim YS, Park IK, Kim WJ et al (2011) SPION nanoparticles as an efficient probe and carrier of DNA to umbilical cord blood-derived mesenchymal stem cells. J Nanosci Nanotechnol 11:1507–1510

125. Gagne MB, Pothier F, Sirard MA (1991) Electroporation of bovine spermatozoa to carry foreign DNA in oocytes. Mol Reprod Dev 29:6–15
126. Hoelker M, Mekchay S, Schneider H et al (2007) Quantification of DNA binding, uptake, transmission and expression in bovine sperm mediated gene transfer by RT-PCR: effect of transfection reagent and DNA architecture. Theriogenology 67:1097–1107
127. Shen W, Li L, Pan Q et al (2006) Efficient and simple production of transgenic mice and rabbits using the new DMSO-sperm mediated exogenous DNA transfer method. Mol Reprod Dev 73:589–594
128. Spadafora C (2007) Sperm-mediated gene transfer: mechanisms and implications. Soc Reprod Fertil Suppl 65:459–467
129. Lavitrano M, Busnelli M, Cerrito MG et al (2006) Sperm-mediated gene transfer. Reprod Fertil Dev 18:19–23
130. Kim TS, Lee SH, Gang GT et al (2010) Exogenous DNA uptake of boar spermatozoa by a magnetic nanoparticle vector system. Reprod Domest Anim 45:e201–e206
131. Campos VF, Komninou ER, Urtiaga G et al (2011) NanoSMGT: transfection of exogenous DNA on sex-sorted bovine sperm using nanopolymer. Theriogenology 75: 1476–1481
132. Perry AC, Rothman A, de las Heras JI et al (2001) Efficient metaphase II transgenesis with different transgene archetypes. Nat Biotechnol 19:1071–1073
133. Lee SH, Gupta MK, Han DW et al (2007) Development of transgenic chickens expressing human parathormone under the control of a ubiquitous promoter by using a retrovirus vector system. Poult Sci 86:2221–2227
134. Uhm SJ, Gupta MK, Das ZC et al (2009) Effect of transgene introduction and recloning on efficiency of porcine transgenic cloned embryo production in vitro. Reprod Domest Anim 44:106–115
135. Uhm SJ, Gupta MK, Kim T et al (2007) Expression of enhanced green fluorescent protein in porcine- and bovine-cloned embryos following interspecies somatic cell nuclear transfer of fibroblasts transfected by retrovirus vector. Mol Reprod Dev 74:1538–1547
136. Uhm SJ, Kim NH, Kim T et al (2000) Expression of enhanced green fluorescent protein (EGFP) and neomycin resistant (Neo(R)) genes in porcine embryos following nuclear transfer with porcine fetal fibroblasts transfected by retrovirus vector. Mol Reprod Dev 57:331–337
137. Symens N, Soenen SJ, Rejman J et al (2012) Intracellular partitioning of cell organelles and extraneous nanoparticles during mitosis. Adv Drug Deliv Rev 64:78–94
138. Symens N, Walczak R, Demeester J et al (2011) Nuclear inclusion of nontargeted and chromatin-targeted polystyrene beads and plasmid DNA containing nanoparticles. Mol Pharm 8:1757–1766
139. Hsu SH, Ho TT, Tseng TC (2012) Nanoparticle uptake and gene transfer efficiency for MSCs on chitosan and chitosan-hyaluronan substrates. Biomaterials 33:3639–3650
140. Rupprecht S, Lipps HJ (2009) Cell cycle dependent histone dynamics of an episomal non-viral vector. Gene 439:95–101
141. Baus J, Liu L, Heggestad AD et al (2005) Hyperactive transposase mutants of the Sleeping Beauty transposon. Mol Ther 12:1148–1156
142. Wang Y, Li Z, Han Y et al (2010) Nanoparticle-based delivery system for application of siRNA in vivo. Curr Drug Metab 11:182–196
143. Woltjen K, Michael IP, Mohseni P et al (2009) PiggyBac transposition reprograms fibroblasts to induced pluripotent stem cells. Nature 458:766–770
144. Liu J, Jeppesen I, Nielsen K et al (2006) Phi c31 integrase induces chromosomal aberrations in primary human fibroblasts. Gene Ther 13:1188–1190
145. Olivares EC, Hollis RP, Chalberg TW et al (2002) Site-specific genomic integration produces therapeutic factor IX levels in mice. Nat Biotechnol 20:1124–1128
146. Riu E, Chen ZY, Xu H et al (2007) Histone modifications are associated with the persistence or silencing of vector-mediated transgene expression in vivo. Mol Ther 15:1348–1355

147. Bell AC, West AG, Felsenfeld G (2001) Insulators and boundaries: versatile regulatory elements in the eukaryotic genome. Science 291:447–450
148. Jackson DA, Juranek S, Lipps HJ (2006) Designing nonviral vectors for efficient gene transfer and long-term gene expression. Mol Ther 14:613–626
149. Papapetrou EP, Ziros PG, Micheva ID et al (2006) Gene transfer into human hematopoietic progenitor cells with an episomal vector carrying an S/MAR element. Gene Ther 13:40–51
150. Rejman J, Oberle V, Zuhorn IS et al (2004) Size-dependent internalization of particles via the pathways of clathrin- and caveolae-mediated endocytosis. Biochem J 377:159–169
151. Becker S, Soukup JM, Gallagher JE (2002) Differential particulate air pollution induced oxidant stress in human granulocytes, monocytes and alveolar macrophages. Toxicol In Vitro 16:209–218
152. Schubert D, Dargusch R, Raitano J et al (2006) Cerium and yttrium oxide nanoparticles are neuroprotective. Biochem Biophys Res Commun 342:86–91
153. Tzeng SY, Guerrero-Cazares H, Martinez EE et al (2011) Non-viral gene delivery nanoparticles based on poly(beta-amino esters) for treatment of glioblastoma. Biomaterials 32:5402–5410

Adv Polym Sci (2013) 254: 87–134
DOI: 10.1007/12_2012_196

Published online: 15 March 2013

Engineering of Polysaccharides via Nanotechnology

Joydeep Dutta

Abstract The exploitation of nanotechnology to engineer polysaccharides is undoubtedly very innovative in the context of nanobiopolymers. The polysaccharides are naturally occurring biopolymers and can potentially be used for diverse applications. Interest in the study of nanotechnology to produce nanoscaled polysaccharides is gradually increasing. Nanotechnology not only changes the structures of polysaccharides but also changes the functionality of the materials. Many properties of polysaccharides are perhaps still undiscovered. The use of nanotechnology to engineer polysaccharides may open up a new horizon that can noticeably change every aspect of human life. Chitosan is an important polysaccharide and is found in the exoskeleton of crab, shrimp, prawn, lobster, squid pen, etc. Despite its abundant availability in nature, it has not yet been completely cherished by the scientific community. The properties of chitosan vary from source to source, which could enable the efficient use of chitosan for a specific application. This chapter deals with various sources of chitosan and discusses its physical, chemical, and biological properties; synthesis of chitosan-based nanoparticles, nanospheres, and nanogels; characterization; and biomedical applications.

Keywords Chitosan · Engineering · Nanotechnology · Polysaccharide

Contents

J. Dutta (✉)
Department of Applied Sciences & Humanities, Institute of Engineering & Technology, ITM University, Uparwara, Raipur, Chattrisgarh 493661, India
e-mail: dutta_joy@yahoo.co.in

1 Introduction

The advent of nanotechnology has given tremendous impetus to the development of polysaccharide-based nanomaterials. There are huge differences between the bulk polysaccharides and nanopolysaccharides in every aspect. No vast amount of research has yet been done on engineering of polysaccharides via nanotechnology. This chapter will mainly deal with chitosan rather than other polysaccharides, namely cellulose, starch, gelatin, alginate, pullulan, carrageenan, and pectin to name a few. Chitin exists widely in cell walls of some microorganisms such as fungi, molds, and yeasts [1] and in the cuticular and exoskeletons of invertebrates such as crab, shrimp, prawn, lobster, squid pen and insects (for example, beetles) [2]. Chitosan exists naturally in only a few species of fungi. Chitosan, a fundamental derivative of chitin, is composed of glucosamine and *N*-acetyl glucosamine units [3–5]. Despite its abundant availability in nature, its exploitation is not very pronounced. Due to the presence of amino groups as well as hydroxyl groups, chitosan can be easily tailored to synthesize different derivatives of interest [6]. Over the last two decades, the studies on chitin and chitosan have intensified as a consequence of their excellent biological properties such as being biodegradable in the human body, biocompatibility, nontoxicity, immunological activity, antibacterial activity, and wound-healing activity [7]. In recent studies, chitosan has been shown to be a very good biomedical candidate for applications ranging from drug delivery to cell delivery to gene delivery to wound healing [8–12]. But, the solubility of chitosan has been a major constraint for its versatile application. Chitosan is not soluble in conventional organic solvents; it is only soluble in acetic acid and a few inorganic acids such as hydrochloric acid [13–15]. Therefore, it has been a major challenge to improve the solubility of chitosan. It is believed that application of nanotechnology to engineer chitosan may not only boost its solubility properties but also change its material characteristics, which would be beneficial for further processing for its commercialization in various products. This belief has led the chitin and chitosan researchers not only to escalate fundamental research on chitosan but also to work aggressively on applied research to develop chitosan-based nanoproducts by

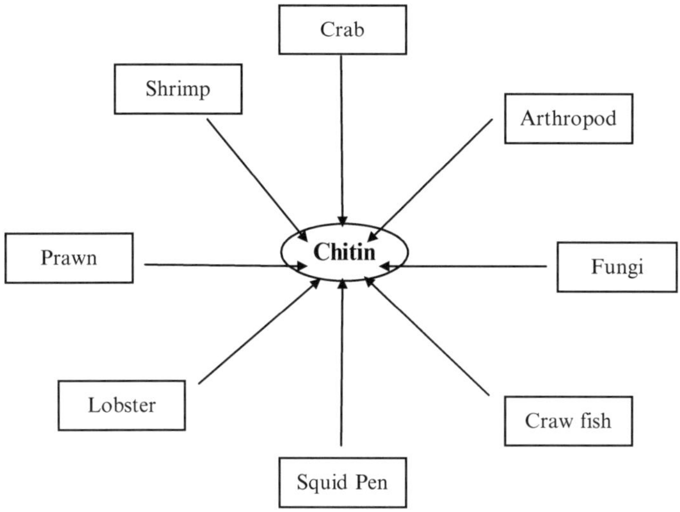

Fig. 1 Various sources of chitin

utilizing its chemical, physical, and biological properties. Chemical modifications of chitosan and its subsequent nanotechnological application are hardly seen. Similarly, the variations in the properties of chitosan obtained from various sources are not yet well documented. This chapter will describe the various sources of chitosan and its multifaceted characteristics and the synthesis, characterization, and biomedical applications of chitosan-based nanocomposites in the form of nanoparticles, nanogels, and nanospheres.

2 Sources of Chitin and Chitosan

There are numerous sources of chitin (Fig. 1) and chitosan is obtained by deacetylating chitin with a hot alkali solution. Chitin has been found in a wide array of natural sources, namely crustaceans, fungi, yeasts, insects, annelids, nematodes, mollusks, coelenterate, marine diatoms, squid pens, etc. [16–19]. However, chitosan is primarily manufactured from the exoskeleton of crustaceans (crab, shrimp, prawn, lobster, krill, and crayfish) because of its abundant availability as a by-product of food processing [20, 21].

3 Isolation of Chitin and Chitosan

In the past, a small quantity of shell waste was utilized for animal feed or chitin isolation [19]. Thus, the processing of shellfish was a major concern for environmental pollution. Nowadays, this problem has been overcome to a certain extent because of an increased desire to explore underutilized resources (mainly shell wastes) for the isolation of chitin and make it suitable for various biomedical applications. The basic approach is more or less same for the isolation of chitin from various sources because, typically, the exoskeleton of crustaceans consists of minerals (especially calcium carbonate), protein, chitin, pigments, etc. No economical and appropriate solvent is, however, known for the dissolution of chitin after isolation of chitin from these exoskeletons. The current industrial method for obtaining chitin from the exoskeletons is to make the calcium carbonate and proteins in the exoskeletons soluble in water and then to remove them from the exoskeletons to obtain chitin. Specifically, the exoskeletons of crabs, lobsters, or shrimps are immersed in a dilute aqueous alkaline solution and heated, after which the degraded proteins are washed off with water. Insoluble matter with chitin contained therein is immersed in a dilute aqueous solution of hydrochloric acid to convert the calcium carbonate (which is still contained in the insoluble matter) into calcium chloride, which is soluble in water. The insoluble matter is then washed with water to remove the calcium chloride, so that chitin is obtained as insoluble matter. The various sources of chitin differ somewhat in their structure and percentage chitin content. Usually, chitin isolation consists of several steps, i.e., demineralization, deproteinization, and decoloration. The first two steps can be reversed according to the need for recovery of carotenoids and protein and for further chitin application. Among other crustacean shells, crab and shrimp shells have received significant attention for the production of chitin and chitosan [22, 23]. The readers can find out about traditional isolation of chitin in many reviews, research articles, patents, and books [24–33]. Succinctly, demineralization is usually achieved by using diluted hydrochloric acid (1–8%) at room temperature [34, 35]. To prevent chitin depolymerization, ethylenediamine tetraacetic acid (EDTA) can be used for removal of mineral salts [36, 37]. On the other hand, deproteinization involves the usage of aqueous sodium or potassium hydroxide solution, and decoloration is usually carried out by a bleaching treatment with NaOCl or H_2O_2 solutions [38].

The most exploited sources of chitin are shrimp and crab shells. Lertsutthiwong et al. [28] has studied the effect of chemical treatment on the characteristics of chitosan obtained from fresh local black tiger shrimp shells. In this study, they treated the shrimp shells by reversing the conventional order of extraction of chitin from crustacean shells. The differences in the characteristics of chitin produced by different serial treatments are shown in Table 1. Further, they showed the effect of these different serial treatments on the quality of chitosan obtained from chitin after deacetylation. If process 2 is followed by subsequent deacetylation then there is a possibility of obtaining chitosan with high viscosity because the protein layer becomes protected from more hydrolysis, and vice versa in the case of process 1 (see Table 2). To obtain chitosan with a high viscosity and a high degree of deacetylation at low temperature, the process should be initiated with decalcification and requires multi-deacetylation.

There are many standard operating procedures available for extraction of chitin and chitosan because of their wide ranges of sources [39–45]. Although lots of work

Table 1 Characteristics of chitin produced by different serial treatments

Process 1	Process 2
Deproteinization precedes decalcification	Decalcification precedes deproteinization
Chitin became whiter with increasing concentration of NaOH (1–4%)	Higher concentration of NaOH (4%) gave more efficient removal of protein and colored matter
Decalcification with 4% HCl for only 2 h was sufficient to remove minerals	Decalcification first with 4% HCl for either 2 or 12 h, gave colored matter and the protein still remained bound to the solid matrix
The yield of chitin was 15–20%. In this case, the protein layer was unprotected because the deproteinization precedes the decalcification. That is why during demineralization more hydrolysis and loss of material in the solid chitin fraction occurred	The yield of chitin was 20–27%, which is attributed to the strong adherence of chitin to protein, leading to less hydrolysis of the backbone because the decalcification precedes the deproteinization
Ash content was less than 1%	Ash content was 1–2%

Table 2 Variations on quality of chitosan with respect to various serial treatments (adapted from [28])

	Parameter			
Treatments	Moisture content (%)	Ash content (%)	Degree of deacetylation (%)	Viscosity (cps)
1% NaOH, 21 h and 4% HCl, 2 h	6.71 ± 0.10	0.51 ± 0.04	75.9 ± 0.35	830 ± 30
4% HCl, 2 h and 1% NaOH, 21 h	9.52 ± 0.16	0.89 ± 0.27	74.3 ± 0.07	5,268 ± 146
2% NaOH, 21 h and 4% HCl, 2 h	6.96 ± 0.04	0.52 ± 0.04	75.7 ± 0.71	486 ± 14
4% HCl, 2 h and 2% NaOH, 21 h	8.13 ± 0.09	1.04 ± 0.08	75.5 ± 0.21	6,370 ± 254
4% NaOH, 21 h and 4% HCl, 2 h	8.36 ± 0.06	0.50 ± 0.08	75.7 ± 0.28	435 ± 13
4% HCl, 2 h and 4% NaOH, 21 h	7.70 ± 0.14	1.01 ± 0.001	75.7 ± 0.85	5,238 ± 190
1% NaOH, 21 h and 4% HCl, 12 h	7.44 ± 0.21	0.84 ± 0.08	73.2 ± 2.05	160 ± 9
4% HCl, 12 h and 1% NaOH, 21 h	12.6 ± 0.15	0.87 ± 0.04	74.8 ± 0.57	3,420 ± 174
2% NaOH, 21 h and 4% HCl, 12 h	8.11 ± 0.15	0.71 ± 0.01	73.8 ± 3.0	110 ± 4
4% HCl, 12 h and 2% NaOH, 21 h	9.38 ± 0.04	0.96 ± 0.01	76.3 ± 0.07	2,919 ± 93
4% NaOH, 21 h and 4% HCl, 12 h	7.97 ± 0.11	0.68 ± 0.02	74.1 ± 1.91	106 ± 5
4% HCl, 12 h and 4% NaOH, 21 h	8.33 ± 0.09	0.97 ± 0.01	76.1 ± 0.07	4,470 ± 115

Fig. 2 Flow chart of the revised process for the production of chitin and chitosan [27]

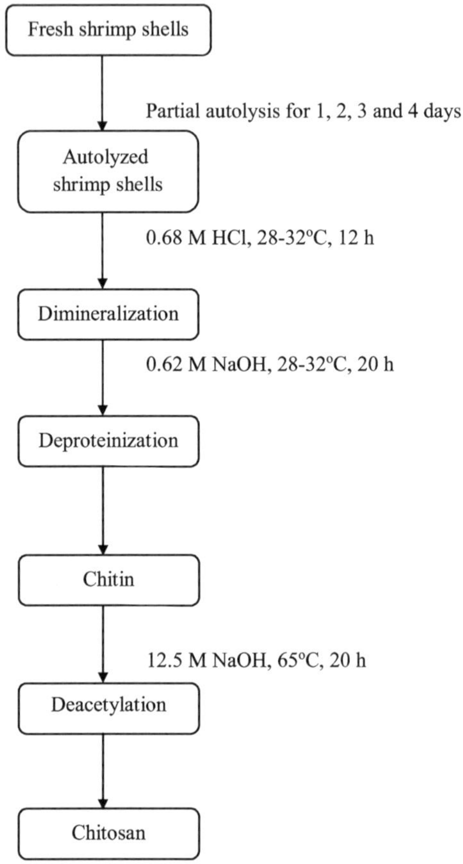

has already been done on extraction of chitin and chitosan from various sources, significant research is still required to improve the production process as well as the quality of chitosan, especially from stored material sources. Toan [27] investigated the effect of partial autolysis during storage of shrimp biowaste on the quality of extracted chitin and chitosan. In this study, a modified method was used in which the shrimp shell was allowed to undergo partial autolysis before it was used to produce chitin and chitosan. This revised process for the production of chitin and chitosan is shown in Fig. 2.

In terms of improving the quality of chitosan, partial autolysis plays a crucial role because it facilitates exposure of $CaCO_3$ directly to the chemicals used during the demineralization process. For the industrial practice of chitin extraction, this means that a shorter time and lower concentration of HCl may be adequate to demineralize and thus avoid damage due to the hydrolytic action of HCl on the polymeric backbone structure of chitin. However, it was found that the protein content after treatment of autolyzed biomaterial with 2% NaOH was similar to the protein content after standard 4% NaOH treatment. Obviously, this study could help

Table 3 Various sources of chitin with their different treatment processes

Source of chitin	Demineralization process	Deproteinization process	References
Crab	HCl	NaOH	[46]
		KOH	[47]
Shrimp	HCl	NaOH	[48]
	CH_3COOH	KOH	[46, 49]
Prawn	HCl	NaOH	[50, 51]
Krill	HCl	$NaHCO_3$	[52]
		KOH	[52]
		NaOH	[53]
Lobster	HCl	NaOH	[54]
	CH_2O_2	NaOH	[55]
Crawfish	HCl	NaOH	[51]
Squid pen	Omitted	NaOH	[20]

Table 4 Ash, moisture, protein, nitrogen, and acetyl content of selected chitosans (adapted from [20])

Sample	Source	% Ash	% Moisture	% Protein	% Nitrogen[a]	% Acetyl
Squid pen chitosan	Squid (IRL)	0.17	2.1	1.3	7.5 (7.8)	49.4
Seacure 443 chitosan	Crab/shrimp (Pronova)	0.58	11.2	1.3	7.2 (8.2)	27.2
Sigma chitosan	Crab (Sigma)	0.51	4.8	1.3	7.1 (8.2)	27.2
Profloc 340 chitosan	Crab/shrimp (Pronova)	0.40	12.9	1.4	7.1 (8.2)	27.2

[a]Numbers in parentheses indicate % nitrogen after adjusting for ash and moisture content

reduce the cost of deproteinization in industry and could help prevent pollution of the environment. The production of chitin from various sources with their different treatment processes are summarized in Table 3.

Keeping in mind the importance of chitosan, as well as its economic value as an industrial product, we must pay attention to its key physical parameter, i.e., turbidity. Depending on source, a marked difference is observed in aqueous solutions of chitosan and its derivatives in terms of their turbidity [33]. Turbid aqueous solutions of chitosan and chitosan-derived products greatly lose their commercial value. Such chitosan cannot be used as a commercial product and in some cases may have to be discarded. Therefore, the selection of source plays a pivotal role in the production of chitin and chitosan. Shepherd et al. [20] reported the production of chitosan from New Zealand Arrow squid (*Notodarus sloani*) pens as well as the evaluation of the functional properties of this squid chitosan compared with chitosan extracted from crustacean sources. Squid pen chitin and chitosan were visibly cleaner than chitin and chitosan obtained from crab and crayfish. In addition, due to the lower mineral content of squid pen as compared to crustacean shells, the demineralization process can be skipped to extract chitin, which also makes the production cheaper. As shown in Table 4, the squid pen chitosan is similar in

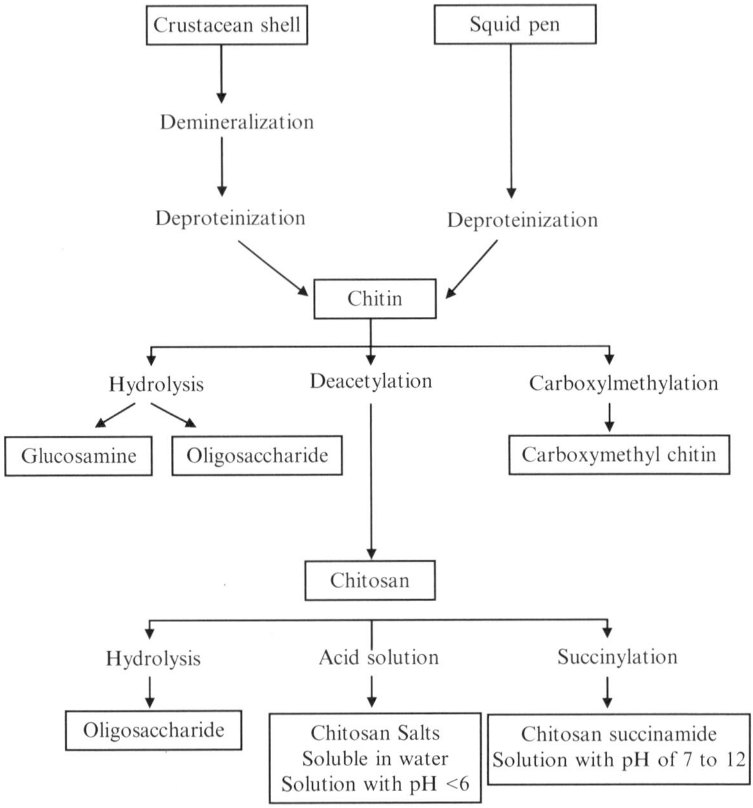

Fig. 3 Chitin and chitosan manufacturing process (adapted from [56])

composition to commercial chitosan. Further, there is a substantially lower amount of ash contained in squid chitosan that makes the aqueous solution noticeably cleaner than samples of chitin and chitosan from crab and crayfish. The manufacturing process of chitin and chitosan from squid pen as well as crustacean shell is shown in Fig. 3. The traditional and commercial chitosan production process has a number of unfavorable characteristics because the process requires expensive heat energy and caustic alkali, which is a potential health hazard. The process also produces large amounts of waste, thereby necessitating significant disposal costs. In addition, the supply of shrimp or crab shells is highly dependent upon seasonal and environmental factors, leading to unpredictable limitations on production capacity [57].

In recent years, chitin derived from fungal mycelia has gained tremendous importance. Fungal mycelia can be cultivated throughout the year by a fermentation that is rapid, synchronized, and can be organized in a closed or semi-closed technological circuit to comply with modern ecological requirements. Moreover, fungal mycelia are relatively consistent in composition and are not associated with inorganic materials. Therefore, no demineralization treatment is required to recover fungal chitin [58]. Teng et al. [29] investigated the concurrent production of chitin from shrimp shells and fungi by placing shrimp shells in direct contact with the fermentation of filamentous fungi in a reactor so as to avoid the use of conventional chemicals, which can lead to undue deacetylation of chitin or chain degradation of the isolated chitin. Three proteolytic *Aspergillus niger* strains (0576, 0307, and 0474) were selected from a screening for protease activity from among 34 zygomycete and deuteromycete strains. The proteolytic enzymes released by the fungi facilitate the deproteinization of shrimp powder and the release of hydrolyzed proteins. The hydrolyzed proteins in turn act as a source of nitrogen for further fungal growth, leading to a lowering of the pH of the fermentation medium, and thereby further enhancing the demineralization of the shrimp-shell powder. Chitin from non-fermented, fermented, and fungal mycelia were directly extracted with 5% lithium-chloride–*N,N*-dimethylacetamide (5% LiCl in DMAc), a non-degradative solvent for chitin [36]. In this process, two different problems are solved simultaneously without using any harsh chemicals. Fungal fermentation is a cost-effective method compared to commercial protease enzymes that deproteinize but not demineralize the shrimp shells.

Demineralization is also an important step in chitin purification from crabs. The chemical method of demineralization includes the use of strong acid (HCl) that harms the physicochemical properties of chitin. For the extraction of chitin from trash crabs (*Podophthalmus vigil*), Das et al. [30] reported the use of organic acids (lactic acid or other organic acids) produced by *Lactobacillus plantarum* as a replacement for HCl for demineralization, and proteolytic enzymes produced by *A. niger* as a substitute for alkaline aqueous solution for deproteinization. Although the removal of protein entirely from crustacean shells has not yet been achieved using biological treatment during isolation of chitin, the quality of chitin obtained by this process compares well with chitin obtained by conventional chemical treatment. The results of chitin derived from the trash crab are shown in Table 5. The study showed that the effectiveness of lactic acid for the demineralization of crustacean shells was virtually comparable to that of HCl. For effective removal of minerals from crab shells using lactic acid, a shell to acid ratio of 1:25 and temperature of 40°C were found to be satisfactory. The use of chemicals causes depolymerization of chitin to a certain extent and, thus, its molecular weight as well as viscosity are affected after solubilization [43] But, in this case, the combination of lactic acid and *A. niger* not only produces partially soluble chitin, which is anticipated to increase its applications in biomedicine and pharmacy, but also reduces the production cost of chitin.

Chitosan is less commonly found in living organisms than chitin but it can be found in the cell walls of certain groups of fungi. The traditional process of obtaining chitosan is the deacetylation of chitin using strong caustic alkali but no significant progress has yet been made in establishing new technologies for the large-scale controlled production of chitosan. The recognition of chitin deacetylase in several fungi and insects has given new momentum to the conversion of chitin into chitosan. In contrast to the currently used chemical procedure, the use of chitin deacetylase offers the possibility of a controlled non-degradable process, resulting in the production of novel, well-defined chitosan oligomers and polymers.

Table 5 Physicochemical parameters for extraction of chitin by chemical and biological processes (adapted from [30])

Parameters	Chemical treatment			Biological treatment		
Concentration of acid	2 N HCl	2 N HCl	2 N HCl	69.5 g/L lactic acid[a]	69.5 g/L lactic acid[a]	69.5 g/L lactic acid[a]
Temperature	RT	RT	RT	40°C	40°C	40°C
Shell to acid ratio	1:15	1:25	1:35	1:15	1:25	1:35
Concentration of deproteinizer	1 N NaOH	1 N NaOH	1 N NaOH	5 g wet weight[b]	5 g wet weight[b]	5 g wet weight[b]
Product appearance	White	Super-white	Super-white	Slightly brownish	White	White
Yield (g)	5.62	5.47	5.41	4.52	4.17	4.03
Yield (%)	28.1	27.35	27.05	22.6	20.85	20.15
Solubility in water	Insoluble	Insoluble	Insoluble	Partially soluble	Partially soluble	Partially soluble
pH	7.2	7.2	7.2	6.9	6.9	6.9

RT room temperature

[a] Lactic acid was produced by *Lactobacillus plantarum*

[b] Deproteinization was done using *Aspergillus Niger*, which produces proteolytic enzymes

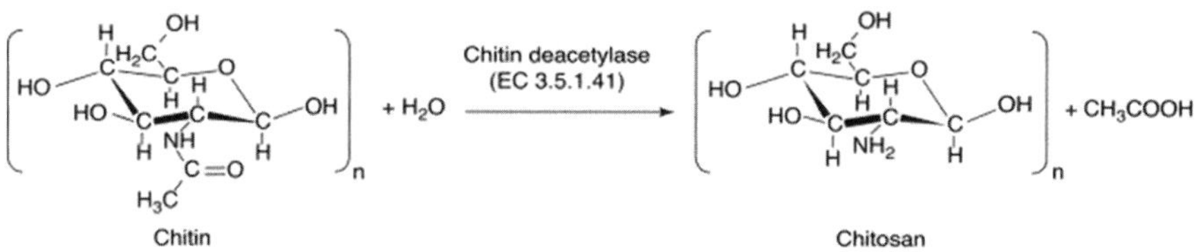

Fig. 4 Conversion of chitin to chitosan under the catalytic action of chitin deacetylases [59]

Table 6 Chitosan yield by various endophytic fungal species (adapted from [18])

Species	Yield of chitosan (%)
Fusarium sp.	–
Aspergillus flavus	5.7
Cladosporium cladosporioides	2.52
Phoma sp.	3.11
Botryodiplodia theobromae	–

Indeed, the use of biological treatment instead of chemical treatment for the isolation of chitin from crustacean shells as well as for production of chitosan will substantially reduce environmental pollution. The use of chitin deacetylase for the preparation of chitosan polymers and oligomers offers the possibility to develop an enzymatic process that could potentially overcome most of the drawbacks discussed earlier [59]. As shown in Fig. 4, chitin deacetylase (CDA; EC 3.5.1.41) catalyzes the hydrolysis of *N*-acetamido bonds in chitin to produce chitosan. The presence of this enzyme has been reported in several fungi and insect species [60–68].

Chen et al. [57] have patented an alternative method for production of chitin and chitosan that does not rely on environmentally harmful chemicals or the variable abundance of the crustacean crop. They demonstrated high-yield production of chitosan or chitin from a culture containing a *Rhizopus azygosporus* or *Actinomucor taiwanensis* fungus. Further, they have shown that a medium containing corn steep liquor, glucose, yeast extract, and ammonium sulfate is capable of increasing the output of chitin and chitosan from a fungal culture. Although the isolation of chitin from fungi is environmentally safe, the enzymes used for the bioconversion of chitin into chitosan require further purification, which could subsequently increase the purification cost. Some fungal species for the isolation of chitosan are shown in Table 6. Therefore, research is being carried out to find corresponding enzymes in bacteria for the biotransformation of chitin to chitosan because bacteria are easier and faster than fungi to grow in a large-scale fermentation system. In addition, bacteria can be utilized without the need to purify the enzyme.

Srinivasan et al. [69] have patented a method for isolation of an unknown bacterium from municipal sewage. This bacterium can deacetylate chitin to the more useful, soluble chitosan. Biotransformation of chitin to chitosan by bacteria can be used in an economical and environmentally friendly process.

In another study, Yong et al. [70] reported a novel chitinase-producing bacterium strain (C4) from soil samples. Molecular identification confirmed that strain C4

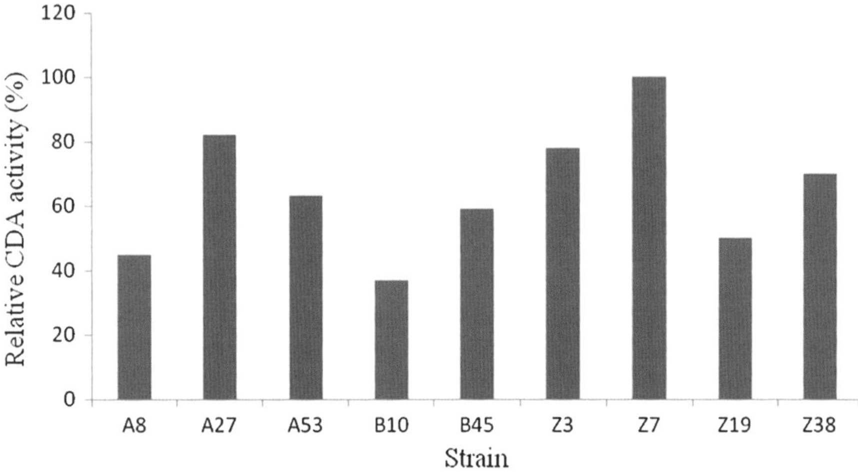

Fig. 5 The relative chitin deacetylase (CDA) activity of nine isolated bacterial strains (adapted from [71])

belongs to the genus *Sanguibacter*. The enzyme extracted from the bacterium was able to degrade polymeric chitin from crab and shrimp shells, and from insect puparium. The enzyme also displayed weak activity against chitosan. Bacterial growth is faster than fungal growth and the chitosan deacetylase (CDA)-producing capabilities of most fungal strains are low and their fermentation requirements are complicated. Zhou et al. [71] reported the isolation of CDA-producing bacteria from soil and optimized its fermentation process. Among 208 bacteria isolated from fresh soil samples, nine were CDA-producing strains of bacteria, namely A8, A27, A53, B10, B45, Z3, Z7, Z29, and Z38. The relative CDA activity of these strains is shown in Fig. 5. As can be seen from this figure, the Z7 strain shows the highest CDA activity, which was determined by enzyme assay screening. To enhance CDA production, various carbon sources (glucose, lactose, sucrose, starch, corn meal, and bran), nitrogen sources (beef extract, yeast extract, casein, proteose peptone, soybean meal, urea, ammonium sulfate, and sodium nitrate), and inorganic salts ($MgSO_4$, $FeSO_4$, $ZnCl_2$, $GuSO_4$, $MnSO_4$, and K_2HPO_4) were separately added to the basal formulated production medium at concentrations of 2%, 1% and 0.04%, respectively. Based on its culture, morphological and physiological characteristics, and molecular identification, strain Z7 was recognized as *Bacillus amyloliquefaciens*. It was concluded that yeast extract at a concentration of 1% was an ideal nitrogen source, starch at a concentration of 2% functioned best as the major carbon source, and magnesium sulfate at a concentration of 0.04% was an ideal inorganic salt. Ideal pH and temperature for optimum production of CDA by this strain were pH 6 and 37°C. Further research is continuing to find a suitable strain of CDA-producing bacteria for the bioconversion of chitin to chitosan to circumvent environmentally related problems caused during the traditional deacetylation of chitin using the harsh chemical NaOH pyrolysis method.

Table 7 Approximate composition of crustacean shell wastes as a percentage of dry weight [47, 73–75]

Source of chitin		Parameter			
		% Ash content	% Protein	% Chitin	% Lipids
Shrimp	*Penaeus monodon*	23.0	47.4	40.4	1.3
	Pandalus borealis	34.2	41.9	17.0	5.2
	Crangon crangon	27.5	40.6	17.8	9.9
Crab	*Collinectes sapidus*	58.6	25.1	13.5	2.1
	Chinoecetes opilio	40.6	29.2	26.6	1.3
Craw fish	*Procamborus clarkia*	46.6	29.8	13.2	5.6
Krill	*Euphausia superba*	23.0	41.0	24.0	11.6
Prawn	–	29.4	61.6	33.0	1.4

In another study, Kaur et al. [72] reported the isolation of 20 strains of bacteria from soil samples collected from different beaches in Chennai, India. Of these 20 bacterial strains, only two strains (S3, S14) are potent degraders of chitin and they are also good producers of the enzyme chitin deacetylase that catalyzes release of chitosan.

However, crustacean shell wastes are extensively used for the production of chitosan because the percentage availability of chitin is higher compared to other sources. Therefore, the approximate composition of different crustacean shell wastes are given in Table 7.

4 Characterization

Because the properties of chitin straightforwardly and inherently depend on the source of chitin, it is necessary to characterize the properties of chitin before it is further utilized for a specific application. In fact, not every chitin or chitosan sample can be used for the same applications. That is why a complete characterization of the samples is mandatory. The physicochemical characteristics of chitin and chitosan and the determination methods are summarized in Table 8.

In the context of characterization of chitin and chitosan, it is noteworthy that the results may vary depending on the method used. Therefore, it is essential to indicate the characterization method. As chitosan is a natural biodegradable polymer with a wide range of biomedical applications, its biodegradation rate also plays a crucial role. But, there is a dilemma regarding the biodegradation rate of chitosan. Some researchers report that it depends on the distribution of the acetamide groups in the chitosan molecule [92]. However, chitosan with a low degree of deacetylation (DD) tends to degrade more rapidly and vice versa. Therefore, it can be presumed that it is impossible to determine the biodegradation rate from the DD alone.

Table 8 Physicochemical characteristics of chitin and chitosan and the determination methods

Physicochemical characteristic	Determination methods
Degree of deacetylation	Infrared spectroscopy [76–78], first derivative UV-spectrophotometry [79, 80], nuclear magnetic resonance spectroscopy (^{1}HNMR) and (^{13}CNMR) [81–83], conductometric titration [83], potentiometric titration [84], differential scanning calorimetry [85]
Average M_w and/or M_w distribution	Viscometry [86], gel permeation chromatography [87], light scattering [88]
Crystallinity	X-ray diffraction [89]
Moisture content	Gravimetric analysis [90]
Ash content	Gravimetric analysis [90]
Protein	Bradford method [91]

5 Structure of Chitin and Chitosan

Chitin is the second-most abundant natural biopolymer (just after cellulose) on earth and it is found in the structure of a wide number of invertebrates (crustacean exoskeletons, insect cuticles) and the cell walls of fungi, among others [93–95]. Depending on source, chitin is available in three different polymeric configurations, namely α-chitin, β-chitin, and γ-chitin. Usually, α-chitin is extracted from the exoskeletons of crustaceans, particularly shrimps and crabs [96, 97]. β-Chitin can be extracted from squid pens, and γ-chitin from fungi and yeast [98]. α-Chitin is the most common form, whereas β-chitin is more reactive [99] and shows a higher affinity for solvents [100]. β-Chitin is easily converted to α-chitin by alkaline treatment followed by flushing in water [101]. Basically, chitin and chitosan are described as a family of linear polysaccharides consisting of varying amounts of β-(1 → 4)-linked residues of *N*-acetyl-2-amino-2-deoxy-D-glucose (A residues) and 2-amino-2-deoxy-D-glucose residues (D residues). The structure of chitin and chitosan is shown in Fig. 6. Chitosan samples have a low amount of D units and hence the polymer is insoluble in acidic aqueous media. On the other hand, the amount of D units in chitosan samples is high enough to allow the polymer to dissolve in acidic aqueous media.

6 Properties of Chitin and Chitosan

Due to the presence of variable compounds in the natural sources from which chitin and chitosan are extracted, the physical and chemical properties of both chitin and chitosan vary remarkably. Chitosan has been described as a nontoxic, biodegradable, and biocompatible polymer with very interesting biological properties, such as permeation-enhancing and mucoadhesive properties, anticoagulant and antimicrobial activity and so on. The properties of chitosan are greatly affected by the conditions under which it is processed, because it is the process conditions that

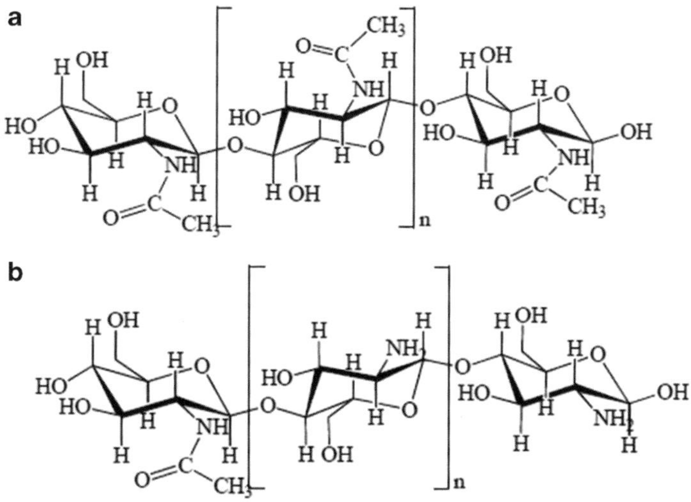

Fig. 6 Chemical structure of 100% acetylated chitin (**a**) and chitosan (**b**)

control the amount of deacetylation that occurs. The degree of deacetylation controls the amount of free amino groups in the polymer chain. The free amino groups give chitosan its positive charge. The amino groups along with the hydroxyl groups gives chitosan its functionality, which allows it to be a highly reactive polysaccharide. Chitosan's positive charge allows it to have many electrostatic interactions with negatively charged molecules. The processing conditions as well as the amount of functional groups created by deacetylation allow for side-group attachment, which then effects crystallinity. This is directly related to the ability of chitosan to solubilize in acidic aqueous solutions, which is an important aspect of its processability [102–104]. The readers are directed to other reviews or articles for more information on the various properties of chitosan [13, 25, 105–109]. A few physical, chemical, and biological properties are listed here:

Physical properties

- Amorphous nature
- Crystallinity
- Viscosity

Chemical properties

- Linear polyamine
- Reactive amino groups
- Reactive hydroxyl groups
- Chelates many transitional metal ions

Biological properties

- Nontoxic
- Biodegradable

- Biocompatible
- Cytocompatible
- Antimicrobial
- Anticholesterolemic
- Anti-oxidant
- Anti-inflammatory
- Analgesic
- Hemostatic
- Fungistatic
- Spermicidal
- Antitumorigenic
- Mucoadhesive
- Stimulates angiogenesis
- Activates macrophages
- Promotes granular tissue and scar formation
- Enhances adsorption
- Binds to mammalian and microbial cells aggressively
- Has a regenerative effect on connective gum tissue
- Accelerates the formation of osteoblasts responsible for bone formation
- Central nervous system depressant
- Immunoadjuvant

7 Limitations of Using Chitin

In spite of having abundant availability of chitin in nature, it is extremely insoluble in dilute aqueous or conventional organic solvents due to its strong intra- and interhydrogen bonds. Therefore, chitin has yet to find large-scale industrial uses. No economical and suitable solvent is known for the dissolution of chitin upon isolation of chitin from the exoskeletons of crustacean shells. The weak solubility of chitin is the main constraint for its usage in biomedicine and biotechnology, even though it possesses versatile biological activities.

8 Chemical Modifications of Chitosan

Over decades, much emphasis has been given to chemical modifications of chitosan, bearing in mind that chitosan is normally insoluble in solutions above pH 6 and requires acid to be protonated [110]. Chitin has two hydroxyl groups, whereas chitosan has one reactive amino group as well as two hydroxyl groups in the repeating *N*-acetyl glucosamine and glucosamine residues, respectively [111]. Therefore, various chemical modifications have been introduced to increase not only the water solubility of chitosan but also to chemically modify these groups for the regeneration reactions, which give rise to various novel biofunctional macromolecular products having the original or new types of organization [24]. The chemical modification of chitosan is a powerful tool for controlling the interaction of the polymer with drugs and thus to enhance the load capability to tailor the release profile of the particles [112]. Chemically modified chitosan improves its bulk properties for the preparation of sustained release systems. There are different ways to chemically modify chitosan. Among them, chemical modification via graft copolymerization seems promising as it can provide a wide variety of molecular characteristics. In view of increasing the water solubility of chitosan, amphiphilic derivatives of chitosan have received much attention by researchers. To prepare such types of compound, hydrophobic moieties are introduced to the chitosan backbone using methodologies such as alkylation, acylation, and graft copolymerization. Many works related to chemical modification of chitosan have been reported in the literature. This section will describe the procedures for some of the chemically modified chitosans.

8.1 Chitosan-g-PEG Copolymer [113]

Under normal conditions, neither the hydroxyl groups nor amino groups of chitosan react with the hydroxyl end groups of poly(ethylene glycol) (PEG). Therefore, the monomethyl ether of PEG (mPEG) is used. Firstly, mPEG is converted into PEG-aldehyde (mPEG-CH=O) by oxidizing it with DMSO and acetic anhydrate [114]. Then, mPEG-CH=O is reacted with chitosan in the presence of $NaCNBH_3$ in order to prepare the grafted materials. In this case, Schiff base is produced first by the reaction of an amino group of chitosan with the aldehyde group of mPEG-CH=O, followed by reductive amination to obtain the copolymer of chitosan and PEG (CS-*g*-PEG). In brief, chitosan (0.5 g) was dissolved in a mixture of aqueous 2% acetic acid (40 mL) and methanol (20 mL) and then a previously prepared aqueous solution of mPEG-CH=O of molecular weight (M_w) 2,000 g/mol (2.9 g) was added dropwise under stirring for 30 min at room temperature [115]. After that, the pH of the chitosan/PEG-aldehyde solution was gradually increased by adding Na_2CO_3 until pH 6. After 1 h, $NaCNBH_3$ (0.183 g) was added and the mixture was stirred for 5 h at 55°C. The precipitate was obtained by pouring the reaction mixture into a saturated ammonium sulfate solution [116]. The precipitate was filtered and dialyzed against aqueous 0.05 M NaOH and water alternately for 96 h using a dialysis membrane bag with a molecular weight cut-off of 12,400. The solutions were frequently changed until the pH of the external water phase reached 7. The material from the inner solution was freeze-dried and washed with ethanol and acetone in order to remove unreacted mPEG in the system. Finally, after drying in a vacuum oven, a white powder of CS-*g*-PEG was obtained. Basically, it is a two-step process. A schematic diagram for the synthesis of CS-*g*-PEG copolymer is shown in Fig. 7.

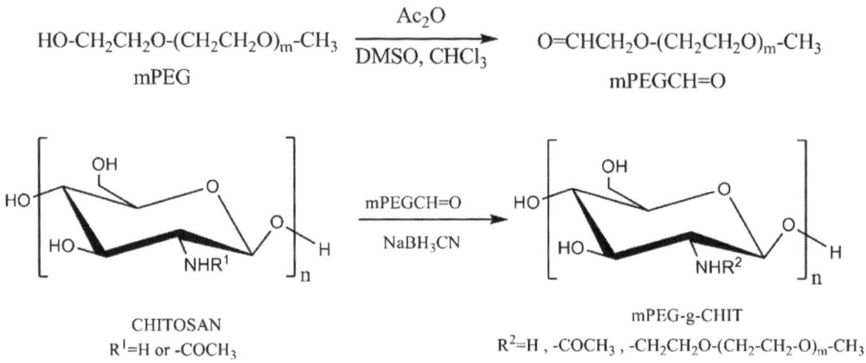

Fig. 7 Synthesis of chitosan-*g*-PEG copolymer [113]

8.2 Chitosan-4-Thiol-Butylamidine [117]

To prepare chitosan-4-thiol-butylamine, chitosan was first selectively depolymerized following the method developed by Huang et al. [118]. Then, the inclusion of thiol groups in the different hydrolyzed chitosans was carried out following the method developed by Bernkop-Schnürch et al. [119]. In brief, 1 g of chitosan was solubilized in 100 mL of acetic acid solution (1%, w/v). The pH of the solution was adjusted to 6.5 with NaOH (1 N). Then, Traunt's reagent (2-iminothiolane) was added in a weight ratio of 5:2 for chitosan:2-iminothiolane. After an incubation period of 24 h at room temperature under continuous stirring, the resulting thiolated polymer was dialyzed using a Spectra/Por 3 membrane with a molecular weight cut-off of 3,500 against different aqueous media. Dialyzed products were freeze-dried and stored at −20°C until use. The resulting polymer was chitosan-4-thiol-butylamidine.

8.3 mPEG-Grafted Phthaloyl Chitosan [120]

Chitosan with a degree of deacetylation of 85% was used for the preparation of phthaloyl chitosan and it was designated as phCS 85. This phCS 85 was further used for the preparation of mPEG-grafted phCS. In brief, phCS 85 (1.0 g) was stirred with mPEG of M_n 2000 (mPEG 2000) (3.04 g, 0.4 mol equivalent to phCS 85) in DMF solution (20 mL). 1-Hydroxy-1*H*-bezotriazole, monohydrate (HOBt) (0.59 g, 3 mol equivalent to mPEG 2000) was added and stirred at room temperature until the solution became clear. 1-Ethyl-3-(3′-dimethylaminopropyl) carbodiimide, hydrochloride (EDC) (0.89 g, 3 mol equivalent to mPEG 2000) was added and stirred overnight. The mixture was dialyzed using a dialysis tubing cellulose membrane with a molecular weight cut-off of 12,400 against distilled water to obtain the product phCS 85-mPEG 2000 in white colloidal solution form. The product was collected in solid form by centrifuging, washing with ethanol, and drying under vacuum.

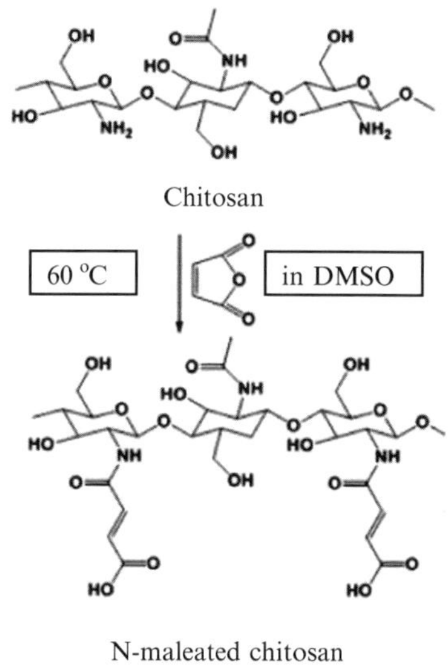

Fig. 8 Preparation of N-maleated chitosan

8.4 *N-Maleated Chitosan*

To synthesize N-maleated chitosan (NMC), Wang et al. [121] adopted the method already published by Lu et al. [122]. In brief, chitosan (2 g) was dispersed in 150 mL of DMSO under stirring, and the maleic anhydride (3.5 g) dissolved in DMSO was added. The reaction was carried out at 60°C for 8 h. The product was precipitated in acetone, filtered, washed with acetone and ether three times, and then dried to obtain NMC. A scheme showing the preparation of NMC is shown in Fig. 8.

9 Importance of Engineering Chitosan-Based Nanoparticles, Nanospheres, and Nanogels

The development of well-organized carrier systems in the nanoscale range for various biomedical applications has been a major challenge, particularly for researchers working in the biomedical domain [123–128]. The development of efficient carrier systems, based on multifunctional polysaccharide-based nanomaterials, combining diagnostic and therapeutic use have recently attracted intensive interest [129]. Polysaccharides are a class of biological macromolecules that play important roles in a wide variety of biochemical and biomechanical functions [130]. Among the several classes of polymeric materials (including polysaccharides) that have become important tools for biomaterial scientists, chitosan possesses unique properties that have recently been used in a wide range of biomedical applications. Unlike other polysaccharides, chitosan is polycationic in nature and, therefore, it is able to bind negatively charged molecules [13]. On the other hand, chitosan offers the possibility of modifying different chemical groups in its structure [131]. In order to realize its full biochemical and biomechanical potential, biomaterial scientists are exploring ways to engineer chitosan and chitosan-based nanostructures that can be used to tune the response of biological systems (namely cells, tissues, etc.) to these materials [132].

9.1 Chitosan-Based Nanoparticles

The importance of chitosan-based nanoparticles is noteworthy in the context of their biomedical applications. Various methods are available for nanoparticles syntheses. In 2007, Zhang et al. [133] reported the synthesis of oleoyl-chitosan (OCH) nanoparticles with an almost spherical shape and a mean diameter of 255.3 nm as drug carriers for doxorubicin (DOX) by oil-in-water emulsification according to a published method [134]. In brief, OCH was dissolved in 0.1 M acetic acid followed by subsequent addition of methylene chloride under stirring. Then, the mixture was homogenized four times. The solution was held under vacuum for 2 h at 20°C to remove methylene chloride and then 1 mL of 0.25% sodium tripolyphosphate solution was added as a crosslinker. DOX-OCH nanoparticles were prepared in a similar way except that DOX was added to the OCH solution before initiating homogenization. The authors evaluated the usefulness of OCH nanoparticles as carriers for DOX by measuring their encapsulation efficiency, drug release profile, and inhibitory rates for different human cancer cells (A549, Bel-7402, HeLa, and SGC-7901) in vitro. The results indicated that the percentage inhibitory rate of DOX solution and of a DOX-OCH nanoparticle suspension for all cancer cells increased with an increase in DOX concentration and extension of treatment, but that DOX-OCH nanoparticles showed better inhibition of cancer cells than DOX itself [133].

In another study, Manaspon et al. [135] reported a drug delivery system using folate-conjugated pluronic F127/chitosan core–shell nanoparticles for delivering DOX to the target cancer cells in a different way. In this case, DOX was encapsulated in pluronic F127 micelle cores in the presence of sodium dodecyl sulfate (SDS) by a self-assembly method, which is a spontaneous process at or above the critical concentration (CMC) of around 0.1% (w/v) [136–139]. This was followed by forming a layer of either chitosan or folate-conjugated chitosan onto the pluronic micelles via an electrostatic interaction. From the study of particle size as well as zeta potential of both micelles, the DOX-loaded micelles were not stable and could easily coagulate, which can be problematic for drug delivery applications. To solve this problem, SDS was used to ameliorate the stability of the micelles. Being an anionic surfactant, SDS can interact with pluronic micelles by binding to the hydrophobic core followed by interaction with hydrophilic corona parts [140]. The initial burst release values of DOX from chitosan and folate-conjugated chitosan nanoparticles were 19.6% and 22.4%, respectively, in the first 24 h followed by a constant sustained release of the drug. The release profiles were quite similar to those obtained for chitosan nanoparticles reported elsewhere [141–143], for which the kinetics often follow the first-order rate law [144]. The cellular uptake rate of the folate-conjugated chitosan nanoparticles was relatively faster than that of the chitosan nanoparticles. This could be due to the more effective folate-mediated endocytosis. Both types of nanoparticles exhibited similar toxicity profiles against MCF-7 cancer cells. Here also, the results indicated that the lower cell viability corresponded to a higher DOX concentration and longer incubation time. These core–shell nanoparticles are very promising as delivery vectors for the treatment of tumors, especially breast cancer, for which folate receptors are overexpressed.

To date, biodegradable thermoresponsive drug carriers are being purposely engineered and constructed with nanometer dimensions [145]. These approaches make it possible to develop smart materials like thermoresponsive drug delivery vehicles. Recently, poly(*N*-vinyl caprolactam) (PNVCL) has superseded the most commonly used pH and temperature-sensitive polymer i.e. poly(*N*-isopropylacrylamide) (PNIPAAm) because of its superior thermoresponsive nature, complexation ability, and biocompatibility [146]. The most important advantages of these thermoresponsive polymeric nanomaterials include their high tunability, through the switch on and off mechanism, and good biodegradability because of the conjugation with a biocompatible chitosan moiety [146]. Surprisingly, at present there are no studies concerning the use of these nanoformulations for cancer therapy. However, Rejinold et al. [146] reported the development of a nanoformulation of 5-fluorouracil (5-FU) with biodegradable thermoresponsive chitosan-*g*-PNVCL nanoparticles (TRC NPs) produced by an ionic crosslinking method. 5-FU, a pyramidine analog that interferes with thymidylate synthesis against solid tumors [147], has a broad spectrum of activity but possesses a short biological half-life due to rapid metabolism, incomplete and non-uniform oral absorption due to rapid metabolism by dihydropyramidine dehydrogenase, and non-selective action against healthy cells. This problem can be overcome by developing appropriate nanocarrier systems that can efficiently release 5-FU to the target sites. Rejinold et al.146 showed that the 5-FU drug release was more prominent above its lower critical solution temperature (LCST) (38°C) compared to release below the LCST. The 5-FU-TRC NPs were toxic to PC3, KB, and MCF7 cancer cells, which indicated the inhibition to cancer cells. Therefore, it can be said that the novel TRC NPs could be used as promising cancer drug delivery vehicles when subjected to a temperature increase using external sources.

The ionic gelation method has received tremendous attention in recent years for the preparation of nanocarriers for low molecular drugs [148]. As far as chitosan-based nanoparticles are concerned, sodium tripolyphosphate (TPP) is usually used as ionic crosslinking agent. Chitosan-based nanoparticles can also be prepared using self-assembly methods. In this case, polyanionic compounds are selected and then allowed to react separately with chitosan and modified chitosan. Electrostatic interactions between oppositely charged polyelectrolytes lead to the formation of nanoparticles. The electrostatic interactions are instrumental in determining the structure and function of living organisms, biopolymers, and drug delivery systems (DDSs). Oppositely charged polyelectrolytes are capable of forming stable intermolecular complexes. The structures formed by opposite charges are normally more stable than neutral block copolymers micelles dissociating upon dilution or with a slight change in the external conditions. Grenha et al. [149] reported the preparation of chitosan-based nanoparticle-loaded mannitol microspheres as a candidate to transport therapeutic protein-loaded nanoparticles to the lungs. In this study, BSA-loaded chitosan nanoparticles were first prepared by the interaction between chitosan and TPP using an ionic gelation method. Next, these nanoparticles were suspended in mannitol for spray-drying to form dry powders of BSA-loaded chitosan/TPP nanoparticles. The distribution of chitosan nanoparticles and mannitol in the microspheres was characterized using confocal laser scanning microscopy (CLSM), X-ray photoelectron spectroscopy, and time-of-flight secondary ion mass spectroscopy. Zhang et al. [150] reported that polyionic hydrogels prepared by ionic gelation have the advantage of creating an environment that favors the stabilization of bioactive agents. In an another study, Fan et al. [151] reported the preparation of monodisperse, low molecular weight (LMW) chitosan nanoparticles by a novel method based on ionic gelation using TPP as a crosslinking agent. The objective of this study was to solve the problem of preparation of chitosan/TPP nanoparticles with a high degree of monodispersity and stability, and to investigate the effect of various parameters on the formation of LMW chitosan/TPP nanoparticles. It was found that the particle size distribution of the nanoparticles could be significantly narrowed by a combination of decreasing the concentration of acetic acid and reducing the ambient temperature during the crosslinking process. The optimized nanoparticles exhibited a mean hydrodynamic diameter of 138 nm with a polydispersity index (PDI) of 0.026 and a zeta potential of +35 mV. The nanoparticles had good storage stability at room temperature for up to at least 20 days. Katas and Alpar [152] prepared chitosan nanoparticles using two methods of ionic crosslinking: simple complexation and ionic gelation using TPP. Both methods produced nanosized particles (<500 nm), depending on the type, molecular weight, and concentration of chitosan. In the case of ionic gelation, the TPP weight ratio and pH also affected the particle size. Papadimitriou et al. [113] demonstrated the production of CS-*g*-PEG nanoparticles by the ionic gelation method using two crosslinking agents, TPP and poly(glutamic acid) (PGA). According to them, no other research has yet been published using PGA as ionic crosslinking agent for CS-*g*-PEG nanomaterials in order to create protein nanocarriers. In brief, aqueous CS-*g*-PEG solutions at different concentrations (0.5, 1.0, 1.5, and 2.0 mg/mL) were first prepared at pH 3.5. Aqueous TPP or PGA (0.5 mg/mL final concentration) were premixed with BSA stock solution (2 mg final drug amount to the samples) and then added into the aqueous CS-*g*-PEG solutions at a rate of 1 mL/min. The obtained nanoparticles were collected by centrifugation at 32,000 rpm for 50 min. Both the crosslinking agents had a great influence on the particle size, although all nanoparticle samples showed a unimodal size distribution. It was found that a CS-*g*-PEG:TPP ratio of 2:1 gave the smaller nanoparticle sizes (177 $\pm$ 7 nm) whereas CS-*g*-PEG: PGA at 2:1 gave the bigger nanoparticle sizes (456 $\pm$ 8 nm). Therefore, the production of nanoparticles with lower particle sizes is attributed to the high reactivity of TPP compared to PGA.

In another study, Yao et al. [153] reported a mild method for the preparation of PEGylated carboxymethylchitosan (CMCTS) nanoparticles via amino groups grafted or crosslinked with mPEG-aldehyde or PEG-bisaldehyde in aqueous media. Synthesis schemes for mPEG-*g*-CMCTS and PEG-CMTCS are depicted in Fig. 9. In brief, mPEG-*g*-CMCTS (Fig. 9a) was synthesized by alkylation of CMCTS followed by Schiff base formation. CMCTS was dissolved in 100 mL water and mPEG-aldehyde added to the solution. The solution was then adjusted to pH 7 with saturated sodium carbonate. The reaction mixture was stirred at about 5°C for 24 h. The solution containing CMCTS nanoparticles was purified by dialysis for 3 days against distilled water and then freeze-dried. PEG-CMCTS was prepared by a similar method to mPEG-*g*-CMCTS, as shown in Fig. 9b. The differences were that PEG was used instead of mPEG and it was crosslinked with CMCTS. The particle size of mPEG-*g*-CMCTC measured by transmission electron microscopy (TEM) varied in the range of 300 nm to 1.1 μm, depending on the graft volume of mPEG 2000. The average size of PEG-CMCTS nanoparticles measured by DLS was 122–500 nm, depending on the molecular weight of PEG. Schiff bases contain a chemical bond that is acid-sensitive at pH 5.4, which is consistent with the pH of tumor tissue. These nanoparticles might give the antitumor drug passive targeting after combining. Therefore, they have great potential as antitumor drug carriers, especially for drug delivery and release.

It is known that exposure of chitosan to γ-irradiation reduces the molecular weight of chitosan, but there has been no report about nanoscale chitosan products being created by γ-irradiation. Pang et al. [154] reported the preparation of deoxycholate-chitosan with the size of 200–600 nm and showed that it had the ability to encapsulate tocopherol acetate and steric methyl ester. As nanoscale material is defined as being smaller than 100 nm, a systematic protocol to reduce and control the particle size of chitosan still requires further studies to achieve not only a simple and effective preparation protocol but also a superior potential product. Pasanphan et al. [155] investigated the systematic preparation of chitosan nanoparticles in the potential range 1–100 nm using γ-irradiation. They showed the effect of irradiation conditions in terms of the physical form of chitosan (i.e., flake, colloidal, and acidic solution) and γ-ray dose. Interestingly, heterogeneous chemical conjugation of deoxycholic acid onto 10 kGy-irradiated colloidal chitosan resulted in particle sizes as small as 50 nm.

Ultrasonication is a common tool for the preparation and processing of polymer nanoparticles. It is particularly effective in breaking up aggregates and in reducing the size and polydispersity of nanoparticles. The physical stability and in vivo distribution of nanoparticles are affected by their mean size, polydispersity, and surface charge density. Despite the widespread applications of ultrasonication in nanotechnology, its effects on chitosan nanoparticles are not well understood.

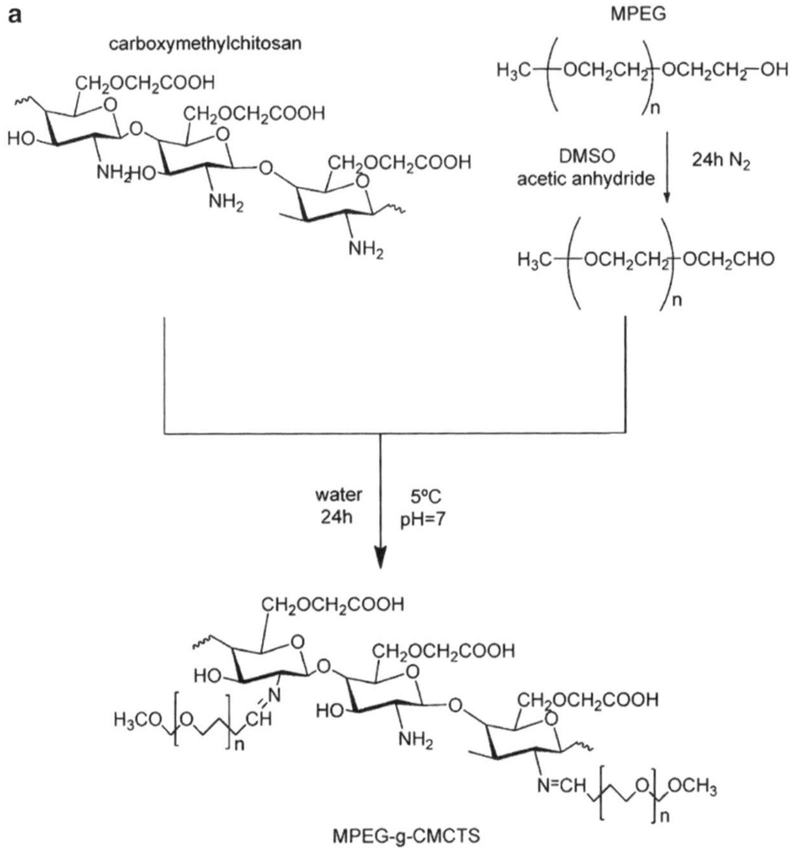

Fig. 9 (continued)

As far as ultrasonication is concerned for the preparation and processing of chitosan nanoparticles for drug delivery applications, high-intensity ultrasonication is not recommended [156].

9.2 Chitosan-Based Nanospheres

Natural polysaccharides such as chitosan and alginate are promising materials for their applications, particularly in drug delivery, due to their favorable biocompatibility, biodegradability, pH sensitivity, and mucoadhesive properties [157, 158]. Covalently crosslinked polysaccharide-based nanospheres are quite stable compared with other several other types of polysaccharide-based nanospheres. The conventional methods for preparation of covalently crosslinked polysaccharide-based nanospheres include emulsion crosslinking, solvent evaporation, spray

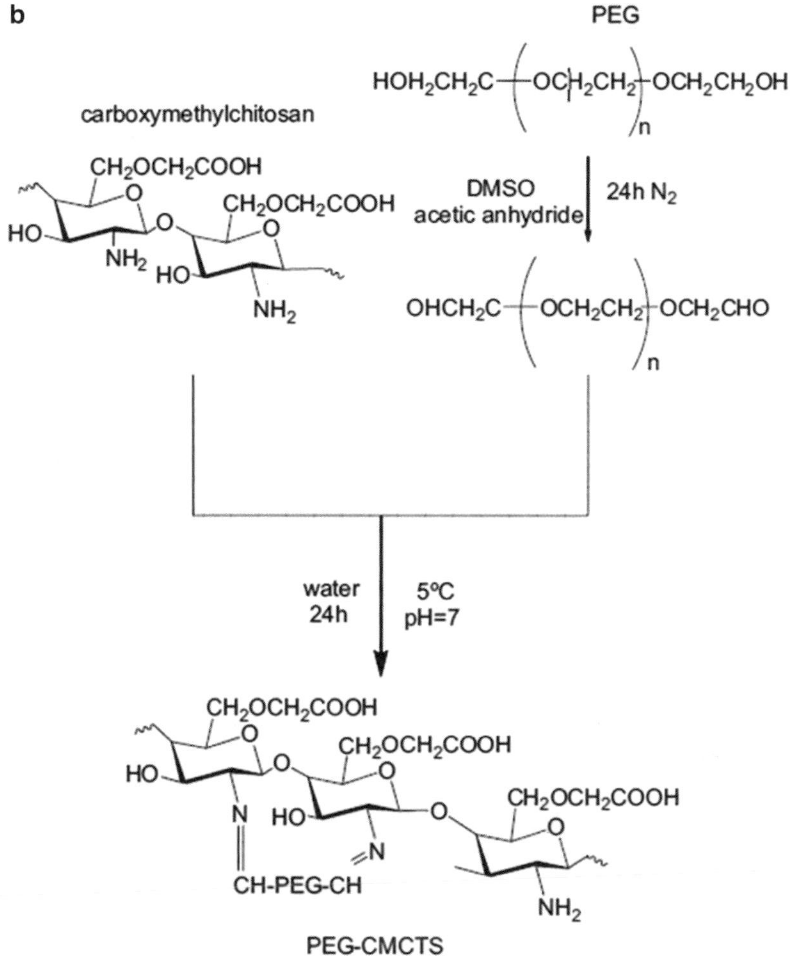

Fig. 9 (**a**) Synthesis of mPEG-*g*-carboxymethylchitosan [153]. (**b**) Synthesis of PEG-*g*-carboxymethylchitosan [153]

drying, and thermal crosslinking [159]. These methods usually involve the use of organic solvents, surfactants, and an elevated preparation temperature, which may lead to loss or reduction in bioactivity of some drugs and cause continuing toxic effects [160]. Therefore, to avoid the use of organic solvents, researchers are trying to prepare crosslinked polysaccharide-based nanospheres with narrow size distribution in aqueous media at room temperature. Wang et al. [121] reported the preparation of chitosan–cyclodextrin nanospheres through in situ formation during Michael addition in an aqueous solution at room temperature. Before preparing these nanospheres, both chitosan and cyclodextrin had been chemically modified.

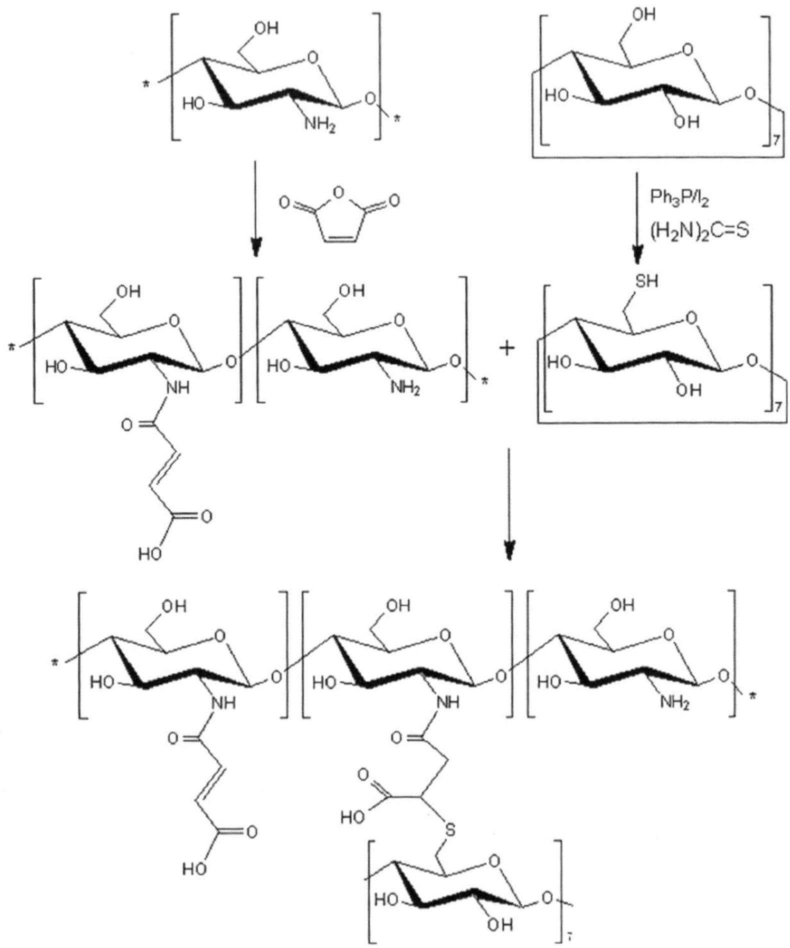

Fig. 10 Synthetic route for the preparation of chitosan–cyclodextrin nanospheres [121]

The synthesis part pertaining to chemical modification of chitosan (i.e., NMC) has been described in Sect. 8.4. On the other hand, thiol-functionalized β-cyclodextrin has been used to crosslink the NMC. Apart from the biodegradability and biocompatibility of β-cyclodextrin, it has the ability to form inclusion complexes with a variety of neutral and charged molecules in aqueous solutions. Particularly for drug delivery applications, the β-cyclodextrin moieties could effectively encapsulate drug molecules and thus improve the solubility of hydrophobic drugs, and enhance the stability and bioavailability of the drugs [161, 162]. A synthetic route for the preparation of chitosan–cyclodextrin nanospheres is shown in Fig. 10. As measured by particle sizer, under neutral and basic conditions, the nanospheres have a negative surface charge with a zeta potential of 34.6 mV due to deprotonated carboxyl groups, which give good dispersibility and high colloidal stability in an aqueous solution.

Further, the negatively charge nanospheres with a size less than 300 nm could provide attractive forces for positively charged drugs. Since many anticancer drugs are positively charged at physiological conditions, the nanospheres prepared in this study are suitable for loading of these drugs. In another study, Choochottiros et al. [120] showed that the surface charge of amphiphilic chitosan nanospheres was negative, resulting in good dispersion at neutral to high pH, but significant precipitation at low pH.

Aiping et al. [163] demonstrated the synthesis of a novel biocompatible *N*-succinyl chitosan (NSCS) with well-designed structure. NSCS can self-assembly in regular nanospheres in distilled water. The NSCS nanospheres are obtained spontaneously under very mild conditions without the need for high temperature, organic solvent, surfactant, or special experimental technology. From X-ray diffraction analysis, it can be known that the crystal behavior of NSCS is weaker than that of chitosan, which is due to the decrease in the intermolecular hydrogen bonding. As a result, NSCS is easy to disperse in distilled water to obtain a stable and transparent colloidal dispersion. TEM was used to check the morphology of the colloid of NSCS and the result shows that NSCS colloids possess regular spherical morphology. The size of the NSCS nanospheres was in the range of 50–100 nm. The NSCS nanospheres have great potential to be used as a novel drug matrix for controlled-release drug delivery.

In recent years, magnetic DDSs have received increasing attention from researchers due to their ability to deliver drugs to the targeted site with the help of an external magnetic field [164–167]. Current approaches to magnetic DDSs usually contain three main steps. First, superparamagnetic nanoparticles are manufactured and second the magnetic nanoparticles are dispersed in shell monomer or polymer. Finally, the DDS is formed after polymerization of shell monomer or crosslinking of shell polymer via chemical bonds. However, surfactants and organic solvents are inevitably used during the synthesis, which brings potential toxicity to the synthesized magnetic DDSs. Therefore, Wang et al. [168] adopted a facile method for the synthesis of chitosan–(acrylic acid) magnetite nanospheres (CS–AA-Fe_3O_4). In this method, CS–AA-Fe_3O_4 is formed in a completely aqueous system and no surfactant is used. Moreover, the CS–AA polyelectrolyte complex was formed via electrostatic interactions, rather than chemical bonds. The size of the CS–AA-Fe_3O_4 nanosphere as determined by laser particle size analyzer was 130 nm before swelling in PBS solution. After swelling, the hydrodynamic size of CS–AA-Fe_3O_4 was increased to 330 nm, which makes it potentially interesting as a DDS.

Of late, significant interest has been paid to hollow polymeric nanostructures (HPNSs) because of the new functionalities and unique physiochemical properties anticipated from polymeric materials at the nanoscale. The development of HPNSs is reflected by the rapid increase in the number of scientific publications and patents on this subject in recent years. HPNSs in the size range of 1–1,000 nm with spherical or cylindrical geometries are of special interest. Spherical HPNSs generally include hollow polymeric nanospheres (HPNSPs), vesicles, or nanopolymersomes. Cylindrical HPNSs are basically polymeric nanotubes (PNTs). In view of possible biomedical applications, HPNSPs are potentially useful as encapsulates

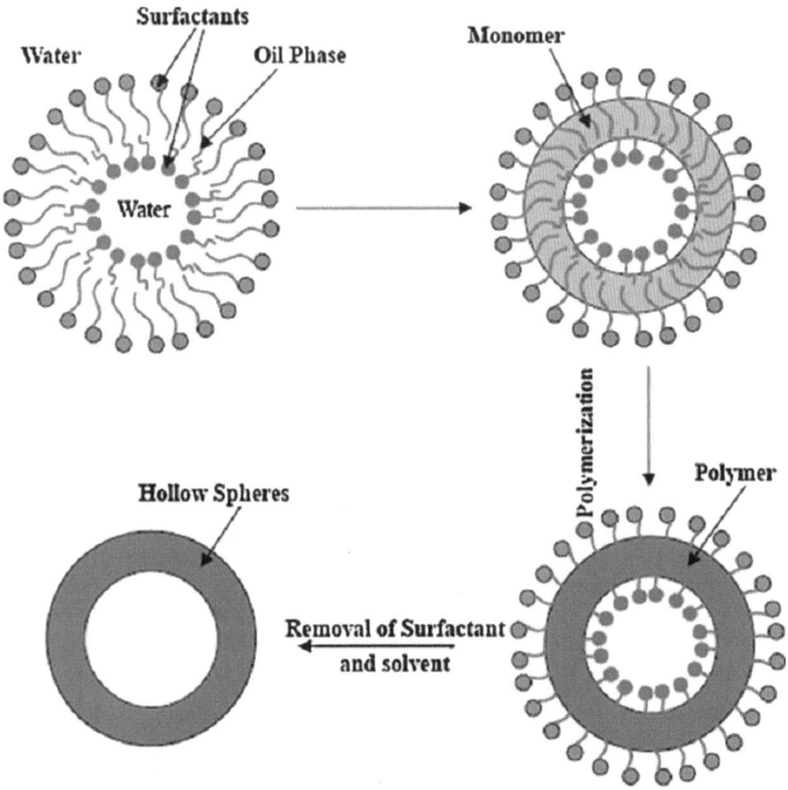

Fig. 11 Preparation of hollow polymeric nanospheres via water–oil–water emulsion polymerization [175]

for drugs, enzymes, protein, genes, and catalysts [169–174]. HPNSPs have a large surface-to-volume ratio, low effective density, and many other advantageous properties in comparison to cylindrical HPNSs, which makes them suitable for applications in encapsulation, catalysis, controlled drug delivery, and imaging. HPNSPs from various polymeric materials have been developed. Here, some hollow chitosan-based nanospheres will be described. HPNSPs have been prepared by suspension polymerization and emulsion polymerization, and from core–shell nanoparticles, self-assembly, template-directed synthesis, and electrospinning. Among them, water–oil–water (W/O/W) emulsion polymerization is a convenient and economical means for the preparation of hollow polymeric micro- or nanospheres [175]. However, this process is less suitable for preparing monodisperse particles with particle size of less than 100 nm. A schematic representation of the preparation of HPNSPs via W/O/W emulsion polymerization is depicted in Fig. 11. The pH-sensitive chitosan/poly(acrylic acid) (PAAC) hollow nanospheres with diameters ranging from 70 to 500 nm were prepared by emulsion polymerization of acrylic acid (AAC) in the presence of chitosan [176]. The AAC molecules are bound electrostatically to chitosan to form the shell of the micelles. Upon polymerization, PAAC interacts with chitosan at the interface of the micelles. The shrinkage of spheres brought about by electrostatic attraction between chitosan and PAAC, as well as the expansion of the spheres due to electrostatic repulsion among the chitosan molecules, causes phase separation between the polymer and the solvent resulting in the formation of HPNSPs.

The template method is also frequently used for the fabrication of hollow spheres. The appropriate selection of template materials plays a crucial role in controlling the core size of the hollow spheres. Further, hollow structures can be easily obtained after removing the template by dissolution, evaporation, or thermolysis. This method requires large amount of surfactants, which can modify the surface of the template to make the template active to interact with shell substance. Wang et al. [177] reported the production of chitosan hollow nanospheres (HNPs) by employing uniform poly (D,L-lactide)–poly(ethylene glycol) (PELA) nanoparticles as templates. Chitosan was adsorbed onto the surface of PELA nanoparticle templates through the electrostatic interaction between the sulfuric acid groups from SDS on the templates and the amino groups of the chitosan. Subsequently, the core-coated structure of chitosan–PELA nanospheres was obtained, with the adsorbed chitosan layer being further crosslinked with glutaraldehyde. After the removal of the PELA cores with acetone, chitosan HNPs were achieved. The mean size of initial PELA nanospheres was 180 ± 2 nm whereas that of chitosan–PELA nanospheres was 280 ± 12 nm and the zeta potential changed from -31.5 to 41.3 mV. The change in zeta potential and size were noteworthy, which indicated that chitosan was absorbed on the surface of PELA nanospheres. Thus, it is possible to form hollow chitosan nanospheres by removing the PELA core.

In order to avoid the post-treatment processes, including organic solvent dissolution or calcination at elevated temperature, for the production of hollow spheres, Deng et al. [178] reported a simple method for fabrication of monodispersed silica hollow nanospheres in a one-step, one-medium process. The nanospheres were then decorated with chitosan using a crosslinking reaction with (3-glycidyloxypropyl) trimethoxysilane (GTPMS) to produce pH-sensitive chitosan–silica hollow nanospheres (CS–SiO_2 HNPs). GPTMS was first used to modify the SiO_2 HNPs in an acidic ethanol medium in which GPMTS reacts with the silanol groups on the SiO_2 surface via the formation of Si–O–Si bonds. The chitosan solution was then added to accomplish crosslinking on the surface of the SiO_2 HNPs [179, 180]. The hydrodynamic diameters (D_h) of SiO_2 and CS–SiO_2 HNPs determined from the dynamic light scattering measurement were 242.8 and 263.7 nm respectively. The zeta potential measurement showed that the SiO_2 HNPs had a negative potential of about -35 mV, whereas the zeta potential of the CS–SiO_2 HNPs increased to +24.7 mV due to the cationic polysaccharide-chitosan decoration. These results demonstrate that chitosan had been successfully introduced to the SiO_2 HNPs.

9.3 Chitosan-Based Nanogels

Polysaccharide-based nanogels are more advantageous than bulk polysaccharide-based hydrogels as a consequence of their high water content, functionality, biocompatibility, tunable size from submicrons to tens of nanometers, large surface area for multivalent bioconjugation, and interior network for the incorporation of therapeutics. Nanogels can be defined as internally crosslinked macromolecular systems wherein the size of hydrogel nanoparticles ranges between 10 and 200 nm [181]. Some reviews have mainly focused on the structure and function of other polysaccharides and give little emphasis to chitosan and their engineered nanoscale assembly for biomedical materials [132, 182, 183]. In this section, only chitosan-based nanogels will be discussed.

Current approaches for the preparation of crosslinked micro- or nanogels can be largely categorized into physical crosslinking, chemical crosslinking, and heterogeneous controlled/living radical crosslinking polymerization (CRCP) [182]. Several methods such as heterogeneous polymerization, continuous extrusion for starch-based biolatex nanoparticles, precipitation in water, micromolding and microfluidic preparation, spray drying, and supramolecular self-assembly and self complexation have been developed for the successful preparation of micro- and nanogels. One general method involves aqueous homogeneous gelation based on chemical and physical crosslinking in water; the gelation is generally conducted in dilute solution in order to prevent macroscopic gelation that leads to bulk hydrogels. As far as physical crosslinking is concerned, it has primarily two advantages, i.e., reversibility and no requirement for other chemicals for the connectivity of chains. In many experiments, the electrostatic interactions of polysaccharide-based polymers with various polyelectrolytes in water have been explored. These polyelectrolytes include polyethyleneimine (PEI), sodium tripolyphosphate (TPP), poly(ethylene oxide) (PEO)-*b*-poly(*N*,*N*,-dimethylaminoethyl methacrylate) block copolymer, and biopolymer-based polyelectrolytes to yield chitosan-based nanogels [184–189]. To prevent dissolution of the nanogel in the aqueous environment, chemical crosslinks involving the formation of covalent bonds are preferred over physical crosslinking. Chitosan-based nanogels were prepared by a covalent chemical crosslinking, based on a carbodiimide coupling of chitosan with an oligo(ethylene glycol) dicarboxylic acid as a water-soluble crosslinker, in water. The resulting nanogels had a diameter of 4–24 nm, as shown by TEM [190]. Chitosan nanogels with a diameter of 70–80 nm were also prepared by the reaction of chitosan with ethylenediaminetetraacetic dianhydride (EDTAA) in water. The stable pH-responsive nanogels enabled a reversible switch of their surface charge upon pH change so that it was positive at pH <4.8 and negative at pH >5.2. These characteristics suggest that the nanogels can be effective candidates for the encapsulation of pH-sensitive anticancer agents such as highly water-insoluble camptothecin [191].

In a recent study, Maggi et al. [192] reported the use of polyion complex micelles (PIC) as nanoreactors for template crosslinking for the preparation of chitosan nanogels. In brief, PIC were formed by interacting chitosan with a block copolymer poly(ethylene oxide)-*block*-poly[sodium 2-(acrylamido)-2-methyl propane sulfonate] (PEO-*b*-PAMPS) synthesized by single-electron transfer living radical polymerization (SET-LRP). Numerous studies have shown that the size and shape of the formed PIC nanoparticles significantly depend on the polymerization degree of PEO block, medium conditions, and molar ratio between the polyelectrolyte blocks. By controlling these parameters, small and highly uniform in size PIC can be obtained with a diameter of 20–50 nm and a coacervate core of 5–15 nm [193–195]. In this study, PIC with a small size of about 50 nm and low polydispersity were obtained up to 5 mg/mL. After crosslinking of chitosan with genipin, the nanoreactors were dissociated by increasing the ionic strength by adding NaCl in order to obtain the free nanogels. The size of the smallest nanogels was about 50 nm in the swollen state and 20 nm in the dry state. The amount of genipin used during reticulation was an important parameter in modulating the size of the nanogels in solution.

Shen et al. [196] reported the synthesis of chitosan-based luminescent/magnetic (CLM) nanomaterials by direct gelation of chitosan, CdTe, and supermagnetic iron oxide into hybrid nanogels for insulin delivery, cell imaging, and research into antidiabetic dietary supplements. The influence of various process parameters was studied to see their effect on the size of the hybrid nanogels. It was found that with an increase in reaction temperature, the size of the resulting hybrid nanogels decreased because of the increasing mobility, weakened entangling degree, and decreasing friction force between the chitosan molecules. However, the rigidity of the product synthesized at high temperature was not enough to maintain a stable shape due to the weak entangling degree. In addition, the photoluminescence intensity of the resulting hybrid nanogels decreased due to changes in the surface trap sites of CdTe quantum dots at high temperature [197].

Oridonin (ORI), extracted from the plant *Rabdosia rubescens*, is a potent anticancer agent in Chinese traditional medicine. Experimentally and clinically, ORI has proven that it is effective against a variety of tumors and cancer cell types, namely liver, prostate, breast, and cervical cancer cells; non-small lung cancer cells; acute promyelocytic leukemia, and glioblastoma multiforme [198–200]. However, its clinical applications against cancers have been impeded by its low therapeutic index induced by the poor solubility and nonspecific systematic distribution. Therefore, it is of great necessity to develop an alternative carrier of ORI. Duan et al. [201], reported the development and characterization of pH-responsive and biocompatible ORI-loaded chitosan-*g*-PNIPAAm-based nanogels for a tumor extracellular targeting DDS using a self-assembly method. In brief, copolymers (500 mg) and ORI (50 mg) were dispersed in 50 mL distilled water and the dispersion was sonicated using a probe-type sonifier at 100 W in an ice bath for 2 min. The resultant ORI-loaded nanogels were collected by centrifugation and lyophilized for storage and use. The drug-encapsulation efficiency and drug-loading efficiency were 86.3% and 5.34%, respectively. From the particle size analysis, it was found that the average hydrodynamic diameter (R_d) of the ORI-loaded nanogels was 99.6 nm, whereas the R_d of the blank nanogels was 81.2 nm, indicating that ORI molecules were successfully encapsulated and that these entrapped ORI increased the size of the ORI-loaded nanogels. In vitro drug release studies indicated that ORI-loaded nanogels exhibited a pH-triggered fast drug release under slightly acidic conditions. Moreover, the MTT assay and cellular morphological analysis demonstrated that the ORI-loaded nanogels could enhance the antitumor activity in an acidic environment. No significant cytotoxicity, however, was observed with the blank carriers themselves. Therefore, this nanogel-based delivery system might be exploitable in tumoral acidic extracellular pH targeting for hydrophobic anticancer drugs.

To deliver anticancer drugs at the targeted sites, the development of both biodegradable and biocompatible thermoresponsive carriers has received significant attention. Although, various polymers like gelatine [202], poly(vinyl alcohol) [203], PNIPAAm, poly(*N*-isopropylmethacrylamide) [204, 205], hydroxypropyl cellulose [206], etc. have been explored for use as temperature-triggered DDSs. Among them, PNIPAAm seems to be a promising thermosensitive DDS because its transition temperature (approximately 32–34°C) can be modulated near human body temperature [206–208]. However, one of the key challenges with the use of thermoresponsive PNIPAAm-based hydrogels is to modulate its LCST to around body temperature and preferably up to 42–43°C, a temperature used for hyperthermia treatment of cancer [209]. But, there is a issue with PNIPAAm in that it is not biodegradable. Therefore, PNIPAAm gel particle surfaces need to be modified or conjugated with some other biodegradable polymers in order to develop biodegradable DDSs [210]. In an attempt, Jaiswal et al. [211] demonstrated the synthesis and characterization of temperature-optimized and magnetically modalized PNIPAAm–chitosan-based nanohydrogels for possible application in hyperthermia for cancer treatment. The particle size analysis data showed that the room temperature hydrodynamic diameters for all the hydrogel particles synthesized in different weight proportions of NIPAAm and chitosan were in the size range of 100–300 nm. The SEM and AFM micrographs of nanohydrogels and magnetic nanohydrogels showed that there was a reasonable decrease in hydrogel size with an increase in the chitosan to NIPAAm ratio, which was consistent with the hydrodynamic diameter measurements obtained by DLS. This decrement might be attributed either to the crosslinking of chitosan to PNIPAAm or due to its role as a surfactant, or both [212, 213]. The observed change in LCST was attributed to hydrophilic modifications of PNIPAAm by water-soluble *N,N′*-methylene-bis-acrylamide (MBAAm) [213] and it was conceived that the associated shift in LCST was due to a decrease in mobility of water molecules trapped in crosslinked chain pores, defined as bound water molecules [214, 215]. Upon reduction in hydrogel size, pores of the mesh diminish in size and, in turn, bound water volume decreases. But, they are held there for a longer time with respect to temperature and that requires a higher amount of heat energy to expel them. Consequently, the LCST shifted in an upward direction. So, it is suggested that chitosan also plays the role of crosslinker as well as serving as a surfactant. Further, an increase in LCST due to the presence of iron oxide MNPs is a very important result and it can be said that dipole–dipole interactions might prevent the collapse of crosslinked polymer segments.

The inclination of the researchers during the last two decades towards the development and implementation of new environmentally responsive smart materials, biomimetics, biosensors, artificial muscles, DDSs, and chemical separations to name a few has given tremendous importance to nanogels instead of bulk hydrogels [216]. Hydrogel transitions, i.e., volume changes that occur in response to changing environmental conditions such as temperature, pH, ionic strength, solvent composition, electric stimuli, and γ-radiation have been studied. Therefore, the nanogels can be designed to undergo large changes in their chemical, mechanical, optical, and electrical make-up in response to a chemical stimulus, biomolecular interaction, or electromagnetic field. Natural biopolymers have been widely used to prepare responsive hydrogels for biomedical application due to their biocompatibility, low toxicity, and a high content of functional groups [217–222]. Although significant benefits have been achieved, few studies have yet been conducted on preparation of natural biopolymer-based bionanomaterials for simultaneous sensing, imaging diagnosis, and therapy. One study describes a multifunctional system on the basis of immobilization of quantum dots (QDs) into the hydroxypropyl cellulose-polyacrylic acid (HPC-PAA) semi-interpenetrating (semi-IPN) nanogel for simultaneous optical pH-sensing, tumor cell imaging, and pH-regulated delivery of anticancer drugs [223]. Despite the exciting conception, several key issues remain unaddressed in this system. To address the key issues, Wu et al. [224] further reported the development of a new class of chitosan-based responsive hybrid nanogels ($R_h < 100$ nm) with CdSe QDs (3.2–3.8 nm) in-situ immobilized in chitosan–poly(methacrylic acid) (chitosan–PMMA) semi-IPN networks for integration of optical pH-sensing, tumor cell imaging, and controlled drug delivery. The mean diameter of chitosan–PMAA nanogels varies from 174.5 to 74.7 nm when the MAA:chitosan ratio changes from 2.10 to 0.53. These hybrid nanogels exhibited marked swelling at both high and low pH values, but a minimum size at pH 5.0–5.5. The swelling of the hybrid nanogels containing pH-responsive groups is governed by the internal osmotic pressure due to the mobile counter-ions contained within the particles, which balance the internal electrostatic repulsion. At high pH values (e.g., >8.0), the nanogels remained nearly at a maximum swelling degree due to the high degree of ionization of the $–COO^-$ groups of the PMMA chains. With a decrease in pH, the $–COO^-$ groups gradually protonate to –COOH, which not only enhance the hydrogen bonding interactions between the chitosan and PMMA chains, but also reduce the Coulombic repulsion within the particles, resulting in a gradual decrease in the size of hybrid nanogels. It is worth mentioning that the fluorescence of the hybrid nanogels does not completely quench, even at extreme pH environments; thus, the hybrid nanogels can be used for cell labeling under physiologically important pH conditions.

10 Biomedical Applications of Chitosan

The vast commercial potential for applications using chitin and chitosan biopolymers has only recently been understood by the scientific community. Commercial uses for chitin and chitosan for biomedical applications include pharmaceuticals and other aspects of health care such as wound care products, medical implants, etc. Applications that take advantage of the biocompatibility and bioactivity of chitin and chitosan represent the numerous useful applications of these biopolymers. For instance, chitosan prevents the formation of scar tissue by inhibiting the formation of fibrin strands in wounds. As a consequence of this property, chitosan, along with chitin, can be used to form sutures, dressings, and other healing agents with properties not found in competing products. In addition, because lysozyme present in wounds breaks down chitin, the need to remove sutures and wound dressings made of chitin after application is avoided. Further, the advent of nanotechnology widens the scope for using chitin and chitosan as wound dressing materials.

Recently, biopolymer-based nanomaterials in the form of microgels, nanogels, and nanoparticles have received a great deal of interest in tissue engineering and drug delivery applications [225–227]. The ability of nanoparticles to manipulate the molecules and their structures has revolutionized the conventional DDS. Chitosan is an excellent biopolymer for preparation of micro- and nanoparticles [228]. Chitosan nanoparticles, because of their biodegradability, biocompatibility, better stability, low toxicity, and simple and mild preparation methods, offer a valuable tool for developing novel DDSs. Li et al. [229] reported an innovative smart DDS based on magnetic and fluorescent multifunctional chitosan nanoparticles, which combined magnetic targeting, fluorescent imaging, and stimulus-responsive drug release properties into one DDS. Water-soluble superparamagnetic Fe_3O_4 nanoparticles, CdTe QDs and pharmaceutical drugs were simultaneously incorporated into chitosan nanoparticles by crosslinking the composite particles with glutaraldehyde. The system showed superparamagnetic and strong fluorescent properties, and was used as a controlled drug release vehicle, which showed pH-sensitive drug release for a prolonged period of time. In another study, Wilson et al. [230] demonstrated tacrine-loaded chitosan nanoparticles prepared by spontaneous emulsification. The prepared particles showed good drug-loading capacity. The in vitro release studies showed an initial burst release followed by a continuous and slow release of the drug. Coating of nanoparticles with Polysorbate 80 slightly reduced the drug release from the nanoparticles. The chitosan nanoparticles have also been reported to have major applications in parentral drug delivery, per-oral administration of drugs, nonviral gene delivery, vaccine delivery, ocular drug delivery, electrodeposition, brain-targeting drug delivery, stability improvement, mucosal drug delivery, controlled drug delivery of drugs, tissue engineering and for the effective delivery of insulin [231]. Recently, chitosan-based nanoparticles for oral controlled delivery of insulin and other therapeutic agents have been reviewed elsewhere [232]. A snapshot of chitosan-based nanomaterials and their biomedical applications is shown in Table 9.

Table 9 Chitosan-based nanomaterials and their biomedical applications

Chitosan-based nanomaterial	Method of preparation	Application	Reference
Chitosan-*g*-PEG nanoparticles	Ionic gelation and self-assembly	Protein delivery	[113]
Heparinized chitosan/poly(γ-glutamic acid) nanoparticles	Self-assembly	Multifunctional delivery of fibroblast growth factor and heparin	[233]
Chitosan–dextran sulfate nanoparticles	Self-assembly	Insulin delivery to protect protein in an oral delivery application	[234]
Chitosan–cyclodextrin nanoparticles	–	To facilitate nasal drug delivery	[235]
Chitosan–heparin polyelectrolyte complex nanoparticles	–	Protein delivery	[236]
Chitosan–DNA nanoparticles	–	Liver-targeted gene delivery	[237]
Monodisperse chitosan nanoparticles	–	Mucosal delivery	[238]
Folate-conjugated pluronic F127/chitosan core–shell nanoparticles	Self-assembly	Breast cancer treatment	[135]
N-Succinyl chitosan nanospheres	Self-assembly	Sustained drug delivery system	[163]
Magnetic chitosan nanospheres	Co-precipitation followed by ionic gelation	Bioseparation	[239]
PEI-PEG-chitosan copolymer-coated iron-oxide nanoparticles	Chemical co-precipitation	Gene delivery	[240]
Folic acid-modified carboxymethyl chitosan nanoparticles	Sonication	Drug carrier for tumor-specific drug delivery	[241]
Chitosan/chondroitin sulfate/nano-SiO_2 composite scaffold	Lyophilization	Bone tissue engineering	[242]
Chitosan-coated ZnS and Zns:Mn^{2+} quantum dots	Radiation technology	Low cytotoxicity and high biocompatibility mean that they could be useful for further biomedical application	[243]

(continued)

Table 9 (continued)

Chitosan-based nanomaterial	Method of preparation	Application	Reference
Chitosan-coated magnetic nanoparticles for high-efficient cellular imaging	Chemical co-precipitation	These new magneto-fluorescent nanoagents have the potentiality for future medical use	[244]
Both folic acid- and mPEG-conjugated chitosan nanoparticles	Ionic gelation and chemical crosslinking	For targeted and prolonged anticancer drug delivery systems	[245]
Carboxymethylchitosan nanoparticles	Ionic gelation	To inhibit the proliferation of keloid fibroblasts	[246]
Galactosylated low molecular weight chitosan nanoparticles	Ionotropic gelation	To selectively deliver doxorubicin to hepatic parenchymal cells	[247]
Water-soluble chitosan nanoparticles	Ionic gelation	Protein delivery	[248]
Hydralazine-loaded chitosan nanoparticles	Ionic gelation	To provide neuroprotection	[249]
N-Hexanoyl chitosan-stabilized magnetic nanoparticles	–	For cellular labeling and magnetic resonance imaging	[250]
Lectin–Gd-loaded chitosan hydrogel nanoparticles	Microemulsion	Excellent candidates for controlled delivery of bioactive compounds to molecular targets and as biospecific diagnostic tools in MRI	[251]
Fibrin gel loaded with chitosan nanoparticles	Ionotropic gelation	Suitable for topical applications at the site of wounds such as ocular surface destructions, burns, fissures, ulcers, blisters, bone breaks, etc., in which the release of growth factors are of crucial importance in the promotion and speed of the healing process. Potential application in tissue engineering	[252]

(continued)

Table 9 (continued)

Chitosan-based nanomaterial	Method of preparation	Application	Reference
Chitosan/PLGA nanoparticles	Emulsion–diffusion–evaporation technique	Cell delivery vehicle	[253]
Chitosan nanogel-associated recombinant NcPDI	Ionic gelation	Chitosan-based nanogels represent an innovative platform for both intranasal (i.n.) and intraperitoneal (i.p.) vaccination approaches to limit the disease caused by *Neospora caninum* infection	[254]

11 Conclusions

Despite the abundant availability of chitosan in nature, it has not yet been completely accepted by the scientific community. This may be due to the wide range of sources of chitosan that gives rise to various types of chitosan with different properties and, hence, to specific applications. In other words, one particular chitosan (depending on its source) used for a particular application may not be suitable for other applications and vice versa. Therefore, it should be mandatory to mention the source of chitin and chitosan when using chitosan for a particular purpose so as to avoid confusion. In view of this, various sources of chitin and chitosan and their extraction processes are discussed here. Chitosan not only possesses versatile properties but also offers a wealth of biological, biochemical, and biomechanical functionality that can be used to develop new biomaterials. In order to realize its full functional potentiality, various ways of engineering chitosan-based nanostructures are being explored along with their possible nanotechnological applications. Therefore, it is hoped that this review will help the reader to accumulate information for further research in new directions with the help of nanotechnology for the betterment of mankind.

References

1. Zheng H, Du Y, Yu J et al (2001) Preparation and characterization of chitosan/poly(vinyl alcohol) blend fibers. J Appl Polym Sci 80:2558–2565
2. Knaul JZ, Hudson SM, Creber KAM (1999) Improved mechanical properties of chitosan fibers. J Appl Polym Sci 72:1721–1732
3. Jiu H, Du Y, Wang X et al (2004) Chitosan kill bacteria through cell membrane damage. Int J Food Microbiol 95:147–155
4. Dong Y, Wang H, Zheng W et al (2004) Liquid crystalline behavior of chitooligosaccharides. Carbohydr Polym 57:235–240

5. Dutta J, Dutta PK, Rinki K et al (2008) Current research on chitin and chitosan for tissue engineering applications and future demands on bioproducts. In: Jayakumar R, Prabaharan M (eds) Current research and developments on chitin and chitosan in biomaterials science. Research Signpost, Trivandrum
6. Tangpasuthadol V, Pongchaisirikul N, Hoven VP (2003) Surface modification of chitosan films. Effects of hydrophobicity on protein adsorption. Carbohydr Res 338:937–942
7. Zhang J, Xia W, Liu P et al (2010) Chitosan modigfications and pharmaceutical/biomedical applications. Mar Drugs 8:1962–1987
8. Khor E, Lim LY (2003) Implantable applications of chitin and chitosan. Biomaterials 24:2339–2349
9. Riva R, Ragelle H, Rieux AD et al (2011) Chitosan and chitosan derivatives in drug delivery and tissue engineering. Adv Polym Sci 244:19–44
10. Malafaya PB, Silva GA, Reis RL (2007) Natural-origin polymers as carriers and scaffolds for biomolecules and cell delivery in tissue engineering applications. Adv Drug Deliv Rev 59:207–233
11. Alves NM, Mano JF (2008) Chitosan derivatives obtained by chemical modifications for biomedical and environmental applications. Int J Biologic Macromol 43:401–414
12. Liu X, Ma L, Mao Z et al (2011) Chitosan-based biomaterials for tissue repair and regeneration. Adv Polym Sci 244:81–128
13. Dutta PK, Dutta J, Tripathi VS (2004) Chitin and chitosan: chemistry, properties and applications. J Sci Ind Res 63:20–31
14. Pillai CKS, Paul W, Sharma CP (2009) Chitin and chitosan: chemistry, solubility, and fiber formation. Prog Polym Sci 34:641–678
15. Guibal E (2005) Heterogeneous catalysis on chitosan-based materials: a review. Prog Polym Sci 30:71–109
16. Franco LO, Thayza CM, Newton P et al (2005) Cunningamella elegans (IFM 46109) como fonte de quitina e quitosana. Rev Anal 4:40–44
17. Muzzarelli RAA (1990) Encyclopedia of polymer science and engineering, 3rd edn. Wiley, New York, p 430
18. George TS, Guru KSS, Sankaranarayanan N et al (2011) Extraction, purification and characterization of chitosan from endophytic fungi isolated from medical plants. World J Sci Technol 1:43–48
19. Synowiecki J, Al-Khateeb NA (2003) Production, properties and some new applications of chitin and its derivatives. Crit Rev Food Sci Nutr 43:145–171
20. Shepherd R, Reader S, Falshaw A (1997) Chitosan functional properties. Glycoconj J 14:535–542
21. Rasmussen RS, Morrissey MT (2007) Chitin and chitosan. In: Shahidi F, Barrow C (eds) Marine nutraceuticals and functional foods. CRC Press, New York
22. Shahidi F, Arachchi JKV, Jeon YJ (1999) Food applications of chitin and chitosan. Trend Food Sci Technol 10:37–51
23. Islam MM, Masum SM, Mahbub KR et al (2011) Antibacterial activity of crab-chitosan against Staphylococcus aureus and Escherichia coli. J Adv Scient Res 2:63–66
24. Dutta PK, Dutta J, Chattopadhyaya MC et al (2004) Chitin and Chitosan: novel biomaterials waiting for future development. J Polym Mater 21:321–333.
25. Kumirska J, Weinhold MX, Thöming J et al (2011) Biomedical activity of chitin/chitosan based materials-influence of physicochemical properties apart from molecular weight and degree of N-acetylation. Polymers 3:1875–1901
26. Limam Z, Selmi S, Sadok S et al (2011) Extraction of chitin and chitosan from crustacean by-products: Biological and physicochemical properties. Afr J Biotechnol 10:640–647
27. Toan NV (2009) Production of chitin and chitosan from partially autolyzed shrimp shell. Open Biomater J 1:21–24
28. Lertsutthiwong P, How NC, Chandrkrachang S et al (2002) Effect of chemical treatment on the characteristics of shrimp chitosan. J Met Mater Miner 12:11–18

29. Teng WL, Khor E, Tan TK et al (2001) Concurrent production of chitin from shrimp shells and fungi. Carbohydr Res 332:305–316
30. Das S, Ganesh EA (2010) Extraction of chitin from trash crabs (Podophthalmus vigil) by an eccentric method. Curr Res J Biol Sci 2:72–75
31. Okada T, Hartsdale NY, Kubo I et al (1999) Fungus useful for chitin production. US Patent 5,905,035
32. Wooten J, Singer NS (2003) Methods of extracting chitin from the shells of exoskeletal animals. US Patent 6,632,941 B2
33. Sannan T, Tsuchida S, Yoshinaga S et al (2006) Purified chitins and production thereof. US Patent 6,989,440 B2
34. Hackman RH, Goldberg M (1965) The studies on chitin. VI. The nature of α- and β-chitins. Aust J Biol Chem 18:941–965
35. No HK, Meyers SP, Lee KS (1989) Isolation and characterization of chitin from crawfish shell waste. J Agric Food Chem 37:138–144
36. Austin PR, Brine CJ, Castle JE et al (1981) Chitin: new facts of research. Science 212:749–753
37. Roberts G (1997) Chitosan production routes and their role in determining the structure and properties of the product. In: Domard A, Roberts GAF, Varum KM (eds) Advances in chitin science. Jaques Andre, Lyon, pp 22–31
38. Roberts GAF (1992) Chitin chemistry. Macmillan, London
39. Hayes M, Carney B, Slater J et al (2008) Mining marine shellfish waste for bioactive molecules: chitin and chitosan. Part A: Extraction methods. Biotechnol J 3:871–877
40. Kjartansson GT, Zivanovic S, Kristbergsson K et al (2006) Sonication-assisted extraction of chitin from shells of fresh water prawns (Macrobrachium rosenbergii). J Agr Food Chem 54:3317–3323
41. Kim SK, Rajapakse N (2005) Enzymatic production and biological activities of chitosan oligosaccharides (COS): a review. Carbohydr Polym 62:357–368
42. Percot A, Viton C, Domard A (2003) Optimization of chitin extraction from shrimp shells. Biomacromolecules 4(1):12–18
43. Waldeck VJ, Daum G, Bisping B et al (2006) Isolation and molecular characterization of chitinase-deficient *Bacillus licheniformis* strains capable of deproteinization of shrimp shell waste to obtain highly viscous chitin. Appl Environ Microbiol 72:7879–7885
44. Mahmoud NS, Ghaly AE, Arab F (2007) Unconventional approach for demineralization of deproteinized crustacean shells for chitin production. Am J Biochem Biotechnol 3:1–9
45. Rødde RH, Einbu A, Varum KM (2008) A seasonal study of the chemical composition and chitin quality of shrimp shells obtained from northern shrimp (*Pandalus borealis*). Carbohydr Polym 71:388–393
46. No HK, Lee MY (1996) Isolation of chitin from crab shell waste. J Korean Soc Food Nutr 24:105
47. Shahidi F, Synowiecki J (1991) Isolation and characterization of nutrients and value-added products from snow Crab (*Cinoecetes opilio*) and shrimp (*Pandalus borealis*) processing discards. J Agric Food Chem 39:1527
48. Tsaih ML, Chen RH (2003) The effect of reaction time and temperature during heterogeneous alkali deacetylation on degree of deacetylation and molecular weight of resulting chitosan. J Appl Polym Sci 88:2917
49. Gagne N, Simpson BK (1993) Use of proteolytic enzymes to facilitate the recovery of chitin from shrimp wastes. Food Biotechnol 7:253
50. No HK, Meyers SP (1997) Preparation of chitin and chitosan. In: Muzzarelli RAA, Peter MG (eds) Chitin handbook. European Chitin Society, Grottammare, Atec, Grottammare, Italy pp 475–489
51. Acosta N, Jimenez C, Borau V et al (1993) Extraction and characterization of chitin from crustaceans. Biomass Bioeng 5:145

52. Synowiecki J, Sikorski ZE, Naczk M (1981) The activity of immobilized enzymes on different krill chitin preparations. Biotechnol Bioeng 23:2211
53. No HK, Meyers SP (1995) Preparation and characterization of chitin and chitosan – a review. J Aq Food Prod Technol 4:27
54. Whistler RS, BeMiller JN (1962) Alkaline degradation of amino sugars. J Org Chem 27:1161
55. Horowitz ST, Roseman S, Blumental HJ (1957) The preparation of glucosamine oligosaccharides separation. J Am Chem Soc 79:5046
56. Refer website: www.france-chitine.com/fab.e.htm
57. Chen MH, Chan HY, Wu CL et al (2002) Production of chitosan and chitin. US Patent 6,485,946 B1
58. Feofilova EP, Nemtsev DV, Tereshina VM et al (1996) Appl Biochem Microbiol 32:437–445
59. Tsigos I, Martinou A, Kafetzopoulos D et al (2000) Chitin deacetylases: new versatile tools in biotechnology. Tibtech 18:305–312
60. Araki Y, Ito E (1975) A pathway of chitosan formation in *Mucor rouxii*: enzymatic deacetylation of chitin. Eur J Biochem 189:249–253
61. Kafetzopoulos D, Martinou A et al (1993) Bioconversion of chitin and chitosan: purification and characterization of chitin deacetylase from *Mucor rouxii*. Proc Natl Acad Sci USA 90:2564–2568
62. Martinou A, Kafetzopoulos D et al (1993) Isolation of chitin deacetylase from *Mucor rouxii* by immunoaffinity chromatography. J Chromatogr 644:35–41
63. Tsigos I, Bouriotis V (1995) Purification and characterization of chitin deacetylase from *Colletotrichum lindemuthianum*. J Biol Chem 270:26286–26291
64. Gao XD, Katsumoto T et al (1995) Purification and characterization of chitin deacetylase from *Absidia coerulea*. J Biochem 117:257–263
65. Tokuyasu K, Ohnishi-Kameyama M et al (1996) Purification and characterization of extracellular chitin deacetylase from *Colletotrichum lindemuthianum*. Biosci Biotechnol Biochem 60:1598–1603
66. Zhao Y, Park RD et al (2010) Chitin deacetylases: properties and applications. Mar Drugs 8:24–46
67. Song YS, Seo DJ, Kim KY et al (2012) Expression patterns of chitinase produced from *Paenibacillus chitinolyticus* with different two culture media. Carbohydr Polym 90:1187–1192
68. Zhao Y, Jo GH, Ju WT et al (2011) A highly N-glycosylated chitin deacetylase derived from a novel strain of Mortierella sp. DY-52. Biosci Biotechnol Biochem 75:960–965
69. Srinivasan VR (1998) Biotransformation of chitin to chitosan. US Patent 5,739,015
70. Yong T, Hong J, Zhangfu L et al (2005) Purification and characterization of an extracellular chitinase produced by bacterium C4. Ann Microbiol 55:213–218
71. Zhou G, Zhang H, He Y et al (2010) Identification of a chitin deacetylase producing bacteria isolated from soil and its fermentation optimization. Afr J Microbiol Res 4:2597–2603
72. Kaur K, Dattajirao V, Shrivastava V et al (2012) Isolation and characterization of chitosan-producing bacteria from beaches of Chennai, India. Enzyme Res. doi:10.1155/2012/421683
73. Muzzarelli RAA, Mattioli-Belmonte M, Muzzarelli B et al (1997) Medical and veterinary applications of chitin and chitosan. In: Domard A, Roberts GAF, Varum KM (eds) Advances in chitin Science. Jacques Andre, Lyon, pp 580–589
74. Naczk M, Synowiecki J, Sikorski ZE (1981) The gross chemical composition of Antarctic krill shell waste. Food Chem 7:175–179
75. Synowiecki J, Al-Khateeb NA (2000) The recovery of protein hydrolysate during enzymatic isolation of chitin from shrimp *Crangon crangon* processing discards. Food Chem 68:147–152
76. Kassai M (2008) A review of several reported procedures to determine the degree of N-acetylation for chitin and chitosan using infrared spectroscopy. Carbohydr Polym 71:497–508
77. Brugnerotto J, Lizardi J, Goycoolea FM et al (2001) An infrared investigation in relation with chitin and chitosan characterization. Polymer 42:3569–3580

78. Van de Velde K, Kiekens P (2004) Structure analysis and degree of substitution of chitin, chitosan and dibutyrylchitin by FT-IR spectroscopy and solid state 13C NMR. Carbohydr Polym 58:409–416
79. Muzzarelli RAA, Rocchetti R (1985) Determination of the degree of acetylation of chitosans by first derivative ultraviolet spectrophotometry. Carbohydr Polym 6:61–72
80. Wu T, Zivanovic S (2008) Determination of the degree of acetylation (DA) of chitin and chitosan by an improved first derivative UV method. Carbohydr Polym 73:248–253
81. Duarte ML, Ferreira MC, Marvao MR et al (2001) Determination of the degree of acetylation of chitin materials by 13C CP/MAS NMR spectroscopy and solid state 13C NMR spectroscopy. Int J Biol Macromol 28:359–363
82. Varum KM, Antohonsen MW, Grasdalen H et al (1991) Determination of the degree of N-acetylation and the distribution of N-acetyl groups in partially N-deacetylated chitins (chitosans) by high-field n.m.r. spectroscopy. Carbohydr Res 211:17–23
83. Raymond L, Morin FG, Marchessault RH (1993) Degree of deacetylation of chitosan using conductometric titration and solid-state NMR. Carbohydr Res 246:331–336
84. Jiang X, Chen L, Zhong W (2003) A new linear potentiometric titration method for the determination of deacetylation degree of chitosan. Carbohydr Polym 54:457–463
85. Guinesi L, Cavalheiro E (2006) The use of DSC curves to determine the acetylation degree of chitin/chitosan samples. Termochim Acta 444:128–133
86. Rinaudo M, Milas M, Le Dung P (1993) Characterization of chitosan: Influence of ionic strength and degree of acetylation on chain expansion. Int J Biol Macromol 15:281–285
87. Brugnerotto J, Desbrieres J, Roberts G et al (2001) Characterization of chitosan by steric exclusion chromatography. Polymer 42:09921–09927
88. Terbojevich M, Cosani A (1997) Molecular weight determination of chitin and chitosan. In: Muzzarelli RAA, Peter MG (eds) Chitin handbook. European Chitin Society, Grotammare, pp 87–101
89. Yen M, Yang J, Mau J (2009) physicochemical characterization of chitin and chitosan from crab shells. Carbohydr Polym 75:15–21
90. ASTM Standard F2103-01 (2001) Standard guide for characterization and testing of chitosan salts as starting materials intended for use in biomedical and tissue-engineered medical product applications. ASTM International, West Conshohocken
91. Bradford M (1976) A rapid and sensitive method for the quantitation of microgram quantities of protein utilizing the principle of protein-dye binding. Ann Biochem 72:248–254
92. Shigemasa Y, Saito K, Sashiwa H et al (1994) Enzymatic degradation of chitins and partially deacetylated chitins. Int J Biol Macromol 16:43–49
93. Ben AC (2011) Chitin and chitosan: Marine biopolymers with unique properties and versatile applications. Global J Biotech Biochem 6:149–153
94. Khoushab F, Montarop Y (2010) Chitin research revisited. Mar Drugs 8:1988–2012
95. Chen JK, Shen CR, Liu CL (2010) N-acetylglucosamine: production and applications. Mar Drugs 8:2493–2516
96. Kumirska J, Czerwicka M, Kaczyński Z et al (2010) Applications of spectroscopic methods for structural analysis of chitin and chitosan. Mar Drugs 8:1567–1636
97. Tolaimate A, Desbrieres J, Rhazi M et al (2003) Contribution to the preparation of chitins and chitosans with controlled physicochemical properties. Polymer 44:7939–7952
98. Campana-Filho SP, De Britto D, Curti E et al (2007) Extraction, structures and properties of α- and β-chitin. Quim Nova 30:644–650
99. Kurita K, Ishii S, Tomita K et al (1994) Reactivity characteristics of squid β-chitin as compared with those of shrimp chitin: high potentials of squid chitin as starting material for facile chemical modifications. J Polym Sci Part Polym Chem 32:1027–1032
100. Sannan T, Kurita K, Iwakura Y (1976) Studies on chitin 2. Effect of deacetylation on solubility. Makromol Chem 177:3589–3600
101. Noishiki Y, Takami H, Nishiyama Y et al (2003) Alkali-induced conversion of β-chitin to α-chitin. Biomacromolecules 4:896–899

102. Dvir T, Tsur-Gang O, Cohen S (2005) Scaffolds for tissue engineering and regeneration. Isr J Chem 45:487–494
103. Senel S, McClure SJ (2004) Potential applications of chitosan in veterinary medicine. Adv Drug Deliv Rev 56:1467–1480
104. Martino AD, Sittinnger M, Risbud MV (2005) Chitosan: a versatile biopolymer for orthopaedic tissue-engineering. Biomaterials 26:5983–5990
105. Mahae N, Chalat C, Muhamud P et al (2011) Antioxidant and antimicrobial properties of chitosan-sugar complex. Int Food Res J 18:1543–1551
106. Li Q, Dunn ET, Grandmaison EW et al (1992) Applications and properties of chitosan. J Bioact Compat Polym 7:370–397
107. New N, Furuike T, Tamura H (2009) The mechanical and biological properties of chitosan scaffolds for tissue regeneration templates are significantly enhanced by chitosan from *Gongronella butleri*. Materials 2:374–398
108. Hwang JK, Shin HH (2000) Rheological properties of chitosan solutions. Korea Aust Rheol J 12:175–179
109. Dhawan S, Singla AK, Sinha VR (2004) Evaluation of mucoadhesive properties of chitosan microspheres prepared by different methods. AAPS PharmSciTech 5:122–128
110. Trapani A, Sitterberg J, Bakowsky U et al (2009) The potential of glycol chitosan nanoparticles as carrier for low water soluble drugs. Int J Pharm 375:97–106
111. Sajomsang W, Tantayanon S, Tangpasuthadol V et al (2009) Quaternization of N-aryl chitosan derivatives: synthesis, characterization, and antibacterial activity. Carbohydr Res 344:2502–2511
112. Prabhakaran M, Mano JF (2005) Chitosan-based particles as controlled drug delivery systems. Drug Deliv 12:41–57
113. Papadimitriou SA, Achilias DS, Bikiaris DN (2012) Chitosan-g-PEG nanoparticles ionically crosslinked with poly(glutamic acid) and tripolyphosphate as protein delivery systems. Int J Pharm 430:318–327
114. Harris JM, Struck EC, Case MG et al (1984) Synthesis and characterization of poly(ethylene glycol) derivatives. J Polym Sci Polym Chem Ed 22:341–352
115. Sugimoto M, Morimoto M, Sashiva H et al (1998) Preparation and characterization of water-soluble chitin and chitosan derivatives. Carbohydr Polym 36:49–59
116. Gorochovceva N, Naderi A, Dedinaite A et al (2005) Chitosan-N-poly(ethylene glycol) brush copolymers: synthesis and adsorption on silica surface. Eur Polym J 41:2653–2662
117. Bravo-Osuna PG, Vauthier C (2007) Tuning of shell and core characteristics of chitosan-decorated acrylic nanoparticles. Eur J Pharm Sci 30:143–154
118. Huang M, Khor E, Lim LY (2004) Uptake and cytotoxicity of chitosan molecules and nanoparticles: effects of molecular weight and degree of deacetylation. Pharm Res 21:344–353
119. Bernkop-Schn rch A, Kast CE, Guggi D (2001) Permeation enhancing polymers in oral delivery of hydrophilic macromolecules: thiomer/GSH systems. J Control Release 93:95–103
120. Choochottiros C, Yoksan R, Chirachanchai S (2009) Amphiphilic chitosan nanospheres: factors to control nanosphere formation and its consequent pH responsive performance. Polymer 50:1877–1886
121. Wang J, Zong JY, Zhao D et al (2011) In-situ formation of chitosan-cylodextrin nanospheres for drug delivery. Colloids Surf B Biointerfaces 87:198–202
122. Lu B, Xu XD, Zhang XZ et al (2008) Low molecular weight polyethylenimine grafted N-maleated chitosan for gene delivery: properties and in vitro transfection studies. Biomacromolecules 9:2594–2600
123. Yateen SP, Saikishore V, Srokanth K et al (2012) Drug delivery systems using chitosan nanoparticles. Am J PharmTech Res 2:1–19
124. Jong WHD, Borm PJA (2008) Drug delivery and nanoparticles. Int J Nanomedicine 3:133–149
125. Singh M, Manikandan S, Kumaraguru AK (2010) Nanoparticles: a new technology with wide applications. Res J Nanosci Nanotechnol 1–12. doi:10.3923/rjnn.2010

126. Arayne MS, Sultana N, Sabah NS (2007) Fabrication of solid nanoparticles for drug delivery. Pak J Pharm Sci 20:251–259
127. Sahoo SK, Parveen S, Panda JJ (2007) The present and future of nanotechnology in human health care. Nanomedicine 3:20–31
128. Diebold Y, Calonge M (2010) Applications of nanoparticles in opthamology. Prog Polym Sci 29:596–609
129. Wu W, Aiello M, Zhou et al (2010) In-situ immobilization of quantum dots in polysaccharide-based nanogels for integration of optical pH-sensing, tumor cell, imaging, and drug delivery. Biomaterials 31:3023–3031
130. Meyers MA, Chen PY, Lin AYM et al (2008) Biological materials: structure and mechanical properties. Prog Mater Sci 53:1–206
131. Andrade F, Goycoolea F, Chiapptta DA et al (2011) Chitosan-grafted copolymers and chitosan-ligand as matrices for pulmonary drug delivery. Int J Carbohydr Chem 1–14. doi:10.1155/2011/865704
132. Boddohi S, Kipper MJ (2010) Engineering nanoassemblies of polysaccharides. Adv Mater 22:2998–3016
133. Zhang J, Chen XG, Li YY et al (2007) Self-assembled nanoparticles based on hydrophobically modified chitosan as carriers for doxorubicin. Nanomedicine 3:258–265
134. Liu CG, Desai KGH, Chen XG et al (2005) Linolenic acid-modified chitosan for formation of self-assembled nanoparticles. J Agric Food Chem 53:437–441
135. Manaspon C, Viravaidya-Pasuwat K, Pimpha N (2012) Preparation of folate-conjugated pluronic F127/chitosan core-shell nanoparticles encapsulating doxorubicin for breast cancer treatment. J Nanomater 1–11. doi:10.1155/2012/593878
136. Sharma PK, Bhatia SR (2004) Effect of anti-inflammatories on Pluronic® F127: micellar assembly, gelation and partitioning. Int J Pharm 278:361–377
137. Escobar-Chàvez JJ, López-Cervantes M, Naïk A et al (2006) Applications of thermore-versible pluronic F-127 gels in pharmaceutical formulations. J Pharm Pharm Sci 9:339–358
138. Kwon SH, Kim SY, Ha KW et al (2007) Pharmaceutical evaluation of genestein-loaded pluronic micelles for oral delivery. Arch Pharm Res 13:1138–1143
139. Kabanov AV, Lemieux P, Vinogradov S et al (2002) Pluronic block copolymers: novel functional molecules for gene therapy. Add Drug Deliv Rev 54:223–233
140. Hecht E, Mortensen M, Gradzielski M, Hoffmann H (1994) Interaction of ABA block copolymers with ionic surfactants in aqueous solution. Langmuir 10:86–91
141. Janes KA, Fresneau MP, Marazuela A et al (2001) Chitosan nanoparticles as delivery systems for doxorubicin. J Control Release 73:255–267
142. Papadimitriou S, Bikiaris D, Avgoustakis K et al (2008) Chitosan nanoparticles loaded with dorzolamide and pramipexole. Carbohydr Polym 73:44–54
143. Su ZQ, Zhanh HL, Wu SH et al (2010) Preparation and characterization of water-soluble chitosan nanoparticles as protein delivery system. J Nanomater. doi:10.1155/2010/898910
144. Srinatha A, Pandit J, Singh S (2008) Ionic cross-linked chitosan beads for extended relaese of ciprofloxacin: in vitro characterization. Indian J Pharm Sci 70:16–21
145. Jayakumar R, Deepthy M, Manzoor K et al (2010) Biomedical applications of chitin and chitosan based nanomaterials – a short review. Carbohydr Polym. doi:10.1016/j.carbpol.2010.04.074
146. Rejinold NS, Chennazhi KP, Nair SV et al (2011) Biodegradable and thermo-responsive chitosan-g-poly(N-vinylcaprolactum) nanoparticles as a 5-fluorouracil carrier. Carbohydr Polym 83:776–786
147. Choi IS, Oh DY, Kim BS et al (2007) 5-FU, folinic acid as first-line palliative chemotherapy in elderly patients with metastatic or recurrent gastric cancer. Cancer Res Treat 39:99–103
148. Kouchak M, Avadi M, Abbaspour M et al (2012) Effect of different molecular weights of chitosan on preparation and characterization of insulin loaded nanoparticles by ion gelation method. Int J Drug Dev Res 4:271–277

149. Grenha A, Seijo B, Serra C et al (2007) Chitosan nanoparticle-loaded mannitol microspheres: structure and surface characterization. Biomacromolecules 8:2072–2079
150. Grenha A, Seijo B, Remunan-Lopez C (2005) Microencapsulated chitosan nanoparticles for lung protein delivery. Eur J Pharm Sci 25:427–437
151. Fan W, Yan W, Xu Z et al (2012) Formation mechanism of monodisperse, low molecular weight chitosan nanoparticles by ionic gelation technique. Colloids Surf B Biointerfaces 90:21–27
152. Katas H, Oya Alpar H (2006) Development and characterisation of chitosan nanoparticles for siRNA delivery. J Control Release 115:216–225
153. Yao RS, Liu L, Deng SS et al (2011) Synthesis and characterization of PEGylated carboxymethylchitosan nanoparticles. Carbohydr Polym 85:809–816
154. Pang HT, Chen XG, Park HJ et al (2007) Preparation and rheological properties of deoxycholate-chitosan and carboxymethyl-chitosan in aqueous systems. Carbohydr Polym 68:419–425
155. Pasanphan W, Rimdusit P, Choofong S et al (2010) Systematic fabrication of chitosan nanoparticle by gamma radiation. Radiat Phys Chem 79:1095–1102
156. Tang ESK, Huang M, Lim LY (2003) Ultrasonication of chitosan and chitosan nanoparticles. Int J Pharm 265:103–114
157. Patel MP, Patel RR, Patel JK (2010) Chitosan mediated targeted drug delivery system: a review. J Pharm Pharm Sci 13:536–557
158. Aranaz I, Harris R, Heras A (2010) Chitosan amphiphilic derivatives: chemistry and applications. Curr Org Chem 14:308–330
159. Bodnar M, Hartmann JF, Borbely J (2005) Preparation and characterization of chitosan-based nanoparticles. Biomacromolecules 6:2521–2527
160. Lees-Haley PR, Williams CW (1997) Neurotoxicity of chronic low-dose exposure to organic solvents: a skeptical review. J Clin Psychol 53:699–712
161. Oda Y, Miura M, Hattori K et al (2009) Syntheses and doxorubicin-inclusion abilities of beta-cyclodextrin derivatives with a hydroquinone alpha-glycoside residue attached at the primary side. Chem Pharm Bull 57:74–78
162. Prabhakaran M, Mano JF (2006) Stmuli-responsive hydrogels based on polysaccharides incorporated with thermo-responsive polymers as novel biomaterials. Macromol Biosci 8:991–1008
163. Aiping Z, Tian C, Lanhua Y et al (2006) Synthesis and characterization of N-succinyl-chitosan and its self-assembly of nanospheres. Carbohydr Polym 66:274–279
164. Gu FX, Karnik R, Wang AZ et al (2007) Targeted nanoparticles for cancer therapy. Nano Today 2:14–21
165. Arruebo M, Rodrigo FP, Ibarra MR et al (2007) Magnetic nanoparticles for drug delivery. Nano Today 2:22–32
166. Ke JH, Lin JJ, Carey JR et al (2010) A specific tumor-targeting magnetofluorescent nanoprobe for dual-modality molecular imaging. Biomaterials 31:1707–1715
167. Lacava LM, Gareia VAP, Kuckelhaus S et al (2004) Long-term retention of dextran-coated magnetite nanoparticles in the liver and spleen. J Magn Magn Mater 272–276:2434–2435
168. Wang Y, Li B, Zhou Y et al (2011) New generation of chitosan-(acrylic acid)-magnetite nanospheres: synthesis, characterization and cell viability test in vitro. J Control Release 152: e245–e246
169. Rösler A, Vandermeulen GWM, Klok HA (2001) Advanced drug delivery devices via self-assembly of amphiphilic block copolymers. Adv Drug Deliv Rev 53:95–108
170. Turner JL, Chen Z, Wooley KL (2005) Regiochemical functionalization of a nanoscale cage-like structure: robust core–shell nanostructures crafted as vessels for selective uptake and release of small and large guests. J Control Release 109:189–202
171. Mu B, Shen RP, Liu P (2009) Crosslinked polymeric nanocapsules from polymer brushes grafted silica nanoparticles via surface initiated atom transfer radical polymerization. Colloid Surf B 74:511–515

172. Gill I, Ballesteros A (1998) Encapsulation of biologicals within silicate, siloxane, and hybrid sol–gel polymers: an efficient and generic approach. J Am Chem Soc 120:8587–8598
173. Lee JM, Bermudez H, Discher BM et al (2001) Preparation, stability, and in vitro performance of vesicles made with diblock copolymers. Biotechnol Bioeng 73:135–145
174. Li W, Szoka FS Jr (2007) Lipid-based nanoparticles for nucleic acid delivery. Pharm Res 24:438–449
175. Fu GD, Li GL, Neoh KG et al (2011) Hollow polymeric nanostructures-synthesis, morphology and function. Prog Polym Sci 36:127–167
176. Hu Y, Chen Y, Chen Q et al (2005) Synthesis and stimuli-responsive properties of chitosan/poly(acrylic acid) hollow nanospheres. Polymer 46:12703–12710
177. Wang W, Luo C, Shao S et al (2010) Chitosan hollow nanospheres fabricated from biodegradable poly-D,L-lactide-poly(ethylene glycol) nanoparticle template. Eur J Pharm Biopharm 76:376–383
178. Deng Z, Zhen Z, Hu X et al (2011) Hollow chitosan-silica nanospheres as pH-sensitive targeted delivery carriers in breast cancer therapy. Biomaterials 32:4976–4986
179. Wu J, Sailor M (2009) Chitosan hydrogel-capped porous SiO_2 as a pH responsive nano-valve for triggered release of insulin. Adv Funct Mater 19:733–741
180. Liu YL, Su YH, Lai JY (2004) In situ crosslinking of chitosan and formation of chitosan-silica hybrid membranes with using gamma-glycidoxypropyltrimethoxysilane as a crosslinking agent. Polymer 45:6831–6837
181. Vinogradov SV et al (2002) Nanosized cationic hydrogels for drug delivery: preparation, properties and interactions with cells. Adv Drug Deliv Rev 54:135–147
182. Oh JK, Lee DI, Park JM (2009) Biopolymer-based microgels/nanogels for drug delivery applications. Prog Polym Sci 34:1261–1282
183. Oh JK (2010) Engineering of nanometer-sized cross-linked hydrogels for biomedical applications. Can J Chem 88:173–184
184. Zhou X, Liu B, Yu X et al (2007) Controlled release of PEI/DNA complexes from mannose-bearing chitosan microspheres as a potent delivery system to enhance immune response to HBV DNA vaccine. J Control Release 121:200–207
185. Ko JA, Park HJ, Hwang SJ et al (2002) Preparation and characterization of chitosan microparticles intended for controlled release of drug delivery. Int J Pharm 249:165–174
186. Shu XZ, Zhu KJ (2000) A novel approach to prepare tripolyphosphate/chitosan complex beads for controlled release drug delivery. Int J Pharm 201:51–58
187. Xu Y, Du Y (2003) Effect of molecular structure of chitosan on protein delivery properties of chitosan nanoparticles. Int J Pharm 250:215–226
188. Mincheva R, Bougard F, Paneva D et al (2009) Polyelectrolyte complex nanoparticles from N-carboxymethylchitosan and polycationic double hydrophilic diblock copolymers. J Polym Sci Part A Polym Chem 47:2105–2117
189. Boddohi S, Moore N, Johnson PA et al (2009) Polysaccharide-based polyelectrolyte complex nanoparticles from chitosan, heparin, and hyaluronan. Biomacromolecules 10:1402–1409
190. Bodnar M, Hartmann JF, Borbely J (2006) Synthesis and study of cross-linked chitosan-N-poly(ethylene glycol) nanoparticles. Biomacromolecules 7:3030–3036
191. Shen X, Zhang L, Jiang X et al (2007) Reversible surface switching of nanogel triggered by external stimuli. Angew Chem Int Ed 46:7104–7107
192. Maggi F, Ciccarelli S, Diociaiuti M et al (2011) Chitosan nanogels by template chemical cross-linking in polyion complex micelle nanoreactors. Biomacromolecules 12:3499–3507
193. Voets IK, Keizer DA, Cohen Stuart MA (2009) Complex coacervate core micelles. Adv Colloid Interface Sci 147–148:300–318
194. Lee Y, Kataoka K (2009) Biosignal-sensitive polyion complex micelles for the delivery of biopharmaceuticals. Soft Matter 5:3810–3817
195. Cohen Stuart MA, Hofs B, Voets IK et al (2005) Assembly of polyelectrolyte-containing block copolymers in aqueous media. Curr Opin Colloid Interface Sci 10:30–36

196. Shen JM, Xu L, Lu Y et al (2012) Chitosan-based luminescent/magnetic hybrid nanogels for insulin delivery, cell imaging, and antidiabetic research of dietary supplements. Int J Pharm 427:400–409
197. Biju V, Makita Y, Sonoda A (2005) Temperature-sensitive photoluminescence of CdSe quantum dot clusters. J Phys Chem B 109:13899–13905
198. Li XT, Lin C, Li PY et al (1985) The comparisons of sensibility of seven kinds of human carcinoma cell lines to the oridonin. Acta Pharm Sci 20:243–246
199. Zhang J, Wang Q, Wang A (2007) Synthesis and characterization of chitosang-poly(acrylic acid)/attapulgite superabsorbent composites. Carbohydr Polym 68:367–374
200. Zhang DR, Ren TC (2003) Pharmaceutical progress of oridonin. Chin Pharm J 38:817–820
201. Duan C, Zhang D, Wang FET et al (2011) Chitosan-g-poly(N-isopropylacrylamide) based nanogels for tumor extracellular targeting. Int J Pharm 409:252–259
202. Hu HS, Liu TY, Liu DM et al (2007) Controlled pulsatile drug release from a ferrogel by a high-frequency magnetic field. Macromolecules 40:6786–6788
203. Tang YF, Du YM, Hu XW et al (2007) Rheological characterisation of a novel thermosensitive chitosan/poly(vinyl alcohol) blend hydrogel. Carbohydr Polym 67:491–499
204. Leon TL, Elaissari A, Vinuesa JLO et al (2007) Hofmeister effects on poly(NIPAM) microgel particles: macroscopic evidence of ion adsorption and changes in water structure. Chem Phys Chem 8:148–156
205. Berndt I, Popescu C, Wortmann FJ et al (2006) Mechanics versus thermodynamics: swelling in multiple-temperature-sensitive core–shell microgels. Angew Chem Int Ed 45:1081–1085
206. Liu TY, Hu SH, Liu DM et al (2009) Biomedical nanoparticle carriers with combined thermal and magnetic response. Nano Today 4:52–65
207. Zhang J, Misra RDK (2007) Magnetic drug-targeting carrier encapsulated with thermosensitive smart polymer: core-shell nanoparticle carrier and drug release response. Acta Biomater 3:838–850
208. Schmaljohann D (2006) Thermo- and pH-responsive polymers in drug delivery. Adv Drug Deliv Rev 58:1655–1670
209. Jordan A, Scholz R, Wust P et al (1999) Magnetic fluid hyperthermia (MFH): cancer treatment with AC magnetic field induced excitation of biocompatible supermagnetic nanoparticles. J Magn Magn Mater 201:413–419
210. Kawaguchi H (2000) Functional polymer microspheres. Prog Polym Sci 25:1171–1210
211. Jaiswal MK, Banerjee R, Pradhan P et al (2010) Thermal behavior of magnetically modalized poly(N-isopropylacrylamide)-chitosan based nanohydrogel. Colloids Surf B Biointerfaces 81:185–194
212. Gao J, Frisken BJ (2003) Cross-linker-free N-isopropylamide gel nanospheres. Langmuir 19:5212–5216
213. Lee CF, Wen CJ, Chiu WY (2003) Synthesis of poly(chitosan-N-isopropylacrylamide) complex particles with the method soapless dispersion polymerization. J Polym Sci Part A: Polym Chem 41:2053–2063
214. Sierra-Martin B, Choi Y, Romero-Cano MS et al (2005) Microscopic signature of a microgel volume phase transition. Macromolecules 38:10782–10787
215. Okada Y, Tanaka F (2005) Cooperative hydration, chain collapse, and flat LCST behavior in aqueous poly(N-isopropylacrylamide) solutions. Macromolecules 38:4465–4471
216. Lee IS, Akiyoshi K (2004) Molecular mechanics of a cholestrol-bearing pullulan nanogel at the hydrophobic interfaces. Biomaterials 25:2911–2918
217. Janczewski D, Tomczak N, Han MY et al (2009) Stimulus responsive PNIPAM/QD hybrid microspheres by copolymerization with surface engineered QDs. Macromolecules 42:1801–1804
218. Zhang J, Xu S, Kumacheva E (2004) Polymer microgels: reactors for semiconductor, metal, and magnetic nanoparticles. J Am Chem Soc 126:7908–7914
219. Wu W, Zhou T, Shen J et al (2009) Optical detection of glucose by CdS quantum dots immobilized in smart microgels. Chem Commun 4390–4392

220. Chen YF, Ji TH, Rosenzweig Z (2003) Synthesis of glyconanospheres containing luminescent CdSe-ZnS quantum dots. Nano Lett 3:581–584
221. Hasegawa U, Nomura SM, Kaul SC et al (2005) Nanogel-quantum dot hybrid nanoparticles for live cell imaging. Biochem Biophys Res Commun 331:917–921
222. Tan WB, Jiang S, Zhang Y (2007) Quantum-dot based nanoparticles for targeted silencing of HER2/neu gene via RNA interference. Biomaterials 28:165–1571
223. Bhang SH, Won N, Lee T et al (2009) Hyaluronic acid-quantum dot conjugates for in vivo lymphatic vessel imaging. ACS Nano 3:1389–1398
224. Wu W, Shen J, Banerjee P et al (2010) Chitosan-based responsive hybrid nanogels for integration of optical pH-sensing, tumor cell imaging and controlled drug delivery. Biomaterials 31:8371–8381
225. Liu Z, Jiao Y, Wang Y et al (2008) Polysaccharides-based nanoparticles as drug delivery systems. Adv Drug Deliv Rev 60:1650–1652
226. Agnihotri SA, Mallakarjuna NN, Aminabhavi TM (2004) Recent advances on chitosan-based micro- and nanoparticle in drug delivery. J Control Release 100:5–28
227. Nanjawade BK, Manvi FV, Manjappa AS (2007) In-situ forming hydrogels for sustained ophthalmic drug delivery. J Control Release 122:119–134
228. Sundar S, Kundu J, Kundu SC (2010) Biopolymeric nanoparticles. Sci Technol Adv Mater 11:014104–014114
229. Li L, Chen D, Zhang Y et al (2007) Magnetic and fluorescent multifunctional chitosan nanoparticles as a smart drug delivery systems. Nanotechnology 18:405102. doi:10.1088/0957-4484/18/40/405102
230. Wilson B, Samanta MK, Santhi K et al (2010) Chitosan nanoparticles as a new delivery system for the anti-alzheimer drug tarcine. Nanomedicine 6:144–152
231. Nagpal K, Singh SK, Mishra DN (2010) Chitosan nanoparticles: a promising system in novel drug delivery. Chem Pharm Bull (Tokyo) 58:1423–1430
232. Chauddhury A, Das S (2011) Recent advancement of chitosan-based nanoparticels for oral controlled delivery of insulin and other therapeutic agents. AAPS PharmSciTech 12:10–20
233. Tang DW, Yu SH, Ho YC et al (2010) Heparinized chitosan/poly (γ-glutamic acid) nanoparticles for multi-functional delivery of fibroblast growth factor and heparin. Biomaterials 31:9320–9332
234. Sarmento B, Rebeiro A, Veiga F et al (2007) Oral bioavailability of insulin contained in polysaccharide nanoparticles. Biomacromolecules 8:3054–3060
235. Teijeiro-Osorio D, Remunan-Lopez C, Alonso MJ (2009) New generation of hybrid polyoligosaccharide nanoparticels as carriers for the nasal delivery of macromolecules. Biomacromolecules 10:243–249
236. Liu ZG, Jiao YP, Liu F et al (2007) Heparin/chitosan nanoparticle carriers prepared by polyelectrolyte complexation. J Biomed Mater Res A 83A:806–812
237. Dai H, Jiang X, Tan GCY et al (2006) Chitosan-DNA nanoparticles delivered by intrabiliary infusion enhance liver-targeted gene delivery. Int J Nanomedicine 1:507–522
238. Zhang H, Oh M, Allen C et al (2004) Monodosperse chitosan nanoparticles for mucosal drug delivery. Biomacromolecules 5:2461–2468
239. Hrictu D, Popa MI, Popa N et al (2009) Preparation and characterization of magnetic chitosan nanospheres. Turk J Chem 33:785–796
240. Kievit FM, Veiseh O, Bhattarai N et al (2009) PEI-PEG-Chitosan copolymer coated iron-oxide nanoparticles for safe gene delivery: synthesis, complexation, and transfection. Adv Funct Mater 19:2244–2251
241. Tan YL, Liu CG (2011) Preparation and characterization of self-assemblied nanoparticles based on folic acid modified carboxymethyl chitosan. J Mater Sci Mater Med 22:1213–1220
242. Kavya KC, Dixit R, Jayakumar R et al (2012) Synthesis and characterization of chitosan/chondroitin sulfate/nano-SiO_2 composite scaffold for bone tissue engineering. J Biomed Nanotechnol 8:149–160

243. Chang SQ, Kang B, Dai YD et al (2011) One-step fabrication of biocompatible chitosan-coated ZnS and ZnS:Mn^{2+} quantum dots via a γ-radiation route. Nanoscale Res Lett 6:591–597
244. Ge Y, Zhang Y, He S et al (2009) Fluorescent modified chitosan-coated magnetic nanoparticles for high-efficient cellular imaging. Nanascale Res Lett 4:187–295
245. Hou Z, Zhan C, Jiang Q et al (2011) Both FA- and mPEG-conjugated chitosan nanoparticles for targeted cellular uptake and enhanced tumor tissue distribution. Nanoscale Res Lett 6:563–573
246. Feng C, Chen X, Zhang J et al (2011) The effect of carboxymethylchitosan nanoparticles on proliferation of keloid fibroblast. Front Chem China 5:31–37
247. Jain NK, Jain SK (2010) Development and in vitro characterization of galactosylated low molecular weight chitosan nanoparticles bearing doxorubicin. AAPS PharmSciTech 2:686–697
248. Chun W, Xiong FU, LianSheng Y (2007) Water-soluble chitosan nanoparticles as a novel carrier system for protein delivery. Chin Sci Bull 52:883–889
249. Cho Y, Shi R, Borgens RB (2010) Chitosan nanoparticle-based neuronal membrane sealing and neuroprotection following acrolein induced cell injury. J Biol Eng 4:2, http://www.jbioleng.org/content/4/1/2
250. Bhattarai SR, Kc RB, Kim SY et al (2008) N-hexanoyl chitosan stabilized magnetic nanoparticles: Implication for cellular labeling and magnetic resonance imaging. J Nanobiotechnol 6:1. doi:10.1186/1477-3155-6-1
251. Pashkunova-Matric I, Kremser C, Galanski M et al (2011) Lectin–Gd-loaded chitosan hydrogel nanoparticles: a new biospecific contrast agent for MRI. Mol Imaging Biol 13:16–24
252. Zhou W, Zhao M, Zhao Y et al (2011) A fibrin gel loaded with chitosan nanoparticles for local delivery of rhEGF: preparation and in vitro release studies. J Mater Sci Mater Med 22:1221–1230
253. Nafee N, Schneider M, Schaefer U et al (2009) Relevance of the colloidal stability of chitosan/PLGA nanoparticles on their cytotoxicity profile. Int J Pharm 381:130–139
254. Debache K, Krope C, Schütz CA et al (2011) Vaccination of mice with chitosan nanogel-associated recombinant NcPDI against challenge infection with Neospora caninum. Parasite Immunol 33:81–94

Adv Polym Sci (2013) 254: 135–168
DOI: 10.1007/12_2012_197

Published online: 15 March 2013

Hydroxyapatite-Packed Chitosan-PMMA Nanocomposite: A Promising Material for Construction of Synthetic Bone

Arundhati Bhowmick, Subhash Banerjee, Ratnesh Kumar, and Patit Paban Kundu

Abstract Natural bone is a composite material composed of organic compounds (mainly collagen) and inorganic compounds (minerals). The inorganic or mineral component of bone primarily consists of crystalline hydroxyapatite (HAp), $[Ca_{10}(PO_4)_6(OH)_2]$, which is present in bones as a major component (60–65%). In recent years, bioactive polymer/HAp nanocomposite materials have been investigated extensively as templates for the regeneration and reconstruction of bone tissue. The advantages of using polymeric biomaterials include ease of manufacturing the components with various simple as well as complex shapes, reasonable cost, and their ability to possess a wide range of physical and mechanical properties. Among other biomaterials, chitosan, a natural polysaccharide composed of glucosamine and *N*-acetylglucosamine units linked in β-(1–4) manner and derived from chitin by partial or complete deacetylation, has played a major role in bone tissue engineering over the last two decades. However, the poor mechanical strength of chitosan has restricted its practical application in bone tissue engineering. Thus, enhancement of the mechanical properties of chitosan is highly desirable. Blending of chitosan with a synthetic polymer having good mechanical properties (e.g. poly methylmethacrylate) could be helpful in increasing the mechanical properties of chitosan. Moreover, the biodegradability and biocompatibility of chitosan makes HAp-packed chitosan nanocomposite materials attractive for the

A. Bhowmick and P.P. Kundu (✉)
Department of Polymer Science and Technology, University of Calcutta, 92 Acharya Prafulla Chandra Road, Kolkata 700009, WB, India
e-mail: ppk923@yahoo.com

S. Banerjee
Department of Chemistry, Guru Ghasidas Vishwavidyalaya, Koni, Bilaspur 495009, CG, India

R. Kumar
National Instutute of Orthopaedic Hadicapped, B.T. Road, Bon-Hoogly, Kolkata 700 090, India

building of synthetic bone. This chapter describes natural bone structure, polymer blends, and recent advances in polymer/(nano)HAp nanocomposites for application as synthetic bone

Keywords Chitosan · Nanocomposite · Nanohydroxyapatite · Synthetic bone · Synthetic polymer

Contents

Abbreviations

μm	Micro meter
BMA	Bone marrow aspirate
BMU	Bone modeling unit
CTS	Chitosan
DP	Degree of polymerization
HAp	Hydroxyapatite
MW	Molecular weight
NP	Nanoparticle
PA	Polyamide
PBMA	Poly(*n*-butyl methacrylate)
PC	Polycarbonate
PCM	Phase change material
PE	Polyethylene
PEEK	Poly(ether ether ketone)
PET	Poly(ethylene terephtalate)
PHEMA	Poly(hydroxyethyl methacrylate)
PMMA	Poly(methyl methacrylate)
POM	Polyoxymethylene

PP	Polypropylene
PSU	Polysulfone
PTFE	Polytetrafluoroethylene
PUs	Selected polyurethanes
PVC	Poly(vinyl chloride)
UHMWPE	Ultrahigh molecular weight polyethylene

1 Introduction

Bone is one of the most commonly replaced tissues, and bone grafts can efficiently repair fractures in bones of the human body. Small defects are usually repaired by the growth of natural bone that is even stronger than the original bone. The bone is naturally regenerated within a few weeks. However, problems arise with major fractures, i.e., large defects that are not repaired by the body. Spinal fusion (arthrodesis), fracture non-unions, skeletal deformities caused by infection, trauma, or tumor resection are examples of such cases where bone healing needs to be improved [1]. Bone grafts are generally used in such cases. Basically, bone grafting is a surgical process that replaces missing bone with the help of bone graft materials in order to repair complex bone fractures in normal healthy bone without damaging living tissue. Bone graft requirements truly depend on the complication of the bone defects. There are a number of bone grafting methods available for the repair of bone defects. Among these, autografting and allografting are commonly used methods. In the autograft technique, bone from a different part of the body is transferred. However, there are some practical complications associated with this method, such as complications in wound healing, extra surgery, donor pain, and limited supply of bone [2]. In the allograft technique, bones from living donors or cadavers are used and the risks associated with this method include rejection and disease transmission, along with a significant structural failure rate due to poor tissue integration [3–6] These limitations and concerns have attracted significant interest in the development of synthetic bone grafts that are equivalent to natural bone tissue in all aspects [7]. A biodegradable and bioactive material is one of the essential components for the development of synthetic bone grafting and a limited number of such materials are available. Among others, chitosan (CTS), a natural polysaccharide derived from chitin, and hydroxyapatite (HAp) serve as the best bioactive biomaterials in bone grafting and are well known for their excellent biocompatibility with the human body environment [8]. Recently, polymer/ceramic composites and nanocomposites consisting of a polymer matrix and bioactive micro- or nanofiller have been utilized as novel bone substitutes. The advantages offered by using polymers in this regard include structural stability, biocompatibility, and preferred shape, which are combined with the characteristics of ceramics that mimic those of bone structure. One of the most important groups of polymer/ceramic composites are polymer/HAp materials. HAp has attracted much attention as a biomaterial because its chemical composition is identical to that of human bone. HAp is a highly biocompatible, biostable, and bio-adoptable natural ceramic without viral effects [9, 10]. Moreover, it has a unique characteristic, i.e., osteoconduction [11–14]. HAp also facilitates new bone formation without resorption and interaction with the living system. Moreover, nanostructured HAp has high surface area and enhanced bioactivity. For this reason, it interacts with the lowest hierarchical levels of bone structure more intensively. As a result, several polymer/HAp nanocomposites have been employed to resemble the structure and properties of human bone tissue. In this chapter, the structure of natural bone, polymer blends, and recent advances in polymer/HAp nanocomposites for application as synthetic bone are described.

2 Structure of Natural Bone

Bone is a vascularized, dense connective tissue that forms the endoskeleton of vertebrates. There are 206 bones in adult humans [15]. The extracellular matrix of natural bone is composed of two major phases: organic, consisting mainly of collagen; and inorganic, consisting of minerals (mainly HAp). The inorganic phase accounts for about 70% of the dry weight of bone and the organic matrix makes up the rest [16]. Collagen fibers are seen at the nanoscale level and are surrounded and infiltrated by minerals. The hierarchical structure of bone has been constructed from these nanostructured building blocks. However, compositional differences in bone exist between different species and age groups. The composition of the two phases of bone are discussed in Sects. 2.1 and 2.2.

2.1 Organic Phase of Bone

The organic matrix mostly consists of type I collagen (90%). Bone collagen is a fibrous molecule of 100–2,000 nm length. The collagen fibers provide the framework and architecture of bone, while the HAp crystals are located in the fibers and between the fibers. The noncollagenous proteins of the organic matrix consist of proteoglycans, glycoproteins, and γ-carboxyglutamic-acid-containing proteins [17, 18]. Type I collagen is synthesized by osteoblasts (bone-forming cells). These molecules have a triple helical structure. After secretion, the globular ends are cleaved off by enzymes and the triple-helical molecules undergo a self-assembly process. In general, there are 12 types of collagen found in the body and each collagen comprises three polypeptide chains containing approximately 1,000 amino acid units in each chain. Specifically, type I collagen (MW 139,000 Da) possesses two identical $\alpha 1(I)$ chains and one unique $\alpha 2$ chain; this configuration produces a fairly rigid linear molecule that is 300 nm long [19]. These self-assembled triple helix bundles have a periodicity of 67 nm, with 40 nm gaps between the ends of the

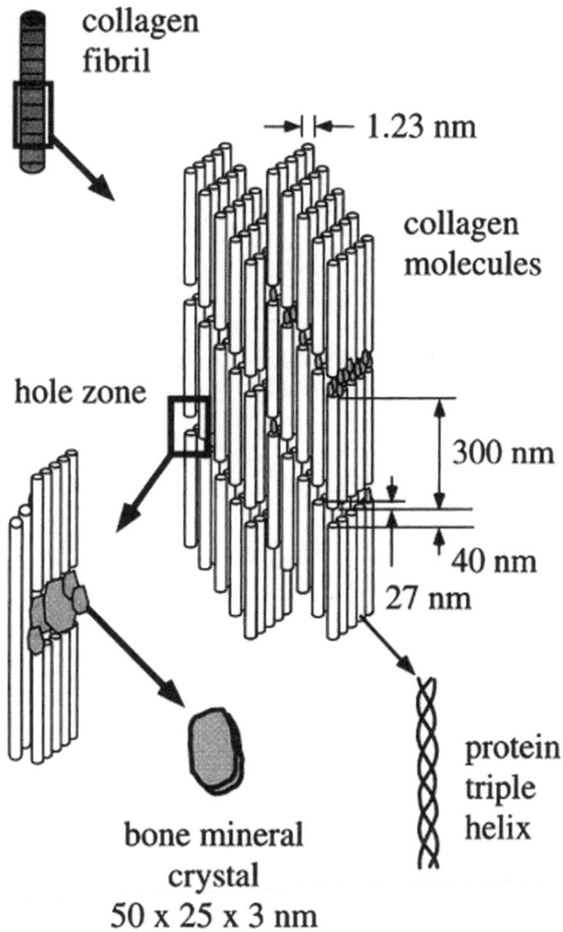

Fig. 1 Assembly of collagen fibers and bone mineral crystals. Reprinted from [24] with permission

molecules and pores between the sides of parallel molecules, which are called hole zones (Fig. 1). The collagen fibers provide the framework and architecture of bone, while the HAp crystals are located in the fibers and between the fibers.

2.2 *Inorganic Phase of Bone*

The mineral phase is mainly formed by calcium and phosphorous, which form spindle- or plate-shaped crystals of HAp [$3Ca_3(PO_4)_2.(OH)_2$]. HAp nanocrystals of bone locate at hole zones within the collagen fibrils. This restricts the possible primary growth of the crystals and forces them to be discrete and discontinuous. The mineral crystals grow with a specific crystalline orientation. The *c*-axes of the

Table 1 Several levels of structural organization of bone, from macro- to sub-nanostructure

Structural level number	Structural organization of bone	Examples
1	Macrostructure	Cancellous and cortical bone
2	Microstructure: 10–500 mm	Haversian system, osteons, single trabeculae
3	Sub-microstructure: 1–10 mm	Lamellae
4	Nanostructure: from a few hundred nanometers to 1 mm	Fibrillar collagen, embedded mineral
5	Sub-nanostructure: below a few hundred nanometers	Mineral, collagen, non-collagenous organic proteins

crystals are parallel to the long axes of the collagen fibrils [20]. The average size of plates is 50 × 25 nm and the crystal thickness is 2–3 nm [21]. The mineralized apatite contains small amounts of impurities, e.g., phosphate, Na, Mg, K, citrate, and carbonate [20], which change certain physical properties (such as solubility) and some important biological features of bone that are vital to normal bone function. For instance, the presence of magnesium in the mineralized matrix may improve cellular activity and promote growth of HAp crystals, followed by new bone formation [22].

2.3 Physical and Mechanical Characteristics of Bone

The hierarchical structure of bone extends over various levels ranging from macrostructure to sub-nanostructure (see Table 1 [49] and Fig. 2) [23]. This hierarchically organized structure has an irregular but optimized arrangement and orientation of the components, which makes bone heterogeneous and anisotropic.

Trabecular or cancellous bone is spongy in nature and occupies about 20% of the total bone. Cancellous bone is lighter, less dense, has higher porosity (pores diameter varies from a few micrometers to millimeters), and a higher concentration of blood vessels than compact bone (also called cortical or dense bone) (Fig. 2). The porous architecture of cancellous bone is easily visible under the microscope or even with the naked eye because it contain very large pores. Cortical bone, which has less porosity and thus a lower concentration of blood vessels, occupies about 80% of the total bone. Due to its lower porosity, its porous architecture is not visible to the naked eye. The diameters of pores are 10–20 μm and mostly separated by 200–300 μm intervals. Spongy bone acts mainly in compression, whereas compact bone acts mechanically in torsion, tension, and compression.

Cortical bone is mechanically stronger than cancellous bone because it is denser than cancellous bone. The relative density and some mechanical properties of bone are shown in Table 2. However, these properties changes with sex, age, dietary history, health status, and anatomical location. Generally, lower density and weaker mechanical properties are observed for diseased bone.

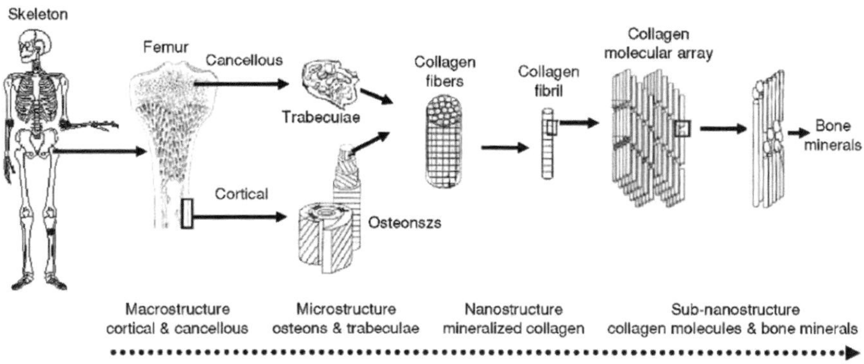

Fig. 2 The hierarchical structure of bone, from macro- to nanoassembly. Reprinted from [34] with permission

Table 2 Relative density and mechanical properties of human bone (Reprinted from [24] with permission)

Property	Cancellous bone	Cortical bone (longitude)	Cortical bone (transverse)
Relative density	0.05–0.7	0.7–1.8	
Elongation (%)	5–7	1–3	
Ultimate tensile strength (MPa)	2–20	130–150	50–60
Elastic modulus (GPa)	0.1–0.5	17–30	7–13

At the microstructural level, the Haversian system or osteon is the repeated structural unit of compact bone, which acts as a weight-bearing pillar. In contrast, spongy bones are made of an interconnecting framework of trabeculae. There are three types of cellular structures in the trabeculae, (1) plate/platelike, (2) plate/barlike, and (3) bar/barlike. The bone tissue primarily consists of collagen nanofibers and bone minerals crystals (particularly HAp) at the nanostructural level.

2.4 *Bone Remodeling and Bone Cells*

Bone has the capability to repair or modify with change of loading conditions of the damaged bone. For instance, bone mineral content can be increased by exercise, which makes bones less likely to fracture [25]. Therefore, bone has the capability of self-repair through the formation of a bone-modeling unit (BMU) under excessive mechanical stress. In this process, three types of bone cells are involved: osteoblasts (bone-forming cells), osteocytes (bone-maintaining cells), and osteoclasts (bone-resorbing cells). The bone cells functioning during the bone remodeling process are represented in Fig. 3 [26]. Osteoclasts are stimulated by cytokines, and the proteins

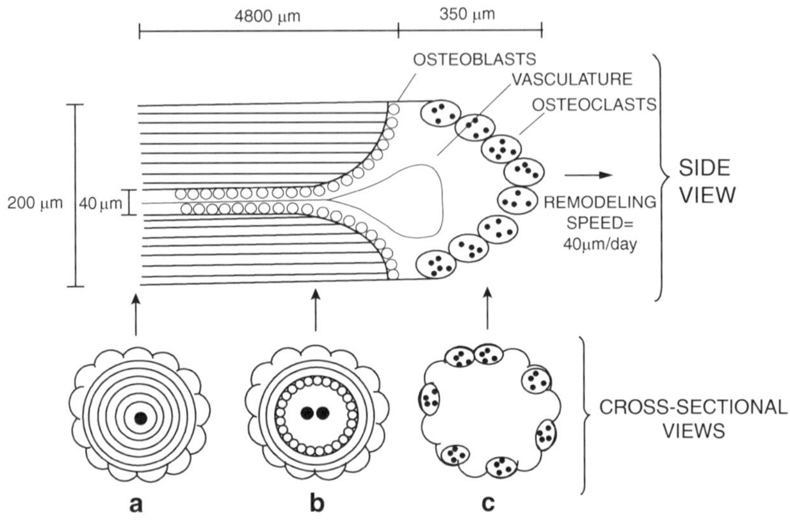

Fig. 3 Coordinated bone cell functions that maintain homeostasis during bone remodeling. Reprinted from [24] with permission

present in the bone matrix are induced to resorb old bone. Osteoblasts are stimulated by insulin-like growth factors I and II to deposit calcium-containing minerals secreted by osteoclasts and/or osteocytes. New bone formation is controlled by osteocytes by modulation of osteoblast differentiation from non-calcium-depositing cells to calcium-depositing cells through secretion of growth factors such as insulin-like growth factor-I and tissue growth factor β [27].

Bone, like other connective tissue in the embryo, is derived from mesenchymal cells. These cells have the ability to proliferate and differentiate into bone cells, which are known as osteoprogenitor cells. They are also called bone-precursor cells.

2.4.1 Osteoblasts

Osteoblast cells, found on the surface of periosteal and endosteal surfaces, are responsible for the formation of new bone tissue. They has an average diameter of 10–50 μm and produce a matrix of osteoid, which is composed mainly of type I collagen. Osteoblasts are also responsible for mineralization of this matrix. Osteoblasts synthesize and deposit bone sialoprotein, osteocalcin (a calcium-binding protein), and other matrix proteins at the beginning of the mineralization process. Osteocalcin interacts with HAp and is thought to mediate pairing to bone resorption by osteoclasts and bone formation by osteoblasts and/or osteocytes.

2.4.2 Osteocytes

Osteocytes are mature cells derived from the osteoblasts implanted in mineralized bone matrix. They have a minor contribution to new bone formation compared to osteoblasts. Osteocytes are arranged around the central lumen of an osteon and between lamellae (Fig. 1). Osteocytes have an interconnecting three-dimensional (3D) network. They are linked with adjacent osteocytes through small channels called canaliculi. Osteocytes are responsive to physiological stress and strain signals in bone tissue and also help to balance osteoblastic and osteoclastic activity to deposit new bone and to dissolve old bone, respectively. Additionally, they act as transporting agents of minerals between bone and blood.

2.4.3 Osteoclasts

Osteoclasts are the large cells (up to 100 μm in diameter) that are found at the surface of bone mineral next to the resorbing bone. Osteoclasts are primarily responsible for bone resorption ("clast" means to break; osteoclasts break down bone). Osteoclasts have multiple nuclei, up to 100 per cell. To dissolve the minerals and collagen from the mature bone, they use acids or enzymes. The dissolved minerals then re-enter the bloodstream and are carried to different parts of the body. Bone-lining cells are responsible for regulating the transportation of bone tissue minerals. They are found along the surface of the mature bone. Osteoclasts can be activated by some exclusive proteins. All these types of cells are together responsible for building the bone matrix with hierarchical self-assembly, maintenance, and remodeling and these processes must be in equilibrium to ensure healthy bone tissue. Importantly, the extent of bone remodeling that occurs at an implant surface will determine the fate of the prosthetic device. For example, loosening and failure of the implant may result from either (1) little or no remodeling in the bone surrounding an implant, which may lead to malnourished juxtaposed bone, or (2) too much remodeling in the bone surrounding an implant, which may lead to excessive bone resorption, or osteolysis.

2.5 *Current Scenario of Bone Grafting*

Bone is the second-most transplanted tissue in humans. Bone grafting is a surgical process that repairs or replaces the defective bones with healthy bones with the help of bone graft materials taken either from the patient's own body or from synthetic sources without damaging living tissues. The bone grafting technique is a widely applied and universally accepted technique in several types of surgery, such as orthopedic surgery, plastic surgery, oral and maxillofacial surgery, and dental surgery. Along with filling defective gaps by bone

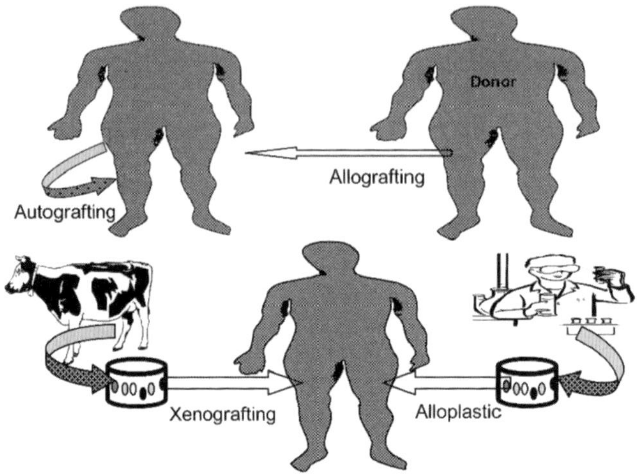

Fig. 4 Bone grafting strategies. Reprinted from [34] with permission

grafting, mechanical and structural support is also provided, which eventually enhances the growth of bone tissue. Using this method, recovery time is reduced and formation of new bone strengthens the defective area by linking implanted bone grafts with host bone. A number of methods are available for the repair of bone defects, such as autografting, allografting, xenografting, and alloplastic or synthetic bone grafting. Each of the methods has its own advantages and limitations [28–33]. A scheme for bone grafting strategies in humans is shown in Fig. 4 [34]. Depending on the nature and complexity of the bone defect, the grafting method is often chosen. Ultimately, the key decision is taken by the surgeon on the basis of experience.

2.5.1 Autografts

In the autograft method, bone tissue is removed from one portion of the skeleton and transplanted to another site of the same body that requires bone regeneration. It is generally taken in the form of cancellous bone from the patient's iliac crest; however, compact bone can also be used for this purpose [35]. There is another newly developed source of autograft, known as bone marrow aspirate (BMA). BMA is osteoblastic stem cells found in bone marrow. In this case, the defective site can be transplanted without complex surgery because these osteoblastic stem cells can be taken by syringe needle from the patient's own body. Additionally, BMA provides osteogenic and osteoinductive factors, which are necessary for fast bone regeneration. Also, delivering cells without using a suitable carrier system increases the chances of migration of cells from the implanted site. The autografting method has the best clinical success rate because it provides osteogenic cells with essential osteoinductive factors for the activation of bone growth; thereby autografts have been the gold standard of bone replacement for long time [36, 37]. Also, this is the safest transplant because of the compatibility of the grafts and, therefore, the chance of rejection is negligible. However, there are a few limitations in various clinical situations, i.e., (a) insufficient amount of grafts when dealing with large bone defects, (b) donor site morbidity, (c) complexity in making the required shape, (d) complications in wound healing, and (e) extra surgery.

2.5.2 Allografts

This is another method for bone grafting, whereby tissue is transplanted between genetically non-identical members of the same species. In 1908, the first clinical use of an allograft was carried out by Lexer [38]. Generally, cancellous, cortical, or both in combination can be used as allograft materials. Cadaver's bones, which are generally frozen or freeze-dried, can be stocked in the bone banks. However, their strength is less after sterilization and they are not resorbed completely after implantation. As a result, they often remain as dead tissue that gradually becomes fragile and can cause medical complications with neighboring tissues. Allografting eliminates harvesting a surgical site, post-operative pain, and the expense of additional surgery. However, allografts are always associated with disease transmission such as hepatitis B, hepatitis C, immune deficiency syndrome (AIDS), and lower effectiveness because of the cleansing and disinfecting process during which the bone growth cells and proteins are removed [39, 40]. These restrict its applications. However, in the case of a large bone defect or when the defect is not repairable by autografting because of insufficiency of the graft materials, this method has particular importance.

2.5.3 Xenografts

Xenografting is a method of transplanting tissue between members of different species; bone from animal to human, for example. Xenografting comes from the Greek word *xenos*, which means strange. In the early 1960s, bone transplantations were mainly performed by xenografts. The bone banks usually stock a plentiful quantity of graft materials because of their simple harvesting compared to bone from humans. Currently, multiple xenografts are being processed, such as frozen calf bone, freeze-dried calf bone, decalcified ox bone, and deproteinized bovine bone [41]. A commercially available xenografts is kiel bone, which comprises deproteinized bone from freshly sacrificed calves [42]. Unlike autografting or allografting, xenografting is not very successful, mainly because of the risk of antigenicity. However, organ scarcity has encouraged continued research in this field.

2.5.4 Synthetic Bone Grafting

Recently, the use of synthetic substances as bone graft materials has increased in importance as an alternative to all other bone grafting methods. Synthetic bone grafting is a surgical process that uses synthetic substances (often called synthetic bone grafts) to repair or redevelop defective bone tissue. It is also known as alloplastic bone grafting. To date, a number of synthetic bone grafts have been developed to eliminate or minimize the complications associated with autografts, allografts, and xenografts [43–46]. Each synthetic material has different characteristic functions either in vitro or in vivo, which makes difficult to judge the best system for bone grafting. There are many advantages and disadvantages associated with each of these materials. Synthetic grafts are considered a good choice for bone grafting because synthetic grafts are abundantly available, sterilizable, reproducible, shapeable, and cost-effective. Also, some of the limitation of autografts or allografts such as donor shortage and the chance of rejection or transmission of infectious disease can be eliminated by synthetic bone grafts. Currently, none of the synthetic grafts resembles natural bone tissue in many aspects because of the lack of proper designing techniques. The biomimetic approach is considered to be promising for the design of such a bone-resembling graft. Research is now progressing in this direction and the next decade may witness many breakthroughs in the fascinating field of biomimetics.

3 Hydroxyapatite

Hydroxyapatite, $Ca_{10}(PO_4)_6(OH)_2$, is the main inorganic component found in hard human tissues such as bone and teeth and is the most extensively used bioceramic in bone tissue engineering. The other materials used for this purpose include alumina, zirconia, titania phosphates, and calcium phosphates [such as calcium tetraphosphate ($Ca_4P_2O_9$) and tricalcium phosphate $Ca_3(PO_4)_2$] and derivatives [47, 48]. The chemical structure of HAp is presented in Fig. 5 [49].

Most of the biological HAps are weakly crystalline and non-stoichiometric. HAp contain various foreign ions, primarily carbonate ion (CO_3^{2-}) and small amounts of sodium (Na^+), magnesium (Mg^{2+}), ferrous (Fe^{2+}), chloride (Cl^-), fluoride (F^-), and hydrogen phosphate (HPO_4^{2-}) ions. CO_3^{2-} consist of 3–8 wt% of the calcified tissue, which varies with bone age [50–52]. Thus, this ion is a major ion in bone metabolism. Bone mineral is a calcium-deficient HAp with a Ca:P ratio of about 1.5. Bone mineral and β-tricalcium phosphate (β-TCP) are chemically and compositionally similar; however, it is structurally similar to HAp (Ca:P = 1.67). Calcium-deficient HAps [$Ca_{10-x}(HPO_4)_x(PO_4)_{6-x}(OH)_{2-x}$] with Ca:P ratios varying from 1.67 to 1.33 are formed by the loss of Ca^{2+} ions from the unit cell [8, 53].

Various methods have been developed for the synthesis of HAp. For example, Hao et al. have synthesized HAp nanocrystals by the hydrothermal method [54, 55].

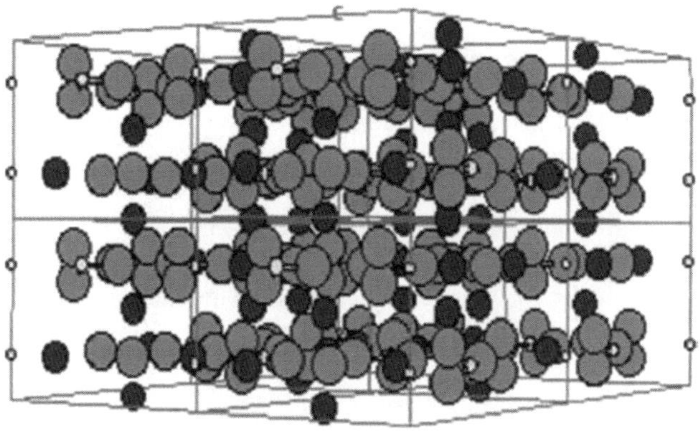

Fig. 5 Atomic structure of HAp; smallest *white* atoms, phosphorus; largest *gray* atoms, oxygen; medium *black* atoms, calcium. Reprinted from [49] with permission

In this method, an aqueous solution of $(NH_4)_2HPO_4$ was slowly added drop by drop into aqueous $Ca(NO_3)_2$ solution under stirring at room temperature. The resultant precipitate was put into an autoclave and then treated hydrothermally at 140°C for 5 h under a pressure of 0.3 MPa. It was then washed with deionized water by a centrifugal method. Finally, the white slurry was dried to produce HAp [55].

Needle-like HAp nanocrystals having a similar morphology and crystal structure to natural apatite (average size of 50 × 25 nm and thickness of 2–3 nm) have been synthesized by hydrothermal methods [54].

Slosarczyk et al. have used a wet method to obtain carbonated HAp powders [56, 57]. Calcium oxide (CaO), calcium nitrate, calcium tetrahydrate [$Ca(NO_3)_2{\cdot}4H_2O$] or calcium acetate [$Ca(CH_3COO)_2{\cdot}H_2O$] were used as the calcium source. As the phosphorous source, phosphoric acid (H_3PO_4) or di-ammonium phosphate [$(NH_4)_2HPO_4$] were used. The molar ratio of Ca:P was 1.67. Ammonium bicarbonate (NH_4HCO_3) or sodium bicarbonate ($NaHCO_3$) were used as reactants to introduce CO_3^{2-} groups. Biological apatites in natural bone, dentin, and enamel contain 7.4, 5.6, and 3.5 wt% of carbonate, respectively. In this method, some of the PO_4^{3-} and OH^- groups are replaced by CO_3^{2-} groups. This carbonated HAp powder has proved to be a promising material for bioresorbable bone substitution, among other applications [56].

Template technology has been used by Zhang et al. to synthesize HAp single-crystal nanowires, and the product was confirmed by using different analytical techniques, e.g., X-ray diffraction (XRD), transmission electron microscopy (TEM), and X-ray photoelectron spectroscopy (XPS). The crystalline order of the HAp precursors was maintained in the electrodeposited nanowires, which were confirmed by the characterization. These HAp single-crystal nanowires displayed structural similarity to the natural HAp found in bone [58].

The precipitation method was adopted by Siddharthan et al. to synthesize calcium-deficient HAp (cdHAp) with a Ca:P ratio of 1.5 [53] using calcium nitrate tetrahydrate and phosphoric acid under accelerated microwave irradiation. The synthesis of cdHAp took a shorter time than by other available methods. The authors have confirmed the formation of 16–39 nm long and 7–16 nm thick needle-like nanocrystallites.

Murugan et al. [51] also applied microwaves for the synthesis of carbonated HAp (cHAp) with structural and chemical similarities to biological apatite. Its bioresorbable nature was determined by its in vitro solubility under physiological conditions and was found to be higher than that of HAp.

cHAp nanospheres for tissue engineering scaffolds have been prepared by Zhou and his group [59] using a nanoemulsion method. Nanoemulsions are a new class of emulsions having very fine and uniform droplet sizes, usually in the range of 20–200 nm [60]. Using this method, microemulsion-like dispersions can be obtained using very low surfactant concentrations, or even without any surfactant. Using this method, nanosized, B-type cHAp particles of spherical shape have been synthesized [59].

Landi et al. [61] have synthesized Mg-doped HAp by a wet-chemical method and 5.7 mol% Mg-doped HAp showed the best result in biological applications. No genotoxicity, carcinogenicity, or cytotoxicity was observed using these materials, which makes them biocompatible. Mg-doped HAp also showed superior material resorption and better osteoconductivity than stoichiometric HAp.

HAp has excellent biocompatibility, osseointegration, and biostability [9, 10]. Additionally, its unique characteristic of osteoconduction [11–14] in host osseous materials has been employed in dental implants [62–66] and in bone cement applications [67–69]. HAp-based implants are made both from natural products like animal bones or corals, and also from synthetic HAp. The use of HAp materials made from animal bones are advantageous because they inherit some properties such as chemical composition and structure from the raw material [70–72]. In spite of these advantages, HAp-based implants from synthetic HAp are preferable because of their uniform composition, high biocompatibility, overall safety, and their completely controllable microstructure [73]. The limitations of an HAp implant include its hardness and brittleness, which makes it difficult to shape in the complex forms required for bone treatment, and its easy migratory tendency from the implanted sites. To overcome these limitations, many efforts have been made towards development of novel HAp/organic polymer composites [74]. In the early 1980s, Bonfield introduced the concept of using bioactive HAp particles/modified polymer composites as implant materials for bone replacement [75]. Since then, HAp-modified polymer-based biocomposites have been widely studied for bone tissue replacement. In the next section, we discuss the use of polymers and give a brief introduction to these polymers and different methods for their preparation.

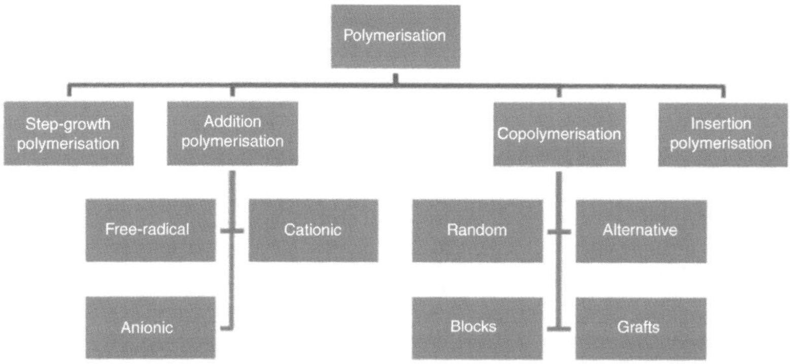

Fig. 6 Different types of polymerization

4 Polymers

Polymers are molecules consisting of a long chain of repeating smaller units called monomers. Polymers have the highest molecular weight of any molecules and may consist of billions of atoms. Human DNA is a polymer with over 20 billion constituent atoms. Proteins, or the polymers of amino acids, and many other molecules that make up life are polymers. Polymers are the largest and most diverse class of known molecules. The length of the polymer chain is specified by the number of repeat unit in the chain and depends on the degree of polymerization (DP). The molecular weight of the polymer is the product of the molecular weight of the repeat unit and the DP. In the next section, different methods for the synthesis of polymers are described.

4.1 Synthesis of Polymers

A number of different methods have been used to synthesize polymers from suitable monomers, such as condensation or step-growth polymerization, addition polymerization, copolymerization, and insertion polymerization (Fig. 6).

4.1.1 Condensation or Step-Growth Polymerization

Condensation polymerization or step-growth polymerization is a process by which two molecules join together, resulting in loss of small molecules, which are often water. The type of end product resulting from a condensation polymerization is dependent on the type of reactive functional groups at the monomeric end, such as –OH, –COOH, –COOR, –COCl, $-NH_2$, –CHO, and –NCO. The reaction is usually a non-catalyzed chemical condensation reaction and continues until the entire amount of one reagent is used up. The equilibrium for these reactions can be controlled. An example of condensation polymerization is given by:

$$CH_3 - NH_2 + HOOC - CH_2 - HC_3 \rightarrow CH_3 - NH - CO - CH_2 - CH_3 + H_2O$$
$$\text{Amine} \quad + \quad \text{Acid} \quad = \quad \text{Amide} \quad + \quad \text{Water}$$

4.1.2 Addition Polymerization

In addition polymerization, simple molecules or monomers are added to each other to form long-chain molecules (polymers) without by-products, thus yielding a polymeric product in which the molecular formula of the repeating unit is identical to that of the monomer. The molecular weight of the polymer so formed is thus the total of the molecular weights of all of the combined monomer units. There are three commonly used types of addition polymerization: free-radical polymerization, cationic polymerization, and anionic polymerization, which are described below. An example of addition polymerization is given by:

$$CH_3 - N = C = O + HO - CH_2 - CH_3 \rightarrow CH_3 - NH - CO - O - CH_2 - CH_3$$
$$\text{Isocyanate} \quad + \quad \text{Alcohol} \quad = \quad \text{Urethane}$$

Free- radical polymerization

A free radical is usually generated by the decomposition of a relatively unstable material called a radical initiator. This process includes initiation, chain propagation, and termination. Examples of this kind of polymerization include polymerization of ethylene to polyethylene, vinyl chloride to poly(vinyl chloride), and vinyl acetate to poly(vinyl acetate).

Cationic polymerization

In this polymerization, the active site has a positive charge, i.e., the process goes through formation of a carbocation intermediate. A monomer containing an electron-donating group is the most appropriate for this type of polymerization. Polymerization initiation can be achieved using a mineral acid or Lewis acid. However, Lewis acid-catalyzed polymerization generally takes place in water or methanol solvent.

Anionic polymerization

In anionic polymerization, the active site on the end of a growing chain is negatively charged, i.e. carboanion. Certain types of ring-opening polymerizations can proceed by this mechanism. In inert solvent without any contaminants, carboanion groups might be present in the system due to the absence of termination reactions. In protic solvents, the chain reaction can be stopped by transfer to solvent.

4.1.3 Copolymerization

In copolymerization, a mixture of more than one monomer is allowed to polymerize and co-polymer is formed. Co-polymers contains multiple units of each monomer used for polymerization in the same polymeric chain. The copolymer with a relatively random distribution of the different monomer in its structure is referred to as a random copolymer. For example, a random copolymer with two different monomers **A** and **B** can be depicted as **AABBBABAAAABBA**. The other types of copolymers are alternating, block, and graft. In the alternating copolymer, two monomers alternate in a regular fashion, e.g., **ABABABABABABAB**. In a block copolymer, one or more long uninterrupted sequence of each monomer can be seen, e.g., **AAAAAABBBBBBBB**; it is a linear copolymer. A graft copolymer is a branched copolymer with a backbone of one type of polymer to which one or more side chains of another type of polymer are attached:

AAAAAAAAAAAAAA.
B
B
B
B
B

An example of copolymerization is given by:

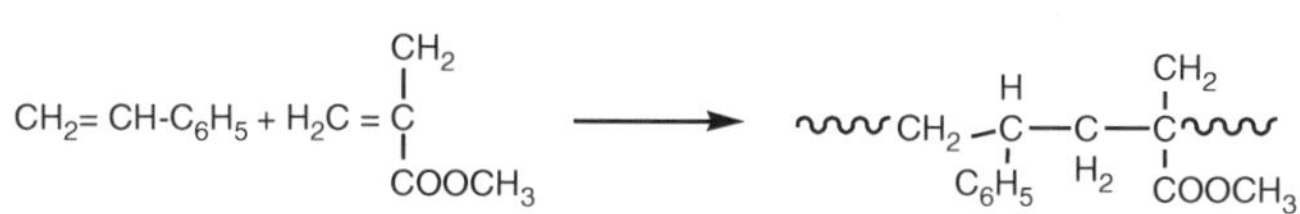

Styrene Methyl methacrylate Copolymer

4.1.4 Insertion Polymerization or Coordination Polymerization

Insertion polymerization is usually catalyzed by an organometallic compound and a monomer is inserted between the metal ion and the carbon atom adjacent to the metal (carboanion). As a result, the polymer chain is pushed out from the solid catalyst surface. Usually, addition polymerizations are involved in the polymerization of olefinic monomers. Polymerization occurs via an insertion of a monomer at the end of the growing chain, mediated by a catalyst. The catalyst stays at the end of the growing chain. Polymers synthesized by insertion polymerization are typically characterized by a very high stereo-regularity. An example of such a polymerization technique is the Ziegler-Natta-catalyzed polymerization.

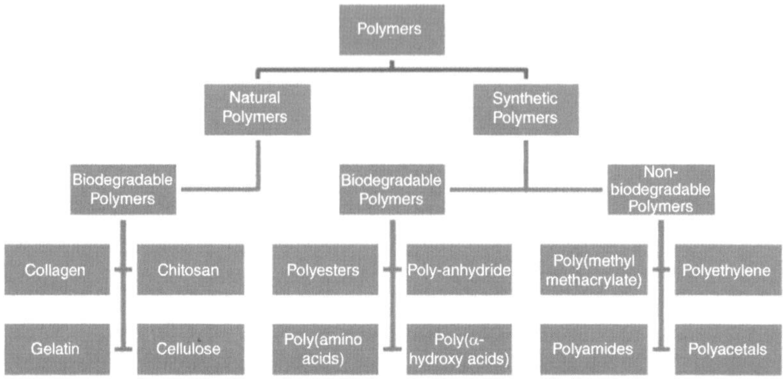

Fig. 7 Classes of polymers

4.2 Classification of Polymers

Polymers can be classified into two main categories, namely, natural and synthetic polymers. The detailed sub-classes of these two categories are given in Fig. 7.

5 Polymers in Bone Tissue Engineering

Metallic biomaterials have their main applications in load-bearing systems such as hip and knee prostheses and for the fixation of internal and external bone fractures. The metallic implants most widely used in orthopedic surgery are stainless steel, titanium and titanium-base alloy, and cobalt–chromium alloys. However, stress-shielding problems can be associated with these materials due to their high elastic modulus [76]. When stiff metal or ceramic implant is placed in bone, it will resorb because of the reduced mechanical environment. For total hip replacement, bone resorption in the proximal femur creates problems like aseptic loosening of the prosthesis, caused by the stress and strain in the femoral cortex after the metallic femoral hip replacement is implanted [8]. There are many other disadvantages of using metallic and ceramic materials for tissue engineering, such as the lack of biodegradability and biocompatibility and limitations in their processability [77–80]. Polymer biomaterials have been using clinically since the 1960s. They offer several advantages like the ease of manufacture of products with diverse and complex shapes, reasonable cost, availability, and the wide range of physical and mechanical properties compared to metallic or ceramic materials [81, 82]. The stiffness of polymeric materials is close to the stiffness of bone, in contrast to metals or ceramics, which gives an additional advantage for certain applications [83, 84]. Polymers can be applied as bone implants because of their tensile strength, the elastic modulus (Fig. 4), and biostability. The most important synthetic nondegradable polymers in bone tissue engineering are polyethylene (PE), polypropylene (PP), selected polyurethanes (PUs), polytetrafluoroethylene (PTFE), poly(vinyl chloride) (PVC), polyamides (PAs), poly (methyl methacrylate) (PMMA), polyoxymethylene (POM, polyacetal resin), polycarbonate (PC), poly(ethylene terephtalate) (PET), poly(ether ether ketone) (PEEK), and polysulfone (PSU). These polymers have several applications in the medical field, ranging from PTFE vascular grafts to ultrahigh molecular weight polyethylene (UHMWPE) acetabular cups [8, 85, 86].

6 Polymer/HAp Composites in Bone Tissue Engineering

Nanocomposites with bone-like properties depend on two main factors: (a) interfacial adhesion should be good between organic polymers and inorganic HAp, and (b) the dispersion of HAp should be uniform at the nanolevel in the polymer matrix [87]. Poor interface between phases is due to the lack of adhesion between HAp particles and the polymer matrix. Additionally, HAp might agglomerate if it is not evenly dispersed in the polymer matrix through proper processing methods, which causes poor mechanical properties. During the past few decades, significant efforts have been made to synthesize polymer/nanoHAp systems that resemble bone structure. Generally, both nondegradable and degradable polymers are used for fabrication of polymer/HAp composite materials in bone tissue engineering.

6.1 Nondegradable Polymer/HAp Nanocomposites

Nondegradable polymers have good mechanical properties and chemical stability. Therefore, they are widely used in bone tissue engineering. However, improvement of their biocompatibility and performance are highly desirable in terms of clinical applications. The nondegradable polymer/HAp systems are generally applied for tissue that cannot be regenerated due to its large losses and, in the case of elderly patients, with a less effective tissue self-healing ability. The use of PMMA as nondegradable polymer is described below.

6.1.1 Poly(methyl methacrylate)

PMMA and its derivatives have been used most often in the bone cements used for fixation in orthopedic surgeries. PMMA is an amorphous thermoplastic polymer. It gives excellent optical clarity, weather resistance, surface hardness, good chemical resistance, rigidity, dimensional stability, low mold shrinkage, good impact strength, nontoxicity, and tastelessness. PMMA powders can be injected and compression molded, extruded, etc. Liquids can be cast into sheets, lenses, and rods [88, 89]. Even though the mechanical characteristics of PMMA are rather poor, it allows even distribution of implant loads and forms a strong mechanical bond with implants. However, there are several limitations to the use of PMMA, such as brittleness and shrinkage of PMMA, void production in processing [90–92], lack of adherence to the bone [93, 94], and the exothermic polymerization reaction, which can damage bone tissue [95, 96].

To minimize this problem, several techniques have been developed. De Santis et al. [97] have prepared a blend of PMMA and paraffin-based phase change material (PCM) that is able to store the thermal energy produced during polymerization of methyl methacrylate [98]. Migliaresi and coworkers have used acryl polymers such as poly(*n*-butyl methacrylate) (PBMA), for orthopedic applications. These polymers are characterized by lower exothermic effect, higher fracture toughness, and superior fatigue life along with lower toxicity to soft tissue and dental pulp [99]. Use of poly(hydroxyethyl methacrylate) (PHEMA) also showed improved biocompatibility [100].

Besides having an inert material for fibroblastic cells observed at the bone–cement interface, PMMA is still the current standard for cement-held prostheses. Vallo and his group measured the flexural, compressive, and fracture properties of PMMA bone cements that incorporated various amounts of HAp. They found that addition of up to 15 wt% HAp increases the flexural modulus and fracture toughness [89]. It was also observed that, with increasing HAp incorporation into PMMA, the biological responses of the bone cement increased and thus increased osteoblast adhesion and response [88]. Kwon and coworkers reported that a PMMA/HAp composite with 30 wt% of HAp increased the interfacial shear strength at the bone–implant interface 6 weeks after implantation in rabbits [101]. Several other research studies revealed that, depending on the type of bone cement, the addition of HAp can improve the mechanical properties of bone cements [102].

Moursi et al. [103] studied the response of osteoblasts to PMMA/HAp materials. They found that osteoblasts attached equally well to PMMA/HAp. In contrast, the increase of osteoblasts on PMMA/HAp was significantly improved, compared to PMMA, after 8 days in culture.

In addition, to improve adhesion between polymer matrix and HAp, several copolymer such as PMMA, PBMA, and PHEMA were grafted onto the surface of HAp using isocyanatoethyl methacrylate (ICEM) or hexamethylene diisocyanate (HMDI)/hydroxyethyl methacrylate (HEMA) [100]. These three methacrylate polymers were coupled to HAp particles via covalent bonding with isocyanate groups.

Skrtic et al. [104] studied methacrylate conversion and volumetric contraction in photo-polymerized composites with amorphous calcium phosphate (ACP). ACP was used because it is highly soluble in aqueous media and can be quickly transformed to HAp. Ideally, on polymerization, the resin systems used in composites should achieve the contradictory goals of both high vinyl conversion and minimal volumetric contraction. A high degree of vinyl conversion, with homogeneous network formation, is desirable because it maximizes the glass transition temperature of the matrix; thereby rendering it less susceptible to the softening effects of the aqueous oral environment. Several researchers have investigated the degree of vinyl conversion and volumetric contraction of different types of resin composites formulated with three types of ACP. They observed that the conversion of methacrylate functional groups in the resin matrix of the various types of composites is independent of the type of filler phase, but dependent on the monomer and the compositional factors of the resin matrix. It was also observed that volumetric contraction of these experimental composites appeared to depend not only on the type of resin, but also on the type of ACP. Kim and coworkers [105] proposed bioactive bone cement (BBC), composed of natural bone powder (HAp), chitosan, and commercially available PMMA-based bone cement. The investigators obtained three types of BBCs with different composition ratios: BBC I, BBC II, and BBC III with 10 wt% of chitosan and 40, 50, and 60 wt% of HAp, respectively. Observation of the interfacial area between the host bone and the bone cement indicated that the BBC II composite has numerous pores that could be expected to afford space for bone ingrowth. However, after 4 weeks, the gaps between the host bone and the BBC II became narrower and PMMA exhibited undesirable cleavage at the interfacial area; simultaneously, histological examinations of the interfaces at 4 weeks post-implantation demonstrated more new bone formations in the BBC II implant than in pure PMMA. In addition, the exothermic effects in the BBCs were considerably lower than for pure PMMA.

6.2 *Degradable Polymer/HAp Nanocomposites*

For the last 40 years, the need for so-called "biodegradable" therapeutic material systems has received much attention in order to replace the use of biostable (or long-lasting) materials. This evolution is aimed at helping injured or diseased tissue to self-repair because living systems have outstanding healing ability. Among these domains, one can distinguish surgery with sutures, osteosynthesis devices (screws, plates, staples, etc.), pharmacology with drug delivery systems, and tissue engineering [106]. Both synthetic and naturally derived biodegradable polymeric materials have been investigated extensively as biomaterials for bone tissue regeneration and reconstruction. An overview of degradable polymeric materials (both synthetic and natural) that are used as polymer matrices for HAp is presented in Table 2.

6.2.1 Synthetic Biodegradable Polymers

Synthetic biodegradable polymers have attracted much interest for short-term medical applications like sutures, drug delivery devices, orthopedic fixation devices, wound dressings, temporary vascular grafts, stents, different types of tissue engineered grafts, etc. [107]. The hydrolytically (synthetic in most cases) degradable polymers are usually preferred as implants compared to enzymatically (natural) degradable polymers due to their minimal site-to-site and patient-to-patient variations [108]. Polymers of glycolic acid and lactic acid have been using in the medical industry since the 1960s, beginning with biodegradable sutures [109]. Other materials, such as poly(dioxanone), poly(trimethylene carbonate) copolymers, and ϵ-caprolactone homo- and copolymers have also been used as medical devices [82, 110, 111]. Functional groups that are easy to hydrolyze, such as ester, phosphazene, anhydride, carbonate, amide, and urethane [112], are logically the most important for the synthetic biodegradable polymers [polyesters, polyphosphazenes, polyanhydrites, polyurethanes, and poly(amino acids)] currently used in biomedical science.

Synthetic biomaterials are generally biologically inert and have more predictable properties than natural polymers. Thus, for biological applications these materials requires a certain level of biological activity. However, strategies have only been developed to incorporate biological motifs (e.g., HAp for bone tissue regeneration) onto synthetic polymers to generate hybrid materials. However, only a few novel procedures are currently being introduced [113].

6.2.2 Natural Degradable Polymers

Polysaccharides

Polysaccharides contain monosaccharide units joined together by glycosidic linkages. Polysaccharides display unique biological functions ranging from cell signaling to immune recognition. Combined with new synthetic routes currently available for synthesis or modification of polysaccharides, their biodegradability and ability to form specific interactions make them one of the most important and extensively investigated groups of natural biomaterials.

Chitin and Chitosan

Chitosan (CTS) is a natural copolymer derived from chitin, which is a homopolymer comprising 2-acetamido-2-deoxy-β-D-glucopyranose units. The deacetylated form, namely 2-amino-2-deoxy-β-D-glucopyranose, is the major composition of CTS chains. With a minimum 50% deacetylation of chitin, it becomes soluble in dilute acids and is referred to as CTS. The main sources exploited are two marine crustaceans, shrimp and crab [114–119]. The structures of chitin and CTS are shown in Fig. 8. Recently CTS has attracted much attention due to its wide range of applications in medicine, drug delivery, waste-water treatment, biomembranes, and hydrogel development [114, 120–122]. CTS is also a potential candidate for pharmaceutical and cosmetic applications due to its biodegradability, biocompatibility, high charge density, nontoxicity, and absorption [123]. Furthermore, the

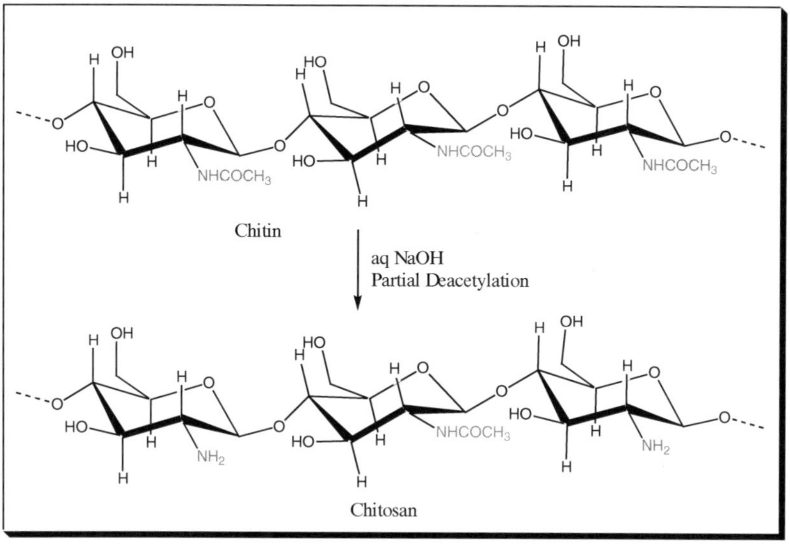

Fig. 8 Structure of chitin and chitosan

macromolecular chain of CTS can be stiff as well as having the ability to stabilize a liquid crystalline phase in acetic acid solution [124]. The stiffness of the macromolecular chain of CTS is very sensitive to pH and a small change in pH can change the properties of the chain. The application of CTS and its derivatives to nonviral gene delivery has been described in many papers, and Gerrit gives an overview of transfection studies that have been performed recently using CTS as transfection agent [125]. Many researchers have also attempted to modify the properties of CTS. The novel properties of CTS make it a versatile biomaterial for cell therapy, tissue engineering, and gene therapy. It is believed that these diverse approaches for regenerative medicine will produce materials with the required properties for the future [126]. For biomedical applications, CTS hydrogels and networks (for instance, interpenetrating polymer networks) formed by aggregation or complexation have been developed [127]. CTS-based products have also been used for the delivery of chemotherapeutics such as antibiotics, antiparasitics, anesthetics, painkillers, and growth promotants to mucosal epithelium for absorption for local or systemic activity [128].

CTS-derived products have found wide application in cosmetic formulations and these products are highly commercialized. Interesting characteristics that render CTS suitable for biomedical applications are a minimal foreign body reaction, an intrinsic antibacterial nature, and the ability to be molded into various geometries and forms such as porous structures that are suitable for cell ingrowth and osteoconduction. Due to its favorable gelling properties, CTS can deliver morphogenic factors and pharmaceutical agents in a controlled fashion. Its cationic nature allows it to form a complex with DNA molecules, making it an ideal candidate for gene delivery strategies [118, 129–131]. Covalent crosslinking of CTS leads to the formation of hydrogels with a permanent network structure because irreversible chemical links are formed. This type of linkage allows absorption of water and/or bioactive compounds without dissolution, and permits drug release by diffusion under pH-controlled conditions. Ionic crosslinked CTS hydrogels exhibit a higher swelling sensitivity to pH changes than covalently crosslinked CTS hydrogels. This fact extends the potential application of ionic crosslinked CTS since dissolution can occur in extreme acidic or basic pH conditions [132].

Wan et al. [116] have investigated the interactions of chitin with calcium species. The strategy for the dispersion of HAp in chitin to produce intimately blended materials has also been reported. Preliminary mechanical tests revealed a reduction in strength for the more highly filled composites, but retention of the plastic properties of the polymer was observed that might be favorable for bone-substitute applications [133]. Chitin/HAp composites with 25, 50, and 75% (w/w) of HAp fractions have also been investigated by Ge et al. [134] and processed into air- and freeze-dried materials. The cell culture study and implantation into the intramusculature of a rat model revealed that these materials are non-cytotoxic and degradable in in vivo systems. The presence of the HAp filler enhanced calcification as well as accelerated degradation of the chitin matrix. CTS/HAp composites with various ratios were obtained by Yamaguchi and coworkers [74] using the co-precipitation method. In these composite materials, calcium phosphate formed crystalline HAp in an acetic acid–lactic acid solvent system for CTS. The amorphous calcium phosphate was formed in organic acids containing more than two carboxyl groups, which resulted in aggregations (length 230 nm) of HAp nanocrystals aligned along the CTS molecules. This is possibly due to the fact that the calcium of HAp nanocrystals formed a complex with the amino groups in CTS. The composites were found to be mechanically flexible and could easily be formed into any desired shape. The mechanical strength could be enhanced by heat treatment in a saturated steam, and this effect was ascribed to the formation of hydrogen bonds between CTS macromolecules. Biodegradable CTS/Gel/HAp composites were prepared as 3D biomimetic scaffolds using phase separation by Zhao et al. [135] and obtained similar compositions to that of normal human bone. The pore size distribution and density of these composite materials was controlled by changing the content and the composition of the variables. Histological and immunohistochemical staining and SEM observations indicated that the osteoblasts attached to and proliferated on the scaffolds. The presence of HAp in the CTS/Gel composite also promoted initial adhesion of human mesenchymal stem cells (hMSC) and supported long-term growth in 3D porous CTS/Gel/HAp scaffolds [136].

A series of CTS/Gel/nanoHAp composites have been fabricated by Li et al. [137] by depositing nanoHAp on the surface of CTS/Gel network films. The amount of polymers in the CTS/Gel networks greatly influenced nucleation and the development of the nanoHAp crystalline phase. Organic functionalities such as –COOH, =C=O, or $-NH_2$ groups can be active sites for the coordination of calcium ions, leading to the formation of complexes that initiate and control formation of nanoHAp nuclei. Therefore, the size of nanoHAp can be controlled by changing the ratio of CTS and Gel to imitate the structure of natural bone.

Zhang et al. [138] fabricated CTS/HAp composites by a co-precipitation method and obtained excellent miscibility with no phase separation between CTS and nanoHAp in the composite. This could be due to strong specific interactions between the components, which endows these composites with good mechanical strength. The maximum compressive strength was about 120 MPa for the 30/70 wt% CTS/nanoHAp composite. In vitro tests showed that degradation of CTS in this composite takes place and that a layer of bone-like apatite forms on the surface of the composite, which is a sign of its high bioactivity.

The mechanical resistance of CTS-based composites with HAp was addressed by Hu et al. [139]. Investigators reported CTS/HAp multilayer nanocomposites with high strength and bending modulus, rendering the materials suitable for possible application in internal fixation of long bone fractures.

A wet-chemical method for production of CTS/HAp nanocrystal composites at low temperature has been demonstrated by Murungan and coworkers [140]. Limitations associated with nanoHAp, such as bioresorption and particle migration, can be alleviated by this approach. The rate of bioresorption of nanoHAp was improved by the addition of CTS, and the viscoelastic nature of the composites prevented the migration of particulate matter into the surrounding tissue upon post-implantation. In addition to these, the smoothness of the composites prevents the soft tissues around the implant from damage.

Injectable bone substitute material consisting of CTS, citric acid, and glucose solution as the liquid phase, and tricalcium phosphate powder as the solid phase, was developed by Liu and coworkers [141]. Four types of cements have been used to investigate the mechanical properties and in vitro biocompatibility of the material. In the presence of citric acid, tricalcium phosphate partially transformed into HAp and dicalcium phosphate.

Kong et al. [142] have investigated CTS/nanoHAp scaffolds for bone tissue regeneration. The bioactivity of the composite scaffolds has was examined by the incubation of apatite formed on the scaffolds in simulated body fluid (SBF), followed by study of the activity of preosteoblasts cultured on them. After incubation in SBF, carbonated HAp was formed on both CTS and CTS/nanoHAp scaffolds. Furthermore, with increasing nanoHAp content in the composite, the quantity of apatite formed on the scaffolds increased. Here, CTS/nanoHAp scaffolds formed more than the pure CTS scaffolds during the bimimetic process. Thus, the cells proliferated better on apatite-coated scaffolds than only on CTS scaffolds.

CTS/nanoHAp microsphere-based scaffolds were produced by Chesnutt et al. [143] using a co-precipitation method. The surface area and surface roughness of composite scaffolds were significantly greater than that of CTS scaffolds. Interestingly, composite scaffolds swelled marginally compared to CTS scaffolds, where significant swelling was observed, and were thus expected to maintain their shape intact in vivo. The compressive modulus of composite scaffolds was higher than the modulus of CTS scaffolds, and both were higher than previous CTS scaffolds fabricated by other techniques. Osteoblast proliferation was also significantly increased on composite scaffolds compared to CTS scaffolds. Recently, Kim et al. [144] obtained a highly flexible HAp/CTS composite film using homogeneously dispersed solutions of microscaled HAp powder and CTS at various concentrations. These films can contain up to 70% HAp without microcracks, phase separation, or brittleness. The HAp powders were well dispersed in the film without clustering or aggregation. These observations were explained by the presence of CTS, which retains the physico-mechanical properties of the film and captures the HAp powders in the matrix. It is anticipated that these films will be potentially useful for the direct or indirect delivery of HAp materials in localized areas of osteoporosis via a patch on skin or bone. Hybrid CTS/HAp composite materials were also developed by Araujo et al. [145]. The effects of the concentration of a crosslinking agent (genipin), concentration of lactic acid, and presence of HAp powder on the evolution of rheological properties have been studied. It was found that the concentration of lactic acid has a marginal influence on the rheological properties, affecting mostly the final microstructure. The pore size of composites decreases with increasing concentration of lactic acid. The microstructure of CTS scaffolds generally consists of large and interconnected pores, the size of which also tends to decrease with increasing amounts of genipin. Moreover, the CTS and CTS/HAp scaffolds exhibited an excellent bioactivity in vitro in the presence of SBF, showing the formation of an apatitic layer, even in the early stages of immersion.

7 Conclusions

Nature has set a high standard for the scientists and engineers who design bone grafts to assist in the repair or regeneration of the functions of defective bone tissues into normal healthy tissues. In this regard, the most preferable goal is to design and produce bone grafts that mimic all the functional physical, mechanical, chemical, and biological properties of natural bone. In this chapter, a detailed account of the structure of natural bone and the different types of bone grafting strategies are described. Recently, polymer/HAp-based nanocomposite materials have received much attention in biomimetic bone grafting because the interfacial adhesion between organic polymers and inorganic nanoHAp is good and also the dispersion of HAp is uniform at the nanolevel. NanoHAp also provides a large surface area and surface reactivity and thus the composites are compositionally and structurally similar to host bone and able to promote bone-related cellular functions extensively. A number of methods have been reported and presented in this chapter for bone grafting using nanoHAp organic polymers. Among several organic polymers, the use of nonbiodegradable and biodegradable CTS/nanoHAp has been described in detail. However, no synthetic graft is yet able to match the performance of natural bone tissue, even with the advances in science and technology. It is believed that a combination of osteoconductive matrix with osteogenic cells and osteoinductive growth factors creates an ideal bone graft. However, the design of bone grafts is still at the laboratory research level, and thus development of novel bone grafting techniques is highly desirable and is a big task for scientists.

Considering the advancements in nanotechnology and tissue engineering in recent years, there is a bright chance in the near future to formulate nanocomposite-based bone grafting methods to replace autogenic bone grafts.

References

1. Einhorn TA, Lee CA (2001) Bone regeneration: new findings and potential clinical applications. J Am Acad Orthop Surg 9:157–165
2. Wagh A (2004) Chemically bonded phosphate ceramics: twenty-first century materials with diverse applications. Elsevier Science, New York
3. Blokhuis TJ, Lindner T (2008) Allograft and bone morphogenetic proteins: an overview. Injury 39:S33–S36
4. Bostrom MP, Seigerman DA (2005) The clinical use of allografts, demineralized bone matrices, synthetic bone graft substitutes and osteoinductive growth factors: a survey study. HSS J 1(1):9–18
5. Eagan MJ, McAllister DR (2009) Biology of allograft incorporation. Clin Sports Med 28 (2):203
6. Goldberg VM, Stevenson S (1994) Bone-graft options – fact and fancy. Orthopedics 17 (9):809
7. Giannoudis P, Dinopoulos H, Tsiridis E (2005) Bone substitutes: an update. Injury 36:20–27
8. Wang M (2003) Developing bioactive composite materials for tissue replacement. Biomaterials 24:2133–2151
9. Martin RB (1999) Bone as a ceramic composite material. Mater Sci Forum 293:5–16
10. Nunes CR, Simske SJ, Sachdeva R, Wolford LM (1997) Long-term ingrowth and apposition of porous hydroxylapatite implants. J Biomed Mater Res 36(4):560–563
11. Yoshikawa H, Myoui A (2005) Bone tissue engineering with porous hydroxyapatite ceramics. J Artif Organs 8(3):131–136
12. Monroe EA, Votava W, Bass DB, Mullen JM (1971) New calcium phosphate ceramic material for bone and tooth implants. J Dent Res 50(4):860–861
13. Tamai N, Myoui A, Tomita T, Nakase T, Tanaka J, Ochi T, Yoshikawa H (2002) Novel hydroxyapatite ceramics with an interconnective porous structure exhibit superior osteoconduction in vivo. J Biomed Mater Res 59(1):110–117
14. Levitt SR, Crayton PH, Monroe EA, Condrate RA (1969) Forming method for apatite prostheses. J Biomed Mater Res 3(4):683–684
15. Steele DG, Bramblett CA (1998) The anatomy and biology of the human skeleton. Texas A&M University Press, TX, p 4
16. Kaplan FS, Hayes WC, Keaveny TM, Boskey A, Einhorn TA, Iannotti J (1994) Form and function of bone. In: Simon SR (ed) Orthopaedic basic science. American Academy of Orthopaedic Surgeons, Columbus, OH, pp 127–185
17. Gehron Robey P, Bianco P, Termine JD (1992) The cellular and biology and molecular biochemistry of bone formation. In: Favus MJ, Coe FL (eds) Disorders of bone and mineral metabolism. Raven, New York
18. Lian JB, Stein GS, Canalis E, Gehron Robey P, Boskey AL (1999) Bone formation: osteoblast lineage cells, growth factors, matrix proteins, and the mineralization process. In: Favus MJ (ed) Primer on the metabolic bone diseases and disorders of mineral metabolism. Lippincott Williams & Wilkins, Philadelphia, PA, pp 14–29
19. Webster TJ (2001) Nanophase ceramics: the future orthopedic and dental implant material. In: Ying JY (ed) Advances in chemical engineering, Nanostructured materials, vol 27. Academic, San Diego, CA, pp 126–160

20. Kuhn-Spearing L, Rey C, Kim HM, Glimcher MJ (1996) Carbonated apatite nanocrystals of bone. In: Bourell DL (ed) Synthesis and processing of nanocrystalline powder. The Minerals, Metals and Materials Society, Warrendale, PA
21. Rho JY, Kuhn-Spearing L, Zioupos P (1998) Mechanical properties and the hierarchical structure of bone. Med Eng Phys 20:92–102
22. Smith R (2004) Bone health and osteoporosis: a report of the surgeon general. U.S. Department of Health and Human Services, Public Health Service, Office of the Surgeon General, Rockville, MD, pp 68–70
23. Cowin SC, van Buskirk WC, Ashman RB (1987) In: Skalak R, Chien S (eds) Handbook of bioengineering. McGraw-Hill, New York
24. Liu H, Webster TJ (2007) Bioinspired nanocomposites for orthopedic applications. In: Webster TJ (ed) Nanotechnology for the regeneration of hard and soft tissues. World Scientific, Singapore, pp 1–52
25. Guyton AC (1991) Textbook of medical physiology. Saunders, Philadelphia, PA, pp 868–881
26. Marin BR, Burr DB (1989) Structure function and adaptation of compact bone. Raven, New York, pp 106–107
27. Trippel SB (1998) Potential role of insulin like growth factors in fracture healing. Clin Orthop Relat Res 355S:S301–S313
28. Damien CJ, Parsons JR (1991) Bone graft and bone graft substitutes: a review of current technology and applications. J Appl Biomater 2:187–208
29. Murugan R, Ramakrishna S (2004) Nanostructured biomaterials. In: Nalwa HS (ed) Encyclopedia of nanoscience and nanotechnology, vol 7. American Scientific, California, pp 595–613
30. Welch RD, Zhang H, Bronson DG (2003) Experimental tibial plateau fractures augmented with calcium phosphate cement or autologous bone graft. J Bone Joint Surg Am 85:222–231
31. Myerson MS, Neufeld SK, Uribe J (2005) Fresh-frozen structural allografts in the foot and ankle. J Bone Joint Surg Am 87:113–120
32. Li XD, Hu YY (2001) The treatment of osteomyelitis with gentamicinreconstituted bone xenograft-composite. J Bone Joint Surg Br 83:1063–1068
33. Cypher TJ, Grossman JP (1996) Biological principles of bone graft healing. J Foot Ankle Surg 35:413–417
34. Murugan R, Ramakrishna S (2007) Nanoengineered biomimetic bone-building blocks. Top Appl Phys 109:301–352
35. Salgado AJ, Coutinho OP, Reis RL (2004) Bone tissue engineering: state of the art and future trends. Macromol Biosci 4(8):743–765
36. Gazdag AR, Lane JM, Glaser D, Forster RA (1995) Alternatives to autogenous bone graft: efficacy and indications. J Am Acad Orthop Surg 3:1–8
37. Asahina I, Sato I, Oda M, Marukawa E, Imranul AM, Enomoto S (1999) In: Bone engineering, 1st edn. Em Squared, Toronto, p 526
38. Lexer E (1908) Substitution of whole or half joints freshly amputated extremities by free plastic operation. Surg Gynecol Obstet 6:601–607
39. Mankin HJ, Gebhardt MC (1996) Long-term results of allograft replacement in the management of bone tumours. Clin Orthop Relat Res 324:86–97
40. Simonds RJ, Holmberg SD, Hurwitz RZ (1992) Transmission of human immunodeficiency virus type 1 from a seronegative organ and tissue donor. N Engl J Med 326:726–732
41. Salama R (1983) Xenogeneic bone grafting in humans. Clin Orthop Relat Res 174:113–121
42. Salama R (1973) Recombined grafts of bone and marrow. J Bone Joint Surg Br 55:402–417
43. Mears DC (1977) Metals in medicine and surgery. Int Metals Rev 218:119–155
44. De Boer HH (1988) The history of bone grafts. Clin Orthop 226:292–298
45. Perry CR (1999) Bone repair techniques, bone graft, and bone graft substitutes. Clin Orthop Relat Res 360:71–86
46. Bohner M (2000) Calcium orthophosphates in medicine: from ceramics to calcium phosphate cements. Injury 31:SD37–SD47

47. Tadic D, Epple M (2004) A thorough physicochemical characterisation of 14 calcium phosphate-based bone substitution materials in comparison to natural bone. Biomaterials 25(6):987–994
48. Ramay HRR, Zhang M (2004) Biphasic calcium phosphate nanocomposite porous scaffolds for load-bearing bone tissue engineering. Biomaterials 25(21):5171–5180
49. Pielichowska K, Blazewicz S (2010) Bioactive Polymer/Hydroxyapatite (Nano)composites for Bone Tissue Regeneration. Adv Polym Sci 232: 97–207
50. Murugan R, Ramakrishna S (2005) Aqueous mediated synthesis of bioresorbable nanocrystalline hydroxyapatite. J Cryst Growth 274:209–213
51. Murugan R, Ramakrishna S (2006) Production of ultra-fine bioresorbable carbonated hydroxyapatite. Acta Biomater 2:201–206
52. Nihouannen DL, Saffarzadeh A, Gauthier O (2008) Bone tissue formation in sheep muscles induced by a biphasic calcium phosphate ceramic and fibrin glue composite. J Mater Sci Mater Med 19:667–675
53. Siddharthan A, Seshadri SK, Sampath Kumar TS (2004) Microwave accelerated synthesis of nanosized calcium deficient hydroxyapatite. J Mater Sci Mater Med 15:1279–1284
54. Hao J, Liu Y, Zhou S (2003) Investigation of nanocomposites based on semiinterpenetrating network of [L-poly(ϵ-caprolactone)]/[net-poly(ϵ-caprolactone)] and hydroxyapatite nanocrystals. Biomaterials 24:1531–1539
55. Deng X, Hao JY, Wang C (2001) Preparation and mechanical properties of nanocomposites of poly(D, L-lactide) with Ca-deffcient hydroxyapatite nanocrystals. Biomaterials 22:2867–2873
56. Slosarczyk A, Paszkiewicz Z, Paluszkiewicz C (2005) FTIR and XRD evaluation of carbonated hydroxyapatite powders synthesized by wetmethods. J Mol Struct 744–747:657–661
57. Rapacz-Kmita A, Cz P, Slosarczyk A (2005) FTIR and XRD investigations on the thermal stability of hydroxyapatite during hot pressing and pressureless sintering processes. J Mol Struct 744–747:653–656
58. Zhang Y, Zhou L, Xue N (2003) Oriented nano-structured hydroxyapatite from the template. Chem Phys Lett 376:493–497
59. Zhou WY, Wang M, Cheung WL (2008) Synthesis of carbonated hydroxyapatite nanospheres through nanoemulsion. J Mater Sci Mater Med 19:103–110
60. Solans C, Izquierdo P, Nolla J (2005) Nano-emulsions. Curr Opin Coll Interface Sci 10:102–110
61. Landi E, Logroscino G, Proietti L (2008) Biomimetic Mg-substituted hydroxyapatite: from synthesis to in vivo behaviour. J Mater Sci Mater Med 19:239–247
62. Ogiso M (1998) Reassessment of long-term use of dense HA as dental implant: case report. J Biomed Mater Res 43(3):318–320
63. Chu TMG, Orton DG, Hollister SJ, Feinberg SE, Halloran JW (2002) Mechanical and in vivo performance of hydroxyapatite implants with controlled architectures. Biomaterials 23 (5):1283–1293
64. Ayers RA, Simske SJ, Nunes CR, Wolford LM (1998) Long-term bone ingrowth and residual microhardness of porous block hydroxyapatite implants in humans. J Oral Maxillofac Surg 56 (11):1297–1301
65. Le Guθhennec L, Soueidan A, Layrolle P, Amouriq Y (2007) Surface treatments of titanium dental implants for rapid osseointegration. Dent Mater 23(7):844–854
66. Fathi MH, Salehi M, Saatchi A, Mortazavi V, Moosavi SB (2003) In vitro corrosion behavior of bioceramic, metallic, and bioceramic-metallic coated stainless steel dental implants. Dent Mater 19(3):188–198
67. Schmitz JP, Hollinger JO, Milam SB (1999) Reconstruction of bone using calcium phosphate bone cements: a critical review. J Oral Maxillofac Surg 57(9):1122–1126

68. Mazor Z, Peleg M, Garg AK, Chaushu G (2000) The use of hydroxyapatite bone cement for sinus floor augmentation with simultaneous implant placement in the atrophic maxilla. A report of 10 cases. J Periodontol 71(7):1187–1194
69. Deb S, Braden M, Bonfield W (1995) Water absorption characteristics of modified hydroxyapatite bone cements. Biomaterials 16(14):1095–1100
70. Haberko K, Bucko M, Brzezinska-Miecznik B (2006) Natural hydroxyapatite – its behavior during heat treatment. J Eur Ceram Soc 26:537–542
71. Gao Y, Cao WL, Wang XY (2006) Characterization and osteoblast-like cell compatibility of porous scaffolds: bovine hydroxyapatite and novel hydroxyapatite artificial bone. J Mater Sci Mater Med 17:815–823
72. Ruksudjarit A, Pengpat K, Rujijanagul G (2008) Synthesis and characterization of nanocrystalline hydroxyapatite from natural bovine bone. Curr Appl Phys 8:270–272
73. Ono I, Tateshita T, Nakajima T (2000) Evaluation of a high density polyethylene fixing system for hydroxyapatite ceramic implants. Biomaterials 21:143–151
74. Yamaguchi I, Tokuchi K, Fukuzaki H (2001) Preparation and microstructure analysis of chitosan/hydroxyapatite nanocomposites. J Biomed Mater Res 55:20–27
75. Bonfield W, Grynpas MD, Tully AE (1981) Hydroxyapatite reinforced polyethylene – a mechanically compatible implant material for bone replacement. Biomaterials 2:185–186
76. Fan JP, Tsui CP, Tang CY (2004) Influence of interphase layer on the overall elastoplastic behaviors of HA/PEEK biocomposite. Biomaterials 25:5363–5373
77. Liu X, Ma PX (2004) Polymeric scaffolds for bone tissue engineering. Ann Biomed Eng 32:477–486
78. Yaszemski MJ (1996) The evolution of bone transplantation: molecular, cellular, and tissue strategies to engineer human bone. Biomaterials 7:175–185
79. Crane GM, Ishaug SL, Mikos AG (1995) Bone tissue engineering. Nat Med 1:1322–1324
80. Chapekar MS (2000) Tissue engineering: challenges and opportunities. J Biomed Mater Res 53:617–620
81. Kenny SM, Buggy M (2003) Bone cements and fillers: a review. J Mater Sci Mater Med 14:923–938
82. Middleton JC, Tipton AJ (2000) Synthetic biodegradable polymers as orthopedic devices. Biomaterials 21:2335–2346
83. Burg KJL, Porter S, Kellam JF (2000) Biomaterial developments for bone tissue engineering. Biomaterials 21:2347–2359
84. Eschbach L (2000) Nonresorbable polymers in bone surgery. Injury 31:SD22–SD27
85. Mano JF, Sousa RA, Boesel LF (2004) Bioinert, biodegradable and injectable polymeric matrix composites for hard tissue replacement: state of the art and recent developments. Comp Sci Technol 64:789–817
86. Blazewicz S (2006) Non-metalic multifunctional composites in biomaterials engineering. In: Nadolny AJ (ed) Biomaterials in regenerative medicine. Polish Academy of Sciences, Warszawa
87. Lee HJ, Kim SE, Choi HW (2007) The effect of surface-modified nano-hydroxyapatite on biocompatibility of poly(e-caprolactone)/hydroxyapatite nanocomposites. Eur Polym J 43:1602–1608
88. Dalby MJ, Silvio LD, Harper EJ (2002) Increasing hydroxyapatite incorporation into poly (methylmethacrylate) cement increases osteoblast adhesion and response. Biomaterials 23:569–576
89. Vallo CI, Montemartini PE, Fanovich MA (1999) Polymethylmethacrylate-based bone cement modified with hydroxyapatite. J Biomed Mater Res (Appl Biomater) 48:150–158
90. Lewis G (1997) Properties of acrylic bone cement: state of the art review. J Biomed Mater Res 38(2):155–182
91. Harper EJ, Bonfield W (2000) Tensile characteristics of ten commercial, acrylic bone cements. J Biomed Mater Res 53(5):605–616

92. Barralet JE, Gaunt T, Wright AJ, Gibson IR, Knowles JC (2002) Effect of porosity by compaction on compressive strength and microstructure of calcium phos-phate cement. J Biomed Mater Res B Appl Biomater 63(1):1–9
93. Zambonin G, Colucci S, Cantatore F, Grano M (1998) Response of human osteoblasts to polymethylmetacrylate in vitro. Calcif Tissue Int 62(4):362–365
94. Cunin G, Boissonnet H, Petite H, Blanchat C, Guillemin G (2000) Experimental vertebroplasty using osteo-conductive granular material. Spine 25(9):1070–1076
95. Lu JX, Huang ZW, Tropiano P (2002) Human bio-logical reactions at the interface between bone tissue and polymethylmethacrylate cement. J Mater Sci Mater Med 13(8):803–809
96. Heini PF, Walchli B, Berlemann U (2000) Percutaneous transpedicular vertebroplasty with PMMA: operative technique and early results. Eur Spine J 9(5):445–450
97. De Santis R, Ambrogi V, Carfagna C (2006) Effect of microencapsulated phase change materials on the thermo-mechanical properties of poly(methyl-methacrylate) based biomaterials. J Mater Sci Mater Med 17:1219–1226
98. Pielichowski K, Flejtuch K (2002) Differential scanning calorimetry studies on polyethylene glycol with different molecular weights for thermal energy storage materials. Polym Adv Techn 13:690–696
99. Migliaresi C, Fambri L, Kolarik J (1994) Polymerization kinetics, glass transition temperature and creep of acrylic bone cements. Biomaterials 15:875–881
100. Liu Q, de Wijn JR, Blitterswijk CA (1998) Covalent bonding of PMMA, PBMA, and poly (HEMA) to hydroxyapatite particles. J Biomed Mater Res 40:257–263
101. Kwon SW, Kim YS, Woo YK (1997) Hydroxyapatite impregnated bone cement: in vitro and in vivo studies. Biomed Mater Eng 7:129–140
102. Harper EJ, Behiri JC, Bonfield W (1995) Flexural and fatigue properties of a bone cement based upon polyethylmethacrylate and hydroxyapatite. J Mater Sci Mater Med 6:799–803
103. Moursi AM, Winnard AV, Winnard PL (2002) Enhanced osteoblast response to a polymethylmethacrylate–hydroxyapatite composite. Biomaterials 23:133–144
104. Skrtic D, Stansbury JW, Antonucci JM (2003) Volumetric contraction and methacrylate conversion in photopolymerized amorphous calcium phosphate/methacrylate composites. Biomaterials 24:2443–2449
105. Kim SB, Kim YJ, Yoon TL (2004) The characteristics of a hydroxyapatite–chitosan–PMMA bone cement. Biomaterials 25:5715–5723
106. Vert M (2007) Polymeric biomaterials: strategies of the past vs. strategies of the future. Prog Polym Sci 32:755–761
107. Ratner BD, Hoffman AS, Schoen FJ (1996) Biomaterials science. In: Lemons JE (ed) An introduction to materials in medicine. Academic, San Diego, CA
108. Katti DS, Lakshmi S, Langer R (2002) Toxicity, biodegradation and elimination of polyanhydrides. Adv Drug Deliv Rev 54:933–961
109. Gilding DK, Reed AM (1979) Biodegradable polymers for use in surgery – polyglycolic/poly (lactic acid) homo- and copolymers. Polymer 20:1459–1464
110. Barrows TH (1986) Degradable implant materials: a review of synthetic absorbable polymers and their applications. Clin Mater 1:233–257
111. Gunatillake P, Mayadunne R, Adhikari R (2006) Recent developments in biodegradable synthetic polymers. Biotechnol Annu Rev 12:301–347
112. Li S (1999) Hydrolytic degradation characteristicts of aliphatic polyesters derived from lactic and glycolic acids. J Biomed Mater Res 48:342–353
113. Nair LS, Laurencin CT (2007) Biodegradable polymers as biomaterials. Prog Polym Sci 32:762–798
114. Rinaudo M (2006) Chitin and chitosan: properties and applications. Prog Polym Sci 31:603–632
115. Khor E, Lim LY (2003) Implantable applications of chitin and chitosan. Biomaterials 24:2339–2349

116. Wan ACA, Khor E, Hastings GW (1997) Hydroxyapatite modified chitin as potential hard tissue substitute material. J Biomed Mater Res 38:235–241
117. Kim IY, Seo SJ, Moon HS (2008) Recent developments in biodegradable synthetic polymers. Biotechnol Adv 26:1–21
118. Di Martino A, Sittinger M, Risbud MV (2005) Chitosan: a versatile biopolymer for orthopaedic tissue-engineering. Biomaterials 26:5983–5990
119. Sionkowska A (2011) Current research on the blends of natural and synthetic polymers as new biomaterials: review. Prog Polym Sci 36:1254–1276
120. Struszczyk MH (2002) Chitin and chitosan. Part II. Applications of chitosan. Polimery 47:396–403
121. Muzzarelli R, Baldassarre V, Conti F, Ferrara P, Biagini G, Gazzanelli G, Vasi V (1988) Biological activity of chitosan: ultrastructural study. Biomaterials 9:247–252
122. Majeti N, Kumar R (2000) A review of chitin and chitosan applications. React Funct Polym 46:1–27
123. Hudson SM, Smith C (1998) Polysaccharide: chitin and chitosan: chemistry and technology of their use as structural materials. In: Kaplan DL (ed) Biopolymers from renewable resources. Springer, New York, pp 96–118
124. Terbojevich M, Cosani A, Conio G, Marsano E, Bianchi E (1991) Chitosan: chain rigidity and mesophase formation. Carbohydr Res 209:251–260
125. Gerrit B (2001) Chitosans for gene delivery. Adv Drug Deliv Rev 52:145–150
126. Chunmeng S, Ying Z, Xinze R, Meng W, Yongping S, Tianmin C (2006) Therapeutic potential of chitosan and its derivatives in regenerative medicine. J Surg Res 133:185–192
127. Berger J, Reist M, Mayer JM, Felt O, Gurny R (2004) Structure and interactions in chitosan hydrogels formed by complexation or aggregation for biomedical applications. Eur J Pharm Biopharm 57:35–52
128. Sevda S, McClure SJ (2004) Potential applications of chitosan in veterinary medicine. Adv Drug Deliv Rev 56:1467–1480
129. Kim TH, Jiang H, Jere D, Park IK, Cho MH, Nah JW (2007) Chemical modification of chitosan as a gene carrier in vitro and in vivo. Prog Polym Sci 32:726–753
130. Dodane V, Vilivalam VD (1998) Pharmaceutical applications of chitosan. Pharm Sci Technol Today 1:246–253
131. Sinha VR, Singla AK, Wadhawan S, Kaushik R, Kumria R, Bansal K, Dhawan S (2004) Chitosan microspheres as a potential carrier for drugs. Int J Pharm 274:1–33
132. Berger J, Reist M, Mayer JM, Felt O, Peppas NA, Gurny R (2004) Structure and interactions in covalently and ionically crosslinked chitosan hydrogels for biomedical applications. Eur J Pharm Biopharm 57:19–34
133. Wan ACA, Khor E, Hastings GW (1998) Preparation of a chitin-apatite composite by in situ precipitation onto porous chitin scaffolds. J Biomed Mater Res 41:541–548
134. Ge Z, Baguenard S, Lim LY (2004) Hydroxyapatite–chitin materials as potential tissue engineered bone substitutes. Biomaterials 25:1049–1058
135. Zhao F, Yin Y, Lu WW (2002) Preparation and histological evaluation of biomimetic three-dimensional hydroxyapatite/chitosan-gelatin network composite scaffolds. Biomaterials 23:3227–3234
136. Zhao F, Grayson WL, Ma T (2006) Effects of hydroxyapatite in 3-D chitosan–gelatin polymer network on human mesenchymal stem cell construct development. Biomaterials 27:1859–1867
137. Li J, Chen Y, Yin Y (2007) Modulation of nano-hydroxyapatite size via formation on chitosan–gelatin network film in situ. Biomaterials 28:781–790
138. Zhang L, Li Y, Yang A (2005) Preparation and in vitro investigation of chitosan/nanohydroxyapatite composite used as bone substitute materials. J Mater Sci Mater Med 16:213–219
139. Hu Q, Li B, Wang M (2004) Preparation and characterization of biodegradable chitosan/hydroxyapatite nanocomposite rods via in situ hybridization: a potential material as internal fixation of bone fracture. Biomaterials 25:779–785

140. Murugan R, Ramakrishna S (2004) Bioresorbable composite bone paste using polysaccharide based nano hydroxyapatite. Biomaterials 25:3829–3835
141. Liu H, Li H, Cheng W (2006) Novel injectable calcium phosphate/chitosan composites for bone substitute materials. Acta Biomater 2:557–565
142. Kong L, Gao Y, Lu G (2006) A study on the bioactivity of chitosan/nano-hydroxyapatite composite scaffolds for bone tissue engineering. Eur Polym J 42:3171–3179
143. Chesnutt BM, Viano AM, Yuan Y (2009) Design and characterization of a novel chitosan/nanocrystalline calcium phosphate composite scaffold for bone regeneration. J Biomed Mater Res 88A:491–502
144. Kim S-H, Lim B-K, Sun F (2009) Preparation of high flexible composite film of hydroxyapatite and chitosan. Polym Bull 62:111–118
145. Araujo AB, Lemos AF, Ferreira JM (2009) Rheological, microstructural, and in vitro characterization of hybrid chitosan-polylactic acid/hydroxyapatite composites. J Biomed Mater Res 88A:916–922

Adv Polym Sci (2013) 254: 169–202
DOI: 10.1007/12_2012_198

Published online: 15 March 2013

Biodegradable Polymers for Potential Delivery Systems for Therapeutics

Sanjeev K. Pandey, Chandana Haldar, Dinesh K. Patel, and Pralay Maiti

Abstract Biodegradable polymers are being extensively used with great interest in areas of nanobiotechnology such as drug delivery, diagnostics, and other applications for clinical and biomedical research covering cardiovascular diseases, diabetes, osteogenesis, cancer, and tissue engineering. Various biodegradable polymers such as poly(lactic acid), poly(lactic-co-glycolic acid), poly (ϵ-caprolactone), chitosan, gelatin, and poly(alkyl cyanoacrylates) have been extensively utilized as polymeric materials and devices for targeted cellular and tissue-specific clinical applications to achieve maximal therapeutic efficacy with minimal or no side effects. Recently, polymeric nanoparticles have revolutionized the area of nanobiotechnology by creating new opportunities for advancing medical science and disease treatment. Polymeric nanoparticles have the potential to act as a carrier of drugs and active constituents to targeted sites, protecting them from the environment and controlling their release rates, thereby enhancing their biological activity and decreasing the adverse side effects. This article compiles updated information regarding various biodegradable polymers, methods of preparation of biodegradable polymeric nanoparticles, and their application in therapeutic and diagnostic strategies for various diseases. This article will support research scientists and clinical physicians who are interested in the development and application of biodegradable polymeric nanoparticles as potential delivery systems for therapeutics.

Keywords Biodegradable polymers · Drug delivery · Nanoparticles

---

S.K. Pandey and C. Haldar
Department of Zoology, Faculty of Science, Banaras Hindu University, Varanasi 221005, India

D.K. Patel and P. Maiti (✉)
School of Materials Science and Technology, Indian Institute of Technology, Banaras Hindu University, Varanasi 221005, India
e-mail: pmaiti.mst@itbhu.ac.in

Contents

1 Introduction

Biodegradable polymers have been considered with great interest as polymeric materials and devices for an enormous number of biomedical applications due to their biocompatibility with tissue and cells together with minimal side effects. Synthetic as well as natural polymers have been extensively used as biodegradable polymeric biomaterials. Polymeric materials are degraded through hydrolytic or enzymatic cleavage of sensitive bonds present in the polymer [1]. During the past two decades, development of biodegradable polymeric materials has been achieved especially for biomedical applications. Biodegradable polymeric biomaterials are preferred for therapeutic devices such as scaffolds for tissue engineering and vehicles for controlled release of drugs. A considerable amount of work has been carried out on drug delivery using biodegradable polymeric materials. Various natural and synthetic biodegradable polymers have been employed for sustained release of drugs. Natural biodegradable polymers and proteins like bovine serum albumin (BSA), human serum albumin (HSA), collagen, gelatin, and hemoglobin have also been used for drug delivery, but the use of natural polymers is much lower due to their high cost of processing and tedious purification [2]. Synthetic biodegradable polymers have been used progressively in last two decades to deliver drugs, proteins, and bioactive molecules by using various devices like microspheres, microcapsules, nanoparticles, pellets, implants, films, and scaffolds.

Polyamides, poly(amino acids), poly(alkyl cyanoacrylates), polyesters, polyorthoesters, polyurethanes, and polyacrylamides have been used to develop various drug-loaded systems. Among them, thermoplastic aliphatic polyesters like poly(lactic acid) (PLA), poly(glycolic acid) (PGA), and especially their copolymer poly(lactic-co-glycolic acid) (PLGA), have generated great interest due to their excellent biocompatibility, bioresorbability, and biodegradability [3, 4]. The most commonly and extensively used biodegradable polymers are PLA, PLGA, poly (ϵ-caprolactone) (PCL), chitosan, gelatin, and poly(alkyl cyanoacrylates), which are usually polymerized through condensation or other conventional polymerization reactions. Hydrolytically degradable polymers have esters, orthoesters, anhydrides, carbonates, amides, urethanes, ureas, etc. in their backbone [5]. Biodegradable polymers are chosen on the basis of various characteristics such as degradability, solubility, permeability, and mechanical stability, which are adequate to influence the manufacturing and performance of polymeric devices. The entrapment efficiency and drug release from the polymeric devices are the important parameters for the polymers to be used as vehicle [6]. Biodegradable polymers are also highly preferred for the majority of biomedical applications in drug delivery, preventive medicine, clinical inspection, surgical treatments of various diseases, and tissue engineering scaffolds [7, 8]. The objective of this article is to summarize the current status of biodegradable polymers for various biomedical applications including transient implants, drug delivery vehicles, and tissue engineering scaffolds.

2 Polymer-Based Nanotechnology

Nanotechnology is a multidisciplinary area of applied science and technology intended to create and fabricate devices or materials that lie within the nanometer dimension. It has significant impact on the development of novel products at the molecular or submicron level [9]. The unique properties of nanomaterials arise from the nanoscale dimensions, which provide them with novel mechanical, thermal, and catalytic properties that are highly desirable for applications in commercial, medical, and environmental sectors. The application of nanotechnology for the prevention and treatment of diseases and for the diagnosis, monitoring, and control of biological systems is referred to as nanomedicine. Recently, the main goal of nanomedicine is controlled delivery and targeting of pharmaceutical, therapeutic, and diagnostic agents [10] with unique physicochemical and biological properties that might be used to overcome the current limitations of molecular imaging and gene/drug delivery [11, 12]. Significant efforts have already been employed in three main areas: (1) drug delivery systems for active molecules, proteins, peptides, and genes for localized or targeted delivery to the tissue; (2) application of nanobiotechnology to the molecular imaging and therapy of the cardiovascular system [8, 13]; and (3) targeted functional nanoparticles for cancer therapy and diagnosis [14, 15]. Nanomaterial devices used for drug delivery should have certain characteristics such as biocompatibility, drug compatibility, suitable biodegradation kinetics, suitable mechanical properties, and ease of processing [16, 17]. In the last two decades, a wide variety of synthetic biodegradable polymers have been used to create therapeutic delivery devices for drugs and genes due to their biocompatibility and biodegradability [18]. Several synthetic biodegradable and biocompatible polymers have been approved by the US FDA for drug delivery devices. While different aspects of polymeric devices have been reviewed in detail elsewhere, this article summarizes the more recent successful evidence for applying polymeric materials and tools to various health complications like cardiovascular disease, cancer, and immunological problems.

3 Biodegradation in Polymers

The term 'biodegradation' defines the gradual breakdown of a material by certain biological activity [19]. Biodegradable polymers have the tendency to degrade into smaller molecules and by-products under a controlled mechanism; the degradation products are then easily eliminated from the body through metabolic pathways. The use of biodegradable polymeric devices in drug delivery applications requires the polymer to be able to degrade under physiological conditions. Knowledge of the degradation mechanisms, degradation kinetics, evolution of mechanical properties, and identification of the degradation products of biomaterials are the basic requirements in selecting and designing polymeric materials for specific applications. Depending on the mode of degradation, polymeric biomaterials are classified into hydrolytically and enzymatically degradable polymers. Most of the naturally occurring polymers undergo enzymatic degradation. Hydrolytically degradable polymers contain chemical bonds such as esters, orthoesters, anhydrides, carbonates, amides, urethanes, ureas, etc. that are susceptible to hydrolysis [5]. The following processes and mechanisms play an important role in the degradation of polymeric materials.

3.1 Oxidation

Polymers degrade in the body fluids and tissues by chemical and/or enzymatic oxidation. It is well known that during the inflammatory response to foreign materials, inflammatory cells particularly leukocytes and macrophages produce highly reactive oxygen-containing species such as superoxide (O_2^-) hydrogen peroxide (H_2O_2), nitric oxide (NO), and hypochlorous acid (HOCl) [20, 21]. The oxidative effect of these reactive oxygen species causes polymer chain scission and ultimately contributes towards the degradation of polymeric materials. Superoxides accelerate the degradation of aliphatic polyesters by the cleavage of ester bonds through nucleophilic attack of O_2^- [22].

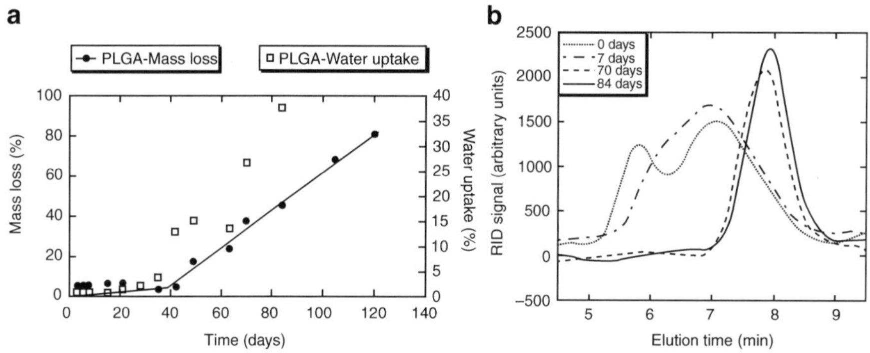

Fig. 1 (**a**) Degradation of PLGA as a function of time, water uptake, and mass loss for PLGA samples showing percolation behavior. (**b**) Gel permeation chromatography traces of PLGA at different degradation times, exhibiting lower molecular weight after longer degradation times [27]

3.2 Hydrolysis

Hydrolysis reactions are usually catalyzed by acids, bases, salts, or enzymes [19]. Polymer hydrolytic degradation is defined as the breakdown of chemical bonds of the polymer backbone by the water molecules to form oligomers or monomers. After implantation, the biomaterial absorbs water and swells and, thereby, degradation starts from the exterior part of the material toward its interior. The hydrophilic and hydrophobic nature of polymeric materials influences their degradation rate because they contain hydrolyzable bonds such as glycosides, esters, orthoesters, anhydrides, carbonates, amides, urethanes, ureas, etc. [19, 23]. Hydrolysis reactions may be catalyzed by enzymes known as hydrolases, which include proteases, esterases, glycosidases, and phosphatases. This class of enzymes comprises cell-derived proteins that play an important role in the degradation of biomaterials by catalyzing the hydrolysis [24]. Aliphatic polyesters such as PLA, PGA, and their copolymer PLGA are biodegradable and are successfully used in medical applications [25]. Degradation of PLA, PGA, and PLGA occurs in an aqueous environment through simple hydrolysis of ester bonds that are auto-catalyzed by carboxylic groups. The rate of hydrolysis increases exponentially with degradation time [26]. The initial degradation starts in the hydrated region of the polymer through ester bond hydrolysis. The changes in mass loss, water uptake, and molecular weight of multilayered PLGA films after hydrolytic degradation are been presented in Fig. 1a, b and show a threshold value of 40 days for mass loss and a gradual decrease of molecular weight (as demonstrated by gel permeation chromatography) [27]. Based on the combined effects of UV treatment and crystallization, the hydrolytic degradation rate of PLA has been shown to be higher for the crystallized films in the early stage, whereas the phenomena reverse in the late stage (as demonstrated by monitoring weight loss and molecular weight) [28]. The nanoparticle-induced hydrolytic degradation of PCL in compost media is presented in Fig. 2 and shows higher biodegradation rates in the

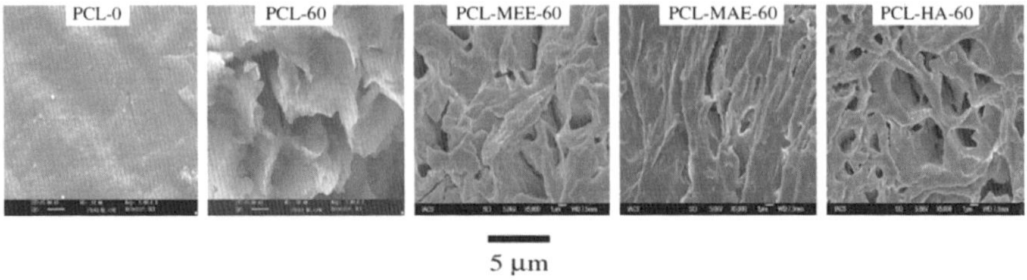

Fig. 2 FE-SEM images of PCL and its indicated nanocomposites before and after soil burial (compost). The *numbers* after the specimen name represent the biodegradation time in days. *0* indicates the sample before biodegradation. Nanoparticles were ion-exchanged with di-methyl di-tallow ammonium (*MAE*), di-poly oxy ethylene alkyl methyl ammonium (*MEE*), and calcium phosphate hydroxide [hydroxyapatite (*HA*)] [29]

presence of various nanoparticles, as evident from the relative erosion caused by biodegradation after soil burial and subsequent observation under SEM [29]. The enzymatic (*Pseudomonas* lipase) degradation of PCL nanocomposites shows a higher rate than degradation of pure polymer, demonstrating the similar nature of the relative biodegradation rate to that of compost. However, the absolute biodegradation is very high vis-a-vis compost media, as shown both by kinetics as well as by the confocal images taken before and after degradation (Fig. 3) [30]. Further, nanoparticles often show decreased and/or increased biodegradation rates vis-a-vis pure poly(hydroxybutyrate-co-valerate) (PHBV), depending upon the nature of the nanoparticles or, more precisely, the chemical modifications on the surface of the nanoparticles (Fig. 4). The phenomena has been explained by the different depolymerase activity of the enzyme in different pH environments arising from the presence of varying nanoparticles (Fig. 5) [31]. The biodegradation of poly (hydroxybutyrate) (PHB)/layered silicate nanocomposites shows a strong nucleating agent effect of the nanoclay for the crystallization of the matrix polymer, which also enhances the rate of biodegradation. Biodegradation in compost media shows a remarkable enhancement of biodegradation rate in the presence of clay (Fig. 6) as a result of an alteration in crystallinity, as evident from lower crystalline or lower spherulitic dimensions in the nanocomposites as compared to pure polymer, resulting in faster biodegradation [32].

4 Polymeric Nanoparticles

The process of making polymer nanoparticles has many useful applications. As described in the literature, nanoparticles from PLA and PLGA have successfully been developed to deliver drugs like peptides, proteins, and vaccines for prolong periods of time at a controlled rate [33]. For larger sized the particles, it is difficult to deliver the drug to targeted tissues via the systemic circulation or across the mucosal membrane [34]. During oral administration, particles less than 500 nm can

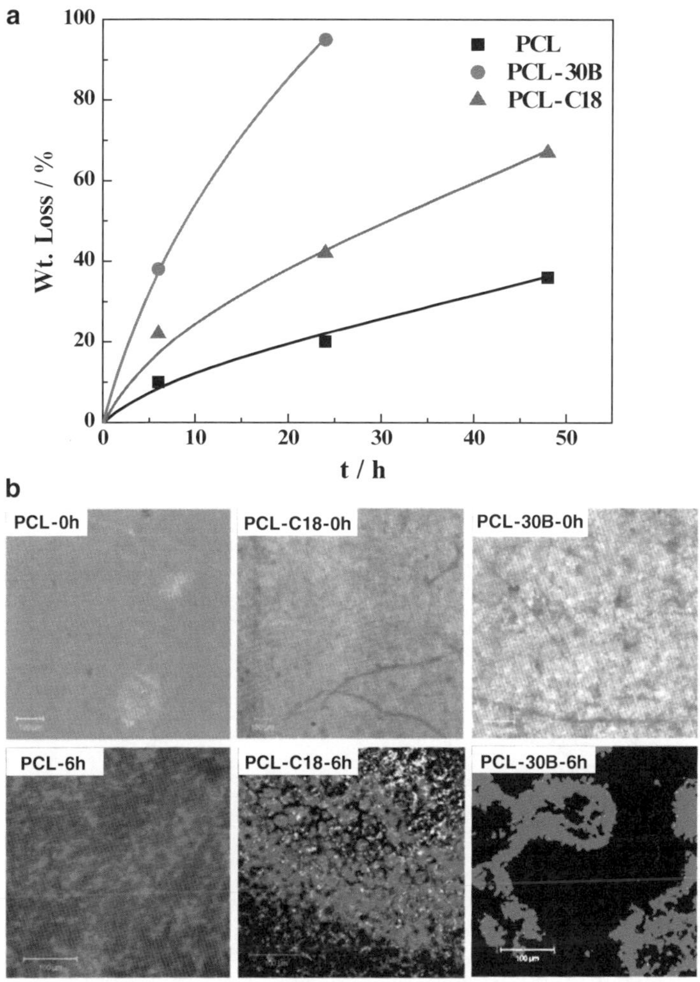

Fig. 3 (**a**) Percentage weight loss of PCL and its indicated nanocomposites during enzymatic (lipase from *Pseudomonas* sp., Type XIII) degradation at 37°C. (**b**) Confocal images of PCL and its indicated nanocomposites before and after enzymatic degradation. Nanoclays were ion-exchanged with methyl tallow bis-hydroxyethyl quaternary ammonium cation (*30B*) and dimethyl-octadecylamine (*C18*) [30]

cross the M cells in the Peyer's patch and the mesentery on the surface of the gastrointestinal mucosa for delivering the drug to the systemic circulation [34]. Accordingly, particles larger than 200 nm can be mechanically filtered in the spleen whereas those smaller than 100 nm leave the blood vessels through fenestrations in the endothelial lining. Therefore, nanoparticles with relatively small size (in the range of 100–200 nm) are desirable for delivery of drugs to tumors [35, 36].

Nanoparticles can be prepared by the same methods as those described for microparticles, but by using a comparatively small ratio of dispersed phase:

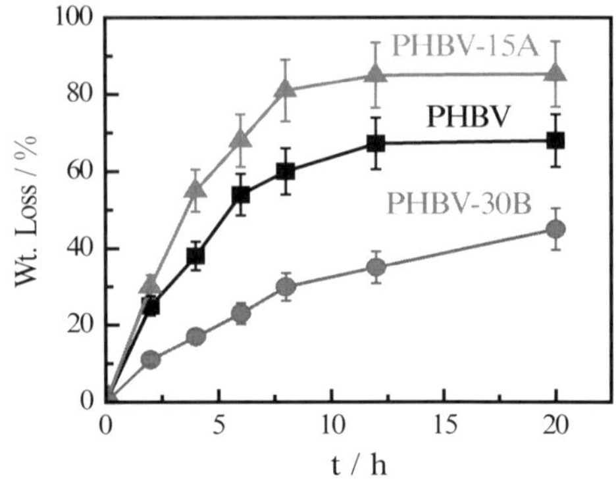

Fig. 4 Percentage weight loss of PHBV and its indicated nanohybrids during biodegradation in enzyme extracted from *Pseudomonase stutzeri* at 37°C. Nanoclays were ion-exchanged with methyl tallow bis-hydroxyethyl quaternary ammonium cation (*30B*) and di-methyl dihydrogenated tallow ammonium cation (*15A*) [31]

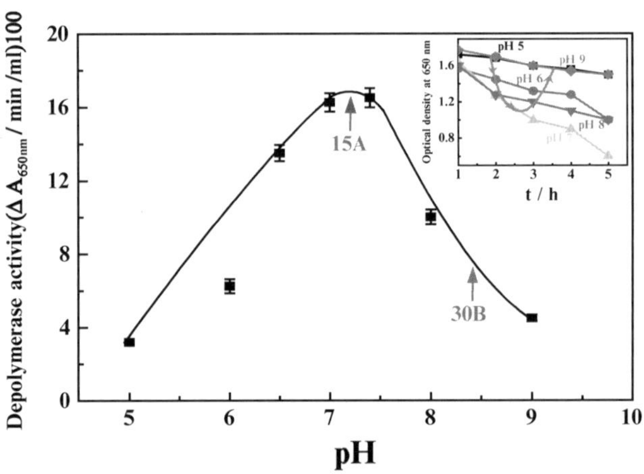

Fig. 5 Effect of pH on *Pseudomonase stutzeri* PHBV-depolymerase activity. The *inset* shows the optical density of PHBV suspension of varying pH as a function of time (in hours). The depolymerase activity was calculated from the initial slope of the curves [31]

dispersion medium and with substantially higher stirring conditions [37]. PLGA nanoparticles have been developed by using organic/water emulsification/solvent evaporation techniques to produce nanoparticles having a size of 150 nm and drug loading of 15.5% (w/w) [38]. Several groups have applied different process variables to get appropriate nanoparticles based on their size and encapsulation efficiency. The nanoparticles were also prepared by using the oil–water (o/w) emulsification technique to determine the effect of some variables on the size distribution of PLGA nanoparticles [39]. The homogenization pressure and PLGA concentration have a linear and predictable effect on both size and polydispersity of the particles. The concentrations of Tween 80 and dichloromethane (DCM) (organic phase) had a large effect on the diameter and polydispersity. Magnetite-containing PLA/PLGA nanoparticles were prepared by high-pressure homogenization of the constituents in aqueous poloxamer-188 solutions [40]. Magnesium salts containing

Fig. 6 Polarizing optical images of PHB and PHBCN2 (2 wt% nanoclay) before and after 6 weeks of biodegradation. The samples were crystallized at 100°C prior to composting to generate different spherulite microstructures [32]

PLA nanoparticles have been prepared by a reversible salting-out process using a cross-flow filtration technique by which 90% drug entrapment was achieved [41]. Niwa et al. prepared nanoparticles for indomethacin (a water-insoluble drug) and 5-fuorouracil (a water-soluble drug) by a novel spontaneous emulsification/solvent diffusion method [42]. The drug and PLGA were dissolved in acetone–DCM mixture and subsequently emulsified in an aqueous PVA solution using a high-speed homogenizer (o/w emulsification). The rapid diffusion of acetone in the aqueous phase resulted in faster deposition and development of polymeric nanoparticles of size less than 500 nm. The preparation of PLGA particles may also be based on double coacervation of PVA and PLGA, whereby PLGA was first dissolved in a mixture of acetone and DCM, ethanol, or methanol and then dispersed in aqueous PVA solution with stirring [43]. The dispersed solution was freeze-dried to obtain powder having size of approximately 300 nm. The coacervated PVA molecules help prevent the aggregation of PLGA nanoparticles due to steric hindrance. A dual-capillary electrospray system was developed to synthesize budesonide-loaded PLGA particles with varying sizes by using different concentrations of PLGA (Fig. 7). Budesonide is a lipophilic, glucocorticoid steroid for the treatment of asthma and non-infectious rhinitis. It has also been studied in the chemoprevention of lung cancer [44]. Electrospray operating in the cone-jet mode generally produces monodisperse particles in sizes ranging from nanometers to micrometers. The investigated dual-capillary electrospray system demonstrated the production of uniform PLGA-coated particles in one step without any later separation or purification steps [45]. Evaluation of in vitro and in vivo anticancer

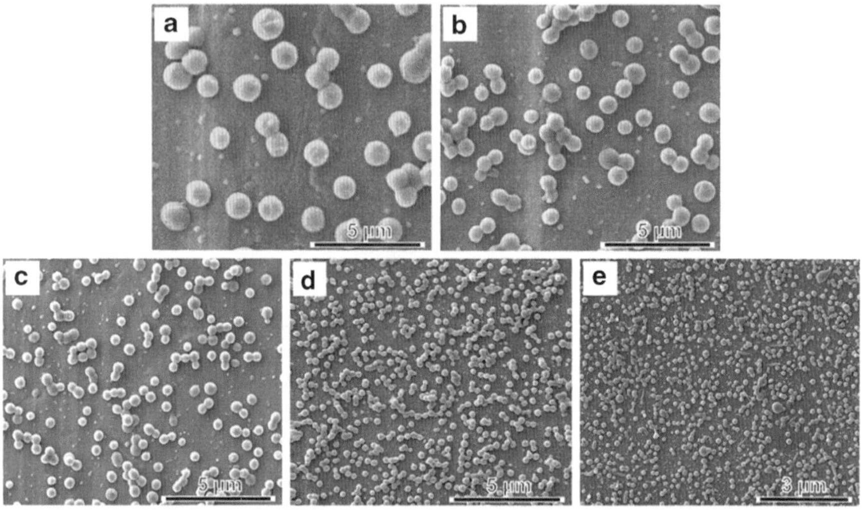

Fig. 7 SEM images of budesonide-loaded PLGA particles of varying sizes produced by tuning the concentration and conductivity of PLGA solutions: (**a**) 5 wt% (1,200 nm), (**b**) 3 wt% (800 nm), (**c**) 0.5 wt% (400 nm), (**d**) 0.2 wt% (289 nm), and (**e**) 0.5 wt% (with 10 mM KCl, 165 nm) [45]

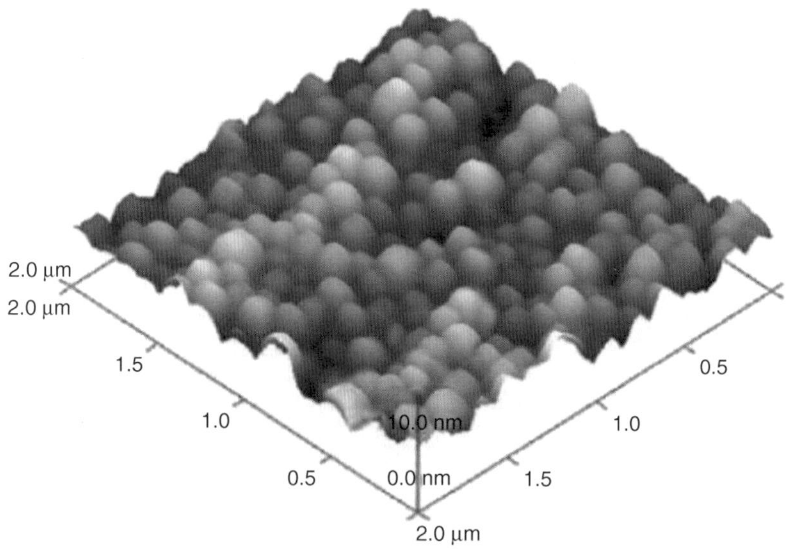

Fig. 8 AFM image of doxorubicin-loaded PLGA nanoparticles [46]

efficacy of orally administered doxorubicin-loaded PLGA nanoparticles (Dox-NPs) has been performed in breast cancer induced in an animal model. Dox-NPs were prepared with an entrapment efficiency and particle size of 55% and 160 nm, respectively (Fig. 8) [46]. Fagui and Amiel studied the preparation and

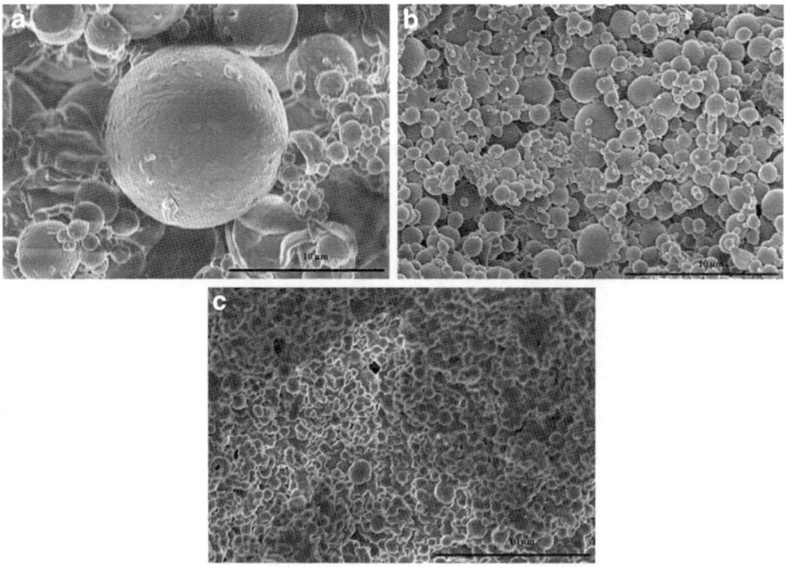

Fig. 9 SEM images of α-tocopherol-loaded PCL nanoparticles [48]

physicochemical properties of benzophenone (hydrophobic)-loaded PLA nanoparticles in the range of 150 nm. Elucidation of its release mechanisms revealed the retention of nanoparticle integrity during the process and the dilution-induced total release of encapsulated benzophenone with slow kinetics [47]. The preparation of α-tocopherol-loaded PCL nanoparticles has been achieved using o/w emulsion solvent evaporation and ultrasonification (Fig. 9). Investigations showed the significant effect of PCL concentration, solvent in the oil phase, and ultrasonification time on the encapsulation efficiency. In contrast, solvent in the oil phase and ultrasonification time did not much affect α-tocopherol loading, and PCL concentration did not affect particle size [48]. A nanoformulation of curcumin, (a low molecular weight hydrophobic drug) loaded in dextran sulfate–chitosan nanoparticles has been developed and characterized by dynamic light scattering (DLS), scanning electron microscopy (SEM), atomic force microscopy (AFM), Fourier transform infrared spectroscopy (FT-IR), X-ray diffraction (XRD) and differential thermal analysis (DTA). The nanoparticles showed an average size of 200–220 nm with a zeta potential value of −30 mV and ~74% drug entrapment efficiency (Fig. 10) [49].

5 Preparation of Polymeric Nanoparticles

Several techniques have been developed and reported for the preparation of polymeric nanoparticles. The selection of the technique depends on the nature of the polymer, drug to be incorporated, proposed applications, and the duration of

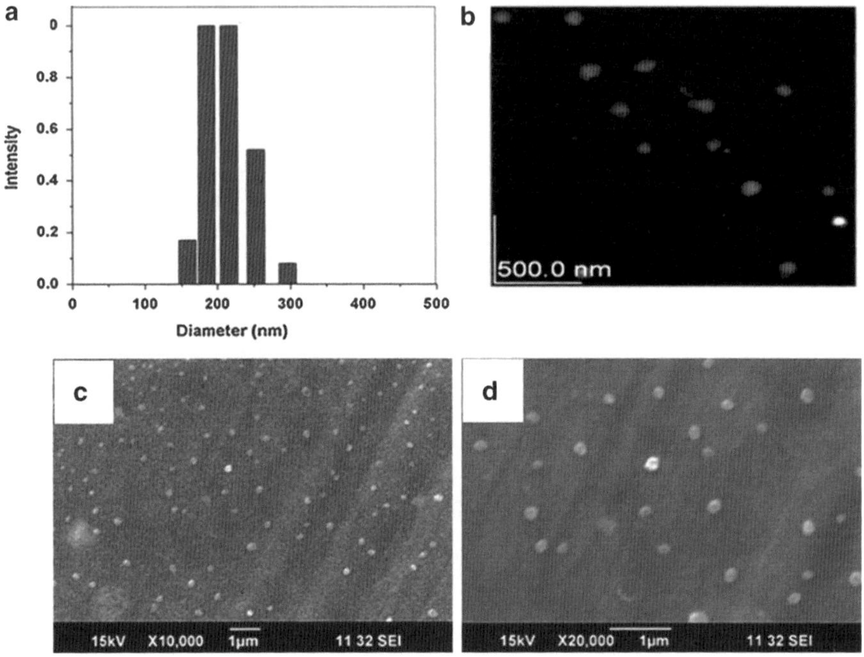

Fig. 10 (**a**) Size distribution of curcumin-loaded dextran sulfate–chitosan nanoparticles, as measured using dynamic light scattering. (**b**) AFM image of curcumin-loaded dextran sulfate–chitosan nanoparticles. (**c**, **d**) SEM images of curcumin-loaded dextran sulfate–chitosan nanoparticles at different magnifications [49]

the therapy [2, 3]. During the process of preparing nanoparticles, the stability and biological activity of the drug should be maintained during entrapment. Further, the nanoparticles should have the required size range without any aggregation or adherence, with high entrapment efficiency. Polymeric nanoparticles can be conveniently prepared either from preformed polymers or by direct polymerization of monomers using classical polymerization [50, 51]. Various methods such as solvent evaporation, salting-out, dialysis, and supercritical fluid technology are being utilized for the preparation of polymeric nanoparticles. Polymeric nanoparticles can also be synthesized by the polymerization of monomers using various techniques such as microemulsion, miniemulsion, surfactant-free emulsion, and interfacial polymerization. Selection of preparation method is made on the basis of a number of factors such as the type of polymer, site of application, size requirement, etc. A polymeric system that is developed for biomedical or environmental application should be completely free from additives, reactants such as surfactants, and traces of organic solvents, etc. Some commonly used methods for the preparation of polymeric nanoparticles are given below, together with a flow diagram to provide a general idea of the preparation of nanoparticles (Scheme 1).

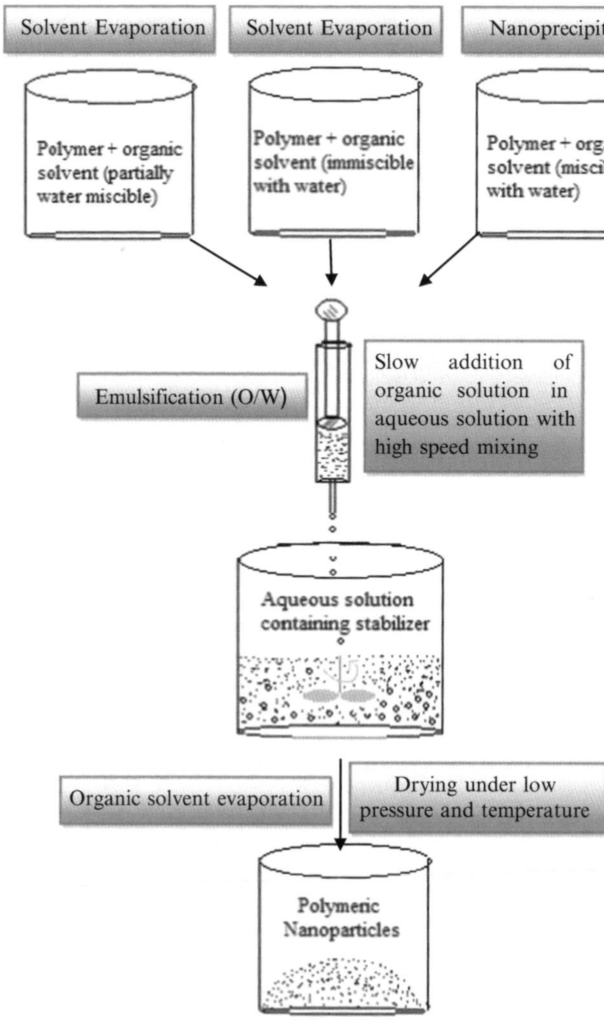

Scheme 1 Method for the synthesis of polymeric nanoparticles

5.1 Emulsification/Solvent Evaporation Method

Solvent evaporation is the most widely used and accepted method for the development of polymeric nanoparticles using a dispersion of preformed polymers [50]. In this method, polymer solutions are prepared in solvents that boil at low temperatures, such as dichloromethane, chloroform, and ethyl acetate, and then emulsification is carried out into an aqueous phase in the presence of stabilizer. The emulsion is converted into a nanoparticle suspension on evaporation of the solvent, which is allowed to diffuse through the continuous phase of the emulsion [52]

(Scheme 1). After evaporation of solvent, the settled solidified nanoparticles are collected by ultracentrifugation and washed with distilled water to remove additives and remaining solvent. PLGA nanoparticles were prepared with a diameter of 130 nm using this method [53]. Nanoparticles based on poly(organophosphazene) derivatives were developed with a particle size of around 200 nm using acetone and poloxamine 908 as the solvent and stabilizing agent, respectively [54]. PLGA nanoparticles have been prepared with a typical particle size of 60–200 nm by employing dichloromethane and acetone (8:2 v/v) as the solvent and PVA as the stabilizing agent [55]. PLGA nanoparticles of about 200 nm were developed by using dichloromethane 1% (w/v) as the solvent and PVA or Span 40 as the stabilizing agent [56]. Both the single-emulsion and double-emulsion methods have been applied for preparation of the nanoparticles and the effect of the emulsion method and the experimental parameters on the particle size have been studied. In the case of the single-emulsion method, an increase in the PVA (stabilizer) concentration results in a significant decrease in the nanoparticle size, while the homogenizer speed, the evaporation rate and freeze-drying only slightly influence the particle size. In the case of the double-emulsion method, PVA concentration in the internal aqueous phase decreases the nanoparticle size as compared to the PVA concentration in the external aqueous phase. The homogenizer speed and freeze-drying only slightly influence the nanoparticle size. The influence of experimental parameters such as temperature, solvent evaporation method, surfactant concentration and internal aqueous phase volume and the influence of the molecular mass of the polymer on the particle size, zeta potential, residual surfactant percentage, and the polydispersity index of the PLA nanoparticles have been reported using the double-emulsion method [57]. Poly(ethylene glycol) (PEG)-coated nanoparticles have been also prepared with a mean diameter of about 200 nm [58]. The influence of the sonication process has been studied on the characteristics of PLGA nanoparticles prepared using a water-in-oil/water solvent evaporation method [59]. The effect of solvent was investigated using two organic solvents (methylene chloride and ethyl acetate) as the dispersed phase to prepare nanoparticles of PLGA and the nanoparticles were found to be larger in methylene chloride than in ethyl acetate [60].

5.2 Salting-Out

This process is based on the separation of a water-soluble solvent from aqueous solution through the salting-out effect. Initially, polymer and drug are dissolved in a solvent such as acetone and then the polymer solution is emulsified into an aqueous solution containing electrolytes (such as magnesium chloride, calcium chloride and magnesium acetate), non-electrolytes (such as sucrose), and a colloidal stabilizer such as poly(vinyl pyrrolidone) or hydroxyethylcellulose. Finally, the oil/water emulsion is diluted with a sufficient volume of water to enhance the diffusion of acetone into the aqueous phase by inducing the formation of nanospheres. Both the solvent and the salting-out agents are then eliminated by cross-flow filtration [61]. Selection of the salting-out agent plays an important role in the encapsulation efficiency of the drug. To overcome the harmful environmental effects, a modified version of the emulsion process was evolved by linking it to a salting-out process [62] to avoid surfactants and chlorinated solvents. An electrolyte-saturated or nonelectrolyte-saturated aqueous solution containing PVA (stabilizer) was added to an acetone solution of the polymer under continuous stirring. The saturated aqueous solution prevented the acetone from mixing with water through the salting-out process [63]. After formation of an oil-in-water emulsion, water was added in an adequate amount to allow the complete diffusion of acetone into the aqueous phase, resulting in formation of nanospheres. PLA nanoparticles were developed by adding an aqueous gel containing magnesium acetate tetrahydrate and PVA to an acetone solution of the polymer to form a water-in-oil emulsion [41]. During the mixing of water with acetone, a liquid–liquid two-phase system was formed due to the presence of the salting-out agent. An oil-in-water emulsion was obtained on further addition of the aqueous solution followed by the addition of sufficient water to allow diffusion of acetone into the aqueous phase, resulting in the formation of nanoparticles with an average size of 295 nm [41]. Recently, poly (trimethylene carbonate) nanoparticles of size 183 and 251 nm were prepared by single emulsion and salting-out methods, respectively [64].

5.3 Nanoprecipitation

Nanoprecipitation is also called solvent displacement and is used for the preparation of polymeric nanoparticles [65]. It is a simple, fast and reproducible method that is broadly employed for the development of both nanospheres and nanocapsules. The method is based on the interfacial deposition of a polymer after displacement of a semipolar solvent from a lipophilic solution [66]. Organic solvents such as ethanol, acetone, hexane, methylene chloride, dioxane, or binary solvent blend, which are miscible in water and easy to remove by evaporation, are selected as solvent [67]. The polymers commonly used are biodegradable polyesters, especially PCL [68, 69], PLA [70], and PLGA [71, 72]. Polymeric nanoparticles are prepared by slow addition of the organic phase to the aqueous phase under stirring. In reverse, the addition of the aqueous phase to the organic phase also leads to the formation of nanoparticles. The physicochemical properties of polymeric nanoparticles are associated with conditions such as organic phase injection rate, aqueous phase agitation rate, and the organic phase to aqueous phase ratio. Surfactant plays an important role in the formation of polymeric nanoparticles by protecting from agglomeration for nanoprecipitation [68].

5.4 Dialysis

Dialysis provides an easy and efficient way for the preparation of small, finely distributed polymeric nanoparticles [73, 74]. In this process, polymer is dissolved in an organic solvent and placed inside a dialysis tube with proper molecular weight cutoff and dialysis performed against a non-solvent that is miscible with the solvent. The displacement of the solvent inside the dialysis membrane causes the progressive aggregation of polymer due to a loss of solubility and the construction of a homogeneous suspension of nanoparticles. The mechanism of the dialysis process is based on similar phenomena to that of nanoprecipitation [65].

6 Biodegradable Polymers and Their Applications

The most extensively used biodegradable polymers are PGA, PLA,PLGA, PCL, poly(alkyl cyanoacrylates), chitosan, and gelatin. The therapeutic applications and preparation methods of these polymeric nanoparticles, with entrapment and release of various drugs are described below and summarized in Table 1.

6.1 Poly(glycolic acid)

PGA was the first biodegradable synthetic semicrystalline polymer (glass transition temperature, T_g ~40°C; melting temperature, T_m >200°C) to be considered for biomedical applications. PGA has been used in bone internal fixation devices due to its good mechanical properties. The good fiber-forming ability and scaffolding ability of PGA, together with cell viability, were used to develop resorbable devices and for tissue regeneration [119]. PGA is being used as a biocompatible durable substitute and as an excellent skin-closing suture material. It also has the ability to help regenerate biological tissue [120]. However, PGA is hydrolyzed in few months, indicating a high degradation rate. It is broken down into glycine under physiological conditions, which can be excreted from the body or converted into carbon dioxide and water via the citric acid cycle [121]. Due to the high rate of degradation and low solubility of PGA, its biomedical and pharmaceutical applications are limited. Therefore, several copolymers containing glycolide units are being developed to overcome the disadvantages of pure PGA.

Table 1 Various biodegradable polymeric nanoparticles used for delivery of different drugs

Method of nanoparticle preparation	Polymer	Encapsulated drug	Particle size (nm)	References
Solvent displacement	PLGA	Indomethacin	~168	[75]
		Doxorubicin	274	[76]
		Cyclosporin A	~170	[75]
		Valproic acid	~166	[75]
		Ketoprofen	~167	[75]
		Vancomycin	~187	[75]
		Insulin	~105–170	[75]
Solvent evaporation or solvent extraction	PLGA	Taxol	~300	[77]
Emulsification/solvent evaporation	PLGA	DNA	~100	[78]
		Cyclosporin A	~300	[79]
		Dexamethasone	~230	[80]
		2-Aminochromone	~100 ± 39	[81]
		Thymopentin	~250	[82]
		Etanidazole	90–190	[83]
Interfacial deposition	PLGA	Paclitaxel	<200	[84]
Nanoprecipitation	PLGA	Taxol	<190	[85]
		9-Nitrocamptothecin	207 ± 26	[86]
Solvent evaporation	PLA	Testosterone	<1,000	[87]
		Albumin	100 or 120	[88]
		Tetanus toxoid	150	[89]
		Loperamide	~300	[90, 91]
		Haloperidol	275 ± 25	[92]
		Progesterone	<335	[93]
		Zidovudine	265.8 ± 15.3	[94]
Solvent displacement	PLA	Doxorubicin	270	[76]
		Taxol	~260	[65]
		Dexamethasone	~300	[65]
		Vitamin K	~270	[65]
Emulsion/solvent diffusion	PLA	p-THPP	125	[95]
		DNA	<300	[96]
		Oridonin	137.3	[97]
		Ellagic acid	120–290	[98]
Double emulsion method	PLA	Hemoglobin	100–200	[99]
		Neurotoxin-I	192.6 ± 15.2	[100]
Emulsion/solvent diffusion	PLGA	p-THPP	117–118	[95]
		Doxorubicin	<1,000	[101]
		Estradiol	116.0–279.3	[102]
		Docetaxel	200	[103]
Solvent displacement technique	PLGA	Xanthones	<300	[104]
		Docetaxel	200	[103]
Salting-out	PLA	Savoxepin	<1,000	[105]
Interfacial deposition	PLA	Indomethacin	230	[65, 106, 107]
Nebulization techniques	PLA	Insulin	400–600	[108]

(continued)

Table 1 (continued)

Method of nanoparticle preparation	Polymer	Encapsulated drug	Particle size (nm)	References
Solvent displacement	PCL	Cyclosporin A	~100–200	[109]
		Tamoxifen	150–250	[110]
		Saquinavir	~200	[111]
Nanoprecipitation method	PCL	Docetaxel	~100 nm	[112]
Emulsion method	PCL	Vinblastine	213–227	[113]
Desolvation method	Gelatin	Paclitaxel	600–1,000	[114]
		Didanosine	140 ± 19	[115]
Solvent evaporation	Gelatin	Chloroquine Phosphate	100–300	[116]
		Sulfamethaxazole	100–300	[117]
Ionotropic gelation	Gelatin	Insulin	~750	[118]

6.2 Poly(lactic acid)

PLA is linear aliphatic thermoplastic polyester (T_g ~55°C, T_m ~175°C), which is being extensively considered for biomedical applications due to its biocompatibility, biodegradation, excellent thermal and mechanical properties, and easy processability [122, 123]. PLA is easily metabolized into lactic acid, a natural component in the body. Lactic acid is a chiral molecule and exists in two stereoisomeric forms (D- and L-lactide) [2]. L-PLA exists in semicrystalline form due to high regioregularity, whereas D,L-PLA is an amorphous polymer with varying D-content because of irregularities in the polymer chain structure [4]. Therefore, the use of D,L-PLA is preferred over L-PLA for homogeneous dispersion of drug in the polymer matrix [124]. Savoxepine is a psychotic drug that has been loaded into PLA nanoparticles with high encapsulation efficiency (95%), which extended its delivery to more than 1 week [125]. However, the semicrystalline L-PLA is preferred for biomaterial devices and controlled delivery systems to help cure various health dysfunctions. Progesterone-loaded PLA nanoparticles have been prepared, with entrapment efficiencies of around 70% and particle size around 260–320 nm, for embryogenesis and gestation in females [93]. Oridonin-loaded PLA nanoparticles have been prepared by a modified spontaneous emulsion/solvent diffusion method [97], with entrapment efficiency of around 92%, for the treatment of esophageal and hepatic carcinoma. Quercitrin-embedded poly (D, L-lactide) nanoparticles have been prepared by a solvent evaporation method, with encapsulation efficiency of 40% and nanoparticles size around 195 nm, for antimalarial and anti-inflammatory activities [126]. The breast cancer drug tamoxifen was loaded into magnetite/poly(L-lactic acid) composite nanoparticles of 200 nm with encapsulation efficiency of 81% using a solvent evaporation method [127]. On the other hand, PLA is preferred for applications related to orthopedic surgery due to its high mechanical strength and toughness [128, 129]. PLA scaffolds satisfy several requirements for use as a suitable material for bone tissue engineering [130].

6.3 Poly(lactic acid-co-glycolic acid)

Polymeric materials based on PLGA, a copolymer of lactide and glycolide, are widely used for biomedical and pharmaceutical applications. PLGA nanoparticles have been considered for sustained and targeted delivery of different compounds, including drugs, proteins, bioactive materials, and plasmid DNA [131, 132]. PLGA is easily biodegraded into lactic acid and glycolic acid in the physiological system as result of hydrolysis of the ester linkages in the presence of water; these are then excreted from the body naturally [133]. The biodegradation rate, crystallinity, and mechanical properties of PLGA can be manipulated by controlling the unit ratio of lactide to glycolide and the overall molecular weight of the copolymer. Surface modifications, additives, molecular weight, and processing conditions of PLGA also influence the effective response to formulated nanomedicines [134]. The composition ratio affects the binding capability of drugs and other active compounds with PLGA due to physicochemical properties [100]. Several cancer-related drugs such as paclitaxel, doxorubicin, 5-fluorouracil, 9-nitrocamptothecin, and cisplatin have been incorporated in PLGA nanoparticles and successfully delivered in vivo [135]. The surface charge of the nanoparticles is important for the cellular internalization of nanoparticles, clustering in blood flow, adherence, and interaction with oppositely charged cell membranes [136]. Formulations using some additive to the PLGA nanoparticles have shown reasonably good activity and much faster administration in comparison to traditional formulations [84]. The incorporation of paclitaxel in the PLGA nanoparticles strongly enhances its antitumoral efficacy as compared to free drug. The controlled delivery of cisplatin in encapsulated PLGA–methoxypoly (ethylene glycol) (mPEG) nanoparticles to tumor cells significantly reduces drug toxicity and improves its therapeutic index. Xanthone, an inhibitory drug for cancer cell lines, has been loaded into PLGA nanoparticles and shown to be physically stable for 3–4 months with sustained release of xanthone [104]. Rose Bengal, which is useful for the treatment of melanoma cancer cells, has been successfully entrapped into PLGA nanoparticles and about 50% of drug release was measured within 30 min in serum [137]. Triptorelin-loaded PLGA nanospheres have been prepared with an encapsulation efficiency varying from 4 to 83% for the treatment of sex-hormone-dependent tumors [138]. High encapsulation efficiency was explained by an ionic interaction occurring between the peptide and the copolymer. Dexamethasone, which reduces the inflammatory response of the body, has been incorporated into PLGA (75:25) nanoparticles and was completely released from this formulation after 4 h of incubation at 37°C in vitro [80]. Similarly, insulin-, haloperidol-, and estradiol-loaded PLGA nanoparticles have been prepared with high encapsulation efficiency and used for therapy of the relevant diseases [92, 134].

6.4 Poly(ϵ-caprolactone)

PCL is a semicrystalline polyester (T_g about −60°C, T_m ~60°C) that is synthesized by ring-opening polymerization (ROP) of ϵ-caprolactone and is degraded through hydrolysis of its ester linkages at physiological conditions. PCL has been considered for long-term drug or vaccine delivery devices. A contraceptive device known as Capronor, composed of PCL, has been used for the controlled release of levonorgestrel [139]. Due to the excellent biocompatibility of both pure PCL and its composites, it has widely been used as scaffolds for tissue engineering [140]. The copolymer of ϵ-caprolactone and glycolide results in less rigidity than polymers of polyglycolide and has been developed as a drug delivery device (SynBiosys). Poly(ethylene oxide) (PEO)-modified PCL nanoparticles prepared by a solvent displacement method were used to embed tamoxifen in polymeric nanoparticles, which gave significantly increased levels of accumulation of tamoxifen within the tumor [141]. Taxol-loaded PEG-PCL nanospheres have been reported as having promising anticancer activity [142]. Nanoparticles prepared by blending PCL and a polycationic nonbiodegradable acrylic polymer have been used as a drug delivery carrier for oral administration of insulin, with entrapment efficiency of around 96% [143], confirming the mucoadhesive properties of polycationic polymers that allow the intestinal uptake of insulin [143]. Poly (*n*-isopropylacrylamide)-*b*-poly(3-caprolactone) (PNPCL) block copolymers have been used to encapsulate 70–90% of the hydrophobic drug clonezepam to enhance the effect of GABA in brain [144]. Further, PCL nanoparticles have been developed in order to study their ability to improve the anti-leishmanial action of Amphotericin B (AmB), with concomitant reduction in the toxicity associated with it [145].

6.5 Poly(alkyl cyanoacrylate)

Poly(alkyl cyanoacrylates) (PACs) are prepared through anionic polymerization of alkyl cyanoacrylic monomers with a trace amount of moisture as the initiator. PAC has been used to prepare excellent synthetic surgical glue, skin adhesive, and an embolic material. PAC is one of the fastest degrading polymers, having degradation times from a few hours to a few days depending on the length of the alkyl side groups. The lower alkyl derivatives degrade within hours in an aqueous environment and release toxic degradation products such as cyanoacetic acid and formaldehyde. Most of the research in this area has therefore been concentrated on higher alkyl derivatives such as octyl and isobutyl cyanoacrylates. PAC nanoparticles have several advantages over other polymeric nanoparticles with their easy preparation, high utility size ranges, absence of solvent residues, ability to form appropriate nanoparticles, and ability to absorb or encapsulate a wide range of drug or protein molecules. Nanoprecipitation has been adopted to prepare pure and drug-loaded PAC nanoparticles for the entrapment of highly sensitive reactive drugs [146, 147]. Ampicillin-loaded poly(isohexyl cyanoacrylate) (PIHCA) nanoparticles have been found to increase the antibiotic efficacy by 120-fold in experimental salmonellosis [148]. Other antibiotics have also been incorporated into various PAC-based nanocarriers [146, 149]. Ciprofloxacin-loaded PAC nanoparticles have been developed that allowed pH-controlled drug release, showing drug-loaded PAC nanoparticles to be as active as the free drug [146]. In vivo experiments with mice have been performed to demonstrate the high efficiency of antibiotic-loaded PACA nanoparticles for the treatment of intracellular infections [150]. Various bioactive compounds (e.g., cytostatics, hormones, peptides, and nucleic acids) have also been loaded into PAC nanocarriers. Although the drug-loaded nanoparticles do not directly reach the cancer cells, the phagocytes in the organs where the nanoparticles become entrapped may serve as drug reservoirs. In vivo studies showed that PAC nanocapsules enhanced the accumulation of drug in the lymph nodes compared to other carriers such as emulsions and liposomes [151]. Drug-loaded PAC nanoparticles have been applied to the brain by modifying their surface with the surfactant polysorbate 80 [152, 153]. It has been proposed that these nanocarriers adsorb apolipoproteins E from blood plasma and then cross the brain endothelium via receptor-mediated endocytosis and transcytosis through the endothelial cells [153]. These routes play an important role in the treatment of brain tumors using drug-loaded PAC nanoparticles. The indomethacin inhibits the production of prostaglandin in the stomach and intestines, which maintains the mucous lining of the gastrointestinal tract. Indomethacin has been encapsulated into poly (isobutyl cyanoacrylate) (PIBC) nanoparticles and found to be stable for up to 12 months [154]. However, PIBC nanocapsules also showed an inhibition of platelet aggregation because of their isobutyl cyanoacrylate content [154]. Doxorubicin-loaded poly(butyl cyanoacrylate) (PBC) nanoparticles have also been reported to increase 60-fold in the brain after being coated with polysorbate 80 [152].

6.6 Chitosan

Chitosan is a naturally occurring biodegradable cationic polymer derived from chitin, which is a fully acetylated polymer that is used in many biomedical applications such as wound dressings and drug delivery vehicles [155, 156]. Chitosan is a linear polysaccharide consisting of β-(1-4)-linked D-glucosamine with randomly located *N*-acetyl glucosamine groups, depending upon the degree of deacetylation of the polymer. Chitosan is degraded by a few enzymes (such as chitosanase, lysozyme, and papain) in vitro and is degraded and hydrolyzed into acetylated residues by lysozyme in vivo [157]. The rate of degradation of chitosan inversely depends on the degree of acetylation and crystallinity of the polymer [158]. The fast degradation rate of chitosan has been attributed to the deformation of strong hydrogen bonds present in chitosan. Due to strong positive charges on chitosan, it can strongly interact with the negatively charged mucous membrane as an effective mucoadhesive [159]. It can be fabricated into various devices because of its solubility in water and can form microsphere and nanosphere formulations without using any organic solvents, thus maintaining the immunogenicity of the antigens [160]. Several chitosan–drug conjugates have also been used for cancer therapy due to its high chemical reactivity [161]. Insulin-loaded chitosan nanoparticles enhanced intestinal absorption of insulin in comparison to oral administration of free insulin due to the mucoadhesive properties of chitosan [118]. Cyclosporin-A-loaded chitosan nanoparticles can make contact very well with the corneal and conjunctiva surfaces, thereby increasing delivery to external ocular tissues, which helps to supply long-term drug levels at target tissues [162]. Chitosan nanoparticles have revealed an excellent capability for the association of ammonium glycyrrhizinate [163]. Addition of PEG results in the reduction of encapsulation efficiency and also reduces the positive charge, which has been achieved by co-synthesis of chitosan polymer along with tripolyphosphate anions (TPP) and PEG using ionic gelation methods. Release of ammonium glycyrrizinate from chitosan nanoparticles follows a slow and continuous release profile, which may improve the oral absorption of ammonium glycyrrhizinate [163].

6.7 Gelatin

Gelatin, a natural polymer derived from collagen, is commonly used for pharmaceutical and biomedical applications due to its biodegradability [164, 165] and biocompatibility in physiological environments [166]. It is also used in food and medical products and is attractive for controlled release due to its nontoxic, biodegradable, and bioactive properties. Gelatin is a polyampholyte having both cationic and anionic groups along with hydrophilic groups. The mechanical and thermal properties and swelling behavior of gelatin significantly depend on the degree of crosslinking. The efficiency of gelatin carriers to encapsulate charged biomolecules such as proteins and plasmid DNA occurs through polyion complexation. Acidic gelatin with an isoelectric point of 5.0 should be used as a carrier for basic proteins in vivo, whereas basic gelatin with an isoelectric point of 9.0 should be used for the sustained release of acidic proteins under physiological conditions [167]. Nanoparticles of gelatin and BSA can absorb 51–72% of water and release BSA from the nanoparticulate matrix following a diffusion-controlled mechanism [168]. Paclitaxel-loaded gelatin nanoparticles have been used against human bladder transitional cancer cells [114]. Didanosine was localized to a greater extent in the spleen lymph nodes and brain after administration of mannan-coated gelatin nanoparticles as compared to its localization after injection in PBS [115]. The toxic effect of the antimalarial chloroquine is reduced by encapsulating it in gelatin nanoparticles [116].

6.8 Polyurethane

Polyurethanes (PUs) are synthesized using diisocyanate, polyol, and a chain extender. PUs consist of alternating hard and soft segments, which control most

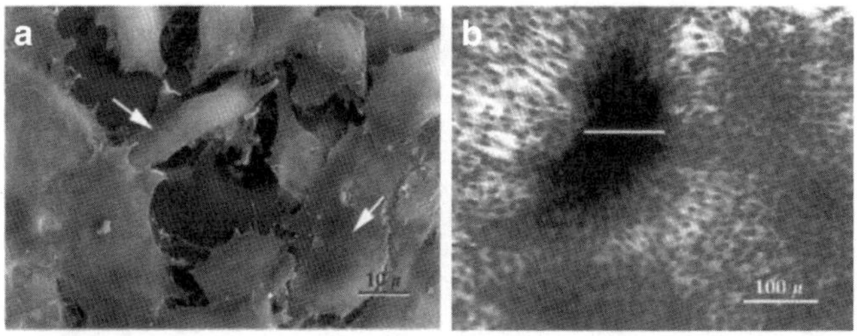

Fig. 11 (**a**) Scanning micrograph of PDI–sucrose polymer exhibiting the attachment and spreading of BMSCs (*arrows*) in vitro after 10 days of culture. (**b**) Light microscopy of PDI–sucrose foam surface exhibiting proliferation of BMSCs (*red arrows*) over a period of 14 days. The *white bar* indicates a large pore into which cells are infiltrating [170]

of the physical properties including mechanical strength and biological activity. The rate of degradation of PU depends on the molar mass, degree of crystallinity, the presence of enzymes, as well as the pH of the medium. PEG is a comonomer constituent used for the preparation of PUs with excellent properties such as hydrophilicity, absence of antigenicity, immunogenicity, and nontoxic degradation products. PUs are frequently used in cell culture, carrier, and biomedical applications due to their biocompatible and cytocompatible nature. PU shows the best biocompatibility, durability, and thrombogenic-resistant properties for use in artificial heart valves [169]. Pentane diisocyanate (PDI)-based PUs are also used in cell culture and drug delivery. Culture of bone marrow stromal cells (BMSCs) on the PU pores showed that the BMSCs are suitable for cell proliferation and retain their morphology in a similar way to cells grown on tissue culture polystyrene (TCPS) (Fig. 11) [170].

7 Polymeric Devices

Development of drug-loaded devices using PLA and PLGA has been reported by several groups [171, 172]. Nail-like ganciclovir-incorporated PLGA implants have been prepared for intraocular drug delivery to treat cytomegalovirus retinitis [173]. The development of 5-fuorouracil-entrapped PLGA subconjunctival coated and uncoated implants/matrices using drug:PLGA ratios of 9:1, 8:2, and 7:3 have been reported [174]. Lin et al. and others have investigated the performance of various antibiotic-loaded PLA and PLGA implants, beads, and cylinders [175, 176]. In vitro and in vivo performances of drug-loaded rods of various polyesters have also been prepared through a melt extrusion process for drug release studies [177]. In order to study the effect of hydrophilic excipient on the drug release from a hydrophobic PLGA (50:50) film, Song et al. designed double-layer films (150 μm thickness) in which the drug-releasing layer consisted of drug/hydrophilic additive/ PLGA (10:10:80 ratio) and the protecting layer consisted of PLGA only [178].

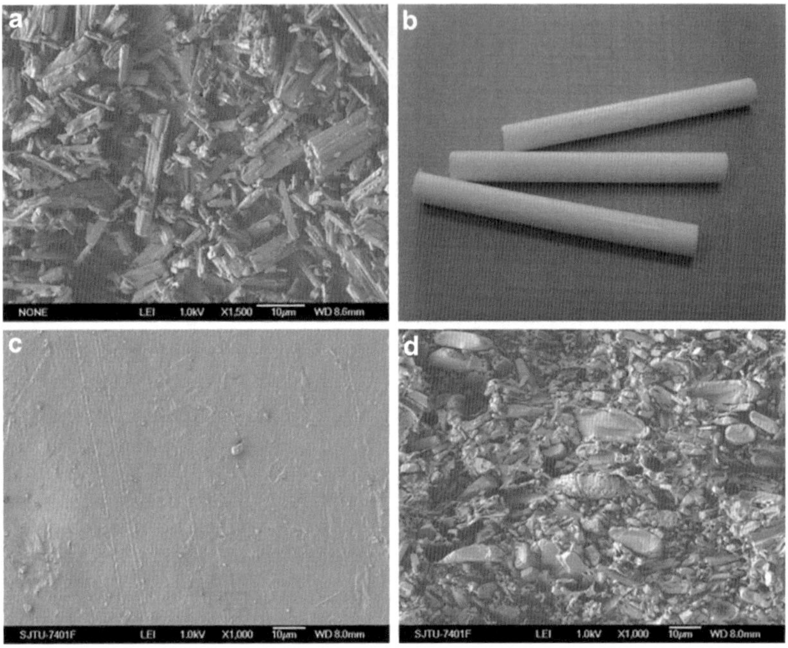

Fig. 12 Praziquantel particles and implants: (**a**) SEM image of praziquantel particles; (**b**) outward appearance of the fabricated implants, taken by digital camera; (**c**) SEM image of the representative surface of the implants; and (**d**) SEM image of the representative cross-section of the implants [180]

Preparation of low density PLGA (85:15) foams having high interstitial void volume has also been reported [179]. A series of praziquantel (anthelmintic)-loaded implants (shown in Fig. 12) based on PEG/PCL blends, fabricated by a combination of twin-screw and hot-melt extrusion, have been reported for in vitro drug release from these implants. The performance of the implants after implantation in rats has been evaluated [180]. Pandey and Banik developed an aqueous two-phase system using biodegradable polymers like dextran T500 and PEG 4000 for better partitioning and extractive fermentation of alkaline phosphatase. An aqueous two-phase system, composed of 9.0% (w/v) PEG 4000 and 9.6% (w/v) dextran T500, was found to be a suitable system for alkaline phosphatase production on the basis of a greater partition coefficient ($k = 5.23$) than various other aqueous two-phase systems [181]. PUs are frequently used in cell culture, carrier, and biomedical applications due to their biocompatible and cytocompatible nature. Out of different materials available for heart valves like collagen, silicon rubber, and polytetrafluoroethylene (PTFE), the PUs show the best biocompatibility, durability, and thrombogenic-resistant properties (Fig. 13) [182, 183]. Subdermal PDI–sucrose implants exhibit no sign of tissue necrosis around the implantation sites; instead, blood vessels were observed around the matrix indicating the biocompatible nature of PU. Additionally, the PDI–sucrose matrix shows an extensive network of fine

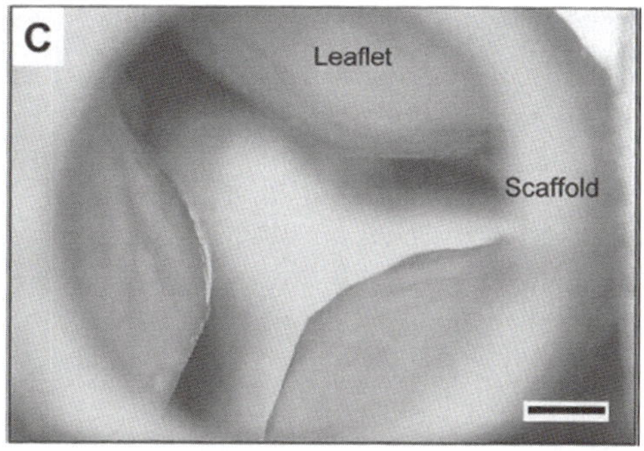

Fig. 13 Macroscopic observation of a biovalve. Bottom section of the boundary between the polyurethane scaffold and the leaflet tissues. Scale bar: 1 mm [182]

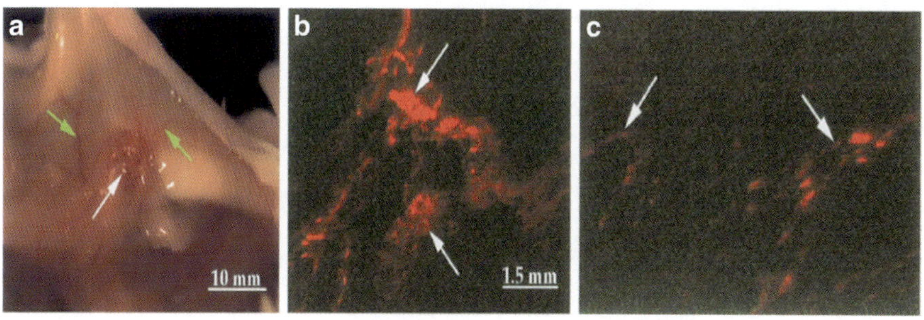

Fig. 14 (**a**) Gross examination of subdermally implanted PDI–sucrose (*white arrow*) exhibiting absence of tissue necrosis, redness, and edema around the polymer 6 weeks after implantation. *Green arrows* indicate capillaries growing towards the polymers. Confocal microscopic examination of vascularization (*white arrows*) in the (**b**) superficial layers and (**c**) deeper layers of subdermally implanted polymer after 3 weeks. The polymer is seen as *green* because of autofluorescence and capillaries are stained *red* because of the intravenous injection of 0.2 μm of fluorospheres containing the fluorescent red dye rhodamine isothiocyanate [170]

capillaries penetrating within the pores located in the deeper layers of the porous polymer (Fig. 14). Thus, the ability to permit vascularization is important for biomaterials that are to be used for tissue engineering. There are a few biodegradable polymer-based devices used in animal and human bodies.

8 Conclusions

The main objective of this article is to summarize information about various biodegradable polymers, including preparation methods for their nanoparticles with/without drugs or other active biological molecules and their biomedical applications. Biodegradable polymers are the most versatile and promising class of biomaterials that can be engineered to fulfill specific requirements in biomedical and pharmaceutical applications such as transient implants, sustained drug delivery, and tissue engineering. At present, several methods are available for the preparation of polymeric nanoparticles suitable for encapsulation of biologically active molecules. It is now possible to select the best method of preparation using an appropriate polymer to achieve efficient entrapment of the drug, depending on the physicochemical characteristics of drug. The selected method should reduce loss of the drug or its pharmacological activity. Nanoparticle drug delivery systems have advantages over conventional drug delivery systems because they can increase the bioavailability, solubility, and permeability of many potent drugs. These advantages include reduced dose, reduced frequency of application, better therapeutic control, and less side effects.

Acknowledgements The authors are thankful to the University Grants Commission (UGC), New Delhi, India, for providing research facilities and financial support to Dr. S.K. Pandey as UGC-Kothari postdoctoral fellow (No.F.4-2/2006(BSR)/13-449/2011) in the Department of Zoology, Faculty of Science, Banaras Hindu University, Varanasi, India. The authors also acknowledge the receipt of research funding from Department of Biotechnology (DBT), New Delhi, Ministry of Science and Technology, Government of India (Project No. BT/PR-6929/BCE/08/433/2005).

References

1. Katti DS, Lakshmi S, Langer R et al (2002) Toxicity, biodegradation and elimination of polyanhydrides. Adv Drug Deliv Rev 54:933–961
2. Jalil R, Nixon JR (1990) Biodegradable poly (lactic acid) and poly (lactide-*co*-glycolide) microcapsules: problems associated with preparative techniques and release properties. J Microencapsul 7:297–325
3. Tice TR, Tabibi ES (1991) Parenteral drug delivery: injectables. In: Kydonieus A (ed) Treatise on controlled drug delivery: fundamentals optimization, applications. Dekker, New York, pp 315–339
4. Wu XS (1995) Preparation, characterization, and drug delivery applications of microspheres based on biodegradable lactic/glycolic acid polymers. In: Wise DL et al (eds) Encyclopedic handbook of biomaterials and bioengineering. Dekker, New York, pp 1151–1200
5. Li S (1999) Hydrolytic degradation characteristics of aliphatic polyesters derived from lactic and glycolic acids. J Biomed Mater Res 48:342–353
6. Orive G, Hernandez RM, Rodriguez Gascon A et al (2003) Drug delivery in biotechnology: present and future. Curr Opin Biotechnol 14:659–664
7. Langer R (1990) New methods of drug delivery. Science 249:1527–1533
8. Kayser O, Lemke A, Hernandez-Trejo N (2005) The impact of nanobiotechnology on the development of new drug delivery systems. Curr Pharm Biotechnol 6:3–5
9. Williams D (2004) Nanotechnology: a new look. Med Device Technol 15:9–10
10. Cheng MMC, Cuda G, Bunimovich YL et al (2006) Nanotechnologies for biomolecular detection and medical diagnostics. Curr Opin Chem Biol 10:11–19
11. Shaffer C (2005) Nanomedicine transforms drug delivery. Drug Discov Today 10:1581–1582
12. Moghimi SM, Hunter AC, Murray JC (2005) Nanomedicine: current status and future prospects. FASEB J 19:311–330
13. Jain KK (2003) Nanodiagnostics: application of nanotechnology in molecular diagnostics. Expert Rev Mol Diagn 3:153–161
14. Kubik T, Bogunia-Kubik K, Sugisaka M (2005) Nanotechnology on duty in medical applications. Curr Pharm Biotechnol 6:17–33

15. Emerich DF (2005) Nanomedicine-prospective therapeutic and diagnostic applications. Expert Opin Biol Ther 5:1–5
16. Norakankorn C, Pan Q, Rempel GL et al (2010) Factorial experimental design on synthesis of functional core/shell polymeric nanoparticles via differential microemulsion polymerization. J Appl Polym Sci 116:1291–1298
17. Rao JP, Geckeler KE (2011) Polymer nanoparticles: preparation techniques and size-control parameters. Prog Polym Sci 36:887–913
18. Chenga J, Teplya BA, Sherifia I et al (2007) Formulation of functionalized PLGA-PEG nanoparticles for in vivo targeted drug delivery. Biomaterials 28:869–876
19. Williams DF, Zhong SP (1994) Biodeterioration/biodegradation of polymeric medical devices in situ. Int Biodeter Biodegrad 34:95–130
20. Coleman JW (2001) Nitric oxide in immunity and inflammation. Int Immunopharmacol 1:1397–1406
21. Labow RS, Tang Y, McCloskey CB et al (2002) The effect of oxidation on the enzyme-catalyzed hydrolytic biodegradation of poly (urethan)s. J Biomater Sci Polym Ed 13:651–665
22. Lee KH, Chu CC (2000) The role of superoxide ions in the degradation of synthetic absorbable sutures. J Biomed Mater Res 49:25–35
23. Gopferich A (1996) Mechanisms of polymer degradation and erosion. Biomaterials 17:103–114
24. Shalaby W, Park H (1994) Chemical modification of proteins and polysaccharides and its effect on enzyme-catalyzed degradation. In: Shalaby S (ed) Biomedical polymers. Hanser Publishers, Munich, pp 213–258
25. Zhang Y, Zale S, Sawyer L et al (1997) Effects of metal salts on poly (DL-lactide-co-glycolide) polymer hydrolysis. J Biomed Mater Res 34:531–538
26. Hakkarainen M, Albetsson AC, Karlsson S (1996) Weight losses and molecular weight changes correlated with the evolution of hydroxy-acids in simulated in vivo degradation of homo- and copolymers of PLA and PGA. Polym Degrad Stabil 52:283–291
27. Loo SCJ, Tan WLJ, Khoa SM et al (2008) Hydrolytic degradation characteristics of irradiated multi-layered PLGA films. Int J Pharm 360:228–230
28. Tsuji H, Shimizu K, Sato Y (2012) Hydrolytic degradation of poly (L-lactic acid): combined effects of UV treatment and crystallization. J Appl Polym Sci 125:2394–2406
29. Singh NK, Banik RM, Kulriya PK et al (2009) Nanoparticle-induced biodegradation of poly (ϵ-caprolactone). Nanosci Nanotechnol Lett 1:52–56
30. Singh NK, Purkayastha BD, Roy JK et al (2010) Nanoparticle-induced controlled biodegradation and its mechanism in poly (ϵ-caprolactone). Appl Mater Interfaces 2:69–81
31. Singh NK, Purkayastha BD, Roy JK et al (2011) Tuned biodegradation using poly (hydroxybutyrate-co-valerate) nanobiohybrids: emerging biomaterials for tissue engineering and drug delivery. J Mater Chem 21:15919–15927
32. Maiti P, Batt CA, Giannelis EP (2007) New biodegradable polyhydroxybutyrate/layered silicate nanocomposites. Biomacromolecules 8:3393–3400
33. Yeh M-K, Davis SS, Coombes AGA (1996) Improving protein delivery from microparticles using blends of poly(D, L lactide-co-glycolide) and poly(ethylene oxide)-poly(propylene oxide) copolymers. Pharm Res 13:1693–1698
34. Brannon-Peppas L (1995) Recent advances on the use of biodegradable microparticles and nanoparticles in controlled drug delivery. Int J Pharm 116:1–9
35. Tabata Y, Ikada Y (1990) Phagocytosis of polymer microspheres by macrophages. Adv Polym Sci 94:107–141
36. Stolnik S, Illum L, Davis SS (1995) Long circulating microparticulate drug carriers. Adv Drug Deliv Rev 16:195–214
37. Arshady R (1991) Preparation of biodegradable microspheres and microcapsules: 2. Polylactides and related polyesters. J Control Release 17:1–22
38. Song C, Labhasetwar V, Guzman L et al (1995) Dexamethasone-nanoparticles for intra-arterial localization in restenosis in rats. Proc Int Symp Control Release Bioact Mater 22:444–445

39. Dawson GF, Halbert GW (2000) The in vitro cell association of invasin coated polylactide-co-glycolide nanoparticles. Pharm Res 17:1420–1425
40. Muller RH, Maaben S, Weyhers H et al (1996) Cytotoxicity of magnetite-loaded polylactide, polylactide/glycolide particles and solid lipid nanoparticles. Int J Pharm 138:85–94
41. Leroux JC, Allemann E, De Jaeghere F et al (1996) Biodegradable nanoparticles-from sustained release formulations to improved site specific drug delivery. J Control Release 39:339–350
42. Niwa T, Takeuchi H, Hino T et al (1993) Preparations of biodegradable nanospheres of water-soluble and insoluble drugs with D, L-lactide/glycolide copolymer by a novel spontaneous emulsification solvent diffusion method, and the drug release behavior. J Control Release 25:89–98
43. Murakami H, Kobayashi M, Takeuchi H et al (1999) Preparation of poly (D, L-lactide-co-glycolide) nanoparticles by modified spontaneous emulsification solvent diffusion method. Int J Pharm 187:143–152
44. Wattenberg LW, Wiedmann TS, Estensen RD et al (1997) Chemoprevention of pulmonary carcinogenesis by aerosolized budesonide in female A/J mice. Cancer Res 57:5489–5492
45. Lee YH, Mei F, Bai MY et al (2010) Release profile characteristics of biodegradable-polymer-coated drug particles fabricated by dual-capillary electrospray. J Control Release 145:58–65
46. Jain AK, Swarnakar NK, Das M et al (2011) Augmented anticancer efficacy of doxorubicin-loaded polymeric nanoparticles after oral administration in a breast cancer induced animal model. Mol Pharm 8:1140–1151
47. Fagui AE, Amiel C (2012) PLA nanoparticles coated with a β-cyclodextrin polymer shell: preparation, characterization and release kinetics of a hydrophobic compound. Int J Pharmaceut 436:644–651. doi:10.1016/j.ijpharm.2012.07.052
48. Byun Y, Hwang JB, Bang SH et al (2011) Formulation and characterization of α-tocopherol loaded poly ϵ-caprolactone (PCL) nanoparticles. LWT Food Sci Technol 44:24–28
49. Anitha A, Deepagan VG, Divya Rani VV et al (2011) Preparation, characterization, in vitro drug release and biological studies of curcumin loaded dextran sulphate–chitosan nanoparticles. Carbohydr Polym 84:1158–1164
50. Vanderhoff JW, Mohamed S, Aasser El et al (1979) Polymer emulsification process. US Patent 4,177,177
51. Couvreur P, Dubernet C, Puisieux F (1995) Controlled drug delivery with nanoparticles: current possibilities and future trends. Eur J Pharm Biopharm 41:2–13
52. Anton N, Benoit JP, Saulnier P (2008) Design and production of nanoparticles formulated from nano-emulsion templates—a review. J Control Release 128:185–199
53. Julienne VMC, Benoit JP (1996) Preparation, purification and morphology of polymeric nanoparticles as drug carriers. Pharm Acta Helv 71:121–128
54. Vandorpe J, Schacht E, Stolnik S et al (1996) Poly(organo phosphazene) nanoparticles surface modified with poly(ethylene oxide). Biotechnol Bioeng 52:89–95
55. Song CX, Labhasetwar V, Murphy H et al (1997) Formulation and characterization of biodegradable nanoparticles for intravascular local drug delivery. J Control Release 43:197–212
56. Lemoine D, Preat V (1998) Polymeric nanoparticles as delivery system for influenza virus glycoproteins. J Control Release 54:15–27
57. Zambaux MF, Bonneaux F, Gref R et al (1998) Influence of experimental parameters on the characteristics of poly(lactic acid) nanoparticles prepared by a double emulsion method. J Control Release 50:31–40
58. Quellec P, Gref R, Dellacherie E et al (1999) Protein encapsulation within poly(ethylene glycol)-coated nanospheres. II. Controlled release properties. J Biomed Mater Res A 47:388–395
59. Bilati U, Allemann E, Doelker E (2003) Sonication parameters for the preparation of biodegradable nanocapsules of controlled size by the double emulsion method. Pharm Dev Technol 8:1–9

60. Mainardes RM, Evangelista RC (2005) Praziquantel-loaded PLGA nanoparticles: preparation and characterization. J Microencapsul 22:13–24
61. Quintanar-Guerrero D, Allemann E, Fessi H et al (1998) Preparation techniques and mechanism of formation of biodegradable nanoparticles from preformed polymers. Drug Dev Ind Pharm 24:1113–1128
62. Bindschaedler C, Gurny R, Doelker E (1990) Process for preparing a powder of water-insoluble polymer which can be redispersed in a liquid phase, the resulting powder and utilization thereof. US Patent 4,968,350
63. Allemann E, Gurny R, Doelker E (1992) Preparation of aqueous polymeric nanodispersions by a reversible salting-out process: influence of process parameters on particle size. Int J Pharm 87:247–253
64. Zhang Z, Grijpma DW, Feijen J (2006) Poly (trimethylene carbonate) and monomethoxy poly (ethylene glycol)-block-poly(trimethylene carbonate) nanoparticles for the controlled release of dexamethasone. J Control Release 111:263–270
65. Fessi H, Puisieux F, Devissaguet JP et al (1989) Nanocapsule formation by interfacial polymer deposition following solvent displacement. Int J Pharm 55:1–4
66. Mishra B, Patel BB, Tiwari S (2010) Colloidal nanocarriers: a review on formulation technology, types and applications toward targeted drug delivery. Nanomedicine 6:9–24
67. Thioune O, Fessi H, Devissaguet JP et al (1997) Preparation of pseudolatex by nanoprecipitation: influence of the solvent nature on intrinsic viscosity and interaction constant. Int J Pharm 146:233–238
68. Moinard-Checot D, Chevalier Y, Briancon S et al (2008) Mechanism of nanocapsules formation by the emulsion–diffusion process. J Colloid Interface Sci 317:458–468
69. Kim E, Yang J, Choi J et al (2009) Synthesis of gold nanorod-embedded polymeric nanoparticles by a nanoprecipitation method for use as photothermal agents. Nanotechnology 20:365602
70. Legrand P, Lesieur S, Bochot A et al (2007) Influence of polymer behaviour in organic solution on the production of polylactide nanoparticles by nanoprecipitation. Int J Pharm 344:33–43
71. Nehilla BJ, Bergkvist M, Popat KC et al (2008) Purified and surfactant free coenzyme Q10-loaded biodegradable nanoparticles. Int J Pharm 348:107–114
72. Yallapu MM, Gupta BK, Jaggi M et al (2010) Fabrication of curcumin encapsulated PLGA nanoparticles for improved therapeutic effects in metastatic cancer cells. J Colloid Interface Sci 351:19–29
73. Jeong YI, Cho CS, Kim SH et al (2001) Preparation of poly (D-L-lactide-co-glycolide) nanoparticles without surfactant. J Appl Polym Sci 80:2228–2236
74. Kostog M, Kohler S, Liebert T et al (2010) Pure cellulose nanoparticles from trimethylsilyl cellulose. Macromol Symp 294:96–106
75. Barichello JM, Morishita M, Takayama K et al (1999) Encapsulation of hydrophilic and lipophilic drugs in PLGA nanoparticles by the nanoprecipitation method. Drug Dev Ind Pharm 25:471–476
76. Nemati F, Dubernet C, Fessi H et al (1996) Reversion of multidrug resistance using nanoparticles in vitro: influence of the nature of the polymer. Int J Pharm 138:237–246
77. Mu L, Feng SS (2003) A novel controlled release formulation for the anticancer drug paclitaxel (Taxol): PLGA nanoparticles containing vitamin E TPGS. J Control Release 86:33–48
78. Prabha S, Zhou W-Z, Panyam J et al (2002) Size-dependency of nanoparticle-mediated gene transfection studies with fractionated nanoparticles. Int J Pharm 244:105–115
79. Sanchez A, Vila Jato JL, Alonso MJ (1993) Development of biodegradable microspheres and nanospheres for the controlled release of cyclosporine. Int J Pharm 99:263–273
80. Gomez-Gaete C, Tsapis N, Besnard M et al (2007) Encapsulation of dexamethasone into biodegradable polymeric nanoparticles. Int J Pharm 331:153–159

81. Labhasetwar V, Song C, Humphrey W et al (1998) Arterial uptake of biodegradable nanoparticles: effect of surface modifications. J Pharm Sci 87:1229–1234
82. Yin Y, Chen DW, Wei XY et al (2007) Lectin-conjugated PLGA nanoparticles loaded with thymopentin: ex vivo bioadhesion and in vivo biodistribution. J Control Release 123:27–38
83. Jin C, Bai L, Wu H et al (2008) Cellular uptake and radiosensitisation off SR-2508 loaded PLGA nanoparticles. J Nanopart Res 10:1045
84. Fonseca C, Simoes S, Gaspar R (2002) Paclitaxel-loaded PLGA nanoparticles: preparation, physicochemical characterization and in vitro anti-tumoral activity. J Control Release 83:273–286
85. Danhier F, Lecouturier N, Vroman B et al (2009) Paclitaxel-loaded PEGylated PLGA-based nanoparticles: in vitro and in vivo evaluation. J Control Release 133:11–17
86. Derakhshandeh K, Erfan M, Dadashzadeh S (2007) Encapsulation of 9- nitrocamptothecin, a novel anticancer drug, in biodegradable nanoparticles: factorial design, characterization and release kinetics. Eur J Pharm Biopharm 66:34–41
87. Gurny R, Peppas NA, Harrington DD et al (1981) Development of biodegradable and injectable lattices for controlled release potent drugs. Drug Dev Ind Pharm 7:1–25
88. Landry FB, Bazile DV, Spenlehauer G et al (1996) Influence of coating agents on the degradation of poly(D, L-lactic acid) nanoparticles in model digestive fluids (USP XXII). Biomaterials 6:195–202
89. Tobio M, Gref R, Sanchez A et al (1998) Stealth PLA-PEG nanoparticles as protein carriers for nasal administration. Pharm Res 15:270–275
90. Ueda H, Kreuter J (1997) Optimization of the preparation of loperamide-loaded poly (L-lactide) nanoparticles by high pressure emulsification solvent evaporation. J Microencapsul 14:593–605
91. Ueda M, Iwara A, Kreuter J (1998) Influence of the preparation methods on the drug release behavior of loperamide-loaded nanoparticles. J Microencapsul 15:361–372
92. Budhian A, Siegel SJ, Winey KI (2005) Production of haloperidol-loaded PLGA nanoparticles for extended controlled drug release of haloperidol. J Microencapsul 22:773–785
93. Matsumoto J, Nakada Y, Sakurai K et al (1999) Preparation of nanoparticles consisted of poly (L-lactide)–poly (ethylene glycol)–poly (L-lactide) and their evaluation in vitro. Int J Pharm 185:93–101
94. Mainardes RM, Gremiao MPD, Brunetti IL et al (2009) Zidovudine-loaded PLA and PLA-PEG blend nanoparticles: influence of polymer type on phagocytic uptake by polymorphonuclear cells. J Pharm Sci 98:257–267
95. Konan YN, Berton M, Gurny R et al (2003) Enhanced photodynamic activity of meso-tetra (4-hydroxyphenyl)porphyrin by incorporation into sub-200 nm nanoparticles. Eur J Pharm Sci 18:241–249
96. Perez C, Sanchez A, Putnam D et al (2001) Poly(lactic acid)-poly(ethylene glycol) nanoparticles as new carriers for the delivery of plasmid DNA. J Control Release 75:211–224
97. Xing J, Zhang D, Tan T (2007) Studies on the oridonin-loaded poly (D, L-lactic acid) nanoparticles in vitro and in vivo. Int J Biol Macromol 40:153–158
98. Sonaje K, Italia JL, Sharma G et al (2007) Development of biodegradable nanoparticles for oral delivery of ellagic acid and evaluation of their antioxidant efficacy against cyclosporine A-induced nephrotoxicity in rats. Pharm Res 24:899–908
99. Sheng Y, Yuan Y, Liu C et al (2009) In vitro macrophage uptake and in vivo biodistribution of PLA-PEG nanoparticles loaded with hemoglobin as blood substitutes: effect of PEG content. J Mater Sci Mater Med 20:1881–1891
100. Cheng FY, Wang SF, Su CH et al (2008) Stabilizer-free poly(lactide-co-glycolide) nanoparticles for multimodal biomedical probes. Biomaterials 29:2104–2112
101. Yoo HS, Oh JE, Lee KH et al (1999) Biodegradable nanoparticles containing PLGA conjugate for sustained release. Pharm Res 16:1114–1118

102. Sahana DK, Mittal G, Bhardwaj V et al (2008) PLGA nanoparticles for oral delivery of hydrophobic drugs: influence of organic solvent on nanoparticle formation and release behavior in vitro and in vivo using estradiol as a model drug. J Pharm Sci 97:1530–1542
103. Esmaeili F, Ghahremani MH, Ostad SN et al (2008) Folate-receptor-targeted delivery of docetaxel nanoparticles prepared by PLGA–PEG–folate conjugate. J Drug Target 16:415–423
104. Teixeira M, Alonso MJ, Pinto MM et al (2005) Development and characterization of PLGA nanospheres and nanocapsules containing xanthone and 3-methoxyxanthone. Eur J Pharm Biopharm 59:491–500
105. Allemann E, Leroux JC, Gurny R et al (1993) In vitro extended release properties of drug-loaded poly(D, L-lactic acid) nanoparticles produced by salting-out procedure. Pharm Res 10:1732–1737
106. Ammoury N, Fessi H, Devissaguet J-P et al (1990) In vitro release pattern of indomethacin from poly(D, L-lactide) nanocapsules. J Pharm Sci 79:763–767
107. Ammoury N, Fessi H, Devissaguet J-P et al (1991) Jejunal absorption, pharmacological activity, and pharmacokinetic evaluation of indomethacin-loaded poly(D, L-lactide) and poly (isobutylcyanoacrylate) nanocapsules in rats. Pharm Res 8:101–105
108. Elvassore N, Bertucco A, Caliceti P (2001) Production of insulin-loaded poly(ethylene glycol)/poly(L-Lactide) (PEG/PLA) nanoparticles by gas antisolvent techniques. J Pharm Sci 90:1628–1636
109. Molpeceres J, Guzman M, Aberturas MR et al (1996) Application of central composite designs to the preparation of polycaprolactone nanoparticles by solvent displacement. J Pharm Sci 85:206–213
110. Shenoy DB, Amiji MM (2005) Poly (ethylene oxide)-modified poly (epsiloncaprolactone) nanoparticles for targeted delivery of tamoxifen in breast cancer. Int J Pharm 293:261–270
111. Shah LK, Amiji MM (2006) Intracellular delivery of saquinavir in biodegradable polymeric nanoparticles for HIV/AIDS. Pharm Res 23:2638–2645
112. Zheng D, Li X, Xu H et al (2009) Study on docetaxel-loaded nanoparticles with high antitumor efficacy against malignant melanoma. Acta Biochim Biophys Sin (Shanghai) 41:578–587
113. Prabu P, Chaudhari AA, Dharmaraj N et al (2008) Preparation, characterization, in-vitro drug release and cellular uptake of poly (caprolactone) grafted dextran copolymeric nanoparticles loaded with anticancer drug. J Biomed Mater Res 90:1128–1136
114. Lu Z, Yeh TK, Tsai M et al (2004) Paclitaxel-loaded gelatin nanoparticles for intravesical bladder cancer therapy. Clin Cancer Res 10:7677–7684
115. Kaur A, Jain S, Tiwary AK (2008) Mannan-coated gelatin nanoparticles for sustained and targeted delivery of didanosine: in vitro and in vivo evaluation. Acta Pharm 58:61–74
116. Bajpai AK, Choubey J (2006) Design of gelatin nanoparticles as swelling controlled delivery system for chloroquine phosphate. J Mater Sci Mater Med 17:345–358
117. Bajpai AK, Choubey J (2005) Release study of sulphamethoxazole controlled by swelling of gelatin nanoparticles and drug–biopolymer interaction. J Macromol Sci 42:253–275
118. Sarmento B, Ribeiro A, Veiga F et al (2007) Alginate/chitosan nanoparticles are effective for oral insulin delivery. Pharm Res 12:2198–2206
119. Gunatillake P, Mayadunne R, Adhikari R (2006) Recent developments in biodegradable synthetic polymers. Biotechnol Annu Rev 12:301–347
120. Terasaka S, Iwasaki Y, Shinya N et al (2006) Fibrin glue and polyglycolic acid nonwoven fabric as a biocompatible dural substitute. Neurosurgery 58:134–139
121. Maurus PB, Kaeding CC (2004) Bioabsorbable implant material review. Oper Tech Sports Med 12:158–160
122. Urayama H, Kanamori T, Kimura Y (2002) Properties and biodegradability of polymer. Macromol Mater Eng 287:116–121
123. Sawai D, Takahashi K, Imamura T et al (2002) Preparation of oriented β-form poly(L-lactic acid) by solid-state extrusion. J Polym Sci B Polym Phys 40:95–104

124. Cohen S, Alonso MJ, Langer R (1994) Novel approaches to controlled release antigen delivery. Int J Technol Assess Health Care 10:121–130
125. Jean-Christophe L, Eric A, Fanny DJ et al (1996) Biodegradable nanoparticles-from sustained release formulations to improved site specific drug delivery. J Control Release 39:339–350
126. Kumari A, Yadav SK, Pakade YB et al (2011) Nanoencapsulation and characterization of *Albizia chinensis* isolated antioxidant quercitrin on PLA nanoparticles. Colloids Surf B Biointerfaces 82:224–232
127. Hu FX, Neoh KG, Kang ET (2006) Synthesis and in vitro anti-cancer evaluation of tamoxifen-loaded magnetite/PLLA composite nanoparticles. Biomaterials 27:5725–5733
128. Leenstag JW, Pennings AJ, Bos RRM et al (1987) Resorbable materials of polyl-lactides VI. Plates and screws for internal fracture fixation. Biomaterials 8:70–73
129. Pathiraja AG, Raju A (2003) Biodegradable synthetic polymers for tissue engineering. Eur Cell Mater 5:1–16
130. Sheridan MH, Shea LD, Peters MC (2000) Bioabsorbable polymer scaffolds for tissue engineering capable of sustained growth factor delivery. J Control Release 64:91–102
131. Panyam J, Labhasetwar V (2003) Dynamics of endocytosis and exocytosis of poly (L-lactide-co-glycolide) nanoparticles in vascular smooth muscle cells. Pharm Res 20:212–220
132. Prabha S, Labhasetwar V (2004) Critical determinants in PLGA/PLA nanoparticle-mediated gene expression. Pharm Res 21:354–364
133. Athanasiou KA, Niederauer GG, Agrawal CM (1996) Sterilization, toxicity, biocompatibility and clinical applications of polylactic acid/polyglycolic acid copolymers. Biomaterials 17:93–102
134. Mittal G, Sahana DK, Bhardwaj V et al (2007) Estradiol loaded PLGA nanoparticles for oral administration: effect of polymer molecular weight and copolymer composition on release behavior in vitro and in vivo. J Control Release 119:77–85
135. Hans ML, Lowman AM (2002) Biodegradable nanoparticles for drug delivery and targeting. Curr Opin Solid State Mater Sci 6:319–327
136. Feng SS (2004) Nanoparticles of biodegradable polymers for new-concept chemotherapy. Expert Rev Med Devices 1:115–125
137. Redhead HM, Davis SS, Illum L (2001) Drug delivery in poly (lactide-co-glycolide) nanoparticles surface modified with poloxamer 407 and poloxamine 908: in vitro characterisation and in vivo evaluation. J Control Release 70:353–363
138. Nicoli S, Santi P, Couvreur P et al (2001) Design of triptorelin loaded nanospheres for transdermal iontophoretic administration. Int J Pharm 214:31–35
139. Nair LS, Laurencin CT (2006) Polymers as biomaterials for tissue engineering and controlled drug delivery. In: Kaplan D, Lee K (eds) Tissue engineering I, Advances in biochemical engineering/biotechnology. Springer, Berlin, pp 47–90
140. Mondrinos MJ, Dembzynski R, Lu L, Byrapogu VK, Wootton DM, Lelkes PI, Zhou J (2006) Porogen-based solid freeform fabrication of polycaprolactone-calcium phosphate scaffolds for tissue engineering. Biomaterials 27:4399–4408
141. Gao H, Wang YN, Fan YG et al (2005) Synthesis of a biodegradable tadpole-shaped polymer via the coupling reaction of polylactide onto mono(6-(2-aminoethyl)amino-6-deoxy)-beta-cyclodextrin and its properties as the new carrier of protein delivery system. J Control Release 107:158–173
142. Kim SY, Lee YM (2001) Taxol-loaded block copolymer nanospheres composed of methoxy poly (ethylene glycol) and poly (epsilon-caprolactone) as novel anticancer drug carriers. Biomaterials 22:1697–1704
143. Damge C, Maincent P, Ubrich N (2007) Oral delivery of insulin associated to polymeric nanoparticles in diabetic rats. J Control Release 117:163–170
144. Changyong C, Chae SY, Jae-Won N (2006) Thermosensitive poly(nisopropylacrylamide)-b-poly(ϵ-caprolactone) nanoparticles for efficient drug delivery system. Polymer 47:4571

145. Espuelas MS, Legrand P, Loiseau PM et al (2002) In vitro antileishmanial activity of amphotericin B loaded in poly (epsilon-caprolactone) nanospheres. J Drug Target 10:593–599
146. Yordanov G, Abrashev N, Dushkin C (2010) Poly(*n*-butylcyanoacrylate) submicron particles loaded with ciprofloxacin for potential treatment of bacterial infections. Prog Colloid Polym Sci 137:53–59
147. Yordanov R, Skrobanska A, Evangelatov A (2012) Entrapment of epirubicin in poly (butyl cyanoacrylate) colloidal nanospheres by nanoprecipitation: formulation development and in vitro studies on cancer cell lines. Colloids Surf B Biointerfaces 92:98–105
148. Balland O, Pinto-Alphandary H, Pecquet S et al (1994) The uptake of ampicillin-loaded nanoparticles by murine macrophages infected with *Salmonella typhimurium*. J Antimicrob Chemother 33:509–522
149. Briones E, Colino CI, Lanao JM (2008) Delivery systems to increase the selectivity of antibiotics in phagocytic cells. J Control Release 125:210–227
150. Forestier F, Gerrier P, Chaumard C et al (1992) Effect of nanoparticle-bound ampicillin on the survival of *Listeria monocytogenes* in mouse peritoneal macrophages. J Antimicrob Chemother 30:173–179
151. Nishioka Y, Yoshino H (2001) Lymphatic targeting with nanoparticulate system. Adv Drug Deliv Rev 47:55–64
152. Gulyaev AE, Gelperina SE, Skidan IN et al (1999) Significant transport of doxorubicin into the brain with polysorbate 80-coated nanoparticles. Pharm Res 16:1564–1569
153. Ramge P, Unger RE, Oltrogge J et al (2000) Polysorbate-80 coating enhances uptake of polybutylcyanoacrylate (PBCE)-nanoparticles by human and bovine primary brain capillary endothelial cells. Eur J Neurosci 12:1931–1940
154. Gursoy A, Eroglu L, Ulutin S et al (1989) Evaluation of indomethacin nanocapsules for their physical stability and inhibitory activity on inflammation and platelet aggregation. Int J Pharm 52:101–108
155. Abraham DJ (2006) Polyionic hydrocolloids for the intestinal delivery of protein drugs: alginate and chitosan. J Control Release 114:1–14
156. Baruch L, Machluf M (2006) Alginate-chitosan complex coacervation for cell encapsulation: effect on mechanical properties and on long-term viability. Biopolymers 82:570–579
157. Nordtveit RJ, Varum KM, Smidstrod O (1996) Degradation of partially N-acetylated chitosans with hen egg white and human lysozyme. Carbohydr Polym 29:163–167
158. Shi C, Zhu Y, Ran X et al (2006) Therapeutic potential of chitosan and its derivatives in regenerative medicine. J Surg Res 133:185–192
159. Martinac A, Filipovi J, Voinovich D et al (2005) Development and bioadhesive properties of chitosan-ethylcellulose microspheres for nasal delivery. Int J Pharm 291:69–77
160. Illum L, Jabbal-Gill I, Hinchcliffe M et al (2001) Chitosan as a novel nasal delivery system for vaccines. Adv Drug Deliv Rev 51:81–96
161. Onishi H, Takahashi H, Yoshiyasu M et al (2001) Preparation and in vitro properties of N-Succinylchitosan or carboxymethylchitin-mitomycin C conjugate microparticles with specified size. Drug Dev Ind Pharm 27:659–667
162. De Campos AM, Sanchez A, Alonso MJ (2001) Chitosan nanoparticles: a new vehicle for the improvement of the delivery of drugs to the ocular surface, Application to cyclosporin A. Int J Pharm 224:159–168
163. Wu Y, Yang W, Wang C et al (2005) Chitosan nanoparticles as a novel delivery system for ammonium glycyrrhizinate. Int J Pharm 295:235–245
164. Yamamoto M, Ikada Y, Tabata Y (2001) Controlled release of growth factors based on biodegradation of gelatin hydrogel. J Biomater Sci Polym Ed 12:77–88
165. Balakrishnan B, Jayakrishnan A (2005) Self-cross-linking biopolymers as injectable in situ forming biodegradable scaffolds. Biomaterials 26:3941–3951
166. Yao CH, Liu BS, Hsu SH et al (2004) Biocompatibility and biodegradation of a bone composite containing tricalcium phosphate and genipin crosslinked gelatin. J Biomed Mater Res 69A:709–717

167. Ikada Y, Tabata Y (1998) Protein release from gelatin matrices. Adv Drug Deliv Rev 31:287–301
168. Li JK, Wang N, Wu XS (1998) Gelatin nanoencapsulation of protein/peptide drugs using an emulsifier-free emulsion method. J Microencapsul 15:163–172
169. Wheatley DJ, Raco L, Bernacca GM et al (2000) Polyurethane: material for the next generation of heart valve prostheses? Eur J Cardiothorac Surg 17:440–448
170. Ganta RS, Piesco PN, Long P (2003) Vascularization and tissue infiltration of a biodegradable polyurethane matrix. J Biomed Mater Res 64A:242–248
171. Gangadharam PRJ, Kailasam S, Srinivasan S et al (1994) Experimental chemotherapy of tuberculosis using single dose treatment with isoniazid in biodegradable polymers. Br J Antimicrobial Chemotherapy 33:265–271
172. Kailasam S, Wise DL, Gangadharam PRJ (1994) Bioavailability and chemotherapeutic activity of clofazimine against *Mycobacterium avium* complex infections in beige mice following a single implant of a biodegradable polymer. J Antimicrob Chemother 33:273–279
173. Kunou N, Ogura Y, Hashizoe M et al (1996) Biodegradable scleral implant for controlled intraocular delivery of ganciclovir. Proc Int Symp Control Release Bioact Mater 23:711–712
174. Wang G, Tucker IG, Roberts MS et al (1996) In vitro and in vivo evaluation in rabbits of a controlled release 5-fluorouracil subconjunctival implant based on poly (D, L-lactide-co-glycolide). Pharm Res 13:1059–1064
175. Ramchandani M, Robinson D (1998) In vitro and in vivo release of ciprofloxacin from PLGA 50:50 implants. J Control Release 54:167–175
176. Lin SS, Ueng SW, Liu SJ et al (1999) Development of biodegradable antibiotic delivery system. Clin Orthop 5:240–250
177. Lemmouchi Y, Schacht E, Kageruka P et al (1998) Biodegradable polyesters for controlled release of trypanocidal drugs: in vitro and in vivo studies. Biomaterials 19:1827–1837
178. Song CX, Labhasetwar V, Levy RJ (1996) Modulation of release rates from double-layer poly(lactic-*co*-glycolic acid) films for periadventital implants in restenosis. Proc Int Symp Control Release Bioact Mater 23:473–474
179. Hsu Y-Y, Gresser JD, Stewart RR et al (1996) Mechanisms of isoniazid release from poly (D, L-lactide-co-glycolide) matrices prepared by dry-mixing and low density polymeric foam methods. J Pharm Sci 85:706–713
180. Cheng L, Lei L, Guo S (2010) In vitro and in vivo evaluation of praziquantel loaded implants based on PEG/PCL blends. Int J Pharm 387:129–138
181. Pandey SK, Banik RM (2011) Extractive fermentation for enhanced production of alkaline phosphatase from *Bacillus licheniformis* MTCC 1483 using aqueous two-phase systems. Bioresour Technol 102:4226–4231
182. Hayashida K, Kanda K, Yaku H et al (2007) Development of an in vivo tissue-engineered, autologous heart valve (the biovalve): preparation of prototype model. J Thorac Cardiovasc Surg 134:152–159
183. Daebrtiz HS, Sachweh SJ, Hermanns B et al (2003) Introduction of a flexible polymeric heart valve prothesis with special design for mitral position. Circulation 108:II-134–II-139

Adv Polym Sci (2013) 254: 203–240
DOI: 10.1007/12_2012_195

Published online: 3 January 2013

Phytomedicine-Loaded Polymeric Nanomedicines: Potential Cancer Therapeutics

S. Maya, M. Sabitha, Shantikumar V. Nair, and R. Jayakumar

Abstract Cancer continues to be one of the leading causes of death worldwide. Repeated treatment with chemotherapeutics has resulted in tumors that are resistant to these agents. So, it is becoming necessary to identify natural products that target multiple signaling pathways and cause growth inhibitory effects on human cancer cells without resulting in toxicity issues in normal cells. Curcumin, epigallocatechingallate (EGCG; green tea extract), resveratrol, saponins, silymarin, and grape seed extract (GSE) are some of the phytochemicals with significant anticancer potential that we will be focusing on in this review. Curcumin, a natural diphenolic compound derived from turmeric *Curcuma longa*, has proven to be a modulator of intracellular signaling pathways that control cancer cell growth, inflammation, invasion and apoptosis, revealing its anticancer potential. EGCG and GSE are two popular plant extracts that have attracted much attention in recent years due to their antioxidant, antimicrobial, anticarcinogenic, and anti-inflammatory properties. Saponins are a group of naturally occurring plant glycosides, of which at least 150 kinds of natural saponins have been found to possess significant anticancer properties. Silymarin, a mixture of mainly three flavonolignans (silybin, silychristin and silydianin), is extracted from the milk thistle and possesses potential biological properties. Even though these agents are potent anticancer agents, they are limited by their solubility, hydrophobicity, and low bioavailability. Polymeric nanocarriers provide an efficient platform for overcoming the factors that limit application of phytochemicals as therapeutic agents. This review focuses on the development of phytochemical-loaded polymeric nanoparticles and their application as potential anticancer therapeutic agents.

S. Maya, S.V. Nair, and R. Jayakumar (✉)
Amrita Centre for Nanosciences and Molecular Medicine, Amrita Institute of Medical Sciences and Research Centre, Amrita Vishwa Vidyapeetham University, Kochi 682041, India
e-mail: rjayakumar@aims.amrita.edu; jayakumar77@yahoo.com

M. Sabitha
Amrita School of Pharmacy, Amrita Institute of Medical Sciences and Research Centre, Amrita Vishwa Vidyapeetham University, Kochi 682041, India

Keywords Cancer therapeutics · Curcumin · Epigallocatechin gallate · Grape seed extract · Nanomedicine · Phytochemicals · Polymeric nanoparticles · Resveratrol · Saponin · Silymarin

Contents

1 Introduction

Cancer prevention is one of the most important priorities in public health. The economic impact that cancer treatment has in public health is well known. As a matter of improving the quality of life and to reduce costs, the real importance of prevention is recognized. It is here that the importance of adopting a healthy diet is emphasized. Natural products are valuable sources of bioactive compounds. Researchers have been putting efforts into evaluation of the chemopreventive role of natural products [1–3]. Studies have demonstrated that several vegetables, fruits and herbs are sources of potential molecules for treating several malignancies. Thus, our diet has a direct impact on public health [4–6]. Chemoprevention, especially through the use of naturally occurring phytochemicals that interfere with the process of carcinogenesis, is a promising approach for cancer management. The safe toxicological profiles associated with the natural products suggests that specific concentrations of phytochemicals from plant sources may produce cancer chemopreventive effects without causing significant levels of toxicity. Thus, a diet rich in fruits and vegetables with a high content of phytochemicals (polyphenols, tannins and anthocyanins) has the potential for chemoprevention [7, 8].

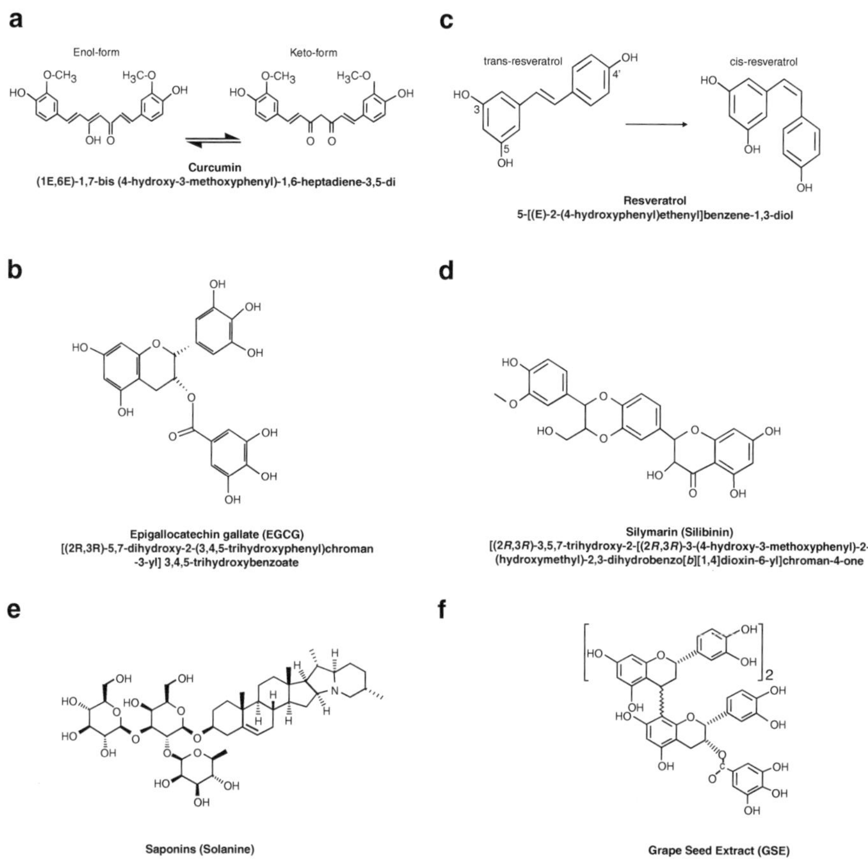

Fig. 1 Structure of different phytochemicals: (**a**) curcumin, (**b**) EGCG, (**c**) resveratrol, (**d**) silymarin, (**e**) saponin and (**f**) GSE

Several thousand polyphenolic molecules have been identified in higher plants and several hundred are found in edible plants. The phytochemicals are mainly plant polyphenols (chemical substances characterized by more than one phenol unit per molecule) composed of flavonoids and non-flavonoids. One or more hydroxyl groups binding to one or more aromatic rings is the structural characteristic of these polyphenols, which are classified according to the number of phenolic rings and the structural elements linking to these rings [9–12]. Flavonoids share a common flavan core formed with 15 carbon atoms, and this class can be divided into flavanols (e.g. catechins), anthocyanidins (e.g. cyaniding, delphinidin), flavones (e.g. apigenin, diosmin), flavanones (e.g. naringenin, hesperitin) and chalcones (e.g. phloretin). The non-flavonoids contain an aromatic ring with one or more hydroxyl groups. Non-flavonoids include stilbene (e.g. *trans*-resveratrol), phenolic acids (e.g. caffeic acid, gallic acid), saponin (e.g. ginsenoside), and other polyphenols such as curcumin and tannins [13]. Figure 1 shows the structure of the different phytochemicals with potential therapeutic effects that are discussed in this review.

1.1 Curcumin

Curcumin (diferuloylmethane with molecular weight of 368.38 g/mol) is a polyphenol derived from the turmeric (*Curcuma longa*) plant. Curcumin (Fig. 1a) has been used extensively in ayurvedic medicine for centuries because it is nontoxic and has a variety of therapeutic properties, including anti-oxidant, analgesic, anti-inflammatory, antimicrobial, and antiseptic activities. Curcumin is reported to act on several biological pathways involved in mutagenesis, oncogene expression, cell cycle regulation, apoptosis, tumorigenesis, and metastasis and, hence, is a potential anticancer agent. Curcumin has shown an antiproliferative effect in multiple cancers, and is an inhibitor of the transcription factor nuclear factor-κβ (NF-κβ) and downstream gene products including c-myc, Bcl-2, cyclooxygenase-2 (COX-2), inducible nitric oxide synthase (iNOS), cyclin D1, tumor necrosis factor-α (TNF-α), interleukins, and matrix metalloproteins (e.g. matrix metalloproteinase 9, MMP-9) [14–18].

1.2 Epigallocatechin-3-Gallate

The habitual consumption of green tea (*Camellia sinensis*), a popular beverage worldwide, has been associated with chemopreventive efficacy. Green tea contains high amounts of the potent antioxidants polyphenols, which may reduce the risk of many chronic diseases including cardiovascular disease and cancer [19–24]. The cancer chemopreventive effects of green tea are reported to be mediated by its polyphenols, known as catechins. The major catechins in green tea are (−)-epigallocatechin-3-gallate (EGCG), (−)-epicatechin-3-gallate, (−)-epigallocatechin, and (−)-epicatechin. EGCG (Fig. 1b) is the major catechin in green tea and accounts for 50–80%, representing 200–300 mg per brewed cup of green tea [4]. Green tea polyphenols have been shown to be chemopreventive for several types of cancers, including those of the liver, prostate, stomach and breast [25–28].

1.3 Resveratrol

Resveratrol (3,5,4′-trihydroxystilbene: $C_{14}H_{12}O_3$, molecular weight 228.25 g/mol) is a non-flavonoid polyphenolic compound abundant in grapes, peanuts, and other foods that are commonly consumed as part of the human diet [29]. The compound was first isolated from the root of *Polygonum cuspidatum*, a plant used in traditional Chinese and Japanese medicine [30]. Resveratrol (Fig. 1c) exists as two structural isomers: *cis*-(*Z*) and *trans*-(*E*). The *trans*-isomer is biologically more active than the *cis*-isomer, probably due to its non-planar conformation [31, 32]. Despite its poor water solubility, resveratrol exhibits high membrane permeability and can be considered a class-II compound in the Biopharmaceutical Classification System [33]. Resveratrol has a broad range of desirable biological actions, including cardioprotection, cancer prevention, and prolongation of lifespan in several species [34–37].

1.4 Silymarin

Silymarin (Fig. 1d) is an active extract from the seeds of the plant milk thistle *Silybum marianum* (L.) Gaertn (Asterceae), and contains approximately 65–80% silymarin flavonolignans (silymarin complex) with small amounts of flavonoids and approximately 20–35% fatty acids and other polyphenolic compounds. The major component of the silymarin complex is silybin. Silymarin has been used for more than 2,000 years as a natural remedy for treating hepatitis and cirrhosis and to protect the liver from toxic substances. Silymarin acts by antioxidative, antilipid peroxidative, antifibrotic, anti-inflammatory, membrane-stabilizing, immunomodulatory, and liver regenerating mechanisms in experimental liver diseases. Furthermore, silymarin has been extensively studied, both *in vivo* and *in vitro,* for its cancer chemopreventive potential against various cancers [38–40].

1.5 Saponins

Saponins are a group of naturally occurring plant glycosides, characterized by their strong foam-forming properties in aqueous solution. The presence of saponins has been reported in more than 100 families of plants, out of which at least 150 kinds of natural saponins have been found to possess significant anticancer properties. Saponins (Fig. 1e) are a vast group of glycosides and are widely distributed in higher plants. Their surface-active properties are what distinguish these compounds from other glycosides. They dissolve in water to form colloidal solutions that foam upon shaking. Saponin-containing plants are sought after for use in household detergents. Saponins have also been sought after in the pharmaceutical industry because some can form the starting point for the semi-synthesis of steroidal drugs. Many have pharmacological properties and are used in phytotherapy and in the cosmetic industry. They are believed to form the main constituent of many plant drugs and folk medicines, and are considered responsible for numerous pharmacological properties [41–44].

1.6 Grape Seed Extract

Grape seed extract (GSE) (Fig. 1f) is a by-product derived from the grape seeds (*Vitis vinifera*) (from grape juice and wine processing). It is extracted, dried, and purified to produce a polyphenolic-rich extract. Standardized grape seed extracts contain 74–78% oligomeric proanthocyanidins and less than approximately 6% of free flavanol monomers on a dry-weight basis. The red color and astringent taste of the GSE can be attributed to polyphenol-rich compounds, especially proanthocyanidins. It also has well-documented antioxidant, antimicrobial, and anti-inflammatory properties. Proanthocyanidins in GSE exhibit protective properties against free radicals and oxidative stress, making it an excellent anti-oxidant in addition to other beneficial biological functions such as cardio- and neuro-protection and antibacterial activity [45].

2 Biological Roles of Phytochemicals

Dietary polyphenols have become potential therapeutic and chemopreventive agents, and chemoprevention of cancer through the use these natural agents has received an increasing interest. Chemoprevention, by definition, is a means of therapy for precancerous lesions, which are called preinvasive neoplasia, dysplasia, or intraepithelial neoplasia, depending on the organ system [8, 46]. Polyphenols have been demonstrated to act on multiple key elements in intracellular signal transduction pathways related to cell proliferation, differentiation, apoptosis, inflammation, angiogenesis, and metastasis. Phytochemicals possess favorable potential therapeutic effects including anti-oxidant, antiviral, antimicrobial, anti-inflammatory, anticancer, cardioprotective, and neuroprotective effects [13, 14, 40, 45]. Many of these phytochemicals exert their biological properties via common mechanisms and this section describes the actions that are very specific. Figure 2 summarizes the major biological properties (antioxidant, anti-inflammatory, and anticancer) of these phytochemicals, with their mechanisms.

2.1 Antioxidant Effect

Phytochemicals exert antioxidant activity and support redox homeostasis in several *in vitro* and *in vivo* models. The antioxidant properties of the polyphenols come by virtue of their chemical structure.

The curcuminoids consist of two methoxylated phenols connected by two α,β-unsaturated carbonyl groups that exist in a stable enol form [47]. Curcumin promotes neutralization of the lipid radicals and acts as a chain breaking antioxidant at the $3'$ position, resulting in an intramolecular Diels–Alder reaction [48]. Curcumin possesses free-radical scavenging activity. Using rat peritoneal macrophages as a model, studies demonstrated that curcumin scavenged macrophage-generated reactive oxygen species (ROS) both *in vitro* and *in vivo* [49, 50]. Curcumin

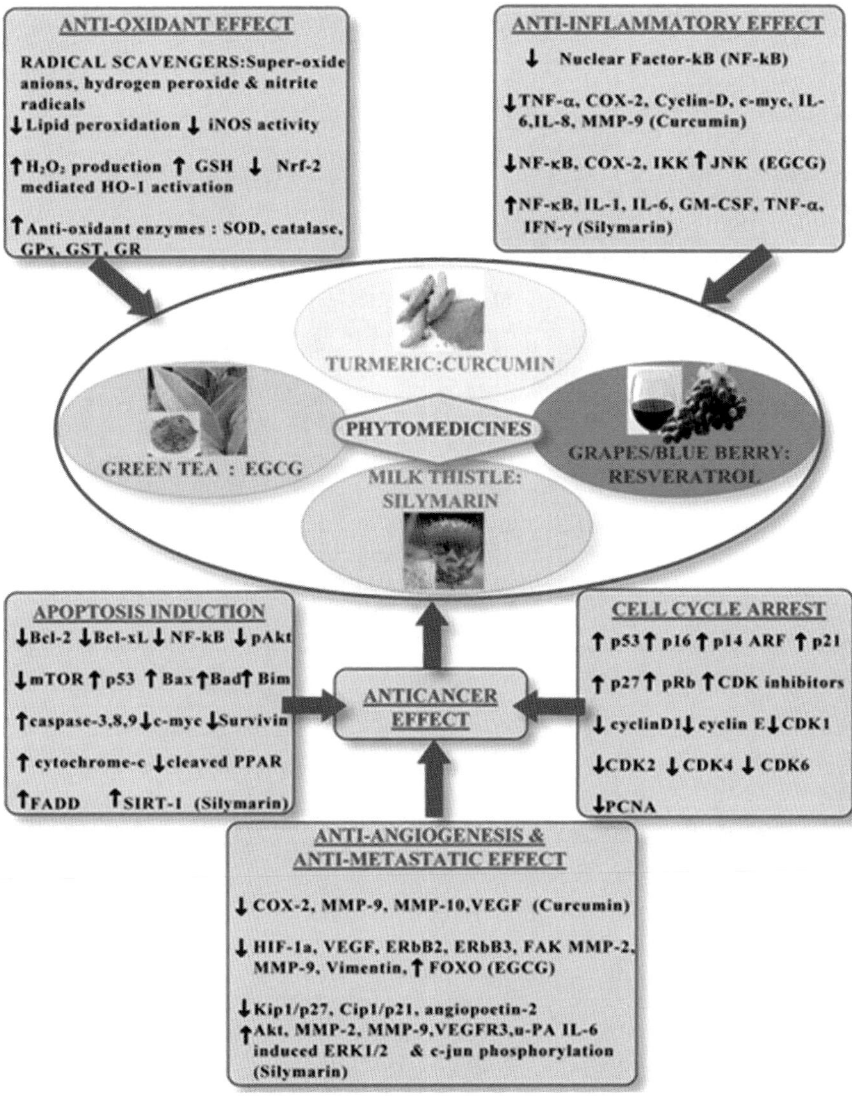

Fig. 2 Biological properties of phytochemicals: scheme summarizing the major genes involved in the pathways that elicit the biological properties of the described phytochemicals

downregulated the activity of iNOS in macrophages, thus reducing the amount of ROS generated in response to oxidative stress [51, 52].

EGCG acquires its antioxidant potential from the phenolic group, which is sensitive to oxidation and generates quinine after oxidation, and also from the trihydroxyl structure in the D ring [53, 54]. EGCG is a powerful radical scavenger; it protects neurons from the oxidative damage induced by commonly used prooxidants such as tertbutylhydroperoxide [55], inhibits expression of the multidrug resistance gene, and can modulate the topoisomerase activity associated with tumor growth. Many studies have provided evidence for the interference of EGCG in different metabolic pathways [56]. It selectively inhibits specific enzymes involved in cancer-developing activities such as DNA methyltransferases and repairs DNA aberrations [57]. EGCG can induce apoptosis by an ROS-dependent mechanism [54]. EGCG also induced H_2O_2 formation in human lung adenocarcinoma (H661) and in Ha-ras gene-transformed human bronchial (21BES) cells, but exogenously added catalase prevented EGCG-induced cell apoptosis, which suggests that H_2O_2 is involved in the apoptotic process provoked by EGCG [58]. Studies suggested that EGCG-mediated ROS production may underlie its ability to induce endogenous antioxidants, at least *in vitro*. Treatment of Hepa1c1c7 human hepatoma cells with EGCG resulted in dose-dependent increases in NADPH:quinone reductase-1 and glutathione gene expression through the electrophile response element. Liquid chromatography–mass spectrometry (LC–MS) analysis of the cell culture medium revealed the presence of EGCG-20-glutathione [59].

Silymarin is also reported to be a strong scavenger of free radicals. In erythrocytes exposed to H_2O_2, silymarin increases the activities of anti-oxidant enzymes like superoxide dismutase, catalase, glutathione peroxidase, glutathione reductase, and glutathione-*S*-transferase. Silymarin reduces glutathione depletion and ROS production [60]. One of the *in vivo* studies reported dose-dependent inhibition of malondialdehyde formation in epidermal microsomes and lipid peroxidation caused by 12-*O*-tetradecanoylphorbol-13-acetate (TPA) and benzoyl peroxide in mouse skin epidermis by silymarin [61]. GSE is also a strong radical scavenger and studies have reported the antioxidant potential of GSE [62].

2.2 Anti-inflammatory Effect

The anti-inflammatory effects of curcumin are related to the inhibition of transcription factor NF-κB, which regulates and coordinates the expression of various genes involved in inflammation, cell survival, differentiation, and growth and also regulates the free-radical scavenging activity of curcumin by decreasing the amount of oxidative stress that can trigger the inflammatory cascade [14]. Curcumin has been shown to suppress the activation of NF-κB, an inducible transcription factor that regulates the expression of a host of genes involved in inflammation, cellular proliferation, and cell survival [13, 63–68]. Genes regulated by NF-κB include those encoding for COX-2, the NF-κB inhibitor IκBα, TNF-α, cyclin D1, ICAM-1, c-myc, Bcl-2, MMP-9, iNOS, and interleukins (including IL-6 and IL-8) [13]. Curcumin blocked the IκK-mediated phosphorylation and degradation of IκBα, thus NF-κB remains bound to IκBα in the cytoplasm and is not able to enter the nucleus to activate transcription [65, 69, 70]. Suppression of NF-κB activity subsequently downregulated COX-2, iNOS and inflammatory markers [71, 72]. The inflammatory cytokines are secreted by activated monocytes and macrophages and include COX-2, lipoxygenase (LOX), iNOS, monocyte chemotactic protein-1 (MCP-1), monocyte inflammatory protein-1α (MIP-1α), and interleukins (including IL-1,

IL-2, IL-6, IL-8, and IL-12). This secretion is inhibited by curcumin. The arachadonic acid pathway for eicosanoid biosynthesis is an important participant in the inflammatory response, generating a host of reactive lipid products including leukotrienes, prostaglandins, prostacyclins, and thromboxanes. Curcumin has been shown to decrease the metabolism of arachadonic acid by downregulating the activity of LOX and COX-2, both at the transcriptional level as well as via post-translational enzyme inhibition. Curcumin inhibited prostaglandin E2 biosynthesis through direct inhibition of the microsomal prostaglandin E2 synthase-1 enzyme.

Studies have demonstrated that silymarin is a potent inhibitor of NF-κB activation in response to TNF-α. This effect was mediated through the inhibition of phosphorylation and degradation of IκB [73]. Silymarin dose-dependently inhibited both cytokine-induced NO production and cell death in RINm5F cells, and prevented IL-1β- and IFN-γ-induced NO production and β-cell dysfunction in human pancreatic islets [74].

In vivo studies on saponins isolated from *Bupleurum rotundifolium* L. (Apiaceae) were reported to have anti-inflammatory activity against both TPA-induced ear oedema and chronic skin inflammation [75]. Kim et al. suggested that the anti-inflammatory activity of these saponins is related to anticomplementary action through the classical inflammation pathway [76]. Tumor promoters induce TNF-α, which stimulates the production of cell adhesion molecules. EGCG inhibited NF-κB and expression of TNF-α, thereby reducing cancer promotion [77].

2.3 Anticancer Effect

Curcumin has been studied in multiple human carcinomas including melanoma, head and neck, breast, colon, pancreatic, prostate, and ovarian cancers [13, 17]. Curcumin has been shown to inhibit the cyclins involved in cell cycle regulation. Specifically, curcumin upregulates the expression of apoptotic gene p53, thus inducing apoptosis at the G2 phase; the expression of Bax, thus initiating the mitochondrial apoptotic pathway; and also promotes binding of TNF-α and Fas ligand (death activators) to their corresponding cell surface receptors, thus stimulating extrinsic apoptotic pathways [78].

Regulation of the cell cycle is one of the mechanisms of action of silymarin in the prevention and therapeutic intervention of cancer. Silymarin has been reported to suppress the proliferation of tumor cells in various cancers, including prostate [79–81], ovarian [82], breast [83], lung [84], skin [73], and bladder [85]. Studies demonstrate that silymarin arrests the cells at the G1 and G2/M phase, induces CDK inhibitors (Cip1/p21 and Kip1/p27) and alters the expression of cyclin D kinases [73]. Silymarin treatment has been shown to inhibit the growth of androgen-dependent (LNCaP) and androgen-independent (PC3 and DU145) prostate cancer cells [39]. Apoptosis or programmed cell death that occurs in various physiological and pathological conditions is one of the hallmarks of cancer. Several phytochemicals that are known to inhibit NF-κB, and AP-1 activation can suppress cell proliferation and sensitize cells to apoptosis induction. Silymarin is reported to induce apoptosis in malignant melanoma cells by increasing the expression of Fas-associated proteins with death domain (FADD) followed by cleavage of procaspase-8 that induces apoptosis, cleavage of poly(ADP-ribose) polymerase (PARP), caspase-9, and caspase-3, and decreasing survivin levels in human prostate cancer. Silymarin also inhibited Akt activity, which was associated with activation of caspase-9 and caspase-3 as well as PARP cleavage, in leukemic cells [86]. It has been shown that dietary supplementation with silymarin inhibits 3,20-dimethyl-4-aminobiphenyl-induced prostate carcinogenesis in male F344 rats by increasing apoptosis and modification of cell proliferation [87].

Many isolated steroidal saponins have been shown to be either cytostatic or cytotoxic to HL-60 human leukemia cell lines [43]. Lee et al. reported the *in vitro* tumor activity of saponins in different cancer cell line and suggested that it could be a good candidate for treatment of pulmonary cancer cells [88].

EGCG inhibits cancer-associated stages and exerts an inhibitory effect on DNA methylation via blocking the performance of DNA methyltransferases, strong free-radical scavenging, and antioxidant activities. Recently, EGCG has been shown to inhibit lipopolysaccharide-induced NO production and iNOS gene expression in isolated peritoneal macrophages by decreasing the activation of NF-κB [89]. EGCG inhibited platelet-derived growth factor (PDGF)-induced apoptosis and cell cycle-regulating pathways of vascular smooth muscle cells, resulting in inhibition of tumor growth, metastasis, and angiogenesis *in vivo* [77]. In a study that tested the effect of EGCG on oral cancer cell lines along with curcumin, EGCG blocked cell division in G1, whereas curcumin blocked cell division in S/G2M. EGCG has antiproliferative activities on tumor cells through the blockage of growth factor binding to the receptor and the suppression of mitogenic signal transduction [89]. Apoptosis induction by EGCG is more prominent in many cancer cells without affecting normal cells because NF-κB is activated in cancer cells. EGCG-induced apoptosis in tumor cells may be mediated through direct inhibition of cyclin-dependent kinases and NF-κB inactivation [24]. EGCG also induces the expression of p21 and p27 while decreasing the expression of cyclin D1 and the phosphorylation of retinoblastoma. However, EGCG inhibited lipopolysaccharide-induced phosphorylation of IκB kinase complex (IKK), but failed to affect NF-kB luciferase reporter gene activation in human colon cancer (HT-29) cells, suggesting that EGCG modulation of NF-κB transcriptional activity is not necessarily dependent on IkBα degradation and subsequent release of NF-κB proteins. EGCG downregulates NF-κB, inducing kinase expression in human lung cancer (PC-9) cells. Activation of NF-κB promotes transcriptional upregulation of Bcl-2 and Bcl-XL. Negative regulation of NF-κB by EGCG decreases expression of the proapoptotic protein Bcl-2 [90]. In human prostate carcinoma LNCaP cells, treatment with EGCG induced apoptosis and was associated with stabilization of p53 and also with a downregulation of NF-κB activity, resulting in a decreased expression of the anti-apoptotic protein Bcl-2 [89]. EGCG (70% lethal dose) at the supra-pharmacological concentration of 10 mg/mL increased the proportion of HNSCC cells in the G1 phase, decreased cyclin D1 protein expression and increased the levels of p21WAF1/CIP1 and p27KIP1 proteins [52]. Many recent studies have demonstrated that EGCG triggers cell growth arrest pathways at the G1 stage of the cell cycle through regulation of cyclin D1, CDK4, CDK6, p21/WAF1/CIP1, and p27/KIP1, and induced apoptosis through generation of ROS and activation of caspase-3 and caspase-9 [89]. Treatment of human colorectal carcinoma HT-29 cells with EGCG resulted in nuclear condensation, DNA fragmentation, caspase activation, disruption of mitochondrial membrane potential, and cytochrome c release, which all appeared to be mediated by the JNK pathway [91].

It has been pointed out that the ability of resveratrol to be proapoptotic, antiproliferative and anti-inflammatory make it a potent anticancer agent [92]. GSE is also reported to have anticancer potential against an array of cancers such as prostate, breast, oral, colorectal cancers, leukemia, and glioblastoma multiforme [93, 94].

2.4 Anti-angiogenic Effect

Anti-angiogenesis is one of the fundamental methods of cancer treatment. Angiogenesis is regulated by a variety of pro-angiogenic genes and signaling molecules, including vascular endothelial growth factor (VEGF), basic fibroblast growth factor (bFGF), epidermal growth factor (EGF), platelet-derived growth factors (PDGFs), hypoxia-inducible factors, angiopoetin-1 and angiopoetin-2, and matrix metalloproteinases [95]. Curcumin has demonstrated to have an anti-angiogenic effect *in vivo* in xenograft models of various tumors including glioblastoma, hepatocelluar carcinoma, and prostate and ovarian carcinomas. Curcumin has been shown to regulate a variety of pro-angiogenic growth factors, enzymes, and transcription factors (including bFGF, VEGF, angiopoetins 1 and 2, COX-2, MMP-9, AP-1, and NF-κB), inhibit the angiogenic response to FGF-2 stimulation in mouse endothelial cells, and decrease the expression of MMP-9 [13].

Silymarin is reported to upregulate Kip1/p27, Cip1/p21, and p53; cause mitochondrial apoptosis and caspase activation; downregulate survivin and inhibit Akt and NF-κB signaling; and induce MMP-2 secretions. These reactions affected the growth and survival of human umbilical vein endothelial cells (HUVECs) by inhibiting capillary tube formation and inducing cell cycle arrest, apoptosis, invasion, and migration [66, 73, 80]. Silymarin reduced VEGF levels in prostate (DU145), breast (MCF) and MDA-MB-468 cancer cells and downregulated MMP-2 and CD34 in human hepatoma cell lines [96], causing anti-angiogenic effects.

EGCG inhibits angiogenesis by enhancing FOXO transcriptional activity. Inhibition of AKT and MEK kinases synergistically induced FOXO transcriptional activity, which was further enhanced in the presence of EGCG. Phosphorylation-deficient mutants of FOXO induced FOXO transcriptional activity and inhibited HUVEC cell migration and capillary tube formation. Inhibition of FOXO phosphorylation also enhanced the anti-angiogenic effects of EGCG through transcriptional activation of FOXO [77, 89].

2.5 *Antimetastatic Effect*

Cancer metastasis, which depends on the motility and invasiveness of cancer cells, can complicate clinical management and is a primary cause of cancer deaths. A study of curcumin in metastatic melanoma demonstrated a dose-dependent reduction in binding to extracellular matrix proteins, decreased expression of $\alpha5\beta1$ and $\alpha(V)\beta3$ integrin receptors, and increased expression of various antimetastatic proteins including tissue inhibitor metalloproteinase 2 (TIMP-2), nonmetastatic gene 23, and E-cadherin [13]. E-cadherin expression is important in maintaining the integrity of intercellular adhesion though binding to various catenins (including β-catenin), and loss of E-cadherin is associated with an increased tendency for tumor metastasis. Antimetastatic effects of curcumin have also been demonstrated in the MDA-MB-231 breast cancer cell line, resulting in decreased expression of MMPs, ICAM-1, and chemokine receptor 4, and suppressed cell migration and invasion [97, 98]. Silibinin has the potential to reduce MMP-2 and urokinase-type plasminogen activator (u-PA) expression through reducing ERK1/2 and Akt phosphorylation, which in turn leads to reduced invasiveness of cancer cells [99]. EGCG blocks urokinase, an enzyme that is essential for cancer growth and metastasis formation, by interfering with the enzyme's ability to recognize its substrates [100].

3 Significance of Polymeric Nanocarriers

Conventional natural products have inherent problems of stability, solubility, toxicity, and bioavailability. Chemopreventive treatments require the administration of low doses of chemopreventive agents to avoid toxic side effects and, hence, the bioavailability of these agents becomes a crucial factor [101]. The chemical structure of the phytochemicals determines their gut absorption, and the urine and plasma levels represent good markers for establishing the bioavailability of polyphenols and their metabolites. Generally, less than 10% of polyphenols, or their metabolites, ingested are found in urine and plasma [102]. Another issue is the loss of activity after metabolism because the metabolites may not be biologically active. Only a small proportion of the molecules remain available following oral administration due to insufficient gastric residence time, low permeability, solubility within the gut, and conditions in the gastrointestinal tract (pH, enzymes, presence of other nutrients), all of which limit the activity and potential health benefits of these nutraceutical components [103]. The major factors that limit the therapeutic application of these phytochemicals include low bioavailability, low tissue distribution, fast metabolism, and short half life [104].

The effectiveness of nutraceutical products in preventing disease depends on preserving the bioavailability of the active ingredients. Novel drug delivery systems have been developed for the potent plant extracts and their compounds. Nanophytomedicines include polymeric nanoparticles, nanocapsules, liposomes, phytosomes, nanoemulsions, microspheres, transferosomes, and ethosomes loaded with bioactive plant extracts. The nanocarriers selected for the phytodrugs should channel the active entity to the site of action and also should deliver them at a rate directed by the need of the body. Thus, the intention of the nanosized delivery systems of herbal drugs is to enhance their activity and overcome the problems associated with plant medicines. The novel nanoformulated phytodrugs are reported to have remarkable advantages over conventional formulations of plant actives and extracts and include enhancement of solubility and bioavailability, improved half life, protection from toxicity, enhancement of pharmacological activity, enhancement of stability, improved tissue macrophage distribution, sustained delivery, and protection from physical and chemical degradation. This review highlights and summarizes the anticancer potency of the available nanophytomedicine formulations (see also [105]).

4 Phytochemicals Encapsulated in Polymeric Nanoparticles

4.1 Curcumin

Despite its promising anticancer properties, the extremely low water solubility of curcumin limits its bioavailability and clinical efficacy; for example, its serum concentrations and tissue distributions are low and, furthermore, it is rapidly metabolized and thus has a short half-life [104]. Curcumin has been encapsulated in liposomes [106, 107], polymeric nanoparticles [108], lipid-based nanoparticles [109], biodegradable microspheres [110], cyclodextrin [111], and hydrogel [112] with the aim of overcoming the issues associated with the conventional form of the drug. Nanoparticles have been used as effective delivery tools to enhance the tumoricidal activities of anticancer drugs. Such delivery systems offer many benefits due to their tumor-targeting abilities, internalization efficiencies and, sometimes, escape from multidrug resistance [113, 114].

Nanoscaled particles with a disk-shaped lipid bilayer stabilized by an apolipoprotein scaffold (curcumin nanodisks, NDs, with <50 nm size) have been used for encapsulating curcumin. NDs self-assemble in solution upon presentation of the scaffold protein to a phospholipid vesicle substrate to which an appropriate hydrophobic bioactive agent has been introduced. The incorporation of curcumin into NDs enhanced growth inhibition of human hepatocellular carcinoma cells (HepG2) and induced apoptosis in mantle cell lymphoma (Jeko cells) [115, 116]. Natthakitta et al. reported the development of mucoadhesive curcumin nanospheres. Ethyl cellulose

(EC) and methylcellulose (MC) were used for efficient curcumin nanoencapsulation by the self-assembly of curcumin into monopolymeric carrier (EC) and dipolymeric carrier (ECMC, a blend of EC and MC) with 50% loading. The nanospheres showed a dose-dependent cytotoxicity and free radical scavenging activity towards human breast adenocarcinoma (MCF-7) and HepG2 cells *in vitro*. The nanosystems were evaluated *in vivo* for their adherence to stomach mucosa and their ability to release curcumin into the circulation via oral administration into mice. The authors reported that curcumin sustainability in blood was improved by the nanoencapsulation [117]. The aqueous solubility and stability over a wide pH range of curcumin was enhanced by its nanoencapsulation in a biodegradable polymeric methoxyPEG-palmitate amphiphilic conjugate. The study also reported that nanoencapsulated curcumin inhibited human cervical cancer (Hela) cell proliferation (half-maximal inhibitory concentration, $IC_{50} = 15.58\ \mu M$) comparably to free curcumin ($IC_{50} = 14.32\ \mu M$), thereby showing its anticancer activity in Hela cells *in vitro* [118].

It was reported that curcumin has been encapsulated in nontoxic and non-immunogenic human serum albumin (HSA) nanoparticles. Albumin-bound nanoparticle technology is an easy fabrication process that does not use any toxic surfactants and is well tolerated *in vivo* [119, 120]. Curcumin-HSA NPs (130–150 nm) improved the water solubility of curcumin by 300-fold and enhanced its *in vivo* antitumor activity (50–66% tumor growth inhibition) in a tumor xenograft HCT-116 model. Kim et al. suggested that this potent antitumor activity of curcumin-HSA NPs could be due to the enhanced water solubility, increased accumulation in tumors, and an ability to traverse vascular endothelial cells [121]. In another study, fibrinogen was used as a nanocarrier for curcumin. Curcumin-loaded fibrinogen nanoparticles (CRC-FNPs) were synthesized via a co-acervation technique with a size distribution of 150–200 nm and 90% loading efficiency. The *in vitro* cytotoxicity assay performed using L929 (mouse fibroblast), PC3 (prostate) and MCF-7 (breast) cancer cell lines confirmed that CRC-FNPs were comparatively nontoxic to the L929 cell line but toxic to PC3 and MCF-7 cancer cells. The study also proved the apoptotic potential of CRC-FNPs in MCF-7 cells. Significant internalization of CRC-FNPs was observed in MCF-7 and PC3 cells, as monitored by fluorescent microscopy and flow cytometry-based uptake studies. These results indicated that CRC-FNPs could be a promising therapeutic agent for prostate and breast cancer treatment [122].

A variety of biodegradable polymers have been studied for nanoencapsulation of curcumin and the nanoformulations have been evaluated in several *in vitro* and *in vivo* models. Based on the safety profile of poly(lactic-*co*-glycolic) acid (PLGA), researchers are interested in nanoformulations based on PLGA. A simple solid-oil–water solvent evaporation method has been used to prepare curcumin-loaded PLGA nanospheres. Curcumin nanospheres (mean particle size 45 nm) exerted a pronounced effect on prostate cancer cells (LNCaP, PC3 and DU145) *in vitro*. Figure 3 illustrates the curcumin-PLGA nanospheres *in vitro* in prostate cancer cells [123]. Grabovac and Bernkop [124] studied the effect of surface functionalization on curcumin-PLGA NPs using thiolated chitosan and Thamake et al. [125] studied surface-functionalized curcumin-PLGA NPs with a homo-bifunctional spacer, bis(sulfosuccinimidyl) suberate (BS3), that facilitated

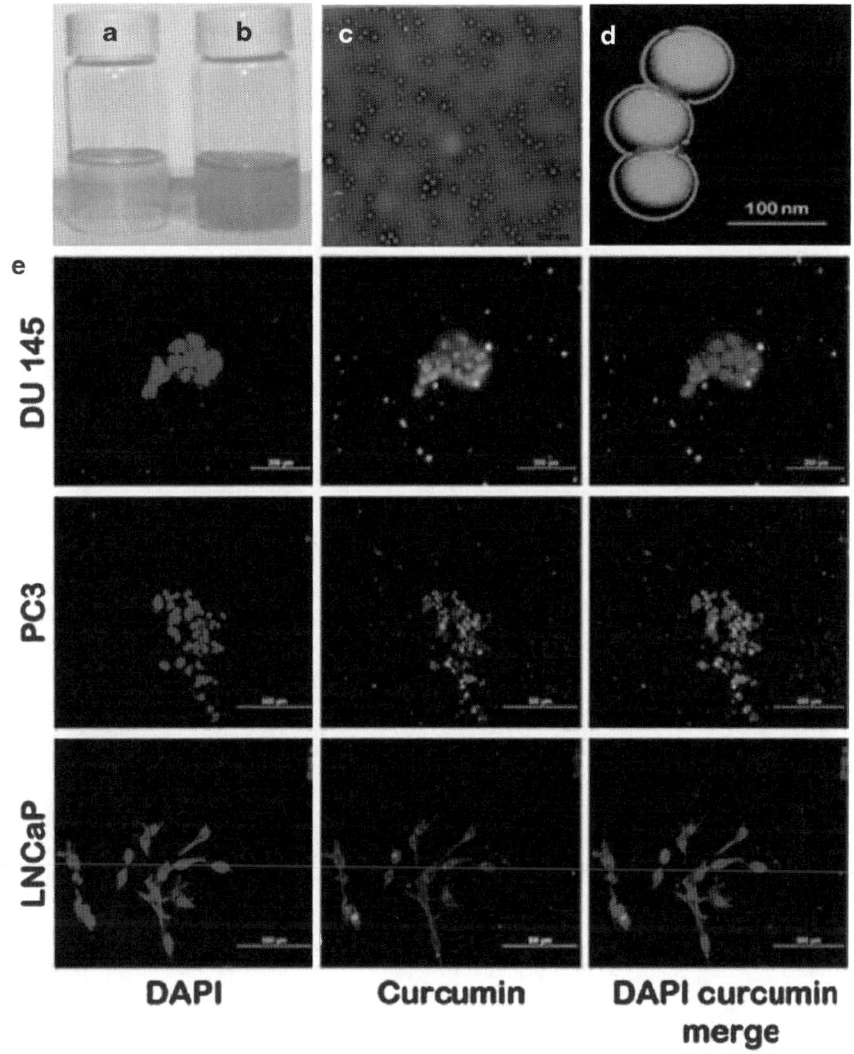

Fig. 3 (**a**) Curcumin-PLGA nanospheres in water and (**b**) free curcumin in water. (**c**) TEM image of curcumin-PLGA nanospheres. (**d**) Confocal microscopy scan illustrating the encapsulation and distribution of curcumin (green fluorescence) within the PLGA nanospheres. (**e**) Confocal images of different cell lines incubated with curcumin-PLGA nanospheres: cell nuclei stained with DAPI, cells showing curcumin in PLGA nanospheres, and merged images of DAPI and curcumin in PLGA nanospheres [123]

conjugation of Annexin A2 and resulted in an efficient target therapy of curcumin to Annexin-A2-positive MDA-MB-231 cancer cells [125]. Curcumin-PLGA NPs stabilized with PEG-5000 exhibited enhanced cellular uptake and increased bioactivity *in vitro* using KBM-5 (human chronic myeloid leukemia), Jurkat (human T cell leukemia), DU145 (prostate carcinoma), MDA-MB-231 (breast adenocarcinoma),

HCT116 (human colon adenocarcinoma), and SEG-1 (human esophageal adenocarcinoma) cell lines and showed superior bioavailability *in vivo*. NF-kB involved in cell proliferation (cyclin D1), invasion (MMP-9), and angiogenesis (VEGF) was inhibited by the nanocurcumin. The study also demonstrated that the nanoformulation increased the half life of curcumin and made it more bioavailable *in vivo* [126]. The effectiveness of curcumin encapsulated in PLGA NPs (NanoCurc) against hepatocellular carcinoma (HCC) has been evaluated using a rat model with oral diethylnitrosamine (DEN)-induced HCC. The study reported that this nanocurcumin was more effective than free curcumin as a free-radical quencher in combating the oxidative damage of hepatic cells and prevented DEN-induced hyperplastic nodule formation. It also promoted apoptosis and thereby eliminated DEN-induced HCC in rats [127]. Yallapu et al. have also reported a curcumin-PLGA nanoformulation (nano-CUR6) with enhanced cellular uptake in cisplatin-resistant A2780CP ovarian and metastatic MDA-MB-231 breast cancer cells. The feasibility of conjugating anticancer antibody [transferrin or anti-TAG-72 (CC49)] to the nanoformulation was analyzed and CC49 antibody-coupled nanoparticles showed intense localization in TAG-72-positive (HPAFII) cells compared to TAG-72-negative (SKOV-3) cancer cells [128]. Curcumin and doxorubicin were co-encapsulated in PLGA NPs and treated multidrug-resistant chronic myeloid leukemia (CML) blast-like cancer cells (K-562 cells) more effectively. Initially, expression of multidrug resistance protein 1 (MDR1) and Bcl-2 were downregulated by curcumin and inhibited the nuclear efflux mechanism; the subsequent release of doxorubicin stimulated cancer cell death [129].

Researchers have also focused on chitosan-based derivatives for formulating nanoparticles for the delivery of curcumin. Curcumin-loaded dextran sulfate-chitosan NPs (200–220 nm) showed preferential killing of cancer cells, thereby sparing the normal cells. Figure 4 shows the cellular uptake of curcumin encapsulated in dextran sulfate-chitosan NPs using fluorescent microscopy in human breast adenocarcinoma (MCF-7), human prostate cancer (PC-3), and human osteosarcoma (MG-63) cell lines and also in normal cells (L929). The cell viability assay showed enhanced cancer cell death (MCF-7, 40.5%; MG 63, 34%; PC3, 34%; and L929, 18%) and the ability of the nanoformulation to induce apoptosis was confirmed in MCF-7 cells [130]. Anitha et al. have also loaded curcumin onto water-soluble chitosan derivatives *O*-carboxymethyl chitosan (*O*-CMC) and *N,O*-carboxymethyl chitosan (*N,O*-CMC). Curcumin-loaded *O*-CMC and *N,O*-CMC NPs (15–200 nm) were found to be biocompatible and induced apoptotic cancer cell death in MCF-7 and PC-3 in a very effective way, thereby retaining its safety for normal cells [131, 132].

Das et al. reported a nanoformulation with a tripolymeric composite carrier (alginate, chitosan and Pluronic) by ionotropic pre-gelation followed by polycationic crosslinking for the delivery of curcumin to cancer cells. The IC_{50} values for free curcumin and encapsulated curcumin on Hela cells were found to be 13.28 and 14.34 μM, respectively. The green fluorescence inside the Hela cells visualized using fluorescent microscopy confirmed the cellular internalization of curcumin-loaded composite NPs [133]. A simple nanoprecipitation method resulted

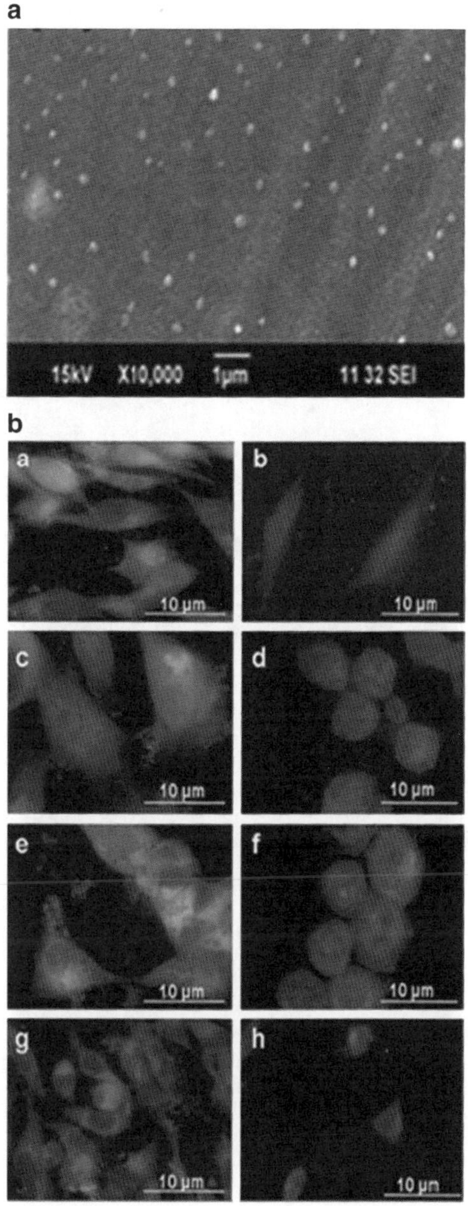

Fig. 4 (**a**) SEM image of curcumin-loaded dextran sulfate–chitosan NPs. (**b**) Fluorescent images of L929 (*a*, *b*), PC3 (*c*, *d*), MCF-7 (*e*, *f*) and MG 63 (*g*, *h*) taken after 1 and 24 h of incubation, respectively [130]

in cationic nanoparticles of curcumin with chitosan and poly(ε-caprolactone) (PCL), which were found to enhance the cellular uptake by Hela cells and human choroidal melanoma (OCM-1). The study suggested that the blending of PCL nanoparticles with the bio-adhesive polymer chitosan resulted in favorable drug loading and release profiles as well as good cellular interaction and *in vitro* anticancer efficacy, thereby retaining the activity of curcumin towards the cancer cells [134].

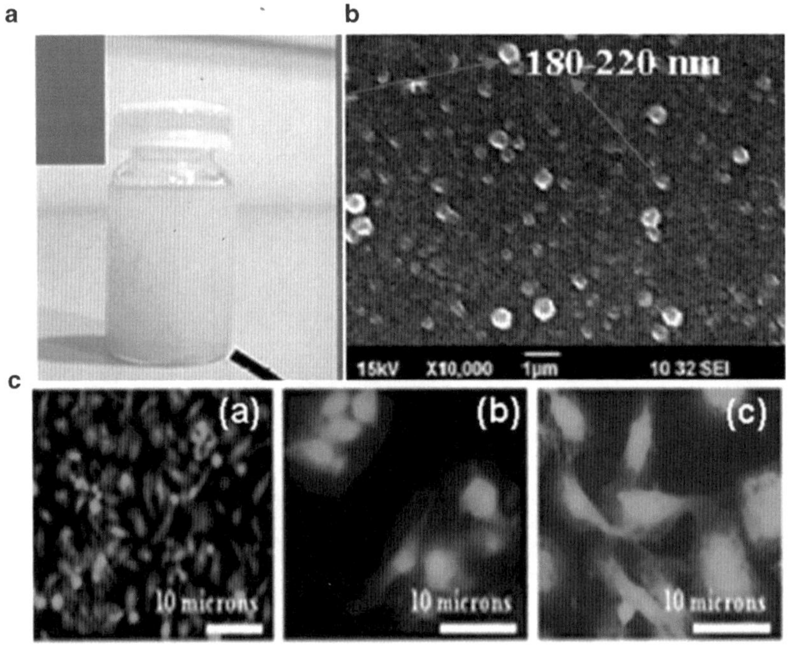

Fig. 5 (**a**) Stable curcumin-loaded TRC-NPs and (**b**) SEM image of curcumin-loaded TRC-NPs. (**c**) Fluorescent images showing cellular uptake of curcumin-loaded TRC-NPs: (*a*) displays the fluorescence of cells treated with bare curcumin; (*b*, *c*) curcumin-loaded TRC-NPs [138]

Chitosan-coated cationic poly(butyl)cyanoacrylate (PBCA) nanoparticles (200 nm) were developed for improving the solubility and bioavailability of curcumin. Nanocurcumin demonstrated comparable *in vitro* toxicity and proapoptotic effects to free curcumin against series of human hepatocellular cancer cell lines (HepG2, Bel7402, and Huh7 cells). HCC growth in murine xenograft models was suppressed by the curcumin nanoparticles and also inhibited tumor angiogenesis by suppressing COX-2 and VEGF expression [135]. Another study reported that the co-encapsulation of curcumin with doxorubicin in PBCA NPs (size 133 ± 5.3 nm) achieved a high reversal efficacy and downregulation of phosphorylated glycoprotein in an adriamycin-resistant breast cancer cell line (MCF-7) [136]. Curcumin loaded in crosslinked nanoparticles based on *N*-isopropylacrylamide, *N*-vinyl-2-pyrrolidinone, and PEG acrylate showed more cytotoxicity in pancreatic cancer cell lines (BxPC3, AsPC1, MiaPaca, XPA-1, XPA-2, PL-11, PL-12, PL-18, PK-9, and Panc 2.03) and enhanced the bioavailability (ninefold higher). The authors demonstrated that curcumin nanoparticles were more active than free curcumin in inhibiting some inflammatory mediators (i.e. TNF-α-induced NF-κB activation) and in suppressing NF-κB-regulated proteins involved in cell proliferation, invasion, and angiogenesis [137].

Sanoj et al. have investigated the *in vitro* anticancer potential of curcumin loaded onto biodegradable thermosensitive nanocarriers (Fig. 5) on oral cancer (KB), prostate cancer (PC3), and breast cancer (MCF-7) cell lines. They formulated curcumin onto chitosan-*g*-poly(*N*-isopropylacrylamide) and chitosan-*g*-poly (*N*-vinylcaprolactam) and showed the *in vitro* cellular uptake by the cancer cells by virtue of the intrinsic fluorescence property of curcumin. The cytotoxicity assay was carried out at the lower critical solution temperature (LCST) of 38°C and below LCST (35°C) and showed enhanced cancer cell death at LCST, indicating the temperature-sensitive release of the drug, inducing toxicity to breast and prostate cancer cells [138, 139].

Curcumin encapsulated in β-cyclodextrin (CD30) was evaluated for its intracellular uptake and anticancer activity towards prostate cancer cells (C4-2 and DU145). The IC_{50} of CD30 is 16.8 in C4-2 cancer cells and 17.6 μM in DU145 cancer cells, values that are comparable to that of free curcumin. Cells treated with 2–10 μM CD30 formed fewer colonies (as shown by colony formation assays) over a 10-day period. Cell proliferation and clonogenic assays demonstrated that β-cyclodextrin–curcumin self-assembly enhanced curcumin delivery and improved its therapeutic efficacy in prostate cancer cells compared to free curcumin [140]. Dhule et al. reported cyclodextrin-based liposomes for the delivery of curcumin against osteosarcoma after evaluation against cancer models of mesenchymal (OS) and epithelial origin (breast cancer). Curcumin encapsulated in cyclodextrins followed by a second encapsulation in liposomes (i.e. 2-hydroxypropyl-γ-cyclodextrin/curcumin–liposome complex) shows promising anticancer potential both *in vitro* and *in vivo* against a KHOS OS cell line and a MCF-7 breast cancer cell line. The curcumin solubilization and entrapment was enhanced by cyclodextrin complexation. Liposomal curcumin initiates the caspase cascade that leads to apoptotic cell death *in vitro*. The efficiency of the liposomal curcumin formulation was also confirmed *in vivo* using a xenograft OS model [141].

The free network structures throughout the crosslinked copolymer nanogels offer the opportunity to load various types of lipophilic drug molecules [142]. Bisht et al. [143] tested the effects of hydrogel nanocurcumin prepared by the redox free-radical polymerization of *N*-isopropylacrylamide and *N*-vinyl-2-pyrrolidone in the presence of PEG monoacrylate on pancreatic cancer cell lines. Nanocurcumin blocked the activation of NF-kB, downregulated steady-state transcripts of multiple pro-inflammatory cytokines and inhibited IL-6 synthesis. The tumor growth was significantly inhibited by parenteral administration of the hydrogel nanocurcumin formulation in both subcutaneous and orthotopic settings in xenograft models of human pancreatic cancer in athymic mice. Also, treatment with gemcitabine combined with hydrogel promoted an enhanced tumor growth inhibition capacity [143]. The hydrogel formulation was found to be effective in inhibiting the proliferation and clonogenicity of medulloblastoma and glioblastoma cell lines. The formulation has achieved a dose-dependent antiproliferation effect in embryonal tumor-derived lines DAOY and D283Med, and the glioblastoma neurosphere lines HSR-GBM1 and JHH-GBM14 [144]. A series of curcumin-loaded dextran-modified hydrogel NPs were studied by Goncalves et al. [145]. A water-dispersible hybrid nanogel for intracellular delivery of curcumin was developed by Wu et al. [147]. The core–shell structured hybrid nanogels were synthesized by coating the Ag/Au bimetallic NPs with a hydrophobic polystyrene gel layer as inner shell, and a subsequent thin hydrophilic nonlinear PEG-based gel layer as outer shell. The cell viability evaluation in mouse melanoma cells (B16F10) demonstrated that curcumin-loaded hybrid nanogels maintain the high anticancer potential of curcumin, resulting in 80–90% of cancer cell death. In the hybrid nanogel system, the Ag/Au core NPs emit strong fluorescence for imaging and also exhibit strong absorption in the near-infrared (NIR) region for photothermal conversion. The inner polystyrene gel layer is introduced to provide strong hydrophobic interactions with curcumin for high drug loading yields, whereas the external nontoxic and thermoresponsive PEG analog gel layer is designed to trigger release of the pre-loaded curcumin, either by variation of surrounding temperature or by exogenous irradiation with NIR light. The curcumin-loaded hybrid nanogels exhibit potent cytotoxicity against B16F10 cells by combined chemo- and photothermal treatment, thus providing higher therapeutic efficacy than chemo- and photothermal treatments alone or their additive efficacy [146]. Sabitha et al. developed biodegradable pH-sensitive chitin nanogels loaded with curcumin (CCNGs) and used them for the transdermal delivery of curcumin to melanoma. They reported that the CCNGs possess excellent skin penetration and retention properties. Due to their efficiency of skin penetration, these nanogels would be a novel approach for trafficking of curcumin through the different layers of skin (Fig. 6); they also showed selective toxicity towards melanoma cell lines A375 [147].

Table 1 shows different polymers used as nanocarriers for curcumin and the cancer cell lines that are affected by the formulations. The nanoencapsulation of curcumin within polymeric nanoparticles improves its therapeutic effect by overcoming the problems of low aqueous solubility and poor bioavailability. Due to its better therapeutic potential, nanocurcumin could be effectively used as a modality for cancer treatment.

4.2 Epigallocatechin-3-gallate

The potential health benefits of tea consumption are attributed to the sensitive water-soluble nutraceutical compound EGCG. But, EGCG faces the issue of instability in neutral and alkaline pH and fast degradation with oxygen, temperature, and pH changes [148]. This necessitates a suitable vehicle for the controlled release of EGCG, thereby enabling it to carry out its therapeutic activities.

The possibility of using thermally modified β-lactoglobulin to form co-assembled nanovehicles (50 nm) for the protected delivery of EGCG against oxidation and degradation has been studied [149]. The study suggested that the approach could be used for encapsulating important polyphenolic nutraceutical compounds and could be used as a preventive medicine. Shao et al. [150] reported electrospinning of the biodegradable polymer PCL with multi-walled carbon nanotubes (MWCNTs) and green tea polyphenol (GTP), as a model drug resulting in the formation of nanofiber meshes. This process imparted controlled release

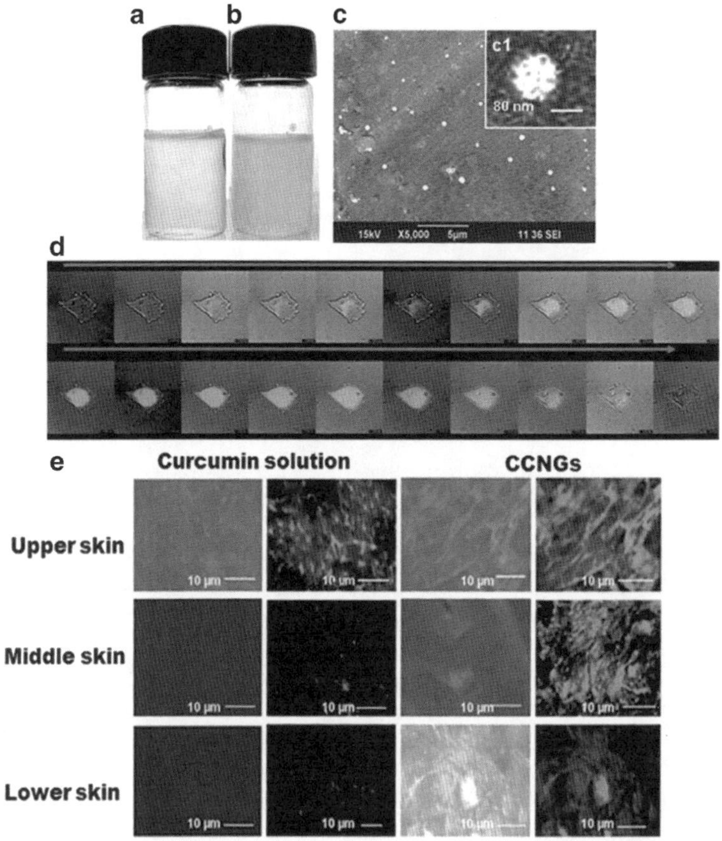

Fig. 6 (**a**) Stable control chitin nanogel and (**b**) curcumin-loaded chitin nanogel (CCNG) at pH 5.0. (**c**) Size distribution of CCNGs by SEM. (**d**) Cellular localization of CCNGs on A375 after 6 h of exposure. (**e**) Skin localization of curcumin by fluorescent microscopy for control curcumin and CCNGs after 6 h of exposure [147]

function and a longer stability of the chemical structure of GTP. GTP-loaded PCL/MWCNT composite nanofibers inhibited cancer cell growth of human hepatoma cells (Hep G2) and human non-small-cell lung carcinoma cells (A549) but had no obvious effect on osteoblasts. The A549 cells have an elliptical structure and conglutinated on nanofiber meshes. Figure 7 displays the fluorescence images and typical SEM photos of cells cultured for 3 days on GTP-loaded NCF-5 and CF-5 composite nanofiber meshes. In accordance with the toxicity results, they also observed that the cancer cells were subjected to apoptosis, which could be due to the released GTP from the nanofibers [150]. A combination of EGCG and caffeic acid was encapsulated within electrospun PCL nanofibers and its *in vitro* anticancer potential evaluated in cultured gastric cancer cell lines (MKN-28). MKN-28 cells were subjected to apoptosis via activated caspase-3 due to the H_2O_2 generated by

Table 1 Polymeric nanoparticles used for encapsulating curcumin and their anticancer properties towards different cancer cells *in vitro* and *in vivo*

Nanocarrier	Size (nm)	Type of cancer	Reference
Nanodisk: apolipoprotein scaffold	<50	HepG2 and Jeko cells	[115, 116]
Nanospheres: ethyl cellulose and methyl cellulose	117.5 ± 92.5	MCF-7 and HepG2	[117]
Micelles: methoxyPEG–palmitate amphiphilic conjugate	41.23	Hela	[118]
Human serum albumin NPs	130–150	HCT-116 tumor xenograft	[121]
Fibrinogen	150–200	PC3 and MCF-7	[122]
Nanosphere: PLGA	45	LNCaP, PC3, and DU145	[123]
PLGA NPs surface functionalized with bis (sulfosuccinimidyl) suberate (BS3)	184.1 ± 17.9 to 193.8 ± 11.34	MDA-MB-231	[125]
PLGA–PEG-5000	80.9	KBM-5, Jurkat, DU145 MDA-MB-231, HCT116, and SEG-1	[126]
PLGA NPs (oral route)	14	*N*-Nitrosodiethylamine-induced hepatocellular carcinoma in rats	[127]
PVA-stabilized PLGA NPs	76.2	Cisplatin-resistant A2780CP and MDA-MB 231	[128]
(Curcumin + doxorubicin)-PLGA NPs	248 ± 1.6	K-562	[129]
Chitosan-dextran sulfate NPs	200–220	MCF-7, PC-3, and MG 63	[130]
O-Carboxymethyl chitosan (*O*-CMC) and *N*,*O*-carboxymethyl chitosan (*N*,*O*-CMC)	150–200	MCF-7	[131, 132]
Tricomposite: alginate, chitosan and pluronic (PF127)	100 ± 20	Hela	[133]
Chitosan/PCL	220–360	Hela, OCM-1	[134]
Chitosan-coated PBCA NPs	200	HepG2, Bel7402, and Huh7 *in vitro* and *in vivo* murine xenograft models	[135]
(Curcumin + doxorubicin)-PBCA NPs	133 ± 5.3	MCF-7	[136]
Micelles: *N*-isopropylacrylamide, *N*-vinyl-2-pyrrolidinone, and PEG acrylate	50	BxPC3, AsPC1, MiaPaca, XPA-1, XPA-2, PL-11, PL-12, PL-18, PK-9, and Panc 2.03	[137]
Chitosan-*g*-poly (*N*-isopropylacrylamide) and chitosan-*g*-poly (*N*-vinylcaprolactam)	150–230	KB, PC3, and MCF-7	[138, 139]
β-Cyclodextrin	100–500	C4-2 and DU145	[140]

(continued)

Table 1 (continued)

Nanocarrier	Size (nm)	Type of cancer	Reference
Liposome: 2-hydroxypropyl-γ-cyclodextrin	98.2 ± 30.1	KHOS and MCF-7	[141]
Hybrid nanogels: polystyrene inner shell and PEG outer shell with bimetallic Ag/Au NP core	50	B16F10	[146]
Nanogels: chitin	70–80	A375	[147]

PBCA poly(butyl cyanoacrylate), *PCL* poly(ε-caprolactone), *PEG* poly(ethylene glycol), *PLGA* poly(lactic-*co*-glycolic) acid, *PVA* polyvinyl alcohol

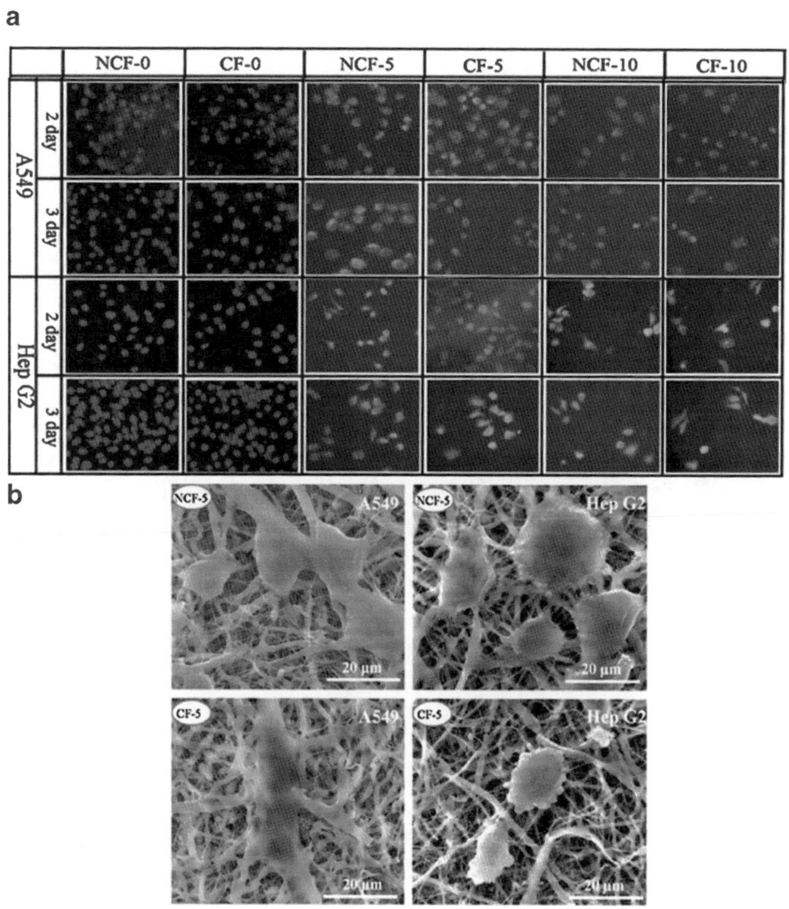

Fig. 7 (**a**) Fluorescence microscope images of A549 and Hep G2 cells cultured on NCF-0, NCF-5, NCF-10, CF-0, CF-5, and CF-10 PCL/MWCNT composite nanofiber meshes for 2 and 3 days (200×). (**b**) Typical SEM images of A549 and Hep G2 cells cultured on NCF-5 and CF-5 nanofiber meshes for 3 days [150]. (NCF-0, NCF-5 and NCF-10 corresponded to pure PCL nanofiber meshes with GTP content of 0%, 5% and 10% and CF-0, CF-5 and CF-10 for PCL/MWCNTs composite nanofiber meshes with GTP content of 0%, 5% and 10%, respectively.)

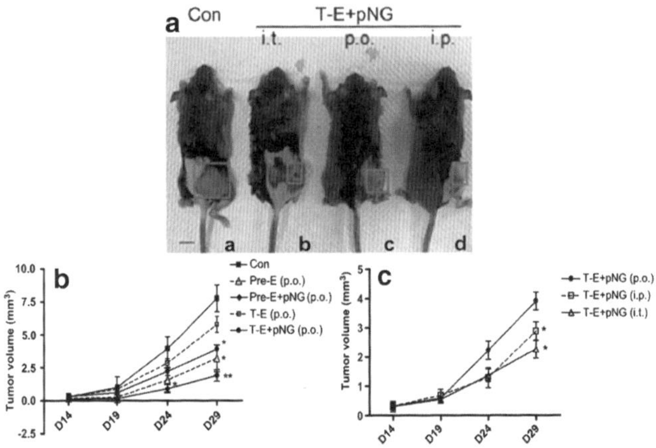

Fig. 8 a) Tumor growth after treatment with EGCG alone or combined with Png *in vivo*. Mice were given oral (p.o.), intraperitoneal (i.p.), or intratumor (i.t.) injection of water (Con), EGCG alone (T-E) or EGCG combined with pNG (T-E þ pNG) every other day on day 14 after tumor implantation. b) The tumor volume in mice orally treated with EGCG and/or pNG. c) The tumor volume in mice from different drug (EGCG combined with pNG, EGCG+NG) delivery strategies

the plant polyphenols released from the nanofibers. The study concluded that nanofibers loaded with these potent polyphenols could offer long-term cancer therapy and could also prevent a recurrence of cancer after surgical operations [151].

The use of polysaccharide nanoparticles for delivering natural antioxidants capable of inhibiting tumorigenesis was highlighted by studies on EGCG encapsulated in carbohydrate particles to preserve its antioxidant properties and improve its bioavailability. Polysaccharide-based NPs (maltodextrin-gum arabic matrix) were fabricated and the presence of EGCG within the carbohydrate matrix demonstrated by the intermolecular dipole–dipole interactions, maintaining its integrity after the spray-drying process. The study also proved the ability of NPs to retain the inhibitory effect of EGCG on the proliferative activity of prostate cancer cells (Du145). The nanoencapsulated EGCG enhanced its inhibitory effect on cell proliferation (10–20%) at lower concentrations (1–2 μM) compared with free EGCG [152, 153].

Hsieh et al. [154] investigated the anticancer activity of EGCG physically adsorbed on gold nanoparticles (pNGs) in C_3H/HeN mice subcutaneously implanted with MBT-2 murine bladder tumor cells. A synergistic growth inhibitory effect was observed *in vitro* on MBT-2 cells and *in vivo* by the treatment of EGCG-loaded pNGs. The study concluded that pNG altered the Bax:Bcl-2 ratio and mitochondrial membrane integrity, thereby activating caspase signaling. Inhibition of tumor cells by the EGCG–pNG (Fig. 8) complex is achieved via the mechanism of cell apoptosis. The EGCG–pNG complex also lowered the level of VEGF and exerted a synergistic effect to outperform the individual components in anti-tumorigenesis [154].

Nanocomplexes assembled from bioactive peptides, caseinophosphopeptides (CPPs), and chitosan were used for the encapsulation of EGCG. EGCG-loaded chitosan-CPP nanoparticles (150 nm) showed a dose-dependent cellular internalization inside human colon carcinoma (Caco-2) cell lines. Furthermore, the intestinal permeability of EGCG using a Caco-2 monolayer was enhanced significantly as delivered by nanoparticles, which indicated the promising elevation of EGCG bioavailability [155]. To address the issue of poor absorption followed by the oral administration of EGCG, EGCG has been encapsulated in chitosan-tripolyphosphate nanoparticles and the oral absorption of EGCG evaluated in Swiss Outbred mice. The study observed an enhanced exposure of EGCG to the jejunum, resulting in increased plasma concentrations of EGC and suggesting that chitosan NPs could be useful for enhancing oral delivery, and therapeutic application, of EGCG in a number of disease conditions. [156].

EGCG encapsulated by poly(lactic acid)–poly(ethylene glycol) (PLA–PEG) nanoparticles was evaluated against prostate cancer cells. They observed that 72% of apoptosis was induced in prostate cancer cell lines (PC-3) when treated with 2.74 μmol/L nano-EGCG but only with 40 μmol/L of non-encapsulated EGCG, indicating the ability of nano-EGCG to induce enhanced cancer cell death with lower doses. Also, fibroblast growth factor (FGF)-induced angiogenesis was more effectively inhibited with nano-EGCG (3 μg EGCG) than with free EGCG. Thus, thc study proved that PLA–PEG nanoparticles retained the biological effectiveness of EGCG with over tenfold dose advantage for exerting its proapoptotic and angiogenesis inhibitory effects [157, 158]. Researchers tried to explore the concept of nanochemoprevention using EGCG NPs functionalized with the small targeting molecule prostate-specific membrane antigen (PSMA) that exhibits a selective *in vitro* efficacy against PSMA-expressing prostate cancer cells. PLGA–PEG–COOH was conjugated with a glutamate-containing urea-based inhibitor (DCL, which contains a primary amine that allows amide bond formation with the carboxyl terminal groups of PEG monomer and has a high affinity for PSMA) [159]. Singh et al. investigated the effect of EGCG and theaflavin loaded within PLGA nanoparticles alone and in combination with the anticancer drug cisplatin in several human cancer lines: A549 (lung carcinoma), HeLa (cervical carcinoma) and THP-1 (acute monocytic leukemia). They showed that the nanoformulation in combination with cisplatin exerted an enhanced anticancer potential by inhibiting cell proliferation, metastasis, angiogenesis, and apoptosis biomarkers, resulting in a 20-fold dose advantage over the free drug [160]. Gelatin can act as a good carrier for the natural polyphenols. EGCG was encased using a layer-by-layer (LbL) technique into 200-nm sized gelatin nanoparticles consisting of a soft gel-like interior with or without a surrounding LbL shell of polyelectrolytes (polystyrene sulfonate/polyallylamine hydrochloride), polyglutamic acid/poly-L-lysine, dextran sulfate/protamine sulfate, or carboxymethyl cellulose/gelatin. Nanoencapsulated EGCG retained its biological activity and blocked hepatocyte growth factor

Table 2 Polymeric nanoparticles used for encapsulating the nutraceutical EGCG and their anticancer properties towards different cancer cells *in vitro* and *in vivo*

Nanocarrier	Size (nm)	Type of cancer	Reference
Nanofibers: PCL/MWCNTs	200–600	Hep G2 and A549	[150]
(EGCG + caffeic acid)-PCL nanofibers	–	MKN-28	[151]
Polysaccharide NPs: (maltodextrin-gum arabic matrix)	400	Du-145	[152, 153]
Gold NPs	20–1200	MBT-2 murine bladder tumor cells *in vivo*	[154]
Caseinophosphopeptides and chitosan	150 ± 4.3	Caco-2	[155]
PLA–PEG	285	PC-3	[157, 158]
DCL–PLGA–PEG	80.5 ± 15	PSMA-expressing LNCaP	[159]
Gelatin NPs	200	MBA-MD-231	[161]
Folate-targeted BSA NPs	200	PC-3	[162]

EGCG epigallocatechingallate, *PLA* poly(lactic acid), *PEG* poly(ethylene glycol), *PLGA* poly (lactic-*co*-glycolic) acid, *PCL* poly(ε-caprolactone), *MWCNT* multiwalled carbon nanotube, *PSMA* prostrate-specific membrane antigen

(HGF)-induced intracellular signaling in the breast cancer cell line MBA-MD-231 as potently as free EGCG [161]. EGCG-loaded bovine serum albumin (BSA) NPs were targeted to prostate cancer cells (PC-3) via folate-mediated targeting: 18.363 μg of folate was conjugated per mg BSA and the uptake of folate-EGCG-BSA NP by cultured PC-3 cells was 23.65 times the uptake of EGCG-BSA NPs in a concentration-dependent manner. The cytotoxicity induced in PC-3 cells by the targeted folate-EGCG-BSA NPs was higher (82.8%) than that induced by free EGCG (58.6%) and non-targeted EGCG-BSA NPs (55.1%) [162].

All the above-mentioned studies demonstrate that nanoencapsulation into polymeric carriers could enhance the therapeutic effectiveness of natural agents such as EGCG in the prevention and treatment of complex diseases like cancer. Table 2 summarizes the cancers affected by nanoencapsulated EGCG and the polymeric carriers used for the encapsulation.

4.3 Resveratrol

With the aim of stabilizing and protecting resveretrol, Nam et al. [163] applied monodisperse functionalized porous polymeric microspheres as support material and Shi et al. [164] demonstrated the successful preparation of yeast-encapsulated resveratrol using *Saccharomyces cerevisiae* as encapsulating wall material. For improving aqueous solubility, researchers have formulated resveratrol complexed with derivatives of β-cyclodextrin (β-CD). A resveratrol nanoemulsion was formulated using poly(oxyethylene)-hydrogenated castor oil as surfactant, ethanol as co-surfactant, and isopropyl myristate as the oil phase. β-CD nanosponges were prepared by crosslinking with carbonyldiimidazole and they were reported to be effective for oral and topical delivery of resveratrol. Polymeric lipid-core nanocapsules have been loaded with *trans*-resveratrol to improve its biodistribution and decrease its very fast metabolism. Nanocapsules (size, 240 nm) of *trans*-resveratrol were prepared and their *in vivo* biodistribution in rats evaluated, showing improved gastro-intestinal safety [165]. In order to enhance the oral bioavailability, *trans*-resveratrol NPs were prepared by temperature-controlled antisolvent precipitation with hydroxypropyl methyl cellulose as the stabilizer. The mean particle size was well controlled, ranging between 232 and 560 nm by adjusting the precipitation temperature and was decreased by lowing the precipitation temperature during the antisolvent process. The study concluded that that the temperature-controlled antisolvent precipitation technique can be considered promising for obtaining nanosized particles to enhance the saturation solubility and dissolution rate, and could be an effective method for achieving therapeutic effects for poorly water-soluble food ingredients and nutraceuticals [166].

The capability of resveratrol to elicit many cellular responses, including cell cycle arrest, differentiation, and apoptosis makes it a potential anticancer agent. Resveratrol is incorporated into methoxyPEG-PCL-based nanoparticles with high encapsulation efficiency and its *in vitro* cytotoxic effect was analyzed in glioma cells (C6). Intracellular ROS levels were shown to exploit the possible antiglioma mechanisms of resveratrol-loaded NPs. Resveratrol NPs demonstrated better efficacy against glioma cells due to the superior cell membrane penetrating ability of the NPs and the ROS scavenging potential of released resveratrol [167]. Wang et al. proposed a strategy of targeting cancer cell mitochondria by resveratrol liposomes [169]. They synthesized a targeting material, dequlinium–poly(ethylene glycol)–distearoylphosphatidylethanolamine (DQA–PEG2000–DSPE), which was used as mitochondriotropic molecule for modifying the surface of liposomes, and evaluated its apoptotic potential in resistant lung cancer cells. Mitochondrial-targeting resveratrol liposomes (size 70–80 nm) exhibited a strong inhibitory effect on the proliferation of A549 cells. Subcellular localization of the drug in A549 and A59/cDDP cells using confocal laser microscopy clearly showed the mitochondrial targeting of the NPs. The inhibitory effect of mitochondrial-targeting resveratrol liposomes on resistant lung cancer xenografts was studied and superior antitumor effects by inducing apoptosis via the mitochondria signaling pathway were achieved with the NPs compared with free resveratrol. The study proved that mitochondrial-targeting resveratrol liposomes modified with the targeting DQA–PEG2000–DSPE conjugate could provide a potential strategy for treating the intrinsic resistant lung cancers [168].

The potential of transferosomes and ethanol-containing vesicles were evaluated for their delivery of resveratrol through skin. Transferosomes (size 83 nm) encapsulating *trans*-resveratrol were prepared with surfactant polysorbate 80 (Tw80), sodium cholate (SC), and sodium deossicholate (SDC) and 116-nm ethanol-containing vesicles of different lipid composition, namely soy phosphatidylcholine (SPC) and cholesterol. Cytotoxicity and the inhibition of ROS production and lipid peroxidation were evaluated on H_2O_2-stimulated human keratinocytes (HaCaT). The presence of *trans*-resveratrol can have a protective effect, reducing

Table 3 Polymeric nanoparticles used for encapsulating the nutraceutical resveratrol and their anticancer properties towards different cancer cells *in vitro*

Nanocarrier	Size (nm)	Type of cancer	Reference
Resveratrol-methoxyPEG-PCL NPs	87.5 ± 9.5	C6	[167]
DQA–PEG2000–DSPE–liposomes	70–80	A549	[168]
Tranferosomes: surfactant polysorbate 80 (Tw80), sodium cholate, and sodium deossicholate	83	Skin cancer	[169]
Resveratrol-BSA NPs	400–500	Ovarian cancer cells (SKOV3) implanted in nude mice	[170]

PEG polyethylene glycol, *PCL* poly(ε-caprolactone), *DQA* dequlinium, *DSPE* distearoylphosphatidylethanolamine, *BSA* bovine serum albumin

ROS formation as well as formation of lipid peroxides. Ethanol-containing vesicles based on SPC were able to promote *trans*-resveratrol permeation through the porcine skin. The nanosystem could be potentially exploited for the transdermal delivery of nutraceuticals with anticancer potential specifically for skin cancer [169]. Guo et al. investigated the antitumor effects and functional mechanism of resveratrol-BSA nanoparticles on human primary ovarian carcinoma cells (SKOV) in nude mice. The administration of resveratrol-BSA NPs significantly retarded the growth of carcinomas in nude mice from the third week onwards, and the inhibition rate was markedly higher than in mice treated with free resveratrol (52.43% vs. 46.34%, $P < 0.05$), without causing weight loss ($P > 0.05$). Apoptotic and necrotic morphological characteristics were observed with electron microscopy in the tumor tissues of treated mice and also revealed that part of the mechanism might be mediated by triggering the release of cytochrome c from the intermembrane space and upregulating the expression of caspase-9 and caspase-3, suggesting that the mitochondrial apoptotic pathway was being activated [170]. Table 3 summarizes the nanoencapsulated resveratrol and its *in vitro* anticancer effects.

4.4 Silymarin

Silymarin encapsulated in PLGA NPs (Fig. 9) were developed by Snima et al. and their potential evaluated for anticancer application *in vitro*. About 60% of silymarin was encapsulated into PLGA nanocarrier via a single-step emulsion technique. The study proved time- and dose-dependent cytotoxic effects towards prostate cancer cells (PC-3). They also suggested that the ability of silymarin-PLGA NPs to decrease the migration of PC-3 could be due to the downregulation of transcription factors such as Snail-1 and SLUG, resulting in upregulation of E-cadherin (a cell adhesion protein) and simultaneous downregulation of vimentin (a cytoskeletal protein) [171].

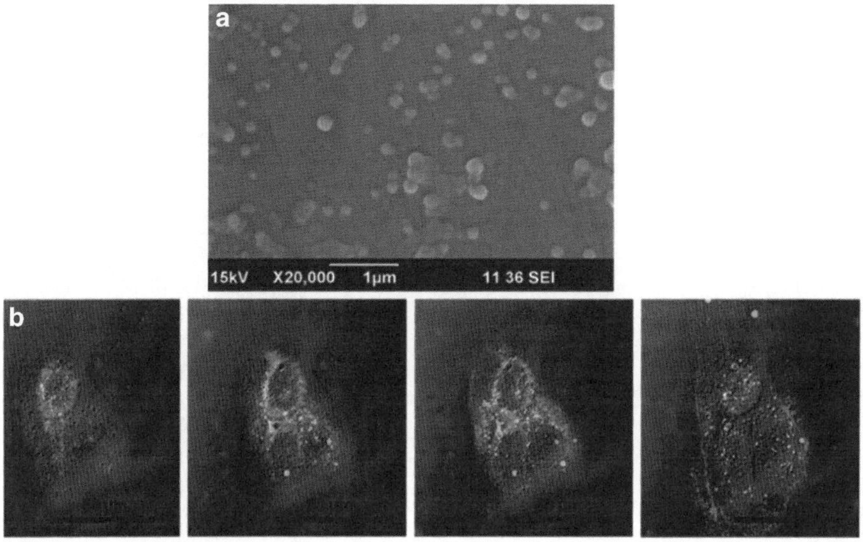

Fig. 9 (**a**) SEM image of silymarin-PLGA NPs. (**b**) Confocal microscopic images showing cellular uptake of rhodamine-123-tagged silymarin-PLGA NPs (0.625 mg/mL) by PC-3 cells. (Z-stacked merged images of PC-3 showing F-actin and rhodamine-123-tagged NPs) [171]

4.5 Saponin

Sanoj et al. formulated NPs of saponin with biocompatible and biodegradable chitosan and reported this nanoformulation to be a better therapeutic agent. The saponin-loaded chitosan nanoparticles (Fig. 10), "nanosaponin", were prepared via a simple crosslinking reaction with sodium tripolyphosphate (TPP) for enhanced, controlled and sustained delivery to prostate cancer cell lines. The cellular uptake studies of nanosaponin using L929 and PC3 cells demonstrated significant internalization and retention of the NPs inside the cells, suggesting that these nanoparticle systems can be used for delivering saponin directly into the cells. The nanosaponin has showed specific toxicity to cancer cells and are nontoxic to normal cells [172].

4.6 Grape Seed Extract

Narayanan et al. reported targeted NP-mediated delivery to enhance the efficacy of the well-known nutraceutical GSE. NanoGSE was prepared by a nanoprecipitation technique whereby GSE was encapsulated in PLGA nanoparticles. The potential cancer-targeting ligand folic acid was conjugated to GSE-PLGA NPs and tested

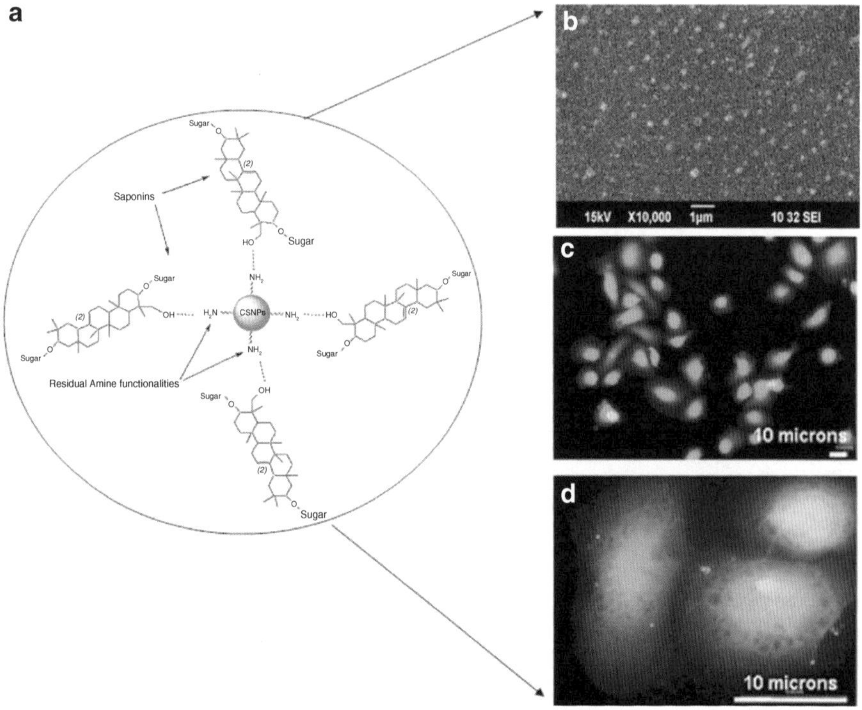

Fig. 10 (**a**) Hypothesized interactions of saponin within chitosan nanoparticles. (**b**) SEM image of nanosaponin. (**c**, **d**) Fluorescent microscopy images of PC-3 cells treated with rhodamine-123-labeled saponin nanoparticles [172]

against folate-receptor-overexpressing cancers (MCF-7). The IC_{50} values were lowered by a factor of about three for folic acid-loaded NanoGSE compared to the free drug, indicating substantially enhanced bioavailability to the tumor cells. Receptor targeting of folic acid-loaded NanoGSE also resulted in a significant increase in apoptotic index [173] (Fig. 11).

5 Conclusions

Nanoformulated nutraceuticals such as polyphenols and flavanoids, which are derived from medicinal plants, provide a new insight for cancer treatment. The mechanism of action of these dietary agents in chemoprevention has gained considerable interest in the area of cancer research. This review has presented several mechanisms explaining the chemopreventive potentials of curcumin, EGCG, resveratrol, saponin, silymarin, and GSE. Their ability to target specific

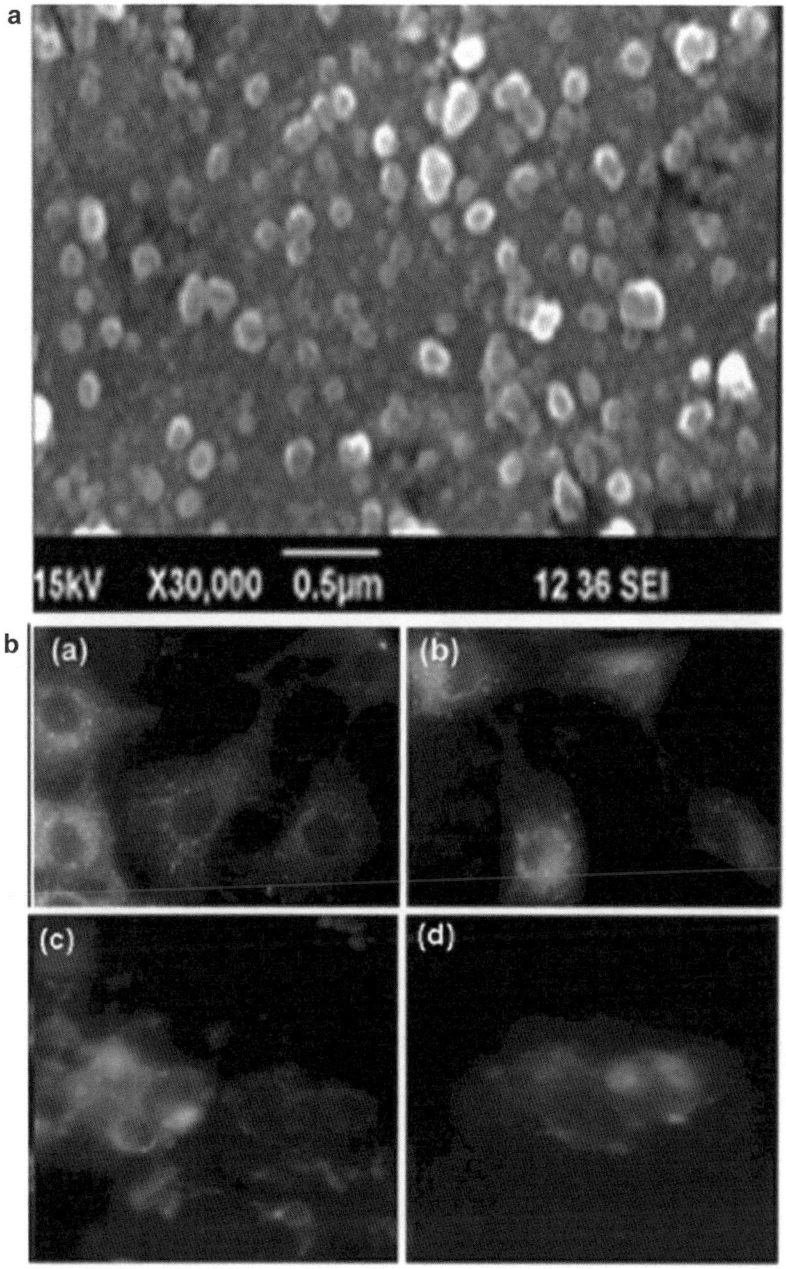

Fig. 11 (**a**) SEM image of NanoGSE. (**b**) Cellular uptake of folic-acid-modified NanoGSE, as shown by fluorescence microscopy after incubation for 2 h: (*a*) KB, (*b*) A549 and (*c*, *d*) KB cells with apoptotic features dually stained using Annexin V and propidium iodide after 24 h of incubation [173]

cell signaling pathways has received considerable attention for regulating cellular proliferation and apoptosis, as has their broad pharmacologic activities.

Several types of polymeric nanocarriers have been found to be suitable for the encapsulation or loading of these phytochemicals to improve their bioavailability and potential as cancer therapeutics. The characteristics of these polymeric phytonanoformulations can be tailored by various mechanisms according to the specific requirement for inducing cellular death. It can be concluded that utilization of encapsulated polyphenols can enhance the stability and bioavailability of the phytodrugs both *in vitro* and *in vivo* and optimize routes for their administration. Overall, the use of phytochemical-loaded polymeric nanoparticles in chemotherapy for cancer treatment improves existing therapies by targeting tumors and by reducing the dose required. Safe toxicological profiles of the biocompatible and biodegradable polymeric carriers with the loaded phytodrugs and their efficacy in the cell-line models highlight their potential for evaluation in *in vivo* models. Future studies have to be conducted clinically after the *in vitro* and *in vivo* evaluations for the use of these phytodrug nanoformulations in effective cancer treatment.

Acknowledgements The authors are thankful to the Department of Biotechnology (DBT), Government of India, for their financial support for this work under the Nanoscience and Nanotechnology Initiative program (Ref. No. BT/PR10882/NNT/28/142/2008). This work was also partially supported by Nanomission, Department of Science and Technology, India under the Theragnostic grant. The author S. Maya also acknowledges CSIR for providing a Senior Research Fellowship (SRF Award No. 9/963(00172)2K11-EMR-I).

References

1. Hostanska K, Jürgenliemk G, Abel G, Nahrstedt A, Saller R (2007) Cancer Detect Prev 31:129
2. Espín JC, García-Conesa MT, Tomás-Barberán FA (2007) Phytochemistry 68:2986
3. Larrosa M, Tomas-Barberan FA, Espin JC (2006) J Nutr Biochem 17:611
4. Faller ALK, Fialho E (2010) J Food Compost Anal 23:561
5. Hervert-Hernández D, García OP, Rosado JL, Goñi I (2011) Food Res Int 44:1182
6. Landete JM (2011) Food Res Int 44:1150
7. Siddiqui IA, Mukhtar H (2010) Pharm Res 27:1054
8. Santos IS, Ponte BM, Boonme P, Silva AM, Souto EB (2012) Biotechnol Adv (in press). doi:10.1016/j.biotechadv.2012.08.005
9. Araujo JR, Goncalves P, Martel F (2011) Nutr Res 31:77
10. Wojdyło A, Oszmiański J, Czemerys R (2007) Food Chem 105:940
11. Manach C, Scalbert A, Morand C (2004) Am J Clin Nutr 79:727
12. Stagos D, Amoutzias GD, Antonios M, Argyris S, Arstides MT, Dimitrios K (2012) Food Chem Toxicol 50:2155
13. Reason W, Veena MS, Marilene BW, Eri SS (2011) Mol Cancer 10:12
14. Murali MY, Meena J, Subhash CC (2012) Drug Discov Today 17:71
15. Purusotam B, Natasa S (2011) Molecules 16:4567
16. Radha KM, Anoop KS, Jaya G, Rikhab CS (2006) Life Sci 78:2081

17. Anand P, Sundaram C, Jhurani S, Kunnumakkara AB, Aggarwal BB (2008) Cancer Lett 267:133
18. Siddiqui IA, Asim M, Hafeez BB, Adhami VM, Tarapore RS, Mukhtar H (2010) FASEB J 25:1198
19. Clement Y (2009) Prev Med 49:83
20. Johnson JJ, Bailey HH, Mukhtar H (2010) Phytomedicine 17:3
21. Katiyar S, Elmets CA, Katiyar SK (2007) J Nutr Biochem 18:287
22. Melgarejo E, Urdiales JL, Sanchez-Jimenez F, Medina MA (2010) Amino Acids 38:519
23. Moore RJ, Jackson KG, Minihane AM (2009) Br J Nutr 102:1790
24. Khan N, Afaq F, Saleem M, Ahmad N, Mukhtar H (2006) Cancer Res 66:2500
25. Khan N, Adhami VM, Mukhtar H (2009) Nutr Cancer 61:836
26. Myung SK, Bae WK, Oh SM, Kim Y, Ju W, Sung J, Lee YJ, Ko JA, Choi HJ (2009) Int J Cancer 124:1496
27. Yang CS, Lambert JD, Ju J, Lu G, Sang S (2007) Toxicol Appl Pharmacol 224:265
28. Ogunleye AA, Xue F, Michels KB (2010) Breast Cancer Res Treat 119:447
29. Amri A, Chaumeil JC, Sfar S, Charrueau C (2012) J Control Release 158:182
30. Vastano BC, Chen Y, Zhu N, Ho CT, Zhou Z, Rosen RT (2000) J Agric Food Chem 48:253
31. Fulda S (2010) Drug Discov Today 15:757
32. Rius C, Abu-Taha M, Hermenegildo C, Piqueras L, Cerda-Nicolas JM, Issekutz AC, Estañ L, Cortijo J, Morcillo EJ, Orallo F, Sanz MJ (2010) J Immunol 185:3718
33. Amidon GL, Lennernäs H, Shah VP, Crison JR (1995) Pharm Res 12:413
34. Das S, Das DK (2007) Cardiovasc Drug Discov 2:133
35. Hung LM, Su MJ, Chen JK (2004) Free Radic Biol Med 36:774
36. Jang M, Cai L, Udeani GO, Slowing KV, Thomas CF, Beecher CW, Fong HH, Farnsworth NR, Kinghorn AD, Mehta RG, Moon RC, Pezzuto JM (1997) Science 275:218
37. Howitz KT, Bitterman KJ, Cohen HY, Lamming DW, Lavu S, Wood JG, Zipkin RE, Chung P, Kisielewski A, Zhang LL, Scherer B, Sinclair DA (2003) Nature 425:191
38. Post-White J, Ladas EJ, Kelly KM (2007) Integr Cancer Ther 6:104
39. Kroll DJ, Shaw HS, Oberlies NS (2007) Integr Cancer Ther 6:110
40. Ramasamy K, Rajesh A (2008) Cancer Lett 269:352
41. Liu J, Henkel T (2002) Curr Med Chem 9:1483
42. Estrada A, Katselis GS, Laarveld B, Barl B (2000) Comp Immunol Microbiol Infect Dis 23:27
43. Sparg SG, Light ME, van Staden J (2004) J Ethnopharmacol 94:219
44. Shuli M, Wenyuan G, Yanjun Z, Luqi H, Changxiao L (2010) Fitoterapia 81:703
45. Perumalla AVS, Hettiarachchy NS (2011) Food Res Int 44:827
46. Sporn MB, Dunlop NM, Newton DL, Smith JM (1976) Fed Proc 35:1332
47. Rao S (1994) J Pharm Pharmacol 46:1013
48. Masuda T, Maekawa T, Hidaka K, Bando H, Takeda Y, Yamaguchi H (2001) J Agric Food Chem 49:2539
49. Joe B, Vijaykumar M, Lokesh BR (2004) Crit Rev Food Sci Nutr 44:97
50. Joe B, Lokesh BR (1994) Biochim Biophys Acta 1224:255
51. Brouet I, Ohshima H (1995) Biochem Biophys Res Commun 206:533–540
52. Chan MM, Huang HI, Fenton MR, Fong D (1998) Biochem Pharmacol 55:1955
53. Mukhtar H, Ahmad N (2000) Am J Clin Nutr 71:698
54. Lambert JD, Elias RJ (2010) Arch Biochem Biophys 501:65
55. Sonee M, Sum T, Wang C, Mukherjee SK (2004) Neurotoxicology 25:885
56. Aggarwal BB, Shishodia S (2006) Biochem Pharmacol 71:1397
57. Fang MZ, Wang Y, Ai N, Hou Z, Sun Y, Lu H (2003) Cancer Res 63:7563
58. Yang TT, Koo MW (2000) Atherosclerosis 148:67
59. Muzolf-Panek M, Gliszczynska-Swiglo A, de Haan L, Aarts JM, Szymusiak H, Vervoort JM (2008) Chem Res Toxicol 21:2352

60. Svobodova A, Zdarilova A, Maliskova J, Mikulkova H, Walterova D, Vostalova J (2007) J Dermatol Sci 46:21
61. Agarwal R, Agarwal C, Ichikawa H, Singh RP, Aggarwal BB (2006) Anticancer Res 26:4457
62. Zhou K, Raffoul JJ (2012) J Oncol 803294. doi:10.1155/2012/803294
63. Chan MM (1995) Biochem Pharmacol 49:1551
64. Singh S, Aggarwal BB (1995) J Biol Chem 270:24995
65. Brennan P, O'Neill LA (1998) Biochem Pharmacol 55:965
66. Aggarwal BB, Shishodia S, Takada Y, Banerjee S, Newman RA, BuesoRamos CE, Price JE (2005) Clin Cancer Res 11:7490
67. Schulze-Tanzil G, Mobasheri A, Sendzik J, John T, Shakibaei M (2004) Ann N Y Acad Sci 1030:578
68. LoTempio MM, Veena MS, Steele HL, Ramamurthy B, Ramalingam TS, Cohen AN, Chakrabarti R, Srivatsan ES, Wang MB (2005) Clin Cancer Res 11:6994
69. Jobin C, Bradham CA, Russo MP, Juma B, Narula AS, Brenner DA, Sartor RB (1999) J Immunol 163:3474
70. Wang D, Veena MS, Stevenson K, Tang C, Ho B, Suh JD, Duarte VM, Faull KF, Mehta K, Srivatsan ES, Wang MB (2008) Clin Cancer Res 14:6228
71. Plummer SM, Holloway KA, Manson MM, Munks RJ, Kaptein A, Farrow S, Howells L (1999) Oncogene 18:6013
72. Han SS, Seo HJ, Surh YJ (2002) J Biochem Mol Biol 35:337
73. Deep G, Agarwal R (2007) Integr Cancer Ther 6:130
74. Matsuda T, Ferreri K, Todorov I, Kuroda Y, Smith CV, Kandeel F, Mullen Y (2005) Endocrinology 146:175
75. Navarro P, Giner RM, Recio MC, Máñez S, Cerdá-Nicolas M, Ríos JL (2001) Life Sci 68:1199
76. Kim DS, Oh SR, Lee IS, Jung KY, Park JD, Kim SI, Lee HK (1998) Phytochemistry 47:397
77. Shankar S, Chen Q, Srivastava RK (2008) J Mol Signal 3:7
78. Sa G, Das T (2008) Cell Div 3:14
79. Singh RP, Agarwal R (2006) Mol Carcinog 45:436
80. Singh RP, Agarwal R (2004) Mutat Res 555:21
81. Zi X, Grasso AW, Kung HJ, Agarwal R (1998) Cancer Res 58:1920
82. Scambia G, De Vincenzo R, Ranelletti FO, Panici PB, Ferrandina G, D'Agostino G, Fattorossi A, Bombardelli E, Mancuso S (1996) Eur J Cancer 32:877
83. Zi X, Feyes DK, Agarwal R (1998) Clin Cancer Res 4:1055
84. Sharma G, Singh RP, Chan DC, Agarwal R (2003) Anticancer Res 23:2649
85. Tyagi A, Raina K, Singh RP, Gu M, Agarwal C, Harrison G, Glode LM, Agarwal R (2007) Mol Cancer Ther 6:3248
86. Zhong X, Zhu Y, Lu Q, Zhang J, Ge Z, Zheng S (2006) Toxicology 227:211
87. Kohno H, Suzuki R, Sugie S, Tsuda H, Tanaka T (2005) Clin Cancer Res 11:4962
88. Lee SJ, Sung JH, Lee SJ, Moon CK, Lee BH (1999) Cancer Lett 144:39
89. Brahma NS, Sharmila S, Rakesh KS (2011) Biochem Pharmacol 82:1807
90. Kim HS, Kim MH, Jeong M, Hwang YS, Lim SH, Shin BA (2004) Anticancer Res 24:747
91. Chan HY, Wang H, Tsang DS, Chen ZY, Leung LK (2003) Nutr Cancer 46:93
92. Aluyen JK, Ton QN, Tran T, Yang AE, Gottlieb HB, Bellanger RA (2012) J Diet Suppl 9:45
93. Kaur M, Singh RP, Gu M, Agarwal R, Agarwal C (2006) Clin Cancer Res 12:6194
94. Zhang F, Yang JY, Mou YH, Sun BS, Ping YF, Wang JM, Bian XW, Wu CF (2009) Chem Biol Interact 179:419
95. Cao Y, Liu Q (2007) Adv Cancer Res 97:203
96. Lah JJ, Cui W, Hu KO (2007) World J Gastroenterol 13:5299
97. Onder TT, Gupta PB, Mani SA, Yang J, Lander ES, Weinberg RA (2008) Cancer Res 68:3645

98. Yodkeeree S, Ampasavate C, Sung B, Aggarwal BB, Limtrakul P (2010) Eur J Pharmacol 627:8
99. Chu SC, Chiou HL, Chen PN, Yang SF, Hsieh YS (2004) Mol Carcinog 40:143
100. Jankun J, Selman SH, Swiercz R, Skrzypczak-Jankun E (1997) Nature 387:561
101. Russo GL (2007) Biochem Pharmacol 74:533
102. Scalbert A, Williamson G (2000) J Nutr 130:2073S
103. Bell LN (2001) Handbook of nutraceuticals and functional foods. CRC, New York, pp 501–516
104. Anand P, Kunnumakkara AB, Newman RA, Aggarwal BB (2007) Mol Pharm 4:807
105. Fang Z, Bhandari B (2010) Trends Food Sci Technol 21:510
106. Li L, Ahmed B, Mehta K, Kurzrock R (2007) Mol Cancer Ther 6:1276
107. Kunwar A, Barik A, Mishra B, Rathinasamy K, Pandey R, Priyadarshini KI (2008) Biochim Biophys Acta 1780:673
108. Sou K, Inenaga S, Takeoka S, Tsuchida E (2007) Int J Pharm 352:287
109. Bisht S, Feldmann G, Soni S, Ravi R, Karikar C, Maitra A, Maitra A (2007) J Nanobiotechnol 5:3
110. Kumar V, Lewis SA, Mutalik S, Shenoy DB, Venkatesh, Udupa N (2002) Indian J Physiol Pharmacol 46:209
111. Salmaso S, Bersani S, Semenzato A, Caliceti P (2007) J Drug Target 15:379
112. Vemula PK, Li J, John G (2006) J Am Chem Soc 128:8932
113. Song G, Mao YB, Cai QF, Yao LM, Ouyang GL, Bao SD (2005) Braz J Med Biol Res 38:1791
114. Byrne JD, Betancourt T, Brannon-Peppas L (2008) Adv Drug Deliv Rev 60:1615
115. Ryan RO (2008) Expert Opin Drug Deliv 5:343
116. Ghosh M, Singh ATK, Xu W, Sulcheck T, Gordon LI, Ryan RO (2011) Nanomed Nanotechnol Biol Med 7:162
117. Natthakitta S, Wijit B, Supason PW, Khajeelak C, Kriengsak L, Jisnuson S (2011) J Control Release 151:176
118. Sahu A, Bora U, Kasoju N, Goswami P (2008) Acta Biomater 4:1752
119. Desai NP, Tao C, Yang A, Louie L, Yao Z, Soon-Shiong P, Magdassi S (1999) Protein stabilized pharmacologically active agents, methods for the preparation thereof and methods for the use thereof. US Patent 5,916,596
120. Segura S, Gamazo C, Irache JM, Espuelas S (2007) Antimicrob Agents Chemother 51:1310
121. Tae HK, Hai HJ, Yu SY, ChanWoong P, Kyung KT, Seulki L, Hyungjun K, Sangyong J, Xiaoyuan C, Kang CL (2011) Int J Pharm 403:285
122. Rejinold NS, Muthunarayanan M, Chennazhi KP, Nair SV, Jayakumar R (2011) J Biomed Nanotechnol 7:521
123. Mukerjee A, Vishwanatha JK (2009) Anticancer Res 29:3867
124. Grabovac V, Bernkop-Schnurch A (2007) Drug Dev Ind Pharm 33:767
125. Thamake SI, Raut SL, Ranjan AP, Gryczynski Z, Vishwanatha JK (2011) Nanotechnology 22:035101
126. Preetha A, Hareesh BN, Bokyung S, Ajaikumar BK, Vivek RY, Rajeshwar RT, Aggarwal BB (2010) Biochem Pharmacol 79:330
127. Debasree G, Somsubhra TC, Swarupa G, Ardhendu KM, Sibani S, Aparajita G, Krishna Das S, Nirmalendu D (2012) Chem Biol Interact 195:206
128. Yallapu MM, Brij KG, Jaggi M, Subhash CC (2010) J Colloid Interface Sci 351:19
129. Misra R, Sahoo SK (2011) Mol Pharm 8:852
130. Anitha A, Deepagan VG, Divya Rani VV, Deepthy Menon, Nair SV, Jayakumar R (2011) Carbohydr Polym 84:1158
131. Anitha A, Maya S, Deepa N, Chennazhi KP, Tamura H, Nair SV, Jayakumar R (2011) Carbohydr Polym 2011(83):452
132. Anitha A, Maya S, Deepa N, Chennazhi KP, Nair SV, Jayakumar R (2011) J Biomater Sci Polym Ed 23(11):1381–1400. doi:10.1163/092050611X581534

133. Das RK, Kasoju N, Bora U (2010) Nanomed Nanotechnol Biol Med 6:153
134. Liu J, Lihu X, Liu C, Zhang D, Wang S, Deng Z, Lou W, Xu H, Bai Q, Ma J (2012) Carbohydr Polym 90:16
135. Jinghua D, Yangde Z, Shiwei H, Yuxiang C, Bo L, Mingmei L, Wei C, Xingming D, Jinfeng Z, Boyun H (2010) Int J Pharm 400:211
136. Duan J, Mansour HM, Zhang Y, Deng X, Chen Y, Wang J, Pan Y, Zhao J (2012) Int J Pharm 426:193
137. Nair HB, Sung B, Yadav VR, Kannappan R, Chaturvedi MM, Aggarwal BB (2010) Biochem Pharmacol 80:1833
138. Sanoj RN, Muthunarayanan M, Divyarani VV, Sreerekha PR, Chennazhi KP, Nair SV, Tamura H, Jayakumar R (2011) J Colloid Interface Sci 360:39
139. Sanoj RN, Sreerekha PR, Chennazhi KP, Nair SV, Jayakumar R (2011) Int J Biol Macromol 49:169
140. Yallapua MM, Jaggi M, Subhash CC (2010) Colloids Surf B 79:113
141. Dhule SS, Patrice P, Trivia F, Ryan W, Joshua F, Grace T, He J, Alina A, Vijay J, Radhika P (2012) Nanomed Nanotechnol Biol Med 8:440
142. Yallapu MM, Vasir JK, Jain TK, Vijayaraghavalu S, Labhasetwar V (2008) J Biomed Nanotechnol 4:16
143. Bisht S, Mizuma M, Feldmann G, Ottendorf NA, Hong SM, Pramanik D et al (2010) Mol Cancer Ther 9:2255
144. Lim KJ, Bisht S, Bar EE, Maitra A, Eberhart CG (2011) Cancer Biol Ther 11:464
145. Gonçalves C, Pereira P, Schellenberg P, Coutinho PJ, Gama FM (2012) J Biomater Nanobiotechnol 3:178
146. Wu W, Shen J, Banerjee P, Zhou S (2011) Biomaterials 32:598
147. Mangalathillam S, Rejinold NS, Nair A, Lakshmanan VK, Nair SV, Jayakumar R (2012) Nanoscale 4:239
148. Zimeri J, Tong CH (1999) J Food Sci 64:753
149. Shpigelman A, Israeli G, Livney YD (2010) Food Hydrocoll 24:735
150. Shijun S, Long L, Guang Y, Jingrong L, Chao L, Tao G, Shaobing Z (2011) Int J Pharm 421:310
151. Young-Jin K, Mi RP, Min SK, Oh HK (2012) J Mater Chem Phys 133:674
152. Peres I, Rocha S, Gomes J, Morais S, Pereira CM, Coelho M (2011) Carbohydr Polym 86:147
153. Rocha S, Generalov R, Pereira CM, Peres I, Juzenas P, Coelho M (2011) Nanomedicine 6:79
154. Hsieh DS, Wang H, Tan SW, Huang YH, Tsai CY, Yeh MK, Wu CJ (2011) Biomaterials 32:7633
155. Bing H, Yuwen T, Xiaoxiong Z, Qingrong H (2011) Carbohydr Polym 89:362
156. Dube A, Nicolazzo AJ, Larson I (2011) Eur J Pharm Sci 44:422
157. Imtiaz AS, Vaqar MA, Dhruba JB (2009) Cancer Res 69:1712
158. Siddiqui IA, Adhami VM, Ahmad N, Mukhtar H (2010) Nutr Cancer 62:883
159. Sanna V, Gianfranco P, Anna MR, Stefania P, Anna MP, Alessandro A, Salvatore M, Pasquale B, Sergio U, Mario S (2011) J Med Chem 54:1321
160. Singh M, Bhatnagar P, Srivastava AK, Kumar P, Shukla Y, Gupta KC (2011) J Biomed Nanotecnol 7:202
161. Shutava TG, Balkundi SS, Vangala P, Steffan JJ, Bigelow RL, Cardelli JA, O'Neal DP, Lvov YM (2009) ACS Nano 3:1877
162. Zu YG, Yuan S, Zhao XH, Zhang Y, Zhang XN, Jiang R (2009) Acta Pharm Sin 44:525
163. Nam JB, Ryu JH, Kim JW, Chang IS, Suh KD (2005) Polymer 46:8956
164. Shi G, Rao L, Yu H, Xiang H, Yang H, Ji R (2008) Int J Pharm 349:83
165. Mérillon JM, Fauconneau B, Waffo P, Barrier L, Vercauteren J, Huguet F (1997) Clin Chem 43:1092
166. Sanggu K, Wai KN, Yuancai D, Surajit D, Reginald BHT (2012) J Food Eng 108:37
167. Shao J, Li X, Lu X, Jiang C, Hu Y, Li Q, You Y, Fu Z (2009) Colloids Surf B 72:40

168. Wang XX, Li LB, Yao HJ, Ju RJ, Zhang Y, Li RJ, Yang Y, Zhang L, Lu WL (2011) Biomaterials 32:5673
169. Scognamiglio I, De Stefano D, Campani V, Mayol L, Carnuccio R, Fabbrocini G, Ayala F, La Rotonda MI, De Rosa G (2012) Int J Pharm (in press). doi:10.1016/j.ijpharm.2012.08.009
170. Guo L, Peng Y, Yao J, Sui L, Gu A, Wang J (2010) Cancer Biother Radiopharm 25:471
171. Snima KS, Arunkumar P, Nair SV, Jayakumar R, Lakshmanan VK (2012) Second international conference on nanotechnology at the bio-medical interface (BioNano2012), Kochi
172. Sanoj RN, Muthunarayanan M, Muthuchelian K, Chennazhi KP, Nair SV, Jayakumar R (2011) Carbohydr Polym 84:407
173. Narayanan S, Binulal NS, Mony U, Manzoor K, Nair SV, Menon D (2010) Nanotechnology 21:285107

Adv Polym Sci (2013) 254: 241–268
DOI: 10.1007/12_2012_199

Published online: 15 March 2013

Proteins and Carbohydrates as Polymeric Nanodrug Delivery Systems: Formulation, Properties, and Toxicological Evaluation

Dhanya Narayanan, J. Gopikrishna, Shantikumar V. Nair, and Deepthy Menon

Abstract Enhancing the clinical benefits of pharmaceuticals at reduced doses with negligible side effects has been an ever-challenging goal in drug delivery. This pursuit towards an optimal formulation is hugely supported through the development of nanoparticulate delivery systems, which encapsulate the drug in its active form. Nanocarriers have emerged as successful candidates in pharmaceutics owing to their multifaceted characteristics such as the capability to entrap pharmaceutical ingredients at therapeutic dose, better biodistribution, size-dependent clearance, targetability, and controlled/sustained release of single or multiple payloads at the site of interest. Additionally, the requirement that the chosen carrier be biocompatible and biodegradable necessitates the use of biopolymers in entrapping payloads. Among the biopolymers, a wide variety of natural and synthetic carbohydrates and proteins (including starch, chitosan, chitin, dextran, alginate, albumin, casein, fibrin/fibrinogen, gelatin, collagen, whey protein, etc.) have evolved as useful materials for delivering hydrophobic or hydrophilic drugs through oral, intravenous, mucosal, ocular, or nasal routes. Targeted-controlled release of pharmaceuticals to diseased sites, environmentally stimulated drug release at desired locations, and many more drug release strategies have been made available by efficient engineering and nanoformulation of carbohydrate and protein biopolymers. This chapter embodies an in-depth discussion of various carbohydrate- and protein-based nanomedicines with respect to the formulation and properties of the nanoconstructs. The manipulation and utilization of these properties for development of better drug delivery devices is described and the toxicological interactions of these nanocarriers with host physiological systems discussed.

D. Narayanan, J. Gopikrishna, S.V. Nair, and D. Menon (✉)
Amrita Centre for Nanosciences and Molecular Medicine, Amrita Institute of Medical Sciences and Research Centre, Kochi 682041, India
e-mail: deepthymenon@aims.amrita.edu; deepsmenon@gmail.com

Contents

1 Introduction

In recent years, nanotechnology has gained incredible attention because of its advantages in the design of drug delivery systems for medical therapeutics. Several therapeutic agents are hydrophobic, low molecular weight compounds that exhibit a nonspecific biodistribution profile, short plasma circulation time, and rapid systemic elimination, with ensuing poor absorption, low bioavailability, and thereby side-effects and poor efficacy [1, 2]. Furthermore, certain drugs demand frequent administration for enhanced therapeutic efficacy, consequently leading to drug-induced toxicity. This is so because a drug is a "potential toxin" when not used within the limits of its "safe" dosage and schedule. Consequently, nearly half of the potentially promising drug candidates with high activities, identified through high-throughput screening, fail to reach the clinical translation stage. The use of nanomaterials to address these downsides [3, 4] can lead to the creation of "smart" drug delivery vehicles and new therapies. Nanoencapsulation is an efficient approach to drug delivery because it evades the issues related to chemical modifications of drug molecules for improved potency [5, 6]. Biodegradable nanocarriers, as drug delivery paradigms, have been extensively explored for loading and delivering a wide range of small molecules as well as macromolecules such as peptides, proteins, and genes [2, 7, 8]. A major goal in designing such pharmaceutical nanoparticles is to control the drug delivery in order to release pharmacologically active agents at the therapeutically optimal rate and site [9]. Use of a biodegradable drug carrier offers numerous advantages: (1) improved solubility and retention time of hydrophilic/hydrophobic drug molecules for enhanced therapeutic value; (2) better drug stability by protection from host enzymes and environment, thereby enhancing cellular uptake; (3) improved bioavailability via delivery of the drug at the site of action at reduced dosage, enabling sustained and controlled release; (4) enhanced biodistribution with reduced toxicity and minimal nonspecific interactions with the host; and ultimately (5) reduced patient expenses and non-compliance by evading repeated administration [10, 11]. In a nutshell, nanoencapsulation of drugs within biodegradable polymers provides enhanced drug efficacy, specificity, tolerability, and a better therapeutic index [12]. The essential hallmarks of the selected drug carrier for attaining these desired outcomes are listed below.

1.1 Loading Capacity

The extent of drug carried by/within a drug carrier is expressed by its loading capacity. Drug loading is denoted by two terms: "entrapment efficiency", which is the drug loaded in a carrier as a fraction of the total drug used in the loading process and "loading efficiency", which is a measure of the amount of drug loaded into a unit value of the drug carrier used, usually expressed as drug-to-polymer ratio. Entrapment and loading efficiencies may vary depending upon the nature of the drug and the method of loading. The higher the loading, the better the clinical therapeutic outcome conferred at low doses.

1.2 Release Characteristics

Delivery of the loaded contents from a drug carrier may occur either via degradation of the drug carrier or by slow diffusion of the drug through the carrier matrix, and is expressed as the release rate. The method of drug incorporation within the nanomatrix influences the release profile considerably. For drugs incorporated within the matrix during its preparation, a relatively small burst effect with better sustained release characteristics is obtained. For nanoparticles stabilized with a polymer coating, release is controlled by diffusion of the drug from the core across the polymeric membrane. Ionic interactions between the drug molecules as well as the addition of surfactant molecules can also drastically affect the release rate. Thus, drug solubility, diffusion, and biodegradation of the matrix materials govern the release process. A slow and sustained release rate at the tissue/organ of therapeutic action will enable improved bioavailability and, thereby, clinical outcomes.

1.3 Stability

One of the main concerns in nanoparticle engineering is to maintain the stability of the primary particles because otherwise the advantages of the small particle size are lost. Physical instability is the major challenge with colloidal nanoparticle systems and, typically, surfactants and/or polymers are used to stabilize the nanosuspensions for better drug efficacy. Biological stability of the nanocarrier is yet another important concern for better efficacy of the drug. This could be conferred by chemical modifications such as crosslinking and surface passivation but, most of all, the stability of the drug carrier is an inherent characteristic of the material used. The better the integrity of the carrier, the better it can hold the drug under storage and host conditions.

1.4 Circulation

Characteristic material properties such as surface charge, size, and hydrophilicity influence the circulation of a drug carrier in the blood stream [13]. These properties of a drug carrier can be engineered for better outcome. The extent of circulation is commonly expressed by the drug "bioavailability". Most of the hydrophobic drugs have a very short circulation time *in vivo* due to their rapid clearance, resulting in low bioavailability. Hence, the longer the circulation time of the drug carrier, the better the availability of the drug at the site of action.

Most of the above-mentioned essential hallmarks are well satisfied by drug carriers made from biopolymers [14]. The diversity of biopolymers, including proteins and carbohydrates, offers a wide range of materials with varied characteristics to choose from viz., degradability, non-antigenicity, nutritional value, abundance, drug-binding capacity, simple chemical structure, etc., all of which provide platforms for an array of manipulations. These exceptional characteristics together with the GRAS (generally regarded as safe) approval of many of the biopolymers make them ideal matrices for drug loading and delivery. Moreover, nanoparticles developed from proteins and carbohydrates can be easily prepared and scaled-up during manufacture, and is also cost-effective, thereby offering tremendous potential for industrial translation. In this chapter, we present an extensive review of various carbohydrates and proteins used to prepare nanomedicines, with a focus on their physicochemical features, formulational aspects, and toxicological behavior. The carbohydrate and protein drug carriers will be assessed on the basis of the above-stated essential hallmarks, their contributing factors, and benefits.

2 Properties of Carbohydrates and Proteins as Drug Delivery Systems

Carbohydrates (the building blocks) and proteins (the working machinery) of living cells have recently been of great interest to the pharmaceutical industry and research for developing drug delivery systems, mainly owing to their abundance, ease of availability, cost-effectiveness, low toxicity, ease of chemical modification due to their complex heterogeneity, and versatile routes of administration [7, 15–18]. Most carbohydrates and proteins have an impeccable capacity to integrate themselves with the host once administered and can be easily absorbed, digested/degraded to release the payload, and forms harmless by-products. Additionally, the availability of diverse functional groups on the surface of nanocarriers developed from carbohydrates and proteins makes them interesting ligands for targeted and site-specific delivery [7]. The possibility to modify the surface helps the nanocarrier to avoid its capture by the host immune or digestive systems, enabling a long sustained release of the encapsulated molecule [19].

The major carbohydrates that are being explored as drug carriers include chitosan and its derivatives, starch, alginate, hyaluronan and dextran. Major protein molecules identified as useful drug carriers are gelatin, albumin, collagen, fibrin/fibrinogen, and milk proteins like casein and whey proteins. Many other proteins like apo-lipoproteins and molecules of plant origin are also being investigated for their utility in drug delivery. Different categories of carbohydrates and protein molecules have varying degrees of drug loading, release, stability, and circulation. These properties also depend upon the method of preparation and the kind of modifications done on their surface. The following section analyzes the utility of the major carbohydrate and protein drug carriers on the basis of the essential hallmarks mentioned earlier.

2.1 Drug Loading

The quantity of drug entrapped within a nanocarrier varies according to the nature of the carrier and the drug used, the method and conditions of particle preparation and its surface modifications. Drug loading in carbohydrate and protein nanocarriers can be accomplished by two different methods: (1) the incorporation/active method, whereby the drug is loaded during preparation of nanoparticles, and (2) the incubation/passive method, whereby the drug is loaded after preparation of particles. The efficiency of drug loading is found to be maximum for the incorporation method, but is also greatly influenced by the method of preparation. For systems requiring faster release, the incubation method would be ideal because this involves only the physical adsorption of drugs to the matrix. Various other processing parameters as well as the presence of additives such as stabilizers, crosslinkers etc., also critically influence drug loading. Both hydrophilic and hydrophobic drugs can be incorporated into any nanoparticulate system but, as mentioned, the loading efficiency and release rate is dictated by the matrix–drug interaction and the nanoparticle preparation route.

2.1.1 Drug Loading Properties of Carbohydrates

Chitosan, a versatile cationic polysaccharide of natural origin, is extensively explored for loading various kinds of drug molecules, genes, etc. For instance, cisplatin, a hydrophilic drug, was loaded into chitosan matrix during the preparation of nanoparticles with an encapsulation efficiency of ~99% [20]. It was observed that neither the amount of crosslinker nor the drug had any effect on drug loading. The major parameter that influenced drug loading was the polymer concentration, which was directly proportional to the percentage drug loading. Likewise, other hydrophobic drugs and drugs that precipitate in acidic pH (such as diclofenac sodium) were also loaded into chitosan via the incubation method. In this case, the influencing parameter was the crosslinker concentration, since the drug was loaded into the polymer during its swelling process. Because higher crosslinking decreased the swelling, the loading efficiency was proportionately reduced [21]. Chitosan as well as derivatives like *O*-carboxymethylchitosan, *N*,*O*-carboxymethylchitosan, chitin, etc. in nanoparticulate form have also demonstrated reasonable success in the encapsulation and delivery of various types of anticancer drugs (curcumin, 5-flurouracil, and polyoxometalates) and antibiotics (tetracyline, etc.) when formulated by ionotropic gelation with the use of sodium tripolyphosphate (TPP) or calcium chloride as the crosslinker [22–24]. However, in certain cases, the drug itself can act as a crosslinker for the polymer, resulting in its high entrapment. For instance, in the case of polyoxometalate (POM)-chitosan nanocomplex, where the anionic POM is crosslinked with cationic chitosan, a good entrapment of ~65% resulted in comparison to ~35% with the use of the routine crosslinker, TPP [25].

Similarly, dextran, a bacterial polysaccharide with neutral charge that is well known for its degradability by dextranase, biocompatibility, and nontoxicity, has been explored for drug delivery wherein active molecules are either incorporated or incubated in the nanoparticles [14, 26, 27]. Furthermore, for improved bioavailability of hydrophobic moieties, an alternative approach of covalent or non-covalent attachment of drug molecules along dextran chains and use of dextran derivatives like carboxymethyl dextran [28] or conjugates of dextran with other carbohydrates like chitosan [29] is found to be useful.

Hyaluronan (HA) is a biodegradable, biocompatible linear polysaccharide that is abundantly present in the extracellular matrix, connective tissues, and organs of all higher animals and is metabolized through enzymatic hydrolysis [30, 31]. However, HA being a carbohydrate with poor biomechanical properties, is less explored for drug delivery in its native form. To facilitate this, various chemical modifications through carboxyl and hydroxyl groups are performed to obtain mechanically and chemically robust HA derivatives that retain the biocompatibility and biodegradability, but differ considerably in their physicochemical properties. Studies performed with hyaluronic acid, the esterified derivative of HA, demonstrate better encapsulation of hydrophilic than hydrophobic drugs due to the partial diffusion of drug molecules from the aqueous phase to the external continuous oil phase during emulsification [32]. Also, the fact that HA can bind to specific cellular receptors such as CD44 makes it promising for targeting antitumor drugs towards various tumors [33].

Starch is a comparatively less explored polymer for drug delivery due to its extreme hydrophilicity. Hence, hydrophobically modified starch derivatives are attracting interest in pharmaceutical research, due to their better entrapment and sustained release. Studies with propyl-, acetyl-, and hydroxyethyl-modified forms of starch have enabled enhanced entrapment through crosslinking and emulsion routes of preparation [34, 35]. Drug loading can also be improved by crosslinking anionic starch with cationic polymers such as chitosan, dextran, etc. [36, 37]. The parameters that crucially influence drug loading include the polymer-to-crosslinker ratio, reaction time, and temperature [38, 39].

Alginate, a linear, unbranched polysaccharide has also earned considerable attention in biomedical research owing to its biocompatibility and abundance. However, this carbohydrate (as in the case of starch) is afflicted by its hydrophilicity. Nonetheless, its ability to form a gel in the presence of cations makes it a potential candidate for drug delivery. The mass relation between calcium (popularly used as crosslinker for alginate) and alginate is the major parameter in determining the encapsulation efficiency [40, 41].

2.1.2 Drug Loading Properties of Proteins

Owing to the differing molecular weights, chemical complexity, and hydrophilicity, each protein exhibits a varying degree of drug loading [4, 42, 43]. Albumin is a nontoxic, biocompatible, and easily metabolizable protein that is widely investigated as a drug carrier, owing to its abundance. It is commercially derived from three major sources, viz., egg white (ovalbumin), bovine serum (BSA), and human serum (HSA). Albumin is an interesting drug delivery agent mainly because of its high affinity to a large number of drugs [42, 44]. It contains a large number of charged amino acids and multiple binding sites, making it ideal for binding both positively and negatively charged molecules. The traditional drug encapsulation strategies for albumin-based drug carriers is to either entrap the drug "inside" the nanocarrier or to attach the drug "onto" the surface of the proteins [7]. Abraxane is a paclitaxel–albumin conjugated chemodrug, prepared by nab (nanoparticle albumin-bound) technology, for treating advanced breast cancer [45]. Nab technology – the technique of conjugating drug molecules onto a protein surface [46, 47] – also serves as the platform for entrapping other anticancer drugs like docetaxel, rapamycine, topoisomerase inhibitors, and other taxanes onto albumin nanoparticles [44, 48]. In a recent study, the antiviral drug Ganciclovir was entrapped in albumin nanocarriers at an efficiency that ranged from 40 to 65% depending on the method of its preparation [49]. A high entrapment efficiency of ~80% was obtained upon loading sodium ferulate in albumin nanoparticles subsequent to crosslinking the protein molecules with glutaraldehyde [50]. Likewise, entrapment and loading efficiencies of ~93% and 27%, respectively, were obtained for paclitaxel in folate-decorated albumin nanoparticles. Such a high entrapment efficiency of paclitaxel can be attributed to the high ionic interactions of the drug with the reactive side chains of albumin [51].

Gelatin, a denaturized form of protein obtained from collagen by either acid or alkaline hydrolysis, has been extensively investigated for its drug bearing and delivering ability [52]. Gelatin is rich in its surface composition with cationic, anionic and hydrophobic domains and, hence, is regarded as a polyampholyte. These surface groups on gelatin are readily available for ligand decoration or drug conjugation [53]. Gelatin nanoparticles have demonstrated success for the delivery of a large category of drugs including anticancer [54–56], anti-inflammatory [57], anti-HIV [58], and antimicrobial drugs [59, 60], as well as activators and growth factors [52, 61, 62]. Two classical methods have been adopted for drug loading in gelatin nanoparticles, i.e., incorporation of the payload within the crosslinked gelatin core, and the use of hybrid nanocarriers by conjugating gelatin with natural or synthetic polymers. Whereas drug loading is typically low and varies between 25 and 40% in the first case, the use of hybrid nanocarriers aids in high drug retention in the gelatin core [58, 62, 63].

Collagen is the parent molecule of gelatin and the most abundant natural protein in the living body, comprising of 20–30% of the total protein pool [64]. It is also widely explored for applications in drug delivery. Biomedical applications of collagen are wide and varied, ranging from implants, shields, particles, drug-eluting gels and sponges to films [64, 65]. A study by Kaufman and Gebhardt, using cyclosporin as a model drug, have identified collagen as an ideal delivery agent for hydrophobic drugs [66]. Entrapment and efficacy of cyclosporine entrapped in collagen nanocarriers were found to be significantly high [66, 67]. The features that make collagen a favorable drug delivery system for hydrophobic payloads are its small size with high surface area, high adsorption capacity, and ability to dispense in water to form a clear colloidal solution [68]. Anticancer drugs such as vinblastin [69], cisplatin [70], 5-flourouracil [71], and paclitaxel have been incorporated within a collagen matrix for localized cancer treatment. In addition to antitumor drugs, antimicrobials, tissue-regulating agents, and growth factors have all been effectively delivered to their site of action by entrapping them in a collagen matrix [64, 65, 72, 73].

Milk proteins routinely transport nutrients to the digestive system in bulk and, hence, are the natural carriers of bioactive molecules in the human body. Two kinds of milk proteins, (flexible caseins and globular whey proteins) have been utilized for drug delivery applications by mimicking their natural transport mechanism [7, 74, 75]. Caseins are exceptionally surface active and stable, and have very high water and ion binding capacity, which makes them favorable candidates for drug delivery [75]. Casein, especially β-casein, is valuable in the oral delivery of hydrophobic bioactives. β-Casein nanovehicles for mitoxantrone, paclitaxel, and vitamin D have been developed [74, 76, 77]. Livney et al. have prepared paclitaxel entrapped in a highly stable formulation of β-casein, with an entrapment efficiency of nearly 100%, and shown its activity on gastric cancer cell lines [77]. Whey proteins have also emerged successful in drug delivery, with an interesting application in the delivery of bioactive bilberry anthocyanins by protein gels [78–80]. Kulozik et al. in 2012 reported the development of a novel whey protein-based aerogel for the delivery of various active ingredients [81].

2.2 Drug Release Properties

The major factors affecting the release of a payload from a polymeric nanocarrier are: (1) degradation rate of the polymer, (2) diffusion of the drug from the matrix, and (3) surface modification of the drug carrier. When nanoparticles come in contact with the dissolution medium, weakly bound molecules on the surface of the matrix desorb into the medium resulting in burst release. Subsequently, carbohydrate nanocarriers become hydrated and swell, forming a gel diffusion layer that hampers the outward transport of the drug within the matrix thereby providing a controlled release effect [17]. Proteins on the other hand, are easily digested by endogenous proteolytic enzymes of the host thereby may result in a faster release. The release kinetics can be altered by the degree of crosslinking, adequate surface modification, and the choice of the matrix. Figure 1 illustrates the different parameters that regulate the burst release or slow and sustained release of drugs from within carbohydrates and polymers.

2.2.1 Drug Release from Carbohydrates

For carbohydrates, the swelling mechanism (i.e., the wetting), which results in loss of integrity of the matrix (mainly for hydrophilic polymers), can be reinforced by aiding proper crosslinking. Since release is carried out in the aqueous media, the hydrophilic drug moieties show a faster release from the matrix than hydrophobic drugs. This is illustrated through the drug release experiment carried out using propyl starch and three model drugs, i.e., flufenamic acid, testosterone, and caffeine (Fig. 2). The hydrophobic drugs flufenamic acid and testosterone showed a sustained release with nearly null burst effect, whereas the hydrophilic drug caffeine showed a much faster, yet linear release within the first 10 h before it reached a plateau phase [35].

The release rate is also observed to be inversely proportional to the molecular weight of the encapsulated drug molecule [82]. The rate of drug release can also be varied by altering the drug–polymer interaction as well as by chemically immobilizing the drug to the polymer backbone using the reactive functional groups, for example, the carboxylate groups in alginate [83, 84]. Other than swelling, degradation can also occur due to the natural enzymatic splitting of polysaccharide bonds. Drug release occurring via matrix degradation can be controlled through appropriate modifications of the native molecule [85]. Surface engineering or modification of nanocarriers with carbohydrates is yet another alternative for obtaining sustained drug release [86]. For certain carbohydrates

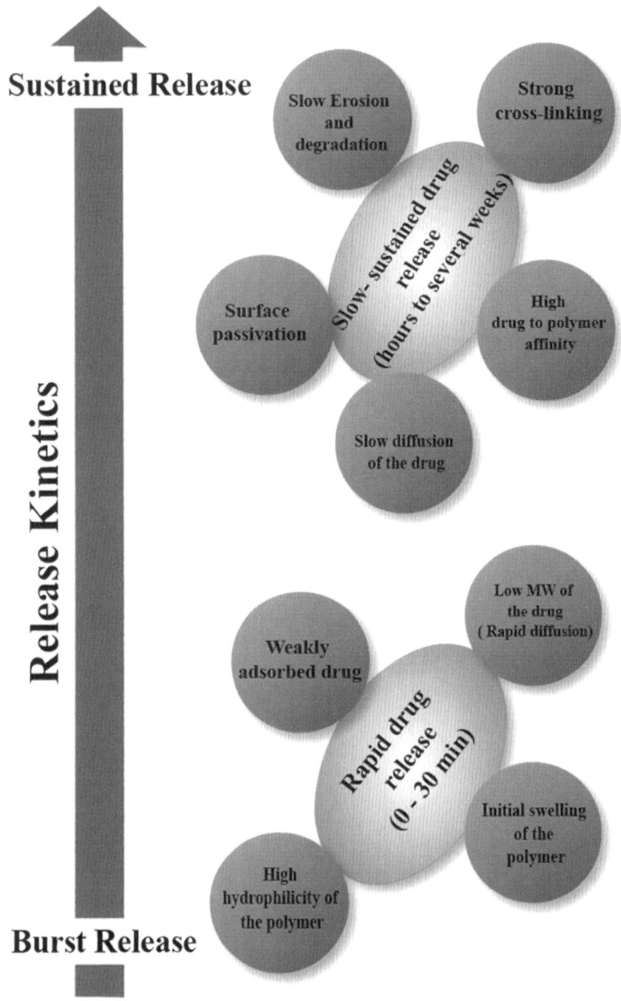

Fig. 1 Factors that regulate drug release from within biopolymeric matrices so that release is rapid or slow and sustained

such as chitosan, its mucoadhesive property plays an important role in attaining controlled release, making it an excellent system for oral, nasal, and ocular drug delivery [87].

Apart from these properties, the release kinetics can be manipulated in such a way that drug release occurs due to a trigger from an environmental stimulus like pH, temperature, etc. [88]. The thermoresponsive graft co-polymeric 5-FU-loaded nanoparticles made from chitosan-*g*-poly(*N*-vinylcaprolactam), prepared by an ionic crosslinking method, showed a lower critical solution temperature (LCST) at 38°C, with a prominent in vitro drug release above LCST [89]. Camptothecin-loaded poly(*N*-isopropylacrylamide) (PNIPAAm)/chitosan nanoparticles were

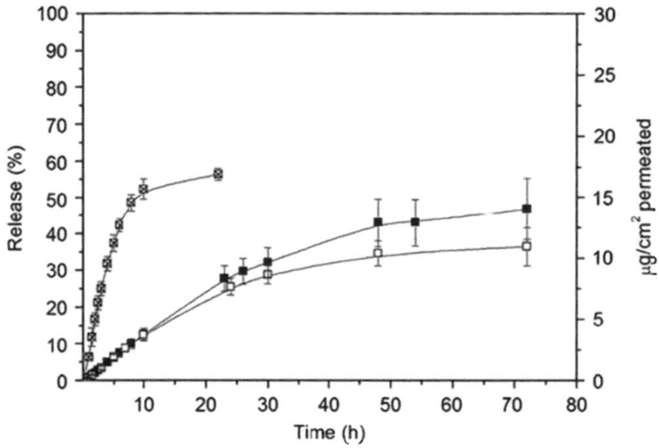

Fig. 2 Release profile of FFA (*filled squares*), testosterone (*open squares*), and caffeine (*crossed squares*) from propyl starch nanoparticles [35]

developed as pH sensitive carriers for specifically targeting tumors, providing an optimum drug release at pH 6.8 and decreased release either below pH 6.5 or above pH 6.9 at 37°C [90].

When sustained drug release using simple dissolution mechanisms such as diffusion, erosion, membrane control, or osmotic systems becomes difficult, retardation mediated by ionic interactions could provide an ideal alternative for attaining controlled release. This can be facilitated by the use of oppositely charged drug and polymeric excipients. For cationic drugs, polymers such as starch, polyacrylates, sodium carboxymethylcellulose, or alginate can be used as the matrix, with chitosan being the best option for anionic drugs. Development of high density complexes using both cationic and anionic polymeric excipients could also slow down the release [91, 92]. Instead of anionic polymers, multivalent inorganic anions such as tripolyphosphate or sulfate can also be used to achieve similar effects [17, 36, 37].

2.2.2 Drug Release from Proteins

In a study that investigates the factors affecting the release of several antimicrobials from albumin microspheres, it was concluded that the denaturation temperature, time, concentration of crosslinking agents, agitation of the release medium, and presence of proteolytic enzymes impact the release profile considerably, with drug leaching from the granular matrix as the key method of drug release [93]. Similar conclusions were drawn in another study investigating the influence of physiochemical parameters on the release of hydrocortisone acetate from albumin microspheres. They observed that the release from the microspheres is initially Fickian, then becomes zero order, and finally is exponentially correlated with time [94].

Gelatin, which forms thicker gel matrices, usually is associated with a much longer and sustained drug release profile. In a phase II clinical trial, an improved angiogenesis was seen in patients with critical limb ischemia, with a sustained release of fibroblast growth factor (FGF) from gelatin carrier gels [95]. The most important parameter that affects drug release from gelatin hydrogels is the degree of crosslinking, which in turn affects parameters such as degradation rate, thickness of the gel, and diffusion of the drug [96, 97]. Another study demonstrated the sustained release of human growth hormone (hGH) from collagen thin films, aiding better wound healing in db/db mice. From the collagen film prepared by air drying, hGH was secreted for 3 days in a sustained manner, resulting in better wound healing with a single administration of the therapeutic [98]. This apart, gelatin and modified gelatin (thiol, PEG) in nanoparticulate form has demonstrated good success in the delivery of various hydrophobic and hydrophilic drugs as well as for intracellular DNA delivery [99–101]. The degree of crosslinking and the hydrophilicity of gelatin regulated the release of the drugs from within the matrix.

Unlike other proteins, caseins and whey proteins are mainly administered orally. Hence, drug release from these carriers mainly depends upon the digestion of the carrier, which is usually a rapid process [74]. Casein and whey proteins form matrices that protect the drugs from the harsh environment of the stomach and intestine.

2.3 *Stability of Nanocarriers*

The most important consideration in developing a pharmaceutical formulation is its long-term stability, which is important from two different viewpoints: (1) stability of the nanocarriers during storage as well as *in vivo* and (2) stability conferred by the nanocarrier to the loaded drug. In both situations, the chemical structure of the parent molecule, the extent of crosslinking, and the physical integrity of the drug and nanocarrier should be intact for a long time during storage and inside the host environment. This is a great obstacle when considering a translation of nanoformulations for clinical applications and, hence, it is essential to devise new strategies to alleviate these issues. The stability of nanoformulations is estimated by assessing the influence of temperature and humidity on various parameters such as particle size, zeta potential, drug content, viscosity, pH, etc. Biological stability, i.e. the ability of the matrix to confer sustained release even when exposed to a harsh environment containing degrading enzymes, is also an important concern that needs to be addressed.

2.3.1 Stability of Carbohydrates

To solve the issues related to the physical instability of carbohydrate nanocarriers, appropriate surface modification, for example glycosylation, seems to be one of the promising strategies [86, 87]. Likewise, for addressing biological instability issues, the commonly adopted strategy is to modify the native material. For example, starch can be easily degraded in presence of the enzyme amylase, which acts on its hydroxyl groups. Hence, in order to obtain a slower release, the hydroxyl groups of starch are modified with propyl, ethyl, or acetyl groups, thereby reducing both its hydrophilicity and degradation rates [34, 35].

A study by Wilson et al. reports that chitosan nanoparticles, intended for the release of the anti-Alzheimer drug tacrine, prepared through spontaneous emulsification are stable at different temperatures (15–20°C, 3–5°C and 37°C), with no significant change in physical appearance, particle size, drug content, or chemical interaction between drug and polymer over a period of 3 months [102]. A study by Prego et al. suggested high stability for ionically crosslinked polyethylene glycol (PEG)-grafted chitosan as a gene carrier [103]. Surface modification of polylactic acid (PLA) nanoparticles was done by using covalent attachment of PEG to PLA and physical adsorption of water-soluble chitosan (WSC) to the particle surface. Two types of WSC, cationic partially deacetylated chitin (PDC) and anionic *N*-carboxy propionyl chitosan sodium (CPCTS), were investigated. The presence of WSC, whether alone or with PEG, highly improved the surface hydrophilicity as well as suspension stability of the nanoparticles [104]. Studies on nanoparticles of dextran esters for drug delivery suggests that better colloidal stability can be attained when a dextran ester of low degree of substitution is used in the aqueous phase, independent of the degree of substitution of the dextran ester present in the core of the particle [27].

2.3.2 Stability of Proteins

Proteins as biomolecules are always liable to enzymatic degradation and digestion as well as to changes in pH and temperature. On the other hand, the use of protein nanocarriers to entrap drug molecules will help evade its uptake by the immune system, prevent opsonization and its subsequent phagocytosis to a great extent, improve circulation of the drug by renal and hepatic re-absorption, and much more. Several techniques have been proposed in the literature to prevent the acid or enzymatic degradation of protein nanocarriers [62, 93, 105, 106].

Physical integrity of the nanocarriers depends upon the preparation route and the kind of crosslinker used. For gelatin, it is found that the use of gluteraldehyde (GA) improves the stability of the nanocarriers. The encapsulated drug also alters the stability of the carrier, as in the case of encapsulated doxorubicin improving the stability of GA crosslinked gelatin nanocarriers. This is believed to be because of the competition between the free amino groups of gelatin and doxorubicin during the crosslinking process [7, 54]. However, the use of GA for controlling nanoparticle resiliency is a concern owing to its cytotoxicity and the fact that it may react undesirably with the therapeutic agents entrapped within the nanoparticles [107]. To address this concern, several cationic polymers such as polyethyleneimine (PEI) and poly-L-lysine (PLL), have been used for crosslinking purpose, thereby enhancing the stability of drug-loaded nanocarriers [108]. In a more recent study, the thermal and mechanical properties of gelatin hydrogels were improved by crosslinking them with biodegradable cellulose nanowhiskers [109].

Albumin nanocarriers are naturally stable, and that is one of the main reasons why they are used for the delivery of nucleotides across the nuclear membrane as well as to deliver drugs across the blood-brain barrier [48, 110]. Albumin nanoparticles showed no change in particle size when stored either in aqueous solution for 6 months at 4°C or as lyophilized powder [111, 112]. Conjugation with carbohydrates or a high degree of crosslinking could further enhance the stability of albumin nanocarriers. A simple self-assembled crosslinking of dextran and albumin with the hydrophobic drug ibuprofen enhanced the stability and drug-loading capacity of the resulting nanocarrier [113]. Cationic biomaterials such as PEI have been used for crosslinking the anionic protein BSA, providing protection against enzymatic degradation [114]. The use of another cationic polymer, PLL, instead of PEI, for coating BSA nanoparticles enhanced its proteolytic resistance, yielding more stable nanoparticles [107]. It was observed that the aqueous solution stability of BSA nanoparticles increased with increasing PLL molecular weight and concentration, as depicted in Fig. 3, wherein a continuous release of FITC–BSA was demonstrated from 0.9 kDa PLL-coated BSA nanoparticles in phosphate buffer (pH 7.4) and almost no release of FITC–BSA from 4.2, 13.8 and 24 kDa PLL-coated BSA nanoparticles after 3 days.

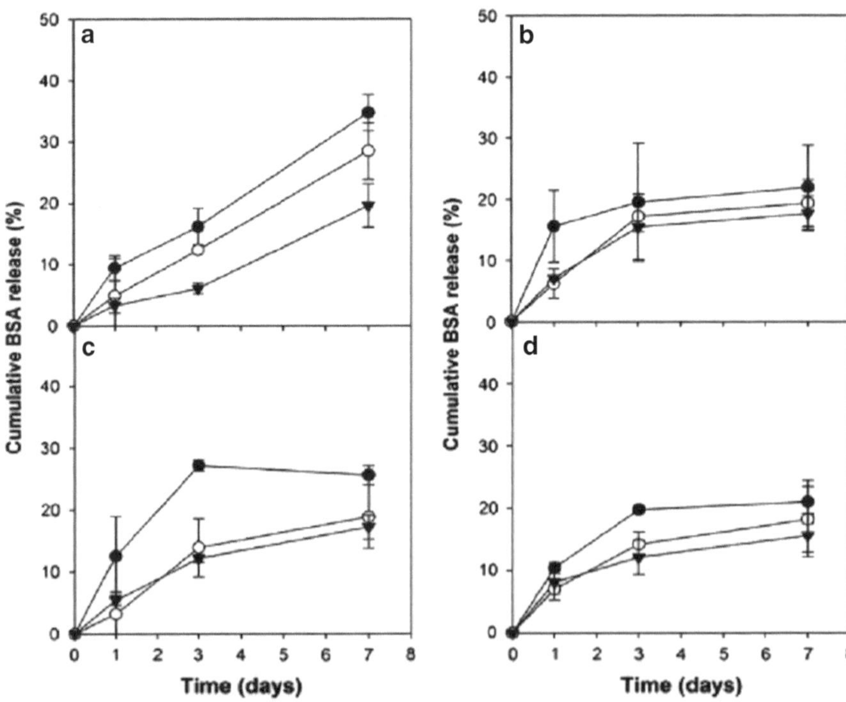

Fig. 3 Stability of PLL-coated BSA nanoparticles based on release profile of FITC–BSA from the nanoparticles in phosphate buffer (pH 7.4). PLL of molecular mass 0.9 kDa (**a**), 4.2 kDa (**b**), 13.8 kDa (**c**), and 24 kDa (**d**) at concentrations of 0.1 mg/mL (*filled circles*), 0.3 mg/mL (*open circles*), and 1.0 mg/mL (*filled inverted triangles*) were used for coating onto BSA nanoparticles [107]

Unlike any other proteins used in drug delivery, milk protein casein is not sensitive to temperature or mild pH changes. In a study investigating the stability of casein nanocarriers over a broad pH range of 6.5–12, it was noted that pH changes do not alter the properties of these nanocarriers [115]. Casein is stable up to 70°C, beyond which it shows considerable degradation [75]. With long-term lyophilized storage, casein itself acts as a cryoprotectant. Celecoxib-loaded β-casein nanoparticles were cryopreserved for 6 months and easily re-suspended on rehydration [116]. Caseins are also well known for protecting the payload even under harsh environments, for example, its effective protection of vitamin D2 and 1,4-dihydroxyanthraquinone (DHA) against UV-induced degradation [117].

2.4 Circulation of Nanocarriers

The activity of a drug in the host is best measured in terms of their plasma half life ($t_{1/2}$) and its plasma concentration. Many drugs have very low $t_{1/2}$ values, once exposed to the host environment because they are metabolized, degraded, or cleared by the local system, thereby limiting their clinical activity to a very short duration, sometimes as low as minutes. Entrapping such fast-clearing drugs within nanocarriers helps to confine the drug in its active state within the matrix, and sustained release from these carriers ensures clinical activity for longer durations. However, a major obstacle in the development of polymeric nanoparticles as effective drug delivery carriers is their rapid clearance from blood. Blood circulation residence time, which ultimately leads to a high therapeutic index, is one of the important factors that determine the clinical success of a pharmaceutical formulation. Surface modification or functionalization of nanoparticles helps in increasing the mean residence time of particles in blood by imparting "stealth" (shielding of the particle surface with a coating material) or sterically stabilized properties to nanoparticles [106].

Carbohydrates and proteins could be potential candidates for successful drug delivery due to their abundance of functional groups that can act as ligands for surface modification. Such surface-modified nanocarriers would block the serum proteins (opsonins) from binding to them and remain camouflaged to phagocytic cells, thereby reducing their rapid clearance from the blood stream. Delays in opsonization can be achieved either through the adsorption of polymers or surfactants on nanocarrier surfaces via covalent or non-covalent attachment of molecules. Although one of the most widely utilized strategies is adsorption or grafting of PEG molecules on the nanocarrier surface, appropriate surface modification of the nanocarriers using carbohydrates can also help evade opsonization [118].

In addition to surface modification, particle size is another key factor that affects the circulation time of nanocarriers. The smaller the particle size, the lesser will be the protein adsorption and slower the clearance. To demonstrate this fact, poly (cyanoacrylate-*co*-*n*-hexadecyl)cyanoacrylate (PHDCA) particles with different sizes, i.e., small (<100 nm), medium (100–200 nm) and large (>200 nm), were incubated with serum proteins for 2 h yielding protein adsorption values of 6%, 23%, and 34%, respectively. A significant correlation between size and protein adsorption was clearly apparent. This was further confirmed by the particle uptake in murine macrophages and its blood clearance kinetics. The experiments helped to conclude that particles with appropriate surface modification and size less than 100 nm showed less protein adsorption, rendering longer circulation time. Apart from hydrophilic coatings and size, surface charge also plays a significant role in the clearance of nanoparticles from circulation. Nonspecific cellular internalization and protein adsorption during circulation can be affected by altering the surface charge of nanocarriers. The particles with higher positive charge are prone to have a rapid uptake by phagocytic cells, as compared to negative or neutral particles, resulting in shorter blood circulation half-lives [13, 119].

2.4.1 Circulation of Carbohydrates

In addition to being a promising matrix for drug delivery, carbohydrates also offer methods for improving the blood residence time of nanocarriers through glycosylation [86]. Glycosylated nanocarriers have the additional attributes of improving the pharmacokinetic profile of drugs, providing carrier stabilization as well as receptor-targeted drug delivery.

Linear dextrans that are frequently used as plasma expanders in medicine can remain in the systemic circulation for extended periods of time that are proportional to their molecular weights. The steric brushes of the dextran macromolecules have shown enhanced stability with reduced protein adsorption for various drugs, nanoparticles (iron oxide, liposomes), etc. [120, 121]. A bacterial exopolysaccharide, i.e., pullulan, which forms conjugates with cholesterol has also been reported to confer protection to liposome surfaces and reduce macrophage uptake *in vivo* [122]. This apart, carbohydrate moieties such as mannose, galactose, etc. have been extensively explored for their potential use as active targeting ligands that bind specifically to cell surface receptors and deliver therapeutic agents at the desired site [123–125].

A combinatorial coating of PEG and cationic WSC, on PLA nanoparticles, has yielded greatly prolonged circulation time for the nanocarriers *in vivo*. In contrast to PLA nanoparticles treated with PEG or WSC alone, PEG and WSC synergistically provided a strong inhibition of macrophage uptake and extended the circulation half-life up to 63.5 h, with concomitant reduced liver sequestration. This $t_{1/2}$ value was much longer than that of control PEG/PVA nanoparticles (1.1 h) as well as that of PLA nanoparticles stabilized with PEG/CPCTS (anionic *N*-carboxy propionyl chitosan sodium) (7.1 h), thereby proving the effect of surface stabilization in developing long-circulating drug delivery carriers [85, 104, 126].

2.4.2 Circulation of Proteins

Long-term retention of protein nanoparticles in host blood circulation is essentially governed by the surface properties, which makes them vulnerable to macrophage uptake, enzymatic degradation, and renal or hepatic clearance. However, by appropriate surface modification and targeting strategies, such uptake and clearance can be alleviated to a certain limit.

In an attempt to demonstrate this, Amiji et al. prepared PEG-modified gelatin nanoparticles and observed that modification of gelatin using hydrophilic PEG molecules helped to effectively evade proteolytic enzymes by steric repulsion [105]. The biodistribution studies using these PEGylated nanoparticles revealed their long circulation, with a half life of 15–39 h [127]. The targetability of PEG-coated gelatin nanocarriers was also assessed in a different study, wherein the surface-modified nanoparticles were found to have the capacity to accumulate at the tumor site and remain at this site up to 12 h [128]. A significant prolongation in blood circulation of ibuprofen sodium (IbS)-loaded gelatin nanoparticles was observed upon PEGylation, providing a sustained release of IbS for ~96 h, thereby improving the bioavailability and pharmacokinetics of the drug when compared to bare IbS and non-PEGylated nanoparticles (Narayanan et al. 2012, unpublished).

Albumin is a natural long-circulating polymer with a plasma half life of 19 days [7, 48, 110]. In a recent study, the plasma half -life of cisplatin was found to be enhanced from 65 min to 24 h, when entrapped within folate-conjugated albumin nanoparticles [129]. Chemical conjugation of PEG to albumin also increases its circulation capabilities at least fivefold, reduces its immunogenicity, and augments the enhanced permeability and retention effect and thereby tumor localization [130]. In another study, PEGylated albumin nanocarriers ensured a slower release of Rose Bengal, even in the presence of proteolytic enzymes, suggesting the effect of steric hindrance offered by PEG [131]. These examples suggest that surface modification of protein nanocarriers either by polymers like PEG, PLL, and PVA or by targeting ligands establishes a barrier between the carrier and its surroundings, thus reducing its chances of detection by the immune system or by enzymes, thereby rendering a longer circulation and targetability for the nanoparticles.

3 Formulational Approach

Hydrophilic and hydrophobic drugs can be encapsulated within carbohydrate or protein nanocarriers in several ways [16, 35, 86, 87]. However, the selection of a specific route of synthesis depends upon various requirements including particle

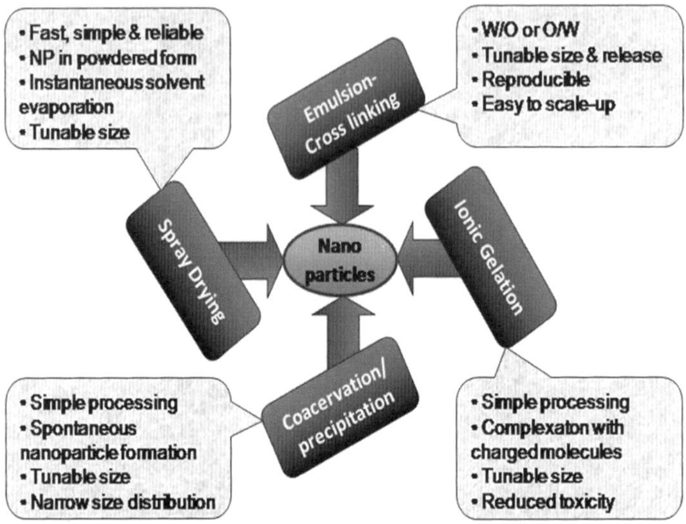

Fig. 4 Representative formulation strategies and their characteristics for preparing nanocarriers

size, stability (both physical and chemical), reproducibility, and release kinetics, with the nature of the drug to be encapsulated (hydrophobic or hydrophilic) and the route of drug administration being the most important criteria. Figure 4 depicts typical formulation strategies adopted for the synthesis of drug-loaded carbohydrate or protein nanocarriers; which are detailed in the following section.

3.1 Emulsion Crosslinking

Emulsion preparation is probably the most important step in the development of nanoparticles. Stable emulsions can be obtained using rudimentary techniques such as stirring and the latest technologies like homogenization and extrusion, but with varying accuracy, efficiency, and properties. Emulsion coupled with crosslinking is reported to provide nanocarriers with higher drug encapsulation efficiency and stability than the emulsion technique alone. Depending upon the nature (hydrophilic/hydrophobic) of the active molecule to be encapsulated, either a water in oil (w/o) or oil in water (o/w) emulsion technique can be adopted [16, 35, 86, 87]. Additionally, the formed emulsions can be stabilized through the use of appropriate surfactants and crosslinking agents. This method helps to control particle size as well as drug release by tuning the solvent:antisolvent ratio, concentrations of polymer and crosslinker, and the stirring speed. The advantages of this technique include high batch-to-batch reproducibility and ease of scaling up. However, this method has a few drawbacks, including incomplete removal of solvents, un-reacted excipients, etc. Overall, the efficacy of this method can be improved via the proper choice of dispersing phase, continuous phase, phase ratio, and surfactants. Chitosan-PEG nanocapsules prepared through the emulsion route for the oral delivery of peptides resulted in stable nanoparticles with improved *in vitro* stability [103]. Likewise, various other carbohydrates such as dextran, hyaluronan, and starch have also been prepared as stable nanoformulations by adopting the emulsion route [32, 132]. Crosslinking of gelatin using transglutaminase is a novel concept. Fuchs et al. detailed the interaction in gelatin–glutaminase crosslinking, and proposed it to be stronger and more biocompatible than the conventionally used glutaraldehyde for gelatin nanoparticle production. [133].

3.2 Ionic Gelation

This technique, which works through the complexation between oppositely charged molecules, has gained wide attention due its simple and mild processing routes. In this method, the polymer solution is added to an oppositely charged ionic solution, resulting in ionic gelation followed by precipitation, yielding spherical particles. Particle size is mainly dictated by the polymer and crosslinker concentrations. One of the main advantages of this technique is its reduced toxicity due to the reversible physical crosslinking by electrostatic interaction rather than chemical crosslinking, which eliminates the use of harsh chemicals. However, the same is also disadvantageous, since this results in particles with poor mechanical stability [16, 35, 86, 87].

Studies have reported a better encapsulation for poorly soluble drugs like 1,4-dihydroxyanthraquinone (DHA) by forming complexes between anionic starch and cationic cyclodextrin derivatives via ionic gelation [36]. Similarly, chitosan nanoparticles prepared using the polyanionic crosslinker tripolyphosphate resulted in stable nanoparticles with better encapsulation efficiency and reduced toxicity, with BSA as a model drug [134]. Zorzi et al. developed new hybrid nanoparticles via ionic gelation between cationized gelatin and the anionic polysaccharides, dextran sulfate and chondroitin sulfate for delivery of plasmid DNA (pEGFP) to the ocular surface [63]. Ionically crosslinked casein nanoparticles loaded with the anti-androgen flutamide, showed satisfactory entrapment efficiency with a positive zeta potential, good colloidal stability, and prolonged *in vitro* drug release. After intravenous administration into rats, pharmacokinetic parameters revealed that flutamide-loaded casein nanoparticles were well tolerated without any side effects and showed a longer circulation time relative to flutamide-free solution, suggesting that ionically crosslinked casein nanoparticles may have a promising future as carriers for hydrophobic drugs [1].

3.3 Precipitation/Coacervation

This method utilizes the physical property of the polymer, whereby the polymer in its solvent is dispersed into a non-solvent (a solvent that precipitates polymer) and the nanoparticles are formed spontaneously during the phase separation, popularly known as the Marangoni effect. The resultant particles can be stabilized by the use of surfactants, and size can be controlled mainly by tuning the solvent:antisolvent ratio. The formed particles are separated and purified by filtration or centrifugation and washing. This route is suitable for encapsulating both hydrophilic and hydrophobic drugs, depending on the choice of solvents. The main advantage of preparing nanoparticles through precipitation is the narrow size distribution of the particles [16, 35, 86, 87]. A study by Zhang et al. suggests that the copolymeric carrier of starch/alginate/chitosan, prepared through coacervation is a better system for the intestinal delivery of proteins and peptides [135]. Proteins such as gelatin, collagen, and albumin are more often prepared by a modified precipitation or coacervation technique [136].

3.4 Spray Drying

This is a fast, simple, and reliable technique for producing drug-loaded nanocarriers in powdered form. The polymer as well as the drug are dissolved in their respective solvents, blended, and then atomized under a stream of hot air, yielding small droplets. During atomization, the solvent evaporates instantaneously and forms a free-flowing powder. Desired particle size can be obtained by varying the processing parameters such as size of the nozzle, spray flow rate, atomization pressure, inlet air temperature, and extent of crosslinking [16, 35, 86, 87].

Alginate nanoparticles were successfully prepared using a spray drying technique, yielding submicron sized particles of mean diameter ~700 nm [135]. Spray drying is the most commonly used encapsulation technique for food products [137]. Among the available biopolymers used for spray drying applications, the most exploited ones for encapsulating food oils and flavors are carbohydrates, including modified and hydrolyzed starches, cellulose derivatives, gums, and cyclodextrins; and proteins such as whey proteins, caseinates, and gelatin [138, 139].In addition, protein-loaded lipid/chitosan nanoparticle complexes were efficiently microencapsulated using the spray drying technique for pulmonary delivery of insulin [140].

Figure 5 illustrates how hydrophilic and hydrophobic drugs can be loaded within biopolymers that are hydrophilic or hydrophobic by adopting these various methods of synthesis.

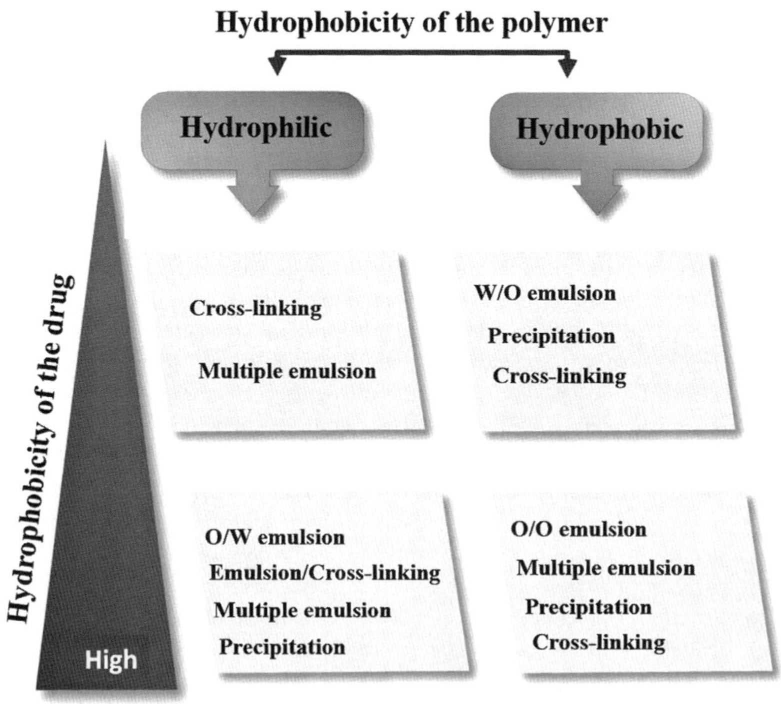

Fig. 5 Loading of hydrophilic or hydrophobic drugs within biopolymers having different hydrophilicities, and the choice of synthesis routes

4 Toxicological Evaluation

Nanomaterials have evolved as a revolutionary technology in various realms of medicine ranging from drug or gene encapsulation and delivery to tissue engineering and medical diagnosis and therapy. This intensive expansion of nanotechnology in the pharmaceutical industries and medical field necessitates an understanding of the potential toxicity of engineered nanomaterials, i.e., their biokinetic evaluation. The toxicity concerns about nanomaterials remain one of the major hurdles in the translation of nanomedicines for human use and this involves physiological, physicochemical, and molecular considerations [141–143].

Material characteristics such as size, shape, chemical composition, surface chemistry, roughness, and surface functionalizations profoundly influence the interactions that occur at the interface between nanomaterials and biological systems (proteins, membranes, endocytic vesicles, organelles, DNA, and biological fluids). Evidence from the literature suggests that the biodistribution of nanosized carriers to specific sites is mostly dictated by the physicochemical characteristics of the nanoparticles as well as the choice of the material used, its biocompatibility, biodegradation [144], etc.

The potential toxicity of nanomaterials is influenced considerably by its biodistribution, phagocytosis, opsonization, and endocytosis, all of which are critically impacted by the nanoparticle size and surface characteristics [145]. Protein opsonization of nanoparticles eventually leads to their clearance, mainly through uptake via the reticulo-endothelial system, resulting in its accumulation. The particles taken up by dendritic cells after opsonization are shown to activate NALP3 inflammasome, which enhances the effects of both innate and antigen-specific cellular immunity. Apart from this, interaction with the vascular components of the blood can also result in local inflammation, which is caused by nanoparticle-mediated oxidative stress. The interaction of nanoparticles with highly oxidatively active mitochondria can lead to an increased release of free radicals as reactive oxygen species (ROS), which depletes the natural defensive antioxidants such as glutathione. The resultant oxidative stress generates a wide variety of cellular events such as cell cycle arrest, apoptosis, inflammation, induction of signaling pathways, etc. Generation of specific antibodies against BSA was observed on administration of BSA-carbon nanotube conjugates [146]. Because the protein nanocarrier is essentially a non-self peptide fragment, generation of antibodies and immune activation is not surprising. Instead of activating the immune system, protein nanoparticles are sometimes also reported to cause downregulation of the immune system. Abraxane, the paclitaxel albumin conjugate drug, resulted in a lower incidence of grade 4 neutropenia (a form of myelosuppression that leads to a decreased number of neutrophils) than the first generation of paclitaxel formulations [147].

For biodegradable polymers such as carbohydrates and proteins, the toxicity associated with their tissue accumulation and metabolism is considerably less. However, upon interaction with blood and its constituents, nanoparticles can induce cyto- or hemotoxicity, inflammation, and oxidative stress. Anionic polymers with sizes greater than 60 nm have been reported to be biocompatible in terms of hemolysis, coagulation, thrombocyte and granulocyte activation, and membrane integrity, as compared to cationic particles with smaller size [148]. In general, cationic carbohydrates such as cellulose and chitosan can cause cytotoxicity by inducing apoptosis and, when in direct contact with blood, can activate the blood coagulation pathway, resulting in clot formation. Even though acute systemic toxicity studies in mice did not show any significant toxicity, chitosan solution even at very low concentrations caused hemeagglutination and platelet aggregation and activation [149–151]. This property of cationic polymers necessitates surface modification or engineering with different molecules such as PEG, dextran, mannose, lactose, galactose etc., with ensuing reduction in toxicity and improvement in the nanoparticle hemocompatibility [103, 151–155]. Figure 6 demonstrates how glycosylation of nanocarriers would be an ideal approach for evading nanoparticle toxicity. Additionally, the unreacted components used in the formulation (crosslinkers, stabilizers, and solvents) can also impart cyto- or hemotoxicity and, hence, complete removal of these components has to be ensured for better performance.

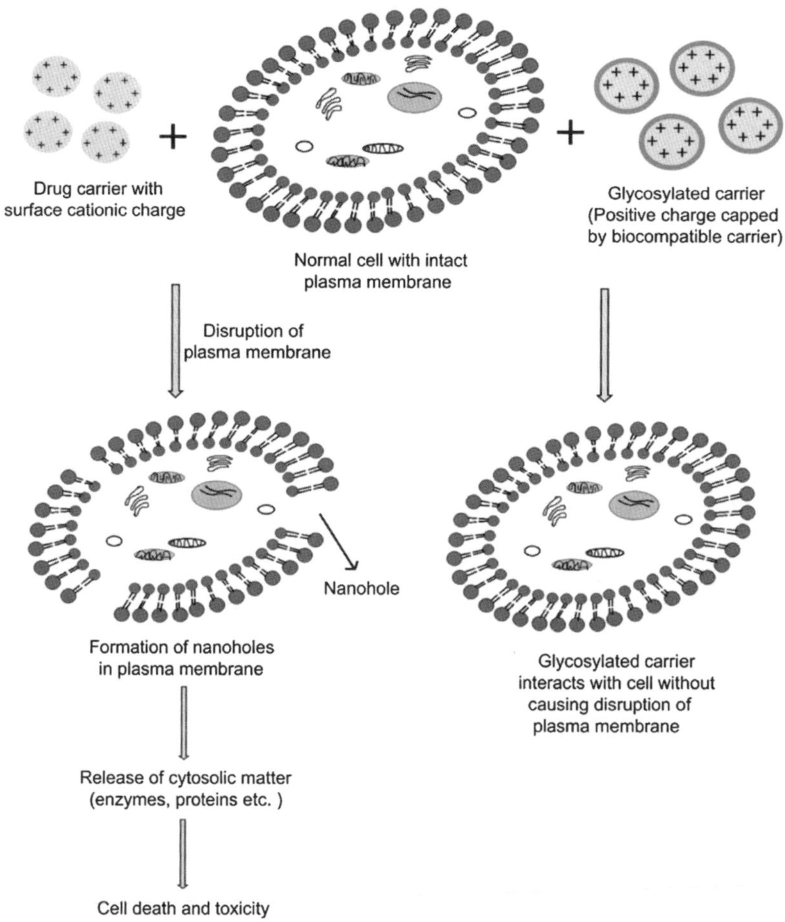

Fig. 6 Glycosylation of nanocarriers can help evade its cellular toxicity [86]

5 Conclusion

In this chapter, we have reviewed different aspects that regulate the use of carbohydrates and proteins for successful drug delivery. Recent research trends in this area have generated a broad spectrum of these biopolymeric nanocarriers offering varying drug incorporation and release profiles, depending on the choice of the matrix and nature of the drug. The ultimate fate of the nanocarriers in terms of their physical and biological stability and *in vivo* circulation are discussed, as well as the design of long-circulating carriers with optimal biological performance through appropriate choice of the material, route of preparation, and surface functionalization. Although certain biopolymers are known to induce blood incompatibilities or immunogenicity, these properties can be manipulated for better performance by varying the processing conditions. In summary, a profound understanding of the properties of the drug and polymer, drug–matrix interactions, and formulational approach would help to improve the biokinetics of active pharmaceutical ingredients through successful nanoencapsulation.

References

1. Elzoghby AO (2012) J Control Release 120:5
2. Kumari A, Yadav SK, Yadav SC (2010) Colloids Surf B Biointerfaces 75:1
3. Menon D, Gopikrishna J, Narayanan D (2012) Hand book of immunological properties of engineered nanomaterials. World Scientific Publishers, Singapore
4. Farokhzad OC, Langer R (2009) Impact of nanotechnology on drug delivery. ACS Nano 3:16
5. LaVan DA, Lynn DM, Langer R (2002) Nat Rev Drug Discov 1:77
6. Torchilin V (2008) Nanotechnology in drugs. Imperial College Press, London
7. Elzoghby AO, Samy WM, Elgindy N (2012) J Control Release 161:38
8. Hans M, Lowman A (2002) Curr Opin Solid State Mater Sci 6:319
9. Sahoo SK, Labhasetwar V (2003) Drug Discov Today 8:1112
10. Suzuki H, Nakai D, Seita T, Sugiyama Y (1996) Adv Drug Deliv Rev 19:335
11. Markovsky E, Baabur-Cohen H, Eldar-Boock A, Omer L, Tiram G, Ferber S, Ofek P, Polyak D, Scomparin A, Satchi-Fainaro R (2012) J Control Release 161:446
12. Miiller R (1991) Colloidal carriers for controlled drug delivery and targeting. Boca Raton, Germany (Wissenschaftliche Verlagsgesellschaft, CRC, Stuttgart)
13. Alexis F, Pridgen E, Molnar LK, Farokhzad OC (2008) Mol Pharm 5:505
14. Domb AJ, Bentolila A, Teomin D (1998) Acta Polymerica 49:526
15. Werz DB, Seeberger PH (2005) Chemistry 11:3194
16. Liu Z, Jiao Y, Wang Y, Zhou C, Zhang Z (2008) Adv Drug Deliv Rev 60:1650
17. Bernkop-Schnürch A, Dünnhaupt S (2012) Eur J Pharm Biopharm 81:463
18. Ghazarian H, Idoni B, Oppenheimer SB (2011) Acta Histochem 113:236
19. Dobrovolskaia MA, Aggarwal P, Hall JB, McNeil SE (2008) Mol Pharm 5:487
20. Akbuğa J, Bergişadi N (2003) J Microencapsul 16:697
21. Agnihotri SA, Mallikarjuna NN, Aminabhavi TM (2004) J Control Release 100:5
22. Maya S, Indulekha S, Sukhithasri V, Smitha KT, Nair SV, Jayakumar R, Biswas R (2012) Int J Biol Macromol 51:392
23. Anitha A, Chennazhi KP, Nair SV, Jayakumar R (2012) J Biomed Nanotechnol 8:29
24. Dev A, Mohan JC, Sreeja V, Tamura H, Patzke GR, Hussain F, Weyeneth S, Nair SV, Jayakumar R (2010) Carbohydr Polym 79:1073
25. Menon D, Thomas RT, Narayanan S, Maya S, Jayakumar R, Hussain F, Lakshmanan V-K, Nair SV (2011) Carbohydr Polym 84:887
26. Thomas H, Tim L, Brigitte H, Stephanie H (2006) In: Klemm D (ed) Polysaccharides II vol 205. Springer, Berlin
27. Kaewprapan K, Inprakhon P, Marie E, Durand A (2012) Carbohydr Polym 88:875
28. Charef S, Papy-Garcia D, Courty J (2010) Biomed Pharmacother 64:627
29. Anitha A, Deepagan VG, Divya Rani VV, Menon D, Nair SV, Jayakumar R (2011) Carbohydr Polym 84:1158
30. Prestwich GD, Marecak DM, Marecek JF, Vercruysse KP, Ziebell MR (1998) J Control Release 53:93
31. Garg HG, Hales CA (2004) Chemistry and biology of hyaluronan. Elsevier Ltd., Oxford
32. Esposito E, Menegatti E, Cortesi R (2005) Int J Pharm 288:35
33. Rivkin I, Cohen K, Koffler J, Melikhov D, Peer D, Margalit R (2010) Biomaterials 31:7106

34. Wöhl-Bruhn S, Bertz A, Harling S, Menzel H, Bunjes H (2012) Eur J Pharm Biopharm 81:573
35. Santander-Ortega MJ, Stauner T, Loretz B, Ortega-Vinuesa JL, Bastos-González D, Wenz G, Schaefer UF, Lehr CM (2010) J Control Release 141:85
36. Thiele C, Auerbach D, Jung G, Qiong L, Schneider M, Wenz G (2011) Polym Chem 2:209
37. Kumari K, Rani U (2011) Adv Appl Sci 2:48
38. Jain AK, Khar RK, Ahmed FJ, Diwan PV (2008) Eur J Pharm Biopharm 69:426
39. Rodrigues A, Emeje M (2012) Carbohydr Polym 87:987
40. Reis Catarina P, Neufeld RJ, Ribeiro António JVF (2005) Chem Ind Chem Eng Quart 12:47
41. Pawar SN, Edgar KJ (2012) Biomaterials 33:3279
42. Jahanshahi M, Babaei Z (2008) Afr J Biotechnol 7:4926
43. Hawkins MJ, Soon-shiong P, Desai N (2008) Adv Drug Deliv Rev 60:876–885
44. Kratz F (2008) J Control Release 132:171
45. Miele E, Spinelli GP, Miele E, Tomao F, Tomao S (2009) Int J Nanomedicine 4:99
46. Vishnu P, Roy V (2010) Womens Health 6:495
47. Cortes J, Saura C (2010) EJC Suppl 8:1
48. Elsadek B, Kratz F (2012) J Control Release 157:4
49. Merodio M, Arnedo A, Renedo MJ, Irache JM (2001) Eur J Pharm Sci 12:251
50. Li F-Q, Su H, Wang J, Liu J-Y, Zhu Q-G, Fei Y-B, Pan Y-H, Hu J-H (2008) Int J Pharm 349:274
51. Zhao D, Zhao X, Zu Y, Li J, Zhang Y, Jiang R, Zhang Z (2010) Int J Nanomedicine 5:669
52. Djagny VB, Wang Z, Xu S (2001) Crit Rev Food Sci Nutr 41:481
53. Zwiorek K, Kloeckner J, Wagner E, Coester C (2005) J Pharm Pharm Sci 7:22
54. Leo E, Angela Vandelli M, Cameroni R, Forni F (1997) Int J Pharm 155:75
55. Fan H, Dash AK (2001) Int J Pharm 213:103
56. Lu Z, Yeh T-K, Tsai M, Au JL-S, Wientjes MG (2004) Clin Cancer Res 10:7677
57. Kumar R, Nagarwal RC, Dhanawat M, Pandit JK (2011) J Biomed Nanotechnol 7:325
58. Jain SK, Gupta Y, Jain A, Saxena AR, Khare P, Jain A (2008) Nanomedicine 4:41
59. Nahar M, Mishra D, Dubey V, Jain NK (2008) Nanomedicine 4:252
60. Saraogi GK, Gupta P, Gupta UD, Jain NK, Agrawal GP (2010) Int J Pharm 385:143
61. Vandervoort J, Ludwig A (2004) Eur J Pharm Biopharm 57:251
62. Kaul G, Amiji M (2005) Pharm Res 2:951
63. Zorzi GK, Párraga JE, Seijo B, Sánchez A (2011) Macromol Biosci 11:905
64. Friess W (1998) Eur J Pharm Biopharm 45:113
65. Lee CH, Singla A, Lee Y (2001) Int J Pharm 221:1
66. Gebhardt BM, Kaufman HE (1995) J Ocul Pharmacol Ther 11:319
67. Gebhardt BM, Varnell ED, Kaufman HE (1995) J Ocul Pharmacol Ther 11:509
68. El-Samaligy MS, Rohdewald P (1983) J Pharm Pharmacol 35:537
69. Sutton R, Yu N, Luck E, Brown D, Conley F (1990) Sel Cancer Ther 6:35
70. Davidson BS, Izzo F, Cromeens DM, Stephens LC, Siddik ZH, Curley SA (1995) J Surg Res 58:618
71. Yu NY, Orenberg EK, Luck EE, Brown DM (1995) Cancer Chemother Pharmacol 36:27
72. Ruszczak Z (2003) Adv Drug Deliv Rev 55:1679
73. Panduranga Rao K (1996) J Biomater Sci Polym Ed 7:623
74. Elzoghby AO, El-Fotoh WSA, Elgindy N (2011) J Control Release 153:206
75. Livney YD (2010) Curr Opin Colloid Interface Sci 15:73
76. Shapira A, Markman G, Assaraf YG, Livney YD (2010) Nanomedicine 6:547
77. Shapira A, Davidson I, Avni N, Assaraf YG, Livney YD (2012) Eur J Pharm Biopharm 80:298
78. Betz M, Kulozik U (2011) Proc Food Sci 1:2047
79. Betz M, Steiner B, Schantz M, Oidtmann J, Mäder K, Richling E, Kulozik U (2012) Food Res Int 47:51
80. Betz M, Kulozik U (2011) Int Dairy J 21:703

81. Betz M, García-González C, Subrahmanyam RP, Smirnova I, Kulozik U (2012) J Supercrit Fluids 72:111
82. Chen L, Pu H, Li X, Yu L (2011) J Control Release 152(Suppl):e51–e52
83. Wee S, Gombotz W (1998) Adv Drug Deliv Rev 31:267–285
84. Morgan SM, Al-Shamkhani A, Callant D, Schacht E, Woodley JF, Duncan R (1995) Int J Pharm 122:121
85. Shukla RK, Tiwari A (2012) Carbohydr Polym 88:399
86. Jain K, Kesharwani P, Gupta U, Jain NK (2012) Biomaterials 33:4166
87. Morris G, Kök S, Harding S, Adams G (2010) Biotechnol Genet Eng Rev 27:257
88. Carreira AS, Gonçalves FAMM, Mendonça PV, Gil MH, Coelho JFJ (2010) Carbohydr Polym 80:618
89. Rejinold NS, Chennazhi KP, Nair SV, Tamura H, Jayakumar R (2011) Carbohydr Polym 83:776–786
90. Fan L, Wu H, Zhang H, Li F, Yang T, Gu C, Yang Q (2008) Carbohydr Polym 73:390
91. Chang PR, Jian R, Yu J, Ma X (2010) Carbohydr Polym 80:420
92. Larsen NE, Balazs EA (1991) Drug delivery systems using hyaluronan and its derivatives. Adv Drug Deliv Rev 7:279
93. Egbaria K, Friedman M (1990) J Control Release 14:79
94. Orienti I, Coppola A, Gianasi E, Zecchi V (1994) J Control Release 31:61–71
95. Marui A, Tabata Y, Kojima S, Yamamoto M, Tambara K, Nishina T, Saji Y, Inui K, Hashida T, Yokoyama S, Onodera R, Ikeda T, Fukushima M, Komeda M (2007) Circ J 71:1181
96. Young S, Wong M, Tabata Y, Mikos AG (2005) J Control Release 109:256
97. Ishikawa H, Nakamura Y, Jo J-I, Tabata Y (2012) Biomaterials 33:9097
98. Maeda M, Kadota K, Kajihara M, Sano A, Fujioka K (2001) J Control Release 77:261
99. Leo E, Cameroni R, Forni F (1999) Int J Pharm 180:23
100. Cascone MG, Lazzeri L, Carmignani C, Zhu Z (2002) J Mater Sci Mater Med 13:523
101. Kommareddy S, Amiji M (2005) Bioconjug Chem 16:1423
102. Wilson B, Samanta MK, Santhi K, Kumar KPS, Ramasamy M, Suresh B (2010) Nanomedicine 6:144
103. Prego C, Torres D, Fernandez-Megia E, Novoa-Carballal R, Quiñoá E, Alonso MJ (2006) J Control Release 111:299
104. Sheng Y, Liu C, Yuan Y, Tao X, Yang F, Shan X, Zhou H, Xu F (2009) Biomaterials 30:2340
105. Kaul G, Amiji M (2002) Pharm Res 19:1061
106. Romberg B, Hennink WE, Storm G (2008) Pharm Res 25:55
107. Singh HD, Wang G, Uludağ H, Unsworth LD (2010) Acta Biomater 6:4277
108. Zhang S, Wang G, Lin X, Chatzinikolaidou M, Jennissen HP, Laub M, Uludağ H (2008) Biotechnol Prog 24:945
109. Dash R, Foston M, Ragauskas AJ (2013) Carbohydr Polym 91:638
110. Elzoghby AO, Samy WM, Elgindy N (2012) J Control Release 157:168
111. Dreis S, Rothweiler F, Michaelis M, Cinatl J, Kreuter J, Langer K (2007) Int J Pharm 341:207
112. Yu S, Yao P, Jiang M, Zhang G (2006) Biopolymers 83:148
113. Li J, Yao P (2009) Langmuir 25:6385
114. Zhang S, Doschak MR, Uludağ H (2009) Biomaterials 30:5143
115. Madadlou A, Mousavi ME, Emam-Djomeh Z, Sheehan D, Ehsani M (2009) Food Chem 116:929
116. Bachar M, Mandelbaum A, Portnaya I, Perlstein H, Even-Chen S, Barenholz Y, Danino D (2012) J Control Release 160:164
117. Zimet P, Rosenberg D, Livney YD (2011) Food Hydrocoll 25:1270–1276
118. Freichels H, Jérôme R, Jérôme C (2011) Carbohydr Polym 86:1093
119. Harush-Frenkel O, Rozentur E, Benita S, Altschuler Y (2008) Biomacromolecules 9:435
120. Pain D, Das PK, Ghosh P, Bachhawat BK (1984) J Biosci 11:1117
121. Moghimi S, Bonnemain B (1999) Adv Drug Deliv Rev 37:295
122. Kang E-C, Aklyoshi K, Sunamoto J (1997) J Bioact Compat Polym 12:14

123. Jayasree A, Sasidharan S, Koyakutty M, Nair S, Menon D (2011) Carbohydr Polym 85:37
124. Oda Y, Yanagisawa H, Maruyama M, Hattori K, Yamanoi T (2008) Bioorg Med Chem 16:8830
125. Sett R, Sarkar HS, Das PK (1993) J Antimicrob Chemother 31:151
126. Owens DE, Peppas NA (2006) Int J Pharm 307:93
127. Kommareddy S, Amiji M (2007) J Pharm Sci 96:397
128. Kaul G, Amiji M (2004) J Drug Target 12:585
129. Chen D, Tang Q, Xue W, Xiang J, Zhang L, Wang X (2010) TJ Biomed Res 24:26
130. Kouchakzadeh H, Shojaosadati SA, Maghsoudi A, Vasheghani Farahani E (2010) AAPS PharmSciTech 11:1206
131. Lin W, Garnett MC, Davis SS, Schacht E, Ferruti P, Illum L (2001) J Control Release 71:117
132. Shi A, Li D, Wang L, Li B, Adhikari B (2011) Carbohydr Polym 83:1604
133. Fuchs S, Kutscher M, Hertel T, Winter G, Pietzsch M, Coester C (2010) J Microencapsul 27:747
134. Gan Q, Wang T (2007) Colloids Surf B Biointerfaces 59:24
135. Zhang L, Liu Y, Wu Z, Chen H (2009) Drug Dev Ind Pharm 35:369
136. Lee EJ, Khan SA, Park JK, Lim K (2012) Bioprocess Biosyst Eng 35:297
137. Jafari SM, Assadpoor E, He Y, Bhandari B (2008) Drying Technol 26:816
138. Hogan SA, McNamee BF, O'Riordan ED, O'Sullivan M (2001) Int Dairy J 11:137
139. Sheu T-Y, Rosenberg M (1998) J Food Sci 63:491
140. Grenha A, Remuñán-López C, Carvalho ELS, Seijo B (2008) Eur J Pharm Biopharm 69:83
141. Vega-Villa KR, Takemoto JK, Yáñez JA, Remsberg CM, Forrest ML, Davies NM (2008) Adv Drug Deliv Rev 60:929
142. Linkov I, Satterstrom FK, Corey LM (2008) Nanomedicine 4:167
143. Aillon KL, Xie Y, El-Gendy N, Berkland CJ, Forrest ML (2009) Adv Drug Deliv Rev 61:457
144. Nel AE, Mädler L, Velegol D, Xia T, Hoek EMV, Somasundaran P, Klaessig F, Castranova V, Thompson M (2009) Nat Mater 8:543
145. Garnett MC, Kallinteri P (2006) Occup Med 56:307
146. Elgrabli D, Abella-Gallart S, Aguerre-Chariol O, Robidel F, Rogerieux F, Boczkowski J, Lacroix G (2007) Nanotoxicology 1:266
147. Green MR, Manikhas GM, Orlov S, Afanasyev B, Makhson AM, Bhar P, Hawkins MJ (2006) Ann Oncol 17:1263
148. Mayer A, Vadon M, Rinner B, Novak A, Wintersteiger R, Fröhlich E (2009) Toxicology 258:139
149. Chou T-C, Fu E, Wu C-J, Yeh J-H (2003) Biochem Biophys Res Commun 302:480
150. Lee D, Powers K, Baney R (2004) Carbohydr Polym 58:371
151. Rao SB, Sharma CP (1997) J Biomed Mater Res 34:21
152. Landesman-Milo D, Goldsmith M, Leviatan Ben-Arye S, Witenberg B, Brown E, Leibovitch S, Azriel S, Tabak S, Morad V, Peer D (2012) Cancer Lett 76:55
153. Sagnella S, Mai-Ngam K (2005) Colloids Surf B Biointerfaces 42:147
154. Benesch J, Tengvall P (2002) Biomaterials 23:2561
155. Mao C, Qiu Y, Sang H, Mei H, Zhu A, Shen J, Lin S (2004) Adv Colloid Interface Sci 110:5

Adv Polym Sci (2013) 254: 269–296
DOI: 10.1007/12_2012_202

Published online: 16 March 2013

Biopolymeric Micro- and Nanoparticles: Preparation, Characterization and Industrial Applications

Anil Kumar Anal and Alisha Tuladhar

Abstract One of the most important industrial applications of the interfacial properties of biopolymeric complexes such as protein–polysaccharide interactions is the formation of microcapsules, microparticles, nanocapsules and/or nanoparticles. Nanogel formation results from the ability of protein–polysaccharide complexes to form a solid film around the droplets containing the product to be encapsulated (microcapsules) and also the possibility of entrapping solvent molecules into the coacervate (microgels). The first, based on the complex coacervation of two oppositely charged biopolymers, consists of the entrapment of a solvent containing the molecule to be encapsulated in the structure of the coacervate. In this case, the microencapsulated product can be either in the core of the complex or in the wall, resulting in some retention problems. The second method of microencapsulation, and the most widespread, is to produce an emulsion stabilized with proteins. The dispersed oil droplets contain the desirable oil-soluble products to be encapsulated (aroma, drugs, cells, etc.). After addition of a polysaccharide, the interfacial coacervation between the protein and the polysaccharide around the oil droplets induces the formation of microcapsules, which are subsequently dried. This type of microencapsulation is referred to as interfacial coacervation. Its main advantage resides in the location of the encapsulated material in the core of the microcapsule.

Keywords Biopolymer · Characterization · Industrial applications · Micro-/nanoparticles · Preparation

A.K. Anal (✉) and A. Tuladhar
Food Engineering and Bioprocess Technology, Asian Institute of Technology, KlongLuang, Pathumthani 12120, Thailand
e-mail: anilkumar@ait.ac.th; anil.anal@gmail.com

Contents

1 Introduction

Biomacromolecules may be either polycations or polyanions, depending on their functional groups. Among various macromolecules, proteins and polysaccharides are often used simultaneously in the food, pharmaceutical and cosmetic industries; knowledge of their interactions is therefore of great importance in controlling the structure and texture of manufactured products. The interactions between proteins and polysaccharides can be either repulsive or attractive. Attractive interactions induce interpolymer complexes, which mostly arise from electrostatic interactions between opposite charges on the polymers.

Polysaccharides are used to modify the flow characteristics of fluids, to stabilize suspensions, to flocculate particles, to produce emulsions, and to encapsulate materials. The majority of polysaccharide additives used by the food industry are extracted from higher plant tissue and seaweeds. Plants synthesize such polymers to serve as structural components of cell walls or as a source of stored energy. However, these polymers have potential problems because of their variations in cost, limited supply and chemical structure.

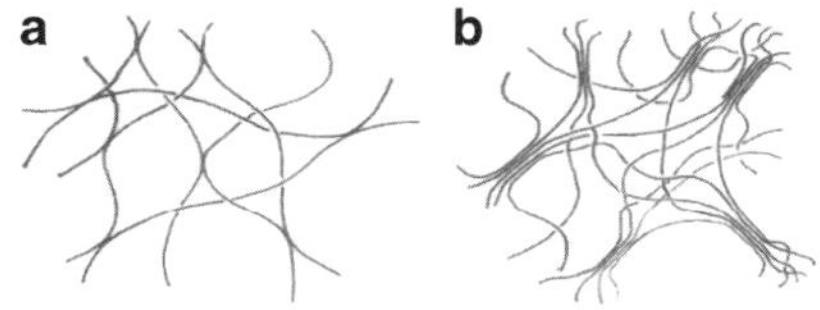

Fig. 1 Gel network: (**a**) molecular network with "point contact"; (**b**) molecular network with "crystalline" junction zones

At present, the discovery of new polysaccharides relies on screening of the extracellular polysaccharides produced by microorganisms. Fungi and yeasts are also potential sources of new polysaccharides. The production of microbial polysaccharide has the advantages of controlled cost, abundant supply and ease of modification of the chemical structure. These new polysaccharides with new properties may generate new market opportunities. Microbial polysaccharides can be classified as extracellular structural or intercellular storage forms. Extracellular polysaccharide can be either exocellular capsules of the cell wall or loose slime components that accumulate outside the cell wall and then diffuse into the medium.

Polysaccharides and polysaccharide mixtures have been used as gelling agents by the food, cosmetic and pharmaceutical industries. A gel is considered to be a coherent colloidal disperse system of at least two components that behaves mechanically as a solid in which the dispersed components form networks interpenetrating and enclosing the solvent phase. Gelation is pictured as the separation of a solid phase (amorphous or crystalline) in such a fashion as to produce a coherent continuous three-dimensional network. An important factor in the demixing of a macromolecular solution is crystallization. Contact points between chains have been taken to be crystallites, although the size of these junction zones may vary from small regions containing few chains to large crystalline regions involving many chains (as shown in Fig. 1) and recognizable as a separate phase. Factors important in gelation are polymer–solvent and polymer–polymer interactions and the effects of preparative conditions on the extent and mechanism of phase separation.

Polysaccharides are found in the ingredients of a wide range of food, cosmetic and pharmaceutical emulsions. The relevant functionalities originate from several molecular properties of the polysaccharides and their interaction with emulsion droplets and other components in the complex food, pharmaceutical and cosmetic systems. Their major functionalities lie in their ability to thicken the emulsion, which is intended to reduce the creaming rate and to improve the texture to the emulsion. Whereas proteins are present primarily as emulsion forming and stabilizing agents, soluble polysaccharides primarily function as thickening and water-holding agents. A wide range of properties are found among the whole group of polysaccharides, varying from insoluble forms (cellulose) to those with high swelling power and solubility (starch, guar gum), low viscosity (gum arabic) to high viscosity (guar gum), and non-gelling (dextran) to gelling (agar). Gel formation is often thermo-reversible and the gel may either melt on heating (alginate, pectin) or set on heating (some cellulose derivatives).

Polysaccharides play a role in emulsions in three main ways:

1. Polysaccharides increase the viscosity of the continuous phase of the emulsion. One of the main functions of polysaccharides in emulsions is to thicken the continuous liquid. The intended effect is usually to impart a desired texture (increase viscosity or stiffness to the system and reduce buoyancy-driven creaming or sedimentation of the emulsion droplets and other particles in the system). Because of their highly swollen molecular structure in solutions, leading to a high effective volume fraction at low concentrations, most polysaccharides are very effective in providing a high viscosity at low concentration.
2. Polysaccharides flocculate the droplets into a gelled droplet network. In a solution of gelling polysaccharide (e.g. alginates, pectin, K-carrageenan), the formation of bonds between segments of the different polysaccharide molecules in solution can lead to formation of a space-filling network of biopolymer molecules. The binding regions in the gelled polysaccharides are often highly structured regular chain conformations, such as double helices, bundles of double helices, egg-box, and double helix-ribbon configurations. A prerequisite for gel formation is that the regular molecular segments are interrupted by nonbonding sequences to form an open, water-holding structure. Such a network can resist a finite applied stress, below which there is no discernable flow. On increasing the applied stress, the structure is broken down, either at a molecular level (related to an apparent yield stress) or at a macroscopic level (related to fracturing of the gel).
3. Some polysaccharides, such as starch granule and inulin, are in particulate form and increase the viscosity of the emulsion by increasing the total volume fraction of dispersed material.

2 Polysaccharides Used in Preparation of Micro- and Nanoparticles

2.1 Chitosan

Chitosan, a biopolymer, has received ample attention and has been widely considered for micro- and nanoparticle preparation. It is the second-most abundant naturally occurring polysaccharide. Properties like biodegradability, low toxicity and good biocompatibility make it appropriate for biomedical and pharmacological inventions and also for use in ophthalmology, anti-diabetic agents, anti-inflammatory drugs and for immobilization of enzymes and proteins.

Generally, chitosan is formed of distributed β-(1–4)-linked D-glucosamine (deacetylated unit) and *N*-acetyl-D-glucosamine (acetylated unit), manufactured by deacetylation of chitin extracted from shells of crabs, shrimps and krill. Chitosan that is accessible commercially is deacetylated 66–95% and has an average molecular weight between 3.8 and 2,000 kDa. It is an admirable biopolymer for research and testing of micro- and nanoparticles due to its excellent biocompatibility and biodegradability. In vivo, it is degraded by lysozyme. In addition, the amino groups give a high charge density to the molecule and are freely available for chemical reactions and salt formation with acids. With wide solubility of chitosan in numerous acids, it also intermingles with polyions to formulate complexes and gels.

2.1.1 Structure of Chitin and Chitosan

In shellfish, including crustaceans, chitin is present in a complex structure with calcium carbonate, forming the rigid skeleton of carpace, shell and tail. Shrimp outer skeleton and crab shells are the most suitable biological sources for the production of chitin [1, 2]. Chitosan is the principal derivative of chitin and is obtained by a deacetylation process. Chitin and its derivatives have widespread applications for biological, physiological and biomedical purposes. Crustacean shell waste mainly consists of protein (30–40% of dry matter), calcium salts (30–50% of dry matter), chitin (20–30% of dry matter), oil (2–3% of dry matter) and astaxanthin (200–250 ppm of dry matter), but the proportions may vary with the species and culture methods [3]. Chitosan can be isolated from crustacean shell, insect cuticles and the membrane of fungi. The properties of chitosan vary with its source. The terms chitin and chitosan do not refer to specific compounds but to two ranges of copolymers, containing the two monomer residues, anhydro-*N*-acetyl-D-glucosamine and anhydro-D-glucosamine, respectively. Chitin is a polymer of β-(1–4)-2-acetamido-2-deoxy-D-glucopyranose. It is one of the most abundant organic materials on earth and second to cellulose and murein, which are the main structural polymers of bacterial cell walls. Chitin is analogous to cellulose and exhibits a highly ordered chain formation. Chitin has the conformation of an extended ribbon and the chains can pack side by side in a crystalline structure. In nature, chitin exists in alpha (sheets are antiparallel and packed), beta (sheets are parallel and packed) and gamma (two parallel sheets are separated by a single antiparallel sheet) conformations (Fig. 2). Shrimp and crab contain α-chitin, which is more difficult to extract as compared with β-chitin in squid and cuttlefish. Chitosan from β-chitin is more hydrophilic and has a more open form. Crustacean chitosan in situ has a molecular weight of over 2 MDa. Fungal chitosan has a molecular weight in the range of 30–60 kDa. The fungal material has the advantage of not containing shrimp protein, which may cause an allergic reaction in 2% of the population. The principal derivative of chitin is chitosan, produced by alkaline deacetylation of chitin. Chitosan is a polymer of β-(1–4)-2-amino-2-deoxy-D-glucopyranose [4].

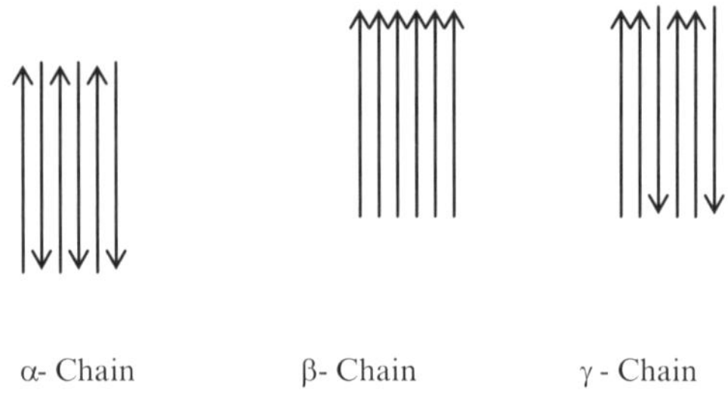

Fig. 2 Arrangement of polymer chains in three forms of chitin

2.1.2 Functional Properties of Chitosan

The three forms of chitin (i.e. non-crystalline, hydrated crystalline and anhydrous crystalline) have been proposed by Ogawa [5]. These types of structure can be easily identified by X-ray crystallography. The type and extent of crystalline fraction are important factors in determining the properties of a chitosan sample and depend both on the treatment of the sample and on the mole fraction of *N*-acetyl-D-glucosamine residues. Chitosan is prepared from chitin by deacetylation with 50% (w/v) sodium hydroxide solution. Because chitosan or partially deacetylated chitin have primary amine groups, they can be used as a biodegradable cationic polymer in aqueous systems [6] The solubility, viscosity and other properties of chitosan depend not only on the degree of deacetylation, but also on the amine groups in a chitosan chain. The solubility of a polymer occurs in two stages. First, solvent molecules slowly diffuse into the polymer to produce a swollen gel, i.e., the polymer–polymer intermolecular forces are high because of crosslinking, crystallinity or strong hydrogen bonding. Chitosan is a hydrophilic polysaccharide, soluble in dilute aqueous organic acid solutions but insoluble in water. The hydrogen bonding leads to a poor solubility in water. During solubilization, the 2-amino group is protonated, thus providing the polycationic nature of chitosan molecules in solutions. Chitosan is not soluble in sulfuric acid solution due to the formation of sulfate salt, which precipitates. Studies on the solution properties of chitosan indicated that the conformation of chitosan is variable in solution, but often has a compact, quasi-globular or flexible chain. From the analysis of physical properties, the structure–property–function relationship can be established and this can provide a scientific basis for the applications of chitin and chitosan. At neutral and alkaline pH, chitosan in free form does not dissolve. At pH 5 and lower, the amine groups in the chitosan become ionized, leading to the perfect dissolution of chitosan in dilute acids. Chitosan solutions of 1–2% are fluid, but at higher concentrations they form a firm gel. Chitin and chitosan are polyglucosides and thus can be hydrolyzed by acids. However, chitosan is more susceptible than chitin in this regard. This difference may be attributed to protonation of the amine groups in acid solution. This may be due to the resultant NH^{3+} groups shielding the glycosidic oxygen from protonation, which is the first stage in hydrolysis of glycosides.

Chemical derivatization of chitosan provides a powerful means to promote new biological activities and to modify its mechanical properties. The primary amino groups on the molecule are reactive and provide a mechanism for side group attachment using a variety of mild reaction conditions. The general effect of addition of a side chain is to disrupt the crystal structure of the material and hence increase the amorphous fraction. This modification generates a material with lower stiffness and often altered solubility, but the precise nature of changes in chemical and biological properties depends on the nature of the side groups. In addition, the characteristic features of chitosan such as being cationic, hemostatic and insoluble at high pH, can be completely reversed by a sulfation process that can render the molecule anionic and water-soluble, and also introduce anticoagulant properties [7].

2.1.3 Biochemical Significance and Degradation of Chitin and Chitosan

Because both chitin and chitosan are biological materials and have excellent biocompatibility, they have been developed as medical materials in recent years. Some enzymes such as lysozyme and chitinase degrade chitin. Plants and insects have chitinase but not mammals except cow, sheep and goat. Chitin and chitosan are thought to be degraded mainly by lysozyme in mammals. Lysozyme in mammalians is present in serum, saliva and other secretary fluids, including those surrounding the cartilage. There are many kinds of glycosidase in the animal body to hydrolyze chitin and its derivatives, including lysozyme. The 6-*O*-acyl, 6-*O*-alkyl and carboxymethyl chitin derivatives have similar enzymatic hydrolysis. The rate of lysozymic hydrolysis of various chitosans are quite low compared with those of other chitin derivatives, probably because of the poor binding ability of chitosan to the active site of lysozyme.

2.1.4 Applications of Chitosan in Biomedical Technology

Biomedical applications relate to the impact of a material, device or procedure in a medical or clinical situation on the health care of humans. The expected outcome should be positive when properly utilized. The application can be a medical device as simple as a syringe or a piece of gauze for cleaning wounds, or as complex as pacemakers, orthopaedic implants and artificial heart valves. For the purpose of this review, we take a focused view by considering the biomedical applications that have been proposed for chitin and chitosan.

The name chitosan covers a large collection of preparations that all have nearly the same natural composition but are different in their characteristics and applicability due to variations in source, charge, charge density, molecular weight and crystallinity. For a specific application, this complicates the search for the most useful form of chitosan, but it presents a large range of chitosan candidates that might serve the specific purpose. Chitosans in diverse physical and chemical forms have been introduced for a variety of applications in the sector of medical technology. However, before medical applications of chitosan can be considered, one has to consider first its safety for application on or in living systems.

Talking about medical application, the first question to address is "is chitosan safe for internal application?" Recent data have confirmed the safety of chitosan. Chitosan has very low toxicity; the LD_{50} of chitosan in laboratory mice is 16 g/kg body weight, which is close to that of sugar or salt. Chitosan is proven to be safe in rats at up to 10% in the diet [8]. There was some concern after some Japanese experiments showed that high doses of chitosan intravenously gave rise to haemorrhage pneumonia, caused by complement activation. However, this should be considered in the light of the law of the medieval scientist Paracelsus, who stated that high concentration makes every compound a toxin. Recent data both in medium-term feeding studies in laboratory animals and humans have confirmed that body and organ weight remained normal, as well as neuro-, cardio-, respiratory, hepato-, digestive, urinary and circulatory systems. There can be some discomfort, such as belching. The only effect is a significantly lower plasma cholesterol level. Various sterilization methods such as ionizing radiation, heat, steam and chemical methods can be adopted for sterilization of chitosan in clinical applications [9].

2.2 *Alginate*

Alginate is a naturally arising, linear unbranched polysaccharide that is soluble in water and extracted from brown seaweed. Its use in numerous biomedical applications, including drug delivery systems, is basically due to its high biodegradable, biocompatible and mucoadhesive polymers. Alginate polymers are also haemocompatible, have not been found to accumulate in any important organs and show indication of in vivo degradation. Sodium alginate is used in a variety of oral and topical pharmaceutical formulations and has been used for the aqueous microencapsulation of drugs, in contrast to more conventional solvent-based systems. Alginate possesses two types of uronic acids, L-guluronic acid and D-mannuronic acid.

Micro- and nanoparticles of alginate are effortlessly achieved by gelation with calcium ions. This property enables creation of a pre-gel consisting of miniature aggregates of gel particles, which can be tracked by the addition of an aqueous polycationic solution to make a polyelectrolyte complex coating. Poly-L-lysine (PLL) is a cationic natural polymer that has been used with alginates to formulate nanoparticles.

The mechanical properties and swelling of alginate beads shaped using cations hinge upon a number of aspects like valence of ions, size of ions, etc. Bajpani et al. [10] describe a comprehensive examination of swelling and degradation performance of alginate beads crosslinked with different metal ions (e.g. Ca^{2+}, Ba^{2+} and Al^{3+}). The results can be explained on the basis of the extent of crosslinking in the beads and the size of the cations involved in the crosslinking process.

Sodium alginate is a sodium salt of alginic acid, a naturally occurring polysaccharide acquired from marine brown algae. Alginate contains two uronic acids, α-L-guluronic acid and β-D-mannuronic acid, and is composed of homopolymeric blocks and blocks with an alternating sequence. Made from a chelated structure with metal ions, the polyguluronate units in the alginate molecules (known as an "egg-box" junction) have apertures that allow the cations to become packed or coordinated [11]. The intersection flanked by the polyguluronate shackles is kinetically stable in the direction of dissociation, whereas the polymannuronate components demonstrate the usual polyelectrolyte characteristics of cation binding [12]. Thus, the two types of connections result in establishment of spherical beads.

The composition of alginate is an important parameter for the formation of alginate particles. The variation in monomer content and viscosity of alginates in brown seaweed with season generally increases from the lowest at the youngest stage of the plant to the maximum at a mature stage. Because sodium and calcium proportions strongly determine the swelling and healing properties of the alginate, this proves to be a significant property [13]. Viscosity classically fluctuates with the percentage of the guluronic content. During alginate gelling, divalent cations bind preferentially to guluronic acid blocks in a highly cooperative manner; the size of the cooperative unit is reported to be more than 20 monomers [14]. A high content of guluronic acid and homopolymer blocks leads to higher interaction between alginate and calcium, which results in a stronger and more stable gel. However, in the emulsification step high guluronic content gives premature gelation, resulting in larger beads with larger dispersions [15] and more porous gels. On the other hand, high mannuronic acid content produces more elastic and weaker gels with good freeze–thaw behaviour. However, at low or very high Ca^{2+} concentrations, high mannuronic alginates produce weaker gels.

3 Proteins Used in Preparation of Micro- and Nanoparticles

3.1 *Albumin*

Albumin is a protein present in blood plasma and is an attractive macromolecular carrier, widely used to prepare nanospheres and nanocapsules due to its availability in pure form and its biodegradability, non-toxicity and non-immmunogenicity. It also acts as an significant extracellular antioxidant [16] and defence mechanism against free radicals and extra-destructive chemical agents [17]. Albumin is easy to formulate in defined sizes, and reactive groups (thiol, amino and carboxylic groups) can be attached to the surface for ligand binding and other surface modifications. Also, albumin nanoparticles have the advantage that ligands can be easily attached by covalent linkage. Drugs captured within albumin nanoparticles can be digested by proteases, and drug loading can be quantified. A number of studies have revealed that albumin gathers in compacted tumours, which makes it a prospective macromolecular transporter for the site-directed delivery of anti-tumour drugs [18]. From the literature, the use of modified serum albumin seems promising as a selective agent for tumour detection or therapy [19] or for delivery of toxic compounds for elimination of *Mycobacterium tuberculosis* via receptor-mediated drug delivery [20]. These exceptional characteristics of albumin have shaped a foremost place for it in drug therapy. Therefore, nanotechnology has correspondingly engaged the well-established assets of albumin (both human serum albumin and bovine serum albumin) for numerous determinations as nanoparticle drug (antibodies, interferon gamma, antiviral compounds) carrier [21–23], therapeutic enhancer of anticancer drugs [24] and as a modified carrier for the provision of drug from the brain to the central nervous system and also across the blood–brain barrier [25, 26].

3.2 Collagen

Collagen is a naturally occurring protein found in animals, especially in the flesh and connective tissues of vertebrates. It is a profusely found mammalian protein making up about 20–30% of total body proteins. Collagen has a distinctive arrangement, size and amino acid sequence that result in the construction of triple helix fibre. It is a beneficial biomaterial due to its tremendous compatibility, biodegradation and availability [27]. Additionally, its responsiveness to alterations have cemented its use in nanoparticle fabrication. Modifications include addition of other proteins, such as elastin, fibronectin and glycosaminoglycans, which results in the improvement of its physicochemical and biological properties [28, 29] as well as control of biodegradability and subsequent release of ligand by use of such crosslinking agents as glutaraldehyde, formaldehyde, ultraviolet and gamma radiation. The biodegradable collagen-based nanoparticles are thermally stable, readily sterilizable, can be uptaken by the reticulo-endothelial system and enable enhanced uptake of drug molecules into cells [30].

3.3 Gelatin

Gelatin is a natural water-soluble macromolecule that results from the heat dissolution and incomplete hydrolysis of collagen. It is a polyampholyte, having both ionic groups. Basically, two types of gelatin are found. Type-A gelatin is achieved by acid treatment of collagen, with isoelectric point (p*I*) being between 7.0 and 9.0. In contrast, type-B gelatin is formed through alkaline hydrolysis of collagen, with p*I* between 4.8 and 5.0. Gelatin has various advantages over other synthetic polymers, i.e., non-irritability, biocompatibility, biodegradability, non-toxicity and non-carcinogenicity, and therefore proves to be one of the most appropriate materials to be used as a carrier molecule [31, 32]. Gelatin has a large number of functional groups on its surface that aid in chemical crosslinking and derivation. These benefits have made it one of the best options for the synthesis of nanoparticles for drug delivery for the last 30 years [33–36].

4 Fabrication Methodologies

Microencapsulation is defined as a technology for packaging solids, liquids or gaseous materials into miniature, sealed capsules that can release their contents at controlled rates under the influences of specific conditions. A microcapsule consists of a semi-permeable, spherical, thin and strong membrane surrounding a solid or liquid core, with a diameter varying from a few microns to 1 mm. In a broad sense, encapsulation can be used for many applications in the food industry, including stabilizing the core material, controlling the oxidative reaction, providing sustained or controlled release (both temporal and time-controlled release), masking flavours, colours or odours, extending the shelf life and protecting components against nutritional loss. Various biopolymers such as alginate, chitosan, CMC, carrageenan, gelatin and pectin are mainly applied, using various microencapsulation technologies. Microcapsules and microspheres can be engineered to gradually release active ingredients. A microcapsule can be opened by many different means, including fracture by heat, solvation, diffusion and pressure. A coating may also be designed to open in specific areas of the body. A microcapsule containing acid-labile core materials that is consumed by gastrointestinal fluids must not be fractured until after it passes through the stomach. A coating can therefore be used that is able to withstand acidic conditions in the stomach and allows those active ingredients to pass through the stomach. Encapsulation technology is widely used in foods and pharmaceutical manufacturing to control the release of active ingredients, protect ingredients from the environment, lower flavour loss during the product shelf-life, extend the flavour perception and mouth-feel over a longer period of time, and enhance the ingredient bioavailability and efficacy.

4.1 Spray Drying

Spray drying is the most commonly used microencapsulation technique in the food industry [37–39]. The process is relatively simple and easily converts liquids to powders and protects volatile compounds against degradation and oxidation [40]. Consequently, spray drying has been used to encapsulate a variety of substances, for example, flavours, vitamins, fish oils and flavours [41].

The principle steps of the process involve preparation and homogenization of the dispersion, atomization of the dispersion and subsequent dehydration of the atomized particles (as depicted in Fig. 3). The dispersion is prepared by mixing the core material into a liquid coating material. This mixture is homogenized and sprayed into a hot chamber, whereby water is evaporated off the surface of the atomized particles to produce capsules [42, 43]. A cyclone separates capsules, which are subsequently collected .

The stability of capsules prepared using spray drying is a consequence of the chemical and physical properties of the coating and core materials, atomization and

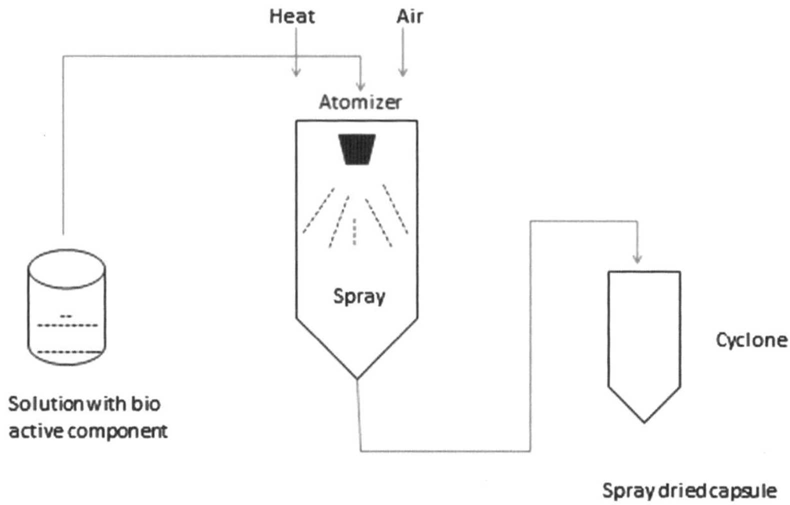

Fig. 3 Schematic presentation of spray drying procedure

drying conditions such as solid content of the dryer and processing temperature. The functionality of the wall material is particularly important and the most effective materials will have high solubility in water, low viscosity at high concentrations and effective emulsifying and film-forming properties [44, 45]. If solubility in water is limited, capsules will adapt a matrix-type structure, whereby small particles of the core material are imbedded throughout a coating material matrix. Even under high viscosity interface during the atomization process, bulky and elongated droplets are formed, which affects adversely at the drying rate [46].

Spray drying is widely used in the food industry due the many advantages offered by the method [47]. The chief advantages are the relatively low process cost, flexibility in allowing various coating materials to be used and adaptability to commonly used equipment. In terms of large-scale production, spray drying can be operated in continuous mode, allowing effortless incorporation into many industrial processes. Furthermore, the method is suitable for heat-labile substances because core materials are exposed to lower temperatures [47].

Spray drying, however, is limited to liquid feeds and therefore coating materials must be water-soluble and have low viscosity at high concentrations. Typical shell materials include gum acacia, maltodextrin and hydrophobic modified starches. Whey proteins and alginate have also been used; however, their solubility in water is much reduced, requiring a greater volume of water to be evaporated. For example, mesquite gum has been shown to impart better stability to oil/water emulsions and provide higher encapsulation efficiencies than gum acacia. Soybean soluble polysaccharide proved to be an advanced emulsifier over gum Arabic in retaining microencapsulated ethyl butyrate during spray drying [48].

4.2 Ionotropic Gelation

Ionotropic gelation is a simple microencapsulation method and can be used to enhance the shelf life and controlled release of the bioactive compounds [49, 50]. Typically, ionotropic gelation is performed with alginate and makes use of the gelation of this hydrocolloid in the presence of divalent metal ions. A major step in ionotropic gelation is the mixing of the wall material and core material together. For example, alginate solution is mixed with oil in water emulsion for about one hour and then the mixture is added drop-wise into a solution containing calcium ions, whereby gelation occurs through a reaction between alginate and the ions in solution. The hardened capsules are subsequently collected and dried. The mechanism of gelation includes the addition of ions, such as calcium or sodium chloride, which shield electrostatic charges on the surface of proteins and allow molecules to move close together. As a result, molecules can aggregate due to the absence of electrostatic repulsion and gelation can occur.

The major advantages of ionotropic gelation are that capsules are easily prepared in aqueous solutions and at room temperature, enabling the method to be used for heat-labile materials. The drawback of this method includes very low payload, unacceptable cost-in-use, large capsules size, limited range of wall materials and undesirable amounts of substantial polymers incorporated into the foods. In order to improve this technique, hydrophobically modified starch, epimerized alginate and polyelectrolyte-based multilayer microcapsules can be used to increase the payload.

4.3 Coacervation

Coacervation is the process by which colloidal particles are separated from solution and deposited around a core material. Coacervation is typically used for the encapsulation of flavours, lipid-soluble vitamins and phytochemicals and certain enzymes. Simple coacervation involves the use of only one hydrocolloid, whereas complex coacervation utilizes two or more polymers. The coacervation process consists of three principle steps: phase separation, deposition and solidification. In the initial step, the coating material, usually consisting of one or more polymers, undergoes a phase separation to form a coacervate. Core materials are suspended or emulsified in the same reaction media and, as particles coalesce, a decrease in surface area occurs. Decreased surface area of the coating particles prompts a decrease in total free interfacial energy of the system, which in turn favours coacervate nuclei adsorption to the core material surface. As a result, a uniform layer of coacervate forms around core particles. In the final step, solidification of the coating material is achieved by crosslinking using chemical, thermal or enzymatic methods. Microcapsules are then collected by filtration or centrifugation before being dried. Spraying a chitosan solution into sodium hydroxide, NaOH–methanol or ethanediamine alkaline solutions using compressed air results in coacervated droplets, forming the nanoparticles [51]. Separation and purification of the particles

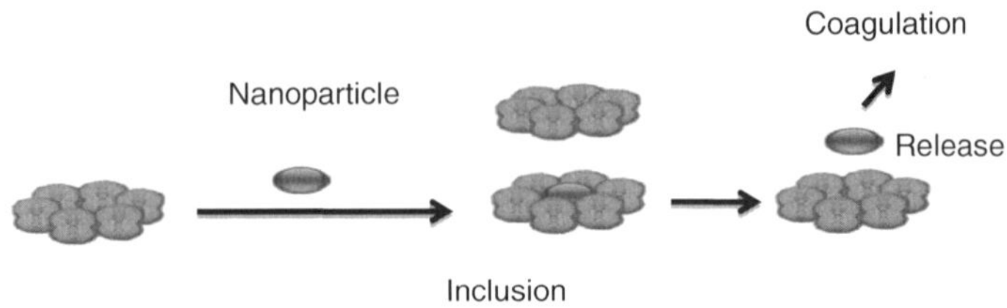

Fig. 4 General overview of molecular inclusion process

is finally achieved by centrifugation, followed by successive washing steps with hot and cold water. In one study, Lin et al. describe a different approach: the formation of particles is obtained after addition of sodium sulfate solution to chitosan solution under mild agitation and continuous sonication for 30 min.

The coacervation technique is analogous to the method of desolvation because it works by mingling of the aqueous protein solution with an organic solvent like acetone or ethanol, which produces minute coacervates. [52]. Numerous parameters such as initial protein concentration, temperature, pH, crosslinker concentration, agitation speed, molar ratio of protein to organic solvent and organic solvent adding rate affect the balance between coacervation and desolvation, which affects the process of providing anticipated assets to the nanoparticles.

4.4 Inclusion Complexation

Inclusion complexation is the entrapment of smaller molecules inside the hollow cavity of a larger molecule [47, 53, 54]. Relatively few food-appropriate molecules exist with such cavities; however, cyclodextrins are commonly used in the food industry.

Cyclodextrins are typically used as coating molecules and are formed by the enzymatic hydrolysis of starch molecules [55]. Glycosyl transferase cleaves starch molecules to form linear fragments, which are then joined to form circular structures with hollow cavities inside. Cyclodextrins can form inclusion complexes with compounds that are small enough to fit inside the cavity. Although the internal diameter varies between cyclodextrins, the typical diameter is 5–8 Å, which is large enough to hold 6–17 molecules of water. Small molecules may displace the water to form thermodynamically stable complexes. The inner cavity is typically hydrophobic, and core molecules are stabilized inside the cavity by van der Waals forces, hydrogen bonding or entropy-driven hydrophobic interactions. The general overview of steps given in Fig. 4 explains the overall procedure.

4.5 Electrospray Drying

The technique of electrospraying creates moderately monodisperse and physically vigorous protein particles. This process encompasses preparation of a protein solution, which is carried out by mixing the dry powder into an electrosprayable solution.

An electric field is used to create acceptable droplets of charged liquid coming out from a thin capillary nozzle. The accumulation of charge on the surface of the liquid is instigated by the electric field between the liquid and counter-electrode. Solvent evaporation leaves a dry residue, which is followed by dispersal of the solution collected on suitable removal substrates. Insulin nanoparticles of size between 88 and 110 nm were prepared in this manner [56]. However, the biological activity of the electrosprayed protein-based nanoparticles is not affected by the process conditions [56]. When the potential difference is sufficient, the electrostatic forces overcome the surface tension, the surface becomes unstable and the meniscus develops a conical shape [57]. If the cone is then disturbed by a further charge, it will break into small charged droplets that will detach from the liquid cone and fly towards a counter-electrode. Several different spraying modes exist [58, 59], but for the production of monodisperse nanosized droplets the cone-jet mode (known as a Taylor cone) is the most desired. Abundant studies have shown that electrospray generated by electrostatic atomization remains functional [60, 61], including an amalgamation of polymeric drug micro- and nanoparticles [62].

5 Characterization of Biopolymeric Micro- and Nanoparticles

5.1 Particle Size

The foremost use of micro- and nanoparticles is their discharge and targeting to the destined site. Particle size plays an important role that adversely affects the targeted delivery of the drug. Smaller particles have a larger surface area, due to which a maximum of the loaded drug will be at the particle surface for quick release. An enlarged surface area of the system indicates augmented surface interactions with the particle environment, which refines, e.g., dissolution to gastrointestinal fluids. Consequently, by decreasing the particle size the dissolution rate can be increased, hence increasing the effectiveness and limiting resistance to the diffusion of substrates. Polymer degradation can also be affected by the particle size. For example, the degradation rate of poly(lactic-*co*-glycolic acid) was found to proliferate with increasing particle size in vitro [63]. Hence, therapeutic nanocarriers or other micro- and nanostructured materials can be potentially beneficial for the engineering of composite drug delivery applications. A major disadvantage is that the smaller particles tend to aggregate during storage and transportation of the nanoparticle dispersion. Hence, there needs to be a balance between minimum size and maximum stability of the nanoparticles.

However, the use of new technologies for determining particle size is in current operation. Techniques such as photon-correlation spectroscopy (PCS) or dynamic light scattering (DLS) are widely used to determine the size of Brownian nanoparticles in colloidal suspensions in the nano- and submicron ranges [64]. With insignificant alteration to PCS, nanoparticle tracking analysis (NTA) is a method industrialized by NanoSight Ltd. to decide the size distribution profile of small particles in a liquid suspension. The method is practiced in conjunction with an ultramicroscope, which permits small particles in liquid suspension to be envisaged moving under Brownian motion.

A few studies have validated the fact that nanoparticles of submicron size have advantages over microparticles as a drug delivery system. Due to their higher intracellular acceptance in comparison to microparticles and their widespread array of biological targets owing to their small size and relative mobility, nanoparticles are considered before any other particle range.

5.2 *Particle Morphology*

The physical and chemical properties of nanoparticles have a deep impact because they govern the relationship of nanoparticles with the environment and biological systems. The techniques used to examine the morphology of nanoparticle are SEM, transmission electron microscopy (TEM) and AFM, along with particle size and distribution analyses. Other parameters like morphology or surface roughness of the nanoparticles are also determined.

For TEM, the sample has to be ultrathin for the electron transmittance, and the dispersion is deposited onto support grids or films. Negative staining material is used to withstand the instrument vacuum and facilitate handling. However, SEM is different in that the solution is primarily changed into a dry powder, which is then stranded on a sample holder and tracked by coating with a conductive metal, such as gold, using a sputter coater. The sample is then scanned with a focused fine beam of electrons.

5.3 *Particle Stability*

Nanoparticle suspensions are moderately stable due to their colloidal nature. In general, colloidal suspensions are unwavering and have very little tendency to separate as a consequence of deliberate deposition, which is a result of the mixing tendencies of diffusion and convection. However, some agglomeration can occur. Colloidal stability can be analyzed through the zeta potential of nanoparticles. The zeta potential reflects the electrical potential of particles and is influenced by the composition of the particle and the medium in which it is dispersed. This potential is an indirect measure of the surface charge. Due to the small particle size and amplified surface area, nanoparticles tend towards a more thermodynamically stable state driven by an increased free energy. Thus, without adequate equilibrium, the accumulation of nanoparticles in the medium will occur shortly soon after particle formation, and at least during long-term storage. Nanoparticles have high particle mobility due to Brownian motion and, thus, the particles collide with each other. The probability of particle–particle interactions proliferates with reduced size. Most colloidal particles have negative zeta potential values ranging from about -100 to -5 mV. Surface charges inhibit the aggregation of nanoparticle polymer dispersions because of strong electrostatic repulsion, thereby augmenting the stability of the nanoparticles. The zeta potential can also provide information concerning the nature of material encapsulated within the nanocapsule or coated onto the surface [65]. In colloids, equilibrium instead of permanent aggregation can be recognized by means of steric (physical barrier) and/or electrostatic stabilization (surface charge) on the surface of the nanoparticles. Ionic or non-ionic surfactants and polymers have been commonly used to provide a barrier against the aggregation of nanoparticles [66, 67]. Steric stabilization can be attained by covering the particles with polymers, which prevent the access of particles to the range of attractive forces by forming a physical barrier. Furthermore, a non-ionic amphiphilic polymer or surfactant with a hydrophobic chain around the hydrophobic drug surface allows the hydrophilic part to project into the water and, thus, increase its hydrophilicity. Steric interactions are more sensitive to temperature fluctuations than electrostatic repulsions [68]. In contrast, ionic stabilizers are very sensitive to changes in pH and ionic strength, and are also incapable of stabilizing nanoparticles in the dry state [69]. Nevertheless, storing of nanoparticles in suspensions for a long period of time leads to many disadvantages. The risk of microbiological adulteration, degradation of premature polymer by hydrolysis, physicochemical instability due to particle aggregation, sedimentation and chemical instability of the drug are a few examples. To avoid such complications, dry formulations are generally favoured over suspensions. To overcome this major drawback, the use of various drying methods have been practiced and the alteration of a liquid samples into a dry product can be accomplished by means of sensitive methods such as freeze-drying or spray-drying processes. Lyophilization (freeze-drying) is considered to be an extremely useful process that is generally functional in augmenting the physicochemical stability of the nanoparticles in order to obtain a pharmaceutically satisfactory product. This is especially useful for cases in which the storage circumstances are unfavourable.

6 Applications of Micro- and Nanoparticles Based on Biopolymers

Drug loading in micro- and nanoparticulate systems is generally carried out by one of two methods, i.e. during the preparation of particles (incorporation) or after their formation (incubation). The drug delivery properties are also essentially dependent on the chemical and textural properties of the matrices, the porosity, wettability, erosion and the surface area. The matrix equally has an impact on the discharge profile for the bioavailability of the entrapped drug. Nanoparticle-based delivery systems have the potential power to improve drug stability, increase the duration of the therapeutic effect and permit enteral or parenteral administration, which may prevent or minimize drug degradation and metabolism as well as cellular efflux [70, 71]. Protein nanoparticles can conveyance medications transversely across the blood–brain barrier that are not usually passed across after injection. A number of authors have demonstrated a considerable tendency for an accumulation of protein nanoparticles in certain tumours. The binding of a variety of cytotoxic drugs, such as 5-fluorouracil, paclitaxel and doxorubicin, to albumin or gelatin nanoparticles significantly enhanced the efficacy against experimental tumours or human tumours transplanted into nude mice in comparison to free drug. Similarly, drug release and polymer biodegradation significantly influence the possibility of obtaining a fruitful nanoparticle system. The factors generally responsible are solubility of drug; desorption of the surface bound or adsorbed drug; drug diffusion through the nanoparticle matrix; nanoparticle matrix erosion or degradation; and combination of erosion and diffusion processes. Accordingly, the solubility, diffusion and biodegradation of the matrix materials control the release practice. When the dispersal of the drug is quicker than matrix erosion, the mechanism of discharge is principally controlled by a diffusion process. When there is a rapid initial release or "burst" it is chiefly accredited to inadequate binding of drug to the large surface of nanoparticles. However, when the drug is loaded by the incorporation method, there is a comparatively slight burst effect and persistent release features. Similarly, when the particle is coated with polymer, the release is then measured by dispersal of the drug from the core across the polymeric membrane.

6.1 Chitosan and Its Derivatives for Pharmaceutical Applications

Recently, the use of chitosan in drug formulation development has increased many fold. Chitosan exhibits excellent compatibility with organic compounds such as cationic dyes and surfactants, starches, quaternary ammonium salts and with most cationic and non-ionic polymers. Multivalent anions can easily crosslink with chitosan to form complexes. Chitosan has many advantages, including its cationic nature that allows for ionic crosslinking with multivalent anions; its mucoadhesive character, which increases residual time at the site of absorption; the lack of need to use hazardous organic solvents for fabricating particles since it is soluble is aqueous acidic solution; it is linear polyamine containing a number of free amine groups that are readily available for crosslinking; and last but not least, its ability to control the release of active agents. Chitosan has been investigated for its possible role in controlling the release of active medicines.

To further enhance the solubility of chitosan and to improve its mucoadhesive and/or permeation enhancing properties, various derivatives such as trimethylated chitosan [72], mono-*N*-carboxymethylchitosan, *N*-sulfo-chitosan and chitosan–EDTA conjugates have been prepared. A further modification is based on the immobilization of thiol-bearing moieties on the polymeric backbone of chitosan.

6.1.1 Chitosan as Binder in Tablets

Chitosan (molecular weight, 19–400 kDa; degree of deacetylation, 75–98%) has been evaluated as a directly compressible vehicle for tablets, but has found limited utility due to the lack of good flow properties and compressibility. The influence of excipient (lactose, sodium lauryl sulfate, sodium alginate, hydroxypropyl methyl cellulose, carbopol and citric acid) on drug release from chitosan matrix tablets was investigated using diltiazem hydrochloride. Upadrastha and coworkers investigated the potential applications of chitosan as a binder. Chitosan, when used at a concentration of 2% of the tablet weight, led to only 30% of the drug being dissolved in the same time period.

6.1.2 Chitosan for Dental Disease

Local delivery of drugs and other bioactive agents directly into the periodontal pocket has received a lot of attention lately. The adherence of oral bacteria on the tooth surface leads to plaque formation. It is believed that the adhesion between the bacteria and tooth surface is due to electrostatic and hydrophobic interactions. These interactions are disrupted by chitosan because of competition by the positively charged amine groups. A cross-over experiment was performed to evaluate the effect of phosphorylated chitosan on plaque formation. The test group of human subjects rinsed with a mouthwash containing 0.5% (w/v) phosphorylated chitosan three times a day. After 3 days of wash out, the same subjects rinsed with a mouthwash without chitosan (placebo group). Following a 4-day rinsing period, the newly formed plaque on the upper front teeth was distinguished from the old plaque by the use of a two-tone disclosing agent (Dis-plaque). In comparison to placebo, the test solution significantly reduced both the newly formed and old plaque on the upper front teeth. The test solution reduced the newly formed plaque by 26% as compared to the placebo, whereas the old plaque was reduced by 61% as compared to placebo [49].

A very intriguing development concerns a drug delivery system for the treatment of periodontitis. It consists of two components. The first is a clay product loaded with tetracycline. The clay product is coated with chitosan to retard drug release. The second component is a thermo-responsive polymer mixed with polyethylene glycol that is syringable at room temperature but solidifies upon warming up to body temperature. The solidified polymer remains 60 min in situ. After that time, it still inhibits the formation of bacterial colonies, and still contains much more of the antibiotic than the control containing no chitosan.

6.2 Chitosan for Buccal Delivery Systems

An ideal buccal delivery system should stay in the oral cavity for at least few hours and release the active ingredients in a unidirectional way towards the mucosa in a controlled or sustained-release fashion. Muco- and bioadhesive polymers are supposed to prolong the residence time of the device in the oral cavity. One of the experiments developed a direct staining method to visualize the polymer adhesion to human buccal cells after exposure to an aqueous dispersion of chitosan. An in vivo study was performed on a test group of human subjects, who rinsed with

30 mL of 0.1% (w/v) of chitosan dispersion for 30 s. Periodically, the buccal cells were harvested from the cheek and suspended in 0.25 M sucrose solution. Buccal cells harvested from the cheeks of human subjects who received no treatment were used as control. The results showed that chitosan formulation was able to adhere to the human buccal cells for at least 1 h.

Chitosan microspheres, prepared by spray-drying, were loaded with cholorohexidine, a drug that is used for treating chronic candida infection. The microspheres along with mannitol and sodium alginate as excipients were compressed into tablets by direct compression. Tablets containing 40 mg of the drug were administered buccally to human subjects. The control group rinsed with a commercial mouthwash (DentosanR) containing 0.2% (w/v) chorohexidine. For the control group, the salivary drug concentrations were very low and no drug was detected after 2 h whereas the drug was still detected even after 3 h in the saliva of the test group [54].

6.2.1 Chitosan Beads, Microcapsules and Microspheres for Gastrointestinal Delivery Systems

Many gastric retentive devices have been developed such as high-density systems, expandable systems, superporous hydrogel systems, intragastric floating systems and mucoadhesive systems. In view of recent problems concerned with *Helicobacter pylori* infection, causing ulcerations, many researchers have been trying to make effective sustained release antibacterial therapy. Tetracycline-loaded chitosan microspheres have been developed for stomach-specific delivery. The microspheres prepared by ionic precipitation with sulfate were spherical in shape with an average diameter of 2–3 μm and were 55% drug-loaded. It was found that only about 60% of the entrapped tetracycline was released in 2 h into the medium of pH 3.5 whereas the microspheres were instantly dissolved at pH 1.2–2.0. Another study examined the gastric residence time of chitosan microspheres following oral administration in fasted gerbils. Because of the instant dissolution of those microspheres at pH 1.2–2.0, the effect of acid suppression on the chitosan microsphere residence time and on the tetracycline concentrations in stomach was also examined. Similar results were obtained as in vitro. As a result of these studies, chemically crosslinked glyoxal chitosan microspheres were developed that showed increased stability even at pH 1.2–2.0. These microparticles were settled and adhere to gastric mucosa by electrostatic interaction. A large fraction of the administered chitosan microsphere dose was found in the colon after 6 h. The tetracycline concentration in the stomach from the crosslinked chitosan microspheres was higher than that of the non-crosslinked microspheres for all times of administration.

Chitosan is a promising polymer for colon delivery because it can be biodegraded by the colonic microflora and it has mucoadhesive character. A thorough study has been carried out on chitosan beads obtained by ionotropic gelation with alginate or tripolyphosphate. The aim was to develop, using natural polymers, a controlled release system that would resist the conditions in the stomach and release its drug content in a slow controlled way, preferably in linear fashion. In a model study, chitosan beads loaded with BSA were obtained by dropping a chitosan albumin mixture in a alginate coagulation bath [48]. Beads were obtained that were amenable to controlled release depending on the fluid in which the beads were incubated. After a successful study with BSA as model compound, these chitosan-alginate beads loaded with ampicillin showed nearly complete drug retention in mimicked gastric conditions in simulated gastric fluid (SGF) but released their content in simulated intestinal fluid (SIF). Bead formation was studied in the presence of competitive cations Ca^{2+} and Ba^{2+} in order to change the release characteristics of the beads [48]. Addition of barium to the system revealed the conditions at which a linear release over about 24 h could be accomplished.

A capsule of chitosan with an enteric coating and filled with the marker carboxy fluoresceine and a low molecular weight enhancer has been studied for uptake in the colon. In preliminary studies, dodecyl maltose was show to be a more effective enhancer than sodium glycocholate, sodium salicylate and sodium caprate. All four enhancers were more active in the colon but the maltose compound was the best. This was tested in the capsule. The capsule containing the maltose compound and the fluoresceine marker was applied orally and gave rise to the highest plasma fluorescence. A solution or gelatin capsules were less effective. The capsule is a candidate for controlled, colon-specific uptake. It should be noted that colon-specific delivery can also be accomplished by coating a capsule with pectin. This is only dissolved in the large intestine. There are interesting possibilities in combination with chitosan.

The *N*-acetyl chitosans are the most interesting for controlled release. Galactosylated chitosan microspheres have been prepared. Production of more specific oligoglycoside-chain chitosan might develop into active targeting of specific tissues. Pectinate gel beads modified with trimethyl chitosan have been prepared and evaluated for colon targeted delivery of macromolecules.

Chitosan beads containing prednisolone were implanted in mouse skin within induced local inflammatory air pouches. These chitosan beads are biocompatible and biodegradable vehicles for treatment. An effective dose can be given in combination with prolonged periods of local drug presence. This compares advantageously with injection of a prednisolone suspension.

6.3 *Chitosan Nanoparticles for Drug Delivery, Gene Therapy and Cancer Therapy*

Among the various drug delivery systems, nanoparticles have several advantages, such as high drug encapsulation efficiency, efficient drug protection against chemical or enzymatic degradation, unique ability to create a controlled release, cell internalization as well as the ability to reverse the multidrug resistance of tumour cells. They could also provide one solution for delivery of new classes of active molecules such as peptides, proteins, genes and oligonucleotides. However, the main drawback of these carriers is their nonspecific interaction with cells and proteins, leading to the drug accumulation in non-target tissues.

Chitosan has been used a carrier for DNA for gene delivery applications. The cationically charged chitosan can form polyelectrolyte complexes with the negatively charged plasmid DNA. Chitosan–DNA complexes could be protected from DNAse to improve the bioavailability of the plasmid DNA delivered into the body for gene therapy. Also, chitosan could be a useful gene carrier because of its mucoadhesive and cohesive properties in the gastrointestinal tract. The advantage of chitosan-based vectors lies not only in avoiding cytotoxicity problems that are inherent in most synthetic polymeric vehicles but also in their unique capability for transcellular transport.

Chitosan–DNA nanoparticles have been prepared and the influence of several parameters on their preparation evaluated. Transfection efficiency of the nanoparticles were cell-type dependent. The study also developed three different schemes to conjugate transferrin or KNOB protein to the nanoparticle surface. The transferrin conjugation only yielded a maximum of fourfold increase in transfection efficiency in HEK293 cells and HeLa cells, whereas KNOB-conjugated nanoparticles could improve the gene expression level in HeLa cells by 130-fold. The usefulness of the complex of chitosan self-aggregates and DNA for transfer of genes into mammalian cells in vitro has also been suggested. Several transfection studies using chemically modified chitosan have been reported. Trimethyl chitosan oligomers, lactosylated chitosan and galactosylated chitosan were examined for their potency as DNA carriers and transfection efficiency in in vitro systems. Incorporation of hydrophobic moieties might considerably increase the transfection efficiency: with longer alkyl side chains, the transfection efficiency was increased and levelled off when the number of carbons in the side chain exceeded eight. The higher transfection efficiency is attributed to increased entry into cells, facilitated by hydrophobic interactions and easier unpacking of DNA from alkylated chitosan carriers due to the weakening of electrostatic attractions between DNA and alkylated chitosan.

Numerous studies have been conducted on prophylactic and therapeutic use of genetic vaccines for combating a variety of infectious diseases in animal models. A human clinical study with the gene gun has validated the concept of direct targeting of dendritic cells in the viable epidermis of the skin. However, it is unclear whether the gene gun technology or other needle-free devices will be commercially viable. A very recent study, has proposed folate–chitosan–DNA nanoparticles as having low cytotoxicity and good DNA condensation, making them a promising candidate for nonviral gene therapy. Chitosan–DNA and folate–chitosan–DNA nanoparticles were prepared using reductive amidation and a complex coacervation process.

The potential of gadolinium-loaded chitosan nanoparticles has been demonstrated. The potential of gadolinium neutron-capture therapy for cancer was evaluated using chitosan nanoparticles as a novel device. Later, the similar gadopentectic acid-loaded chitosan nanoparticles were prepared for gadolinium neutron-capture therapy. Their releasing properties and ability for long-term retention in the tumour indicated that these particles are useful as intratumoural injectable devices for gadolinium neutron capture therapy. The accumulation of drug-loaded chitosan nanoparticles in tumour cells (L929 fibroblast cultured cells as model) was 100–200 times higher in comparison to dimeglumine gadopenteate aqueous solution (a magnetic resonance contrast agent). These findings indicated the higher affinity of chitosan nanoparticles to tumour cells, thus contributing to the long retention of gadolinium in tumour tissue and leading to suppression of tumour growth in in vivo studies. The nanoparticles have recently been regarded as promising carriers in breaching the epithelial and enzymatic barriers to enhance the bioavailability of biomolecules. Surface-modified triglyceride nanoparticles, using chitosan for salmon calcitonin delivery, have been developed and showed that chitosan could be formed around the triglyceride nanoparticles by simple incubation of the lipid cores in chitosan solution, due to the high affinity of chitosan for the lipid core, to provide a continuous and slow release of the associated peptide. Recently, chitosan nanoparticles have been coated to enhance their stability and to prevent immediate desorption in body fluids, including gastrointestinal fluids. Sodium alginate was used as a coating material. In vitro release studies showed that the presence of an alginate layer around the nanoparticles was able to prevent a burst release of loaded model antigen and to improve the stability of the nanoparticles in simulated gastrointestinal fluid. Liposomes have been coated by chitosans to enhance their stability for oral administration of peptide drugs. It was found that the chitosan-coated liposomes are more penetrative into the intestinal mucosa and have higher retentive properties than non-coated liposomes.

The emergence of bioactive food compounds (neutraceuticals) with health benefits provides an excellent opportunity to improve public health. The incorporation of bioactive compounds, such as peptides and vitamins, into food systems holds much promise in the development of innovative functional foods that may have physiological benefits or reduce the risk of diseases. Compared to pharmaceutical applications, little work has been done on the encapsulation properties of chitosan nanoparticles for the oral administration of neutraceuticals in healthy food. More recently, chitosan has been identified as a versatile biopolymer for a broad range of health and food applications because of its safety and non-toxicity in human beings. Because chitosan matrix is not stable at very low pH, modifications to the chitosan micro- or nanoparticles might be able to protect them.

6.4 Applications of Alginate Nanoparticles

6.4.1 Alginate Nanoparticles as Antituberculosis Drug Carriers

An alginate-based nanoparticulate delivery system was developed for frontline antituberculosis drugs (rifampicin, isoniazid, pyrazinamide and ethambutol). Alginate nanoparticles were prepared by controlled cation-induced gelification and administered orally to mice. The drug levels were analysed by high performance liquid chromatography (HPLC) in plasma and tissues. The therapeutic efficacy was evaluated in *M. tuberculosis* H37Rv-infected mice. High drug encapsulation efficiency was achieved in alginate nanoparticles, ranging from 70% to 90%. A single oral dose resulted in therapeutic drug concentrations in the plasma for 7–11 days and in the organs (lungs, liver and spleen) for 15 days. In comparison to free drug (which were cleared from plasma and organs within 12–24 h), there was a significant enhancement in the relative bioavailability of encapsulated drugs. In TB-infected mice, three oral doses of the formulation spaced 15 days apart resulted in complete bacterial clearance from the organs, compared to 45 conventional doses of orally administered free drug.

6.4.2 Sodium Alginate–Chitosan Composite Films for Fabrication of Antibacterial Silver Nanoparticles

A new and simple ecofriendly method for the synthesis of silver nanoparticles (Ag NPs) using a natural biopolymer, sodium alginate, as both reducing and stabilizing agent has been reported. The synthesized NPs were characterized using UV–vis spectroscopy, TEM and selected area electron diffraction (SAED) pattern. The alginate-capped NPs (Alg–Ag NPs) were found to be antibacterial. The Alg–Ag NPs were blended with varying amounts of chitosan to form polyelectrolyte complexes that were cast into stable films. The films were characterized by field emission scanning electron microscopy (FESEM), optical microscopy, Fourier transform infrared spectroscopy (FTIR) and X-ray diffraction (XRD). The water uptake and mechanical properties of the films were also studied. The blended films demonstrated excellent antibacterial activity against both Gram-negative and Gram-positive bacteria, with more activity against Gram-positive bacteria. Thus, the developed films have a potential to be used for various antibacterial applications in biotechnology and biomedical fields.

6.5 Applications of Gelatin Nanoparticles

6.5.1 Nonviral Gene Delivery Vectors

Cationized gelatin nanoparticles have shown the potential of being a new effective carrier for nonviral gene delivery. The major benefit of gelatin nanoparticles is not only the very low cell toxicity, but also their simple production, combined with low cost. Native gelatin nanoparticles were prepared by a two-step desolvation technique. In order to bind DNA by electrostatic interaction onto the surface of the particles, the quartenary amine cholamine was covalently coupled to the particles. The modified nanoparticles were loaded with different amounts of plasmid in varying buffers and compared to polyethyleneimine–DNA complex as a standard. Transfection ability of the loaded nanoparticles was tested on B16F10 cells. Additionally, the cell toxicity of the formulation was monitored. Different setups resulted in efficient gene delivery, displayed by an exponential increase in gene expression. The gene expression itself occurred with a certain delay after transfection. In contrast to polyethyleneimine complexes, cationized gelatin nanoparticles showed almost no significant cytotoxic effects.

7 Conclusions

The progression of micro- and nanoparticle research approaches has remained noticeable for the low amount of contaminated and poisonous components, generalization of the technique to permit financial scale-up, and optimization to expand yield and set-up efficiency. Well-organized drug frame-up and conversion to larger scales are of greatest importance to manufacturing applicability. Micro- and nanoparticles are basically established from a diversity of resources, e.g. synthetic polymers, proteins and polysaccharides. Natural polymeric materials are beneficial in comparison to man-made polymers because they are biodegradable, biocompatible and nontoxic. With abundant groundwork, approaches for manufacturing particles and significant industrial developments have been accomplished. Depending on the physico-chemical characteristics of a drug, superlative techniques for preparing the polymer to accomplish a well-organized entrapment of the drug have also become an easy task. The selected scheme can curtail the loss of medicine or control its pharmacological movement.

References

1. Muzzarelli RAA (ed) (1977) Chitin. Oxford, Pergamon, p 307
2. Stevens WF (2001) Production of chitin and chitosan: refinement and sustainability of chemical and biological processing. In: Uragami et al (eds) Chitin and chitosan: chitin and chitosan in life science. Kodansha Scientific, Tokyo, pp 293–300
3. Joensen O, Villadsen A (1994) Ecological sustainable production of chitin and chitosan. In: Karnicki ZS, Brzeski MM, Bykowski PJ, Wojtasz-Pajak A (eds) Chitin world. Wirtschaftsverlag, Bremerhaven, pp 38–47
4. Roberts AFG (1992) Chitin chemistry. Macmillan, London, pp 85–102
5. Ogawa K (1991) Effect of heating an aqueous suspension of chitosan on the crystallinity and polymorphs. Agric Biol Chem 55:2375–2377
6. Mucha M (1997) Rheological characteristics of semi-dilute chitosan solutions. Macromol Chem Phys 198:471
7. Suh JKF, Matthew HWT (2000) Application of chitosan-based polysaccharide biomaterials in cartilage tissue engineering: a review. Biomaterials 21:2589–2598
8. Arai K, Kinumaki T, Fujita T (1968) Toxicity of chitosan. Bull Tokai Reg Fish Lab 43:89–94
9. Chandy T, Sharma CP (1990) Chitosan – as a biomaterial. Biomater Artif Cells Artif Organs 18:1–24
10. Bajpai SK, Sharma S (2004) React Funct Polym 59:129–140
11. Park K, Sharaby WSW, Park H (1993) Biodegradable hydrogels for drug delivery. Technomic Publishing, Lancaster, pp 99–140
12. Smidsrod O (1974) Chem Soc 57:263–270
13. Sartori C, Finch DS, Ralph AB (1996) Determination of the cation content of alginate thin films by FTIR spectroscopy. Polymer 38:43–51

14. Walsh PK, Isdell FV, Noone SM, O'Donovan MG, Malone DM (1996) Growth patterns of Saccharomyces cerevisiae microcolonies in alginate and carrageenan gel particles: effect of physical and chemical properties of gels. Enzyme Microb Technol 18:366–372
15. Poncelet D (2001) Production of alginate beads by emulsification/internal gelation. Ann N Y Acad Sci 944:74–82
16. Halliwell B (1988) Biochem Pharmacol 37:569
17. Emerson TE (1989) Crit Care Med 17:690
18. Kreuter J, Tuber U, Illi V (1979) Distribution and elimination of poly(methyl-2-14C methacrylate) nanoparticle radioactivity after injection in rats and mice. J Pharm Sci 68:1443–1447
19. Sinn H, Schrenk HH, Friedrich EA, Schilling U, Maier-Borst W (1990) Nucl Med Biol 17:819
20. Majumdar S, Basu SK (1991) Antimicrob Agents Chemother 35:135
21. Nakagawa Y, Takayama K, Ueda H, Machida Y, Nagai T (1987) Drug Des Deliv 2:99
22. Irache JM, Merodio M, Arnedo A, Camapanero MA, Mirshahi M, Espuelas S (2005) Mini Rev Med Chem 5:293
23. Santhi K, Dhanaraj SA, Joseph V, Ponnusankar S, Suresh B (2002) Drug Dev Ind Pharm 28:1171
24. Kreuter J, Hekmatara T, Dreis S, Vogel T, Gelperina S, Langer K (2007) J Control Release 118:54
25. Merodio M, Irache JM, Eclancher F, Mirshahi M, Villarroya H (2000) J Drug Target 8:289
26. Lynn AK, Yannas IV, Bonfield W (2004) J Biomed Mater Res B Appl Biomater 71:343
27. Barbani N, Giusti P, Lazzeri L, Polacco G, Pizzirani G (1995) J Biomater Sci Polym Ed 7:461
28. Lefebvre F, Gorecki S, Bareille R, Amedee J, Bordenave L, Rabaud M (1992) Biomaterials 13:28
29. Ruderman RJ, Wade CWR, Shepard WD, Leonard F (1973) J Biomed Mater Res 7:253
30. Bender A, von Briesen H, Kreuter J, Duncan IB, Rubsamen-Waigmann H (1996) Antimicrob Agents Chemother 40:1467
31. Schwick HG, Heide K (1969) Bibl Haematol 33:111
32. Ward AG, Courts A (1977) The science and technology of gelatin. Academic, New York
33. Bajpai AK, Choubey J (2006) J Mater Sci Mater Med 17:345
34. Stevens KR, Einerson NJ, Burmania JA, Kao WJ (2002) J Biomater Sci Polym Ed 13:1353
35. Marios Y, Chakfe N, Deng X, Marios M, How T, King W, Guidoin R (1995) Biomaterials 16:1131
36. DiSilvio L, Courtney-Harris RG, Downes S (1994) J Mater Sci Mater Med 5:819
37. Anal AK, Singh H (2007) Recent advances in microencapsulation technologies for probiotics for industrial (11–12). Trends Food Sci Technol 18:240–251
38. ChampagneC P, Fustier P (2007) Microencapsulation for the improved delivery of bioactive compounds into foods. Curr Opin Biotechnol 18:184
39. Gharsallaoui A, Roudaut G, Chambin O, Voilley A, Saurel R (2007) Applications of spray-drying in microencapsulation of food ingredients: an overview. Food Res Int 40:1107–1121
40. Gouin S (2004) Microencapsulation: industrial appraisal of existing technologies and trends. Trends Food Sci Technol 15:30–347
41. Gibbs BF, Kermasha S, Alli I, Mulligan CN (1999) Encapsulation in the food industry: a review. Int J Food Sci Nutr 50:213–224
42. Dziezak JD (1988) Microencapsulation and encapsulated ingredients. Food Technol 42:136–151
43. Wantabe Y, Fang X, Adachi S, Matsuno R (2002) Suppressive effect of saturated acyl L-ascorbate on the oxidation of linoleic acid encapsulated with maltodetrin or gum Arabic by spray drying. J Agric Food Chem 50(14):3984–3987
44. Goubet I, Le Quere JL, Voilley A (1998) Retention of aroma compounds by carbohydrates: influence of their physicochemical characteristics and of their physical state. J Agric Food Chem 48:1981–1990
45. Bomben JL, Bruin B, Thijssen HAC, Merson RL (1973) Aroma recovery and retention in concentration and drying of foods. Adv Food Res 20:2–11

46. Rosenberg M et al (1990) Factors affecting retention in spray-drying microencapsulation of volatile materials. J Agric Food Chem 38(5):1288–1294
47. Madene A, Jacquot M, Scher J, Desobry S (2006) Flavour encapsulation and controlled release – a review. Int J Food Sci Technol 41:1–21
48. Anal AK, Stevens WF (2005) Chitosan-alginate multilayer beads for controlled release of ampicillin. Int J Pharm 290:45–54
49. Anal AK, Bhopatkar D, Tokura S, Tamura H, Stevens WF (2003) Chitosan-alginate multilayer beads for gastric passage and controlled intestinal release of protein. Drug Dev Ind Pharm 29:713–724
50. Nishimura K (1986) J Biomed Mater Res 20:1359
51. Lin W, Coombes A, Davies M, Davis S, Illum L (1993) J Drug Target 1:237
52. Hedges A, McBride C (1999) Utilization of β-cyclodextrin in food. Cereal Foods World Microporous Sugars Adsorpt 44:700–704
53. Patiington JS (1986) β-cyclodextrin and its uses in the flavour industry. In: Brich GG, Lindley MG (eds) Developments in food flavour. Elsevier Applied Science, London
54. Gomez A, Bingham D, De Juan L, Tang K (1999) Mater Res Soc Symp Proc 550:101
55. Mandal BB, Kundu SC (2009) Nanotechnology 203:55101
56. Tamada Y, Sano M, Niwa K, Imai T, Yoshino G (2004) J Biomater Sci Polym Ed 15:971
57. Aramwit P, Kanokpanont S, De-Eknamkul W, Kamei K, Srichana T (2009) J Biomater Sci Polym Ed 20:1295
58. Zhang YQ (2002) Biotechnol Adv 20:91
59. Zhang YQ, Ma Y, Xia YY, Shen WD, Mao JP, Xue RY (2006) J Control Release 11:5307
60. Aramwit P, Kanokpanont S, De-Eknamkul W, Srichana T (2009) J Biosci Bioeng 10:7556
61. Dunne M, Corrigan OI, Ramtoola Z (2000) Biomaterials 21:1659
62. Berne BJ, Pecora R (1975) Dynamic light scattering. Wiley, New York
63. Mohanraj VJ, Chen Y (2006) Trop J Pharm Res 5:561
64. Van Eerdenbrugh B, Van den Mooter G, Augustijns P (2008) Top-down production of drug nanocrystals: nanosuspension stabilization, miniaturization and transformation into solid products. Int J Pharm 364:64–75
65. Raghavan SL, Schuessel K, Davis A, Hadgraft J (2003) Formation and stabilisation of triclosan colloidal suspensions using supersaturated systems. Int J Pharm 261:153–158
66. Rabinow BE (2004) Nanosuspensions in drug delivery. Nat Rev Drug Discov 3:785–796
67. Farrokhpay S (2009) A review of polymeric dispersant stabilisation of titania pigment. Adv Colloid Interface Sci 151:24–32
68. Sarmento B, Ribeiro A, Veiga F, Sampaio P, Neufeld R, Ferreira D (2007) Pharm Res 24:2198
69. Pan Y, Li YJ, Zhao HY, Zheng JM, Xu H, Wei G, Hao JS, Cui FD (2002) Int J Pharm 249:139
70. Sieval AB, Thanou M, Kotze AF, Verhoef JC, Brussee J, Junginger HE (1998) Preparation and NMR-characterization of highly substituted N-trimethyl chitosan chloride. Carbohydr Polym 36:157–165
71. Sano T, Mimura R, Ohshima K (2001) Phylogenetic analysis of hop and grapevine isolates of Hop stunt viroid supports a grapevine origin for hop stunt disease. Virus Genes 22:53–59
72. Giunchedi P, Juliano C, Gavini E, Cossu M, Sorrenti M (2002) Formulation and in-vivo evaluation of drug carriers for cerebral tumours. J Microencapsul 17:625–638

Adv Polym Sci (2013) 254: 297–342
DOI: 10.1007/12_2012_208

Published online: 17 May 2013

Applications of Glyconanoparticles as "Sweet" Glycobiological Therapeutics and Diagnostics

Naresh Kottari, Yoann M. Chabre, Rishi Sharma, and René Roy

Abstract Combining nanotechnology with glycobiology has triggered an exponential growth of research activities for the design of novel functional bionanomaterials (glyconanotechnology). More specifically, recent advances in the tailored and versatile synthesis of glycosylated nanoparticles (glyconanoparticles), considered as synthetic mimetics of natural glycoconjugates, have paved the way towards diverse biomedical applications. The accessibility of a wide variety of these structured nanosystems, in terms of shape, size, and organization around stable nanoparticles, have readily contributed to their development and application in nanomedicine. In this context, glycosylated gold nanoparticles, glycosylated quantum dots, fullerenes, single-wall nanotubes, and self-assembled glyconanoparticles using amphiphilic glycopolymers or glycodendrimers have received considerable attention for their application in powerful imaging, therapeutic, and biodiagnostic devices.

This chapter provides an overview of the most recent syntheses and applications of glyconanoparticles that have deepened our understanding of multivalent carbohydrate–protein interactions. Together with sensitive detection devices, inhibitors of bacterial adhesion to host tissue, and cancer vaccines in therapeutic systems (including photosensitizers for photodynamic therapies), these novel bionanomaterials are finding widespread relevance. The recent development of glyconanoparticle-based chemotherapeutic drug delivery agents will also be addressed, together with their molecular imaging properties for MRI or as X-ray contrast agents, as platforms for selective immunolabeling and imaging of cells, and as colorimetric devices in various bioassays.

Keywords Diagnostics · Glycodendrimers · Glyconanoparticles · Quantum dots · Therapeutics

---

N. Kottari, Y.M. Chabre, R. Sharma, and R. Roy (✉)
Pharmaqam – Department of Chemistry, Université du Québec à Montréal, P.O. Box 8888, Succ. Centre-ville, Montréal, QC, Canada H3C 3P8
e-mail: roy.rene@uqam.ca

Contents

1 Introduction

Nanoparticles (NPs) conjugated with sugar moieties (referred to here as glyconanoparticles or GNPs) have been extensively used for multivalent presentation of carbohydrates in biomedical applications. The globular shapes and diversified nanometer sizes of NPs, some of which are comparable to those of biomacromolecules, make them efficient scaffolds for the syntheses of "glycocalyx-like" building blocks exposing multiple copies of oligosaccharide conjugates. By using simple and efficient preparative methods available for the synthesis and functionalization of NPs, a wide variety of biorelevant glycans have been coated onto the NPs. Furthermore, several molecules have been attached to single nanoparticle surfaces to afford hybrid NPs and, by varying the ratios of different ligands, their presentation density can be controlled [1]. A wide variety of linkers have been used for the chemical ligation between the two partners and for better exposure of carbohydrate head groups to their cognate molecular receptors, which is key to achieving high avidity multicontact binding.

By altering experimental conditions, it is possible to tune the dimensions of the NPs, which is directly coupled to their stability, cytotoxicity and electronic, optical, and magnetic properties. The size and shape of NPs have been shown to play several crucial roles in their in vivo characteristics. For instance, renal clearance, liver and mononuclear phagocyte uptake, and intracellular delivery are a few examples wherein the use of NPs could be advantageous. Intracellular delivery is also dictated by the charge of the NP surface groups, and internalization can occur through interaction with specialized cellular receptors or by one of the endocytosis mechanisms. In addition, surface functionalization is also a major factor in determining the stability, toxicity, and long-term circulation abilities of the NPs. Poly(ethyleneglycol) (PEG) and carbohydrates (including mono- and polysaccharides) are often used to reduce the toxicity and to achieve the long-term circulation of NPs in the blood stream.

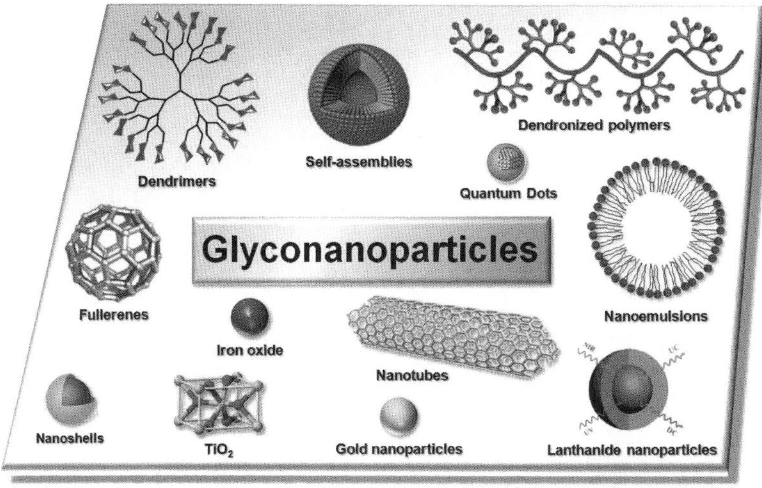

Fig. 1 Representation of accessible multivalent glyconanoparticles reviewed in this chapter

For chemotherapy, NP-based drugs have the additional advantage of being more readily accumulated in tumor tissues than in normal tissues due to the enhanced permeability and retention (EPR) effect of the tumor vasculature. The increased vascular permeability of tumor blood vessels to NPs and ineffective lymphatic drainage due to the poor development of lymphatic vessels in tumor tissue can lead to accumulation of NPs in the tumor surroundings. As such, NP-based drugs conjugated with cellular targeting carbohydrate agents could be effective chemotherapeutic agents for different types of cancers. This subject will be extensively examined throughout the chapter.

Combining nanotechnology with glycobiology has triggered an exponential growth of research activities in the design of novel functional bionanomaterials (glyconanotechnology). More specifically, recent synthetic advances towards the tailored and versatile architectural assembly of glycosylated nanoparticles (glyconanoparticles), considered as synthetic mimetics of natural glycoconjugates, have paved the way towards diverse biomedical applications. The accessibility of a wide variety of these structured nanosystems, in terms of shape, size, and organization around stable nanoparticles, have readily contributed to their development and application in nanomedicine [2]. In this context, glycosylated gold nanoparticles, glycosylated quantum dots (QDs), fullerenes, single-wall nanotubes (SWNTs), and self-assembled glyconanoparticles using amphiphilic glycopolymers or glycodendrimers have received considerable attention for their application in powerful imaging, therapeutic, and biodiagnostic devices (Fig. 1).

This chapter will provide an overview of the most recent syntheses and applications of glyconanoparticles that have deepened our understanding of multivalent carbohydrate–protein interactions. Together with sensitive detection devices, inhibitors of bacterial adhesion to host tissue, and cancer vaccines in therapeutic systems, including photosensitizers for photodynamic therapies (PDT), these novel bionanomaterials are finding widespread relevance. The recent development of

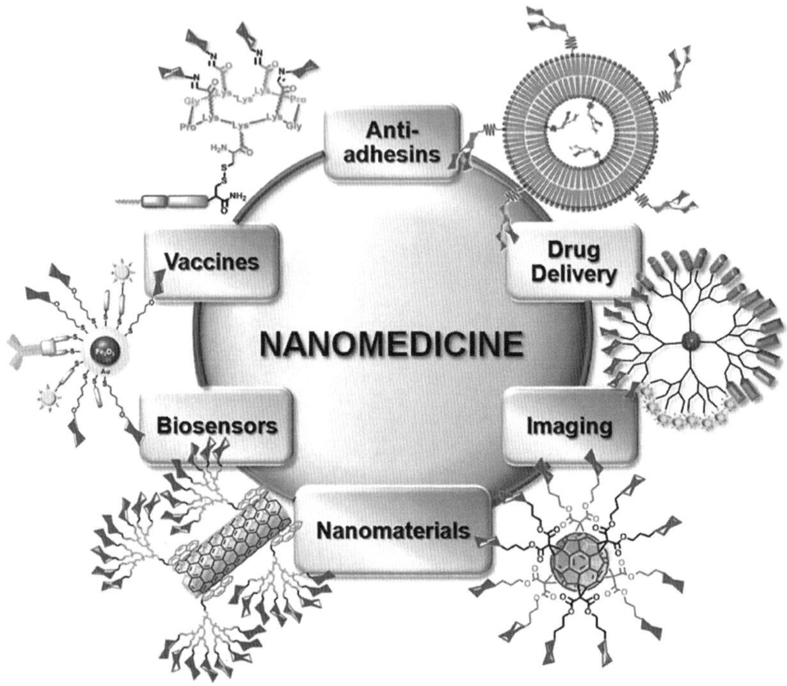

Fig. 2 Applications involving glycosylated nanoparticles

glyconanoparticle-based chemotherapeutic drug delivery agents will also be addressed, together with their molecular imaging properties for magnetic resonance imaging (MRI) or as X-ray contrast agents, as platforms for selective immunolabeling and imaging of cells, and as colorimetric devices in various bioassays (Fig. 2). Several specialized reviews have already dealt with most of these topics [3–9]; however, this chapter will comprehensively disclose a few examples for each of all known nanoparticle carriers (Fig. 1).

2 Glycodendrimers

A dendrimer is a synthetic highly branched monodisperse and polyfunctional macromolecule, which is constituted by repetitive units (so-called "generations") that are chemically bound to each other by an arborescent process around a multifunctional central core [10, 11]. Thus, dendrimers are structurally well defined and can be synthesized from a fully controlled iterative approach. The concept of repetitive and controlled synthetic growth with branching was first reported in 1978 by Vögtle et al., who achieved the construction of a low molecular weight "cascade" polyamine [12], followed by divergent macromolecular synthesis giving birth to the first true well-characterized dendrimers, named "arborols" [13] and "PAMAM" [poly(amidoamine)] dendrimers [14]. Until the mid-1990s, scientific efforts boosted the efficient control of the two main synthetic strategies that were used to construct perfectly branched dendrimers: the divergent and the convergent approaches [10]. Over time, accelerated alternatives have been developed in order to increase the synthetic efficiency and to reach a substantially higher number of peripheral functionalities. Furthermore, clever adaptations and combination of expeditious strategies based on suitably functionalized cores or dendrons have allowed the direct access of complex dendritic structures. The perfect control of tri-dimensional parameters (size, shape, geometry); the covalent introduction of functionalities to the core, branches, or high number of surface groups; or the physical encapsulation of functionalities in the microenvironment created by cavities can all help confer optimal properties of solubility and hydrophilic/hydrophobic balance for the required application. Thus, creativity has allowed these structures to become integrated with nearly all contemporary scientific disciplines.

Undoubtedly, the merging of biology and nanomedicine represents new fields that have generated the highest passion for these architectures. Thus, the very unique structures and properties of glycodendrimers have motivated their use in numerous disciplines including nanomedicine, with biomedical and therapeutic applications such as drug or gene delivery devices with beneficial EPR effect for anticancer therapy [15, 16], and antibacterial, antiviral, or antitumor agents. The use of dendrimers in biological systems, together with systematic studies of the most common dendritic scaffolds to determine their biocompatibility, including in vitro and in vivo cytotoxicity, biostability, and immunogenicity, have been extensively reviewed [17, 18]. One typical example concerns the use of dendrimers as "glycocarriers" for the control of multimeric presentation of biologically relevant carbohydrate moieties, which are useful for targeting modified tissue in malignant diseases for diagnostic and therapeutic purposes. In such molecules (glycodendrimers), the saccharide portions are conjugated according to the principles of dendritic growth or are ligated to pre-existing highly functionalized and repetitive dendritic scaffolds having varied molecular weights and structures. Since they first appeared in the literature in 1993 [19], glycodendrimers and related glycodendrons, with their spheroidal or dendritic wedge structures, have been designed as bioisosteres of cell surface multi-antennary glycans and have stimulated wide interest within the scientific community. Similarly to conventional dendritic structures, glycodendrimers can be obtained as dendrons, as spherical or globular architectures, or as "hybrid dendronized-polymers" according to divergent, convergent, or accelerated approaches. All these original synthetic clusters were constructed in such a way that their number of surface groups, shapes, and carbohydrate contents could be varied at will with a controlled integration of dendritic building blocks. As artificial glycoforms, the elaborated glycodendritic architectures are classically used to study, rationalize, and understand the critical carbohydrate–protein interactions that are reinforced when multivalent ligand copies are presented to clustered biological receptors through the well-known "glycoside cluster effect" [20, 21]. As such, they notably constituted powerful synthetic tools for the development of potent bacterial or viral inhibitors together with efficient inhibitors of mammalian lectins such as galectines, which are involved in cancer progression. Although the applications of glycoclusters and related glycodendrimers as potent biological process inhibitors or effectors have been extensively reviewed [22–31], the panel of related smart dendritic systems has recently been widened to generate promising glycosylated candidates as imaging agents or vaccines for instance.

In this context, Fukase and colleagues described the multivalent presentation of N-glycans containing or lacking sialic acid termini scaffolded around a dendritic poly-L-lysine backbone for the development of in vivo noninvasive imaging devices [32]. Considering the critical role of glycan residues in cell–cell recognition, adhesion, and signal transduction [33] and their implication as anti-inflammatory agents of immunoglobulin G (IgG) through Siglec interactions [34] with antibody-dependent cellular toxicity (ADCC) and/or complement-dependent cytotoxicity (CDC), the dynamics of these multivalent glycans have been systematically studied in vivo. The main goal was to achieve promising sensors with optimized structures and generating organ-specific accumulation. To this end, different generations of polyvalent polylysine-based glycodendrons have been generated by solid-phase synthesis to investigate the cluster effect. The architectures thus consisted in the mixed presentation of four (**1a–e**), eight (**2a–e**) and sixteen (**3a–e**) peripheral N-glycan derivatives, attached via CuAAc (copper-catalyzed azide-alkyne cycloaddition or "click chemistry") and terminal benzylated histidine. The presence of a focal lysine ε-amino group allowed the subsequent incorporation of ^{68}Ga-cryptand (^{68}Ga-DOTA) as positron emission tomography (PET) radiolabel or of Cy5 and NBD as fluorophores through straightforward 6π-azaelectrocyclization (Fig. 3). The administration of the glycoclusters labeled with ^{68}Ga-DOTA (**1a–3e**^{68}Ga-DOTA) to the tail vein of BALB/c nude mice allowed determination of the distribution of the glycosylated derivatives after a whole-body PET scan over 4 h. Remarkably, the in vivo dynamics and biodistribution appreciably differed from the different generations of clusters containing the bis-Neuα(2,6)Gal termini (**1a–3a**). Prolonged in vivo lifetimes and suitable biodistributions were observed for the larger derivative **3a**, suggesting its plausible efficient uptake by hepatocytes through the galactose/*N*-acetylgalactosamine (Gal/GalNAc) lectin receptor.

These observations highlighted the critical role of molecular size, hydrophobic/hydrophilic balance, and multivalency effects on the in vivo dynamics. While 4- and 8-mers (**1a** and **2a**, respectively) were rapidly cleared through the kidney, the larger congener **3a** was retained in the body after 4 h, with a substantial radioactivity detected in the liver, gallbladder, and blood. Importantly, the 16-mer homologue lacking the terminal sialic acid residue (**3b**) or presenting the bis-Neuα(2,3)Gal glycan (**3c**), were more rapidly cleared through the kidney, although **3b** accumulated in the liver due to the presence of asialoglycoprotein receptors in this organ. On the other hand, the variation of the specific sialoside linkage to galactose, i.e., the Neuα(2,6)Gal linkage, was of a crucial importance in the circulatory residence of N-glycans resulting in the uptake of **3a** in the liver, also avoiding excretion by the biofiltration pathway in the kidney, as observed for **3c**.

A biodistribution study was performed on 16-mer homologues (**3a–3e** Cy5) using fluorescence imaging. The importance of at least one specific Neuα(2,6)Gal epitope was reinforced, with higher in vivo stability observed for Cy5-labeled homogeneous **3a**, or heterogeneous **3d** and **3e**, with fluorescence accumulated both in the liver and the spleen after 4 h. Considering that the spleen could be involved in the immune system or in the production of antigen-specific antibodies by the interaction of T cells

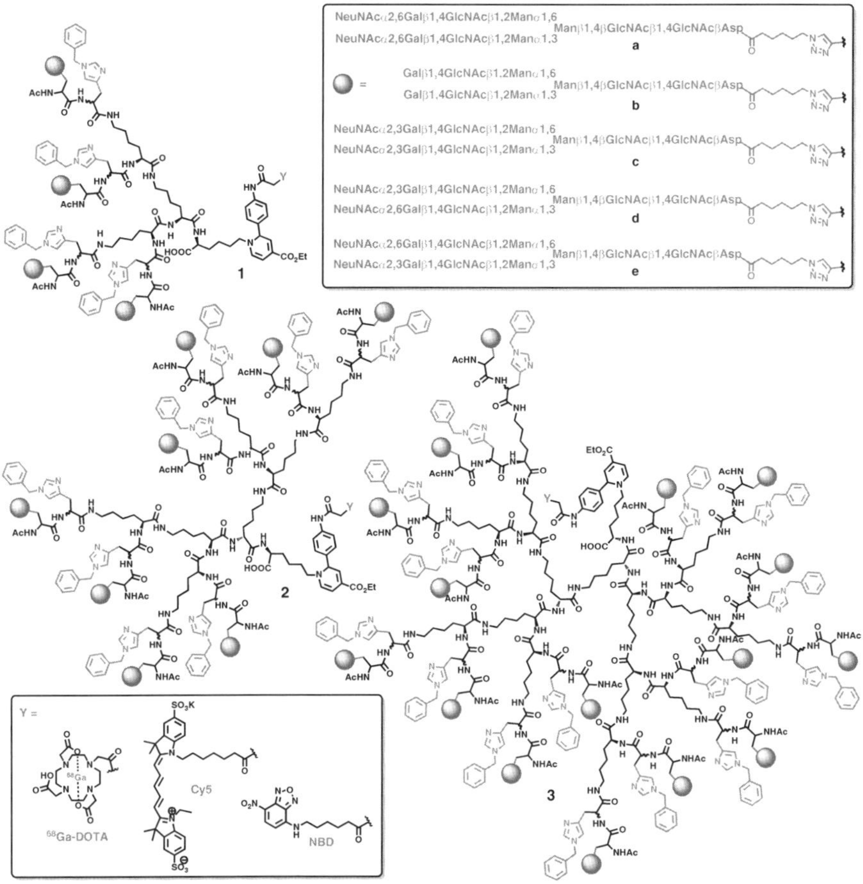

Fig. 3 Glycoclusters proposed by Tanaka and colleagues for in vivo dynamics [32]

with mature B cells [35], the authors investigated the interaction of glycoclusters **3a**, **3d**, and **3e** Cy5 with Siglec 2 (CD22), which is a Neuα(2,6)Gal-specific lectin overexpressed in mature B cells. In this context, the expressed murine of human CD22 only detected weak interactions with the ligands, suggesting their early capture by the reticuloendothelial system (RES) consisting of phagocytic cells such as monocytes and macrophages. An additional examination of the distribution of the ligands in a cancer model was performed after subcutaneous injection of DLD-1 colorectal adenocarcinoma in the dorsal division of BALB/cAJcl-nu/nu mice. Overall, preliminary results indicated that the clusters did not target the tumor tissues. Nevertheless, marked differences were achieved in in vivo dynamics, with accelerated excretion rates in the cancer-affected mice for glycoclusters **3d** and **3e** without accumulation in the spleen, whereas **3b** was efficiently retained, thus validating the proposed methodology as suitable for the development of in vivo molecular imaging in living animals.

Another promising therapeutic application involving glycodendrimers as functional antigens with requisite immunochemical abilities recently arose from their molecularly defined multivalent scaffolds and their abundant architectural variations [36]. Interestingly, although carbohydrates are T-cell-independent antigens with poor immunogenicity, their conjugation to protein carriers serving as T-cell-dependent epitopes generated glycosylated structures capable of acquiring the requisite immunochemical ability. This observation triggered numerous efforts directed towards synthetic carbohydrate-based vaccine candidates. More particularly, early examples of strongly immunogenic glycoclusters and glycodendrimers bearing critical components of tumor-associated carbohydrate antigens (TACAs) such as the immune-dominant T_N and T_F antigens (α-GalNAc-*O*-Ser/Thr and β-D-Gal-(1→3)-α-D-GalNAc-*O*-Ser/Thr, respectively) have been described as promising antitumor vaccines [37]. More recent investigations pointed to four optimum structural requirements for constructing successful glycoconjugate vaccines for cancer immunotherapy:

1. The presence of key TACA as recognition motif, acting as B-cell epitope against which antibodies should be raised.
2. The presence of $CD4^+$ T-cell peptide epitope and $CD8^+$ T lymphocyte (CTL) to trigger both humoral and cytotoxic immunity, respectively.
3. The incorporation of nonimmunogenic and nontoxic lipidic adjuvant targeting the Toll-like receptor 2 (TLR-2) to provide a self-adjuvanting property to the construct. Particularly, lipid-bound antigens can be directly recognized by TLRs, from which they are internalized within the antigen-presenting cells (APCs), processed, and finally exposed on the surface for $CD8^+$ natural killer cell activation [38–40].
4. Multivalency, either insured by the multitask platform used to expose the epitopes, or naturally generated through the propensity of lipid derivatives to form liposomes, particularly if glycolipid QS21 adjuvant is co-injected.

Dumy et al. capitalized on these cumulated observations to achieve a fully synthetic four-component antitumor vaccine, built on a nonimmunogenic cyclodecapeptide template (known as "RAFT" [41] for regioselectively addressable functionalized template) [42]. The optimized multiepitopic construct incorporated four copies of a T_N antigen analogue as TACAs (B-cell epitope), the universal $CD4^+$ helper T-cell peptide from the type I poliovirus protein, together with a $CD8^+$ CTL from ovalbumin ($OVA_{257–264}$ peptide SIINFEKL) to which was covalently added palmitic acid as lipid adjuvant. The synthetic versatility of the pre-activated RAFT platform displaying four glyoxoaldehyde functions allowed the efficient ligation of the aminooxylated sugar (T_N) derivatives on the platform via iterative oxime ligation. The peptidic epitopes were further combined to the scaffold through disulfide bridge formation to afford the desired vaccine NPs composed of ovalbumin glycol-lipopeptide (OVA-GLP) **4** (Fig. 4).

The study highlighted both B- and T-cell antigenicity as well as immunogenicity in vitro and in vivo. The authors first assessed the safety, immunogenicity, and protective efficacy of **4** using a MO5/BALB/c tumor mouse model. The investigations revealed the production of tumor-specific antibodies (Abs) developed against the T_N

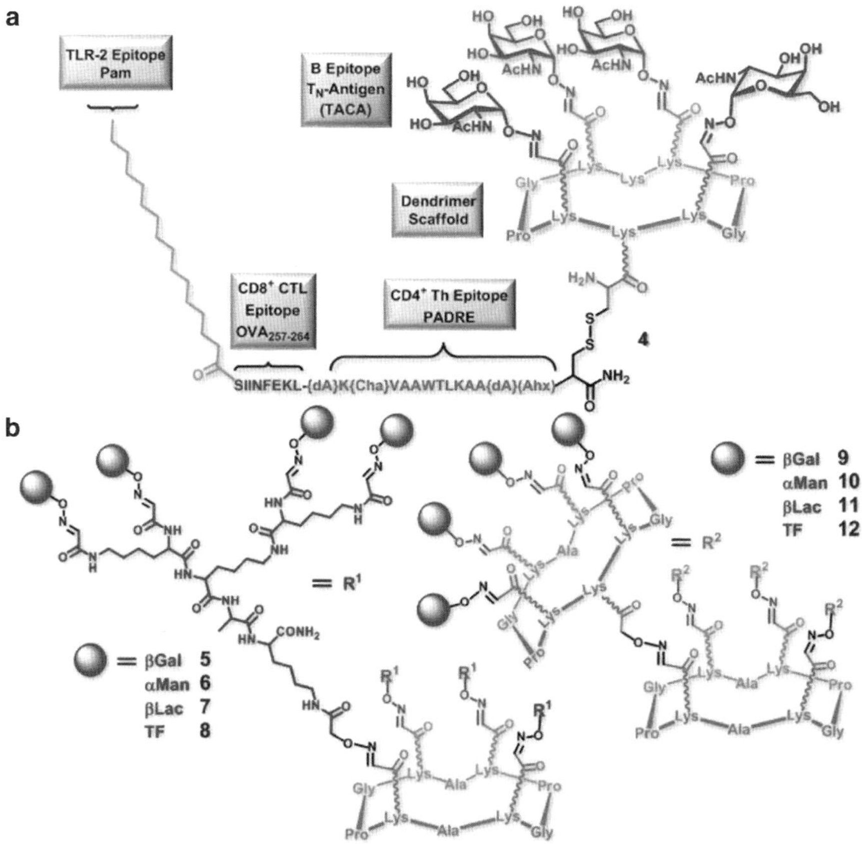

Fig. 4 (**a**) Recent OVA-GLP multicomponent TACA vaccine incorporating the requisite elements to raise both humoral and cytotoxic immunity. (**b**) Hexadecavalent dendri-RAFT structures proposed by Renaudet and colleagues

cluster displayed in the synthetic vaccine that efficiently recognized the native form of T_N antigen presented on human tumor cells, as determined by flow cytometry between the immune serum IgG and the common breast cancer cell line MCF7. The efficient stimulation of T cells was also observed, together with tumor regression and significant increase of survival in mice inoculated with MO5 carcinoma cells in both immunotherapeutic and immunoprophylactic settings [43]. These results clearly evidenced that vaccination with totally synthetic self-adjuvanting GLP prototype nanovaccine **4** induced a strong protection against tumors. For a deeper understanding of the cellular and molecular mechanisms involving GLPs and the influence of the position of the lipid moiety on immunogenicity and protective efficacy of GLPs, the same authors investigated the uptake, processing, and cross-presentation pathways of two different congeners. The main goal of this study was to determine whether the position of the lipid moiety in GLP molecules, positioned either on the N-terminal end or in the middle of the PADRE-CD8$^+$ T-cell epitope sequence, would affect the processing and presentation of T- and B-cell epitopes as well as their in vivo immunogenicity [44]. Towards this goal, two structural analogues bearing a $CD8^+$ CTL peptide taken from the sequence of human epidermal growth factor 2 ($HER_{420-429}$), a transmembrane receptor-like glycoprotein with tyrosine kynase activity, were synthesized and compared. The results indicated that the position of the lipid moiety not only profoundly affected the strength of immunotherapeutic efficacy, the uptake, and cross-presentation pathways for GLP in dendritic cells but also modulated the magnitude of antitumor antibody and $CD8^+$ T-cell protective immunity. Particularly, after challenge inoculation with cancerous MO5 cells, regression of established tumors was induced by both conjugates. However, the attachment of the palmitic acid moiety at the peptide N-terminal afforded the most advantageous results, including (a) stronger and longer lasting $HER_{420-429}$-specific IFN-γ producing $CD8^+$ T-cell response, (b) stronger DC maturation together with the production of potent TLR-2-dependent T-cell activation, and (c) higher inhibition of tumor growth.

These promising results suggest that RAFT scaffolds are suitable tools for engineering other potent synthetic anticancer vaccines with optimized structural variations, including: (a) controlled multi-and heterotopic antigen presentation on a single cyclic framework to mimic the heterogeneity of targeted cancers [45, 46], (b) a rigid structure to provide epitopes with limited conformational freedom towards their optimized and precise presentation for a cluster-recognizing tumor-specific antibody response, and (c) an enhanced number of TACAs through real dendritic structures. Towards this goal, and although immunogenicity of the related architectures have not yet been investigated, second-generation dendri-RAFTs exhibiting various topologies have been elaborated to access higher glycosidic valency, density, and spatial presentation with various levels of rigidity [47]. Hence, new series of molecularly defined hexadecavalent glycoclusters were synthesized in a controlled manner using robust and versatile divergent protocols allowing successful oxime attachment. The iterative strategy thus generated derivatives containing peripheral clustered Gal, Man, Lac, and T_F antigen analogue presented through flexible polylysine dendrons (**5**, **6**, **7**, and **8**, respectively) or constrained glycosylated cyclopeptides (**9**, **10**, **11**, and **12**, respectively) (Fig. 4). Binding assays were performed with model lectins and three human galectins [47, 48] that highlighted the beneficial influence of the enhanced multivalency on their inhibitory potencies and on their ability to protect cells from toxin and/or lectin binding. It can be safely anticipated that the combination of these structural features with those described earlier can provide NPs with enhanced anticancer vaccine properties.

3 Glycofullerenes

Fullerenes, the third allotropic form of carbon along with graphite and diamond, represent a novel class of globular shaped molecules made exclusively of carbon atoms. This intriguing structure has generated enthusiasm and many research efforts over the past few years [49]. Hence, the chemical and physical features of C_{60} (also named Buckminsterfullerene), the most representative structure among fullerenes, have extensively been explored. The intrinsic properties of fullerenes, such as size, hydrophobicity, three-dimensionality, and electronic properties have made them extremely promising nanostructures with interesting features at the interface of material sciences [50] with biological and medicinal chemistry [51–53]. Interestingly, initial investigations indicated that these novel and fascinating architectures were not carcinogenic when applied on skin and did not affect the proliferation and viability of cells when they were internalized. Hence, despite the observed dose-dependent toxicity of some derivatives, early observations suggested great promise for applications in DNA cleavage, photodynamic therapy, and enzymatic inhibition, and for their antiviral, antibacterial, and antiapoptotic activity [53, 54]. However, the total absence of solubility in aqueous or physiological media hampered their quick and efficient development as suitable carriers. The natural repulsion of fullerenes for water was cleverly circumvented by the application of several methodologies comprising their entrapment into tailored microcapsules, their suspension with the help of co-solvents, and their chemical derivatization, including their introduction onto peripheral solubilizing appendages. Furthermore, it has been shown that the multivalent presentation of polar groups around the fullerene spheres could avoid clustering phenomenon in reasonably dilute solutions and consequently increase the hydrosolubility of the resulting conjugates. In this context, a variety of chemical functionalities have been utilized both to increase the hydrophilicity (e.g., OH, COOH, NH_2, quaternary ammonium, cyclodextrin groups) and to prepare original compounds possessing biological and pharmacological activity.

Among the panel of fullerene derivatives proposed through the years, the family of fulleroglyco-conjugates (also called glycofullerenes) includes promising candidates that combine crucial properties related to water solubility with biological relevance. Their emergence particularly arose from the spherical topology of fullerenes that furnished suitable scaffolds for multivalent presentation of peripheral carbohydrate residues, and from the development of adapted chemistry for their efficient and controlled conjugation.

Besides numerous applications validating the beneficial glycoside cluster effect for lectin recognition and inhibition that have been treated elsewhere [22, 55], investigations on multivalent enzyme inhibition with globular glycofullerenes have recently emerged. Most of the enzymes possess a single and deep active site that is usually less accessible than the shallow CRDs present on multimeric lectins. These unfavorable structural features first dissuaded the syntheses of multivalent inhibitors. Additionally, limited success for enzyme inhibition remained disappointing [56, 57]. Nevertheless, two striking examples based on potent fullerene-based glycosylated structures presenting T_h-symmetrical octahedral patterns were recently presented.

The attachment of iminosugar analogues such as 1-deoxynojirimycin, known as glycosidase inhibitors, around the fullerene framework has been described [58]. In the context of carbohydrate-processing enzyme inhibition, biological assays involving a series of glycosidases were used to determine the inhibition profile of polytopic fullerene **13** decorated with twelve peripheral *N*-alkyl analogues of 1-deoxynojirimycin (Fig. 5). The presence of the elongated epitopes located at the extremity of the hexyl spacer and the general poor selectivity of this class of compounds made them attractive as models for the examination of the influence of multivalency on inhibition

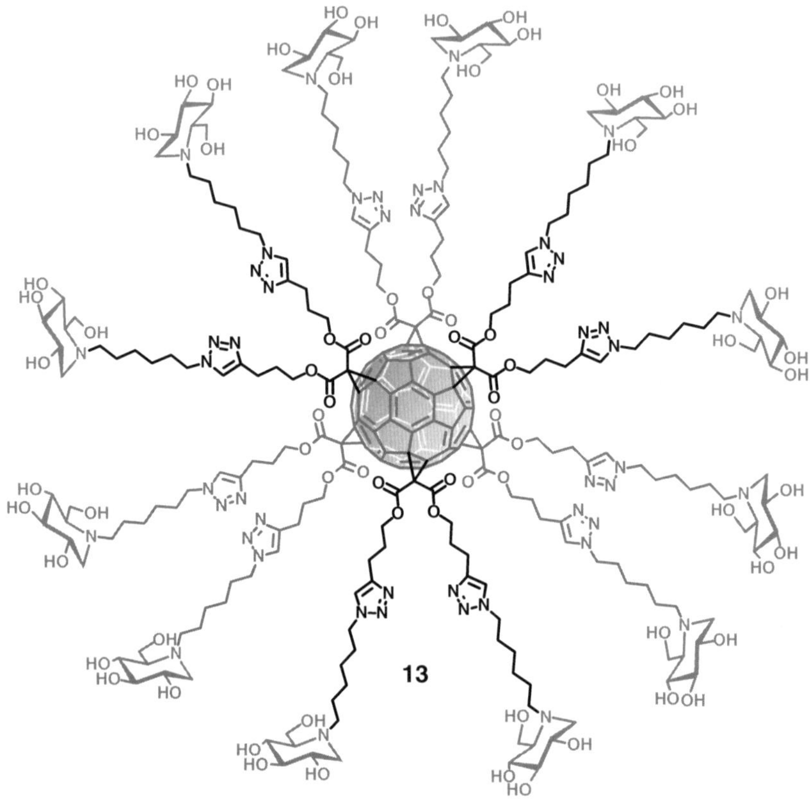

Fig. 5 Dodecavalent fullereno-glycocluster with iminosugars as glycosidase inhibitors

selectivity over a large panel of commercially available glycosidases. Although a significant decrease in affinity compared to the monovalent reference was observed on sweet almond β-glucosidase, investigations on amyloglucosidase and bovine liver β-glucosidase revealed no substantial improvements in using the multivalent approach. More interestingly, better results emerged from the inhibition of green coffee α-galactosidase, Baker's yeast isomaltase, and *Penicillium decumbens* naringinase with inhibition constants of two orders of magnitude higher than that of the corresponding monomer. The most spectacular result was obtained with the Jack bean α-mannosidase, with a 180-fold enhancement in valency-corrected efficiency of the globular dodecavalent conjugate **13** that compared very well with a previously described trivalent congener [59].

These results unequivocally demonstrated the existence of tailored presentation of peripheral recognition moieties insured by the rigid central scaffold. Clear multivalent effects beyond the expected statistical rebinding and local concentration effects was observed. In addition, the plausible implication of a sliding mechanism of the inhibitors in the enzyme active site or an additional binding process at a close allosteric site might explain the distinct specificity enhancement.

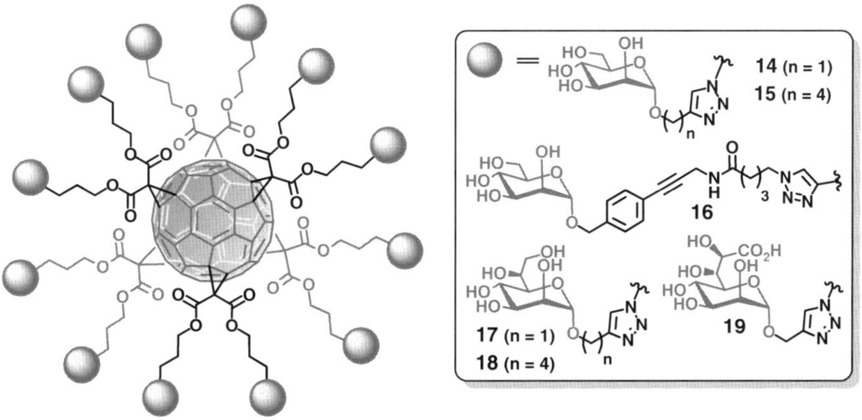

Fig. 6 Dodecavalent fullerene-based mannopyranosides with heptosyltransferase WaaC inhibition properties

This approach was further extended to the mannopyranose moiety of bacterial lipopolysaccharides (LPS) L,D-heptoside constituents [60]. Interestingly, the alteration of LPS structures that are key components of the outer membrane of Gram-negative bacteria can be achieved through glycosyltransferase (GT) inhibition. This structural variation may widely reduce their virulence by rendering them more susceptible to phagocytosis by macrophages and more sensitive towards hydrophobic antibiotics [61, 62]. The GTs involved in LPS biosynthesis thus represent attractive targets for the development of new treatments for infectious diseases, including antibacterial agents. The development of a new class of multivalent inhibitors of biologically relevant bacterial GTs, namely heptosyltransferase WaaC, which triggers the incorporation of the first heptosyl-unit of ADP-L-heptose onto the growing core moiety of LPS, allowed the demonstration of the beneficial implication of multivalent effects. Indeed, the inhibition of WaaC activity by the glycofullerenes was observable and effective. Glycofullerenes were decorated via CuAAc with different unprotected mannosides, equipped with various linkers and terminated with appropriate azido or alkyne functionalities, which were attached to the rigid framework. Hence, a small library of monodisperse dodecavalent conjugates was prepared, containing either simple mannosides with alkyl chains ($n = 1$ and 4 for **14** and **15**, respectively) or more sophisticated linkers (**16**), and (pseudo)mimetics of the ADP-L-heptose termini as heptosides (**17** and **18**) or octosides (**19**). The former were engineered by introduction of an ionic functionality (terminal carboxylate) to provide additional coulombic interactions with specific amino acids in the binding pocket of the WaaC heptosyl-transferase (Fig. 6).

The specific activity of the enzyme was followed by a coupled enzymatic assay involving pyruvate kinase to transform the ADP reaction product to ATP, giving access to IC_{50} values for each candidate. Interestingly, all multivalent derivatives displayed low micromolar inhibition levels, with IC_{50} values ranging from 7 to 47 μM with strong improvements of at least one order of magnitude over the corresponding monovalent references (500 μM) in most cases. As anticipated, the overall results suggested that additional interactions with the transferase were observed with the incorporation of the appropriate functionalization at the 5-position of the carbohydrate unit, considering multivalent analogues with the same spacers (**14**, **17** and **19**). The marked decrease in IC_{50} values along this series when going from the hexose to the octose derivative demonstrated this tendency, with the best inhibition profile for octoside **19**. The presence of shorter spacer and the introduction of aromatic moiety in the aglycone also tended to improve the binding properties.

Although significant levels of multivalent inhibition have been measured for two different enzymes in the presence of dodecavalent glycosylated C60 fullerenes, more investigations have to be performed to rationalize these promising results and optimize the combination of multivalency and glycomimetic for novel antibacterial agents.

4 Glyconanotubes

Carbon nanotubes (CNTs) represent members of the fullerene structural family and consist exclusively of carbon atoms arranged in a series of condensed benzene rings, organized as graphite sheets rolled-up into tubular structures [63]. According to specific nomenclature, single-walled CNTs (SWNTs) are made up of single graphene layer wrapped into cylindrical structures, whereas multi-walled CNTs (MWNTs) are generated from a central tubule of nanometric diameter and surrounded by several graphite layers spaced by a distance of about 0.34 nm. Typically, CNTs form bundles that are entangled together in the solid state giving rise to highly complex "spaghetti" networks [64]. Intrinsically, CNTs possess very interesting and unique physicochemical properties such as high surface area, ordered structure with high aspect ratio, ultralight weight, excellent chemical stability, high electrical conductivity, high thermal conductivity, and metallic or semiconducting behaviors. For these reasons, these nanomaterials have raised great enthusiasm and expectations in a wide range of different applications including material, biological, and medical sciences when suitable functionalizations are realized [65]. In fact, their inherent hydrophobic properties, associated with their natural propensity to form bundles via aggregation through van der Waals forces, constitute major technical barriers for their utilization in medicinal applications. To overcome these drawbacks, the modification of their surface is typically achieved by adsorption, electrostatic interaction, or covalent attachment of various appendages to reach the specific targeted properties [66]. Through such modifications, the water solubility of CNTs is generally improved and the aggregation phenomenon reduced. Furthermore, their biocompatibility and biodistribution profiles are completely transformed and improved. Particularly, recent investigations boosted the emergence of functionalized CNTs as new biologically relevant alternatives for therapeutic and diagnostic applications [67], including their use as vectors for delivery of therapeutic molecules [68] such as plasmid genes [69], peptides/proteins [70], and antibiotics [71]. More interestingly in the context of this chapter, these nanosystems recently acted as useful tridimensional platforms for carbohydrate presentation. The resulting densely oriented sugars as pendant groups

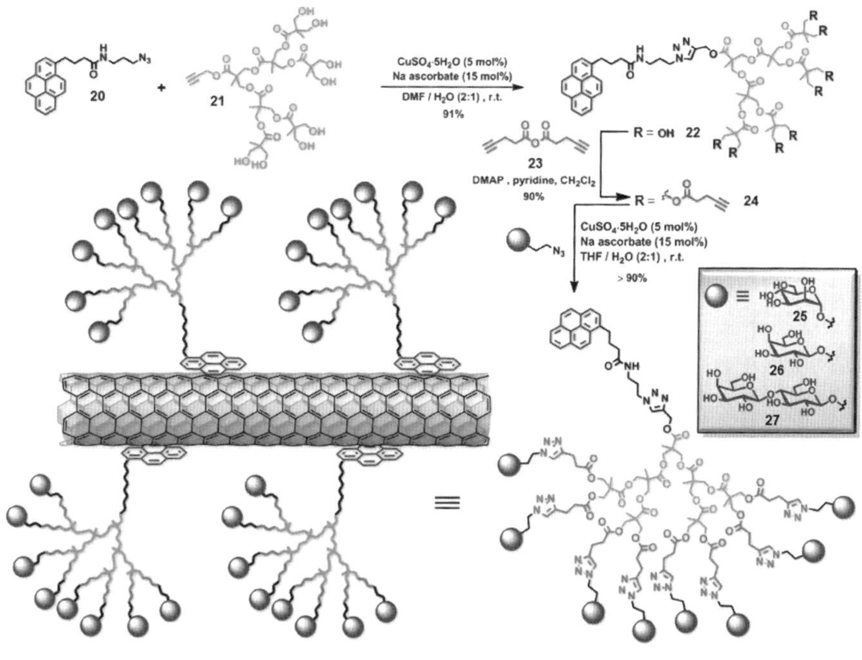

Fig. 7 Synthesis of glycodendrons containing hydrophobic and fluorescent pyrene head group and scheme of the SWNT coating

on the CNT hybrids become multivalent carbohydrate ligands showing potential biological activities as well as glycoconjugate polymers. Hence, multivalent presentation of simple sugar residues or clustered wedges have been described via a non-covalent stabilization process through favorable hydrophobic π–π stacking, or electrostatic interactions in water between the absorbates and CNTs, or in a more controllable fashion involving covalent functionalization [22, 72, 73].

The ability of pyrene-terminated glycodendritic wedges to interact with SWNTs through π–π interactions has been described [74]. The homogeneous bioactive coatings ensured the foreseen biocompatibility and hydrosolubility that made the cell-surface nanosystem mimetics suitable for biological evaluation. The synthesis of the sugar appendages was again initiated with a CuAAC methodology involving azido-terminated pyrene derivative **20** and a dendronized polyol **21** having a propargyl function at the focal point. The quasi-quantitative coupling afforded triazole **22**, which upon subsequent elongation with pent-4-ynoic anhydride **23** generated the G(3)-dendritic precursor **24** exhibiting eight alkyne groups at the periphery (Fig. 7). A similar strategy was also described for the construction of lower generation G(2) wedges containing four alkyne termini with suitable building blocks. The last step of the synthetic protocol involved the attachment of unprotected 2-azidoethyl mono- or disaccharide using CuAAc to give the octavalent glycodendrons **25**–**27** in high yields.

Mixtures of SWNTs suspended in aqueous glycodendron solutions resulted in complete solubilization through specific adsorption of the glycosylated pyrene motif onto the nanotubes, regardless of the nature of the sugars. The suspensions of functionalized SWNTs exhibited remarkable stability in water for several months. Furthermore, unlike thick and heterogeneous coatings generated with the previous use of glycopolymers [75], the well-defined glycodendrons uniformly wrapped the nanostructure to form small bundles, as revealed by scanning electron microscopy (SEM) and transmission electron microscopy (TEM).

The capability of the assemblies to bind to specific proteins was probed with different labeled lectins using fluorescein isothiocyananate (FITC)-conjugated *Canavalia ensiformis* (Con A), *Arachis hypogaea* (PNA) and *Psophocarpus tetragonolobus* (PTA) agglutinins, which recognize α-mannose, lactose, and β-galactose, respectively. As expected, specific recognition-induced fluorescence was generated and measured by fluorescence microscopy when mannosylated **25**-SWNT was treated with Con A, whereas no substantial signal was detected from unrelated PTA and PNA lectins. Similarly, galactosylated **26**-SWNT and lactosylated **27**-SWNT bound to FITC-PTA and both labeled PNA and PTA, respectively. Additional experiments were conducted with mixed glycosylated structures around the SWNT to more accurately model the sugar diversity present on cell surfaces. Hence, SWNTs functionalized with a mixture of mannosylated and lactosylated G(2)-dendrons at various ratios, were incubated with a 1:1 mixture of Texas Red-conjugated PNA and FITC-labeled Con A. Importantly, fluorescence intensities associated with specific recognition events simultaneously increased with the percentage of complementary ligands presented in the initial mixtures. These results strongly suggest that multiple epitopes displayed on SWNTs could bind simultaneously to discrete proteins. The authors took advantage of the functionalized SWNT-associated specific fluorescence to investigate and observe the strong interaction of the labeled nanosystems with Chinese hamster ovary (CHO) cell surfaces. In fact, the remaining binding sites of tetravalent Con A, which partly bound to mannosylated SWNT structures, offered additional opportunities for subsequent complexation with Man residues present on cell-surface glycans. Finally, G(2) and G(3) glycodendrimer-coated SWNTs induced no cytotoxicity in the presence of HEK293 cells whereas unfunctionalized nanotubes hampered their growth.

CNTs also represent attractive candidates for multifunctional carrier systems by virtue of their inherent dual role both as hosts for active payload that can be located in the internal spaces and as active frameworks for subsequent covalent chemical surface modifications. An eye-catching example demonstrating that filled and carbohydrate-functionalized SWNTs could be used as efficient radioprobes with specific ability to target organs in vivo has been published [76]. The authors first filled SWNTs with metal halide such as CuBr and the more relevant $Na^{125}I$, carried out at high temperature. The subsequent cooling process was responsible for closing of the ends of the SWNTs, resulting in the desired encapsulation of high-density radio-emitting nanocrystals by using molten phase capillary wetting.

The successful sealing and filling by metal halides were confirmed by high-resolution TEM and scanning TEM.

The surface of these filled SWNTs was then covalently modified with bi-antennary D-*N*-acetylglucosamine (GlcNAc) dendron to further improve the water dispersibility and the biocompatibility of the nanosystems, and to provide the ability to target lung tissues. For this, mild azomethine ylid 1,3-dipolar cycloaddition (Prato reaction) [67] allowed the attachment of PEGylated glycoconjugate appendages together with the 2,3,5-triiodophenyl motif used as tagging agent, with the preservation of the tubular structure and the closed ends of the filled SWNTs. Considering the fact that iodine radioemitters used for treatment of thyroid cancer are still considered to be among the most effective for systemic radiotherapy, the potential of the resulting functionalized ^{125}I-nanocapsule **28** to act as defined containers capable of localizing radionuclides was investigated (Fig. 8a). Intravenous administration of **28**, in the tail vein of BALB/c mice was tracked in vivo using single-photon emission computed tomography (SPECT). Results suggested strong differences in tissue and organ biodistribution of the filled and glycosylated **28** compared to non-encapsulated Na^{125}I. Particularly, nanoencapsulation of ^{125}I prevented leakage of the radionuclide to organs such as the thyroid and the stomach, known to have innate affinity for iodide, and also avoided urinary or biliary excretion of the iodide, even after 7 days (Fig. 8b, c). In fact, when **28** was administrated, the tissue distribution profile indicated predominantly specific lung accumulation and persistent retention of the construct, while no signals were detected in the thyroid, stomach, or bladder at $t = 0$ or after 4 h, 24 h, or 7 days. This specificity was rationalized by the authors to be attributed to the presence of lung-localized proteins able to bind the GlcNAc ligands presented around the nanotube framework. Results also highlighted the high in vivo stability of the synthetic radioemitter. Additional quantitative study to assess the organ biodistribution and blood-clearance profile of the nanotube derivatives by direct gamma counting reinforced the initial observations for a strong lung accumulation and negligible blood content 3 minutes after administration (1% versus 27% for free Na^{125}I). The percentage of injected dose per gram tissue confirmed that more than 80% of ^{125}I persisted in the lung for 24 h only when inside SWNTs. In addition, histological examination of different organs such as lung, liver, spleen, and kidneys realized after administration of **28** further confirmed lung accumulation and indicated no sign of necrosis or fibrosis. Finally, no cytotoxic response was obtained from a human lung epithelial cell line (A549) following interaction with **28**, even with prolonged exposition times and concentrations up to 125 μg/mL, as determined by in vitro cytotoxicity assay (LDH). In summary, the specific tissue accumulation in the lung, coupled to high in vivo stability, prevented the leakage of radionuclide to high-affinity organs (thyroid and stomach) and excretion, and resulted in ultrasensitive non-invasive imaging devices able to deliver an unprecedented radiodose density.

Those examples thus paved the way for the conceptualization of optimized multi-task organ-specific therapeutics and diagnostics constructed using CNT nanocapsules surrounded by the multivalent presentation of relevant glycoconjugates.

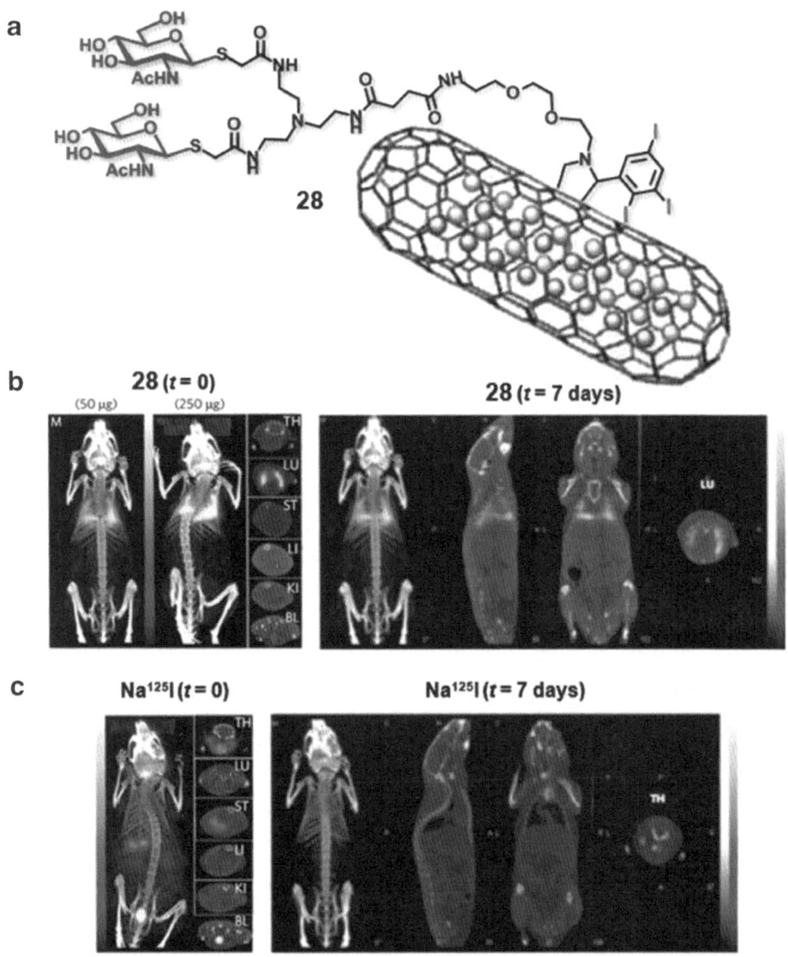

Fig. 8 (**a**) Structure of the SWNT-based ^{125}I-nanocapsule **28** used as radioprobe. (**b**) Whole body SPECT/CT imaging carried out immediately after tail-vein injection (*left*) and 7 days post-injection of **28** (*right*) (50 and 250 μg). (**c**) Whole body SPECT/CT imaging carried out immediately after tail-vein injection (*left*) and 7 days post-injection of Na^{125}I (*right*). Cross-sections of the thyroid (TH), lung (LU), stomach (ST), liver (LI), kidney (KI), and bladder (BL) at equivalent time points are shown. Adapted with permission from Mcmillan Publishers Ltd from [76]. Copyright 2010 Nature Publishing Group

Another elegant application of glycosylated SWNT devices has recently been reported that involved their preparation into electrolyte-gated field-effect transistors (FETs) to afford nanoelectronic detection of lectins [77]. Towards this goal, non-covalently functionalized SWNTs incorporating tetravalent porphyrin-based glycoconjugates $\mathbf{29}_{\mathbf{porphyrin}}$, $\mathbf{30}_{\mathbf{porphyrin}}$, and $\mathbf{31}_{\mathbf{porphyrin}}$ acted as conducting channels that transduced the specific binding between glycoconjugates and lectins into electrical signals. Therefore, specific molecular recognition was efficiently probed between

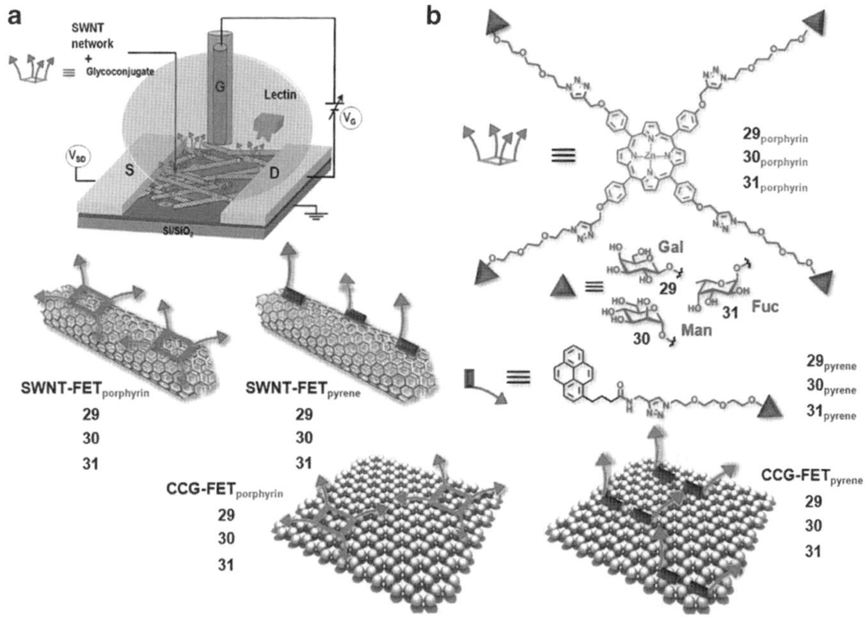

Fig. 9 (**a**) Representation of glycoconjugate-functionalized SWNT-field effect transistor (SWNT-FET) as sensitive lectin nanodetection platforms. (**b**) Structures of porphyrin- and pyrene-based glycoconjugates, together with illustrations of nanocarbon FET devices. Adapted with permission from Mcmillan Publishers Ltd from [77]. Copyright 2010 American Chemical Society

three different porphyrin systems carrying β-D-galactose (**SWNT-FET$_{porphyrin}$ 29-Gal**), α-L-fucose (**SWNT-FET$_{porphyrin}$ 30-Fuc**), and α-D-mannose epitopes (**SWNT-FET$_{porphyrin}$ 31-Man**) displayed on SWNTs and the corresponding bacterial PA-IL, PA-IIL, and plant Con A lectins, respectively (Fig. 9). The results indicated that substantial responses were induced by the specific interactions between the saccharides and their cognate lectins, without nonspecific adsorption on Zn-porphyrins or SWNTs. This observation corroborated that the glycoconjugate functionalization served the dual purpose of selectively detecting lectin binding as well as preventing nonspecific protein binding to SWNTs. Moreover, the sensitivity and the selectivity of the methodology was demonstrated, with detection limit of SWNT-FET devices being comparable to more traditional techniques used for lectin detection such as optical microarray, electrochemical surface plasmon resonance (SPR), electrochemical impedance spectroscopy, or voltammetric and colorimetric protocols. The IC_{50} values compared well with those previously determined by SPR or by hemagglutination inhibition assay (HIA), thus confirming that glycosylated SWNT-FETs could be used for the quantitative determination of affinity constants.

The same group also compared the performance of these SWNT-FET devices with similarly functionalized chemically converted graphene-based FET systems (**CCG-FET$_{porphyrin}$ 29-Gal**, **CCG-FET$_{porphyrin}$ 30-Fuc**, and **CCG-FET$_{porphyrin}$ 31-Man**), also adding pyrene-terminated appendages to extend the applicability of these nanodetection platforms [78]. As evoked earlier, glycosylated pyrene derivatives $\mathbf{29}_{\mathbf{pyrene}}$, $\mathbf{30}_{\mathbf{pyrene}}$, and $\mathbf{31}_{\mathbf{pyrene}}$, synthesized via click chemistry, were able to form π–π stacking interactions with SWNT and graphene surfaces to generate corresponding $\mathbf{SWNT\text{-}FET}_{\mathbf{pyrene}}$ **29–31** and $\mathbf{CCG\text{-}FET}_{\mathbf{pyrene}}$ **29–31**. The general tendencies suggested that SWNT-FET induced a larger response and better selectivity than CCG-FET devices, regardless of the glycoconjugate appendages exhibited through the pyrene or porphyrin framework. Computational modeling revealed that the one-dimensional SWNT systems provided more flexibility for the sugar termini to interact with lectin, leading to higher sensitivity than the two-dimensional CCG systems. More specially, the stacking of pyrene or porphyrin moieties of the glycoconjugates on the SWNT yielded galactose residues that were accessible for PA-IL lectin, whereas CCG congener provoked the epitopes' presentation only in close proximity to the carbon sheet, reducing their accessibility. Atomic force microscopy (AFM) imaging was performed to obtain evidence for the attachment of glycoconjugates onto the carbon nanostructures, and showed substantial and progressive increase in height at different stages of functionalization and during specific lectin interactions. This technique, together with fluorescence microscopy, also allowed visualization of the continuously high coverage on SWNTs whereas the binding sites were scattered randomly on the CCG surface with a reduced number of adsorption sites for lectins, in accordance with the previous computer modeling studies. Finally, these comparative studies confirmed the potential of these carbon nanomaterial-based FET systems as effective bacteria or virus detection devices in water, soil, or human fluids.

5 Gold Nanoparticles

Gold NPs are the most stable and studied of the metal-based nanoclusters. Owing to their exceptional electronic, magnetic, and optical properties that are highly size dependent, they have been extensively studied in material and biomedical applications. Even though a large number of methodologies are known for the preparation of gold NPs, the Brust technique and its variations are still the most popular for preparing a monolayer of coated gold NPs. The technique is based on the high affinity of the thiol groups for the gold atoms because of the soft character of both Au and S atoms [79]. In this method, gold NPs are synthesized through reduction of gold salt ($HAuCl_4$) by $NaBH_4$ in the presence of a thiol. Gold NPs coated with glycan monolayers have been studied extensively in various areas of biomedical research [80–82]. Since the first report in 2001 [83], gold glyconanoparticles (gold GNPs) have received great attention among glycochemists. Here, we discuss a few recent examples of gold GNPs that have been used for biomedical applications related to diagnostics and therapy.

Gold NPs have been used as multivalent and multifunctional platforms for tumor-associated carbohydrate antigens in combination with suitable immunogenic carriers to develop carbohydrate-based vaccines. The pioneering group of Penadés was the first to identify gold NPs as potential scaffolds for vaccine preparation [84].

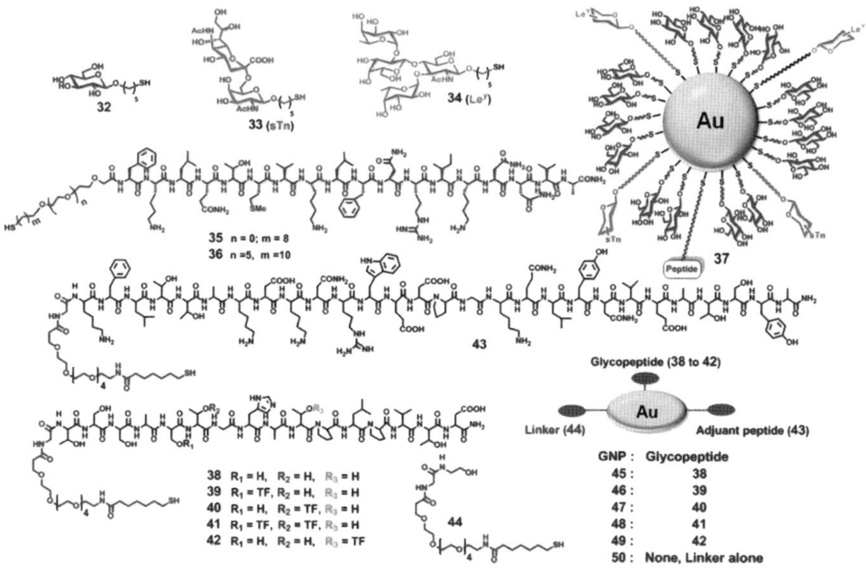

Fig. 10 Molecular structures of the gold NP-based tumor vaccine constructs

Accordingly, in an early report, hybrid NPs **37** have been synthesized by combining the Tn and Lewis oligosaccharides tumor markers (**33** and **34**), a peptide from tetanus toxoid (TT) as T-cell helper (**35** or **36**), and an inert glucose moiety **32** to control the density of the above conjugates (Fig. 10). Preliminary data showed that gold GNPs **37** exhibited potent immunogenic activity. In a very recent work, Barchi Jr. and coworkers reported another potential tumor vaccine construction prepared from gold NPs [83]. Gold GNPs (**45**–**50**) have been loaded with three components that included: (a) glycopeptides **38**–**42** consisting of Mucin-4-based peptides with the Thomsen–Friedenreich antigen (TF) at various positions, (b) a 28-residue peptide from complement-derived protein C3d as a B-cell activating molecular adjuvant **43**, and (c) an inert linker **44** (in Fig. 10, the linker portion is highlighted for glycopeptides and adjutant peptide). In vivo studies on sera from mice immunized with GNPs showed small but statistically significant antibody responses (both IgG and IgM) against the carbohydrate antigen. In addition to TF-functionalized gold GNPs (**46**–**49**), glycoconstructs from the peptides without the TF-antigen **45** or linker alone **50** also showed some immunogenicity. Among the TF-functionalized gold GNPs, the mono-functionalized conjugates (**46**, **47**, and **49**) showed the highest activity of the GNPs with two TF units (**48**).

The respective positioning of the TF antigen along the peptide chain, as in GNPs **49** with the antigen at the 10th position, showed better selectivity toward IgG over IgM, whereas conjugates with the sugar at other positions (**46** and **47**) exhibited similar selectivity for both IgG and IgM. Even though the results are still preliminary, the efficiency of the NPs to acts as scaffolds for vaccine construction was shown successfully in the above applications.

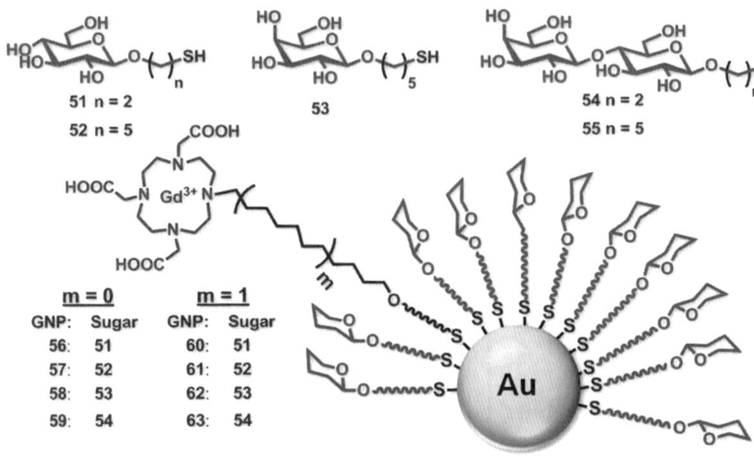

Fig. 11 Molecular structures of gold GNPs **56–63**

Paramagnetic gold NPs functionalized with sugar conjugates and Gd complexes have been studied for MRI applications [86]. A library of GNPs (**56–63**) was synthesized with variations in the sugar residues and spacer (Fig. 11). Thiol-ended sugars (**51–54**) and *N*-alkyl (pentyl or undecanyl)tetraazacyclododecane triacetic acid (DO3A, a chelating agent for Gd) have been coated onto gold NPs and subsequently chelated with the Gd metals, which led to the formation of paramagnetic GNPs. The longitudinal relaxitivity (r_1) values obtained from the relaxation time T_1 have been found to depend on the nature of the sugar and the spacer between the gold NPs and sugar head groups (i.e., the relative position of the sugar with respect to Gd complex). Gold GNPs **62** and **63** exhibited the highest r_1 values of the series (~20 $mM^{-1}\ s^{-1}$), which is about six times higher than that of commercial contrast agent Dotarem (3.1 $mM^{-1}\ s^{-1}$). In vitro imaging studies with lactose-containing GNP **63** showed better contrast abilities than Dotarem. On the other hand, in vivo studies of glucoside-functionalized GNPs **61** in mice gliomas generated from GL261 tumor cells, demonstrated enhanced contrast in tumor zones in comparison to clinically used substances. Lactosylated conjugates **63**, although showing activity outside the brain, were unable to reach the brains, which was thought to be due to the liver uptake of GNPs mediated by the known galactose–a-sialoglycoprotein receptor interactions. The above assays showed the influence of sugar on the in vivo behavior of GNPs.

Russel and coworkers reported other gold GNPs-based colorimetric sensors for the rapid detection of cholera toxin (CT), secreted by the bacterium *Vibrio cholera* [87]. The assay was developed based on the visible color change resulting from the shift in the surface plasmon absorption upon NP aggregation. CT is known to bind to the GM1 ganglioside oligosaccharides present at the surface of the mucosal cells, with the galactoside residues representing the major epitope [88]. Taking the advantage of the galactose specificity of the CT lectin, gold NPs coated with the galactoside-containing lactose disaccharide (**54**) (Fig. 11) have been synthesized

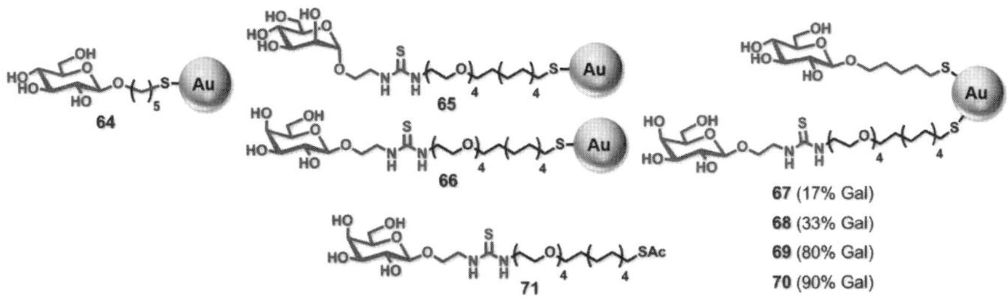

Fig. 12 Molecular structures of gold GNPs with sugar appendages (**64–70**) used as bacterial sensors

for the detection of CT. Upon binding with CT, due to the multiple binding events between the CT lectin and the gold GNPs, the monodispersed GNPs formed aggregates, which resulted in visible color changes from red to deep purple. Upon aggregation, the UV–vis absorption spectra were shifted from 524 to 620 nm and the amount of bound protein could be quantified by measuring the absorbance at 620 nm. To mimic the biological samples, the efficiency of the assay was also assessed in the presence of other ions present in the human stool samples. The developed assay was found to be robust and the added ions had no effect on the aggregation of NPs upon binding to CT lectin. Using the colorimetric assay, it was possible to detect and quantify the very low concentration (110 nM) of the CT lectin within 10 min. The stability of the NPs was assessed with the freeze-drying process, which did not have any effect on the CT detection abilities of gold GNPs.

Perez and coworkers reported the synthesis of gold NP-based multivalent galactoside clusters and studied their interactions with the bacterial *Pseudomonas aeruginosa* PA-IL lectin [89]. To assess the effect of the sugar density upon protein binding, gold GNPs (**66–70**) showing various ratios of "active glycans" were synthesized in combination with inert glucoside residues taken as negative control (Fig. 12). The protein binding studies of the galactoside-coated gold NPs with the lectin PA-IL were assessed using HIA, SPR and ITC techniques. In the HIA assays, galactose gold NPs showed 10- to 100-fold enhanced activity compared to monovalent glycoside **71** in the inhibition of agglutination of rat erythrocytes mediated by the PA-IL lectin. However, there was a difference in the kinetic parameters obtained from ITC and SPR techniques. The binding affinity of the GNPs was very high compared to monovalent glycosides and, for the GNP series, the affinity increased with increasing the sugar density. NPs loaded with glucoside or mannoside (**64** and **65**) did not show any activity with PA-IL, thus confirming the specificity of the interactions. The activity of mannose GNPs **65** with the lectin BC2L-A (a mannose-specific lectin) served as positive control. ITC data showed that the gold GNPs with 100% galactoside (**66**) residues exhibited 2,800-fold enhancement in K_D value (50 nM) compared to monovalent compound **71**. This provided an additional strong demonstration of the classical example of the glycoside cluster effect, and gold GNP **66** is one of the most efficient ligands known against the bacterial PA-IL lectin.

6 Quantum Dots and Quantum Rods

Nanocrystals with semiconducting metals are broadly referred as quantum dots (QDs). QDs are known to exhibit unique optical properties, such as broad absorption, size-dependent photoluminescence with high quantum efficiency, long-term photostability, and low photobleaching. QDs have been synthesized from various combinations of metals to provide selected photoluminescence properties [2]. Due to their inherent photoluminescence properties, QDs coated with glycan units have been used in several biomedical applications, especially as luminescent labels in animal and cell imaging studies and as a fluorescent probes for studying carbohydrate–protein interactions [90–92]. Although the semiconducting metals are generally toxic in nature, their toxicity can be reduced significantly by coating with hydrophilic glycan monolayers for their use in vivo. In this section, we describe a few recent applications of glyco-QDs that have been used as imaging agents, nanosensors, and drug delivery vehicles. Nishimura and coworkers demonstrated the synthesis of multifunctionalized glyco-QDs with various sugars and studied their in vivo imaging applications [93]. QDs (CdSe/ZnS or CDSeTe/CdS) have been functionalized with phosphorylcholine derivative **72** in combination with amine functionalized thiol **73**, which was subsequently used for the incorporation of glycans (**74–80**) (Fig. 13). Further enzymatic modifications of the GlcNAc-QDs **76** afforded the more complex functionalized oligosaccharides (**81–84**) at the surface of the nano-objects. The formation of glyco-functionalized QDs was confirmed by MALDI-ToF MS (matrix-assisted laser desorption/ionization time-of-flight mass spectrometry) analysis. In vivo near-infrared fluorescence imaging studies in mice demonstrated the stability and the non-fouling nature of the glyco-QDs. Most of the glycan-functionalized QDs shown in Fig. 13 (**74–79**, **81** and **83**), except for sialylated oligosaccharides, accumulated rapidly in the liver after 5–10 min of tail vein injection in mice and the fluorescence of the QDs was almost quenched over 1 h in most cases, indicating the degradation of glyco-QDs.

On the other hand, glyco-QDs having terminal sialic acid residues (**82** and **84**) exhibited long-term stability in the body 2 h after injection, without noticeable accumulation into the liver, and were distributed among various tissues. The study emphasized the influence of the neighboring sugar group on the in vivo characteristics of GNPs, such as stability and organ-specific distribution and consequently on target-specific drug delivery applications. Live animal imaging studies for the Lewis X-functionalized nano-objects **83** were conducted for the first time.

Ragusa and coworkers studied the cellular uptake and in vitro drug release abilities of glyco-quantum rods (QRs) (**85** and **86**) functionalized with lactose and dopamine (Fig. 14) [94]. In addition, the fluorescent QRs were used as nanosensors in protein interaction studies. The galactoside ligands were introduced via reductive amination of reducing lactose to an amine-functionalized QR, whereas the dopamine drug was installed through ester bonds with the hydroxyl groups of the galactose. The bioavailability of the galactose groups of the QRs was verified in a protein binding study with *Ricinus communis* (RCA_{120}) lectin. After the lectin binding, the quenching of photoluminescence properties of QRs **85** and **86** were observed. Although the

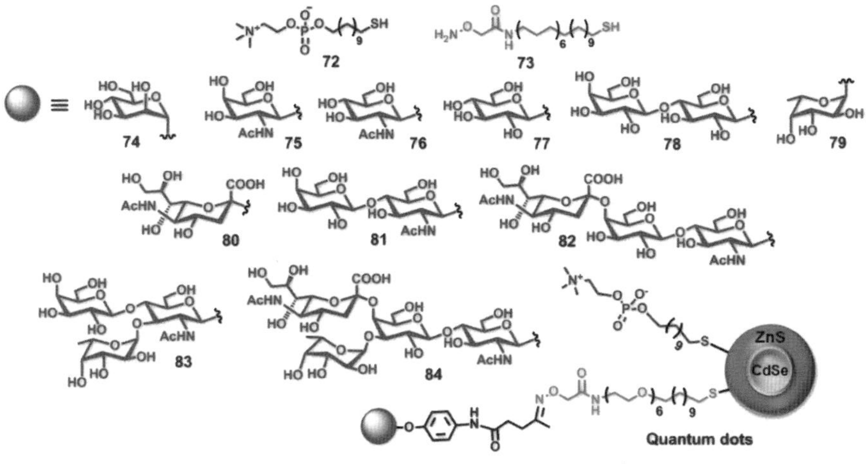

Fig. 13 Molecular structures of glyco-QDs **74–84**

mechanism for the fluorescence quenching was unclear, the dissociation constants of QRs **85** derived from the fluorescence quenching studies was in the order of 10^7, which was three to four orders of magnitude higher than that of the β-D-galactoside monomers, indicating the enhanced interactions of the clustered QRs with lectin. The specificity of the interactions was confirmed by competitive studies through the addition of galactose, which caused complete recovery of the fluorescence intensity of the QRs. In vitro enzyme hydrolysis studies using porcine liver esterase, the regulated release of dopamine, demonstrated the potency of QRs **86** to act as drug delivery agents. In biological studies, glyco-QRs have been selectively recognized by human nasopharyngeal epidermal carcinoma (KB) cells through the GLUT-1 transporter, a glucose receptor found at the outer KB cell membrane. In the assays, cellular uptake of the QRs was confirmed from their high internalization values. The specificities of glyco-QRs in in vivo assays have been confirmed through competitive studies with the free galactose.

QDs functionalized with lactose have been studied for protein interactions, cell imaging, and anti adhesive applications [95]. Lactose-QDs have been prepared with and without PEG spacer (**87** and **88**, respectively). In the case of **88** without spacer, QDs with different lactoside densities were synthesized in combination with ethanethiol (Fig. 14). The amount of sugar loading was determined from NMR studies in conjunction with thermogravimetric analysis (TGA). The protein binding studies of QDs were performed in SPR assays over the surface of the immobilized PNA lectin. Glyco-QDs showed higher affinity than free lactose and, among the glyco-QDs, the affinity was lower for the QDs **87** having PEG spacer as compared to **88** deprived of spacer, thus demonstrating the importance of linker design in multivalent glycoconjugates. The anti-adhesive ability of glyco-QDs to prevent the adhesion of human acute monocytic leukemia cells to human umbilical vein endothelial cells (HUVEC) was measured. These cells play a critical role in the pathogenesis of septic shock. In the assays, glyco-QDs exhibited five to sixfold higher activity than the

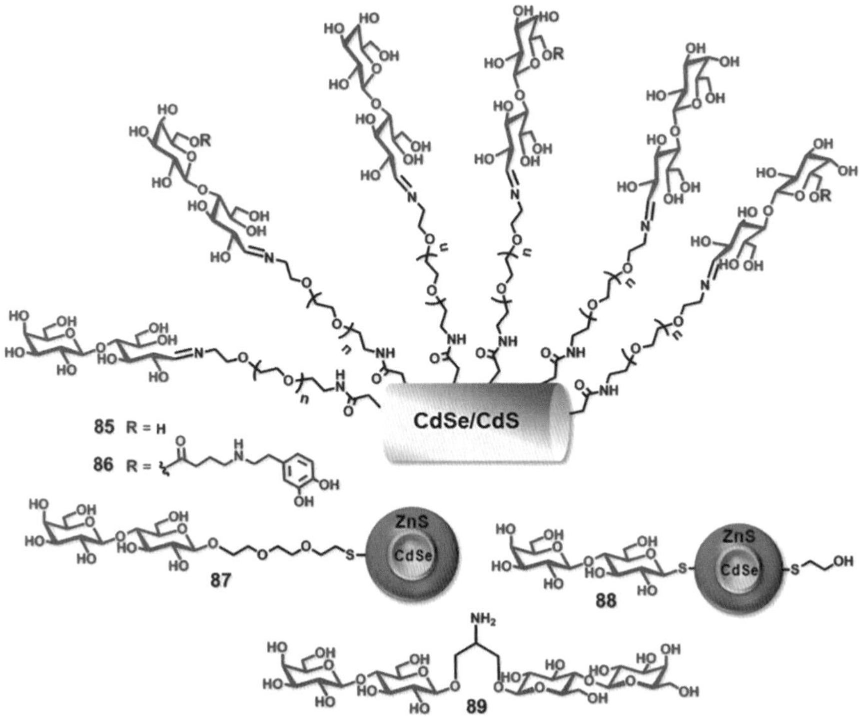

Fig. 14 Molecular structures of glyco-QDs **85–88**

divalent lactoside **89**. The results showed that the QDs without spacer exhibited higher activity, consistent with SPR studies. In fluorescence imaging studies with leukocytes, the potency of the QDs to act as cell-imaging agents was also demonstrated.

7 Iron-Oxide-Based Nanoparticles

Nanoclusters with iron oxides (magnetite Fe_3O_4 and/or maghemite γ-Fe_2O_3) have attracted extensive interest due to their superparamagnetic properties and low cytotoxicity [96]. Due to the superparamagnetic nature of the iron oxide NPs, cell detection through MRI can be achieved without any further labeling. Consequently, iron oxide NPs coated with glycan units have been studied for glycan-directed cell imaging and other biomedical applications. In this section, we will discuss two recent examples of iron-oxide-based GNPs that have been used to label various cancer cells and different human blood cell populations.

Huang and coworkers reported iron-oxide-based magnetic GNPs as sensors for the detection and differentiation of various cancer cells [97]. Silica-coated magnetic NPs have been functionalized with mannoside, galactoside, fucoside, sialic acid, and *N*-acetylglucosamine glycoconjugates (**90–94**, Fig. 15) through amide bond and click chemistry. Protein binding studies of GNPs using a series of lectins have

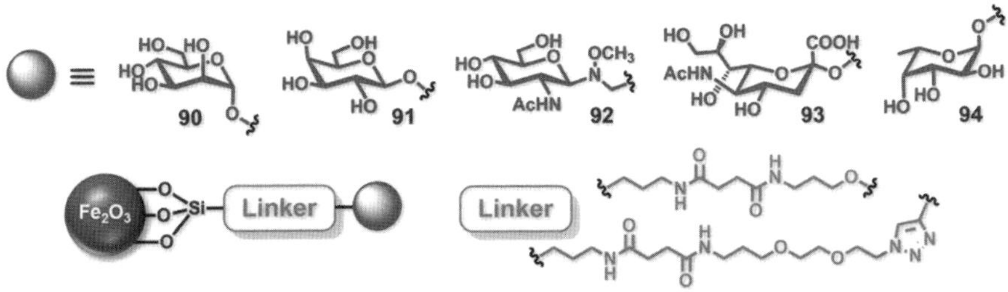

Fig. 15 Molecular structures of iron oxide GNPs

confirmed the bioaccessibility of the sugar head groups for the receptor interactions. In vitro studies on the interaction of nine different cancer cell types showed that the GNPs formed cohesive aggregates, leading to shorter T_2 transverse relaxation times and consequently darkened MRI images. The specificity of the interactions was confirmed through control studies with NPs without sugar, which did not show any activity. In addition to detection, the sugar-binding preferences of the different cancer cell lines were identified. Using linear discriminant analysis (LDA) with five different types of GNPs (**90–94**), it was possible to identify the characteristic MRI signature for each of the nine different cancer cell lines. By using galactose-coated GNPs **91**, it was feasible to differentiate the breast cancer MCF-7/Adr-res cells from normal breast endothelial cells 184B5 with the changes in T_2 values, wherein the former (cancer cells) exhibited larger changes (ΔT_2) in comparison with the latter (normal cells).

It was also demonstrated that the magnetic GNPs were able to differentiate between closely related isogenic cancer cells. For instance, B16F10 and B16F1 mouse melanoma cells have been differentiated using galactosylated GNPs **91** by the changes in T_2 values. Furthermore, cellular uptake assays have shown that the GNPs are internalized by tumor cell lines, which significantly reduces the cancer cell adhesion.

Multifunctionalized magnetic GNPs have been prepared from bimetallic gold-coated ferrite (core Fe_3O_4 and shell Au) and used for the specific labeling of cell populations of human blood using MRI and fluorescence techniques (Fig. 16) [98]. Glycan unit **95** and bifunctional linker **96** were initially attached onto the surface of the gold-coated ferrite. Subsequently, **96** has been used for its functionalization with fluorescent Texas Red dye **97** and protein G. Two different types of antibodies were coated onto the GNPs through binding with protein G for the specific labeling of human blood cells (anti-human CD45 for peripheral blood mononuclear cells, PMBC, and CD235a IgG for red blood cells, RBC). The multifunctionalization of GNPs did not affect the stability, photoluminescence and magnetic properties. Immunomagnetic-fluorescent NPs have been tested for their ability to differentiate the cell populations of the human blood using MRI and fluorescence techniques. These investigations demonstrated that GNPs **98** with anti-human CD45 antibody were able to label PMBC, whereas GNPs **99** selectively recognized RBC. In fluorescence studies, GNPs **98** were capable of specifically detecting a very low

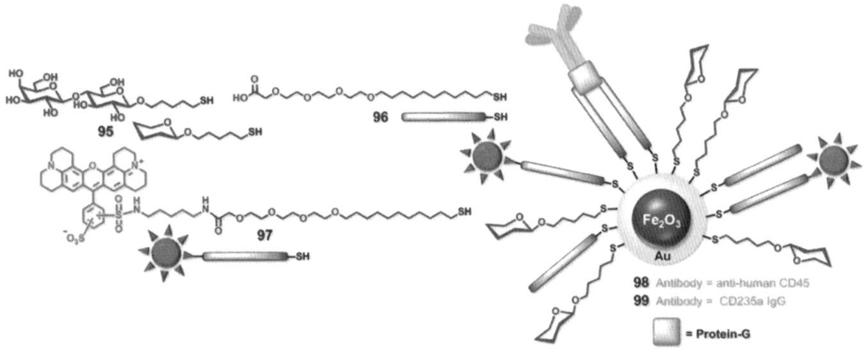

Fig. 16 Molecular structures of gold-coated ferrite GNPs

amount of PMBC (0.01%) in a complex biological media containing a higher population of RBC. The efficient orientation of antibodies obtained through protein G strategy enhanced the selectivity in the detection of different cell populations of human blood samples.

8 Liposomes

Liposomes are bilayer aggregates of amphiphilic lipid molecules. The core and the outer surface of these self-assembled architectures are equipped with hydrophilic head groups. In aqueous medium, liposomes consist of an aqueous internal compartment in addition to bulk aqueous phase. Liposomes can act as delivery vehicles for both hydrophobic and hydrophilic molecules. Liposomes coated with multiple copies of carbohydrates (glycoliposomes) have been used extensively in glycan-directed biomedical applications [80, 99]. The hydrophilic sugar head groups of glycoliposomes consist of two important structural characteristics pertaining to natural cell membranes, namely, the ability to present sugar moieties in multivalent states and the formation of an integral lipidic environment around the glycan units. A few recent examples of glycoliposomes acting as biosensors and as scaffolds for the synthesis of multivalent glycoclusters are discussed herein.

Among various liposomes, the polydiacetylene (PDA) liposomes consisting of ene–yne type linkages are of particular interest (Fig. 17) [100]. PDA liposomes exhibit blue color due to the extended conjugation, and the color of the liposomes is sensitive to external forces such as heat and mechanical stress and to the ligand–receptor interactions at their surfaces. Disturbances in the extended conjugation account for the color change in PDA assemblies. Jiang and coworkers demonstrated the application of PDA glycoliposomes as potent sensors for influenza viruses by targeting their cell surface hemagglutinin receptors (HA1) [101]. Glycoliposomes have been synthesized from the active sialic acid glycoside (**100**) and/or inactive β-lactoside (**101**) derivatives with the combination of DMPC and diacetylene monomer **102** (Fig. 17). Subsequent photopolymerization of the mixed glycoliposomes led to the formation of ene–yne

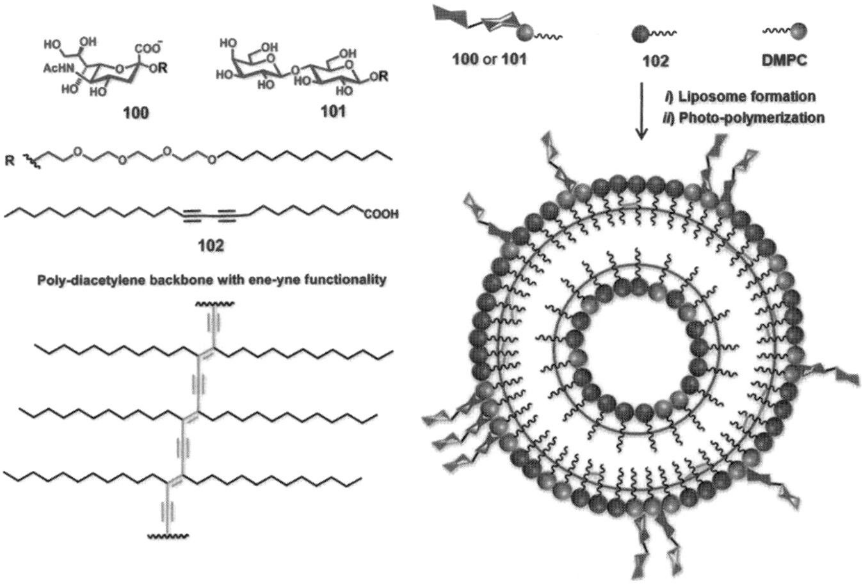

Fig. 17 Molecular structures of sialic and lactoside derivatives (**100** and **101**) and representation of PDA-glycoliposome formation

functionalized glycan-PDA assemblies. Oligosaccharides with tetraethylene glycol spacer have been used in the preparation of the glycoliposomes for better exposure of the sugar head groups at the liposomal surface to facilitate the interactions with the proteins. In biological studies, upon binding associations between the HA1 protein to the sialic acid groups of the liposomal surface, liposomes **100** exhibited the expected color change from blue to red. Liposomes with inactive lactoside head groups **101** did not show any activity, thus confirming the specificity of the interactions. Interestingly, liposomes with the combination of **100** and **101** in a 1:1 molar ratio showed better activity (color changes detected at 10 ng/mL of HA1) than the sialic acid (**100**) alone, resulting from the synergetic effect between the two sugars. The morphology changes of the liposomes with the addition of the protein was identified from TEM studies, which showed a clear transition from spherical to rectangular shape.

Marchant and coworkers reported the synthesis of biomimetic liposomes functionalized with 3′-sulfo-Lewis a (SO_3^--Le^a) oligosaccharides and their binding properties were assessed with P-selectin [102]. P-Selectin is a receptor expressed on activated platelets that binds to the P-selectin glycoprotein ligand 1 (PSGL-1) on leukocytes during inflammation processes [103]. P-Selectin plays a crucial function in the early stages of the inflammatory response to vascular injury by regulating the platelet–leukocyte interactions. The P-selectin binding site of the PSGL-1 is extended (50 nm away) from the leukocyte surface [104]. To mimic the native structures, glycoliposomes have been synthesized with PEG spacer-.functionalized sugar conjugate **103** (Fig. 18). DSPC and cholesterol have been used along with sialylated oligosaccharide **103** for the preparation of glyco-PEG-liposomes.

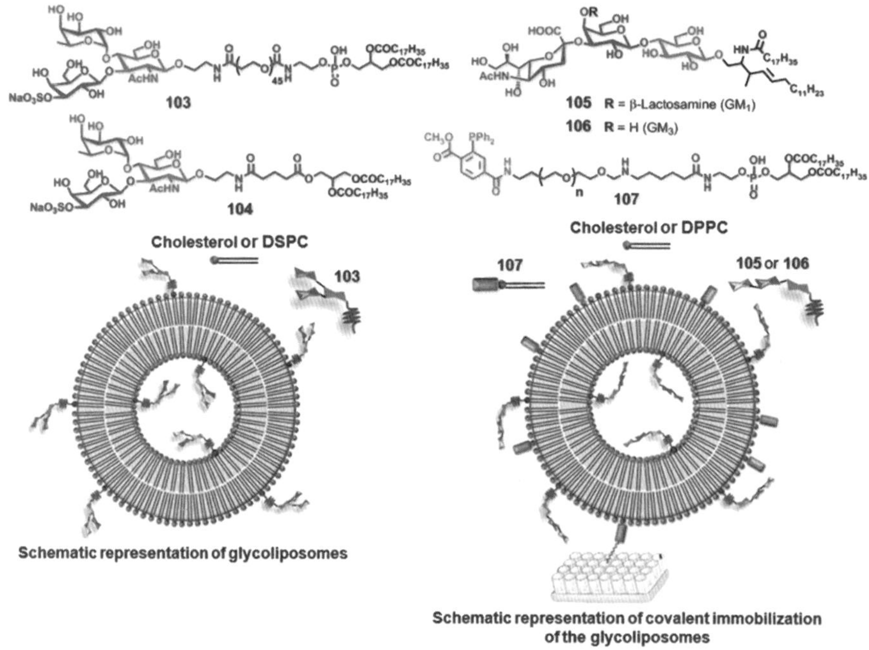

Fig. 18 Molecular structures of glycolipids **103–107** and representation of their liposome formation

Subsequently, studies on protein binding to the glycoliposomes have been performed using ELISA. The ability of the glycoliposomes to inhibit the binding of PSGL-1 to P-selectin was measured. The IC_{50} value for the glyco-PEG-liposomes was found to be 0.13 mM, which is significantly higher in comparison to free SO_3^--Le^a antigen (5.61 mM) or glycoliposomes without PEG spacer (2.92 mM) synthesized from the sialylated oligosaccharide **104**. The higher activity of glyco-PEG-liposomes demonstrated the importance of the PEG spacer in mimicking the extended structure of PSGL-1, thereby facilitating interactions with the lectin. Further, fluorescent glycoliposomes have been synthesized with incorporation of nitrobenzodiazole-modified phosphatidylcholine (NBD-PC) (1%) for their use in interaction studies with activated platelets in vitro. In the assays, glyco-PEG-liposomes showed very high activity compared to the control PEG-liposomes without sugar head groups.

In addition to solution phase studies, glycoliposomes have also been fabricated onto microarray surfaces to study their receptor interactions. In the liposomal microarrays, the multivalent glycans at the solid surface are surrounded by hydrophobic lipid chains similar to those of natural cell membranes, a situation not found when direct immobilization of glycoconjugates onto the solid surfaces occurred. Sun and coworkers synthesized glycoliposomes with ganglioside head groups **105** or **106**, azide-reactive PEG lipid **107** in combinations with DPPC and cholesterol (Fig. 18) [105]. Upon reaction with azide-coated glass surfaces, the lipid portion **107** underwent a chemically selective and biocompatible Staudinger reaction to form an amide bond through which the glycoliposomes were covalently linked to the glass surface. Fluorescent dye release kinetics and fluorescence imaging studies confirmed the immobilization and stability of the liposomes over the solid surface. The bioavailability of the sugar head groups in liposomal microarrays has been ascertained by specific interaction studies with fluorescently labeled lectins from Cholera toxin B (CTB), *Maackia amurensis* agglutinin (MAA), and peanut lectin from *Arachis hypogea* (PNA lectin). Control studies with PSA lectin (which does not have any specificity for **105** and **106**) confirmed the specificities of the interactions at the liposomal surface.

9 Micelles

Micelles are monolayer aggregates of lipids with hydrophobic and hydrophilic groups. For every lipid, there is a certain concentration known as critical micelle concentration (CMC), above which lipids cannot be soluble as individual molecules and are forced to form micelles in aqueous medium to decrease the aqueous contact with hydrophobic groups. In aqueous medium, the hydrophobic groups are located at the core of the micelles, whereas the hydrophilic groups will be in contact with the water at the surface of the micelles. Due to their hydrophobic core, micelles can solubilize hydrophobic molecules and can act as carriers for applications in drug delivery. Self-aggregates of amphiphilic glycolipids have been used in a variety of biomedical applications [106]. Mignet and coworkers designed perfluoroalkylated glycolipids (**108–110**, Fig. 19) for imaging studies [107]. The perfluoroalkyl groups were chosen as the lipid portion due to their low miscibility with natural phospholipids as compared with "hydrocarbons", which minimizes the toxicity of the amphiphilic perfluoro-glycolipids [108]. Glycolipids have been functionalized with chelating agents such as DTPA (diethylenetriaminepentaacetic acid), which was used for complexation to fluorescent or radiolabel markers for biological studies. In aqueous medium, glycolipids form self-aggregates with 10–50 nm diameters. In agglutination studies with the galactoside-binding RCA lectin, the glycolipid micelles formed large aggregates that could be quantified by absorbance at 490 nm, thus confirming the formation of multivalent complexes. In competitive inhibition studies, a decrease in the absorbance of the aggregates with the addition of lactose was evidence for the specificity of the interactions. In vivo scintigraphy imaging studies in mice showed that, at pH 6, glycolipid micelles complexed with radiolabeled ^{99m}Tc accumulated mostly in the liver. Further, fluorescence studies showed that micelles complexed with europium interacted very poorly with blood serum proteins. The poor reactivity with the complex blood serum proteins and organ-specific accumulation of micelles dictated by the sugar head groups of the perfluoro-glycolipids can be used for specific drug delivery applications.

Fort and coworkers studied the aggregation and protein binding properties of amphiphilic rod–coil type of glycoconjugates (**111** and **112**, Fig. 19) synthesized using click chemistry between suitably functionalized precursors [109]. The glycan-functionalized amphiphiles were self-assembled in aqueous medium and their size and morphologies were dictated by the amphiphilic balance (volume fraction of

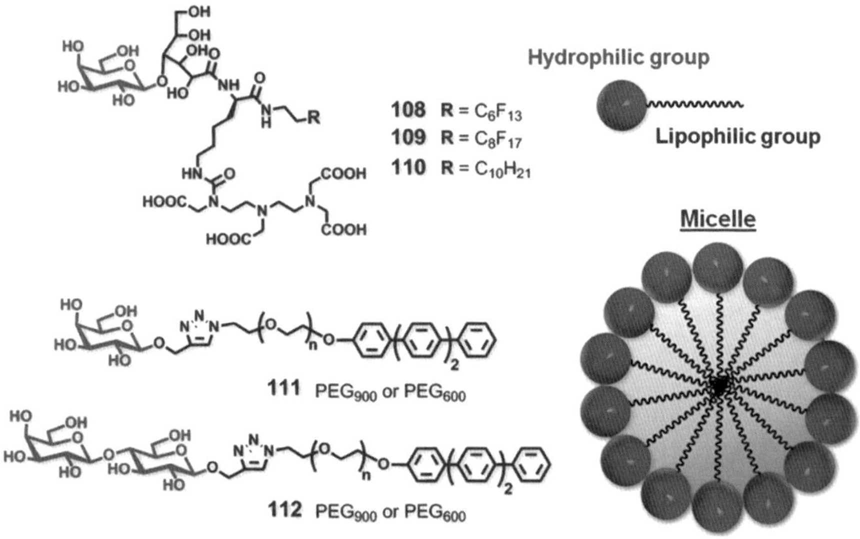

Fig. 19 Molecular structures of glycolipids **108–112** and representation of a micelle

hydrophobic to hydrophilic segments) in the glycolipids. Glycosides with longer PEG_{900} hydrophilic segments formed spherical micelles with 10 nm diameter, whereas shorter PEG_{600} glycosides showed the formation of larger vesicles. The accessibility of the sugar head groups for bio-interactions was verified with protein binding studies with wheat germ agglutinin (WGA) and PNA lectins. NPs with lactoside head groups were found to interact with their cognate PNA lectin, whereas NPs with GlcNAc head groups were recognized by the WGA lectin; in the other cases (WGA with lactose or PNA with GlcNAc) activity was not observed.

10 Glycopolymers

Glycopolymers, a term initially defined by one of the authors [110], constitute an important class of multivalent glycans that have attracted much attention in biological and biomedical areas [3, 111–113]. In glycopolymers, it is possible to trigger the density of the carbohydrate ligands by controlling the degree of polymerization. The required spacer between the sugar head groups can be achieved by choosing appropriate monomers in the glycopolymer synthesis. Here we discuss a few recent examples of glycopolymers that have shown potential applications in biomedical research. Stengel and coworkers studied the antitumor potency of the polymeric version of the carbohydrate-based drug auronafin (**113**, Fig. 20), which is known to be effective against rheumatoid arthritis and a few cancers [114]. In the polymer synthesis, the deacetylated version of the drug was loaded onto acrylate block copolymers (**114**, Fig. 20), known to be equally potent [115]. Upon water addition to the polymer solution of **114** in DMSO, spherical aggregates with 75 nm

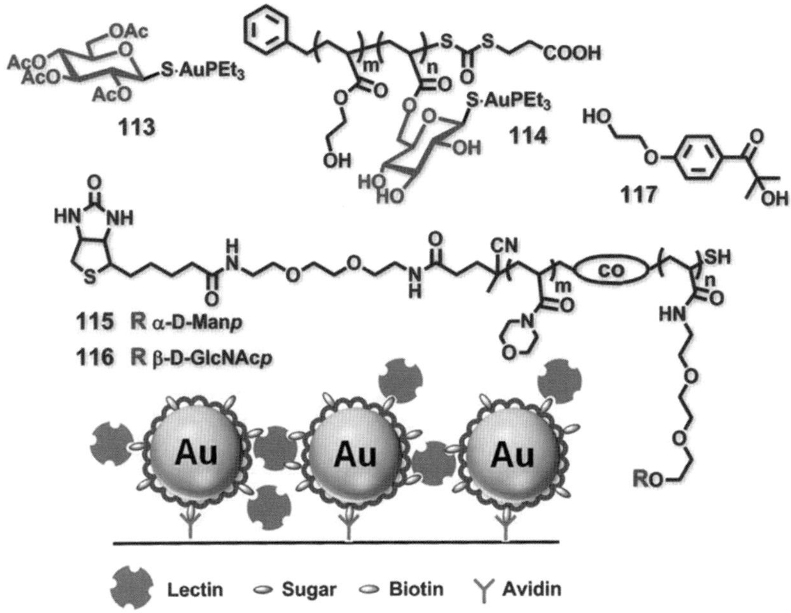

Fig. 20 Molecular structures of the glycopolymers and representation of glycopolymer-NP immobilization

diameter were generated. The activity of the polymeric auronafin micelles against tumor cells was assessed by measuring the proliferation of human ovarian carcinoma cells OVCAR-3. In the studies, it was found that the micelles of **114** showed better efficiency (IC_{50} 0.94 μM) than the drug auronafin alone (IC_{50} 1.51 μM), which was known to be most the potent form. The high efficiency of the micelles was thought to be due to the route of entry via endocytosis rather than via diffusion, as used by the smaller molecule auronafin.

Naraian and coworkers demonstrated the synthesis and protein binding properties of NPs synthesized from mannoside- and GlcNAc-functionalized glycopolymers containing biotin and thiol functionalities (**115** and **116**, Fig. 20) [116]. GNPs have been synthesized from glycopolymers (**115** or **116**) in combination with thiol-functionalized PEG using a recently developed photochemical process, which involves the reduction of gold salts by using photochemical decomposition of benzoin derivative **117** [117]. Glycopolymer-NPs have been immobilized onto avidin-coated surfaces through avidin–biotin interactions (Fig. 20). The bioavailability of the carbohydrate head groups of the NPs over the solid surface was assessed by protein binding studies. The interactions were monitored using diffractive optics technology (DOT)-based sensors. It was found that the glycopolymers were able to detect the corresponding lectins (i.e., Con A with mannose polymer and WGA with GlcNAc polymer). In the other cases, the lectins did not show any cross-activity, thus confirming again the biospecificity of the interactions between the GNPs and their non-cognate lectins.

Li and coworkers reported glucoside- and pH-sensitive block copolymer assemblies for the triggered release of insulin [118]. Triblock amphiphilic copolymer poly(acrylic acid-*co*-acrylamidophenylboronic acid)-*block*-poly(2-acryloxyethyl galactose)-*block*-poly(acrylic acid-*co*-acrylamidophenylboronic acid) {[(PAA-*co*-PAAPBA)-*b*-$]_2$ PAEG} was synthesized using atom radical transfer polymerization, which formed self-assembled structures with particles ranging from 230 to 270 nm. The boronic acid polymer (PBA) of the block copolymer self-assemblies is responsible for its glucose-sensitive nature through complexation with adjacent hydroxyl groups of the carbohydrate moieties [119]. In vitro insulin binding studies revealed that the profiles of NP micelles were dependent on both the pH and the glucose concentration. The self-aggregated NPs of the block copolymer have been shown to release insulin rapidly upon exposure to high concentrations of glucose.

11 Nanogels

Nanogels are colloidal solutions composed of crosslinked synthetic or biopolymer networks. They have been synthesized with both chemically or physically associated crosslinkages between the polymer networks. Recently, nanogels have attracted major attention for the encapsulation and delivery of drugs and biomacromolecules [120, 121]. Narian and coworkers reported the synthesis of carbohydrate-based nanogels (**118** and **119**, Fig. 21) and studied their ability to encapsulate and deliver biomacromolecules such as DNA, enzymes, or proteins in their active forms [122]. The size and morphologies of **118** and **119** were found to be thermoresponsive and, at 37°C, nanogels have been obtained with diameters ranging from 30 to 40 nm. In biological studies, although nanogels **118** and **119** exhibited good complexation properties with DNA, their gene transfection in Hep G2 cells was found to be low. This was attributed to the high hydrophobicity and aggregation of DNA–nanogel complexes. The incorporation of cationic polymer **120** onto the gels led to an increase in cellular uptake and gene expression of DNA–nanogel complexes. In the case of different sugars, the high activity of lactoside-coated nanogels was thought to be due to their higher uptake by Hep G2 cells, which is mediated by galactose–asialoglycoprotein receptor interactions as above. The activity of the biomacromolecules, after encapsulation in cationic nanogels, was studied with β-galactosidase. It was shown that the enzymatic activity remained after complexation with the nanogels. In addition to monomolecular encapsulation, cationic nanogels have also been used for bimolecular delivery applications. The co-delivery of protein (BSA) and DNA in Hep G2 cells was successfully achieved with these cationic nanogels.

Zhou and coworkers recently reported the synthesis of chitosan-based nanogels and their pH-responsive biomedical applications [123]. Nanogels have been prepared from chitosan and poly(methacrylic acid) (PMAA) polymers in two ways: using covalent crosslinking or physical association between the polymers (Fig. 22). Naturally occurring chitin, chitosan, and their derivatives have found widespread applications in various biological settings [124].

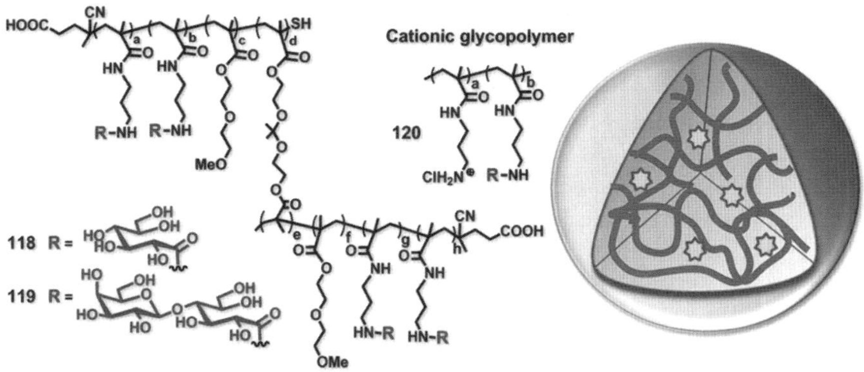

Fig. 21 Molecular structures of crosslinked glycopolymers **118** and **119** used in nanogel preparation and representation of nanogels with encapsulated biomacromolecules

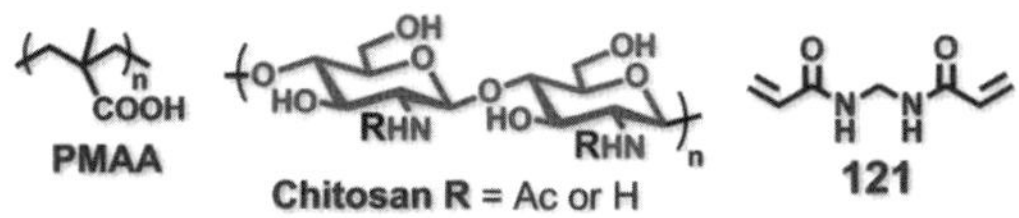

Fig. 22 Molecular structures of the PMAA and chitosan polymers used in gel preparation and the crosslinker **121**

The polymerization of MAA monomers was performed in aqueous solution containing chitosan in the presence or absence of a crosslinker **121**, to afford covalently crosslinked or physically associated gels, respectively. The in situ insertion of CdSe quantum dots during nanogel preparation afforded hybrid nanogels with photoluminescent properties. Under pH variation, the hybrid nanogels having covalent crosslinking showed reversible volume phase transitions that were stable in both structure and composition, whereas the physically associated gel hybrids were degraded under pH variations. In vivo studies with mouse melanoma cells B16F10 showed the potential of the hybrid nanogels to act as a tumor cell imaging agents. In drug delivery studies, nanogels with covalent crosslinkages showed controlled release of the anticancer drug Temozolomide under slightly acidic pH range (5–7.4), which is close to the pH found in pathological tumor sites. Because of their degradation, nanogels formed under physical associations were not successful for drug delivery studies. In cytotoxicity studies, covalently crosslinked nanogels exhibited lower toxicity as compared to physically associated nanogels. This study clearly showed the importance that crosslinking in nanogels has on their properties.

12 Polysaccharides

Polysaccharide nanoparticles have been extensively used in biomedical applications due to their biocompatibility, biodegradability, high abundance, and low toxicity [125, 126]. The wide range of biological activities of polysaccharides arises from the great diversities of their chemical composition. For instance, the glycan-directed binding abilities of polysaccharides to different body tissues can be used in targeted drug delivery applications. In addition, polysaccharides can be readily functionalized through their intrinsic multiple hydroxyl groups. The most commonly used polysaccharides are chitosan, alginate, hyaluronan, dextran, heparin, chondroitin sulfate, amylose, pullulan, arabinogalactan, and cyclodextrins. NPs have been synthesized from polysaccharides using various methods. Typical methodologies include: (a) crosslinking between different polymers (one example is discussed in the section on nanogels); (b) formation of interpolymer complexes through electrostatic interactions between oppositely charged polysaccharides; (c) self-assembly of amphiphilic hydrophobic groups grafted onto polysaccharides; and (d) from polymer–drug conjugates (one example is discussed in the next section). The inherent ionic charges associated with polysaccharides have been successfully used for encapsulation and delivery of oppositely charged drugs and biomacromolecules. Polysaccharides have also been coated to the surface of nanoparticles to achieve increased stability, long-term circulation, and targeting abilities for drug delivery applications.

13 Miscellaneous Examples

Lee and coworkers designed a pH-sensitive chitosan-based drug conjugate **122** for photodynamic therapy (Fig. 23) [127]. Chitosan polymers were functionalized with 3-diethylaminopropyl isothiocyanate (DEAP), chlorine e6 (Ce6, used as a phoιosensitizing model drug) and a PEG block. The drug conjugate was self-assembled in aqueous solution such that the hydrophobic DEAP and Ce6 were located at the hydrophobic core, while the hydrophilic PEG and CSG polymers were exposed at the hydrophilic surface of the self-assemblies. The importance of the drug conjugate **122** resides in its conformational change under pH variation from 7.4 (normal tissue pH) to 6.8 (pH in the tumor microenvironment). With a change in pH from 7.6 to 6.8, the coiled assembly structure (self-quenched state of Ce6, not suitable for singlet-oxygen production) undergoes a conformational change to form uncoiled extended random molecules (de-quenched state of Ce6 suitable for singlet-oxygen formation) (Fig. 23). The protonation of the DEAP block within the conjugate **122** at pH 6.8 was responsible for the conformational change. Fluorescence studies demonstrated that the "singlet-oxygen productivity" of the drug conjugate **122** at pH 6.8 was similar to that of free Ce6, whereas that at pH 7.4 was found to be low, hence confirming the uncoiled structure of **122** at pH 6.8 with the de-quenched state of Ce6 suitable for singlet-oxygen formation.

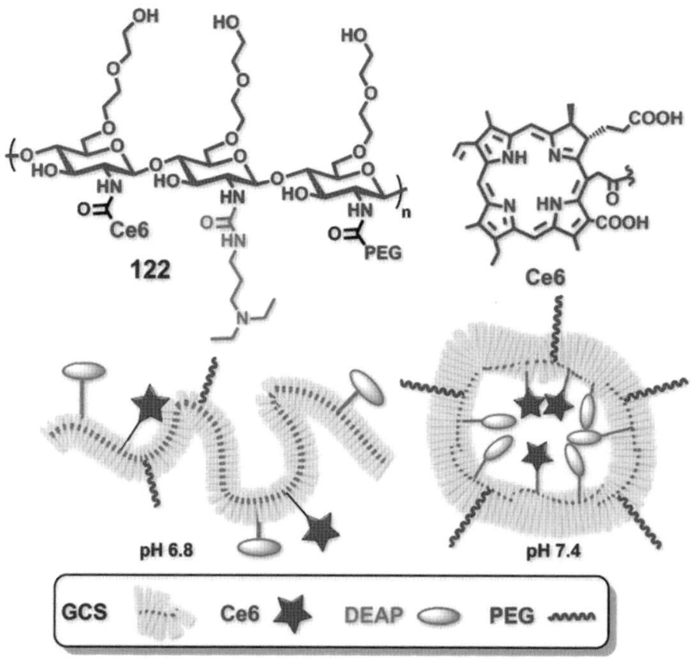

Fig. 23 Molecular structures of drug conjugate **122** and representation of its conformational changes upon pH variation from 7.4 to 6.8

In phototoxicity assays at pH 6.8 and 6.4, the drug conjugate **122** exhibited higher toxicity for human epithelial carcinoma HeLa cells than at pH 7.4 due to higher production of singlet oxygen. This was consistent with fluorescence studies. In vivo imaging studies on mice having HeLa tumor cells showed that the drug–conjugate exhibited strong fluorescence at tumor sites, whereas the control compounds (free Ce6 or drug conjugate without DEAP) showed no or very weak fluorescence. The drug–conjugate **122** exhibited persistent fluorescent signal for 24 h at tumor sites, which demonstrated the efficiency of the drug candidate for anticancer therapy without affecting normal cells.

Our group demonstrated the application of luminescence resonance energy transfer (LRET) as an analytical technique for investigating carbohydrate–protein interactions (Fig. 24) [128]. Lanthanide ($NaGdF_4$:Er^{3+}, Yb^{3+}) NPs have been functionalized with PAMAM dendrimers via ligand exchange and subsequently glycosylated with the dendrimer's amine groups. The formation of lanthanide GNPs **123** was confirmed by IR and TGA analysis. Upon excitation at 980 nm (near IR region), the monodispersible lanthanide GNPs exhibited "upconversion" red and green fluorescence centered at 525, 540, and 600 nm. Lectin binding studies of the glyco-NPs with rhodamine isothiocyanated Con A (RITC-labeled Con A) have been monitored using LRET phenomenon. With the increase in lectin concentration, the unconverted green fluorescence of the GNPs led to an increase in acceptor (RITC) fluorescence at 585 nm, with a corresponding decrease in unconverted green fluorescence of the GNPs at

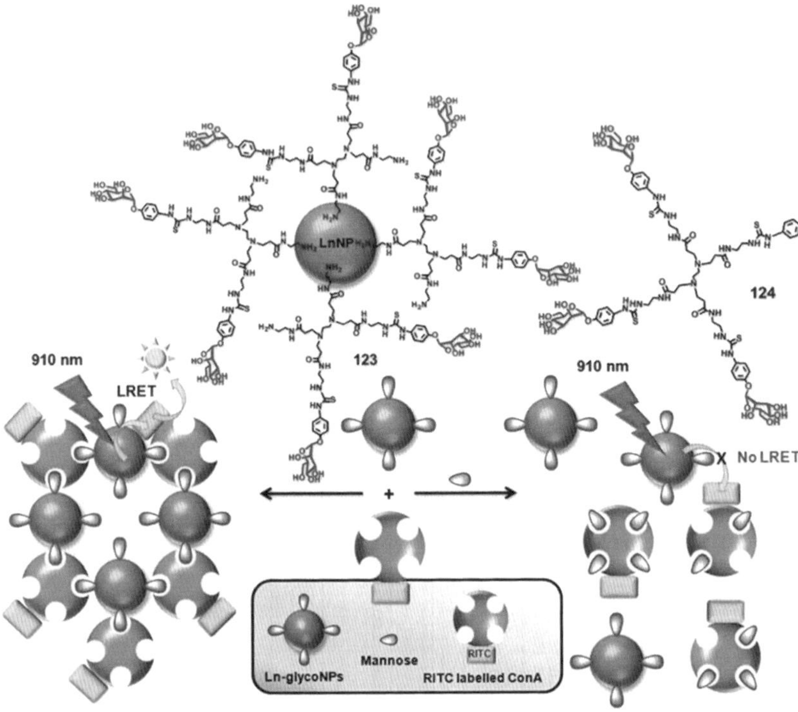

Fig. 24 Molecular structures of lanthanide GNPs and representation of LRET phenomenon

540 nm due to the non-radiative excitation of the RITC fluorophore in Con A (Fig. 24). The absence of LRET phenomenon with galactoside-coated NPs and the reduction of LRET in the presence of mannoside in competitive studies confirmed the specificity of the interactions. The enhancement in the binding constant for the GNPs (7×10^4) in comparison with either Me-Man (7×10^3) or mannose-functionalized PAMAM G0 dendrimer **124** (1.9×10^4) was anticipated to be due to clustered interactions.

Muller and coworkers demonstrated the synthesis of glyconanospheres from the incorporation of glycopolymers onto the methacrylic-group-displaying iron oxide nanospheres (Fig. 25) [129]. The glyco-fluorescent polymer **125** was installed onto the nanospheres **126** through a thiol–ene reaction. The glyconanopheres exhibited superparamagnetic and photoluminiscent properties from the iron core and from the glyco-fluorescent polymer, respectively. After 24 h of incubation with A549 cells, imaging studies with the fluorescent glyconanospheres revealed efficient internalization by the cells. This was particularly evident in close proximity to the cell nucleus and was presumed to be due to the presence of galectins on the nucleus of the cells. In another approach, instead of attaching the glycopolymer onto the nanospheres, the polymerization was carried out at the surface of the nanospheres [130]. Atom transfer radical polymerization of GlcNAc monomer **127** with the initiator-functionalized polystyrene core **128**, followed by deprotection of the ester groups, led to the formation of novel spherical sugar-containing polymer brushes.

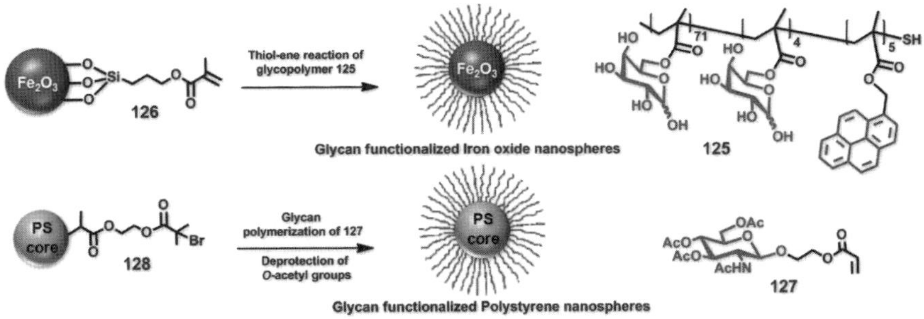

Fig. 25 Representation of glyconanospheres

In protein binding studies with WGA lectin, the glyconanospheres exhibited higher affinity (about six to seven times) than the monomer. The formation of aggregates of multivalent complexes between the glyconanospheres and WGA lectin was also confirmed from TEM studies.

14 Conclusion

This review has clearly illustrated that carbohydrate appendages on nanoparticles of varied genesis offer several advantages for biomedical applications. The mixed expertise in nanosciences, glycochemistry, and glycobiology, coupled to expert knowledge in photophysics, has permitted the frontiers of nanomaterial science to be pushed closer to nanomedicine. Several examples have also been used to demonstrate that, once glycosylated, several hitherto toxic nanoparticle scaffolds acquired cellular safety. The rapidly growing interest in this exciting field is opening the way to new discoveries that are limited only by the creativity of scientists. Better insights into multivalent carbohydrate–protein interactions will further our appreciation of the "glycocodes" [131, 132] complexity, which should undoubtedly allow more precise addressing of targeted drugs and gene delivery.

If one combines state-of-the-art synthetic organic methodologies with the latest developments in carbohydrate chemistry, the next generation of glyconanoparticles will exemplify even more complex saccharide structures that will probably also include glycomimetics. This follows the actual trend in improved design of carbohydrate ligands and inhibitors to encompass more hydrolytically stable analogues, most of which successfully demonstrate much simpler hybrid glyco-pharmacophore structures that better mimic newer drug conceptions.

References

1. Wang X, Ramström O, Yan M (2010) Glyconanomaterials: synthesis, characterization, and ligand presentation. Adv Mater 22:1946–1953
2. Doane TL, Burda C (2012) The unique role of nanoparticles in nanomedicine: imaging, drug delivery and therapy. Chem Soc Rev 41:2885–2911

3. Narain R (ed) (2011) Engineered carbohydrate-based materials for biomedical applications-Polymers, surfaces, dendrimers, nanoparticles, and hydrogels. Wiley, Hoboken
4. Marradi M, Garcia I, Penadés S (2011) Carbohydrate-based nanoparticles for potential applications in medicine. Prog Mol Biol Transl Sci 104:141–173
5. Garcia I, Gallo J, Marradi M et al (2011) Glyconanoparticles: new nanomaterials for biological applications. In: Narain R (ed) Engineered carbohydrate-based materials for biomedical applications-polymers, surfaces, dendrimers, nanoparticles, and hydrogels. Wiley, Hoboken
6. El-Boubbou HX, Huang X (2011) Glyco-nanomaterials: translating insights from the "sugar code" to biomedical applications. Curr Med Chem 18:2060–2078
7. Garcia I, Marradi M, Penadés S (2010) Glyconanoparticles: multifunctional nanomaterials for biomedical applications. Nanomedicine 5:777–792
8. De la Fuente JM, Penadés S (2006) Glyconanoparticles: types, synthesis and applications in glycoscience, biomedicine and material science. Biochim Biophys Acta 1760:636–651
9. Cipolla L, Peri F, Airoldi C (2008) Glycoconjugates in cancer therapy. Anticancer Agents Med Chem 8:92–121
10. Newkome GR, Moorefield CN, Vögtle F (2001) Dendrimers and dendrons: concepts, synthesis, applications. Wiley, New York
11. Fréchet JMJ, Tomalia D (2001) Dendrimers and other dendritic polymers. Wiley, New York
12. Buhleier E, Wehner W, Vögtle F (1978) "Cascade"- and "Nonskid-Chain-like" syntheses of molecular cavity topologies. Synthesis 1978:155–158
13. Newkome GR, Z-q Y, Baker GR et al (1985) Micelles. Part 1. Cascade molecules: a new approach to micelles. A [27]-arborol. J Org Chem 50:2004–2006
14. Tomalia DA, Baker H, Dewald J et al (1985) A new class of polymers: starburst-dendritic macromolecules. Polym J 17:117–132
15. Haag R, Kratz F (2006) Polymer therapeutics: concepts and applications. Angew Chem Int Ed Engl 45:1198–1215
16. Tekade RK, Kumar PV, Jain NK (2009) Dendrimers in oncology: an expanding horizon. Chem Rev 109:49–87
17. Duncan R, Izzo L (2005) Dendrimer biocompatibility and toxicity. Adv Drug Deliv Rev 57: 2215–2237
18. Boas U, Christensen JB, Heegaard PMH (2006) Dendrimers in medicine and biotechnology: new molecular tools. RSC Publishing, Cambridge
19. Roy R, Zanini D, Meunier SJ et al (1993) Solid phase synthesis of dendritic sialoside inhibitors of influenza A virus haemagglutinin. J Chem Soc Chem Commun 1993:1869–1872
20. Lundquist JL, Toone EJ (2002) The cluster glycoside effect. Chem Rev 102:555–578
21. Roy R (1996) Syntheses and some applications of chemically defined multivalent glycoconjugates. Curr Opin Struct Biol 6:692–702
22. Chabre YM, Roy R (2010) Design and creativity in synthesis of multivalent neoglycoconjugates. Adv Cabohydr Chem Biochem 63:165–393
23. Chabre YM, Roy R (2008) Recent trends in glycodendrimer syntheses and applications. Curr Top Med Chem 8:1237–1285
24. Roy R (2003) A decade of glycodendrimer chemistry. Trends Glycosci Glycotechnol 15: 291–310
25. Touaibia M, Roy R (2007) Glycodendrimers as anti-adhesin drugs against type 1 fimbriated *E. coli* uropathogenic infections. Mini Rev Med Chem 7:1270–1283
26. Nepogodiev SA, Stoddart JF (2003) Glycodendrimers: chemical aspects. Adv Macromol Carbohydr Res 2:191–239
27. Röckendorf N, Lindhorst TK (2001) Glycodendrimers. Top Curr Chem 217:201–238
28. Cloninger MJ (2002) Biological applications of dendrimers. Curr Opin Chem Biol 6:42–748
29. Bezouska K (2002) Design, functional evaluation and biomedical applications of carbohydrate dendrimers (glycodendrimers). Rev Mol Biotechnol 90:269–290

30. Turnbull WB, Stoddart JF (2002) Design and synthesis of glycodendrimers. Rev Mol Biotechnol 90:231–255
31. Li Y, Cheng Y, Xu T (2007) Design, synthesis and potent pharmaceutical applications of glycodendrimers: a mini review. Curr Drug Discov Technol 4:246–254
32. Tanaka K, Siwu ERO, Minami K et al (2010) Noninvasive imaging of dendrimer-type N-glycan clusters: in vivo dynamics dependence on oligosaccharide structure. Angew Chem Int Ed 49:8195–8200
33. Kamerling JP, Boons GJ, Lee YC, Suzuki A, Taniguchi N, Voragen AGJ (eds) (2007) Analysis of glycans, polysaccharide functional properties & biochemistry of glycoconjugate glycans, carbohydrate-mediated interactions: comprehensive glycoscience: From Chemistry to Systems Biology, vol. II and III. Elsevier, Amsterdam
34. Kaneko Y, Nimmerjahn F, Ravetch JV (2006) Anti-inflammatory activity of immunoglobulin G resulting from Fc sialylation. Science 313:670–673
35. Crocker PR, Paulson JC, Varki A (2007) Siglecs and their roles in the immune system. Nat Rev Immunol 7:255–266
36. Shiao TC, Roy R (2012) Glycodendrimers as functional antigens and antitumor vaccines. New J Chem 36:324–339
37. Toyokuni T, Singhal AK (1995) Synthetic carbohydrate vaccines based on tumour-associated antigens. Chem Soc Rev 24:231–242
38. Toyokuni T, Dean B, Cai S et al (1994) Synthetic vaccines: synthesis of a dimeric Tn antigen-lipopeptide conjugate that elicits immune responses against Tn-expressing glycoproteins. J Am Chem Soc 116:395–396
39. Kaiser A, Gaidzik N, Becker T et al (2010) Fully synthetic vaccines consisting of tumor-associated MUC1 glycopeptides and a lipopeptide ligand of the toll-like receptor 2. Angew Chem Int Ed 49:3688–3692
40. Kudryashov V, Glunz PW, Williams LJ et al (2001) Toward optimized carbohydrate-based anticancer vaccines: epitope clustering, carrier structure, and adjuvant all influence antibody response to LewisY conjugates in mice. Proc Natl Acad Sci USA 98:3264–3269
41. Dumy P, Eggleston IM, Cervigni S et al (1995) A convenient synthesis of cyclic peptides as regioselectively addressable functionalized templates (RAFT). Tetrahedron Lett 36: 1255–1258
42. Renaudet O, BenMohamed L, Dasgupta G et al (2008) Towards a self-adjuvanting multivalent B and T cell epitope containing synthetic glycolipopeptide cancer vaccine. ChemMedChem 3:737–741
43. Bettahi I, Dasgupta G, Renaudet O et al (2009) Antitumor activity of a self-adjuvanting glycol-lipopeptide vaccine bearing B cell, CD4+ and CD8+ T cell epitopes. Cancer Immunol Immunother 58:187–200
44. Renaudet O, Dasgupta G, Bettahi I et al (2010) Linear and branched glyco-lipopeptide vaccines follow distinct cross-presentation pathways and generate different magnitudes of antitumor immunity. PLoS One 5:e11216
45. Jeon I, Lee D, Krauss IJ et al (2009) A new model for the presentation of tumor-associated antigens and the quest for an anticancer vaccine: a solution to the synthesis challenge via ring-closing metathesis. J Am Chem Soc 131:14337–14344
46. Fiore M, Thomas B, Duléry V et al (2012) Synthesis of multi-antigenic platforms as vaccine candidates against cancers. New J Chem. doi:10.1039/C2NJ40972K
47. Bossu I, Šulc M, Křenek K et al (2011) Dendri-RAFTs: a second generation of cyclopeptide-based glycoclusters. Org Biomol Chem 9:1948–1959
48. André S, Renaudet O, Bossu I et al (2011) Cyclic neoglycodecapeptides: how to increase their inhibitory activity and selectivity on lectin/toxin binding to a glycoprotein and cells. J Pept Sci 17:427–437
49. Kroto HW, Heath JR, O' Brien SCO et al (1985) C60: buckminsterfullerene. Nature 318: 162–163

50. Giacalone F, Martin N (2006) Fullerene polymers: synthesis and properties. Chem Rev 106: 5136–5190
51. Jensen AW, Wilson SR, Schuster DI (1996) Biological applications of fullerenes. Bioorg Med Chem 4:767–779
52. Da Ros T, Prato M (1999) Medicinal chemistry with fullerenes and fullerene derivatives. Chem Commun 1999:663–669
53. Bosi S, Da Ros T, Spalluto G et al (2003) Fullerene derivatives: an attractive tool for biological applications. Eur J Med Chem 38:913–923
54. Marchesan S, Da Ros T, Spalluto G et al (2005) Anti-HIV properties of cationic fullerene derivatives. Bioorg Med Chem Lett 15:3615–3618
55. Chabre YM, Roy R (2011) Solving promiscuous protein carbohydrate recognition domains with multivalent glycofullerenes. In: Renaudet O, Spinelli N (eds) Synthesis and biological applications of glycoconjugates. Bentham Science, Oak Park, pp. 64–77
56. Wennekes T, van den Berg RJBHN, Bonger KM et al (2010) Synthesis and evaluation of dimeric lipophilic iminosugars as inhibitors of glucosylceramide metabolism. Tetrahedron Asymmetry 20:836–846
57. Lohse A, Jensen KB, Lundgren K, Bols M (1999) Synthesis and deconvolution of the first combinatorial library of glycosidase inhibitors. Bioorg Med Chem 7:1965–1971
58. Compain P, Decroocq C, Iehl J et al (2010) Glycosidase inhibition with fullerene iminosugar balls: a dramatic multivalent effect. Angew Chem Int Ed Engl 49:5753–5756
59. Diot J, Garcia-Moreno MI, Gouin SG et al (2009) Multivalent iminosugars to modulate affinity and selectivity for glycosidases. Org Biomol Chem 7:357–363
60. Durka M, Buffet K, Iehl J et al (2012) The inhibition of liposaccharide heptosyltransferase WaaC with multivalent glycosylated fullerenes: a new mode of glycosyltransferase inhibition. Chem Eur J 18:641–651
61. Escaich S (2008) Antivirulence as a new antibacterial approach for chemotherapy. Curr Opin Chem Biol 12:400–408
62. Moreau F, Desroy N, Genevard JM et al (2008) Discovery of new Gram-negative antivirulence drugs: structure and properties of novel *E. coli* WaaC inhibitors. Bioorg Med Chem Lett 18:4022–4026
63. Tasis D, Tagmatarchis N, Bianco A et al (2006) Chemistry of carbon nanotubes. Chem Rev 106:1105–1136
64. Bianco A, Kostarelos K, Prato M (2005) Applications of carbon nanotubes in drug delivery. Curr Opin Chem Biol 9:674–679
65. Martin CR, Kohli P (2003) The emerging field of nanotube biotechnology. Nat Rev Drug Discov 2:29–37
66. Lin Y, Taylor S, Li H et al (2004) Advances toward bioapplications of carbon nanotubes. J Mater Chem 5:27–541
67. Prato M, Kostarelos K, Bianco A (2008) Functionalized carbon nanotubes in drug design and discovery. Acc Chem Res 41:60–68
68. Pastorin G, Kostarelos K, Prato M et al (2005) Functionalized carbon nanotubes: towards the delivery of therapeutic molecules. J Biomed Nanotechnol 1:133–142
69. Pantarotto D, Singh R, McCarthy D et al (2004) Functionalized carbon nanotubes for plasmid DNA gene delivery. Angew Chem Int Ed 43:5242–5246
70. Shi Kam NW, Jessop TC, Wender PA et al (2004) Nanotube molecular transporters: internalization of carbon nanotube-protein conjugates into mammalian cells. J Am Chem Soc 126:6850–6851
71. Bianco A, Kostarelos K, Partidos CD et al (2005) Biomedical applications of functionalised carbon nanotubes. Chem Commun 2005:571–577
72. Gorityala BK, Ma J, Wang X (2010) Carbohydrate functionalized carbon nanotubes and their applications. Chem Soc Rev 39:2925–2934
73. Bandaru NM, Voelcker NH (2012) Glycoconjugate-functionalized carbon nanotubes in biomedicine. J Mater Chem 22:8748–8758

74. Wu P, Chen X, Hu N et al (2008) Biocompatible carbon nanotubes generated by functionalization with glycodendrimers. Angew Chem Int Ed 47:5022–5025
75. Chen X, Lee GS, Zettl A et al (2004) Biomimetic engineering of carbon nanotubes by using cell surface mucin mimics. Angew Chem Int Ed 43:6112–6116
76. Hong SY, Tobias G, Al-Jamal KT et al (2010) Filled and glycosylated carbon nanotubes for in vivo radioemitter localization and imaging. Nat Mater 9:485–490
77. Vedala H, Chen Y, Cecioni S et al (2010) Nanoelectronic detection of lectin-carbohydrate interactions using carbon nanotubes. Nano Lett 11:170–175
78. Chen Y, Vedala H, Kotchey GP et al (2012) Electronic detection of lectins using carbohydrate-functionalized nanostructures: graphene versus carbon nanotubes. ACS Nano 6:760–770
79. Brust M, Walker M, Bethell D et al (1994) Synthesis of thiol-derivatized gold nanoparticles in a 2-phase liquid–liquid system. J Chem Soc Chem Commun 1994:801–802
80. Marradi M, Chiodo F, García I et al (2011) Glycoliposomes and metallic glyconanoparticles in glycoscience. In: Renaudet O, Spinelli N (eds) Synthesis and biological applications of glycoconjugates. Bentham Science, Oak Ridge, pp 164–202
81. Chang-Ming D (2011) Glyconanoparticles for biomedical applications. Combin Chem High Throughput Screen 14:173–181
82. Marradi M, Martin-Lomas M, Penadés S (2010) Glyconanoparticles: polyvalent tools to study carbohydrate-based interactions. Adv Carbohydr Chem Biochem 64:211–290
83. de la Fuente JM, Barrientos AG, Rojas TC et al (2001) Gold glyconanoparticles as water-soluble polyvalent models to study carbohydrate interactions. Angew Chem Int Ed 40: 2257–2261
84. Ojeda R, de Paz JL, Barrientos AG et al (2007) Preparation of multifunctional glyconanoparticles as a platform for potential carbohydrate-based anticancer vaccines. Carbohydr Res 342:448–459
85. Brinãs RP, Sundgren A, Sahoo P et al (2012) Design and synthesis of multifunctional gold nanoparticles bearing tumor-associated glycopeptide antigens as potential cancer vaccines. Bioconjug Chem 23:1513–1523
86. Marradi M, Alcántara D, de la Fuente JM et al (2009) Paramagnetic Gd-based gold glyconanoparticles as probes for MRI: tuning relaxivities with sugars. Chem Commun 2009:3922–3924
87. Schofield CL, Field RA, Russell DA (2007) Glyconanoparticles for the colorimetric detection of cholera toxin. Anal Chem 79:1356–1361
88. Turnbull WB, Precious BL, Homans SW (2004) Dissecting the cholera toxin-ganglioside GM1 interaction by isothermal titration calorimetry. J Am Chem Soc 126:1047–1054
89. Reynolds M, Marradi M, Imberty A et al (2012) Multivalent gold glycoclusters: high affinity molecular recognition by bacterial lectin PA-IL. Chem Eur J 18:4264–4273
90. Weingart JJ, Sun XL (2011) Glyco-functionalized quantum dots. ACS Symp Ser 1091: 105–121
91. Svarovsky SA, Barchi JJ Jr (2007) De novo sunthesis of biofunctional carbohydrate-encapsulated quantum dots. ACS Symp Ser 960:375–392
92. Dong CM (2011) Glyconanoparticles for biomedical applications. Comb Chem High Throughput Screen 14:173–181
93. Ohyanagi T, Nagahori N, Shimawaki K et al (2011) Importance of sialic acid residues illuminated by live animal imaging using phosphorylcholine self-assembled monolayer-coated quantum dots. J Am Chem Soc 133:12507–12517
94. Malvindi MA, Corato RD, Curcio A et al (2011) Multiple functionalization of fluorescent nanoparticles for specific biolabeling and drug delivery of dopamine. Nanoscale 3:5110–5119
95. Yang Y, Yu M, Yan T-T et al (2010) Characterization of multivalent lactose quantum dots and its application in carbohydrate–protein interactions study and cell imaging. Bioorg Med Chem 18:5234–5240

96. Gupta AK, Gupta M (2005) Synthesis and surface engineering of iron oxide nanoparticles for biomedical applications. Biomaterials 26:3995–4021
97. El-Boubbou K, Zhu DC, Vasileiou C, Borhan B et al (2010) Magnetic glyco-nanoparticles: a tool to detect, differentiate, and unlock the glyco-codes of cancer via magnetic resonance imaging. J Am Chem Soc 132:4490–4499
98. Gallo J, García I, Genicio N, Padro D, Penadés S (2011) Specific labelling of cell populations in blood with targeted immuno-fluorescent/magnetic glyconanoparticles. Biomaterials 32:9818–9825
99. Jølck RI, Feldborg LN, Andersen S et al (2011) Engineering liposomes and nanoparticles for biological targeting. Adv Biochem Eng Biotechnol 125:251–280
100. Ahn DJ, Kim J-M (2008) Fluorogenic polydiacetylene supramolecules: immobilization, micropatterning, and application to label-free chemosensors. Acc Chem Res 41:805–816
101. Deng J, Sheng Z, Zhou K et al (2009) Construction of effective receptor for recognition of Avian influenza H5N1 protein HA1 by assembly of monohead glycolipids on polydiacetylene vesicle surface. Bioconjug Chem 20:533–537
102. Zhu J, Xue J, Guo Z et al (2007) Biomimetic glycoliposomes as nanocarriers for targeting P-selectin on activated platelets. Bioconjug Chem 18:1366–1369
103. McEver RP, Cummings RD (1997) Cell adhesion in vascular biology. J Clin Invest 110: 485–491
104. Li F, Erickson HP, James JA et al (1996) Visualization of P-selectin glycoprotein ligand-1 as a highly extended molecule and mapping of protein epitopes form monoclonal antibodies. J Biol Chem 271:6342–6348
105. Ma Y, Sobkiv I, Gruzdys V et al (2012) Liposomal glyco-microarray for studying glycolipid–protein interactions. Anal Bioanal Chem 404:51–58
106. van Dongen SFM, de Hoog H-PM, Peters RJRW et al (2009) Biohybrid polymer capsules. Chem Rev 109:6212–6274
107. Richard C, Chaumet-Riffaud P, Belland A et al (2009) Amphiphilic perfluoroalkyl carbohydrates as new tools for liver imaging. Int J Pharm 379:301–308
108. Chabaud E, Barthélémy P, Mora N et al (1998) Stabilization of integral membrane proteins in aqueous solution using fluorinated surfactants. Biochimie 80:515–530
109. Bó AGD, Soldi V, Giacomelli FC et al (2012) Self-assembly of amphiphilic glycoconjugates into lectin-adhesive nanoparticles. Langmuir 28:1418–1426
110. Roy R (1996) Blue-prints, syntheses and applications of glycopolymers. Trends Glycosci Glycotechnol 8:79–99
111. Ladmiral V, Melia E, Haddleton DM (2004) Synthetic glycopolymers: an overview. Eur Polym J 40:431–449
112. Roy R (1996) Syntheses and some applications of chemically defined multivalent glycoconjugates. Curr Opin Struct Biol 65:692–702
113. Chabre YM, Roy R (2009) The Chemist's way to prepare multivalency. In: Gabius HJ (ed) The sugar code: fundamentals of glycosciences. Wiley-Blackwell, Weinheim, pp 53–70
114. Pearson S, Scarano W, Stenzel MH (2012) Micelles based on gold-glycopolymer complexes as new chemotherapy drug delivery agents. Chem Commun 48:4695–4697
115. Gaujoux-Viala C, Smolen JS, Landewé R et al (2010) Extended report: Current evidence for the management of rheumatoid arthritis with synthetic disease-modifying antirheumatic drugs: a systematic literature review informing the EULAR recommendations for the management of rheumatoid arthritis. Ann Rheum Dis 69:1004–1009
116. Jiang X, Housni A, Gody G et al (2010) Synthesis of biotinylated α-D-mannoside or N-acetyl β-D-glucosaminoside decorated gold nanoparticles: study of their biomolecular recognition with Con A and WGA lectins. Bioconjug Chem 21:521–530
117. McGilvray KL, Decan MR, Wang D et al (2006) Facile photochemical synthesis of unprotected aqueous gold nanoparticles. J Am Chem Soc 128:15980–15981
118. Wang Y, Zhang X, Han Y et al (2012) pH- and glucose-sensitive glycopolymer nanoparticles based on phenylboronic acid for triggered release of insulin. Carbohydr Polym 89:124–131

119. Egawa Y, Seki T, Takahashi S et al (2011) Electrochemical and optical sugar sensors based on phenylboronic acid and its derivatives. Mater Sci Eng C 31:1257–1264
120. Oh JK, Bencherif SA, Matyjaszewski K (2009) Atom transfer radical polymerization in inverse miniemulsion: a versatile route toward preparation and functionalization of microgels/nanogels for targeted drug delivery applications. Polymer 50:4407–4423
121. Ebara M (2011) Carbohydrate-derived hydrogels and microgels. In: Narain R (ed) Engineered carbohydrate-based materials for biomedical applications-polymers, surfaces, dendrimers, nanoparticles, and hydrogels. Wiley, Hoboken
122. Ahmed M, Narain R (2012) Intracellular delivery of DNA and enzyme in active form using degradable carbohydrate-based nanogels. Mol Pharm 9:3160–3170
123. Wu W, Shen J, Banerjee P et al (2010) Chitosan-based responsive hybrid nanogels for integration of optical pH-sensing, tumor cell imaging and controlled drug delivery. Biomaterials 31:8371–8381
124. Dutta PK (ed) (2005) Chitin and chitosan opportunities and challenges. SSM International Publication, Midnapore
125. Mizrahy S, Peer D (2012) Polysaccharides as building blocks for nanotherapeutics. Chem Soc Rev 41:2623–2640
126. Bhaw-Luximon A (2011) Modified natural polysaccharides as nanoparticulate drug delivery devices. In: Narain R (ed) Engineered carbohydrate-based materials for biomedical applications-polymers, surfaces, dendrimers, nanoparticles, and hydrogels. Wiley, Hoboken
127. Park SY, Baik HJ, Oh YT et al (2011) A smart polysaccharide/drug conjugate for photodynamic therapy. Angew Chem Int Ed 50:1644–1647
128. Bogdan N, Vetrone F, Roy R et al (2010) Carbohydrate-coated lanthanide-doped upconverting nanoparticles for lectin recognition. J Mater Chem 20:7543–7550
129. Pfaff A, Schallon A, Ruhland TM et al (2011) Magnetic and fluorescent glycopolymer hybrid nanoparticles for intranuclear optical imaging. Biomacromolecules 12:3805–3811
130. Pfaff A, Shinde VS, Lu Y et al (2011) Glycopolymer-grafted polystyrene nanospheres. Macromol Biosci 11:199–210
131. Gabius HJ, André S, Jiménez-Barbero J et al (2011) From lectin structure to functional glycomics: principles of the sugar code. Trends Biochem Sci 36:288–303
132. Gabius HJ (ed) (2009) The sugar code: fundamentals of glycosciences. Wiley-Blackwell, Weinheim

Index

Printed by Printforce, the Netherlands